METHODS IN MOLECULAR BIOLOGY

AF693923

Series Editor
John M. Walker
School of Life Sciences
University of Hertfordshire
Hatfield, Hertfordshire, AL10 9AB, UK

For further volumes:
http://www.springer.com/series/7651

Pseudomonas Methods and Protocols

Edited by

Alain Filloux

MRC Centre for Molecular Bacteriology and Infection (CMBI), Department of Life Sciences, Imperial College, London, UK

Juan-Luis Ramos

Estación Experimental del Zaidín, Consejo Superior de Investigaciones Científicas, Granada, Spain

Humana Press

Editors
Alain Filloux
MRC Centre for Molecular Bacteriology
and Infection (CMBI)
Department of Life Sciences
Imperial College
London, UK

Juan-Luis Ramos
Estación Experimental del Zaidín
Consejo Superior de Investigaciones Científicas
Granada, Spain

ISSN 1064-3745 ISSN 1940-6029 (electronic)
ISBN 978-1-4939-4416-3 ISBN 978-1-4939-0473-0 (eBook)
DOI 10.1007/978-1-4939-0473-0
Springer New York Heidelberg Dordrecht London

© Springer Science+Business Media New York 2014
Softcover reprint of the hardcover 1st edition 2014
This work is subject to copyright. All rights are reserved by the Publisher, whether the whole or part of the material is concerned, specifically the rights of translation, reprinting, reuse of illustrations, recitation, broadcasting, reproduction on microfilms or in any other physical way, and transmission or information storage and retrieval, electronic adaptation, computer software, or by similar or dissimilar methodology now known or hereafter developed. Exempted from this legal reservation are brief excerpts in connection with reviews or scholarly analysis or material supplied specifically for the purpose of being entered and executed on a computer system, for exclusive use by the purchaser of the work. Duplication of this publication or parts thereof is permitted only under the provisions of the Copyright Law of the Publisher's location, in its current version, and permission for use must always be obtained from Springer. Permissions for use may be obtained through RightsLink at the Copyright Clearance Center. Violations are liable to prosecution under the respective Copyright Law.
The use of general descriptive names, registered names, trademarks, service marks, etc. in this publication does not imply, even in the absence of a specific statement, that such names are exempt from the relevant protective laws and regulations and therefore free for general use.
While the advice and information in this book are believed to be true and accurate at the date of publication, neither the authors nor the editors nor the publisher can accept any legal responsibility for any errors or omissions that may be made. The publisher makes no warranty, express or implied, with respect to the material contained herein.

Printed on acid-free paper

Humana Press is a brand of Springer
Springer is part of Springer Science+Business Media (www.springer.com)

Preface

Pseudomonas aeruginosa is a highly versatile Gram-negative bacterium that can thrive in a wide range of ecological niche including soil or marine habitats. This microorganism has a broad range of metabolic and catabolic activities allowing growth on unusual carbon sources. Further, *P. aeruginosa* is now considered as a deadly super bug, recognized as a major nosocomial agent associated with high levels of morbidity and mortality, deadly for cystic fibrosis patients and thus a major threat for human health.

Progress in bacterial genetics and understanding of basic molecular mechanisms in living cells has for decades largely benefited of the studies performed with *Escherichia coli*, the laboratory model organism. With the emergence of new molecular methodologies, the high-throughput sequencing of bacterial genomes, novel model organisms are emerging and what we can learn from them is invaluable.

P. aeruginosa is one of them, and we described its versatility, which suggests that the organism is able to adapt efficiently to all kinds of environments and is thus equipped with most pathways one may think of. Not only the organism can be used to increase our basic knowledge on bacterial life, but this knowledge may have direct therapeutic applications considering it is a dreadful pathogen. Again it displays versatility in terms of infections since it can develop in most parts of the human body and can infect all kind of non-mammalian organisms such as plants, fishes, nematodes, insects or amoeba. This gives also a unique opportunity to perform high-throughput study on host–bacterium interactions using genetically characterized organisms such as *Caenorhabditis elegans* or *Drosophila melanogaster*.

The complete sequence of the genome of *P. aeruginosa* strain PAO1 was released in Nature in 2000 and since then it has constantly been updated. Genomes from several other isolates have also been posted. This is a large size genome, over 6 Mb, which makes you feeling it contains a gold mine for understanding bacterial adaptation and versatility. The genome sequence has been accompanied with the development of a large tool kit box for studying *P. aeruginosa*, which has been made available to the *Pseudomonas* community.

It is time to summarize all the available methods and tools for the molecular study of this fascinating microorganism. This is the purpose of this book, which convenes the most prominent scientists in the *Pseudomonas* field and presents a series of 60 chapters that covers aspects of simple microbiological methods to fancy and high-throughput molecular techniques that have been developed over the last decade.

Enjoy reading it and help the community to keep going with new and cutting-edge findings.

London, UK *Alain Filloux*
Granada, Spain *Juan-Luis Ramos*

Contents

Part V Lifestyles and Infection Models

Editors and Contributors

Editors

Alain Filloux received his PhD (Microbiology and Molecular Biology) in 1988 from the University of Aix-Marseille (France). From 1988 to 1993, he stayed at the University of Utrecht, (The Netherlands), initially as a postdoctoral fellow and later appointed as assistant Professor. In 1994, he is recruited at the French National Centre of Scientific Research (CNRS) in Marseille (France), where he is appointed as head of a research unit in 2003. In 2008, he is appointed as Professor in Molecular Microbiology at Imperial College London (UK) and is deputy director of the MRC Centre for Molecular Bacteriology and Infection. In 2013, he is appointed as Chief Editor for FEMS Microbiology Reviews. The main focus of his research is on bacterial pathogenesis, molecular mechanisms of protein secretion, biofilm formation, and regulation of gene expression. The system model he used is the bacterial pathogen *Pseudomonas aeruginosa*.

Juan-Luis Ramos is Head of the Biotechnology Unit at Abengoa Research. Dr. Ramos obtained his Bachelor's and Master's Degree in Biology at the University of Seville (Spain). He was awarded a PhD in Biochemistry by the same university in 1981. From 1982 to 1984, he worked at the Unit of Nitrogen Fixation at the University of Sussex through an

EMBO fellowship. After that he joined the Department of Medical Biochemistry at the University of Geneva. In 1987 he returned to CSIC in Granada where he was appointed Director of Estación Experimental del Zaidín and became Full Research Professor. He is Fellow of the European and American Academy of Microbiology. He is King Jaime I Laureate for his contribution to the field of Environmental Protection, and he received FEMS Lowff prize for his work on Environmental Microbiology. He has been Otto Monsted Professor at the DTU Center for Biosustainability in Lyngby, Denmark.

Contributors

AHMAD MOHAMMAD ABDEL-MAWGOUD • *INRS-Institut Armand Frappier, Québec, Canada*

JOHN ALVERDY • *Department of Surgery, Pritzker School of Medicine, University of Chicago, Chicago, IL, USA*

ERIN M. ANDERSON • *Department of Molecular and Cellular Biology, University of Guelph, Guelph, Canada*

GREGORY G. ANDERSON • *Department of Biology, Indiana University Purdue University Indianapolis, Indianapolis, IN, USA*

FREDERICK M. AUSUBEL • *Department of Genetics, Harvard Medical School, Massachusetts General Hospital, Boston, MA, USA; Department of Molecular Biology, Massachusetts General Hospital, Boston, MA, USA*

JOANA AZEREDO • *Centre of Biological Engineering, University of Minho, Campus de Gualtar, Braga, Portugal*

REGINA L. BALDINI • *Departamento de Bioquímica, Instituto de Química, Universidade de São Paulo, São Paulo, Brazil*

GENEVIÈVE BALL • *Laboratoire d'Ingénierie des Systèmes Macromoléculaires, UMR7255, CNRS, Aix Marseille Université, IMM, Marseille, France*

NICOLAS BARRAUD • *Centre for Marine Bio-Innovation and School of Biotechnology and Biomolecular Sciences, The University of New South Wales, Sydney, Australia*

DAVID A. BARRETT • *Centre for Analytical Bioscience, School of Pharmacy, University of Nottingham, Nottingham, UK*

VOLKER BEHRENDS • *Department of Life Sciences, Faculty of Natural Sciences, Imperial College London, London, UK*

ILARIA BENEDETTI • *Systems Biology Program, Centro Nacional de Biotecnologia, Madrid, Spain*

SOPHIE DE BENTZMANN • *Laboratoire d'Ingénierie des Systèmes Macromoléculaires (LISM), UMR7255, CNRS, Aix Marseille Université, Institut de Microbiologie de la Méditerranée, Marseille, France*

SOPHIE BLEVES • *Laboratoire d'Ingénierie des Systèmes Macromoléculaires, UMR7255, CNRS, Aix-Marseille Université, IMM, Marseille, France*

CHRISTOPHE BORDI • *Laboratoire d'Ingénierie des Systèmes Macromoléculaires (LISM), UMR7255, CNRS, Aix Marseille Université, Institut de Microbiologie de la Méditerranée, Marseille, France*

BRIAN BOYLE • *Institut de Biologie Intégrative et des Systèmes (IBIS), Université Laval, Québec, QC, Canada*

ALESSANDRA BRAGONZI • *Infection and Cystic Fibrosis Unit, Division of Immunology, Transplantation, and Infectious Diseases, San Raffaele Scientific Institute, Milan, Italy*
FIONA S.L. BRINKMAN • *Department of Molecular Biology and Biochemistry, Simon Fraser University, Burnaby, BC, Canada*
IWONA BUCIOR • *Department of Medicine, University of California, San Francisco, CA, USA; Microbial Pathogenesis and Host Defense Program, University of California, San Francisco, CA, USA*
JACOB G. BUNDY • *Department of Surgery and Cancer, Faculty of Medicine, Imperial College London, London, UK*
FRÉDÉRIC CADORET • *Laboratoire d'Ingénierie des Systèmes Macromoléculaires, UMR7255, CNRS, Aix Marseille Université, IMM, Marseille, France*
HENRIK CALUM • *Department of Clinical Microbiology, Rigshospitalet, University Hospital of Copenhagen, Copenhagen, Denmark*
MIGUEL CÁMARA • *School of Life Sciences, Centre for Biomolecular Sciences, University of Nottingham, Nottingham, UK*
BRENT O. CEZAIRLIYAN • *Department of Genetics, Harvard Medical School, Massachusetts General Hospital, Boston, MA, USA; Department of Molecular Biology, Massachusetts General Hospital, Boston, MA, USA*
STEVE J. CHARETTE • *Institut de Biologie Intégrative et des Systèmes (IBIS), Pavillon Charles-Eugène-Marchand, Université Laval, Quebec, QC, Canada; Centre de recherche de l'Institut Universitaire de Cardiologie et de Pneumologie de Québec (CRIUCPQ), Quebec, QC, Canada; Département de biochimie, de microbiologie et de bio-informatique, Faculté de sciences et de génieUniversité Laval, Quebec, QC, Canada*
YIMEI CHEN • *Electron Microscopy Core Facility, University of Chicago, Chicago, IL, USA*
CRISTINA CIGANA • *Infection and Cystic Fibrosis Unit, Division of Immunology, Transplantation, and Infectious Diseases, San Raffaele Scientific Institute, Milan, Italy*
BARBARA M. COFFEY • *Department of Biology, Indiana University Purdue University Indianapolis, Indianapolis, IN, USA*
ROBBERT H. COOL • *Department of Pharmaceutical Biology, University of Groningen, Groningen, The Netherlands*
PASCAL COSETTE • *"Polymères, Biopolymères, Surfaces" laboratory, UMR CNRS 6270, PISSARO Proteomic Facility, Institute for Research and Innovation in Biomedicine, University of Rouen, Rouen, France*
JOSÉ ANTONIO REYES DARIAS • *Department of Environmental Protection, Estación Experimental del Zaidín, Consejo Superior de Investigaciones Científicas, Granada, Spain*
ERIC DÉZIEL • *INRS-Institut Armand Frappier, QC, Canada*
GONZALO DURANTE-RODRÍGUEZ • *Department of Environmental Biology, Centro de Investigaciones Biológicas, Madrid, Spain*
JOANNE ENGEL • *Department of Medicine, University of California, San Francisco, CA, USA; Microbial Pathogenesis and Host, Defense Program, University of California, San Francisco, CA, USA; Department of Microbiology and Immunology, University of California, San Francisco, CA, USA*
MARCELLA FACCHINI • *Infection and Cystic Fibrosis Unit, Division of Immunology, Transplantation, and Infectious Diseases, San Raffaele Scientific Institute, Milan, Italy*
DOMINIQUE FERRANDON • *Equipe Fondation Recherche Médicale, UPR 9022 du CNRS, Université de Strasbourg, IBMC, Strasbourg, France*

GENEVIÈVE FILION • *Institut de Biologie Intégrative et des Systèmes (IBIS), Pavillon Charles-Eugène-Marchand, Université Laval, Quebec, QC, Canada; Centre de recherche de l'Institut Universitaire de Cardiologie et de Pneumologie de Québec (CRIUCPQ), Quebec, QC, Canada*

ALAIN FILLOUX • *MRC Centre for Molecular Bacteriology and Infection (CMBI), Department of Life Sciences, Imperial College London, London, UK*

MATTHEW FLETCHER • *School of Life Sciences, Centre for Biomolecular Sciences, University of Nottingham, Nottingham, UK*

EMANUELA FRANGIPANI • *Department of Sciences, University Roma Tre, Roma, Italy*

DARA W. FRANK • *Department of Microbiology and Molecular Genetics, Center of Infectious Disease Research, Medical College of Wisconsin, Milwaukee, WI, USA*

CRISTINA GARCÍA-FONTANA • *Department of Environmental Protection, Estación Experimental del Zaidín, Consejo Superior de Investigaciones Científicas, Granada, Spain*

JEAN-MARC GHIGO • *Unité de Génétique des Biofilms, Institut Pasteur, Paris, France*

MARÍA GÓMEZ-LOZANO • *Department of Systems Biology, Technical University of Denmark, Lyngby, Denmark; Novo Nordisk Foundation Center for Biosustainability, Technical University of Denmark, Hørsholm, Denmark*

DAE-GON HA • *Department of Microbiology and Immunology, Geisel School of Medicine at Dartmouth, Hanover, NH, USA*

SAMANTHA HALLER • *Equipe Fondation Recherche Médicale, UPR 9022 du CNRS, Université de Strasbourg, Strasbourg, France*

NIGEL HALLIDAY • *School of Life Sciences, Centre for Biomolecular Sciences, University of Nottingham, Nottingham, UK*

YOUAI HAO • *Department of Molecular and Cellular Biology, University of Guelph, Guelph, Canada*

ISABEL HARTMANN • *Microbial Proteomics, Helmholtz Centre for Infection Research (HZI), Inhoffenstr, Braunschweig, Germany*

SUSANNE HÄUSSLER • *Twincore, Centre of Experimental and Clinical Infection Research, Hannover, Germany; Helmholtz Centre for Infection Research, Braunschweig, Germany*

RONEN HAZAN • *Institute of Dental Sciences, Hebrew University, Campus Hadassah, Jerusalem, Israel; Department of Surgery, Harvard Medical School, Boston, MA, USA; Massachusetts General Hospital, Boston, MA, USA; Shriners Burns Institute, Boston, MA, USA; Department of Microbiology and Immunobiology, Harvard Medical School, Boston, MA, USA; Shriners Hospitals for Children Boston, Boston, MA, USA; IYAR, The Israeli Institute for Advanced Research, Rehovot, Israel*

FRANÇOISE HOEGY • *UMR 7242, Université de Strasbourg-CNRS, ESBS, Strasbourg, France*

NIELS HØIBY • *Department of Clinical Microbiology, Rigshospitalet, University Hospital of Copenhagen, Copenhagen, Denmark*

PAUL A. HOSKISSON • *Strathclyde Institute of Pharmacy and Biomedical Sciences, University of Strathclyde, Glasgow, UK*

ANNE IMBERTY • *Centre de Recherches sur les Macromolécules Végétales, UPR5301 CNRS, Université Joseph Fourier and ICMG, Grenoble, France*

YASUHIKO IRIE • *Centre for Biomolecular Sciences, School of Molecular Medical Sciences, University of Nottingham, UK*

BÉRENGÈRE IZE • *Laboratoire d'Ingénierie des Systèmes Macromoléculaires, UMR7255, CNRS, Aix Marseille Université, IMM, Marseille, France*

KARL-ERICH JAEGER • *Institute of Molecular Enzyme Technology, Research Centre Juelich Heinrich-Heine-University of Duesseldorf, Juelich, Germany*
DIETER JAHN • *Institute of Microbiology, Technische Universität Braunschweig, Braunschweig, Germany*
LOTHAR JÄNSCH • *Microbial Proteomics, Helmholtz Centre for Infection Research (HZI), Inhoffenstr, Braunschweig, Germany*
KATY JEANNOT • *Laboratoire de Bactériologie, Faculté de Médecine-Pharmacie, Centre Hospitalier Régional Universitaire, Université de Franche-Comté, Besançon cedex, France*
JULIE JEUKENS • *Institut de Biologie Intégrative et des Systèmes (IBIS), Université Laval, Québec, QC, Canada*
POL NADAL-JIMENEZ • *Department of Pharmaceutical Biology, University of Groningen, Groningen, The Netherlands; Bacterial Signalling Group, Instituto Gulbenkian de Ciência, Oeiras, Portugal*
CERITH JONES • *Department of Life Sciences, MRC Centre for Molecular Bacteriology and Infection (CMBI), Imperial College London, London, UK*
THIERRY JOUENNE • *"Polymères, Biopolymères, Surfaces" laboratory, UMR CNRS 6270, PISSARO Proteomic Facility, Institute for Research and Innovation in Biomedicine, University of Rouen, Rouen, France*
SUZANNE E. KERN • *Biology Department, Massachusetts Institute of Technology, Cambridge, MA, USA; Division of Biology and Biological Engineering, California Institute of Technology, Pasadena, CA, USA*
EFRAT KESSLER • *Sackler Faculty of Medicine, Maurice and Gabriela Goldschleger Eye Research Institute, Tel Aviv University, Sheba Medical Center, Tel-Hashomer, Israel*
NATALIA V. KIRIENKO • *Department of Genetics, Harvard Medical School, Massachusetts General Hospital, Boston, MA, USA; Department of Molecular Biology, Massachusetts General Hospital, Boston, MA, USA*
GUDRUN KOCH • *Department of Pharmaceutical Biology, University of Groningen, Groningen, The Netherlands*
FILIP KOVACIC • *Institute of Molecular Enzyme Technology, Research Centre Juelich Heinrich-Heine-University of Duesseldorf, Juelich, Germany*
TINO KRELL • *Department of Environmental Protection, Estación Experimental del Zaidín, Consejo Superior de Investigaciones Científicas, Granada, Spain*
SHERRY L. KUCHMA • *Department of Microbiology and Immunology, Geisel School of Medicine at Dartmouth, Hanover, NH, USA*
IRENA KUKAVICA-IBRULJ • *Institut de Biologie Intégrative et des Systèmes (IBIS), Université Laval, Québec, QC, Canada*
HEMANTHA D. KULASEKARA • *Department of Microbiology, University of Washington, Seattle, WA, USA*
JESÚS LACAL • *Department of Environmental Protection, Estación Experimental del Zaidín, Consejo Superior de Investigaciones Científicas, Granada, Spain*
JOSEPH S. LAM • *Department of Molecular and Cellular Biology, University of Guelph, Guelph, Canada*
AURÉLIE LAUBIER • *Laboratoire d'Ingénierie des Systèmes Macromoléculaires, UMR7255, CNRS, Aix-Marseille Université, IMM, Marseille, France*
FRANÇOIS LÉPINE • *INRS-Institut Armand Frappier, QC, Canada*
ROGER C. LEVESQUE • *Institut de Biologie Intégrative et des Systèmes (IBIS), Université Laval, Québec, QC, Canada*

Stefanie Limmer • *Institut für Neuro- und Verhaltensbiologie, Münster, Germany*
María A. Llamas • *Department of Environmental Protection, Estación Experimental del Zaidín-CSIC, Granada, Spain*
Katherine S. Long • *Department of Systems Biology, Technical University of Denmark, Lyngby, Denmark; Novo Nordisk Foundation Center for Biosustainability, Technical University of Denmark, Hørsholm, Denmark*
Victor de Lorenzo • *Systems Biology Program, Centro Nacional de Biotecnologia, Madrid, Spain*
Andrés Corral Lugo • *Department of Environmental Protection, Estación Experimental del Zaidín, Consejo Superior de Investigaciones Científicas, Granada, Spain*
Ashleigh MacKenzie • *The James Hutton Institute, Dundee, UK; Strathclyde Institute of Pharmacy and Biomedical Sciences, University of Strathclyde, Glasgow, UK*
Zofia Magnowska • *Microbial Proteomics, Helmholtz Centre for Infection Research (HZI), Inhoffenstr, Braunschweig, Germany*
Esteban Martínez-García • *Systems Biology Program, Centro Nacional de Biotecnologia, Madrid, Spain*
Rasmus Lykke Marvig • *Department of Systems Biology, Technical University of Denmark, Lyngby, Denmark*
Damien Maura • *Department of Surgery, Harvard Medical School, Boston, MA, USA; Massachusetts General Hospital, Boston, MA, USA; Department of Microbiology and Immunobiology, Harvard Medical School, Boston, MA, USA; Shriners Hospitals for Children Boston, Boston, MA, USA*
Gaëtan L.A. Mislin • *UMR 7242, Université de Strasbourg-CNRS, ESBS, Strasbourg, France*
Søren Molin • *Department of Systems Biology, Technical University of Denmark, Lyngby, Denmark; Novo Nordisk Foundation Center for Biosustainability, Technical University of Denmark, Hørsholm, Denmark*
Joana A. Moscoso • *MRC Centre for Molecular Bacteriology and Infection (CMBI), Department of Life Sciences, Imperial College London, London, UK*
Claus Moser • *Department of Clinical Microbiology, Rigshospitalet, University Hospital of Copenhagen, Copenhagen, Denmark*
Daniela Muhl • *Department of Life Sciences, MRC Centre for Molecular Bacteriology and Infection (CMBI), Imperial College London, London, UK*
Antje Munder • *Clinical Research Group, Clinic for Pediatric Pneumology, Allergology and Neonatology, Hannover Medical School, Hannover, Germany*
Chandrika Goh Nair • *Department of Life Sciences, Imperial College London, London, UK*
Dianne K. Newman • *Division of Biology and Bioengineering, Division of Geological and Planetary Sciences, Howard Hughes Medical Institute, California Institute of Technology, Pasadena, CA, USA*
George A. O'Toole • *Department of Microbiology and Immunology, Geisel School of Medicine at Dartmouth, Hanover, NH, USA*
Catharine A. Ortori • *Centre for Analytical Bioscience, School of Pharmacy, University of Nottingham, Nottingham, UK*
Matthew R. Parsek • *Department of Microbiology, University of Washington, Seattle, WA, USA*
Diana P. Pires • *Centre of Biological Engineering, University of Minho, Campus de Gualtar, Braga, Portugal*

PATRICK PLÉSIAT • *Laboratoire de Bactériologie, Faculté de Médecine-Pharmacie, Centre Hospitalier Régional Universitaire, Université de Franche-Comté, Besançon cedex, France*
JENNIFER R. POWELL • *Department of Biology, Gettysburg College, Gettysburg, PA, USA*
ALEXA PRICE-WHELAN • *Division of Biology and Bioengineering, California Institute of Technology, Hughes Medical Institute, Pasadena, CA, USA; Department of Biological Sciences, Columbia University, New York, NY, USA*
WIM J. QUAX • *Department of Pharmaceutical Biology, University of Groningen, Groningen, The Netherlands*
YOK AI QUE • *Department of Surgery, Harvard Medical School, Boston, MA, USA; Massachusetts General Hospital, Boston, MA, USA; Department of Microbiology and Immunobiology, Harvard Medical School, Boston, MA, USA; Shriners Hospitals for Children Boston, Boston, MA, USA*
LAURENCE G. RAHME • *Department of Surgery, Harvard Medical School, Boston, MA, USA; Massachusetts General Hospital, Boston, MA, USA; Department of Microbiology and Immunobiology, Harvard Medical School, Boston, MA, USA; Shriners Hospitals for Children Boston, Boston, MA, USA*
JUAN-LUIS RAMOS • *Department of Environmental Protection, Estación Experimental del Zaidín, Consejo Superior de Investigaciones Científicas, Granada, Spain*
MIRIAM RICO-JIMÉNEZ • *Department of Environmental Protection, Estación Experimental del Zaidín, Consejo Superior de Investigaciones Científicas, Granada, Spain*
STEPHANIE L. ROLSMA • *Department of Microbiology and Molecular Genetics, Center of Infectious Disease Research, Medical College of Wisconsin, Milwaukee, WI, USA*
BEN RYALL • *School of Molecular Bioscience, University of Sydney, Sydney, NSW, Australia*
IRINA SADOVSKAYA • *USC ANSES, Université du Littoral-Côte d'Opale, Biochimie des Produits Aquatiques, Boulogne-sur-mer, France*
MARY SAFRIN • *Sackler Faculty of Medicine, Maurice and Gabriela Goldschleger Eye Research Institute, Tel Aviv University, Sheba Medical Center, Tel-Hashomer, Israel*
THIBAULT G. SANA • *Laboratoire d'Ingénierie des Systèmes Macromoléculaires, UMR7255, CNRS, Aix-Marseille Université, IMM, Marseille, France*
ASTRID M. VAN DER SAR • *Department of Medical Microbiology and Infection control, VU University Medical Center, BT Amsterdam, The Netherlands*
ISABELLE J. SCHALK • *UMR 7242, Université de Strasbourg-CNRS, ESBS, Strasbourg, France*
SEBASTIAN SCHULZ • *Twincore, Centre of Experimental and Clinical Infection Research, Hannover, Germany; Helmholtz Centre for Infection Research, Braunschweig, Germany*
MEGHA SHAH • *Department of Surgery, Pritzker School of Medicine, University of Chicago, Chicago, IL, USA; School of Biological Sciences, University of Auckland, Auckland, New Zealand*
SANNA SILLANKORVA • *Centre of Biological Engineering, University of Minho, Campus de Gualtar, Braga, Portugal*
HORTENCIA SILVA-JIMÉNEZ • *Department of Environmental Protection, Estación Experimental del Zaidín, Consejo Superior de Investigaciones Científicas, Granada, Spain*
RAFAEL SILVA-ROCHA • *Faculdade de Medicina, University of Sao Paulo, Sao Paulo, Brazil*
CHANTAL SOSCIA • *Laboratoire d'Ingénierie des Systèmes Macromoléculaires, UMR7255, CNRS, Aix Marseille Université, IMM, Marseille, France*
MELISSA STARKEY • *American College of Physicians, Philadelphia, PA, USA*

CLAUS STERNBERG • *Department of Systems Biology, Technical University of Denmark, Kongens Lyngby, Denmark*
LEWIS STEWART • *Strathclyde Institute of Pharmacy and Biomedical Sciences, University of Strathclyde, Glasgow, UK*
TIM TOLKER-NIELSEN • *Department of International Health, Immunology and Microbiology, Faculty of Health and Medical Sciences, University of Copenhagen, Copenhagen, Denmark*
CINDY TRAN • *Microbial Pathogenesis and Host Defense Program, University of California, San Francisco, CA, USA; Department of Pediatrics, University of California, San Francisco, CA, USA*
NICHOLAS P. TUCKER • *Strathclyde Institute of Pharmacy and Biomedical Science, University of Strathclyde, Glasgow, UK*
BURKHARD TÜMMLER • *Clinical Research Group, Clinic for Pediatric Pneumology, Allergology and Neonatology, Hannover Medical School, Hannover, Germany*
LYNNE TURNBULL • *The ithree institute, University of Technology Sydney, Ultimo, NSW, Australia*
ANNABELLE VARROT • *Centre de Recherches sur les Macromolécules Végétales, UPR5301 CNRS, Université Joseph Fourier and ICMG, Grenoble, France*
VÉRONIQUE VIARRE • *Laboratoire d'Ingénierie des Systèmes Macromoléculaires, UMR7255, CNRS, Aix Marseille Université, IMM, Marseille, France*
ROMÉ VOULHOUX • *Laboratoire d'Ingénierie des Systèmes Macromoléculaires, UMR7255, CNRS, Aix Marseille Université, IMM, Marseille, France*
CYNTHIA B. WHITCHURCH • *The ithree institute, University of Technology Sydney, Ultimo, NSW, Australia*
PAUL WILLIAMS • *School of Life Sciences, Centre for Biomolecular Sciences, University of Nottingham, Nottingham, UK*
HUW D. WILLIAMS • *Department of Life Sciences, Imperial College London, London, UK*
GEOFFREY L. WINSOR • *Department of Molecular Biology and Biochemistry, Simon Fraser University, Burnaby, BC, Canada*
OLGA ZABORINA • *Laboratory for Surgical Infection Research and Therapeutics, Department of Surgery, University of Chicago, Chicago, IL, USA; Department of Surgery, Pritzker School of Medicine, University of Chicago, Chicago, IL, USA*

Part I

Basic Pseudomonas Features

Chapter 1

Gene Transfer: Transduction

Emanuela Frangipani

Abstract

Bacteriophages able to propagate on *Pseudomonas* strains are very common and can be easily isolated from natural environments or lysogenic strains. The development of transducing systems has allowed bacterial geneticists to perform chromosome analyses and mutation mapping. Moreover, these systems have also been proved to be a successful tool for molecular microbiologists to introduce a foreign gene or a mutation into the chromosome of a bacterial cell. This chapter provides a description of the phage methodology illustrated by Adams in 1959 and applicable to strain PAO1 derivatives.

Key words Generalized transduction, Bacteriophage, Gene transfer, MOI, Plaque-forming units

1 Introduction

Transduction represents the process by which DNA is transferred from one bacterium to another, by means of a bacteriophage (phage) [1].

Transduction may occur either through the lytic cycle or through the lysogenic cycle of a phage. For transduction to occur, phage particles must contain fragments of the bacterial genome derived from the host which they have been propagated on. These bacterial DNA fragments are encapsulated into the phage particles, either in combination with the phage genome or alone; in the latter case they replace the entire phage genome. Phages containing bacterial DNA fragments are called transducing phages and are usually noninfectious viral particles, as the bacterial DNA replaces most of the essential phage genes required to complete the life cycle. In many cases transduction is not specific, and virtually any genome fragment of the donor bacterium may be transferred to the recipient infected bacterial cell with similar probabilities (10^{-5} to 10^{-6}) [2]. This mechanism of genetic exchange, called *generalized transduction*, was described for the first time over 60 years ago in *Salmonella typhimurium* [1]. During generalized transduction, the genetic material of the donor bacterium is disrupted while the

Alain Filloux and Juan-Luis Ramos (eds.), *Pseudomonas Methods and Protocols*, Methods in Molecular Biology, vol. 1149, DOI 10.1007/978-1-4939-0473-0_1, © Springer Science+Business Media New York 2014

phages multiply, leading to the accidental incorporation of virtually any bacterial DNA fragment into the phage particles [2].

Another type of phage-mediated transfer of genetic material, called *specialized transduction*, was described for the first time in the case of phage λ in *Escherichia coli* [3]. During specialized transduction the transducing phage particles usually carry a small specific portion of the bacterial genome together with the majority of the phage genome. Specialized transduction only occurs with some temperate lysogenic phages. In contrast to lytic virulent phages, which are only able to undergo the lytic life cycle leading to the immediate death of their bacterial hosts, temperate phages may integrate into the bacterial chromosome and exist as a dormant prophage (lysogeny), prior to entering the lytic phase [3]. Excision of the prophage from the host chromosome is sometimes imprecise, leading to the inclusion of some bacterial genes into the phage particles [4]. Bacterial DNA fragments that might be carried in these transducing phage particles depend on the site of the insertion of the prophage in the host cell.

In both generalized and specialized transduction, once a bacterial cell is infected with a transducing phage, the bacterial DNA fragment carried by the phage can be integrated into the chromosome of the infected bacterium through homologous recombination, generating transductant bacteria containing DNA derived from the donor cell [2].

Bacteriophages have significantly contributed to the understanding of bacterial genetics. Together with conjugation and transformation, transduction has played a key role in determining genome organization in several bacterial species, including *Pseudomonas aeruginosa* [5, 6]. Bacteriophages able to propagate on *Pseudomonas* strains are very common and can be easily isolated from sewage, lysogenic strains or soil [5]. However, while phages able to propagate on *Pseudomonas* species other than *aeruginosa* have been described, the greatest body of information on transduction remains limited to *P. aeruginosa* [5]. In general, bacteriophages are not suitable for studies involving interspecific gene transfer, due to their limited host range that seems to follow the species lines, at least in *Pseudomonas* [5, 6]. The development of transducing systems for genetic analysis or mutation mapping in *P. aeruginosa* PAO1 has been relatively straightforward, due to the abundance of isolated phages suitable for infection of this strain [6]. The most commonly used phages in transduction experiments in *P. aeruginosa* PAO1 are phages F116L [7] and G101 [8], both generalized transducing lysogenic phages of 41 and 38 MDa, respectively, able to transduce markers at a frequency of 1×10^{-7} to 5×10^{-7} per plaque-forming unit [9]. These phages have been extensively used in mapping of the *P. aeruginosa* PAO1 chromosome [10–13]. Interestingly, a mutant of the virulent lytic bacteriophage E79, named E79*tv-2*, capable of mediating generalized transduction in *P. aeruginosa* PAO1 has also been successfully used for linkage studies [14] and even more recently, to transduce mutations among *P. aeruginosa* PAO1 mutant strains [15, 16]. Phage F116L is a pilin-specific phage [17], while phage E79*tv2* is specific for the lipopolysaccharide (LPS) [18]. Their successful use in transduction experiments is, therefore, subject to the expression of these structures by the host strain.

This chapter provides a description of the phage methodology illustrated by Adams in 1959, which includes standard techniques for generalized transduction procedures [2]. The method represents an efficient way to transduce genes and mutations among *P. aeruginosa* strains via generalized transduction and has been recently applied within *P. aeruginosa* PAO1 mutants using phages F116L [7, 11], or E79*tv-2* [14–16].

The procedures described below might be applied to other *Pseudomonas* species, given the existence of transducing phages with suitable host ranges.

2 Materials

Prepare all solutions using double-distilled, deionized water (ddH_2O). Store at room temperature.

2.1 Preparation of Growth Media and Solutions

1. *Nutrient Yeast Broth (NYB)* [19]. Weigh 25 g Oxoid nutrient broth No. 2, 5 g Oxoid yeast extract and transfer to a 1 L graduated cylinder or a beaker. Add ddH_2O to a volume of 900 ml and dissolve using a magnetic stir bar. Check that the pH of the medium is between 7.0 and 7.5 (*see* **Note 1**). Make up to 1 L with ddH_2O, transfer to a glass bottle, and sterilize by autoclaving.
2. *Nutrient Agar (NA)* [19]. Weigh 40 g Oxoid blood agar base, 5 g Oxoid yeast extract and transfer to a 1 L glass bottle. Add ddH_2O to a volume of 1 L and sterilize by autoclaving. Cool the medium to ~50 °C and pour 25 ml medium per plate to obtain ~40 plates/L. Store plates at 4 °C in a plastic bag, to limit plate over-drying.
3. *Semi-solid nutrient layer agar (TLA)*. Proceed as for preparation of NYB. Dispense in five 200 ml glass bottles (*see* **Note 2**), each containing 0.8 % Oxoid agar No. 1 (1.6 g per bottle). Mix and sterilize by autoclaving.
4. *Tris buffer with sodium chloride and magnesium sulfate (TNM)* [13]. Weigh 1.21 g Tris, 8.77 g NaCl and 2.46 g $MgSO_4 \cdot 7H_2O$ and transfer to a 1 L cylinder or a beaker. Add ddH_2O to a volume of 900 ml and dissolve using a magnetic stir bar. Adjust the pH to 7.4 using 1 M HCl. Make up to 1 L with ddH_2O, transfer to a glass bottle, and sterilize by autoclaving.

2.2 Other Materials Required

1. Sterile petri dishes.
2. Disposable spreaders.
3. Sterile centrifuge tubes.
4. Phage stocks, e.g., F116L [7] or E79*tv-2* [14].
5. *P. aeruginosa* donor and recipient strain(s).

3 Methods

Carry out all the procedures under sterile conditions.

3.1 Preparation of Phage Lysates

1. Grow *P. aeruginosa* (or a mutant strain containing the mutation or the phenotype to be transduced) for 12–14 h in NYB at 37 °C, in the presence of the appropriate antibiotic or selective marker, under vigorous shaking.
2. Mix 100 μl of the *P. aeruginosa* culture and 100 μl of the phage stock (~10^5 pfu/ml), with 3–4 ml of warm (~45–50 °C) TLA. Briefly vortex and pour onto an NA plate. Repeat this procedure twice. Close the plates and let them solidify for 10 min (*see* **Note 3**). Incubate for 12–14 h at 37 °C, until semi-confluent plaques appear (Fig. 1).
3. Using a sterile disposable plastic spreader, scrape off the layers of 2 NA plates containing semi-confluent phage plaques and transfer them into a sterile 30 ml centrifuge tube (*see* **Note 4**).
4. Add 2 ml of TNM to the 30 ml centrifuge tube and vortex vigorously for 1 min.

Fig. 1 Semi-confluent plaques produced in a lawn of *P. aeruginosa* PA01 by the replication of phage E79*tv-2* and consequent lysis of the bacterial hosts

5. Centrifuge the mixture at 20,000 × *g* for 10 min and remove the supernatant by using a 2 ml syringe.
6. Filter the phage-containing supernatant through a 0.45 μm filter and store in a screw-cap vial at 4 °C.
7. Determine the phage titre as described below in Subheading 3.2.

3.2 Titration of the Phage Lysates

1. Grow *P. aeruginosa* (or the strain from which the phage lysate has been obtained) for 12–14 h in NYB at 37 °C, under vigorous shaking.
2. Mix 100 μl of the *P. aeruginosa* culture with 3–4 ml of warm (~45–50 °C) TLA in a 15 ml pre-warmed centrifuge tube. Briefly vortex and pour onto an NA plate. Close the plate and let it settle for 10 min (*see* **Note 5**).
3. Make serial 1/10 dilutions of the phage lysate using sterile TNM. For this purpose, prepare 10 sterile 1.5 ml centrifuge tubes each filled with 90 μl of TNM. Add 10 μl of the phage lysate to the first tube and mix well by pipetting. Transfer 10 μl from the first dilution to the second tube and mix well by pipetting. Proceed until you reach the tenth tube (*see* **Note 6**).
4. Spot 5 μl of each dilution, prepared during **step 3**, onto the *P. aeruginosa* lawn prepared in **step 2**. Take care to indicate where each dilution is spotted on the NA plate containing the *P. aeruginosa* overlay. Let dry until the spots are completely absorbed onto the agar (*see* **Note 7**). Incubate for 12–14 h at 37 °C.
5. Count the resulting plaques and calculate the plaque forming units per milliliter (pfu/ml) in the original phage lysate stock by applying the following formula:

pfu/ml = Numbers of pfu counted * dilution factor at which the pfu are counted * 200

3.3 Transduction into the Recipient Strain

1. Grow the recipient *P. aeruginosa* strain for 12–14 h in NYB at 37 °C, in the presence of the appropriate antibiotic, under vigorous shaking.
2. Transfer 500 μl of the *P. aeruginosa* recipient strain culture (~2×10^9 cell/ml) into a microcentrifuge sterile tube.
3. Centrifuge for 3 min at maximum speed in a benchtop centrifuge.
4. Discard the supernatant and resuspend in 500 μl sterile TNM.
5. Add 500 μl of phage suspension (~5×10^9 pfu/ml) and incubate for 20 min at 37 °C to allow phage adsorption.
6. Remove the non-adsorbed phage with two washes in TNM buffer. Centrifuge the 1 ml suspension containing the bacteria–phage mix for 3 min at maximum speed in a benchtop centrifuge.

Discard the supernatant and resuspend the pellet in 1 ml of sterile TNM. Repeat this step twice. Finally resuspend the pellet in 1 ml of sterile TNM (*see* **Note 8**).

7. Spread 200 μl and 800 μl of the washed infected bacteria onto selective plates and let them grow for 12–24 h at 37 °C, until bacterial colonies are visible (*see* **Note 9**).
8. As controls, plate 100 μl of the recipient strain on a selective plate (spontaneous mutation control) and 100 μl of the phage lysate on a selective plate (contamination of the phage lysate control).
9. Record the frequency of transduction by applying the following formula (*see* **Note 10**):

$$\text{Frequency of transduction} = \text{Number of transductants} / \text{Number of pfu of the phage lysate}$$

4 Notes

1. The pH of the medium, prepared as described, should be within the range of 7.0–7.5, so usually there is no need to adjust it. However, a few drops of 1 M NaOH can be used to reach the required pH, if needed.
2. TLA is used to distribute phages or bacteria evenly in a thin layer over the surface of a plate. Before use, melt TLA by heating in a microwave oven on low power (defrost mode), then cool to and hold at 45 °C to 50 °C. Alternatively, TLA bottles can be directly stored in a 55 °C water bath after autoclaving and used within 1 week.
3. In order to obtain a concentrated phage stock of about 10^{13} pfu/ml, it is important to retrieve phages from semi-confluent plaques. Using plates where few plaques are visible or large clear lysis zones appear might result in low phage titre or low transduction efficiency. Increase or decrease the amount of phage to use and repeat the preparation of the phage lysates from Subheading 3.1, **step 1**.
4. This step can be problematic as sometimes it is challenging to separate the TLA layer from the NA bottom of the plate. Start by making a small incision in the layer with a sterile micropipette tip and then proceed by scraping off the TLA layer with the disposable spreader. Do not apply too much pressure while using the plastic spreader, as this might result in the breakage of the underlying agar.
5. In order to obtain a smooth overlay of *P. aeruginosa* cells, it is important that the NA plate is dry and pre-warmed, as otherwise

TLA will clump upon pouring. It is usually sufficient to dry plates for 10–15 min in an oven at 42 °C before pouring the bacteria–TLA mix. It is also desirable that warm TLA is aliquoted in 15 ml sterile tubes (3–4 ml in each tube) and kept in a 55 °C water bath until needed.

6. When working with phages is critical not to contaminate laboratory equipment. To avoid contamination of micropipettes the use of filter tips is advisable.
7. It may take relatively a long time for the spots to adsorb onto the agar. If using a sterile hood, petri dishes might be left to dry with the lid open to speed up the adsorption process.
8. This washing step is very important. Removing as many non-adsorbed phages as possible would prevent the transductants from become lysogens with F116L or from being killed by E79*tv2*. Another washing step might be added or, in some cases, E79*tv2* transducing lysates can be attenuated for virulence by a 15 s treatment with UV, prior to incubation with recipient strain [15].
9. The selective marker on plates depends on mutation or gene that has to be transferred by transduction. If the aim is to transfer a mutation that has been generated in the donor strain by the insertion of an antibiotic resistance cassette, the selective plate must contain the antibiotic for which the cassette confers resistance. Similarly, if the aim is to transduce a specific phenotype, the selective plate should allow a fast screening of this phenotype (e.g., milk agar to identify protease production, minimal medium to screen for complementation of auxotrophic mutants). Make sure that plates are dried for 10–15 min at 42 °C before plating the bacterial cells. Before plating 800 μl of the infected bacteria suspension, it is advisable to concentrate the mix by centrifuging it for 3 min at maximum speed in a benchtop centrifuge and resuspend the pellet in 200 μl of sterile TNM.
10. If the frequency of transduction is very low, one way to increase it is to increase the multiplicity of infection (MOI). MOI is the ratio of infectious agents (i.e., phages) to infection targets (i.e., bacterial cells) present in a defined space. Usually a MOI of 1/2.5. is sufficient to obtain a good transduction frequency ($\sim 10^{-6}$ to 10^{-7}), when using a transducing phage such as F116L, G101, or E79*tv-2*. Many parameters may influence the frequency of transduction (size of fragment to be transduced, type of selection for transducing cells). Therefore, each experiment may need adjustment of the MOI accordingly. Another parameter that might influence the frequency of transduction is the time allowed for phage adsorption on bacterial cells. Usually 15–20 min at 37 °C is sufficient, but

allowing adsorption for longer times (30 min) might be advisable.

Acknowledgements

I wish to thank Dr Áine Fox, with whom I had the pleasure of working on several transduction experiments whilst at the University of Lausanne, for critically reviewing this book chapter and Professor Dieter Haas, my Ph.D. supervisor, for introducing me to the fascinating world of phages.

References

1. Zinder ND, Lederberg J (1952) Genetic exchange in *Salmonella*. J Bacteriol 64: 679–699
2. Adams MH (1959) Bacteriophages. Interscience, New York
3. Morse ML, Lederberg EM, Lederberg J (1956) Transduction in *Escherichia coli* K-12. Genetics 41:142–156
4. Lederberg EM, Lederberg J (1953) Genetic studies of lysogenicity in *Escherichia coli*. Genetics 38:51–64
5. Holloway BW (1969) Genetics of *Pseudomonas*. Bacteriol Rev 33:419–443
6. Holloway BW, Krishnapillai V, Morgan AF (1979) Chromosomal genetics of *Pseudomonas*. Microbiol Rev 43:73–102
7. Krishnapillai V (1971) A novel transducing phage. Its role in recognition of a possible new host-controlled modification system in *Pseudomonas aeruginosa*. Mol Gen Genet 114: 134–143
8. Holloway BW, van de Putte P (1968) Lysogeny and bacterial recombination. In: Peacock WJ, Brock RD (eds) Replication and recombination of genetic material. Australian Academy of Sciences, Canberra, pp 175–183
9. Holloway BW, Krishnapillai V (1975) Bacteriophages and bacteriocins. In: Clarke PH, Richmond MH (eds) Genetics and biochemistry of *Pseudomonas*. John Wiley and Sons, London, pp 99–132
10. Fargie B, Holloway BW (1965) Absence of clustering of functionally related genes in *Pseudomonas aeruginosa*. Genet Res 6: 284–299
11. Fox A, Haas D, Reimmann C et al (2008) Emergence of secretion-defective sublines of *Pseudomonas aeruginosa* PAO1 resulting from spontaneous mutations in the *vfr* global regulatory gene. Appl Environ Microbiol 74: 1902–1908
12. Haas D, Holloway BW, Schomböch A et al (1977) The genetic organization of arginine biosynthesis in *Pseudomonas aeruginosa*. Mol Gen Genet 154:7–22
13. Mee BJ, Lee BTO (1967) An analysis of histidine requiring mutants in *Pseudomonas aeruginosa*. Genetics 55:709–722
14. Morgan AF (1979) Transduction of *Pseudomonas aeruginosa* with a mutant of bacteriophage E79. J Bacteriol 139:137–140
15. Malone JG, Jaeger T, Spangler C et al (2010) YfiBNR mediates cyclic di-GMP dependent small colony variant formation and persistence in *Pseudomonas aeruginosa*. PLoS Pathog 12(6(3)):1000804
16. Valentini M, Storelli N, Lapouge K (2011) Identification of C(4)-dicarboxylate transport systems in *Pseudomonas aeruginosa* PAO1. J Bacteriol 193:4307–4316
17. Pemberton JM (1973) F116: a DNA bacteriophage specific for the pili of *Pseudomonas aeruginosa* strain PAO. Virology 55:558–560
18. Jarrell K, Kropinski AM (1977) Identification of the cell wall receptor for bacteriophage E79 in *Pseudomonas aeruginosa* strain PAO. J Virol 23:461–466
19. Stanisich VA, Holloway BW (1972) A mutant sex factor of *Pseudomonas aeruginosa*. Genet Res 19:91–108

Chapter 2

Gene Transfer: Transformation/Electroporation

Frédéric Cadoret, Chantal Soscia, and Romé Voulhoux

Abstract

Since *Pseudomonas aeruginosa* is a non-naturally competent bacterium, various methods have been developed to transfer exogenous DNA. Alternatively to transduction and conjugation, electroporation can also be used to transfer exogenous DNA molecules into *Pseudomonas*. Electroporation uses an electric field which generates pores in bacterial membranes allowing the entry of the exogenous DNA molecule. In contrast to conjugation which is restricted to the transfer of DNA from one bacterial cell to another, electroporation can be used to transfer all types of DNA resuspended in water.

Key words DNA transfer, *Pseudomonas aeruginosa*, Competent cells, Electroporation, Electro-transformation

1 Introduction

Competence is a natural ability for a wide range of bacteria to easily uptake DNA by horizontal DNA transfer. However, some bacteria such as *Pseudomonas* species which are not naturally competent must use more complex strategies such as conjugation to uptake DNA. Bacterial conjugation (described Chapter 4) is very efficient in *Pseudomonas* but requires the presence of a specific origin of transfer supported by conjugative pili encoded by the donor strain. Alternatively, electroporation procedures of *Escherichia coli* and *P. aeruginosa* have been described by several investigators [1, 2]. This technique allows the efficient transfer of DNA in solution directly into the bacterial cell. This approach can be used to transfer any type of circular or linear, replicative or non-replicative DNA for many applications such as gene expression, gene replacement, site-specific gene integration, homologous recombination, gene integration, or transposon delivery experiment [3].

Electroporation uses a high-intensity electric field that permeabilizes bacterial cell membranes, allowing the entry of exogenous DNA molecules. This protocol can be divided in two steps. The first step consists in the preparation of *P. aeruginosa* electro-competent

Alain Filloux and Juan-Luis Ramos (eds.), *Pseudomonas Methods and Protocols*, Methods in Molecular Biology, vol. 1149, DOI 10.1007/978-1-4939-0473-0_2, © Springer Science+Business Media New York 2014

cells using sucrose treatment to permeabilize membrane. The second step is the electroporation per se and consists in mixing electrocompetent cells with DNA followed by the application of an adapted electric shock. Although a large proportion of the cells are killed during this electrical pulse, many survival cells integrate the foreign DNA. This chapter describes a protocol specifically adapted to *P. aeruginosa* electroporation.

2 Materials

2.1 Bacterial Growing Media

Luria–Bertani rich medium (LB): Bacto Tryptone 10 g/L; Bacto Yeast Extract 5 g/L; NaCl 10 g/L; pH 7.4. Sterilize using an autoclave.

Super Optimal Broth (SOC): Bacto Peptone 20 g/L; Bacto Yeast Extract 5 g/L; NaCl 10 mM; KCl 2.5 mM; $MgCl_2$ 10 mM; $MgSO_4$ 10 mM. Adjust to 1 L with distilled H_2O. Sterilize using an autoclave. Add glucose 2 % to the medium after autoclaving.

Solid selective growth medium: *Pseudomonas* Isolation Agar (PIA) from DIFCO company; Peptone 20 g/L, $MgCl_2$ 1.4 g/L, K_2SO_4 10 g/L, Irgasan™ 25 mg/L, agar 13.6 g/L. Glycerol 4 % final.

2.2 Electroporation Buffer

The 300 mM saccharose (sucrose) solution is obtained by dissolving pure saccharose (in powder) in ultrapure sterile water (*see* **Note 1**) and stored at 4 °C.

2.3 Electroporation Material

Gene Pulser Electroporation System (Bio-Rad). The parameter "time constant" must be selected. For an efficient transformation, the Gene Pulser has to be adjusted at the following settings: 200 Ω, 25 μF, and 2.5 kV for 2 mm gap width cuvettes and 1.8 kV for 1 mm gap width cuvettes.

2.4 Electroporation Cuvettes

Electroporation cuvettes stored at −20 °C, 24 h before experiment.

2.5 Other Material

For liquid culture of bacteria, use a to-and-fro shaking incubator set at 180 rpm and 37 °C (*see* **Note 2**).
To measure Optical Density (OD_{600}) of bacterial cultures, use a spectrophotometer set at 600 nm and appropriate cuvettes.

2.6 Antibiotics

Antibiotics are solubilized in distilled water except for tetracycline which is solubilized in 70 % ethanol. Antibiotic solutions are filtered and used according to the concentrations mentioned Table 1. For streptomycin and kanamycin, fresh solutions must be used.

Table 1
Antibiotic concentrations used for *Pseudomonas aeruginosa* selection. For antibiotics marked with an asterisk, specific adjustment might be necessary

Antibiotic	Concentration for plasmid expression (μg/ml)	Concentration for chromosomal expression (μg/ml)
Carbenicillin	500	300–500
Gentamicin	50–150	50–75
Kanamycin*	1,500	500–1,000
Streptomycin*	2,000	500–750
Tetracycline*	200	50–100

3 Methods

1. From an overnight culture obtained in LB at 37 °C under shaking conditions, inoculate fresh LB rich medium (*see* **Note 3**) at an OD_{600} of 0.05 at 37 °C under shaking conditions.
2. At an OD_{600} of 0.5–0.6, centrifuge the culture for 10 min at 2,300 × *g* (*see* **Notes 4** and **5**).
3. Discard the supernatant and gently resuspend bacterial pellet in an equal volume of a cold 300 mM saccharose solution.
4. Spin down the bacteria for 10 min at 2,300 × *g* in a centrifuge pre-cooled at 4 °C.
5. Discard the supernatant and gently resuspend bacterial pellet in 0.5 volume of a cold 300 mM saccharose solution.
6. Spin down the bacteria for 10 min at 2,300 × *g* in a centrifuge pre-cooled at 4 °C.
7. Discard the supernatant and gently resuspend the bacterial pellet in 0.01 volume of a cold 300 mM saccharose solution.
8. Mix gently 80 μl of cell suspension with up to 10 μl of a DNA solution at 0.1–1 μg/ml (*see* **Note 6**).
9. Chill the DNA–cell suspension on ice for at least 30 min prior to electroporation.
10. Transfer the DNA–cell mix in an electroporation cuvette (*see* **Note 7**).
11. Place the electroporation cuvette containing the DNA–cell mix in the Gene Pulser. Set the parameters as described in Subheading 2 and apply the electric shock (*see* **Note 8**).

12. Transfer the mix in 12 ml falcon tube containing 2 ml of SOC glucose medium.
13. Cultivate the cells for 2 h at 37 °C under shaking conditions (*see* **Note 9**).
14. Display 200 μl of the bacterial culture on selective agar plates (*see* **Note 10**) and incubate at 37 °C for at least 48 h.

4 Notes

1. The saccharose solution can be sterilized by filtration (preferred) or autoclaving at 110 °C maximum in order to avoid saccharose damage.
2. To-and-fro motion incubator supplies a horizontal non-rotative shaking, more efficient for *P. aeruginosa* growth.
3. 10 ml of culture is necessary for each electroporation assay.
4. For easy electroporations (high-copy plasmids) it is possible to centrifuge 6 ml of an overnight culture as described by Choi et al. [3].
5. From this step to **step 12**, it is very important to use precooled solution and material (centrifuge and pipettes) at 4 °C. Electroporation cuvettes must be previously stored overnight at −20 °C.
6. It is very important to use pure DNA resuspended in deionized water since the presence of salt often lead to electric arcs that strongly compromise the transformation. In case of non-replicative DNA, it could be useful to perform a control electroporation with replicative DNA.
7. The mix can directly be made in the cuvette. The presence of bubbles may lead to electric arcs during electroporation which will compromise the experiment. In case of electric arcs, it is possible to continue the experiment since some positive colonies could nevertheless be obtained.
8. For the Bio-Rad Gene Pulser Electroporation system, the pulse is considered efficient when it lasts between 4.5 and 5 ms.
9. This step allows the phenotypic expression of resistance markers. It can be extended to 3 h in the case of non-replicative DNA to allow its insertion on the bacterial chromosome.
10. For low efficiency electroporation (non-replicative DNA) the whole culture should be centrifuged 10 min at 2,300 ×*g*, resuspend in 200 μl of LB medium and plate on selective solid medium.

References

1. Dower WJ, Miller JF, Ragsdale CW (1988) High efficiency transformation of *E. coli* by high voltage electroporation. Nucleic Acids Res 16:6127–6145
2. Smith A, Iglewski BH (1989) Transformation of *Pseudomonas aeruginosa* by electroporation. Nucleic Acids Res 17:10509
3. Choi KH, Kumar A, Schweizer HP (2006) A 10-min method for preparation of highly electrocompetent *Pseudomonas aeruginosa* cells: application for DNA fragment transfer between chromosomes and plasmid transformation. J Microbiol Methods 64:391–397

Chapter 3

Gene Transfer: Conjugation

Thibault G. Sana, Aurélie Laubier, and Sophie Bleves

Abstract

Conjugation is a gene transfer process in which a recipient bacterium receives DNA from a donor bacterium by cell-to-cell contact through conjugative pili. Conjugation is mediated by certain plasmids or transposons. Here, we describe plasmid conjugation.

Key words Conjugation, Gene transfer, Conjugative plasmid, Helper plasmid, Donor/recipient bacterium, *oriT*, Tra, Mob, Mating pair

1 Introduction

Conjugation involves the transfer of DNA from donor bacterium to recipient bacterium by cell contact [1]. The first step in conjugation involves a conjugative pilus (sex pilus or F pilus) on the donor bacterium (called F+) binding a recipient bacterium lacking a conjugative pilus (called F−). Typically the conjugative pilus retracts (depolymerizes) pulling the two bacteria together and the two cells become bound together at a point of direct envelope-to-envelope contact. A series of membrane proteins coded by the conjugative plasmid then forms a bridge and an opening between the two bacteria, now called a mating pair. Using the rolling circle model of DNA replication, a nuclease breaks one strand of the plasmid DNA at the origin of transfer site (*oriT*) of the plasmid and that nicked strand enters the recipient bacterium progressing in the 5′ to 3′ direction. A complex of auxiliary proteins assemble at *ori*T and assist in the nicking process to form this intermediate (called the relaxosome) in the DNA transfer. One DNA strand begins to pass through the cytoplasmic bridge to the F− cell, while the other strand remains in the donor cell. The complementary strands are synthesized in both donor and recipient cells. Both bacteria now possess the conjugative plasmid. The recipient then becomes an F+ male and can make a sex pilus. During conjugation, no cytoplasm or cell material except DNA passes from donor to recipient.

Alain Filloux and Juan-Luis Ramos (eds.), *Pseudomonas Methods and Protocols*, Methods in Molecular Biology, vol. 1149, DOI 10.1007/978-1-4939-0473-0_3, © Springer Science+Business Media New York 2014

Some bacterial plasmids are conjugative plasmids that have the ability of transfer themselves into another host. Mobilizable plasmids are not able to promote their own transfer unless an appropriate conjugation system is provided in *trans*. Mobilizable vectors contain the site for transfer initiation, *ori*T, and have sequences encoding Mob proteins involved in the mobilization of the DNA during the conjugative process. Mob proteins alone are not sufficient to achieve the transfer. Additional proteins for transfer (Tra) are involved in the formation of the pilus through which the DNA passes to the recipient bacteria. Mobilizable plasmids do not encode Tra proteins and for this reason they require a helper plasmid providing the *tra* genes. The *tra* genes can also be present in the genome of the donor strain.

As the broad-host-range plasmids available for *P. aeruginosa* study are mobilizable plasmids, we describe here a triparental mating involving a third strain that contains the helper plasmid. Briefly, in a first step the donor strain is mobilized by the helper strain, and in a second step, the plasmid is transferred to the recipient strain. The conjugant strain is then selected for the presence of the plasmid on an appropriate selective medium.

2 Materials

Prepare all solutions using ultrapure water. Prepare and store liquid medium at room temperature and in sterile conditions.

2.1 Media

1. LB Broth (Luria-Bertani) (BD Difco™): tryptone 10 g, yeast extract 5 g, sodium chloride 10 g per liter. For liquid LB medium, add 25 g of LB Broth in 1 L of distilled water and autoclave it 15 min at 121 °C. For LB agar plates, add 15 g per liter of Bacto Agar (BD Bacto™ Agar) and autoclave.
2. Selection media for *Pseudomonas aeruginosa* conjugants: *Pseudomonas* Isolation Agar (PIA) (BD Difco™): peptone 20 g, magnesium chloride 1.4 g, potassium sulfate 10 g, Irgasan™ 25 mg, agar 13.6 g per liter. For PIA plates, add 45 g of PIA powder in 1 L of distilled water and autoclave it 15 min at 121 °C. Add, as carbon source, 30 ml of glycerol 80 % (V/V) per liter and the appropriate antibiotics (Table 1). The PIA medium is used to counter-select *Pseudomonas* for its intrinsic resistance to Irgasan against *E. coli*. Do not use plates older than 1 week.
3. Isolation media (to maintain conjugants 1 week at 4 °C after selection): LB plates with appropriate concentration of antibiotics (Table 1).
4. Antibiotics (Sigma-Aldrich) are solubilized in distilled water (except for chloramphenicol which is solubilized in ethanol 100 %, and tetracycline which is solubilized in ethanol 70 %)

Table 1
Concentrations of antibiotics used for *E. coli* and *P. aeruginosa* growth

Antibiotic	Initial concentration (mg/ml)	Concentration for *E. coli* strains (μg/ml)	Concentration for *Pseudomonas* selection (μg/ml)	Concentration for *Pseudomonas* isolation (μg/ml)
Ampicillin	25	25–50	/	/
Carbenicillin	25	/	500–750	300–500
Gentamicin	10	10–20	100–150	50–75
Kanamycin	10	50–100	1,500–2,000	1,000–1,500
Streptomycin	100	50–100	1,500–2,000	1,000–1,500
Tetracycline	15	10–15	100–200	50–100

/ : Not concerned

and filtered. Add antibiotics to liquid and solid media according to Table 1.

2.2 Incubators

1. For liquid culture of bacteria, use a shaking incubator (180 RPM) at 37 °C or 42 °C (*see* **Note 1**).
2. For bacteria growth on plates, use a stove at 37 °C.
3. For growth at 42 °C, use a stove or a water bath.
4. To dry plates, use a stove at 55 °C.
5. To measure Optical Density (OD) of bacterial cultures, use a spectrophotometer set at 600 nm and appropriate cuvettes.

2.3 Small Equipment

1. Bacteria growth in sterile glass erlens of 50 ml with stopper.
2. Sterile loop to inoculate liquid media.
3. Sterile eppendorfs of 1.5 ml for bacteria culture before plating.
4. A vortex to resuspend bacteria in eppendorf tubes.
5. Round sterile petri dishes.
6. Sterile tips and Pipetman are used to sample liquid cultures.

3 Methods

Carry out all procedures at room temperature, in sterile conditions, unless otherwise specified.

3.1 Precultures

1. Inoculate 5 ml of sterile liquid LB broth containing the appropriate antibiotic with donor (which contains the plasmid to be mobilized), helper (which contains the helper plasmid), and recipient strains (*see* **Note 2**).
2. Incubate overnight at 37 °C in a shaking incubator.

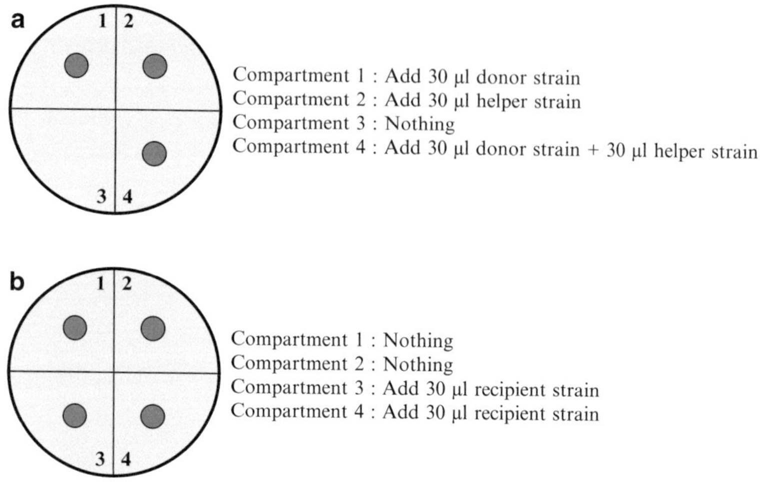

Fig. 1 How to prepare a conjugation plate assay. (**a**) Mobilization step (*t*). (**b**) Conjugation step (*t*+ 2 h)

3.2 Mobilization (See Note 3)

1. Dry a LB plate at 55 °C for 1 h. Divide it into 4 compartments with a pen. The first three compartments serve as controls, while the fourth one is for the conjugation itself.
2. Spot 30 μl of the donor preculture in the first compartment, and 30 μl of the helper preculture in the second one. Mix 30 μl of donor and 30 μl of helper precultures in the fourth compartment (*see* **Note 4**). An example of LB plate is represented in Fig. 1a.
3. Carefully place the plate at 37 °C without shaking during 2 h.
4. At the same time, incubate the recipient preculture at 42 °C without shaking during 2 h (*see* **Notes 5** and **6**).

3.3 Conjugation

5. After 2 h, spot 30 μl of the recipient in the fourth compartment (containing the donor and the helper) to initiate the conjugation (Fig. 1b). Place the LB plate at 37 °C without shaking during 4 h (*see* **Note 7**).
   Spot also 30 μl of the recipient preculture in the third compartment (control).

3.4 Selection

6. Prepare 4 sterile eppendorfs with 400 μl of sterile liquid LB medium. Resuspend each spot with a loop and vortex vigorously for 2–5 s.
7. Place the eppendorfs at 37 °C with shaking during 1 h (*see* **Note 8**).
8. At the same time, dry the selection plates at 55 °C during 1 h (*see* **Note 9**).
9. Spread out 200 μl of each culture (corresponding to each of the four spots) onto selection plates and incubate them at

37 °C without shaking overnight. At this time, there are four plates: three negative controls and one conjugation plate (*see* **Note 10**).

10. 24 h after plating, check for the absence of colonies on the three negative control plates. Select 2–8 colonies on the conjugation plate and isolate them on isolation plates (*see* **Notes 11** and **12**).

4 Notes

1. To-and-fro motion incubators allow for better growth of *P. aeruginosa*.
2. For conjugal transfer, plasmids must contain an origin of transfer, either the *mob* region from RSF1010 (pBBRMCS series, pJN105) or the *oriT* from R6K (pMMB67, pMM190). Those mobilizable plasmids are not capable of self-transfer and rely on mobilization functions provided in *trans*. Biparental mating involves some *Escherichia coli* strains which posses RP4 transfer genes on the chromosome (e.g., *E. coli* SM10 strain, 2). In triparental mating, the transfer functions are provided by a helper plasmid (e.g., pRK2013 helper plasmid carrying the *tra* genes on RK2, 3).
3. In case of biparental mating, the protocol begins at Subheading 3.2, **step 4**.
4. The preculture can be spotted on a sterile filter (1 cm diameter filters from Millipore) instead of directly on the agar.
5. The recipient bacteria are placed at 42 °C to inhibit DNA restriction-modification system that can degrade foreign DNA. In case of low efficiency of conjugation, this step can be realized at 43 °C. However, this condition may induce more mutations in the genome of the recipient strain, so we recommend trying first at 42 °C.
6. Particular case of the MiniCTX plasmids: we find that chromosomal integration of MiniCTX is more efficient when the preculture of the recipient strain is done overnight at 42 °C with shaking, instead of 37 °C. However, this condition may induce mutations in the genome of the recipient strain, so we recommend trying first the preculture at 37 °C and 2 h at 42 °C the day of the conjugation.
7. This step can be extended to 6 h if necessary, and sometimes overnight in some particular cases (e.g., pJN105 vector).
8. This step is required for the expression of the selection marker genes (e.g., antibiotic resistance gene) in the recipient bacteria before the selection step.

   If the gene of interest is cloned on the plasmid under the control of a pLac or a pBAD promoter and, if the coded protein may be toxic for the bacteria, we recommend adding glucose 0.4 % to avoid expression of this gene during this step.

9. Selection media should allow only the growth of conjugants (the recipient bacteria that have received the mobilizable plasmid), whereas the donor, the helper, and the recipient bacteria must not grow. Controls have been done to check this.
10. If there is a lawn of bacteria on the conjugation plate, we recommend diluting the culture before spreading out on the selection plate.
11. If there are colonies on any of the negative controls, first check that media used were sterile. If it is the case, we recommend increasing the antibiotic concentration of on selection plates.

    If there are no colonies after 48 h, ensure that antibiotic concentration is not too high, otherwise no conjugant will grow. Even after modulating antibiotic concentration if no conjugant grows, a selection on LB plates instead of PIA plates can be performed.
12. Presence of the plasmid in conjugants can be checked by colony PCR with specific primers.

References

1. Zechner EL, Lang S, Schildbach JF (2012) Assembly and mechanisms of bacterial type IV secretion machines. Philos Trans R Soc Lond B Biol Sci 367:1073–1087
2. de Lorenzo V, Timmis KN (1994) Analysis and construction of stable phenotypes in gram-negative bacteria with Tn*5*- and Tn*10*-derived minitransposons. Methods Enzymol 235: 386–405
3. Figurski DH, Helinski DR (1979) Replication of an origin-containing derivative of plasmid RK2 dependent on a plasmid function provided *in trans*. Proc Natl Acad Sci USA 76: 1648–1652

Chapter 4

Pseudomonas Bacteriophage Isolation and Production

Joana Azeredo, Sanna Sillankorva, and Diana P. Pires

Abstract

Bacterial viruses or bacteriophages were discovered nearly 100 years ago and are ubiquitous in nature, readily isolated for a variety of bacterial hosts and from a diversity of sources. Here, we describe the methods used to isolate, concentrate, purify, and produce bacteriophages specific for *Pseudomonas* species.

Key words Bacteriophage, Enrichment, Isolation, Titration, Production

1 Introduction

Bacteriophages are bacterial viruses hypothesized to exist wherever their hosts thrive. The vast majority of known bacteriophages, including those targeting pseudomonads, belong to the *Caudovirales* order comprising three families of dsDNA bacteriophages differing in their tail characteristics: long and non-contractile (*Siphoviridae*), long and contractile (*Myoviridae*), and short and non-contractile (*Podoviridae*) [1]. Bacteriophages can have lytic or lysogenic life cycles. Lytic or virulent bacteriophages reproduce within the host and induce lysis resulting in the release of progeny bacteriophages that start another round of infection. Bacteriophages with lysogenic growth (temperate bacteriophages), integrate their prophage into the chromosome or other replicon of the host bacteria which, through cell division, pass the bacteriophage genome (prophage) to all daughter cells [2]. When plated, clear plaques will be formed on the lawns of hosts completely susceptible to the bacteriophages and turbid plaques will arise when hosts are partially resistant to the bacteriophages. The plaque formed can vary in size from barely visible to several millimeters in diameter, depending mainly of some of the bacteriophage characteristics (e.g., adsorption efficiency, latent period, and burst size), the host physiological state, and the growth conditions. Furthermore, plaques can be surrounded by a halo which is representative of the presence of genes encoding for

Alain Filloux and Juan-Luis Ramos (eds.), *Pseudomonas Methods and Protocols:* Methods in Molecular Biology, vol. 1149, DOI 10.1007/978-1-4939-0473-0_4, © Springer Science+Business Media New York 2014 extracellular polymeric substances (EPS) depolymerase within the bacteriophage genome.

Bacteriophages specific for *Pseudomonas* were first described in mid-twentieth century [3, 4] and, to date, are known for only little over a dozen of the more than 230 described *Pseudomonas* species [5]. Bacteriophages specific for this bacterium have been isolated from a wide range of sources including sewage (hospital, wastewater treatment plant) [6–8]; soil [9]; urine [10], and even sporocarps of oyster mushroom [11], and some of their applications described in literature include typing of pseudomonads [12], in vitro efficacy tests [13], and in vivo trials (e.g., application on patients with chronic otitis and venous leg ulcers) [14, 15], among others. Due to its pathogenicity, bacteriophages targeting *P. aeruginosa* have gathered most attention and to date, over 70 % of the pseudomonad bacteriophages sequenced are specific for this species.

In this Chapter we describe all necessary steps involved in the isolation and production of *Pseudomonas* bacteriophages.

2 Materials

Prepare all solutions using distilled water and store at room temperature (unless indicated otherwise). The media used in the procedures described herein is Tryptic Soy Broth (TSB), but alternative media can be used (*see* **Note 1**).

2.1 Bacteriophage Enrichment from Collected Samples

1. Samples for bacteriophage isolation: collect solid samples (e.g., soil, foods) and place in sterile jars and/or liquid samples (e.g., sewage, water) in sterile 1 L bottles.
2. Sterile 50 mL Falcon tubes.
3. Sterile double strength TSB (2× TSB): weigh 48.0 g of commercially available TSB and place in a 1 L bottle and add 800 mL of distilled water. Autoclave at 121 °C for 15 min.
4. Overnight grown bacteria in 100 mL Erlenmeyers containing 25 mL of sterile TSB prepared as follows: Add 24.0 g of commercially available TSB and dissolve in 800 mL of water. Autoclave at 121 °C for 15 min.
5. Sterile 100, and 500 mL bottles.
6. Sterile 500 mL Erlenmeyers.
7. Filters 0.45 μm (Whatman™, USA).

2.2 Preparation of Bacterial Lawns by Pour-Plate Technique

1. Agar plates containing TSB: Add 24.0 g of commercially available TSB and agar 1.2–1.5 % (w/v) in 800 mL of distilled water (*see* **Note 2**). Autoclave at 121 °C for 15 min and pour onto petri dishes.

2. Overnight grown bacteria in 100 mL Erlenmeyers containing 25 mL of sterile TSB.
3. Sterile Molten Top-Agar (MTA): Weigh 15.0 g of TSB, 3.0 g of agar and place in a 500 mL bottle. Adjust to 500 mL with distilled water (*see* **Note 3**). Autoclave at 121 °C for 15 min and store accordingly (*see* **Note 4**).
4. Sterile 15 mL test tubes.

2.3 Bacteriophage Isolation

1. Overnight grown bacteria in 100 mL Erlenmeyers containing 25 mL of sterile TSB, agar plates containing TSB, and sterile MTA (47 °C) for bacterial lawn preparation.
2. Sterile paper strips: Cut a letter size paper sheet (80 g per m^2) in small, approximately 1 cm × 5 cm, rectangles and autoclave at 121 °C for 15 min.
3. Sterile toothpicks.

2.4 Bacteriophage Production

1. Overnight grown bacteria in 100 mL Erlenmeyers, containing 25 mL of sterile TSB.
2. Agar plates with TSB or 100 cm^2 tissue culture flasks containing an agar layer of TSB.
3. Sterile MTA (47 °C).
4. Sterile Saline Magnesium buffer (SM buffer): Prepare 1 M Tris–HCl buffer (pH 7.5) in a 100 mL bottle—Weigh 6.06 g of Tris-Base, add 50 mL of water and adjust the pH of the buffer with HCl to 7.5. Then, to a 1 L bottle, add 5.8 g of NaCl, 2.0 g of $MgSO_4 \cdot 7H_2O$, and 50 mL of the prepared 1 M Tris–HCl (pH 7.5), and make up to 1 L with water. Autoclave at 121 °C for 15 min (*see* **Note 5**).
5. Sterile 50 mL Falcon tubes.
6. Sterile 250 mL Erlenmeyers.
7. Sterile 50, 100, 200 mL bottles.
8. Sterile paper strips (*see* Subheading 2.3).
9. Filters 0.2 μm (Whatman™, USA).
10. Sterile toothpicks.

2.5 Bacteriophage Titration

1. Sterile SM buffer.
2. Sterile MTA (47 °C).
3. Sterile 1.5 mL tubes (*see* **Note 6**).
4. Sterile 15 mL test tubes

2.6 Bacteriophage Concentration and Purification

1. Solid NaCl.
2. Solid Polyethylene Glycol 8000 (PEG 8000).
3. Chloroform (≥99.8 %).

4. Sterile 250 mL bottles.
5. Sterile 500 mL Erlenmeyers.
6. Sterile 15 and 50 mL Falcon tubes.

3 Methods

Carry out all procedures at room temperature unless otherwise specified.

3.1 Bacteriophage Enrichment

1a. For liquid samples (100 mL), pour the sample into two 50 mL Falcon tubes and centrifuge (9,000 × *g*, 10 min, 4 °C).

1b. For solid samples, add to a 500 mL bottle, 100 mL of saline solution and 10 g of solid sample, mix thoroughly, and incubate for at least 1 h at room temperature. Then, pour onto 50 mL Falcon tubes and centrifuge (9,000 × *g*, 10 min, 4 °C).

2. Collect and filter (0.45 μm) the supernatant to 500 mL sterile Erlenmeyers.
3. Add 50 μL of overnight grown bacterial suspension and 100 mL of 2× TSB to the Erlenmeyer containing the filtrated supernatant (*see* **Note 7**).
4. Incubate at the proper growth temperature, under agitation (120–180 rpm), for 24 up to 48 h.
5. Pour the enriched sample onto 50 mL Falcon tubes and centrifuge (9,000 × *g*, 4 °C for 10 min).
6. Collect and filter (syringe filter 0.2 μm) the supernatant to sterile 100 mL bottles.

3.2 Preparation of Bacterial Lawns by Pour-Plate Technique

1. Add to a 15 mL test tube 100 μL of overnight grown bacterial suspension and 3–5 mL of MTA (47 °C) and tap gently.
2. Pour onto an agar plate with TSB and swirl gently.
3. Let the plates dry for 1–2 min.

3.3 Bacteriophage Isolation

3.3.1 Spot Test Verification of the Enriched Samples

1. Add 1–4 drop(s) of 10 μL of the filtrated sample (Subheading 3.1, **step 6**) on a bacterial lawn (prepared as described in Subheading 3.2) of the strain used in the enrichment.
2. Let the plate stand until the drop(s) have completely dried (*see* **Note 8**).
3. Incubate the plate overnight at the proper growth temperature.
4. Check for clear and turbid lysis zones indicative of the presence of bacteriophages.

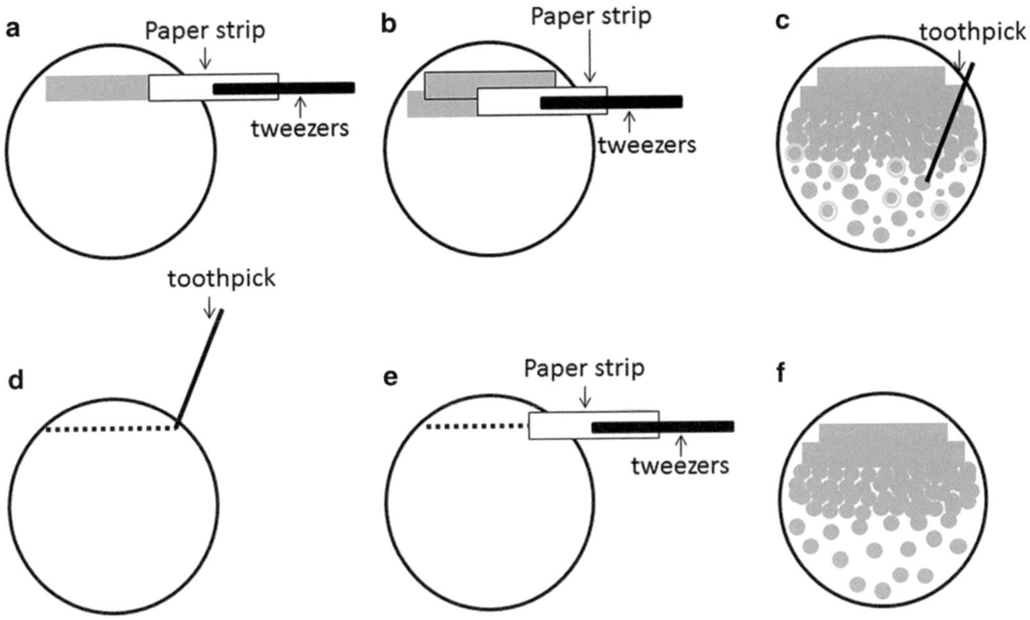

Fig. 1 Isolation of bacteriophages. (**a**) initial streak; (**b**) streaking downwards on a bacterial lawn; (**c**) isolated bacteriophage plaques and picking a single colony with a toothpick; (**d**) making a puncture line with a toothpick; (**e**) streaking with paper strips; (**f**) morphologically identical plaques

3.3.2 Bacteriophage Plaque Isolation [16]

Positive results from Subheading 3.3.1, **step 3** need to be investigated for the presence and further isolation of different bacteriophages in the enriched samples using the procedure as follows:

1. Wet the tip of a sterile paper strip (*see* **Note 9**) in the bacteriophage suspension obtained in Subheading 3.1, **step 6**.
2. Streak once on a petri dish containing a bacterial lawn (prepared as described in Subheading 3.2) of the strain used in the enrichment step (Fig. 1a).
3. Streak downwards, changing the paper strip after every streak, making certain that the paper strip touches the previous streak (Fig. 1b).
4. Incubate the plate overnight at optimum temperature for bacterial growth.
5. Analyze the bacteriophage plaque morphologies to check for differences in size, presence of halo, and turbidity (examples of bacteriophage plaques with varying morphology in Fig. 1c).
6. Pick different plaques with a toothpick and stick the toothpick several times (in a line), in an agar plate with a bacterial lawn prepared as described in Subheading 3.2 (Fig. 1d).

7. Use sterile paper strips to streak the bacteriophages as described above (Subheading 3.3.2, **step 3**) (Fig. 1e).
8. Incubate the plates with different plaque morphologies overnight at the proper temperature and repeat **steps 6–7** until all bacteriophage plaques are uniform (Fig. 1f).

3.4 Preparation of Bacteriophage Stocks and Lysates

3.4.1 Preparation of Bacteriophage Stocks from Single Bacteriophage Plaques Using the Plate Lysis and Elution Procedure

1. Pick with a toothpick a bacteriophage plaque from Subheading 3.3.2, **step 8** and stick it on 1–20 agar plate(s) containing a bacterial lawn prepared as described previously (Subheading 3.2).
2. Without changing the paper strip, pass it through the punctures made, and spread evenly to ensure maximum coverage (maximum replication) by the bacteriophages.
3. Incubate overnight at the proper temperature.
4. Add 3–5 mL of SM buffer to each plate(s) used.
5. Incubate with agitation (50–90 rpm) at 4 °C for 5–18 h.
6. Collect the SM buffer with the eluted bacteriophages to a 50 mL Falcon tubes (*see* **Note 10**) and discard the agar plate.
7. Centrifuge the solution (9,000 ×*g*, 4 °C, and 10 min) to remove all bacteria.
8. Collect and filter (0.2 μm) the supernatant to 100 mL bottles.
9. Store at 4 °C until needed.

3.4.2 Preparation of Bacteriophage Lysate Using a Bacterial Culture

1. Dilute the bacteriophage stock solution to have a concentration of approximately 1×10^5 plaque forming units (PFU) per mL.
2. To a 250 mL Erlenmeyer, add 1–5 mL of the diluted bacteriophage and 25 mL of an overnight grown culture (*see* **Note 11**)
3. Incubate at proper temperature and under agitation (150–200 rpm).
4. When lysates clear (4–6 h), transfer to 50 mL Falcon tubes and centrifuge (9,000 ×*g*, 4 °C, 10 min).
5. Collect and filter (0.2 μm) the supernatant to 100 mL bottles.
6. Store at 4 °C until needed.

3.4.3 Preparation of Lysate Using the Soft-Agar Overlay Technique

1. Add 1 mL of an overnight grown culture and 1 mL of diluted bacteriophage (approximate titre 1×10^5 PFU per mL) in 1–5 tissue culture flask(s) (100 cm^2) containing an agar layer of TSB, and mix gently.
2. Incubate for 15 min at the proper temperature to allow bacteriophages to adsorb to the host bacterium.
3. Add 30 mL of MTA (47 °C) and leave the tissue culture flask(s) still until the MTA has hardened.
4. Incubate overnight, without inverting, at the proper temperature.

5. Add 30 mL of SM buffer to the tissue culture flask(s) (100 cm^2) and incubate overnight at 4 °C.
6. Remove the SM buffer with the eluted bacteriophages to 50 mL Falcon tubes.
7. Centrifuge (9,000 ×*g*, 4 °C, and 10 min), carefully collect the supernatant, and filter (0.2 μm) to sterile 50 mL (1 tissue culture flask used) or 200 mL (5 tissue culture flasks used) bottles.
8. Store at 4 °C until further use.

3.5 Bacteriophage Titration

3.5.1 Bacteriophage Enumeration by Double Agar Overlay

1. Prepare successive serial dilutions (1:10) in SM buffer of the bacteriophage solutions produced in Subheading 3.4 (add 100 μL of bacteriophage solution and 900 μL of SM buffer to a 1.5 mL eppendorfs, or 20 μL of bacteriophage solution and 180 μL of SM buffer if performing the dilutions in 96-well microplates).
2. Add to a test tube 100 μL of bacteriophage solution, 100 μL of overnight grown bacteria, and 3–5 mL of MTA (47 °C) and tap gently.
3. Pour it on an agar plate with TSB and swirl carefully.
4. Let the plates dry for 1–2 min.
5. Incubate overnight under optimal conditions.
6. Count the bacteriophage plaques in the dilution which resulted in 20–200 plaques.
7. Determine the titre of triplicate preparations according to Eq. 1.

$$\text{Bacteriophage titer}\left(\text{PFU per mL}\right)=\frac{\text{Nr. of bacteriophage plaques formed}\times\text{Dilution factor}}{\text{Volume of bacteriophage sample}\left(\text{mL}\right)} \quad (1)$$

3.5.2 Bacteriophage Enumeration Using the Small Drop Plaque Assay [16]

This procedure can only be used with bacteriophages forming large plaques (>2 mm).

1. Serially dilute the bacteriophage sample in eppendorfs or 96-well microplates (*see* Subheading 3.5.1, **step 1**).
2. Add 20 μL of diluted bacteriophage solution to 20 μL of an overnight grown bacterial culture in eppendorfs or 96-well microplates.
3. Incubate for 15 min at the proper growth temperature to allow bacteriophage adsorption to the bacteria.
4. Add a drop of 20 μL of the dilution mixture onto an agar plate with TSB.
5. Let the plates stand still until the drops have completely dried.
6. Incubate overnight at the proper growth temperature.

7. Count the plaques formed in the drop of the dilution with 3–30 bacteriophage plaques.
8. Calculate the titre of the bacteriophage of triplicate preparations using the Eq. 1.

3.6 Concentration and Purification of Bacteriophage Lysates

This protocol was adapted from Sambrook and Russell [17]; however, it includes some minor modifications and does not include an ultracentrifugation step.

1. To a 250 mL bottle, add 100 mL of bacteriophage lysate, and DNase I and RNase A (1 μg/mL each).
2. Incubate for 30 min at room temperature.
3. Add 5.84 g of NaCl.
4. Incubate at 4 °C or on ice under agitation (50–90 rpm) for 1 h.
5. Pour onto 50 mL Falcon tubes and centrifuge (9,000 × *g*, 4 °C, 10 min).
6. Collect the supernatant to a 500 mL Erlenmeyer.
7. Add PEG to the supernatant (10.0 g PEG per 100 mL solution).
8. Incubate 5 h to overnight at 4 °C under gentle agitation (50–90 rpm).
9. Pour onto 50 mL Falcon tubes and centrifuge (9,000 × *g*, 4 °C, 10 min).
10. Discard the supernatant and resuspend the pellet, containing the precipitated bacteriophage particles, in SM buffer containing 2 % gelatin. Use 6 mL of SM buffer for each 50 mL of centrifuged sample.
11. Transfer the bacteriophage solution to 15 mL Falcon tubes and add chloroform (*see* **Note 12**) in a proportion of 1:4 (v/v). Vortex for 30 s.
12. Centrifuge (3,500 × *g*, 4 °C, 5 min).
13. Recover and filter (0.2 μm) the aqueous phase (upper phase) which contains the bacteriophage to 15 or 50 mL Falcon tubes and store at 4 °C.
14. Determine the bacteriophage titre as described previously in Subheading 3.5 (*see* **Note 13**).

4 Notes

1. TSB is a commonly used media for *Pseudomonas* growth; however, other media such as LB or NB can be used instead. TSB is commercially available otherwise prepare as follows: To 1 L of distilled water add 17.0 g of Enzymatic Digest of Casein; 3.0 g of Enzymatic Digest of Soybean Meal; 5.0 g of Sodium

Chloride; 2.5 g of Dipotassium Phosphate; 2.5 g of Dextrose and adjust the pH to have a final concentration of 7.3 ± 0.2 at 25 °C.

2. Alternatively, commercially available TSA which corresponds to TSB plus agar can be used according to the manufacturer's instructions.
3. MTA is generally prepared with 0.6 % of agar; however, other agar percentages such as 0.4–0.7 % can be used. Alternatively, MTA can be prepared with agarose instead of agar.
4. MTA can be stored at 47 °C if used within 1 day or at 4–21 °C. Solid MTA can be melted using a water bath or a microwave oven.
5. Optional: 2 % of gelatin (w/v) can be added to SM buffer. Gelatin is known to preserve bacteriophages and thus can be used in the later steps of bacteriophage purification.
6. In case several different samples are handled, sterile 96-well plates can be used to perform serial dilutions.
7. A loopful of freshly grown host bacterium can be picked from agar plates and suspended in saline solution (0.9 % NaCl) and used instead of overnight grown bacterial suspension.
8. Several different bacteriophage enrichment samples, from different sources and origins, can be spotted on one bacterial lawn.
9. The paper strip has the same functionality for streaking bacteriophages as the inoculating loops have for streaking colonies.
10. The MTA layer can also be collected.
11. For some bacteriophages it is preferable to use cells at a mid-exponential phase instead of overnight cultures.
12. Chloroform is used to extract PEG; however, it should be used with caution as some bacteriophages may be sensitive to it.
13. The titer obtained at the end of the concentration and purification step varies with the burst size of the bacteriophage. For instance, bacteriophages with high burst sizes (e.g., *T7likeviruses*) result in titers above 10^{10} PFU per mL; however, phages with lower burst sizes (e.g., those of the *Myoviridae* family) result in average only in 10^{9} PFU per mL.

References

1. Ackermann HW (2009) Phage classification and characterization. Methods Mol Biol 501: 127–140
2. Summers WC (2012) The strange history of phage therapy. Bacteriophage 2(2):130–133
3. Kellenberg G, Kellenberg E (1957) Electron microscopical studies of phage multiplication. III. Observation os single cell bursts. Virology 3:275–285
4. Holloway BW, Egan JB, Monk M (1960) Lysogeny in Pseudomonas aeruginosa. Aust J Exp Biol Med Sci 38: 321–330
5. Ceyssens PJ, Lavigne R (2010) Bacteriophages of Pseudomonas. Future Microbiol 5(7): 1041–1055
6. Pires D et al (2011) Use of newly isolated phages for control of Pseudomonas aeruginosa

PAO1 and ATCC 10145 biofilms. Res Microbiol 162(8):798–806

7. Sillankorva S, Kropinski AM, Azeredo J (2012) Genome sequence of the broad-host-range Pseudomonas phage Phi-S1. J Virol 86(18): 10239
8. Monson R et al (2011) The Pseudomonas aeruginosa generalized transducing phage phiPA3 is a new member of the phiKZ-like group of "jumbo" phages, and infects model laboratory strains and clinical isolates from cystic fibrosis patients. Microbiology 157: 859–867
9. Adriaenssens EM et al (2012) Complete genome sequence of the giant Pseudomonas phage Lu11. J Virol 86(11):6369–6370
10. Golkar Z, Jamil N (2010) Hepatitis C virus, human immunodeficiency virus and Pseudomonas phage PS5 triad share epitopes of immunogenic determinants. Virol J 7:346
11. Sajben-Nagy E et al (2012) Isolation of new Pseudomonas tolaasii bacteriophages and genomic investigation of the lytic phage BF7. FEMS Microbiol Lett 332(2): 162–169
12. Gould JC, McLeod JW (1960) A study of the use of agglutination sera and phage lysis in the classification of strains of Pseudomonas aeruginosa. J Pathol Bacteriol 79:295–311
13. Sillankorva S, Neubauer P, Azeredo J (2010) Phage control of dual species biofilms of Pseudomonas fluorescens and Staphylococcus lentus. Biofouling 26(5):567–575
14. Rhoads DD et al (2009) Bacteriophage therapy of venous leg ulcers in humans: results of a phase I safety trial. J Wound Care 18:237–238
15. Wright A et al (2009) A controlled clinical trial of a therapeutic bacteriophage preparation in chronic otitis due to antibiotic-resistant Pseudomonas aeruginosa; a preliminary report of efficacy. Clin Otolaryngol 34:349–357
16. Mazzocco A, Waddell TE, Lingohr E, Johnson RP (2008) Enumeration of Bacteriophages Using the Small Drop Plaque Assay System. Methods Mol Biol 501:81–85. doi:10.1007/978-1-60327-164-6_9
17. Sambrook J, Russell DW (2001) Molecular cloning—a laboratory manual, 3rd edn. Cold Spring Harbor Laboratory, Cold Spring Harbor

Chapter 5

Genotyping Methods

Burkhard Tümmler

Abstract

Genotyping allows for the identification of bacterial isolates to the strain level and provides basic information about the evolutionary biology, population biology, taxonomy, ecology, and genetics of bacteria. Depending on the underlying question and available resources, *Pseudomonas aeruginosa* strains may be typed by anonymous fingerprinting techniques or electronically portable sequence-based typing methods such as multiple locus variable number tandem repeat (VNTR) analysis (MLVA), multilocus sequence typing, or oligonucleotide microarray.

Macrorestriction fragment pattern analysis is a genotyping method that is globally applicable to all bacteria and hence has been and still is the reference method for strain typing in bacteriology. Agarose-embedded chromosomal DNA is cleaved with a rare-cutting restriction endonuclease and the generated 20–70 fragments are then separated by pulsed-field gel electrophoresis. The chapter provides a detailed step-by-step manual for SpeI genome fingerprinting of *Pseudomonas* chromosomes that has been optimized for SpeI fragment pattern analysis of *P. aeruginosa*.

Key words *Pseudomonas aeruginosa*, Genotyping, Random amplified polymorphic DNA, RAPD, Pulsed-field gel electrophoresis, PFGE, Genome fingerprint, Multiple locus variable number tandem repeat analysis, VNTR, MLVA, Multilocus sequence typing, MLST, Whole-genome sequencing

1 Introduction

The identification of bacterial isolates to the strain level is often necessary for studies in ecology, epidemiology, taxonomy, or biotechnology. Typical applications are the control of the identity of strains deposited in culture collections or used in biotechnology. The population structure, natural history, and microevolution of taxa may be tracked by genotyping of strain collections. The molecular epidemiology, surveillance, and control of infections are the major themes in medicine where genotyping of bacterial isolates has become an indispensible tool. Within the Pseudomonas field genotyping has predominantly been utilized to monitor the epidemiology and outbreaks of infections with the opportunistic pathogen *Pseudomonas aeruginosa*.

Alain Filloux and Juan-Luis Ramos (eds.), *Pseudomonas Methods and Protocols*, Methods in Molecular Biology, vol. 1149, DOI 10.1007/978-1-4939-0473-0_5, © Springer Science+Business Media New York 2014

The currently used genotyping methods are either anonymous fingerprinting techniques or sequence-based typing methods. The underlying principles are described in the following paragraphs.

1.1 Anonymous Fingerprinting Techniques

The related methods Arbitrarily Polymerase Chain Reaction (AP-PCR) and Random Amplified Polymorphic DNA (RAPD) are based on the observation that PCR performed at relatively low stringency (in the case of AP-PCR) or with low selectivity primers at high stringency (in the case of RAPD) yield a collection of products that depend on the template and primer sequences. Arrayed on an agarose or polyacrylamide gel, this collection of products can be viewed as a "fingerprint."

The number and size of products generated by AP–PCR are affected by numerous factors such as Mg^{2+} concentration, nucleotide concentration, primer and DNA template concentrations, and the physical conditions of amplification [1]. Consequently, this method is only applicable for the comparative genotyping of *Pseudomonas* strains that are processed in parallel under standardized conditions at a single site [1].

In the case of RAPD a collection of oligonucleotides is first tested to determine whether PCR amplification generates stable and reproducible patterns from the taxon of interest. Eight of 100 decamers, for example, were found to amplify reproducible polymorphisms suitable for strain differentiation of *P. aeruginosa* [2]. RAPD has turned out to be a robust and rapid technique for strain genotyping. Its ability to detect unique and clonal strains in a collection of *P. aeruginosa* strains has been determined to be about 80 % and 70 %, respectively [3]. The relatively low predictive value of RAPD to differentiate between clonally related and unrelated strains infers that RAPD cannot be recommended to study the population genetics of pseudomonads.

Macrorestriction fragment pattern analysis is a genotyping method that is globally applicable to all bacteria and hence has been and still is the reference method for strain typing in bacteriology [4]. Agarose-embedded chromosomal DNA is cleaved with a rare-cutting restriction endonuclease and the generated 20–70 fragments are then separated by pulsed-field gel electrophoresis (PFGE). Since the recognition sites of most restriction endonucleases are randomly distributed in the chromosome, fragment patterns represent unbiased genome fingerprints and can thus be exploited to assign clonal relationships between strains. SpeI is the optimal enzyme for *Pseudomonas* chromosomes to produce informative fingerprints of multiple nonoverlapping fragments [5]. SpeI fragment patterns of completely sequenced *Pseudomonas* genomes serve as size markers. Minor differences between fragment band patterns are typical for members of the same clonal complex. These shifts are not caused by de novo point mutations in SpeI recognition sites which only occur every 10–20 years in a non-mutator Pseudomonas clone but represent genome rearrangements by insertions, deletions or inversions [6]. The accessory genome of pseudomonads is highly dynamic in the uptake and loss of mobile genetic elements such as phages, transposons, and genomic islands which manifests in changes of SpeI fragment length [7].

1.2 Sequence-Based Typing Methods

Anonymous fingerprinting techniques are poorly portable because they index variation that is difficult to compare among laboratories. In contrast, sequence-based typing methods generate data that are electronically portable and can be reliably compared using online databases.

Multiple locus variable number tandem repeat (VNTR) analysis (MLVA) is a PCR-based typing method that relies on the inherent variability found in some regions of repetitive DNA. Typing schemes of different marker sets have been developed for *P. aeruginosa* [8] and *P. syringae* [9]. Searches of the completely sequenced *P. aeruginosa* PAO1 and PA14 genomes identified polymorphic tandem repeats, each of which having 4–12 alleles with different numbers of repeats. The repeat length varies between 6 and 129 bp. The amplicons of each VNTR locus are generated by monoplex [8] or—more recently—dye-colored multiplex [10] PCR and separated by agarose or capillary electrophoresis. Genotyping of strains by MLVA was consistent with that by macrorestriction fragment pattern analysis. The low level of congruence between various MLVA typing schemes, however, indicates that the MLVA method is not suitable to infer clustering or evolutionary relationships, at least in *P. aeruginosa* [10].

Multilocus sequence typing (MLST) utilizes nucleotide sequence data of internal fragments of, typically, seven housekeeping genes [11]. MLST is a portable, universal and definitive method for the genotyping of bacterial strains. The target genes are amplified by PCR and the amplicons are sequenced. To allow high-throughput multiplex sequencing of numerous bacterial isolates at the same time by next generation sequencing technology, the amplicons of each strain are provided with a unique barcode [12]. Within the Pseudomonas community accepted MLST schemes exist for the genotyping of *P. syringae* [13, 14] and *P. aeruginosa* [15] strains. MLST and PFGE have a comparably high predictive value and ability to detect unique and clonally related *P. aeruginosa* strains [3, 16, 17].

MLST only scans the genetic diversity of the core genome and relies on the comparably slow DNA-sequencing technology of purified PCR products, which still cannot be performed under point-of-care or rural field conditions. Recently a robust and rapid oligonucleotide microarray has been developed that types *P. aeruginosa* strains in both the conserved core and the flexible accessory genome [18]. DNA, amplified from the bacterial colony by cycles of multiplex primer extension reactions, is hybridized onto a microarray to yield an electronically portable binary multimarker genotype that represents the core genome by single nucleotide polymorphisms (SNPs) and the accessory genome by markers of genomic islets and islands. The hexadecimal code of the SNP genotypes assigns the strain to a clonal complex. This classification has meanwhile been validated by whole-genome sequencing [7, 19, 20]. Strains sharing the same hexadecimal code that were isolated from unrelated sources differ by some dozen to a few hundred SNPs from each other, whereas sequence variation between strains with unequal hexadecimal code ranges from 10,000 to 100,000 SNPs.

The whole genome represents the ultimate genotype of a strain. Since the already existing sequencing technologies HiSeq and SOLiD5500 allow the simultaneous sequencing of more than a hundred *Pseudomonas* strains with an average 50-fold genome coverage in one run [21], whole-genome genotyping becomes an attractive alternative to the typing methods mentioned above. Current bottlenecks to make whole-genome typing a routine procedure are the comparably tedious protocols to prepare the bar-coded genome libraries and the challenges to build up a pipeline for the rapid and reliable processing and evaluation of the generated sequence reads.

Table 1 gives an overview of the pros and cons of the typing methods presented in this Introduction.

The author's laboratory developed two methods for the genotyping of *Pseudomonas* bacteria, the globally applicable macrorestriction fragment pattern analysis [4] and the genotyping of *P. aeruginosa* by low-resolution multimarker microarray [18]. To learn about genotyping by microarray, the reader is referred to a 24-page step-by-step manual that is freely available for download as a pdf file at the Web site of the sole supplier (http://alere-technologies.com/en/products/lab-solutions/ps-aeruginosa.html). The manual contains a hands-on protocol for genotyping, provides hints for troubleshooting, describes the image analysis and guides the user of how SNP-patterns obtained from the microarray are converted to a hexadecimal code. Further open access protocols and practical hints are provided in the Supporting Information of ref. 18.

Easily accessible step-by-step manuals exist for genotyping by microarray, but not for genome fingerprinting by PFGE. Hence, a hands-on protocol for the genotyping of *Pseudomonas* strains by macrorestriction fragment pattern analysis [4] is provided below.

2 Materials

All buffers should be autoclaved before use unless otherwise stated.

1. Broth for growing *Pseudomonas* bacteria. Tryptone broth: 10 g casein hydrolysate, 5 g NaCl, add 1 L H_2O, adjust to pH 7.5.

Table 1
Features of current genotyping methods

Topic	AP-PCR	RAPD	PFGE (SpeI fingerprints)	MLVA	MLST	GeneChip	Whole-genome sequencing
Performance							
Ability to detect unique strains	Low	High	High	High	High	High	100 %
Predictive value thereof	Low	Moderate	High	High	High	High	100 %
Ability to detect clonal strains	Poor	Moderate	High	Low	High	High	100 %
Predictive value thereof	Poor	Low	High	Low	High	High	100 %
Portable	No	No	Difficult	Yes	Yes	Yes	Yes
Electronically portable	No	No	Difficult	Yes	Yes	Yes	Yes
Online database available	No	No	No	Yes	Yes	Yes	Yes
Applicability							
Point-of-care analysis	No	Perhaps	No	No	No	Yes	No
Local infection control	Yes	Yes	Yes	Yes	Yes	Yes	Yes
Domestic or international surveillance	No	No	No	(Yes)	Yes	Yes	Yes
Field study, microbial ecology	No	No	Yes	No	Yes	Yes	Yes
Microevolution	No	No	Yes	No	No	Yes	Yes
Population structure	No	No	Yes	No	Yes	Yes	Yes
Features of method							
Robustness	Poor	Low	Moderate	High	High	High	High
Turnover of specimens	High	High	Moderate	Moderate	Moderate	High	Low
Time spent per assay	½ day	½ day	3–7 day	1–2 day	2–7 day	2–4 h	Weeks
Requested skills							
Manual skills	Unimportant	Unimportant	Necessary	Desirable	Unimportant	Unimportant	Unimportant
Hands-on experience	Desirable	Desirable	Necessary	Desirable	Unimportant	Desirable	Unimportant
Level of knowledge in Statistics/bioinformatics	Low	Low	Low	Low	Intermediate	Intermediate	High
Costs							
Consumables	Low	Low	Moderate	Low	Moderate	Moderate	High
Equipment	Low	Low	Moderate	Low	High	Moderate	Very high

2. SE buffer: 75 mM NaCl, 25 mM EDTA, pH 7.5.
3. ES: 0.5 M EDTA, 1 % (w/v) *N*-lauroylsarcosine, pH 9.5. Sterilize by filtration. Add 0.5 mg/ml proteinase K before use.
4. 50 mg/ml proteinase K (dissolve in sterile aqua bidest.)
5. Lysozyme buffer: 100 mM NaCl, 25 mM Tris, pH 5.0. Add 10 mg/ml lysozyme before use.
6. 10 mg/ml lysozyme (dissolve in sterile aqua bidest.)
7. TE buffer: 10 mM Tris, 10 mM EDTA, pH 7.5.
8. 10× TPE buffer: 0.9 M Tris, 15.5 ml 85 % (v/v) phosphoric acid, 0.02 M EDTA, pH 8.3–8.6.
9. 10× TBE buffer: 0.9 M Tris, 0.9 M boric acid, 1 mM thiourea, 0.02 M EDTA, pH 8.3–8.6.
10. 2 % (w/v) low-melting-point agarose (Type VII, Sigma) in SE buffer.
11. SpeI-buffer: 50 mM NaCl, 10 mM $MgCl_2$, 10 mM Tris, pH 7.5 at 37 °C.
12. 10 mg/ml bovine serum albumin (dissolve in sterile double-distilled water).
13. 0.5 M dithiothreitol (DTT, in sterile double-distilled water), sterilize by filtration.
14. 0.1 M ß-mercaptoethanol (freshly prepared in sterile double-distilled water).
15. 1 mg/ml ethidium bromide (dissolve in aqua bidest.)
16. Agarose (ultrapure agarose, Life Technologies).
17. CHEF (contour-clamped homogeneous electric field) instrument: CHEF-DRII system or CHEF-DRIII system or CHEF mapper XA system (all from BIO-RAD).

3 Methods

To generate high resolution macrorestriction fingerprints, one has to adhere to general measures of hygiene and handling that are listed in *see* **Notes 1–4**.

3.1 Preparation of Pseudomonas DNA Embedded in Agarose Plugs (See Note 5)

1. Transfer a single colony of the bacterial strain into culture broth (culture volume dependent on your needs of how many agarose plugs you would like to prepare). Incubate about 18 h at 37 °C.
2. Dilute 0.1 ml of the overnight culture 1:10 with SE. Measure OD_{570} (0.6 OD 570 ~ 1×10^9 cells/ml).
3. Centrifuge the overnight culture at 1,200 × *g* at room temperature for 10 min. Resuspend the cells in SE (the same volume as

before). Centrifuge again at 1,200 × *g* at room temperature for 10 min.

4. Resuspend the cells to a final concentration of 3×10^9 cells/ml.
5. Boil 2 % (w/v) low-melting-point agarose in SE and then maintain at 45 °C.
6. Incubate 300 μl of the bacterial suspension shortly at 45 °C.
7. Mix 300 μl 2 % (w/v) low-melting-point agarose with 300 μl bacteria suspension thoroughly and transfer the mixture immediately into molds (volume of molds: 0.6 cm (length) × 1.0 cm (height) × 0.1 cm (width)).
8. Store at 4 °C for 15–30 min. Agarose blocks will harden within the molds within this time period (*see* **Note 6**).
9. Carefully transfer up to five blocks into one 1.5 ml Eppendorf tube containing 1 ml ES + proteinase K.
10. Incubate overnight at 56 °C. Shake the tube gently several times in the first few hours of incubation.
11. Discard the ES + proteinase K solution the next day.
12. Add 1 ml TE buffer.
13. Store the DNA at 4 °C. The DNA can be kept for up to 10 years (*see* **Note 7**).

3.2 SpeI Digestion of DNA Embedded in Agarose Blocks (See Note 8)

1. Equilibrate half a block first for 30 min with 1 ml fresh TE buffer and then three times for 30 min each with 1 ml of the SpeI restriction buffer. Remove the buffer each time using a Pasteur pipet (*see* **Note 9**).
2. Add: 60 μl SpeI restriction buffer, 1 μl DTT (0.5 M), 1 μl bovine serum albumin (BSA, 10 mg/ml), 10 U SpeI.
3. Incubate overnight (16 h) at 37 °C.
4. Stop the reaction with 1 ml TE buffer.
5. Store the tubes at 4 °C. Plugs should be used for PFGE within the next 24 h (*see* **Note 10**).

3.3 Gel Casting

1. Boil the ultrapure agarose (1.5 % w/v) in 0.5 TBE on a heater with stirring (*see* **Note 11**).
2. Let the agarose cool until the flask is hand-warm.
3. Put the gel casting chamber on a level surface.
4. Place the comb for 20 gel slots about 1 cm from the edge of the frame. Make sure that the comb is perpendicular to the gel plate.
5. Pour the agarose in the gel chamber. The 20 × 20 cm square gel should be about 0.4 cm thick. Keep the rest of the agarose in the flask in a hot water bath so that the agarose remains fluid.

6. When the agarose becomes cloudy, the gel is solidified (about 20 min).

3.4 Loading of Samples into Gel Molds

1. Fill molds with TE-buffer.
2. Cut the blocks to the size of the gel slots with a scalpel.
3. Load the blocks into the slots with the help of a scalpel and a plastic spatula. As size standard a total SpeI digest of a completed sequenced *P. aeruginosa* strain, e.g., strains PAO1 [22] or PA14 [23], should be applied to the two outermost and the innermost lanes (*see* **Note 12**).
4. Seal all slots with warm agarose (*see* Subheading 3.3, **step 5**).
5. Equilibrate the gel for 30 min in the electrophoresis chamber.
6. Start pulsed-field gel electrophoresis in a CHEF apparatus.

3.5 PFGE

1. PFGE should be performed with a CHEF instrument. High-resolution SpeI fragment patterns of *Pseudomonas* DNA are achieved with the following conditions:
   - 20 cm × 20 cm agarose gels
   - Electrophoresis buffer: 0.5 TBE (*see* **Note 13**)
   - Temperature: 10 °C ± 0.5 °C (*see* **Note 14**)
   - 120° reorientation angle
   - Electric field: 6 V/cm (*see* **Note 15**)
   - Pulse times are linearly increased in three ramps from 8 s to 50 s for 24 h, from 12 s to 25 s for 22 h, and from 1 s to 14 s for 14 h (*see* **Notes 16** and **17**).
2. After PFGE the gel is transferred into a container and stained for 45 min with ethidium bromide (1,000-fold diluted aqueous solution of the ethidium bromide stock solution) and then destained by three washes for 30 min each with distilled water.
3. Upon exposure with UV-light in a cabinet or dark room the SpeI restriction fragment patterns on the gel are visualized by the emission of fluorescent light and documented by taking digital pictures with a camera (*see* Table 2 for troubleshooting of CHEF gels in case of unsatisfactory quality). Photos are processed by imaging software. Any commercial gel documentation system is suitable for this job.

3.6 Evaluation of SpeI Macrorestriction Fingerprints

Strains with identical or almost identical fragment patterns on the gel are easily recognized by visual inspection. If strains separated on different gels are suspected to share highly related fragment patterns, it is strongly recommended to rerun the SpeI digests of the two strains in adjacent lanes in a further PFGE experiment.

Since SpeI cleaves 30–60 times in the 4.5–7.5 Mb large Pseudomonas chromosomes, highly informative fingerprints with

Table 2
Troubleshooting [4]

Problem	Probable cause	Remedy (rule of thumb)
(a) No bands after ethidium bromide staining	No cell disruption by ES/proteinase K treatment	Add proteinase K, check solutions
	Proteinase K and detergent are not sufficient	Use lysozyme treatment (*see* **Note 1**) prior to proteinase K treatment to lyse the cells
	Restriction digest failed (*see* point d)	
	Inhibition of the restriction enzyme by acetylated BSA	Try BSA from other suppliers or test non-acetylated batches of BSA
(b) Faint bands after ethidium bromide staining	Ethidium bromide concentration too low	Increase ethidium bromide concentration of the staining solution or extend staining
	DNA concentration of agarose blocks too low	Prepare new agarose blocks
	High background of ethidium bromide fluorescence	Extend destaining
	Improper digestion of cell protein	Shake blocks during incubation, use more proteinase K or try other proteases
(c) Smeared or diffuse bands	Buffer concentration too low or too high	Check buffer composition
	Gel made up of water instead of buffer	Prepare new gel and try again
	Degradation of embedded DNA by forces especially when loading the gel (*the most common problem of the beginner*)	Use more care or another technique in handling agarose blocks
(d) Smeared or diffuse bands after use of restriction enzymes		First check the quality of an undigested agarose block (*see* also points above)
	Enzyme partially degraded	Check storage conditions and age of enzyme
	Incomplete removal of proteinase K after block preparation	Change TE more frequently
	Insufficient equilibration with restriction buffer	Prolong equilibration or change buffer more often
	Improper restriction buffer	Check recipe, check calibration of pH-meter, test buffer recipes from different suppliers
	Restriction buffer, BSA, DTT, etc. with DNase contamination	Incubate agarose blocks with the respective solutions omitting the restriction enzyme
	Improper pH at incubation temperature	Consider the temperature dependence of pH of Tris buffers and choose the correct one for incubation temperature

(continued)

Table 2 (continued)

Problem	Probable cause	Remedy (rule of thumb)
	Enzyme batch with DNAse contamination	Reduce incubation time, check batches from several suppliers
	Star digestion	Reduce incubation time, check buffer recipe
	Restriction enzyme may stick to the DNA	Digest with ES/proteinase K for 1h after restriction enzyme digestion
	Failure in gel cooling	Examine chamber cooling
	Field strength too high for resolution of fragments larger than about 1.5 Mbp	Reduce field strength (voltage)
(e) Poor resolution of bands	DNA concentration in the blocks too high	Make new blocks with lower DNA concentration
	Improper parameters for separation of fragments	Change pulse times and/or voltage
(f) Ethidium bromide staining of slots containing no samples	Microbial growth on the gel matrix(especially if buffers on basis of acetic acid are used)	Disinfect electrophoresis chamber

a low number of overlapping bands are generated by PFGE on a 15–18 cm long lane. Consequently false positive matches of bands with similar size, but unrelated sequence are rare. Hence, the bacterial strains can be quantitatively compared in their fragment length patterns.

The relatedness of the SpeI patterns between two strains is assessed as follows [24]. First, the relatedness between the two SpeI fragment length patterns A and B is scored by the Dice coefficient S_D, i.e., the ratio of twice the number n_i of bands common to their pattern to the sum of all bands in the two patterns $(n+m)$:

$$S_{D(A,B)} = 2n_i / (n+m) \quad (1)$$

If $n \approx m$, two strains are more related than expected from similar genome size and a random restriction site distribution ($P < 0.01$), if

$$S_D > 0.5 + 1.5n^{1/2}/n \quad (2)$$

This nonrandom presence of bands of identical size indicates that the two isolates belong to the same clonal complex.

An example: A *P. aeruginosa* chromosome contains about 50 SpeI sites, 30 of which will be within the range of 40 kb to 800 kb that is highly resolved under the PFGE conditions described above.

Insertion of $n \approx 30$ into Eq. 2 yields that two *P. aeruginosa* isolates are members of a clonal complex if they share more than 78 % of SpeI bands in the 40–800 kb range, i.e., 23 out of 30 bands.

If the CHEF PFGE resulted in gels with collinear lanes, quantitative gel-to-gel comparisons are feasible. The known SpeI fragment lengths of sequenced *Pseudomonas* reference chromosomes are taken as size markers and used to calculate for each gel the nonlinear calibration curve between fragment size and migration distance during PFGE. According to our experience cubic splines as theoretical functions are the adequate interpolation tool to construct the calibration function (http://newtonexcelbach.wordpress.com/2009/07/02/cubic-splines/).

Next, the positions of the individual restriction fragments in each lane of the gels are converted to fragment lengths. Then the number of band matches between each pair of fragment patterns is calculated. If the similarity matrix reveals pairs of related fragment patterns, the respective SpeI digests are rerun on a CHEF gel on adjacent lanes to verify the numeric data analysis.

4 Notes

1. All solutions, glassware and plasticware should be sterilized or sterile filtrated, respectively, to prevent contaminations by DNAse.
2. Wear gloves, whenever you are handling agarose blocks in order to avoid DNase contamination which might lead to the degradation of DNA in a subsequent restriction digest.
3. Always prepare the agarose gel with the appropriate electrophoresis buffer, do not use water.
4. The blocks should not be mechanically stressed especially when loading the gel, since degraded DNA will give a smear after electrophoretic separation.
5. This is a standard protocol. It does not only work with pseudomonads, but with all Gram-negative bacteria. In case of Gram-positive bacteria, the outer cell wall should first be digested with lysozyme prior to incubation with proteinase K. **Step 9** is then modified as follows: Transfer up to five blocks into one 1.5 ml Eppendorf tube containing lysozyme buffer. Incubate 2 h at 37 °C. Discard the lysozyme buffer. Add 1 ml ES + proteinase K and continue with **step 10**.
6. If agarose blocks do not harden in the mold, the agarose concentration is too low. Hence, prepare a new stock solution of 2 % w/v agarose. Moreover, please take care that you harden the blocks at 4 °C. If the temperature drops below 0 °C, the blocks will not harden properly. Never store blocks below 0 °C.

If the agarose hardens already at 45 °C, you ordered an inappropriate type of agarose. Use low-melting-point agarose.

7. Never store the blocks below freezing point.
8. SpeI is the optimal restriction endonuclease to generate informative fragment patterns of *Pseudomonas* chromosomes on PFGE gels. DraI and XbaI are two other restriction enzymes that are often used for strain genotyping of pseudomonads. These enzymes, however, generate too many fragments in the range below 100 kb so that a quantitative fragment length analysis becomes equivocal.

   PFGE-based macrorestriction fragment pattern analysis is globally applicable to the genotyping of bacterial strains [4]. The selection of appropriate restriction enzymes is guided by the G+C-content of the taxon and an estimate of the frequency of the recognition sequence in the genome that is based on the known frequency of the respective oligonucleotide in the genome of a completely sequenced strain of that respective taxon. Recommendations for the selection of restriction enzymes are also available from the literature [25] (still available as an E-book). High-resolution and informative fingerprints are obtained with enzymes that cleave 30- to 70-times in the genome.
9. Some people recommend to treat the agarose plugs with PMSF after DNA isolation in order to inactivate residual amounts of proteinase K. We never detected any proteolytic degradation of a restriction enzyme. So we recommend PMSF treatment only if minute amounts of restriction enzyme are used (1U per digest or less). This may be the case in partial digests that are very sensitive to traces of proteinase K.
10. When small fragments (<20 kbp) are intended to be visualized by PFGE, use the plugs immediately, because these small fragments will diffuse out of the plug within hours (<1 kb) to a few days (20 kb).
11. The resolution of SpeI fragments of *Pseudomonas* chromosomes is higher with 1.5 % (w/v) agarose gels than with the conventional 1 % (w/v) agarose gels, the latter concentration typically being reported in the literature.
12. Linear DNA fragments of the same size should migrate in all lanes at the same rate, but very often some inhomogeneity and/or gradient of the electric field lead to nonidentical mobility of linear DNA of the same length. Application of SpeI digests of a sequenced *Pseudomonas* reference strain of known SpeI fragment lengths to the outermost and innermost lanes of the CHEF gel will allow for normalization of differential mobilities in the lanes of the gel so that accurate size estimates of the SpeI fragments are obtained.

13. The 0.5 TBE buffer contains 50 μM thiourea to protect *P. aeruginosa* DNA from radical-mediated fragmentation. Free radicals are generated during electrophoresis by the Tris molecules. The radicals degrade the SpeI-restricted chromosomal DNA and generate a broad smear in the ethidium bromide stain of PFGE-separated SpeI digests. This phenomenon has been observed in about 4 % of *P. aeruginosa* strains. In our hands the addition of thiourea to the electrophoresis buffer increased the typeability of *P. aeruginosa* by PFGE from 96 % to 100 % [26].
14. The resolution of fragment patterns is critically dependent on PFGE runs at a constant temperature. Separation at room temperature yields broader bands than separation at the optimal temperature of 10 °C. A constant buffer temperature of 10 °C during electrophoresis is achieved by two cooling circuits. A rotary pump circulates the electrophoresis buffer of the first circuit with a pump output of 1–2 L/min. The buffer passes through one to two heat exchangers that are cooled by the second cooling circuit whose lines are connected with a circulation pump and a cryothermostat set to 6 °C. The delivery rates of the two pumps are regulated by hose cocks. Temperature is monitored in the CHEF electrophoresis chamber with a thermometer.
15. Electrophoretic mobility of linear DNA nonlinearly increases with the strength of the electric field. An electric field of 6 V/cm is a compromise of a preferably low electrophoresis run time and a sufficiently high resolution of fragments in the range of 30–800 kb.
16. Three linear ramps from 8 s to 50 s for 24 h, from 12 s to 25 s for 22 h, and from 1 s to 14 s for 14 h have empirically been found to be optimal for the separation of SpeI digests of *P. aeruginosa* DNA by CHEF, but also yield excellent results for other *Pseudomonas* taxa. The total electrophoresis time is 60 h which may be too long for applications in hospital infection control. If shorter run times are requested, we recommend two linear ramps from 5 s to 25 s for 20 h and then from 5 s to 60 s for 17 h. This set-up will still yield high-resolution fingerprints. However, if one needs results within a day, one linear ramp from 5 s to 35 s over 19 h is sufficient for routine applications in hospital infection control.
17. During PFGE electrolysis takes place as is indicated by the production of gas bubbles at the electrodes. Electrolysis varies between electrodes in the hexagonal CHEF field. Users will note that after several PFGE runs some platinum electrodes become thinner whereas other remain unaffected. Thin electrodes are fragile and finally will break. Hence, platinum electrodes need to be regularly checked and exchanged when necessary.

Acknowledgments

PFGE equipment and protocols have initially been set up by Dietmar Grothues, Wilfried Bautsch, Joachim Greipel, Uta Koopmann, and Ute Römling. The protocols were subsequently modified, evaluated, and adapted to the needs of individual applications by (in chronological order) Ute Römling, Rainer Fislage, Thomas Heuer, Jutta Ulrich, Christiane Bürger, Karen Larbig, Stefanie Breitenstein, Peter Gudowius, Christian Weinel, and Jens Klockgether. The protocol for genotyping of *P. aeruginosa* by oligonucleotide microarray has been developed by Lutz Wiehlmann in collaboration with a team at CLONDIAG Chip Technologies GmbH.

References

1. Dabrowski W et al (2003) Optimisation of AP-PCR fingerprinting discriminatory power for clinical isolates of *Pseudomonas aeruginosa*. FEMS Microbiol Lett 218:51–57
2. Mahenthiralingam E et al (1996) Random amplified polymorphic DNA typing of *Pseudomonas aeruginosa* isolates recovered from patients with cystic fibrosis. J Clin Microbiol 34: 1129–1135
3. Waters V et al (2012) Comparison of three typing methods for *Pseudomonas aeruginosa* isolates from patients with cystic fibrosis. Eur J Clin Microbiol Infect Dis 31:3341–3350
4. Römling U et al (1994) Bacterial genome analysis by pulsed field gel eletrophoresis techniques. In: Chrambach A, Dunn MJ, Radola BJ (eds) Advances in electrophoresis, vol 7. Weinheim, VCH Verlagsgesellschaft, pp 353–406
5. Römling U, Greipel J, Tümmler B (1995) Gradient of genomic diversity in the *Pseudomonas aeruginosa* chromosome. Mol Microbiol 17: 323–332
6. Römling U, Schmidt KD, Tümmler B (1997) Large genome rearrangements discovered by the detailed analysis of 21 *Pseudomonas aeruginosa* clone C isolates found in environment and disease habitats. J Mol Biol 271:386–404
7. Klockgether J et al (2011) *Pseudomonas aeruginosa* genomic structure and diversity. Front Microbiol 2:150
8. Vu-Thien H et al (2007) Multiple-locus variable-number tandem-repeat analysis for longitudinal survey of sources of *Pseudomonas aeruginosa* infection in cystic fibrosis patients. J Clin Microbiol 45:3175–3183
9. Gironde S, Manceau C (2012) Housekeeping gene sequencing and multilocus variable-number tandem-repeat analysis to identify subpopulations within *Pseudomonas syringae* pv. maculicola and *Pseudomonas syringae* pv. tomato that correlate with host specificity. Appl Environ Microbiol 78:3266–3279
10. Sobral D et al (2012) A new highly discriminatory multiplex capillary-based MLVA assay as a tool for the epidemiological survey of *Pseudomonas aeruginosa* in cystic fibrosis patients. Eur J Clin Microbiol Infect Dis 31: 2247–2256
11. Maiden MCJ et al (1998) Multilocus sequence tying: a portable approach to the identification of clones within populations of pathogenic microorganisms. Proc Natl Acad Sci USA 95:3140–3145
12. Boers SA, van der Reijden WA, Jansen R (2012) High-throughput multilocus sequence typing: bringing molecular typing to the next level. PLoS One 7:e39630
13. Sarkar SF, Guttman DS (2004) Evolution of the core genome of *Pseudomonas syringae*, a highly clonal, endemic plant pathogen. Appl Environ Microbiol 70:1999–2012
14. Bull CT et al (2011) Multilocus sequence typing of *Pseudomonas syringae* sensu lato confirms previously described genomospecies and permits rapid identification of *P. syringae* pv. coriandricola and *P. syringae* pv. apii causing bacterial leaf spot on parsley. Phyto Pathol 101:847–858
15. Curran B et al (2004) Development of a multilocus sequence typing scheme for the opportunistic pathogen *Pseudomonas aeruginosa*. J Clin Microbiol 42:5644–5649
16. Johnson JK et al (2007) Multilocus sequence typing compared to pulsed-field gel electrophoresis for molecular typing of *Pseudomonas aeruginosa*. J Clin Microbiol 45:3707–3712

17. Kidd TJ et al (2011) Comparison of three molecular techniques for typing *Pseudomonas aeruginosa* isolates in sputum samples from patients with cystic fibrosis. J Clin Microbiol 49:263–268
18. Wiehlmann L et al (2007) Population structure of *Pseudomonas aeruginosa*. Proc Natl Acad Sci USA 104:8101–8106
19. Cramer N et al (2011) Microevolution of the major common *Pseudomonas aeruginosa* clones C and PA14 in cystic fibrosis lungs. Environ Microbiol 13:1690–1704
20. Klockgether J et al (2013) Intraclonal diversity of the *Pseudomonas aeruginosa* cystic fibrosis airway isolates TBCF10839 and TBCF121838: distinct signatures of transcriptome, proteome, metabolome, adherence and pathogenicity despite an almost identical genome sequence. Environ Microbiol 15:191–210
21. Liu L et al (2012) Comparison of next-generation sequencing systems. J Biomed Biotechnol 2012:251364
22. Stover CK et al (2000) Complete genome sequence of *Pseudomonas aeruginosa* PAO1, an opportunistic pathogen. Nature 406:959–964
23. Lee DG et al (2006) Genomic analysis reveals that *Pseudomonas aeruginosa* virulence is combinatorial. Genome Biol 7:R90
24. Grothues D, Tümmler B (1991) New approaches in genome analysis by pulsed-field gel electrophoresis: application to the analysis of *Pseudomonas* species. Mol Microbiol 5:2763–2776
25. Römling U, Fislage R, Tümmler B (1996) Macrorestriction mapping and analysis of bacterial genomes. In: Birren B, Lai E (eds) Nonmammalian genome analysis—a practical guide. Academic Press, San Diego, pp 165–195, http://dx.doi.org/10.1016/B978-01210 1285-4/50007-2
26. Römling U, Tümmler B (2000) Achieving 100 % typeability of *Pseudomonas aeruginosa* by pulsed-field gel electrophoresis. J Clin Microbiol 38:464–465

Chapter 6

Drug Susceptibility Testing by Dilution Methods

Katy Jeannot and Patrick Plésiat

Abstract

Serial twofold dilution methods are widely used to assess the bacteriostatic activities of antibiotics. This can be achieved by dilution of considered drugs in agar medium or in culture broth, and inoculation by calibrated inoculums. Although seemingly simple, these methods are greatly influenced by the experimental conditions used and may lead to discrepant results, in particular with untrained investigators. The present step-by-step protocol has been validated for *Pseudomonas* species, including *P. aeruginosa*. Introduction of appropriate control strains is crucial to ascertain minimal inhibitory concentration values and compare the results of independent experiments.

Key words Antibiotics, Resistance, *Pseudomonas*, Minimum Inhibitory Concentration (MIC), Dilution method

1 Introduction

Highly versatile and opportunistic microorganisms such as *Pseudomonas* have developed sophisticated mechanisms to cope with hostile environments. Appropriately, the poor outer membrane permeability of these species usually potentiates the activity of an impressive set of active, multispecific efflux systems and detoxifying enzymes aimed at preventing the accumulation of inhibitors into the cell interior [1]. Because of their widespread production by some soil bacterial and fungal species, antibiotics represent very common inhibitors in nature though present at rather low concentrations most microbes can adapt to. The massive therapeutic use of some of these molecules by man since the mid-twentieth century has imposed unprecedented selective conditions in some environments (e.g., hospitals, sewage) pushing bacteria to become ever more resistant [2]. A detailed presentation of the numerous strategies by which human and animal pathogens escape the action of antibiotics is clearly beyond the scope of this chapter and can be found elsewhere [3]. In *Pseudomonas* species including the major nosocomial pathogen *P. aeruginosa*, increase in drug resistance can

Alain Filloux and Juan-Luis Ramos (eds.), *Pseudomonas Methods and Protocols*, Methods in Molecular Biology, vol. 1149, DOI 10.1007/978-1-4939-0473-0_6, © Springer Science+Business Media New York 2014

be achieved by both intrinsic (i.e., related to the core genome) and transferable mechanisms (i.e., related to the accessory genome or plasmids) [1]. With the growing emergence of nosocomial strains recalcitrant to more than one antibiotic class, new terms have been proposed to better define the extent of drug resistance in human and animal pathogens, ranging from multidrug (MDR) to extensively (XDR) to extremely to pan-drug resistant (PDR) [4]. These definitions however still await a more general agreement. Moreover, their use requires testing extensive panels of antibiotics, some of which may not be completely available commercially in some countries.

All these definitions refer to a clinical view of bacterial resistance. Indeed, a strain will be categorized as "susceptible," "intermediate," or "resistant" to a given antibiotic molecule depending on whether the minimum inhibitory concentration (MIC) of this agent can be reached or not in the serum of humans by standard doses, as determined by toxicologic and pharmacokinetic studies. Breakpoint concentrations allowing this classification are established and regularly reevaluated by national (e.g., CLSI, http://www.clsi.org) or international (e.g., EUCAST, http://www.eucast.org) scientific societies. These institutions also provide detailed protocols for drug susceptibility testing and results interpretation. It should be recalled here that by definition the MIC represents the smallest drug concentration at which no more growth is visible by the naked eye, irrespective on whether the bacteria are killed or not. MBC which estimates the bactericidal activity of antibiotics (minimum concentration able to kill $\geq$99.99 % of the initial inoculum) is much rarely determined or substituted with time kill experiments. Analysis of the MIC distribution curves for the most pertinent human pathogens has proved essential for the discrimination between wild-type and resistant populations, in other words to characterize those organisms having acquired mutational and/or tansferable resistance mechanisms to a specific agent. The MIC value corresponding to the upper limit of the susceptible population is called ECOFF (Epidemiological Cut-OFF). This parameter is taken into consideration to set up clinical susceptibility breakpoints and to follow the development of antibiotic resistance in surveillance studies.

Because many technical factors may influence MIC values, it is highly recommended to repeat experiments at least twice and to systematically include appropriate reference strains as internal controls of repeatability and reproducibility. A special attention will also be paid to the composition of media used for MIC determinations, as it can significantly vary from one manufacturer to another, or from one batch to another. For instance, concentration of divalent cations Ca^{2+} and Mg^{2+} is known to strongly influence aminoglycoside MICs in *Pseudomonas* species [5].

Small differences in drug susceptibility falling in the twofold range of MIC thus should systematically be confirmed by three determinations to prove they are significant. Because the resistance level of a given strain to a given inhibitor results from interplays of complex mechanisms, the physiological condition of bacteria to be tested may have a dramatic impact on drug MICs. To avoid getting uninterpretable results, bacteria should always be freshly subcultured from a frozen stock at −70/80 °C the day before starting with drug susceptibility experiments. Finally, presentation of drug susceptibility data will benefit of using internationally recognized acronyms for the antibiotic molecules tested.

2 Materials

2.1 Media and Bacterial Strains

1. Mueller-Hinton Agar (MHA, Difco™ Mueller-Hinton agar, Becton Dickinson and company, Maryland, USA) or MHA cation adjusted (BBL™, Mueller-Hinton II Agar) depending on the antibiotic used (*see* **Note 1**).
2. Mueller-Hinton Broth (MHB) cation adjusted (BBL™, Mueller-Hinton II Broth Cation-adjusted, Becton Dickinson and company) (*see* **Note 1**). Add 22 g of powder to 1 L of deionized water and proceed according to manufacturer's instructions. Autoclave (e.g., 121 °C, 15 min, 1 bar) and store at 4–8 °C until use.
3. Sterile petri dishes with cap 14×90 mm (VWR International).
4. Sterile Multipoint inoculator (Boekel Scientific, 48-pin replicator; Mast Group, 36-pin replicator) to deliver 0.001 mL volumes (*see* **Note 2**).
5. Sterile 96-well microtiter plates (microbroth dilution method only).
6. Micropipettes (0.1–1 mL).
7. Sterile disposable pipettes tips.
8. Sterile pipettes of 2 or 5 mL.
9. Sterile tubes with caps (130×100 mm, VWR International).
10. Sterile conic tubes with caps (50 mL) (BD Falcon™).
11. Erlenmeyer flasks, 250 or 500 mL.
12. Tabletop incubator shaker.
13. Sterile deionized water.
14. Spectrophotometer (wavelength set at 625 nm).
15. Polystyrene spectrophotometer cuvettes.
16. Vortex mixer.

17. Bacterial strains stored at −70/80 °C in Brain-Heart broth (BBL™, Brain-Heart Infusion, Becton Dickinson and company) with 20 % (v/v) sterile glycerol, in 1.5 mL screw-capped microtubes.
18. Control *P. aeruginosa* strains such as PAO1 [6], ATCC 27853 (American Type Culture Collection, Manassas, VA 20108, USA), NCTC 10662 (National Collection of Type Cultures, Health Protection Agency Culture Collection, London, UK), CIP 76.110 (Collection de L'Institut Pasteur, Paris, France).
19. Sterile inoculation loops of 10 μL or sterilized cotton swabs (VWR International).

2.2 Antibiotic Stock Solutions

1. Purified antibiotic powders for nonclinical use may be obtained from pharmaceutical companies or suppliers such as Sigma-Aldrich® (St. Louis, MO, USA). The antibiotic powders are stored according to manufacturers' recommendations until expiry date.
2. Technical information about the potency (titer in μg per mg) of powders is needed to prepare stock solutions.
3. Define the range of concentrations to be tested for each antibiotic (*see* Table 1).
4. Prepare stock solutions using the following formula:

$$W = (C \times V \times 1000) / P$$

   W=weight of antibiotic powder (mg) to be dissolved, *C*=final concentration of stock solution (mg/mL), *V*=volume required (mL), *P*=potency (titer) of the powder (μg/mg) (*see* **Note 3**).
5. Dissolve antibiotic powders (*see* **Notes 4** and **5**). Most of the stock solutions can be prepared in advance and stored for several days. Recommended storage conditions can be found in reference [7] (*see* **Note 6**).

3 Methods

There are alternative methods to determine MIC values. The choice depends on the number of strains and antibiotics to be tested. For instance, agar dilution methods are more suitable than broth dilution techniques when large series of strains are considered. In contrast, broth dilution techniques are more convenient to establish drug susceptibility profiles of a limited (e.g., ≤10) number of strains and/or antibiotics.

All the procedures are carried out at room temperature unless specified otherwise. MIC values are obtained within 3 days with fast growing bacteria such as *Pseudomonas*.

Table 1
Antibiotics frequently tested on *Pseudomonas*

Antibiotic	Solvent	Storage 4 °C	Control strains		Clinical interpretive criteria				Acronym
			PA01	ATCC[a] 27853	CLSI[b]		EUCAST[b]		
β-lactams									
Penicillins									
Ticarcillin (sodium)	[c]	1 week	16–32	8–32	≤16	≥128	≤16	>16	TIC
Ticarcillin/clavulanate[d]	[c]	1–3 days	16/2	8/2–32/2	≤16/2	≥128/2	≤16	>16	TIM
Piperacillin (sodium)	Water	2 days	4–8	1–8	≤16	≥128	≤16	>16	PIP
Piperacillin/tazobactam[e]	water	-[f]	4/4	1/4–8/4	≤16/4	≥128/4	≤16	>16	TZP
Cephalosporins									
Cefepime (dihydrochloride)	[c]	–	2	0.5–4	≤8	≥32	≤8	>8	FEP
Ceftazidime (pentahydrate)	[g]	1 day	1–2	1–4	≤8	≥32	≤8	>8	CAZ
Monobactams									
Aztreonam (anhydrous crystalline β form)	[h]	1 day	4	2–8	≤8	≥32	≤1	>16	ATM
Carbapenems									
Imipenem (monohydrate)	[i]	1 day	1	1–4	≤2	≥8	≤4	>8	IPM
Meropenem (trihydrate)	Water	–	0.5	0.25–1	≤2	≥8	≤2	>8	MEM
Doripenem	[j]	–	0.2	0.12–0.5	≤2	≥8	≤1	>4	DOR
Aminoglycosides									
Gentamicin (sulfate)	Water	6 months	1–2	0.5–2	≤4	≥16	≤4	>4	GEN
Amikacin (sulfate)	Water	7 days	4	1–4	≤8	≥32	≤8	>16	AMK
Tobramycin (sulfate)	Water	1 week	0.5	0.25–1	≤4	≥16	≤4	>4	TOB
Netilmicin (sulfate)	Water	6 months	4	0.5–8	≤16	≥64	≤4	>4	NET
Fluoroquinolones									
Ciprofloxacin (hydrochloride monohydrate)	Water	2 weeks	0.1–0.2	0.25–1	≤1	≥4	≤0.5	>1	CIP
Levofloxacin (hemihydrate)	[k]	–	0.25	0.5–4	≤2	≥8	≤1	>2	LVX
Lipopeptides									
Colistin (sulfate)	Water	–	1–2	0.5–4	≤2	≥8	≤4	>4	CST
Polymyxin B (sulfate)	Water	–	1–2	1–4	≤2	≥8	ND	ND	PMB

ND, not determined

[a] ATCC, American Type Culture Collection. The results are identical for strains NCTC 10662 and CIP 76.110

[b] CLSI, Clinical and Laboratory Standards Institute; EUCAST, European Committee on Antimicrobial Susceptibility Testing; breakpoints 2013

[c] Phosphate buffer, pH 6.0, 0.1 mol/L

[d] Fixed concentration of clavulanate (2 μg/mL)

[e] Fixed concentration of tazobactam (4 μg/mL)

[f] No data available on stability, prepare extemporaneously

[g] A solution containing anhydrous sodium carbonate at 10 % (wt/wt) the amount of ceftazidime is prepared extemporaneously

[h] Satured sodium bicarbonate ($NaHCO_3$) solution

[i] Phosphate buffer, pH 7.2, 0.01 mol/L

[j] 0.85 % physiological saline

[k] Add 0.1 mol/L NaOH dropwise to dissolve in 1/2 volume of water, and then complete with water

3.1 Antibiotic Dilution (See Note 7)

1. For each antibiotic tested, choose an appropriate range of concentrations in order to "cover" wild-type and resistant strains (e.g., from 0.125 to 256 μg/mL).
2. Label sterile capped tubes with the desired antibiotic concentrations (e.g., 0.125, 0.25, 0.5, 1, 2, 4, 8, 16, 32, 64, 128, 256) (*see* **Notes 8** and **9**). This step is repeated for each strain and each antibiotic tested when performing the broth macrodilution method.
3. By using a micropipette, dispense 0.5 mL of MHB or MHB II in each tube except those corresponding to the highest final drug concentration (e.g., 256 μg/mL).
4. Introduce an appropriate volume of antibiotic stock solution into the tube corresponding to the highest final concentration (e.g., 256 μg/mL), as calculated with the following equation:

$$\mathrm{Vr} = \left[(\mathrm{Cf} \times \mathrm{Vf}) / \mathrm{Vi} \right] / \mathrm{C}) \times 1.5$$

Vr = volume of stock solution, Cf = final antibiotic concentration in the medium (mg/mL), Vf = final volume of MH in petri dishes (usually 25 mL) or tubes (usually 2 mL) (*see* **Note 10**), Vi = volume of antibiotic solution used to supplement MHA Vi (usually 0.5 mL), C = concentration of the stock solution (mg/mL) (*see* **Notes 11** and **12**).

5. Adjust the volume to 1.5 mL by adding MHB or MHB II with a micropipette or a pipette.
6. Mix thoroughly with a vortex shaker.
7. Transfer 0.5 mL from the starting tube (e.g., 256 μg/mL) to the second tube of the series to obtain a twofold dilution (e.g., 128 μg/mL) (*see* **Note 13**).
8. Vortex vigorously.
9. Repeat **steps 7** and **8** serially with the remaining tubes.

3.2 Agar Dilution Plate (Only for Agar Dilution Method)

1. Prepare MHA or MHA cation adjusted depending on which antibiotic has to be tested (*see* **Note 1**, Table 1). Add 38 g of powder to 1 L of deionized water and proceed according to manufacturer's instructions (*see* **Note 14**).
2. Sterilize in an autoclave (e.g., 121 °C, 15 min, 1 bar).
3. Let the agar medium cool down to about 50 °C in a waterbath. The medium can be kept under these conditions for several hours before use (*see* **Note 15**).
4. Pour 25 mL of melted medium into a sterile graduated container (e.g., 50 mL Falcon tube or 50 mL cylinder).
5. Add 0.5 mL of the less concentrated antibiotic solution (e.g., 0.125 μg/mL) with a micropipette.

6. Invert the tube several times. Do not vortex to avoid forming bubbles in the medium.
7. Pour the medium gently into a labelled (final antibiotic concentration) sterile petri plate and allow it to solidify for 1 h on a horizontal benchtop. Avoid making bubbles (*see* **Note 16**).
8. Repeat **steps 4–7** for each antibiotic concentration.
9. Pour two control, drug-free agar plates.
10. Allow the plates to dry for 30 min with their lids off in a 37 °C incubator or in a laminar airflow hood (*see* **Note 17**).
11. Mark the agar plates to define an orientation and leave them upside down in stacks on the bench.

3.3 Bacterial Suspension

1. Directly inoculate MHA plates with frozen bacteria in order to obtain well separate colonies (*see* **Note 18**).
2. Incubate the plates for 18–24 h at 35–37 °C.
3. With a sterile loop, pick out ≥3 colonies of similar aspect per plate and resuspend them in 5 mL of fresh pre-warmed MHB or MHB-II (*see* **Note 19**).
4. Briefly mix on vortex shaker.
5. Incubate with shaking (250 rpm) at 35–37 °C until the culture reaches an absorbance of 0.08–0.13 at 625 nm, which corresponds to 10^8 cfu/mL (*see* **Note 20**).
6. Invert the spectrophotometer cuvettes several times before measuring optical density (*see* **Notes 21** and **22**).

Agar dilution method **steps 7–9**

7. Transfer 0.01 mL of bacterial culture into a well of a sterile 96-well microtiter plate pre-filled with 0.09 mL of MHB or MHB-II.
8. Place a new tip on the micropipette and repeat **step 7** for each strain to be tested. Be careful not to cross-contaminate the wells with droplets when pipetting off bacterial suspensions. If using a multipoint inoculator, adapt the distribution of individual cultures into the microtiter plate. For up to 48 tests, inoculate within rows A–H, columns 1–6 of the plate.
9. Make a note of the position of each strain in the wells (*see* **Note 23**).

Broth macrodilution method **steps 10–14**

10. Prepare a 1:100 dilution of the culture obtained **step 5** by pipetting 0.1 mL into a sterile tube containing 9.9 mL of pre-warmed MHB or MHB II. If ≥10 drug dilutions need to be tested per strain, repeat **step 10** or adapt the volume of dilution.

11. Repeat **step 10** for each strain to be tested.
12. Dispense 1 mL of the 1:100 suspension in the tubes containing serial antibiotic dilutions to yield ca. 5×10^5 cfu/mL (*see* Subheading 3.1).
13. In parallel, transfer 1 mL of the 1:100 suspension in a control tube containing 1 mL drug-free MHB (control 1). Make one control for each strain tested.
14. Incubate 18–20 h at 35–37 °C.

3.4 Inoculation (Only for Agar Dilution Method)

1. If the agar plates have been stored at 4–8 °C, leave them at room temperature for 1 h or so before the inoculation step.
2. Prepare a microtiter plate containing 95 % (v/v) ethanol to sterilize the inoculator.
3. Use the multipoint inoculator to deposit 0.001 mL volumes of bacterial suspension (ca. 10^4 CFU/mL) onto the agar surface of an antibiotic free MHA plate (control 1).
4. Visually check the size of the spots onto the agar plate after removing the inoculator (diameter of 3–5 mm).
5. Inoculate the drug containing plates starting with the lowest antibiotic concentration.
6. Repeat **step 5** for the different antibiotics tested (*see* **Note 24**).
7. Sterilize the multipoint inoculator by leaving the pins in the wells containing 95 % ethanol for 3 min.
8. At the end of the experiment, inoculate a second MHA plate without antibiotic (control 2).
9. Keep the inoculated plates at room temperature until the spot are dry.
10. Stack up the plates (lids down).
11. Incubate for 18–20 h at 35–37 °C.

3.5 Reading and Interpretation

1. Check all of the strains have grown on control plate 1 or in control tube 1.
2. Check there is no contamination on control plate 2 (*only for agar dilution method*).
3. Determine the MIC for control strain(s) (*see* **Notes 25** and **26**).
4. Determine the MIC for the other strains tested.

4 Notes

1. Most of the MIC determinations can be performed with MHA. However, MIC results for aminoglycosides and polymyxins greatly vary according to Mg^{2+} and Ca^{2+} concentrations.

It is thus recommended to use cation adjusted media in this particular case.

2. A multichannel pipette set at 0.001 mL can be used to dispense accurate volumes of bacterial suspension.
3. For example, amikacin sulfate salt, $V=1$ mL, $C=50$ mg/mL, $P=710$ µg; $W=70.42$ mg of powder.
4. Most of the antibiotics are soluble in distilled or deionized water except the ones listed in Table 1, that require specific solvents.
5. Microbial contamination of antibiotic powders is rather unusual. It therefore is unnecessary to sterilize stock solutions by ultrafiltration. If needed, never sterilize by autoclaving.
6. Do not freeze stock solutions more than once. Some antibiotics like ß-lactams can undergo substantial degradation when thawed and refrozen. Aliquots of stock solutions can be conveniently prepared in advance and individually used where necessary.
7. Twofold serial antibiotic dilutions in liquid or solid MH medium should be prepared and inoculated the same day.
8. In the agar dilution method, MHA plates are each prepared with 0.5 mL of antibiotic solution at 50× the final concentration. This volume may be adapted if necessary (e.g., from 0.1 to 1 mL).
9. In the broth macrodilution method, antibiotic solutions are at 4× the final concentrations.
10. A 25 mL volume gives a 4 mm high agar medium when using petri dishes of 90 mm diameter. This volume will be adapted if larger plates are used.
11. For example, with the agar dilution method: amikacin sulfate, Cf = 0.256 mg/mL, Vf = 25 mL, Vi = 0.5 mL, C = 50 mg/mL, Vr = 0.384 mL.
12. For example, with the broth macrodilution method: amikacin sulfate, Cf = 0.256 mg/mL, Vf = 2 mL, C = 50 mg/mL, Vr = 0.01536 mL.
13. Change the pipette tip before preparing another dilution.
14. The volume of MHA to be prepared depends on the number of strains and antibiotics tested.
15. Below 45 °C the medium will solidify rapidly. Temperatures ≥55 °C may inactivate some antibiotics.
16. The same container can be used serially if the MHA plate with the lowest antibiotic concentration is made first. It is possible to use the same pipette tip to dispense the antibiotic solutions (0.5 mL).

17. Ideally, the plates should be used the same day. They can be stored at 4–8 °C for up to 3 days if containing stable antibiotics.
18. Use a drug free medium for subculture except for maintaining a plasmid.
19. At least three methods can be used for the inoculum preparation. One of them is proposed to be appropriate for *Pseudomonas.*
20. An incubation period of 1.5–4 h is required to obtain adequate amounts of log phase bacteria depending on their growth rates.
21. The relationship between cell density (cfu/mL) and optical density is linear only with cultures of absorbance ≤0.6 U.
22. The bacterial suspension should be used within 30 min.
23. To avoid errors, it is recommended to make an asymetric distribution of the strains in the microtiter plate.
24. Sterilize the inoculator before each new antibiotic and strain series.
25. The MIC is defined as the lowest concentration of antibiotic able to inhibit visible bacterial growth. Ghost spots sometimes appear on MHA plates. Do not consider them as positive culture unless distinct colonies are visible inside.
26. The MIC for the control strain should be equal to the expected MIC (Table 1). A variation of ± one twofold dilution is usually considered as acceptable.

References

1. Lister PD, Wolter DJ, Hanson ND (2009) Antibacterial-resistant *Pseudomonas aeruginosa*: clinical impact and complex regulation of chromosomally encoded resistance mechanisms. Clin Microbiol Rev 22:582–610
2. Davies JA, Davies D (2010) Origins and evolution of antibiotic resistance. Microbiol Mol Biol Rev 74:417–433
3. Plésiat P (2010) Biochemistry of resistance. In: Couvalin P, Leclercq R, Rice LB (eds) Antibiogram. ESKA publishing, Portland (OR), pp 17–23
4. Magiorakos AP, Srinivasan A, Carey RB et al (2012) Multidrug-resistant, extensively drug-resistant and pandrug-resistant bacteria: an international expert proposal for interim standard definitions for acquired resistance. Clin Microbiol Infect 18:268–281
5. Barry AL, Reller LB, Miller GH et al (1992) Revision of standards for adjusting the cation content of Mueller-Hinton broth for testing susceptibility of *Pseudomonas aeruginosa* to aminoglycosides. J Clin Microbiol 30:585–589
6. Stover CK, Pham XQ, Erwin AL et al (2000) Complete genome sequence of *Pseudomonas aeruginosa* PAO1, an opportunistic pathogen. Nature 406:959–964
7. Andrews JM (2001) Determination of minimum inhibitory concentrations. J Antimicrob Chemother 48(S1):5–16

Chapter 7

Plate-Based Assay for Swimming Motility in *Pseudomonas aeruginosa*

Dae-Gon Ha, Sherry L. Kuchma, and George A. O'Toole

Abstract

Swimming motility is a flagellum-dependent form of movement observed in the Gram-negative bacterium *Pseudomonas aeruginosa*. Swimming motility is defined as the movement in liquid or low-viscosity conditions (up to 0.3 % agar concentration). Unlike swarming motility, swimming motility requires a functional flagellum, but neither quorum sensing (QS) systems nor biosurfactants. While swimming motility can also be observed via microscopy, here we describe a reproducible plate-based method.

Key words Swimming motility, Flagellum, Chemotaxis

1 Introduction

Pseudomonas aeruginosa has a polar flagellum that it uses to swim in liquid-to-low viscosity environments. This flagellum is composed of numerous proteins including the cap, filament, hook, and the basal body. Collectively, the hook and the filament are capable of both clockwise (CW) and counterclockwise (CCW) rotation to propel the bacterium, while the basal body (stator and motor parts combined) hold the flagellum in place, as well as conduct ions necessary to power this machine [1]. Similar to ATP synthesis via oxidative phosphorylation, conductance of cations (H^+ for most neutrophiles, and Na^+ for alkalophiles and *Vibrio* species) through the basal body fuels flagellar motility [2–4].

Swimming motility is a unicellular behavior, and requires a functional polar flagellum with its motor-stator complex. The plate-based swimming motility assay described below measures the degree of this flagellar-dependent motility of the bacterium. As a consequence, a mutant lacking a properly functioning flagellum demonstrates decreased motility compared to its wild-type counterpart (*see* Fig. 1). It is important to note that the extent of the bacterium's swimming motility is scored based on its radial migration through the agar.

Alain Filloux and Juan-Luis Ramos (eds.), *Pseudomonas Methods and Protocols*, Methods in Molecular Biology, vol. 1149, DOI 10.1007/978-1-4939-0473-0_7, © Springer Science+Business Media New York 2014

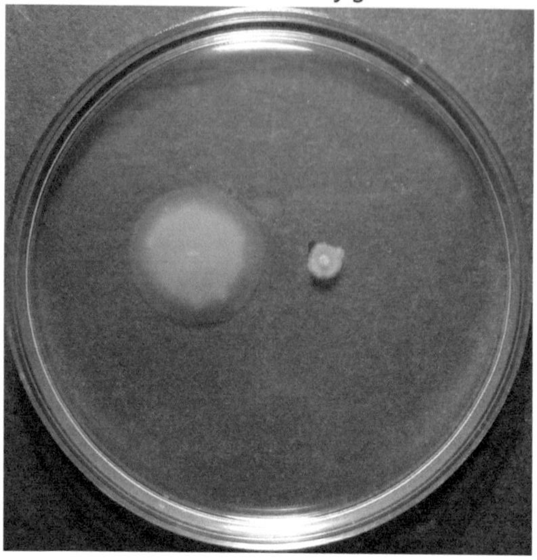

Fig. 1 Swimming motility of *Pseudomonas aeruginosa* PA14 wild type vs. Δ*flgK* mutant

It should also be noted that in addition to a functioning flagellum and the basal body, chemotaxis machinery is also important for proper bacterial motility. Loss of chemotaxis results in a phenotype similar to loss of flagellar function, and thus, this plate-based assay does not readily distinguish between flagellar and chemotactic defects. Therefore, should the investigator be interested in chemotaxis-related phenotypes, microscopy-based observations are a recommended alternative approach to this plate-based assay.

In this section, we describe a reproducible, plate-based method to measure swimming motility in *P. aeruginosa*. Although some reports have documented swimming motility assays conducted on rich media, e.g., lysogeny broth (LB) agar plates [5, 6], we have used and found M8-supplemented plates to be both consistent and reliable. And while we aim for uniformity in these assays, our general protocol should help produce reproducible results regardless of the chosen medium.

2 Materials

Prepare all solutions using ultrapure, distilled deionized water. All solutions used in this assay are sterilized (autoclaved at least 45 min at 121 °C, 15 psi for 1 L solution) prior to use and are kept at room temperature (unless indicated otherwise). Follow all waste disposal procedures when disposing of contaminated solutions at the completion of the experiments.

1. Tube(s) of *P. aeruginosa* strain(s) grown overnight at 37 °C in lysogeny broth (LB: 0.5 g yeast extract, 1 g tryptone, 0.5 g NaCl in 100 mL water) from a fresh streak plate, <7 days old.
2. Petri plates: sterile, 100 mm × 15 mm.
3. Autoclavable bottle: 1 L.
4. Autoclavable flask: 2 L.
5. Automatic pipettor.
6. Disposable, sterile serological pipettes: 10, 25, 50 mL.
7. 20 % Casamino acids solution in water (*see* **Note 1**).
8. 20 % Glucose solution in water.
9. 1 M $MgSO_4$ solution in water.
10. Pipette: P1000.
11. Agar.
12. 1.5 mL disposable, sterile microcentrifuge tubes.
13. Sterile wooden toothpicks or sterile yellow plastic disposable pipette tips.
14. 5× M8 solution: dissolve 64 g $Na_2HPO_4{\cdot}7H_2O$, or alternatively, 30 g Na_2HPO_4; 15 g KH_2PO_4; 2.5 g NaCl in water, and bring the final volume to 1 L (*see* **Note 2**).

3 Methods

All procedures are performed at room temperature (RT) unless specified otherwise. The following protocol is scaled to make 1 L batch of agar. This protocol for 1 L batch should yield roughly 30–36 swimming motility plates. For a different volume, scale accordingly to maintain the appropriate concentration for each component.

1. Add 3 g (for final concentration of 0.3 %) of agar to 800 mL of water. Autoclave sufficiently in 2 L flask to yield a sterile, homogeneous agar suspension; 45 min cycle is sufficient for a 1 L batch (*see* **Note 3**).
2. Autoclave 5× M8 solution in 1 L bottles. Add 200 mL of 5× M8 solution to the melted agar (*see* **Note 4**).
3. Add 10 mL of 20 % glucose to the melted agar; final concentration: 0.2 % glucose.
4. Add 25 mL of 20 % casamino acids to the melted agar; final concentration: 0.5 % casamino acids.
5. Add 1 mL of 1 M $MgSO_4$ to the melted agar; final concentration: 1 mM $MgSO_4$.
6. Mix the agar medium and cool prior to pouring onto petri plates (*see* **Note 5**).

7. Pour thick plates (~25 mL/plate) and let plates solidify at room temperature (RT) for a few hours (*see* **Note 6**).
8. Pipet a small volume (~100 μL) of the overnight bacterial culture into a 1.5 mL microcentrifuge tube.
9. Using a sterile toothpick (or a sterile, disposable yellow pipette tip), dip it in the overnight culture. Use the same toothpick (or pipette tip) to stab *into* the agar layer of the plate, but not all the way to the base of the petri plate (*see* **Notes** 7 and **8**).
10. Incubate plate(s) upright at 37 °C for 16–24 h (*see* **Note 9**) and observe the phenotype. For semi-quantitative analysis, software like ImageJ can be used to compare radial growth of swimming phenotype between strains (*see* **Note 10**).

4 Notes

1. Since 5× M8 solution lacks any source of nitrogen, supplementation with a nitrogen source is key in a swimming motility assay. Therefore, casamino acids can be substituted with a different source of nitrogen depending on the investigator's experimental question. As for 20 % stock solution of casamino acids, precipitates do accrue in older solutions at RT. In our experience, while these precipitates do not redissolve with increased temperature (~50 °C), they do not impact the final outcome of swimming motility phenotypes.
2. Completely dissolving Na_2HPO_4 (both for heptahydrate- and anhydrous-forms) into water takes some time; however, warming the solution is unnecessary. Add reagents to 800 mL of water before bringing up the final volume to 1 L.
3. We find it best to prepare a fresh batch of molten agar on the day of the assay. Addition of a magnetic stir bar to the agar is recommended, as it will assist in mixing other components.
4. Depending on the temperature of the molten agar, addition of the 5× M8 solution could either prematurely solidify the agar, or speed up the cooling process. We find that the best approach is to pre-warm the 5× M8 solution in a warm water bath (~55 °C) prior to adding into the molten agar. If working on a shorter time frame, however, 5× M8 solution at RT can be added to a hot molten agar to decrease the time needed to cool the molten agar prior to pouring plates.
5. While swimming motility measures bacterial movements within the agar layer, and not on top, for clear and easy to image swim plates, we recommend pouring a smooth agar layer. Mix the added components using a magnetic stir bar; aggressive mixing tends to create unnecessary foam. Once poured, any remaining bubbles/foam can be removed by gently running a flame over

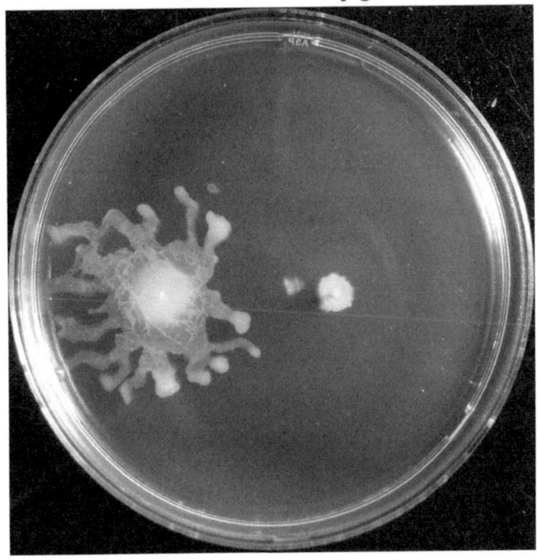

Fig. 2 Occasional swarming motility may interfere with interpreting swimming motility

a poured plate (not recommended on batch volumes >1 L, as the plates begin solidifying rather quickly once poured). Agar suspension that is cool to the touch (5 s to the touch without burning one's hand) is when we generally start pouring plates. We feel this temperature is warm enough to prevent any premature solidifying of the agar, but also cool enough to pour manually without feeling too hot, nor melting the plastic petri plates.

6. Plate thickness does not have to be exact, but the general rule of thumb is that thicker plates will produce better results. The reason being that thicker agar layer retains moisture longer, and prevents premature drying of the plate. Since swimming motility occurs through the agar medium, retention of moisture ensures a constant viscosity. After the plates have been poured, we find it best to solidify them either as single plates or in small stacks (<4 plates/stack) of plates for approximately 3–4 h, as plates on the upper level of the stack tend to dry slower than those at the bottom. When plates are dry, make sure to remove any condensation that has accumulated on the underside of the petri plate cap. Overnight incubation at 37 °C will further increase condensation, which may fall back onto the agar.
7. Whereas swimming motility occurs in the agar medium, any residual inoculum on the surface of the agar can result in unwanted swarming motility. Tendril projections of swarming motility may interfere with estimating the radial growth representing swimming motility (*see* Fig. 2). In order to minimize

residual inoculum on the agar surface, we find it best to preferentially use a pointy-ended toothpick or pipet tip, and also to dip less (do not dunk) into the overnight culture. Prior to inoculating the agar medium, briefly check the toothpick or pipet tip end to ensure there is not any excess inoculum on the sides. Furthermore, it is helpful to keep the toothpick or pipet tip perpendicular to the agar medium to prevent any unintentional inoculation on the agar surface.

8. Should the inoculum pass through the entire layer of the agar medium and come in contact with the basal petri plate surface, twitch motility may occur and may interfere with the interpretation of the radial growth depicting swimming motility. It is recommended that the investigator simply poke part way into the agar, rather than applying force to stab into the agar medium. Considering the possibility of these unintentional errors, we recommend replicate plates per strain tested (at least 3 per strain).
9. Even though the agar solidifies after 3–4 h on the benchtop, due to its low viscosity (0.3 % agar), the plate should be maintained in an upright position. This applies to both during- and post-incubation at 37 °C. Failure to do so and inverting the plate for any extended period of time (>10 min) may result in loss of the integrity of the agar medium, allowing it to detach from the base of the petri plate. Due to lower viscosity and easier detachment of the agar, it is also recommended that the investigator take extra caution when disposing of these plates.
10. While this plate-based assay can be scored qualitatively, software such as ImageJ can aid in semi-quantitatively measuring the swimming motility of each strain. Simply measure the area covered by each strain's swimming motility zone (as depicted by a radial growth), and draw a comparison with other strains in question. In case of an obstructed view due to swarming phenotype, omit the plate from consideration, and if necessary, repeat the assay.

Acknowledgement

This work was supported by the NIH grant R01A1003256 to G.A.O. and the Rosaline Borison predoctoral fellowship awarded to D.G.H.

References

1. Blair DF (2003) Flagellar movement driven by proton translocation. FEBS Lett 545: 86–95
2. Larsen SH, Adler J, Gargus JJ, Hogg RW (1974) Chemomechanical coupling without ATP: the source of energy for motility and chemotaxis in bacteria. Proc Natl Acad Sci U S A 71:1239–1243
3. Manson MD, Tedesco P, Berg HC, Harold FM, Van der Drift C (1977) A protonmotive force drives bacterial flagella. Proc Natl Acad Sci U S A 74:3060–3064

4. Hirota N, Imae Y (1983) Na+-driven flagellar motors of an alkalophilic *Bacillus* strain YN-1. J Biol Chem 258:10577–10581
5. Murray TS, Ledizet M, Kazmierczak BI (2010) Swarming motility, secretion of type 3 effectors and biofilm formation phenotypes exhibited within a large cohort of *Pseudomonas aeruginosa* clinical isolates. J Med Microbiol 59:511–520
6. Soscia C, Hachani A, Bernadac A, Filloux A, Bleves S (2007) Cross talk between type III secretion and flagellar assembly systems in *Pseudomonas aeruginosa*. J Bacteriol 189:3124–3132

Chapter 8

Plate-Based Assay for Swarming Motility in *Pseudomonas aeruginosa*

Dae-Gon Ha, Sherry L. Kuchma, and George A. O'Toole

Abstract

Swarming motility is one of three distinct modes of motility observed in the gram-negative bacterium *Pseudomonas aeruginosa*. Swarming motility is defined as the movement across a semisolid surface, and in *P. aeruginosa* requires flagellar motility and the production of biosurfactants. Swarming motility is thought to occur on gelatinous/viscous surfaces inside a host, such as on epithelial cells. There is currently no standardized in vitro assay to visualize and study swarming motility, and the assays used can vary greatly between laboratory groups. Here, we describe a detailed, reproducible in vitro swarming motility assay for *P. aeruginosa*. While different protocols have previously been reported in the literature, we hope that adopting this method will improve the reproducibility of these swarming motility assays and allow comparisons of swarming motility findings between and among groups.

Key words Swarming motility, Flagellum, Biosurfactants, Surface motility

1 Introduction

In addition to flagellum-dependent swimming and type IV pili-dependent twitching motility, *Pseudomonas aeruginosa* is also capable of swarming motility, which occurs on semisolid surfaces. Unlike swimming motility, swarming motility has been shown to require the production of two biosurfactants—rhamnolipids (RL) and 3-hydroxyalkanoic acids (HAA)—along with a functional flagellum [1–4].

Different strains of *P. aeruginosa* show distinct swarming motility phenotypes, for example, *P. aeruginosa* PAO1 is associated with a circular pattern, whereas *P. aeruginosa* PA14 forms a unique flower-like pattern, with noticeable tendril projections from the point of inoculation (*see* Fig. 1).

At present, one of the major drawbacks with swarming motility assays is the lack of consistency and reproducibility. That is, each swarming assay, even from a single batch of plates, can yield drastically different outcomes based on multiple factors, including

Alain Filloux and Juan-Luis Ramos (eds.), *Pseudomonas Methods and Protocols*, Methods in Molecular Biology, vol. 1149, DOI 10.1007/978-1-4939-0473-0_8, © Springer Science+Business Media New York 2014

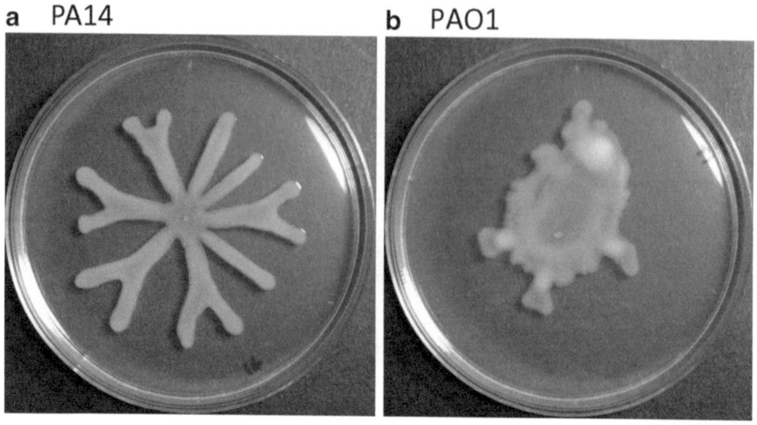

Fig. 1 Swarming motility of *Pseudomonas aeruginosa* (**a**) PA14 vs. (**b**) PAO1

(but not limited to): freshness of the bacterial culture, agar concentration (%), thickness of the plate, dryness of the plate, pH, incubation temperature [5], and location within the incubator. Compounding the issue is the variability in medium composition used among the laboratories investigating swarming motility. To date, however, investigators have yet to systemically investigate the relative merits of each medium, leaving the novice investigators somewhat bewildered when attempting this assay for the first time.

Here, we provide a useful method that has been optimized by our laboratory group through extensive experimentation over the past decade, and yields reproducible swarming motility phenotypes. While our laboratory has adopted the M8-supplemented swarming medium [1], we believe our protocol will also improve the reproducibility and consistency in other swarming media as well [5, 6].

2 Materials

Prepare all solutions using ultrapure, distilled deionized water. All solutions used in this assay are sterilized (autoclaved at least 45 min at 121 °C, 15 psi per 1 L solution) prior to use and are kept at room temperature (unless indicated otherwise). Follow all waste disposal procedures when disposing of contaminated solutions at the completion of the experiments.

1. Tube(s) of *P. aeruginosa* strain(s) grown overnight at 37 °C in lysogeny broth (LB: 0.5 g yeast extract, 1 g tryptone, 0.5 g NaCl in 100 mL water) from a fresh streak plate, <7 days old.
2. Petri plates: sterile, 100 mm × 15 mm.
3. Autoclavable bottle: 1 L.
4. Autoclavable flask: 2 L.
5. Automatic pipettor.
6. Disposable, sterile serological pipettes: 10, 25, 50 mL.

7. 20 % Casamino acids solution in water (*see* **Note 1**).
8. 20 % Glucose solution in water.
9. 1 M $MgSO_4$ solution in water.
10. Pipette: P10, P1000.
11. Agar.
12. 5× M8 solution: dissolve 64 g $Na_2HPO_4 \cdot 7H_2O$, or alternatively, 30 g Na_2HPO_4; 15 g KH_2PO_4; 2.5 g NaCl in water, and bring the final volume to 1 L (*see* **Note 2**).

3 Methods

All procedures are performed at room temperature (RT) unless specified otherwise. The following protocol is scaled to make 1 L batch of agar. In our experience, 1 L batch should yield 30–36 swarm motility plates. For a different volume, scale accordingly to maintain the appropriate concentration for each component.

1. Add 5–8 g (for final concentration of 0.5–0.8 %) of agar to 800 mL of water. Autoclave sufficiently in 2 L flask to yield a sterile, homogeneous agar suspension; 45 min cycle is sufficient for a 1 L batch (*see* **Note 3**).
2. Autoclave 5× M8 solution in 1 L bottles. Add 200 mL of 5× M8 solution to the melted agar (*see* **Note 4**).
3. Add 10 mL of 20 % glucose to the melted agar; final concentration: 0.2 % glucose.
4. Add 25 mL of 20 % casamino acids to the melted agar; final concentration: 0.5 % casamino acids.
5. Add 1 mL of 1 M $MgSO_4$ to the melted agar; final concentration: 1 mM $MgSO_4$.
6. Mix the agar medium and cool prior to pouring onto petri plates (*see* **Note 5**).
7. Pour thick plates (~25 mL/plate) and let plates solidify at RT for a few hours (*see* **Note 6**).
8. Inoculate the center of the plate(s) using the overnight bacterial culture, 2.5 μL per inoculum. Pipette tip should be close to the agar surface during inoculation (*see* **Note 7**).
9. Incubate plate(s) upright at 37 °C for 16–24 h (*see* **Notes 8** and **9**) and observe for phenotype (*see* **Note 10**).

4 Notes

1. Since 5× M8 solution lacks any source of nitrogen, supplementation with a nitrogen source is key in a swarm motility assay. Therefore, casamino acids can be substituted with a different

source of nitrogen depending on the investigator's experimental question. As for 20 % stock solution of casamino acids, precipitates do accrue in older solutions at RT. In our experience, while these precipitates do not redissolve with increased temperature (~50 °C), they also do not impact the final outcome of swarm motility phenotypes.

2. Completely dissolving Na_2HPO_4 (both for heptahydrate- and anhydrous forms) into water takes some time; however, warming the solution is unnecessary. Add reagents to 800 mL of water before bringing up the final volume to 1 L.
3. We find it is best to prepare a fresh batch of molten agar on the day of the assay. Addition of a magnetic stir bar to the agar is recommended, as it will assist in mixing other components.
4. Depending on the temperature of the molten agar, addition of the 5× M8 solution could either prematurely solidify the agar, or speed up the cooling process. We find that the best approach is to pre-warm the 5× M8 solution in a warm water bath (~55 °C) prior to adding into the molten agar. If working on a shorter time frame, however, 5× M8 solution at RT can be added to a hot molten agar to decrease the time needed to cool the molten agar prior to pouring plates.
5. Since swarm motility occurs on the surface of the agar, it is best to eliminate any irregularities on the surface. Aggressive mixing of the components tends to cause foaming, which can later result in unwanted air bubbles on the plate surface. Magnetic stir bars, at low-to-medium rotational speed will achieve both homogeneous mixing and bubble-free medium. Alternatively, remaining air bubbles can be "popped" by gently running a flame over a poured plate (not recommended on batch volumes >1 L, as the plates begin solidifying rather quickly once poured).

   Agar suspension that is cool to the touch (5 s to the touch without burning one's hand) is when we generally start pouring plates. We feel this temperature is warm enough to prevent any premature solidifying of the agar, but also cool enough to pour manually without feeling too hot, nor melting the plastic petri plates.
6. Plate thickness does not have to be exact, but the general rule of thumb is that thicker plates will produce better swarm results. Thicker agar layer retains moisture longer during the incubation step, which prevents premature drying of the plate. When plates have been poured, we find it best to solidify them either as single plates or in small stacks of plates (<4 plates/stack) for approximately 3–4 h, as plates on the upper level of the stack tend to solidify and dry more slowly than those at the bottom. A previous report demonstrated increased reproducibility of swarming motility phenotypes by drying under

laminar flow [5]; however, the uneven distribution of laminar flow resulted in different dryness of the agar based on the plate's location within the flow, which can impact swarming phenotypes. For this reason, and the lack of laminar flow equipment in certain labs, we prefer drying our plates at RT as was described here. When plates are dry, make sure to remove any condensation that has accumulated on the underside of the petri plate cap. Incubation at 37 °C will further increase condensation, which may fall back onto the agar and alter the development of swarming motility.

7. Unless the investigator is performing high-throughput screens of swarm motility mutants, we generally recommend *one* point of inoculation per plate in the center of the plate. As was seen in the example figure, 2.5 μL of *P. aeruginosa* spotted on the center of the plate, then incubated overnight at 37 °C tends to create a robust swarm phenotype spanning almost the entire plate. Two inoculations can be performed when investigating tendril–tendril interactions between two strains/mutants, but this is the maximum we recommend. As for screening purposes, inoculum size should be ≤1 μL with sufficient space between each inoculum to distinguish the phenotypes of each mutant/strain. We recommend 96 well replicators or the like for this purpose.

   Regarding inoculation, it is recommended that pipet tips be as close to the agar surface prior to inoculating. Direct contact with the agar could result in a squirt of bacterial culture over the agar surface, rather than a uniform point of inoculum. Conversely, if the pipet tip is distant from the agar surface while inoculating, multiple inoculation points can inadvertently result in points of inoculation due to splashing.

8. Numerous factors can contribute to variability in swarm phenotypes. To mitigate the effects of these variables, consistency is extremely important. We generally stack the plates of a given strain, with each stack being of same height as other strains being tested within the same batch. Each stack is composed of <8 plates. These stacks of plates are placed near each other within the incubator at the same time, and also removed at the same time. At the end of the 16–24 h of incubation, place them at RT for 1–2 h for coloration of the swarm tendrils. This additional incubation is only for aesthetic purposes as green pigments of *P. aeruginosa* are accentuated during this RT treatment.

9. Controlled humidity in the incubation chamber is key in producing reproducible swarming phenotypes. Ideally, a humidity-controlled incubation chamber will yield the best set of results. Other non-humidity controlled chambers will require investigators to identify region(s) with (relatively) consistent humidity. For example, with incubation chambers utilizing circulating

airflow, preliminary experiments will be necessary to identify regions within the chamber that do not over-dry swarming motility plates during the 16–24 h incubation period. Another solution we have tried and have found success with is covering the stack of swarming motility plates with an inverted tray. It is also important to incubate swarm plates on an even surface as even a slight slant could trigger the flow of the inoculum from its initial point of inoculation. The result will resemble a streak, rather than tendril formations and flower-like pattern, typically associated with positive swarming motility in *P. aeruginosa*. Should this happen, the investigator should not score the streak phenotype as a "positive" swarming phenotype.

10. Swarming motility continues at room temperature post-incubation at 37 °C. We have noticed that prolonged incubation thickens tendrils and accentuates the coloration of green pigments. It has also been observed that continued incubation at room temperature can cause negative-swarming mutants to start to produce tiny tendrils. In our hands, <24 h at room temperature following 37 °C incubation accentuates swarm tendrils for documenting purposes.

Acknowledgement

This work was supported by the NIH grant R01A1003256 to G.A.O., and the Rosaline Borison predoctoral fellowship awarded to D.G.H.

References

1. Caiazza NC, Shanks RM, O'Toole GA (2005) Rhamnolipids modulate swarming motility patterns of *Pseudomonas aeruginosa*. J Bacteriol 187:7351–7361
2. Deziel E, Lepine F, Milot S, Villemur R (2003) *rhlA* is required for the production of a novel biosurfactant promoting swarming motility in *Pseudomonas aeruginosa*: 3-(3-hydroxyalkanoyloxy)alkanoic acids (HAAs), the precursors of rhamnolipids. Microbiology 149:2005–2013
3. Kohler T, Curty LK, Barja F, van Delden C, Pechere JC (2000) Swarming of *Pseudomonas aeruginosa* is dependent on cell-to-cell signaling and requires flagella and pili. J Bacteriol 182:5990–5996
4. Tremblay J, Richardson AP, Lepine F, Deziel E (2007) Self-produced extracellular stimuli modulate the *Pseudomonas aeruginosa* swarming motility behaviour. Environ Microbiol 9:2622–2630
5. Tremblay J, Deziel E (2008) Improving the reproducibility of *Pseudomonas aeruginosa* swarming motility assays. J Basic Microbiol 48:509–515
6. Overhage J, Lewenza S, Marr AK, Hancock RE (2007) Identification of genes involved in swarming motility using a *Pseudomonas aeruginosa* PAO1 mini-Tn5-lux mutant library. J Bacteriol 189:2164–2169

Chapter 9

Motility Assay: Twitching Motility

Lynne Turnbull and Cynthia B. Whitchurch

Abstract

Twitching motility is a mode of solid surface translocation that occurs under humid conditions on semisolid or solid surfaces, is dependent on the presence of retractile type IV pili and is independent of the presence of a flagellum. Surface translocation via twitching motility is powered by the extension and retraction of type IV pili and can manifest as a complex multicellular collective behavior that mediates the active expansion of colonies cultured on the surface of solidified nutrient media, and of interstitial colonies that are cultured at the interface between solidified nutrient media and an abiotic material such as the base of a petri dish or a glass coverslip. Here we describe two methods for assaying twitching motility mediated interstitial colony expansion in *P. aeruginosa*. The first method, the "Macroscopic Twitching Assay," can be used to determine if a strain is capable of twitching motility mediated interstitial colony expansion and can also be used to quantitatively assess the influence of mutation or environmental signals on this process. The second method, the "Microscopic Twitching Assay," can be used for detailed interrogation of the movements of individual cells or small groups of bacteria during twitching motility mediated colony expansion.

Key words Twitching motility, Type IV pili, Surface motility, tfp, Fimbriae, Interstitial colony, Interstitial biofilm

1 Introduction

Pseudomonas aeruginosa undergoes a form of surface translocation known as twitching motility. This motility is independent of the flagellum but is dependent upon the presence of retractile type IV pili (tfp), also called type IV fimbriae. The tfp are polar filamentous appendages of 5–6 nm in diameter and are of average length of 1–4 μm but can be up to 10 μm long. The tfp are composed of several thousand copies of a single subunit protein termed pilin or PilA, arranged in a helical symmetry of 4 subunits per turn. The production of tfp is dependent on the culture conditions, being produced during plate or stationary broth culture but not during vigorously shaken broth culture (for recent reviews *see* refs. 1–3).

When cultured in humid conditions on the surface of solidified nutrient media, colonies of *P. aeruginosa* that are capable of twitching motility grow as flat, rough, spreading colonies with a

Alain Filloux and Juan-Luis Ramos (eds.), *Pseudomonas Methods and Protocols*, Methods in Molecular Biology, vol. 1149, DOI 10.1007/978-1-4939-0473-0_9, © Springer Science+Business Media New York 2014

characteristic "ground glass" edge. Strains that lack functional tfp or are no longer capable of twitching motility produce smooth, domed colonies that do not actively expand across the surface via twitching motility [4, 5]. Whilst the obvious distinction in these phenotypes has been utilized in genetic screens to identify genes involved in twitching motility [6], the observation of differences in colony morphology provides little insight into the process by which twitching motility mediates active colony expansion. More sensitive assays of twitching motility were developed when it was observed that *P. aeruginosa* exhibits extremely active twitching motility mediated colony expansion when cultured at the interstitial surface between agar and the base of a plastic petri plate [4, 7]. When a small inoculum of *P. aeruginosa* strains that are capable of twitching motility are stabbed through a layer of agar to the underlying plastic, a large halo of interstitial colony expansion is obtained after overnight incubation, whereas non-twitching strains produce no such zone of colony expansion ([4]; *see* Fig. 1). This "Macroscopic Twitching Assay" is now used as a de facto measurement of twitching motility activity and can be modified easily to observe the effect of different environmental stimuli by simply adding compounds to the solidified nutrient media.

However, the "Macroscopic Twitching Assay" does not allow detailed analyses of individual cell movements during interstitial colony expansion. High-resolution microscopy of twitching motility mediated colony expansion is possible using a thin layer of solidified nutrient media and a glass coverslip to create the interstitial interface at which active colony expansion occurs [4]. Microscopic observation of the interstitial colony created at the glass–media interface reveals that under these conditions, twitching motility manifests as a highly organized multicellular collective behavior that enables rapid colonization of large areas and involves the formation of an intricate lattice network of cells behind outgoing leading edge rafts of cells ([4]; *see* Fig. 2).

In this chapter we describe these two methods for assaying twitching motility in *P. aeruginosa*. The "Macroscopic Twitching Assay" is suitable for quantitative medium throughput screening of twitching motility ability on a macroscopic level. This method may be used for simple determination of the ability of a strain to demonstrate twitching motility mediated interstitial colony expansion or may be performed with replicates to allow quantitative determination of the effect of mutation or environmental changes. Suitable positive control strains for this assay include PA01, PA103, PA14, or PAK. A suitable negative control strain would be an isogenic mutant that is unable to assemble functional tfp such as a mutant of the major pilin subunit gene *pilA*.

The second method, the "Microscopic Twitching Assay," allows microscopic examination of collective and individual

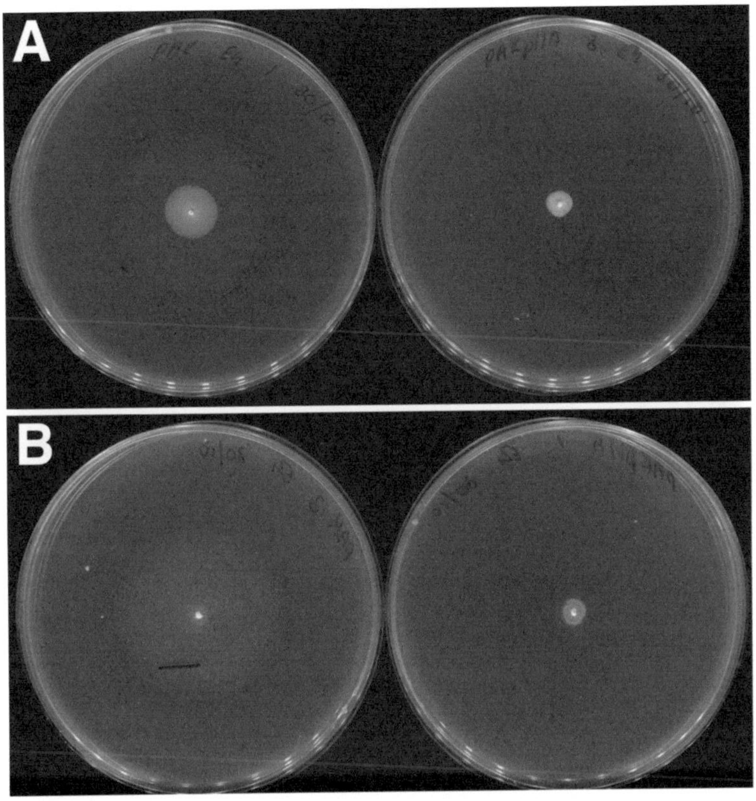

Fig. 1 Macroscopic Twitching Assay. The interstitial biofilm can be visualized by flooding the plates with TM developer solution and recording images of the petri plates on a flatbed document scanner. In these examples, wild-type *P. aeruginosa* strain PAK (*left*) had produced a large interstitial colony that has expanded due to twitching motility whilst the isogenic non-twitching *pilA* mutant (*right*) has produced only a small zone of growth. (**a**) Shows two petri plates that have been marked for measurement but are have not had TM developer solution added. Note that the top colony is clearly visible on both plates and that the interstitial colony on the left plate is quite faint. (**b**) Shows two petri plates after TM developer solution has been applied. A 1 cm mark has been drawn onto one of the plates as a scale bar for quantification and to assist with accurate figure preparation where scale bars are required. Note that the top colonies have been removed and the interstitial colony on the left plate is more easily seen and the small interstitial colony seen on the right plate is attributable to growth only

behaviors of bacteria undergoing twitching motility mediated interstitial colony expansion. The "Microscopic Twitching Assay" described below is optimized for obtaining a layer of single cells at the leading edge of the colony to allow high-resolution microscopy and tracking of individual cell movements when time-series are captured. In this assay the nutrient medium is solidified with a gellan gum, which provides greater optical clarity than agar.

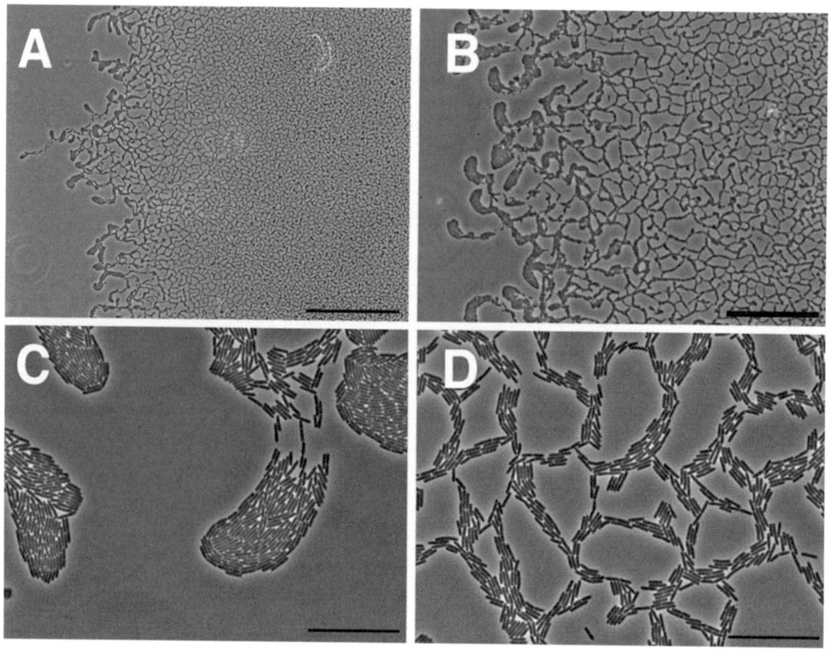

Fig. 2 Microscopic Twitching Assay. Representative images of interstitial colonies formed by wild-type *P. aeruginosa* strain PAK visualized with phase-contrast microscopy. Under the conditions of the assay twitching motility mediates active expansion of the interstitial colony and manifests distinct micro-morphological features including an intricate lattice-like network of cells that forms behind rafts of cells at the leading edge of the colony. *Panels* (**a**) and (**b**) show overviews of the outermost edge of the expanding interstitial colony obtained with 10× (**a**) and 20× (**b**) low magnification dry objectives. *Panels* (**c**) and (**d**) are high-resolution images of the leading edge rafts (**c**) and lattice-like network (**d**) obtained with 100× oil immersion objective. Scale bars: (**a**) = 200 μm; (**b**) = 100 μm; and (**c** and **d**) = 20 μm

2 Materials

Prepare all solutions using ultrapure water (prepared by purifying deionized water to attain a conductivity of 18 M Ω/cm at 25 °C) and analytical grade reagents. Chemicals specific to these protocols are listed below. Prepare and store all reagents at room temperature (unless indicated otherwise).

2.1 Macroscopic Twitching Assay

1. 1 % LB-Lennox agar (1 %LBA): 4 g Tryptone, 2 g yeast extract, 2 g NaCl, 4 g agar (*see* **Note 1**), 400 mL water.
2. Petri plates: sterile, 90 mm × 10 mm (*see* **Note 2**).
3. Autoclavable bottle, minimum volume 500 mL (*see* **Note 3**).
4. Water bath set to 55 °C.
5. Disposable 25 mL sterile serological pipettes.
6. Automatic pipettor.

7. Wooden toothpicks or yellow plastic disposable pipette tips (*see* **Note 4**).
8. Plastic disposable sterile inoculation loop, 10 μL.
9. Humidified chamber: Plastic or glass lunchbox/container with lid containing paper towel in base dampened with sterile water.
10. TM developer solution: 400 mL water, 100 mL glacial acetic acid, 500 mL methanol, store at 4 °C.

2.2 Microscopic Twitching Assay

1. Twitching motility Gellan Gum media (TMGG): 0.8 g gellan gum (*see* **Note 5**), 0.4 g tryptone, 0.2 g yeast extract, 0.2 g NaCl, 0.1 g $MgSO_4{\cdot}7H_2O$, 100 mL water.
2. Dry heating block set to 65 °C with holder for 50 mL tubes or a dry heating bath containing metal beads set to 65 °C (*see* **Note 6**).
3. Heated slide preparation stage set to 65 °C (*see* **Note 7**).
4. Large petri plates (145 mm diameter), sterile.
5. 50 mL disposable plastic tubes, sterile.
6. 5 mL disposable plastic tubes, sterile.
7. Forceps, metal.
8. Microscope slides (75 × 25 mm).
9. Coverslips (22 × 40 mm) (*see* **Note 8**).
10. Spatula, fine.
11. Disposable plastic inoculation loops, 10 μL.
12. Tissues (*see* **Note 9**).
13. Lint free cleaning tissues (Kimwipes).
14. Paraffin film (Parafilm).
15. Humidified chamber: Plastic or glass lunchbox/container with lid containing paper towel in base dampened with sterile water sized to hold large petri dish.

3 Methods

3.1 Macroscopic Twitching Assay

All procedures are performed at room temperature unless otherwise specified. The following protocol will make 40 assay plates. Scale the volume and components accordingly to make the number of plates that you require.

1. Streak each strain to be assayed onto a 1.5 %LBA media plate and incubate overnight at 37 °C.
2. Add tryptone, yeast extract, and NaCl to 400 mL water and mix until fully dissolved.
3. Add agar to autoclavable bottle.

4. Add dissolved medium to agar bottle and gently mix to avoid agar clumping. Agar will not dissolve.
5. Autoclave to yield a sterile homogenous suspension (20 min at 121 °C, 15 psi should be sufficient).
6. Place autoclaved 1 %LBA in heated water bath and allow to cool until medium is at 55 °C.
7. Place 40 petri plates on bench in a single layer.
8. Using a 25 mL disposable serological pipette attached to an automatic pipettor, dispense 10 mL of molten 1 %LBA into each petri plate. If volume does not cover bottom of petri plate, swirl gently to spread evenly.
9. Leave agar to set for 1 h.
10. Invert petri plates on the bench and leave overnight to dry in a single layer (*see* **Note 10**).
11. The following day, label the base of each plate appropriately.
12. Remove inoculation plates from 37 °C incubation. With a plastic inoculation loop take a small portion of the outer edge of the bacterial streak and gently mix in a sterile area of agar until the bacterial culture is smooth (*see* **Note 11**).
13. Use a sterile toothpick or sterile yellow pipette tip to scoop a match-head sized inoculum from the mixed culture. Try to keep the inoculum size similar for each plate.
14. In the center of each 1 %LBA plate, stab the inoculum perpendicular to the agar down to the agar–plastic interface at the bottom of the plate. Jiggle the toothpick/tip up and down a few times to ensure that the inoculum has reached the bottom of plate. Do not move toothpick from side-to-side (*see* **Note 12**).
15. Invert plates into a humidified chamber in stacks of no more than three plates high.
16. Place in standard 37 °C incubator with lid slightly ajar to allow air to enter the humidified chamber. Incubate in humidified environment at 37 °C for 24 h.
17. Remove petri plates from incubator (*see* **Note 13**).
18. There are two types of colony growth on the petri plate. There will be a colony on the surface of the agar around the inoculation point (top colony) and a visible halo of bacteria that have twitched across the plate between the bottom of the agar and the plastic petri plate (interstitial colony). The interstitial colony can be seen by holding the plate up to a light source (*see* **Note 14**). Mark each edge of interstitial colony with a permanent marker ("Sharpie") (*see* Fig. 1a).
19. The surface area of the interstitial colony can be calculated as follows. If the interstitial colony is circular in shape then the

surface area can be calculated using the formula πr^2 where $r = ½$ the diameter of the interstitial colony. If the interstitial colony is oval in shape, measure the longest and shortest diameters and calculate the surface area by using the formula $\pi \times a \times b$ where $a = ½$ longest diameter and $b = ½$ shortest diameter.

20. To visualize the interstitial colonies for image capture or to aid visualization for measurement, flood the plates with a small amount of cold TM developer solution. The solution makes the interstitial colony appear as an opaque white halo (*see* Fig. 1b).
21. Leave the TM developer solution on the petri plates until the interstitial colony is visible as a white halo or for up to 30 min (*see* **Note 15**).
22. Using a plastic disposable loop, gently scrape the top colony away from the agar, being careful not to scratch the surface of the agar.
23. Carefully decant the solution from the plates into a beaker. Make sure that the outsides of the plates are dry and clean by wiping with a tissue if necessary.
24. Images of the interstitial colonies can be recorded by scanning the petri plates using a flatbed document scanner (Fig. 1). It is useful to draw a 1 cm line on the base of the petri plate prior to scanning for use as a scale bar.

3.2 Microscopic Twitching Assay

1. Weigh out each component of TMGG except the gellan gum and add to water in a beaker containing a magnetic stirring bar.
2. Heat the medium with stirring on a magnetic stirring/heating block.
3. Gradually add the gellan gum in small amounts with stirring to prevent clumps forming. Heat the medium to boiling while stirring to dissolve.
4. Once the mixture turns clear, autoclave for 20 min at 121 °C, 15 psi (*see* **Note 16**).
5. Remove sterile TMGG from autoclave and decant into 50 mL tubes that have been preheated to 65 °C either in a dry block heater or in the metal beads of a 65 °C dry bath. Allow to cool to 65 °C. Gellan gum will set below 60 °C, so care must be taken to keep temperature of TMGG above 65 °C. All procedures should be done as quickly as possible to avoid TMGG setting in tubes. Care should also be taken to minimize water loss by steam or condensation and by minimizing the time that lids are removed from containers (*see* **Note 17**).
6. Preheat the slide preparation stage to 65 °C, and ensure that it is level (*see* **Note 18**).
7. Preheat 5 mL tubes to 65 °C in the heating block or dry bath.

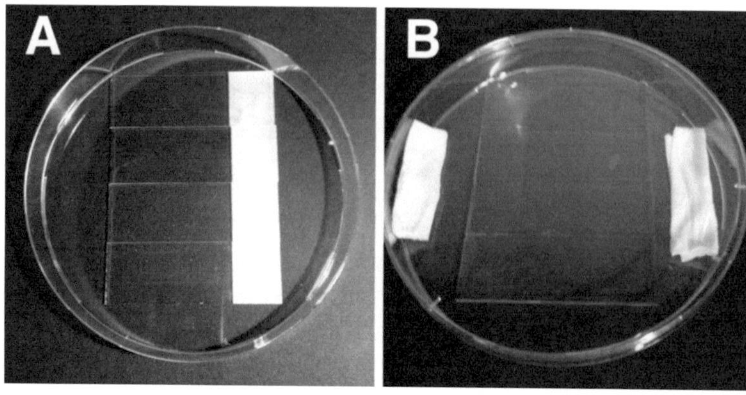

Fig. 3 Microscopic Twitching Assay. Prepare the microscope slides by flame-sterilizing and aligning side-by-side in a large petri plate. Push the slides together tightly until the long edge on one slide is firmly abutted to the rim of the petri plate as shown in (**a**). Pour the TMGG solution and tilt gently to ensure an even surface coverage, add the damp tissues and prepare for 4 °C storage as shown in (**b**)

8. Preheat a large petri plate on the 65 °C warm stage.
9. Flame-sterilize four microscope slides by holding each slide with a pair of forceps, dipping the slide into a glass beaker containing 96 % ethanol, touching the corner of the slide to the edge of beaker to drain off most of the ethanol and passing the slide through the flame of a Bunsen burner to burn the remaining ethanol. Align each slide side by side inside the petri plate using sterile forceps. Push the slides tightly together so that the long edge of one slide is touching the wall of the petri plate and allow to preheat to 65 °C (*see* Fig. 3a).
10. Decant 5 mL of TMGG into the pre-warmed 5 mL tube. Return the 5 mL tube to heating block or dry bath to maintain temperature.
11. Remove lid of petri plate and rest on bench so that you have two free hands.
12. Remove 5 mL tube with TMGG from heating block and pour TMGG over all four slides. Gently tilt the petri dish from side to side so that the TMGG covers all surfaces of the slides. Try to get an even coverage of TMGG on the four slides but not let the TMGG pour off the edges of the slides. Some TMGG may leak between the slides and settle between the slides and the bottom of the petri plate; this is acceptable.
13. Allow TMGG to settle, replace lid and move petri plate carefully from the warm stage to a cool, level area and allow TMGG to set for 15 min.
14. Fold a tissue in half and in half again until you have a small wad approximately 1 × 4 cm in size. You will need two per petri plate.

Wet the tissues with sterile water until the tissues are full but not quite dripping and place at empty ends of the petri dish. Do not let free water run into the base of the petri plate. Be careful not to let the tissues touch the slides (*see* Fig. 3b).

15. Holding the petri plate level, seal the lid to the base using paraffin film and transfer to a 4 °C refrigerator until required (*see* **Note 19**).
16. Streak the desired strain(s) to be assayed onto a 1.5 %LBA plate and incubate overnight at 37 °C.
17. The following morning, remove the pre-poured slides from 4 °C.
18. Remove the two tissues from each end of the plate and keep aside aseptically.
19. Flame-sterilize a fine spatula and use to gently score the TMGG between each slide and separate slides a little using the spatula.
20. Place open large petri plate (lid removed) in a Type 1 or Type 2 Biohazard Cabinet for approximately 20 min to dry slides (use the same cabinet each time for consistency, *see* **Note 20**).
21. Remove petri plate from cabinet and remove excess TMGG from under the slides. To do this, slowly lift out each slide using a spatula and wipe the back of each slide with a lint free cleaning tissue. Also wipe out the petri plate. Replace slides into the large petri plate.
22. Using the edge of a flame-sterilized fine spatula, gently score around the edge of each slide. Remove excess TMGG by lifting gently away with the flat edge of the spatula. Trim 1–2 mm from the long edges of the slide, 5–10 mm from the clear short edges to prevent contact of the TMGG with the microscope stage and all of the white section so that slide can be adequately labeled (*see* Fig. 4a).
23. With a plastic inoculation loop take a small portion of the outer edge of the bacterial streak and gently mix in a sterile area of agar until the bacterial culture is smooth (*see* **Note 11**). Collect a small amount of culture on the tip of the loop and lightly dab onto the center of the TMGG slide, creating a small flat inoculation point of 1–2 mm in diameter.
24. Using 96 % ethanol, flame-sterilize a 22×40 mm coverslip and allow to cool. Placing one short edge of coverslip onto the TMGG, slowly lower coverslip onto the TMGG at an angle using either sterilized metal forceps or spatula, trying to minimize bubble formation between the coverslip and the TMGG (*see* Fig. 4b). Repeat **steps 23** and **24** for all slides (*see* **Note 21**).
25. Place slides back into petri dish along with damp tissues, being careful not to let the tissues touch the slides (to prevent excess moisture being introduced under the coverslip).

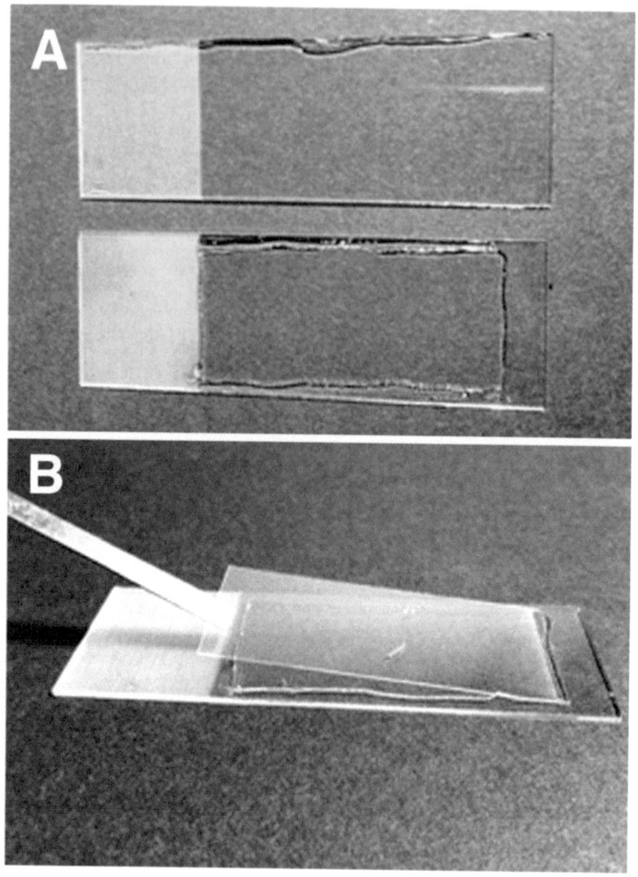

Fig. 4 Microscopic Twitching Assay. After slides are gently separated from one another, the slides are covered in an even layer of TMGG (**a**, *upper slide*). This is then trimmed as described in **step 22** to achieve a smaller rectangle of TMGG with clear edges and an area to label the slide (**a**, *lower slide*). After drying, the slide is inoculated and a glass coverslip gently lowered to create an interstitial interface (**b**)

26. Incubate at 37 °C for at least 4 h, until required for microscopic analysis. The resulting interstitial colony can be visualized by phase-contrast microscopy at high or low magnification (*see* Fig. 2; **Note 22**). Series of time-lapse images can also be captured to create a movie of twitching motility.

4 Notes

1. Agar suppliers: The brand of agar used can significantly affect the results of this assay. It is recommended that you acquire some samples of agar from different suppliers and test in this assay. Some brands of agar may even inhibit twitching motility. For testing with 1 %LBA, you should be able to see a 3–5 cm

diameter interstitial colony at 24 h with a wild-type strain such as PAO1, PAK, or PA103. We use agar No. 1 obtained from Oxoid in our laboratory.

2. Petri plates: We use standard sterile petri plates available from most bulk suppliers, but again some are better than others for enhancing twitching motility. We note that tissue-culture treated petri plates will dramatically enhance twitching motility mediated interstitial colony expansion. If we have an ambiguous result with a strain, we will use a tissue-culture treated petri plate as a follow up to increase the sensitivity of the assay.

3. It is advisable to set aside a stock of bottles that are only used for making TM assay media. Results will be more reproducible if you minimize the possibility of chemical contamination into the bottles.

4. We usually choose the cheapest toothpicks at a supermarket. These tend to be the best for this assay. Some brands are impregnated with antimicrobial compounds and should be avoided.

5. Gellan gum is available under the trade names Gel-Gro, Gelzan, and Gelrite. Gellan gum is optically clearer than agar as a setting agent and does not autofluoresce. This reduces background for both phase-contrast and fluorescence microscopy and enables the addition of fluorescent dyes to the TMGG or to use bacteria expressing fluorescent proteins for the assay.

6. We do not recommend the use of water baths for this method. We have found that the need for rapid movements during this method can lead to contamination. Dry heating is best.

7. We use a heated slide preparation stage for this step. It is the best option if you have access to one. However, you can also use a dry block heater with the blocks turned upside down to maximize the heating surface. It is very important that you can control the temperature so we do not recommend a normal heating plate unless it has a temperature control feature.

8. Coverslip and slides: We have found that some batches of slides and coverslips can be covered in debris. If this happens, we recommend cleaning the slides with a toothbrush and powder cleaner such as Ajax and rinsing well with deionized water. Coverslip can be cleaned also with a powder cleaner by rubbing gently between thumb and forefinger, taking great care not the crack the coverslip or cut the operator. If you are observing the twitching motility with phase-contrast microscopy, you may use #1 coverslips (0.13–0.16 mm thick). However, if you are going to perform high-resolution fluorescence microscopy, we recommend the use of #1.5 coverslips (0.17 mm thick) for optimal imaging.

9. Tissues: We have found that occasionally tissues can be a source of contamination. If you are experiencing contamination in

your assay, make a batch of wet tissues, autoclave and keep aseptically until needed.

10. It is important to dry plates so that no water is visible on the surface but not to over-dry them. If the conditions are too moist, it is possible that the bacteria will swarm or swim instead of twitch and if the plates are too dry, twitching will be inhibited.
11. Homogeneous inoculation culture: as the area of most active twitching motility is located at the outermost edge of the colony, we have found that in order to make the assay as consistent as possible (especially if doing large numbers of assays from the one streak plate), it is best to mix by swirling with a plastic inoculation loop, a small portion of the outer active edge of the plate culture until smooth and homogeneous and use this for all replicate inoculations. We do not recommend using liquid overnight cultures for inoculation as flagellated strains will swim or swarm if too much liquid is deposited in the stab site. Although the liquid will eventually dry out at the inoculation site and the strains will then twitch, the rapid swimming or swarming may give a false positive or a falsely enlarged interstitial twitching colony.
12. It is important to keep the toothpick or pipette tip as upright as possible during the inoculation. Make sure that you have been able to push some bacterial culture to the bottom of the plate by turning the petri plate upside down and visually inspecting. It is also important during this step to not leave a large mound of bacterial inoculation culture on the top of the agar to form a large top colony. If the top colony grows too large, it may inhibit interstitial colony expansion and you will get skewed results. For all initial tests of new strains or environmental stimuli, we recommend using only one inoculation site in the center of the petri plate. Multiple inoculations result in inhibition of twitching motility as the expanding colonies approach each other and this may skew results.
13. If necessary, plates can be stored at 4 °C at this stage for measurement later. However, if you are doing any microscopy or image capture, this needs to be done immediately as the cold storage changes the interstitial colony morphology.
14. It is important to use an appropriate negative control especially if assaying clinical or mutant strains. Some strains may be capable of some, albeit, aberrant twitching motility and therefore show a reduced interstitial colony size. A non-twitching control will allow the assessment of the effects of growth and replication that result in a very small interstitial colony but are not attributable to twitching motility (*see* Fig. 1).

15. Do not leave in the developer solution for more than 30 min. If the solution is left on too long, you will scratch the surface of the agar in **step 22** and this may interfere with being able to see the white halo if it is small. Previous literature reports the use of Coomassie Brilliant Blue to stain the interstitial biofilm [4, 5]. This method involved dehydrating the medium with paper towel, staining and destaining the agar to visualize the interstitial colony. The method described here is simple, rapid, uses fewer consumables and visualizes the interstitial colony without the need for dehydrating the medium.
16. It is important to visually inspect the boiling solution and to make sure that there are no small lumps in the solution. These will interfere with the gellan gum setting.
17. The TMGG solution can be used immediately (after cooling to 65 °C) or it can be allowed to set, stored at room temperature, then melted later when required. When melting TMGG, use a microwave set to low power and monitor the solution to make sure that it does not boil over. Make sure that the solution is completely melted before using, as any small lumps will affect the ability to form flat surfaces for microscopy. Again take care when handling the heated solution.
18. Ensure that the surface is level by using a spirit level. This is important or the slides will have an uneven surface and this makes microscopic inspection very difficult.
19. The slides may be stored for up to 1 week at 4 °C before use. We recommend making all slides needed for 1 week from one batch of molten TMGG and discarding the disused portion.
20. Drying times will vary between different cabinets and will need to be optimized. 15 min drying is a good starting point, but it is useful to come to recognize how much drying a particular set of slides needs through tactile and visual senses. Slides that have been dried the right amount will have a slightly matt sheen over the surface of the TMGG and will be slightly tacky to handle.
21. If the coverslip is too warm, not only will the bacteria be burnt, moisture will be trapped due to condensation between the coverslip and the TMGG. It doesn't matter if there are a small number of bubbles trapped under the coverslip, as some will go away during incubation. You can use areas away from the bubbles for observation. The lowering of the coverslip step is the most difficult and you will get better with practice.
22. We recommend the use of high-quality phase-contrast microscopy rather than bright-field or differential interference contrast (DIC) microscopy to visualize individual cells.

References

1. Burrows LL (2012) *Pseudomonas aeruginosa* twitching motility: type IV pili in action. Annu Rev Microbiol 66:493–520
2. Mattick JS (2002) Type IV pili and twitching motility. Annu Rev Microbiol 56:289–314
3. Whitchurch CB (2006) Type IV pili in Pseudomonas species. In: Ramos J-L, Levesque RC (eds) Pseudomonas. Kluwer, New York, pp 139–188
4. Semmler AB, Whitchurch CB, Mattick JS (1999) A re-examination of twitching motility in *Pseudomonas aeruginosa*. Microbiology 145: 2863–2873
5. Henrichsen J (1983) Twitching motility. Annu Rev Microbiol 37:81–93
6. Hobbs M et al (1993) PilS and PilR, a two-component transcriptional regulatory system controlling expression of type 4 fimbriae in *Pseudomonas aeruginosa*. Mol Microbiol 7(5):669–682
7. Darzins A (1993) The pilG gene product, required for *Pseudomonas aeruginosa* pilus production and twitching motility, is homologous to the enteric, single-domain response regulator CheY. J Bacteriol 175:5934–5944

Chapter 10

Qualitative and Quantitative Assays for Flagellum-Mediated Chemotaxis

José Antonio Reyes Darias, Cristina García-Fontana, Andrés Corral Lugo, Miriam Rico-Jiménez, and Tino Krell

Abstract

A primary driving force during bacterial evolution was the capacity to access compounds necessary for growth and survival. Since the species of the genus *Pseudomonas* are characterized by metabolic versatility, these bacteria have developed chemotactic behaviors towards a wide range of different compounds. The specificity of a chemotactic response is determined by the chemoreceptor, which is at the beginning of the signaling cascade and to which chemoattractants and chemorepellents bind. The number of chemoreceptor genes of *Pseudomonas* species is significantly higher than the average number in motile bacteria. Although some of the receptors have been annotated with a function, the cognate signal molecules for the majority of them still need to be identified. Different qualitative and quantitative methods are presented that can be used to study flagellum-mediated taxis.

Key words Flagellum-mediated taxis, Motility, Chemoreceptor, Agarose plug assay, Quantitative and qualitative capillary assay

1 Introduction

The molecular machinery that mediates chemotaxis in bacteria is significantly more complex than any other bacterial signal transduction system [1]. The core proteins of a chemotactic signaling pathway are the chemoreceptors, the CheA histidine kinase, the CheW coupling protein, the CheY response regulator, the CheR methyltransferase, and the CheB methylesterase. In addition, a number of auxiliary proteins have been identified, like CheD, CheC, CheZ, CheV, and CheX, that are present in fewer chemosensory pathways [1]. The initial step of a typical chemotaxis signaling pathway consists in the binding of signal molecules to the ligand binding region of chemoreceptor, which in turn creates a molecular stimulus that is transmitted across the membrane to the chemoreceptor signaling domain where it modulates the activity of bound CheA. Alteration of the phosphorylation state of CheA will

Alain Filloux and Juan-Luis Ramos (eds.), *Pseudomonas Methods and Protocols*, Methods in Molecular Biology, vol. 1149, DOI 10.1007/978-1-4939-0473-0_10, © Springer Science+Business Media New York 2014

alter the transphosphorylation activity to CheY. Only the phosphorylated form of CheY has the capacity to bind to the flagellar motor, thereby altering its activity [2].

Analysis of sequenced genomes revealed that more than half of the bacterial genomes possess chemotaxis genes [1]. Based on variations in domain architecture and on the participation of different auxiliary proteins in the bacterial genomes analyzed, it was possible to differentiate between 17 different types of chemotaxis pathways [1]. Chemotaxis in this article refers to flagella-mediated taxis since other chemosensory systems were found to mediate type IV pili-mediated taxis. The number of chemoreceptors per genome was found to differ significantly and varies between 1 and 64 receptor genes per genome [3, 4]. *Pseudomonas aeruginosa* and *P. putida* were found to possess 26 chemoreceptors, which is significantly more than the well-studied *E. coli*, which has five chemoreceptors [5]. It is generally assumed that an increased number of chemoreceptors provide the organism with the capacity to respond to a wide range of different signal molecules. In this context, species of the genus Pseudomonas have been found to respond to compounds like amino acids [6], peptides [7], inorganic phosphate [8], glucose [9], the plant hormone ethylene [5], thiocyanates [10], aromatic compounds like benzoate [11], naphthalene [12], toluene [13], polychlorinated biphenylic compounds [14], chloroethylenes [13, 15], benzoxazinoids (plant secondary metabolites present in root exudates) [16], Krebs cycle intermediates [17, 18], the herbicide atrazine [19], cytosine [20], or caffeine [21]. This very large range of compounds is related to the metabolic versatility of *Pseudomonas* and many of these compounds can be used for growth. However, the chemotactic behavior of *Pseudomonas* species is not uniform, which is exemplified by the fact that *P. aeruginosa* is repelled by chloroethylenes [15], whereas *P. putida* is attracted to these compounds [13].

Although the cognate signal molecule for most of the 26 *Pseudomonas* chemoreceptor still needs to be identified, some of the chemoreceptors have been annotated with a function like CtpH and CtpL as receptors for inorganic phosphate [22], PctA, PctB, and PctC for amino acids [23], NahY for naphthalene [24], McpT for mono- and biaromatic compounds [25], McpS for Krebs cycle intermediates [18], or PA2652 for malate [17]. There appears to be a significant diversity amongst *Pseudomonas* species in the type chemoreceptors for a given compound. For example, McpS and PA2652 are the sole chemoreceptors for malate in *P. putida* and *P. aeruginosa* [17, 18]. However, receptors differ entirely in their ligand binding region (LBR). PA2652 has a small cluster I LBR [3] predicted to form a four-helix bundle structure, whereas the McpS-LBR belongs to the larger cluster II family and forms a bimodular structure composed of two 2 long and 4 short helices [18]. Other examples are the receptors NahY and McpT that mediate taxis towards naphthalene in the *P. putida* strains G7 and DOT-T1E, respectively [24, 25]. Although the ligand binding domains of both receptors are predicted to form a four-helix bundle, both domains do not share any significant sequence similarities. These differences in chemoreceptor architecture hamper their functional annotation by sequence comparison, which thus underlines the necessity to identify functions experimentally. In this chapter we summarize experimental approaches to visualize and quantify chemotaxis in *Pseudomonas*. The different methods are of qualitative, semiquantitative, and quantitative nature.

2 Materials

Prepare all solutions using ultrapure water (prepared by purifying deionized water to attain a sensitivity of 18 M Ω cm at 25 °C) and analytical grade reagents.

2.1 Media

1. Minimal saline (MS) medium: 10 mM Na_2HPO_4, 20.6 mM KH_2PO_4, 25 mM NH_4NO_3, 0.8 mM $MgSO_4$, 1.0 ml of trace elements solution per liter.
2. Trace elements solution: 63 mM $FeCl_3$, 3.3 mM $CoCl_2$, 4 mM $CaCl_2$, 1.2 mM Na_2MoO_4, 1.6 mM H_3BO_3, 0.7 mM $ZnSO_4$, 0.7 mM $CuSO_4$, 0.7 mM $MnSO_4$.
3. MS-S medium; MS medium containing 10 mM succinate.
4. MS-Glucose agar: MS medium (see above) supplemented with 10 mM glucose and 0.25 % (w/v) agar.

2.2 Agarose Plug Assays

1. Petri dishes (60 mm diameter).
2. Microscope slides and coverslips.
3. NuSieve GTG low-melting agarose (FMC Bioproducts, Rockland, Maine).

2.3 Capillary Assays

1. Swimming plates: LB medium diluted tenfold, containing 0.25 % (w/v) agarose.
2. Capillary tubes of 1 μl capacity (Drummond Scientific, USA).
3. Microscope slides and coverslips.
4. Bunsen burner.
5. Two pairs of tweezers.
6. 1.5 ml Eppendorf tubes.
7. Bulb for Pasteur pipette and self-made rubber adaptor to connect to capillary.
8. Self-made V-shaped metal rod.
9. MS medium (see above).

10. NuSieve GTG low-melting agarose (FMC Bioproducts, Rockland, Maine).
11. Chemotaxis buffer (CB); 50 mM potassium phosphate buffer, 6 mM $MgCl_2$, 0.1 M EDTA, pH 7.0.

2.4 Assays Using Hydroxypropyl Methylcellulose

1. MS medium.
2. Hydroxypropyl methylcellulose (formulated to give a viscosity of about 4,000 cP in 2 % (w/v) aqueous solution) (Sigma Chemical Co.).
3. Petri dishes (60 mm diameter).
4. Chemotaxis buffer (CB, see above).

2.5 Plate Gradient Assay

Square petri dishes (120 mm × 120 mm) with a grid (Greiner bio one, Cat.-No.: 688102).

3 Methods

3.1 Agarose Plug Assays

The principle of this assay is the immobilization of the chemoattractant into a solid piece of agar, which is then surrounded by a bacterial suspension. Ring formation around the agarose plug is indicative of chemotaxis. Two experimental setups are being used making use of petri dishes or microscope slides and microscope coverslips. This approach is well suited for initial screening, and in case of taxis, a confirmation by other techniques is advised.

3.1.1 Petri-Dish Agarose Plug Assay

1. Cells are grown overnight in MS-S and an aliquot of the resulting culture is diluted with MS-S to reach an OD_{660} of 0.05. The culture is then grown for further 3–5 h until the logarithmic phase (OD_{660} of 0.6–0.8) (*see* **Note 1**).
2. Manipulations are carried out at room temperature. Bacteria are collected by low-speed centrifugation (800 × *g*) on a benchtop microcentrifuge (*see* **Note 2**). Cells are then washed twice with MS medium by consecutive resuspension and low speed centrifugation (800 × *g*).
3. Cells are then resuspended in MS-medium.
4. A 100 μl aliquot of melted agarose (2 %, w/v) containing the chemoattractant is placed into the center of a petri dish (Fig. 1a). In a separate petri dish a 100 μl aliquot of 2 % (w/v) melted agarose containing MS medium is placed as negative control (*see* **Note 3**).
5. After the agarose has solidified (around 15 min) the bacterial suspension (5 ml) is poured around the agarose plug (Fig. 1b, c). Plates should be monitored every 30 min for up to 2 h. Ring formation around the agarose plug is indicative of taxis, no ring formation should be observed for the control experiment.

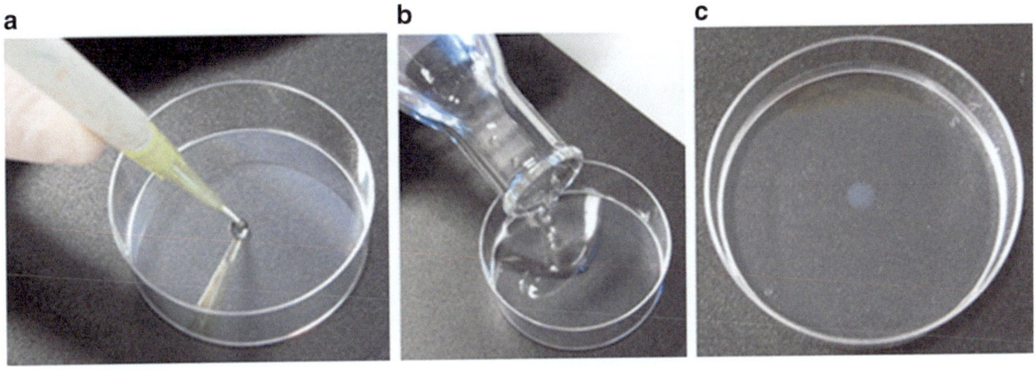

Fig. 1 Experimental setup of the petri-dish agarose plug assay. (**a**) Melted agarose containing the chemoattractant is deposited onto a petri dish. (**b**) After the solidification of the agarose plug the bacterial suspension is added. (**c**) Agarose plug submerged in bacterial suspension immediately after setup

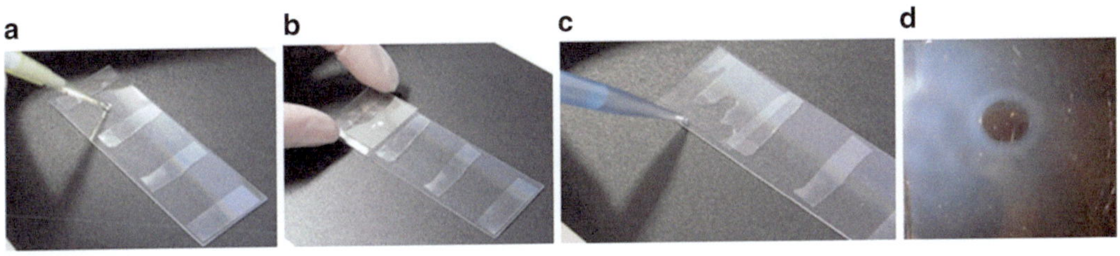

Fig. 2 Experimental setup for microscope slide agarose plug assay. (**a**) Placing the melted agarose drop containing the chemoattractant. (**b**) Covering the field with a microscope coverslip. (**c**) Addition of the bacterial suspension. (**d**) Observation of ring formation indicative of chemoattraction

3.1.2 Microscope Slide Agarose Plug Assay

1. Do **steps 1** and **2** of Subheading 3.1.1 (see above).
2. Using fragments of microscope coverslips a microscope slide is divided in three sections as illustrated in Fig. 2a.
3. Place 5 μl drops of 2 % (w/v) melted agarose containing the chemoattractant onto the fields (Fig. 2a).
4. Cover immediately the corresponding field with a microscope coverslip (Fig. 2b).
5. After solidification of the agarose plug, apply the bacterial suspension (Fig. 2c) to fill the entire space of the chamber.
6. Observe ring formation which typically occurs after several minutes (Fig. 2d).

3.2 Capillary Assays

3.2.1 Optimization of Bacterial Mobility

The need for the optimization of motility is based on empirical observations of several laboratories. The molecular processes that occur during the optimization are to our knowledge not fully understood. Prior to the capillary assays the motility of bacteria was optimized as follows:

1. Swimming plates were inoculated with bacteria by stabbing into the center of the agar followed by an incubation at 30 °C for 24 h.

2. An aliquot of bacteria taken from the outer edge of the swimming zone is used to inoculate another swimming plate in the same manner.
3. After a minimum of three successive passages, bacteria from the outer edge of the last swimming plate are used to inoculate the liquid medium for chemotaxis assays.

3.2.2 Qualitative Capillary Assay

1. Bacteria were grown to an OD_{660} of 0.6–0.8 in MS-S medium at 30 °C. The culture is washed twice with CB (at 30 °C) by consecutive centrifugation at a benchtop microfuge and resuspension.
2. An analysis chamber is made using a V-shaped metal rod, a microscope slide and coverslip (Fig. 3a, b).
3. Capillaries are sealed at one end by melting over a flame (Fig. 3c).
4. The sealed capillary is then warmed over the flame and the open end inserted into the attractant solution, which contains 1 % (w/v) low-melting point agarose in CB (Fig. 3d). As the capillary cools, the chemoattractant solution is sucked into the capillary by capillary force.
5. The filled capillaries are then inserted into the analysis chamber (Fig. 3e). Typically experiments are carried out in triplicates.
6. Freshly grown cells are suspended in CB (*see* **Note 4**) to an OD_{660} of approximately 0.08 and placed into the analysis chamber (Fig. 3f).
7. The tip of the capillary is inspected by a microscope with 40 fold magnification. Cell accumulation at the capillary tip as shown in Fig. 3i is indicative of taxis.

3.2.3 Quantitative Capillary Assay

1. Do the **steps 1–6** of Subheading 3.2.2, except **step 4** in which agarose is omitted from the attractant solution.
2. For quantification purposes the capillaries are removed from the analysis chamber after an exposure for 30 min.
3. The capillary is then rinsed with sterile water or MS medium.
4. To empty the capillary the sealed end of the capillary is broken with a second pair of tweezers (Fig. 3g).
5. The content of the capillary is then transferred into a 1.5 ml Eppendorf tube containing 1 ml of MS medium. Efficient transfer is guaranteed by using a Pasteur pipette bulb connected to the capillary with a self-made rubber adaptor (Fig. 3h).
6. Eppendorf tubes are inverted several times to homogenize contents and 100 μl aliquots are streaked out onto 1.5 % agar plates with M9 medium supplemented with 10 mM of succinate.
7. Plates are incubated for 24–48 h and the number of colony forming units is determined (*see* **Note 5**, Fig. 3j).

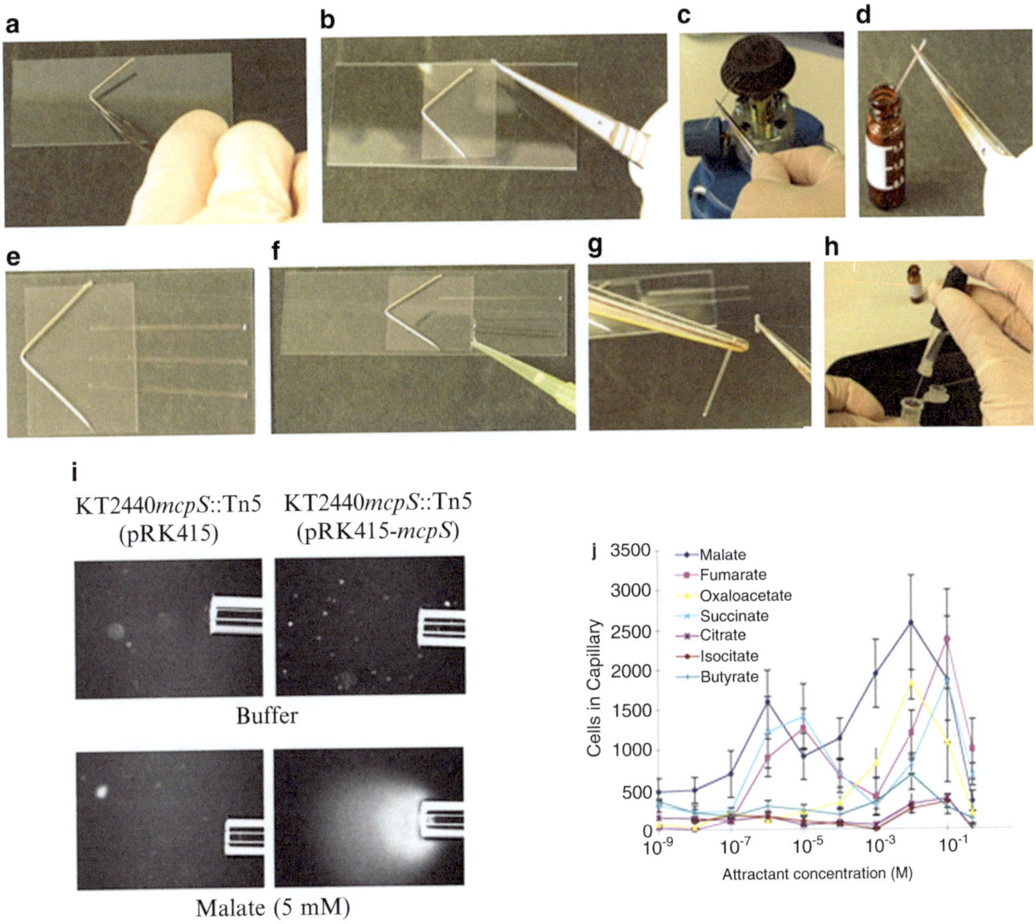

Fig. 3 Experimental setup of capillary assays and final results. (**a**) A V-shaped metal rod is placed onto a microscope slip and (**b**) covered with a coverslip. (**c**) One end of the capillary is sealed on open fire. (**d**) Sealed and warmed capillary is inserted into chemoattractant solution for filling. (**e**) Filled capillaries are inserted into the analysis chamber. (**f**) Bacterial suspension is introduced into the analysis chamber. (**g**) Capillaries are removed from the analysis chamber and the sealed end is broken off. (**h**) Contents of capillary are emptied into an Eppendorf tube containing MS medium. (**i**) Images taken from qualitative capillary assays. Shown is taxis towards buffer (control) and malate of *P. putida* KT2440 Δ*mcpS* and of the same mutant complemented with the *mcpS* gene. (**j**) Results from quantitative capillary assays of *P. putida* KT2440 to different carboxylic acids. Data were taken with permission from ref. 18

3.3 Assay Using Hydroxypropyl Methylcellulose

1. Harvest 40 ml of cells in the logarithmic growth phase (OD_{660} of 0.6–0.8) and resuspend in 12 ml of CB or CB containing 10 mM succinate (*see* **Note 6**).
2. Add to the bacterial suspension 3 ml of a 1 % aqueous solution of hydroxypropylmethylcellulose.
3. The resulting suspension is then placed into petri dishes (60 mm) just enough (5–6 ml) to cover the bottom of the dish (height of around 3 mm).

3. Add a small amount of the chemoattractant to the center of the dish. This can either be solids such as poorly soluble polycyclic aromatic hydrocarbons or 10 μl drops of concentrated soluble chemoattractants. Place MS medium into another petri dish as the negative control.
4. The plates are kept at room temperature and are inspected at regular intervals during around 2 h. A chemotactic response results in the formation of a turbid ring near the center of the petri dish after about an hour.

3.4 Plate Gradient Assays

1. Square petri dishes (120 mm × 120 mm) with vents (Greiner bio one) are filled with 50 ml of semisolid agar containing minimal MS medium, 10 mM glucose (*see* **Note 7**) and 0.25 % (w/v) agar. Plates are cooled at room temperature for at least 1.5 h.
2. At the vertical central line of the plate 10 μl aliquots of chemoattractant solution dissolved in MS are placed at regular distance (Fig. 4a). Plates are incubated for 12–16 h at 4 °C. During this time the chemoattractant will diffuse into the agar and create a concentration gradient.
3. Bacteria are grown overnight in MS medium supplemented with 10 mM succinate to saturation and diluted to an OD_{660} of 0.8–1.
4. Cells are then washed twice with MS medium by consecutive resuspension and centrifugation at 3750 × *g* for 3 min.
5. Two microliter aliquots of bacterial suspension are then placed horizontally to each of the chemoattractant spots but with

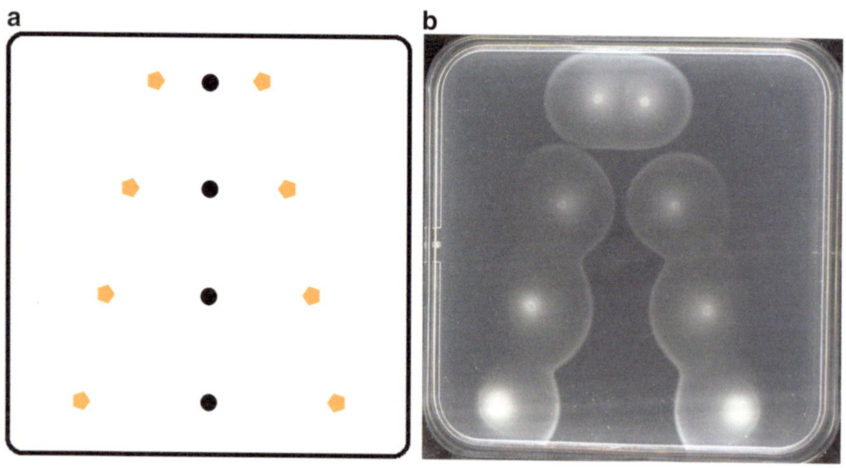

Fig. 4 The plate gradient assay. (**a**) Experimental setup. *Black dots* represent the deposited aliquots of the chemoattractant. Objects in *yellow* are aliquots of the bacterial suspension. (**b**) Typical image of plate gradient assay. Shown is taxis of *P. putida* KT2440R to 10 % (x/v) casamino acids. The image was taken 16 h after the inoculation of the plate

varying distance (0.5–3.5 cm) to the spot (Fig. 4a). This is done on both sides of the chemoattractant spot and different bacterial strains can be used for each side (such as the wild type strain and a mutant). Plates are incubated at 30 °C for 16–30 h.

6. Plates are inspected after 16–30 h and the formation of oval bacterial accumulation zones are indicative of chemotaxis (Fig. 4b).

4 Notes

1. The bacterial culture must achieve the late exponential phase, since the development of the flagella occurs in this growth phase.
2. The cell harvest by centrifugation is done at low speed (5 min in $800 \times g$) to prevent the loss/damage of flagella.
3. There are many different types of taxis (aerotaxis, temperature taxis, pH taxis, etc.). This has to be kept in mind and appropriate controls have to be made. Control experiments using the buffer for cell resuspension are essential. In addition avoid the introduction of air bubbles into the agarose plug (aerotaxis). When analyzing pH-active compounds adjust pH to that of the buffer used for cell resuspension.
4. The presence of EDTA in CB has been found to be essential for chemotaxis in *P. aeruginosa* [26], which indicates that heavy metals inhibit motility and chemotaxis.
5. To normalize for day-to-day differences, the chemotactic response can also be expressed as a ratio of the accumulation in attractant capillaries to that in control capillaries (the relative response). This value shows less variability between different cell preparations than the absolute values. Figure 3j shows two peaks of chemotaxis in function of the chemoattractant concentration, which illustrates the need to conduct chemotaxis assays using a concentration range instead of a single concentration.
6. Because chemotaxis is an energy-requiring process, a potential energy source can be added to the MS medium to stimulate the chemotactic response.
7. We use 10 mM of glucose as carbon source. No significant response to 30 mM glucose was detected in cells grown on succinate or on malate. The lack of response to glucose of *P. aeruginosa* when grown on succinate, citrate or malate is due to the lacking expression of the glucose-binding protein, necessary for the chemotactic response [27]. Well-fed bacteria show reduced responses and the use of soft agar swim plates is recommended if responses to N- or C-sources are to be detected.

Acknowledgements

We acknowledge financial support from the Andalusian regional government Junta de Andalucía (grant P09-RNM-4509 to T.K.) and the Spanish Ministry for Economy and Competitiveness (grant Bio2010-16937 to T.K.). We thank Juan-Luis Ramos for reading the chapter and his continuous support.

References

1. Wuichet K, Zhulin IB (2010) Origins and diversification of a complex signal transduction system in prokaryotes. Sci Signal 3:r50
2. Hazelbauer GL, Falke JJ, Parkinson JS (2008) Bacterial chemoreceptors: high-performance signaling in networked arrays. Trends Biochem Sci 33:9–19
3. Lacal J, Garcia-Fontana C, Munoz-Martinez F, Ramos J-L, Krell T (2010) Sensing of environmental signals: classification of chemoreceptors according to the size of their ligand binding regions. Environ Microbiol 12:2873–2884
4. Alexandre G, Greer-Phillips S, Zhulin IB (2004) Ecological role of energy taxis in microorganisms. FEMS Microbiol Rev 28:113–126
5. Kato J, Kim HE, Takiguchi N, Kuroda A, Ohtake H (2008) *Pseudomonas aeruginosa* as a model microorganism for investigation of chemotactic behaviors in ecosystem. J Biosci Bioeng 106:1–7
6. Kuroda A, Kumano T, Taguchi K, Nikata T, Kato J, Ohtake H (1995) Molecular cloning and characterization of a chemotactic transducer gene in *Pseudomonas aeruginosa*. J Bacteriol 177:7019–7025
7. Kelly-Wintenberg K, Montie TC (1994) Chemotaxis to oligopeptides by *Pseudomonas aeruginosa*. Appl Environ Microbiol 60: 363–367
8. Kato J, Ito A, Nikata T, Ohtake H (1992) Phosphate taxis in *Pseudomonas aeruginosa*. J Bacteriol 174:5149–5151
9. Sly LM, Worobec EA, Perkins RE, Phibbs PV Jr (1993) Reconstitution of glucose uptake and chemotaxis in *Pseudomonas aeruginosa* glucose transport defective mutants. Can J Microbiol 39:1079–1083
10. Ohga T, Masduki A, Kato J, Ohtake H (1993) Chemotaxis away from thiocyanic and isothiocyanic esters in *Pseudomonas aeruginosa*. FEMS Microbiol Lett 113:63–66
11. Harwood CS (1989) A methyl-accepting protein is involved in benzoate taxis in *Pseudomonas putida*. J Bacteriol 171:4603–4608
12. Grimm AC, Harwood CS (1997) Chemotaxis of Pseudomonas spp. to the polyaromatic hydrocarbon naphthalene. Appl Environ Microbiol 63:4111–4115
13. Parales RE, Ditty JL, Harwood CS (2000) Toluene-degrading bacteria are chemotactic towards the environmental pollutants benzene, toluene, and trichloroethylene. Appl Environ Microbiol 66:4098–4104
14. Gordillo F, Chavez FP, Jerez CA (2007) Motility and chemotaxis of Pseudomonas sp. B4 towards polychlorobiphenyls and chlorobenzoates. FEMS Microbiol Ecol 60:322–328
15. Shitashiro M, Kato J, Fukumura T, Kuroda A, Ikeda T, Takiguchi N, Ohtake H (2003) Evaluation of bacterial aerotaxis for its potential use in detecting the toxicity of chemicals to microorganisms. J Biotechnol 101:11–18
16. Neal AL, Ahmad S, Gordon-Weeks R, Ton J (2012) Benzoxazinoids in root exudates of maize attract *Pseudomonas putida* to the Rhizosphere. PLoS One 7:e35498
17. Alvarez-Ortega C, Harwood CS (2007) Identification of a malate chemoreceptor in *Pseudomonas aeruginosa* by screening for chemotaxis defects in an energy taxis-deficient mutant. Appl Environ Microbiol 73: 7793–7795
18. Lacal J, Alfonso C, Liu X, Parales RE, Morel B, Conejero-Lara F, Rivas G, Duque E, Ramos J-L, Krell T (2010) Identification of a chemoreceptor for tricarboxylic acid cycle intermediates: differential chemotactic response towards receptor ligands. J Biol Chem 285: 23126–23136
19. Liu X, Parales RE (2009) Bacterial chemotaxis to atrazine and related s-triazines. Appl Environ Microbiol 75:5481–5488
20. Liu X, Wood PL, Parales JV, Parales RE (2009) Chemotaxis to pyrimidines and identification of a cytosine chemoreceptor in *Pseudomonas putida*. J Bacteriol 191:2909–2916
21. Dash SS, Sailaja NS, Gummadi SN (2008) Chemotaxis of Pseudomonas sp. to caffeine

and related methylxanthines. J Basic Microbiol 48:130–134

22. Wu H, Kato J, Kuroda A, Ikeda T, Takiguchi N, Ohtake H (2000) Identification and characterization of two chemotactic transducers for inorganic phosphate in *Pseudomonas aeruginosa*. J Bacteriol 182:3400–3404
23. Taguchi K, Fukutomi H, Kuroda A, Kato J, Ohtake H (1997) Genetic identification of chemotactic transducers for amino acids in *Pseudomonas aeruginosa*. Microbiology 143(Pt 10):3223–3229
24. Grimm AC, Harwood CS (1999) NahY, a catabolic plasmid-encoded receptor required for chemotaxis of *Pseudomonas putida* to the aromatic hydrocarbon naphthalene. J Bacteriol 181:3310–3316
25. Lacal J, Munoz-Martinez F, Reyes-Darias JA, Duque E, Matilla M, Segura A, Calvo JJ, Jimenez-Sanchez C, Krell T, Ramos J-L (2011) Bacterial chemotaxis towards aromatic hydrocarbons in Pseudomonas. Environ Microbiol 13:1733–1744
26. Moulton RC, Montie TC (1979) Chemotaxis by *Pseudomonas aeruginosa*. J Bacteriol 137: 274–280
27. Stinson MW, Cohen MA, Merrick JM (1977) Purification and properties of the periplasmic glucose-binding protein of *Pseudomonas aeruginosa*. J Bacteriol 131:672–681

Chapter 11

Microscopic Analysis: Morphotypes and Cellular Appendages

Olga Zaborina, John Alverdy, Megha Shah, and Yimei Chen

Abstract

Microscopic analysis is a well-accepted technique in microbiology to characterize a single colony and single cell morphotypes. Although colony morphotype can be imaged by light microscopy, morphotypes of microbial cells are normally viewed by transmission electron microscopy (TEM). The specific location of proteins on microbial cell surface or inside the cells is normally detected using immunoelectron microscopy (IEM). Here we describe the technique of detection of type-specific cellular appendages that are produced by several strains of *Pseudomonas aeruginosa* under conditions of phosphate limitation. The ability to produce the appendages in *P. aeruginosa* herein described is thus far attributable only to certain multidrug-resistant strains isolated from critically ill patients. The appearance of appendages is highly suppressed in phosphate rich media and enhanced during growth in phosphate depleted media. Under these dual conditions, the absence or presence of appendages correlates with an adhesive and virulent phenotype in *P. aeruginosa* strains. The proper technique for appendage visualization is critical to facilitate the necessary studies to further elucidate their structure and function.

Key words Transmission electron microscopy (TEM), Immunoelectron microscopy (IEM), Cellular PstS-containing appendages, *Pseudomonas aeruginosa*

1 Introduction

Outer cell surface appendages play a crucial role in the virulence of bacterial pathogens [1, 2]. Bacteria that possess surface appendages are better able to attach to host tissues. The adhesive capacity (adhesiveness) provides a critical physiologic advantage to bacteria to competitively acquire nutrients from host cells when nutrients are scarce. The capacity of bacteria to adhere is affected by multiple elements within the local ecology one of which our group has identified to be the phosphate concentration [3].

Phosphorus is an essential element that supports the growth of all living organisms [4–7]. Microorganisms are exquisitely sensitive to the concentration of extracellular phosphate and express varying phenotypes when phosphate is limited [8–10]. Owing to

Alain Filloux and Juan-Luis Ramos (eds.), *Pseudomonas Methods and Protocols*, Methods in Molecular Biology, vol. 1149, DOI 10.1007/978-1-4939-0473-0_11, © Springer Science+Business Media New York 2014

the presence of conserved phosphosensory and phosphoregulatory pathways that sense and respond to fluctuations in extracellular phosphate, many microbes express enhanced virulence during phosphate limitation in order to invade host tissues for the purposes of phosphate acquisition [9]. Among such pathogens is *Pseudomonas aeruginosa*, a conditional and opportunistic pathogen that can cause lethal sepsis when it colonizes the gut of critically ill patients [11, 12]. Results from prior studies in our laboratory demonstrate that in mice, local intestinal mucosal phosphate becomes depleted during surgical injury and triggers intestinal *P. aeruginosa* to express a virulent phenotype characterized by enhanced ability to invade mucosal tissues and cause lethal sepsis [13]. The generalizability of phosphate mediated triggering of the virulence of opportunistic intestinal pathogens is wide and involves various pathogens that colonize the gut of critically ill patients [8, 14]. In vitro work in our laboratory has demonstrated that isolation of strains of *P. aeruginosa* from critically ill patients with the potential to cause sepsis can be provoked to express novel and unusual appendages rich in PstS when grown under low phosphate conditions [3]. These studies underscore the value and importance of defining the ultrastructural events in single microbial cells during conditions similar to those encountered in vivo in order to uncover novel virulence traits in key microbes known to cause sepsis.

Transmission and immunoelectron microscopy allows for imaging of single microbial cells and the detection of specific proteins in cells at very high resolution. The technique of transmission electron microscopy applied to microbial structure is based on the transmission and interaction of a beam of electrons with microbial surface structure. TEM was first introduced 80 years ago (http://inventors.about.com/od/mstartinventions/a/microscope_2.htm) [15, 16], and since then, advancement of electron microscopes design has greatly improved its resolution. The critical step in TEM is sample preparation technique. The thickness of a sample cannot exceed nanometer scale, with the optimum of ten nanometer and a greatest thickness of hundreds nanometers. The preparation technique for a given sample is dictated by the sample structure per se. For biological objects, previous knowledge regarding object structure dictates the specificity of the sample preparation. Here, we demonstrate the application of TEM analysis for imaging of outer surface appendages on bacterial cells that are strongly influenced by microenvironmental conditions such as phosphate limitation. We were able to demonstrate that slightly curved extracellular outer surface appendages are formed by multidrug-resistant strains of *P. aeruginosa*. Using immunoelectron microscopy we were able to visualize high phosphate affinity binding proteins, PstS harbored within these appendages [3]. The PstS appendages production and structure stability are strongly affected by the extracellular phosphate concentration, and visualization of the appendages requires a specific approach. This chapter describes the details of electron microscopy technique for appendage visualization.

2 Materials

2.1 Bacterial Plates

1. *TSB plates.* Suspend 30 g of tryptic soy broth media (TSB, Fluka, cat#T8907) and 15 g of agar (Fisher, laboratory grade, cat# A360) in 1 L of distilled water. Autoclave at 121 °C for 20 min. Chill in water bath to 60 °C, add antibiotic if needed, and pour into dishes.
2. *PIA plates.* Use Difco™ Pseudomonas isolation agar (BD Ref #292710, distributed by Spectrum, cat#743-32406) to prepare the plates. Suspend 45 g of PIA in 980 ml of distilled water, add 20 ml of glycerol, and heat with agitation to completely dissolve the powder. Autoclave at 121 °C for 20 min. Chill in water bath to 60 °C, add antibiotic if needed, and pour into dishes (*see* **Note 1**).
3. *10 % glycerol.* Mix 10 ml of glycerol (Fisher, G33-1) and 90 ml of distilled water. Sterilize by filtering via 0.22 μm filter (Whatman, cat# 6780-2502).
4. *Sterile 4.5 mm glass plating beads.* Rattler™ Plating Beads (Zymo Research, cat# S1001) are needed to seed bacterial cells at normalized density.
5. *Lazy-L-spreaders* (Phenix, Fisher, Sigma, Denville Scientific Inc. etc.). Disposable Lazy-L-spreaders are needed to spread bacterial suspension on PIA plates.

2.2 Preparation of Samples for Transmission Electron Microscopy (Negative Staining)

1. 24 and 48 h old *P. aeruginosa* PIA plates (prepared as described in Subheading 3.1, **step 2**).
2. Glow-discharged (using evaporator Edwards Auto 306) 400 mesh formvar/carbon-coated copper grids.
3. Ultrapure distilled water (Milli-Q water purification system, Millipore) passed through 0.22 μm filter (Whatman, cat# 6780-2502).
4. *TE buffer.* 10 mM Tris–HCl, 1 mM EDTA, pH 8.0. Prepare TE buffer using ultrapure distilled water. Purify prepared buffer by passing through 0.22 μm filter (Whatman, cat# 6780-2502). Aliquot and keep at −20 °C.
5. *1 % aqueous solution of uranyl acetate.* Prepare uranyl acetate solution using ultrapure distilled water. Purify prepared solution by passing through 0.22 μm filter (Whatman, cat# 6780-2502).
6. *Forceps.* Needed to manipulate with support grids.

7. *Pieces of Whatman filter paper.* Needed to remove drops of solutions from prepared support grids.
8. Micropipette.

2.3 Preparation of Samples for Immunoelectron Microscopy (Gold-Labeling)

1. *Rabbit anti-PstS antibodies.* The polyclonal rabbit antiserum against 192–212 peptide KEEALCKGDFRPNVNEQPGS of PA5369 from *P. aeruginosa* PAO1 was produced in rabbits (SynPep Corporation, Dublin, CA). Anti-PA5369 antibodies were affinity purified by AminoLink Plus Immobilization Kit (Pierce) using 192–212 peptide to create an affinity column. Given the conservative structure of PstS proteins among strains of *P. aeruginosa*, we assumed that the antibodies raised again the proteins/specific peptides from one strain can be used for different strains of *Pseudomonas* spp. As we demonstrate, these PstS antibodies recognize PstS in different species of *P. aeruginosa* [17].
2. Glow-discharged (using evaporator Edwards Auto 306) 400 mesh formvar/carbon-coated nickel grids.
3. PBS.
4. 1 % BSA.
5. Humidified chamber.
6. Goat anti-rabbit IgG conjugated with 10 nm gold particles (TED PELLA).
7. 1 % glutaraldehyde in PBS.
8. 2.5 % lead citrate.
9. Saturated uranyl acetate.

2.4 Imaging of Bacterial Cell Surface

FEI Tecnai F30 TEM operated at 300 kV equipped with Gatan CCD digital micrograph.

3 Methods

3.1 Preparation of Bacteria for TEM Analysis (Negative Staining)

1. The bacterial strains are stored as 10 % glycerol stocks at −80 °C. Plate *P. aeruginosa* from frozen stock on solidified TSB medium and grow overnight at 37 °C. Suspend the colonies in 10 % glycerol at $OD_{600nm} = 0.5$.
2. Make tenfold dilutions in 10 % glycerol and spread 50 μl of 10^{-3} dilution on PIA plate. Alternatively, pour 50 μl of the 10^{-2} to 10^{-3} dilution on PIA plate and add 6–8 sterile beads/plate. Shake the plate for 30 s and then remove the beads. You can also use overnight grown liquid TSB bacterial culture to inoc-

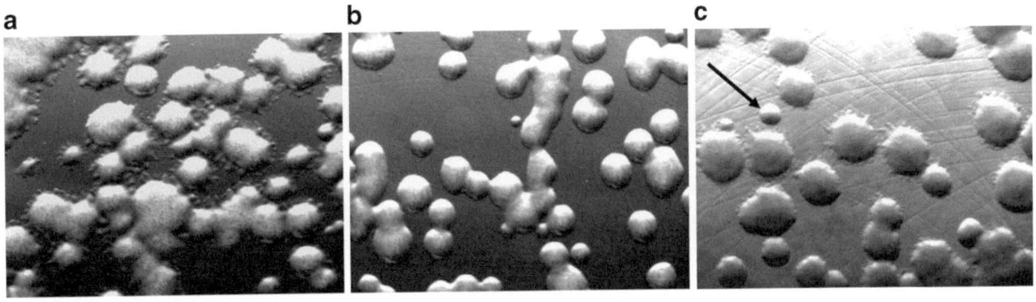

Fig. 1 Colony morphotypes of *P. aeruginosa* MDR25 grown on (**a**) PIA, Gm 50 μg/ml and (**b**) PIA, Gm 50 μg/ml supplemented with 25 mM potassium phosphate buffer, pH 6.0. (**c**) The appearance of small smooth colonies on PIA, Gm 50 μg/ml

ulate PIA plates. In this case, take 5 μl of overnight culture and add it to 995 μl of 10 % glycerol. Vortex the mixture. Take further 2.5 μl from diluted bacterial suspension and add it to 995 μl of 10 % glycerol. Vortex the mixture. Plate 50 μl of the final dilution on PIA plate, add 6–8 sterile beads, and shake the plates. Remove the beads and incubate plates at 37 °C for 24–48 h. The 1- and 2-day-old PIA plates are used for TEM analysis (*see* **Note 2**).

Important:

1. We have noticed that colony density plays an important role in that the highest yield of appendages develops at a confluence of about 70 %. Perhaps this represents a density at which nutrients are most efficiently consumed from the medium but where colonies are still separated from each other (*see* Fig. 1).
2. The addition of phosphate into PIA abrogates the appendage formation by *P. aeruginosa* (*see* Fig. 1).
3. Based on the colony morphology, assumptions can be made regarding which colony cells produce appendages. There is a lower probability of finding appendages in cells from smooth colonies than from wrinkled colonies. Images of *P. aeruginosa* colony phenotypes on PIA supplemented with 50 μg/ml of gentamicin are represented on Fig. 1. Mostly wrinkled colonies are seen in PIA (Fig. 1a) while smooth colonies are seen in PIA supplemented with 25 mM phosphate buffer, pH 6.0 (Fig. 1b). The appearance of small smooth colonies on PIA plates (Fig. 1c, shown by black arrow) correlates with decreased appendages formation. Figure 1c was incorporated from Zaborina et al. [3].

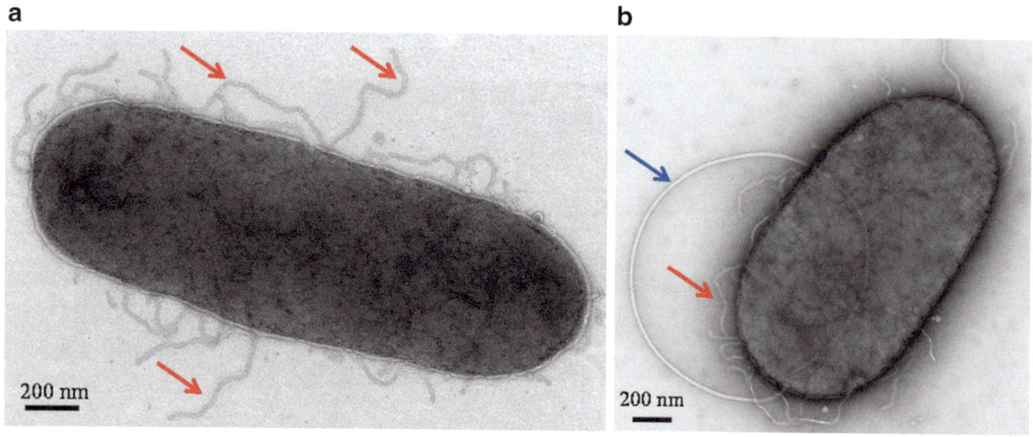

Fig. 2 Visualization of appendages by TEM in *P. aeruginosa*. (**a**) Image of MDR25 producing appendages on PIA, Gm 50 μg/ml. Appendages are pointed with *red arrows*. (**b**) Image of MDR1 producing the appendages (shown by *red arrow*) and flagella (shown by *blue arrow*). Images are reprinted from Zaborina et al., PLoS Pathogens, 2008

3.2 TEM Microscopy (Negative Staining)

Drop 10 μl of ultrapure sterile water on the chosen colony(s) on PIA plate and cover the drop of water with the glow-discharged 400 mesh formvar/carbon-coated copper support grid. In 30 s, remove the support grid, wash several times with TE buffer, and stain briefly with 1 % uranyl acetate. Remove the extra drops of uranyl acetate from the grid, and dry on air. The samples are ready now for viewing with the electron microscope. The sample on the support grids can be kept for at least 2 days.

3.3 Immunoelectron Microscopy (Gold Labeling)

Drop 10 μl of ultrapure sterile water on the chosen colony(s) on PIA plate and cover the drop of water with the 200 mesh formvar-coated nickel grids. In 30 s, remove the support grid, and wash with several drops of TE buffer. Treat with PBS for 20 min, and then with 1 % BSA for 20 min followed by transferring to anti-PA5369 antibodies diluted in 1 % BSA. Incubate in a humidified chamber for 4 h, at room temperature, followed by extensive washing with PBS (six times), blocking with 0.5 % BSA for 25 min, and incubating in the humidified chamber for 1 h with goat anti-rabbit IgG conjugated with 10 nm gold particles (TED PELLA) at 1:10 dilution in 0.5 % BSA. Wash grids with PBS three times, fix with 1 % glutaraldehyde in PBS for 10 min, wash with water three times and stain briefly with saturated uranyl acetate and then with 2.5 % lead citrate. Dry on air at room temperature.

1. Imaging of PstS appendages with FEI Tecnai F30 electron microscope equipped with Gatan CCD digital micrograph.
2. The PstS appendages are clearly distinguishable from flagella and pili.

It is easy to visualize PstS appendages on TEM images. The appendages are slightly curved, have an irregular length that can

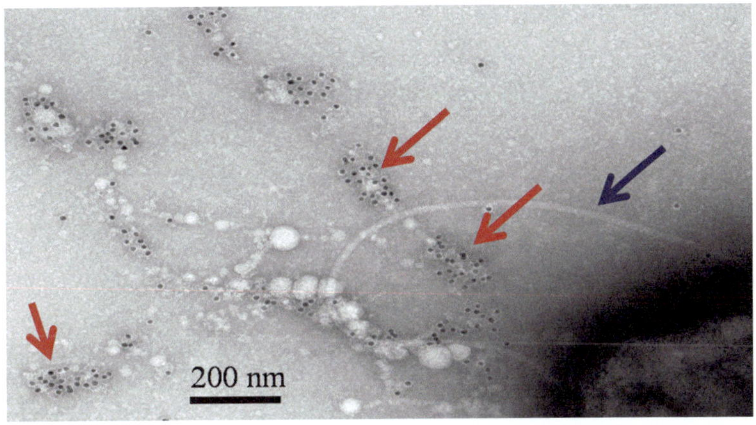

Fig. 3 Immunoelectron microscopy image of *P. aeruginosa* MDR25 grown on PIA, Gm 50 μg/ml. Primary anti-PstS antibodies were used for staining. PstS proteins are depicted by *red arrows*. Flagellum is indicated by *blue arrow*

vary from tens nanometer to hundreds nanometer long, and surround the cells. The diameter of appendages is about 20 nm. The TEM image of *P. aeruginosa* strain MDR25 with appendages is displayed on Fig. 2a (indicated by red arrows). The appendages seem to envelope the cell as seen on Fig. 2a. Although *P. aeruginosa* flagella are also around 20 nm thick, they are very long and have a sharp bend just outside the outer membrane. They project from the cell pole in *P. aeruginosa*. The differences between the flagella (shown by blue arrow) and the appendages (shown by red arrow) are seen in TEM image of *P. aeruginosa* strain MDR1 displayed in Fig. 2b. Pili are also obviously distinguishable from the appendages as they have a smaller width of ~6 nm and extend from a cell for up to 2 μm.

3. PstS localization on the appendages
   Cells of strain MDR25 have been harvested from PIA plates, loaded on grids, and incubated with anti-PA5369 antibody followed by incubation with gold-labeled goat anti-rabbit antibodies (Fig. 3). Gold spots correspond to PstS proteins localized to appendages. Note the fragility of the appendages after the antibody staining. In contradistinction, flagellum is seen to be intact (Fig. 3, shown by blue arrow).

4 Conclusion

Innovative methods in single cell technology are emerging to delineate the structure and function of novel microbial organelles that have evolved under historically unprecedented conditions as a result of the selective pressure of modern medical treatments.

Understanding the molecular mechanisms involved in their formation and function will lead to novel antisepsis strategies against multidrug-resistant pathogens when anti-antibiotic approaches fail.

Although the methods described in this chapter have been developed to visualize the specific PstS-containing appendages in *P. aeruginosa*, these methods are straightforward and can be applied to detect other extracellular components including flagella and pili. Attention should be paid to the specificity and purity of primary antibodies (affinity purified antibodies would be the best) and their concentration. Negative controls (no primary antibodies) for the specificity of secondary gold-labeled antibodies should be included in experiments.

5 Notes

1. BD Company produces different types of Pseudomonas agar. Do not use Difco™ Pseudomonas agar F (Ref#244820), as it contains dipotassium phosphate.
2. Culture media composition and growth conditions such as oxygen concentration (hypoxia vs normoxia), co-incubation with soluble host-stress-derived factors, etc. can vary to elucidate the additional stimulating or inhibiting factors for appendages formation.

Acknowledgments

This study was funded by the National Institute of Health (RO1-GM062344-12).

References

1. Josenhans C, Suerbaum S (2002) The role of motility as a virulence factor in bacteria. Int J Med Microbiol 291:605–614
2. Jonson AB, Normark S, Rhen M (2005) Fimbriae, pili, flagella and bacterial virulence. Contrib Microbiol 12:67–89
3. Zaborina O, Holbrook C, Chen Y, Long J, Zaborin A et al (2008) Structure-function aspects of PstS in multi-drug-resistant *Pseudomonas aeruginosa*. PLoS Pathog 4:e43
4. Rouached H, Arpat AB, Poirier Y (2010) Regulation of phosphate starvation responses in plants: signaling players and cross-talks. Mol Plant 3:288–299
5. Yang XJ, Finnegan PM (2010) Regulation of phosphate starvation responses in higher plants. Ann Bot 105:513–526
6. Lin WY, Lin SI, Chiou TJ (2009) Molecular regulators of phosphate homeostasis in plants. J Exp Bot 60:1427–1438
7. Hanrahan G, Salmassi TM, Khachikian CS, Foster KL (2005) Reduced inorganic phosphorus in the natural environment: significance, speciation and determination. Talanta 66: 435–444
8. Romanowski K, Zaborin A, Valuckaite V, Rolfes RJ, Babrowski T et al (2012) *Candida albicans* isolates from the gut of critically ill patients respond to phosphate limitation by expressing filaments and a lethal phenotype. PLoS One 7:e30119
9. Lamarche MG, Wanner BL, Crepin S, Harel J (2008) The phosphate regulon and bacterial virulence: a regulatory network connecting

phosphate homeostasis and pathogenesis. FEMS Microbiol Rev 32:461–473

10. Rao NN, Gomez-Garcia MR, Kornberg A (2009) Inorganic polyphosphate: essential for growth and survival. Annu Rev Biochem 78:605–647
11. Alverdy J, Holbrook C, Rocha F, Seiden L, Wu RL et al (2000) Gut-derived sepsis occurs when the right pathogen with the right virulence genes meets the right host: evidence for in vivo virulence expression in *Pseudomonas aeruginosa*. Ann Surg 232:480–489
12. Alverdy JC, Chang EB (2008) The re-emerging role of the intestinal microflora in critical illness and inflammation: why the gut hypothesis of sepsis syndrome will not go away. J Leukoc Biol 83:461–466
13. Long J, Zaborina O, Holbrook C, Zaborin A, Alverdy J (2008) Depletion of intestinal phosphate after operative injury activates the virulence of *P. aeruginosa* causing lethal gut-derived sepsis. Surgery 144:189–197
14. Norris FA, Wilson MP, Wallis TS, Galyov EE, Majerus PW (1998) SopB, a protein required for virulence of *Salmonella dublin*, is an inositol phosphate phosphatase. Proc Natl Acad Sci U S A 95:14057–14059
15. Ruska E (1987) Nobel lecture. The development of the electron microscope and of electron microscopy. Biosci Rep 7:607–629
16. Ruska E (1980) The early development of electron lenses and electron microscopy. Microsc Acta (Suppl):1–140
17. Shah M et al (2014) Localization of DING proteins on PstS-containing outer-surface appendages of *Pseudomonas aeruginosa*. FEMS Microbiol Lett 352:54–61

Part II

Proteins and Protein Transport

Chapter 12

Determination of Lipolytic Enzyme Activities

Karl-Erich Jaeger and Filip Kovacic

Abstract

Pseudomonas aeruginosa is a versatile human opportunistic pathogen that produces and secretes an arsenal of enzymes, proteins and small molecules many of which serve as virulence factors. Notably, about 40 % of *P. aeruginosa* genes code for proteins of unknown function, among them more than 80 encoding putative, but still unknown lipolytic enzymes. This group of hydrolases (EC 3.1.1) is known already for decades, but only recently, several of these enzymes have attracted attention as potential virulence factors. Reliable and reproducible enzymatic activity assays are crucial to determine their physiological function and particularly assess their contribution to pathogenicity. As a consequence of the unique biochemical properties of lipids resulting in the formation of micellar structures in water, the reproducible preparation of substrate emulsions is strongly dependent on the method used. Furthermore, the physicochemical properties of the respective substrate emulsion may drastically affect the activities of the tested lipolytic enzymes. Here, we describe common methods for the activity determination of lipase, esterase, phospholipase, and lysophospholipase. These methods cover lipolytic activity assays carried out in vitro, with cell extracts or separated subcellular compartments and with purified enzymes. We have attempted to describe standardized protocols, allowing the determination and comparison of enzymatic activities of lipolytic enzymes from different sources. These methods should also encourage the *Pseudomonas* community to address the wealth of still unexplored lipolytic enzymes encoded and produced by *P. aeruginosa*.

Key words *Pseudomonas aeruginosa*, Lipase, Esterase, Phospholipase A, Lysophospholipase, Agar-plate assay, Colorimetric assay, Fluorometric assay, Titrimetric assay, Lipase fingerprinting, Enantioselectivity assay

1 Introduction

Lipases and phospholipases A (PLA) are carboxylester hydrolases (EC 3.1.1) that catalyze the cleavage of ester bonds for example of triacylglycerol and phospholipid substrates, respectively. Both esterases and lipases hydrolyze triacylglycerols; however, esterases show specificity for water-soluble, short-chain fatty acid (less than ten carbon atoms) substrates, while lipases usually hydrolyze water-insoluble, long-chain fatty acid (more than nine carbon atoms) substrates [1–3]. Most lipases show strongly enhanced activity at the interface between the hydrophobic triacylglycerol

Alain Filloux and Juan-Luis Ramos (eds.), *Pseudomonas Methods and Protocols*, Methods in Molecular Biology, vol. 1149, DOI 10.1007/978-1-4939-0473-0_12, © Springer Science+Business Media New York 2014

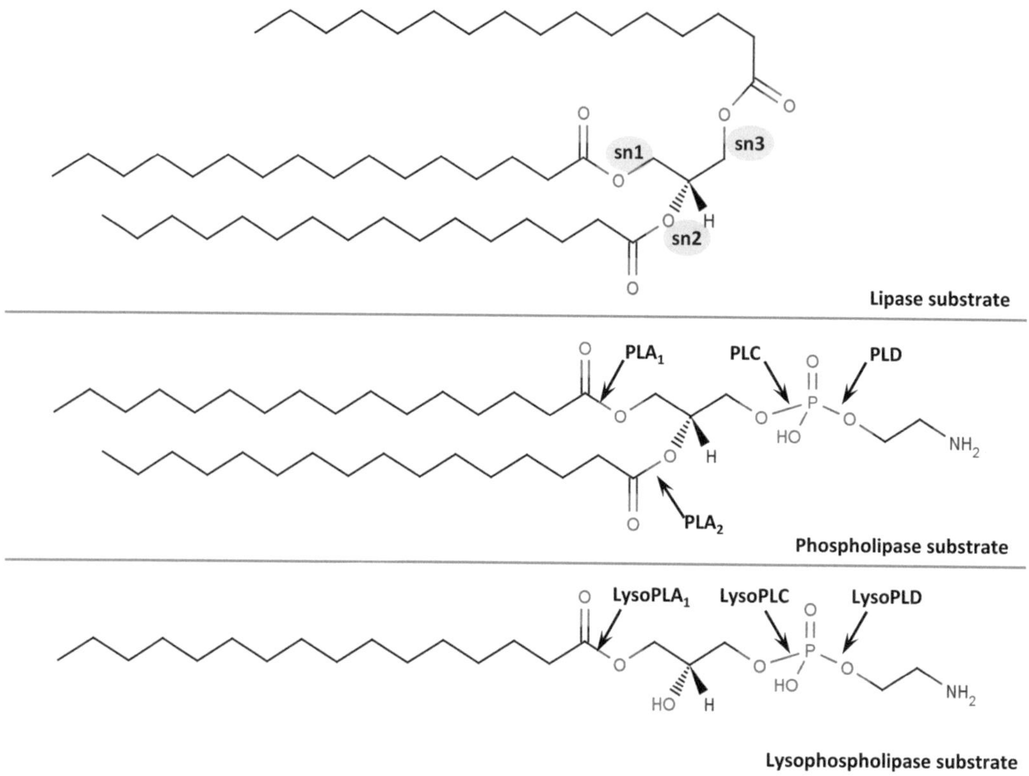

Fig. 1 Tripalmitoylglycerol, dipalmitoylphosphatidylcholine, and monopalmitoyl-phosphatidylcholine as typical lipase, phospholipase, and lysophospholipase substrates, respectively, show structural similarity at *sn*1 and *sn*2 positions which both represent cleavage sites for these three types of lipolytic enzymes. Lipases can hydrolyze ester bonds at all three positions (*sn*1, *sn*2, and *sn*3); cleavage sites of phospholipases A1 (PLA1), A2 (PLA2), C (PLC), and D (PLD), lysophospholipases A1 (LysoPLA1), A2 (LysoPLA2), C (LysoPLC), and D (LysoPLD) are indicated with *arrows*. Oxygen, nitrogen, and phosphorus atoms are indicated in *red*, *blue*, and *brown*, respectively (Color figure online)

substrate and water, this phenomenon called "interfacial activation" was proposed to distinguish between esterases and lipases [4]. Meanwhile, structural and kinetic studies of lipases and esterases have challenged this hypothesis [5–7]. Furthermore, broad substrate specificity and substrate promiscuity were observed among carboxylesterases and a number of enzymes were identified that possess esterase, lipase, and PLA activities at the same time [8, 9]. The structural similarity of typical substrates for all three groups of lipolytic enzymes (Fig. 1) suggests a plausible explanation for the observed substrate promiscuity. In living cells, lipases are thought to catalyze exclusively the hydrolysis of ester substrates; however, they are also capable to catalyze the backwards reaction, i.e., the synthesis of esters at low water conditions. Such esterification reactions play an important role for a variety of biotechnological applications, e.g., in the synthesis of enantiopure compounds [10] and acylated carbohydrates [11, 12], thus making lipases the most important group of biocatalysts for the chemical industries [13–15].

The genus *Pseudomonas* comprises a diverse group of medical, environmental, and biotechnological important bacteria [16–19] known to synthesize a variety of different lipolytic enzymes. *Pseudomonas aeruginosa* represents the best studied member [20] of this group with its genome sequenced and carefully annotated [21]. Here, we focus on lipolytic enzymes from *P. aeruginosa* although the assays described below are certainly applicable for other lipolytic enzymes as well. The systematic sequence analysis of the *P. aeruginosa* PA01 genome [21] using amino acid sequences of known lipases and phospholipases and the keywords "phospholipase," "esterase," and "lipase" revealed a total number of 104 genes encoding putative and hypothetical lipolytic enzymes. Out of these, 88 genes encode enzymes without homology to any previously characterized lipolytic enzymes, thus underlining the biosynthetic capacity of *P. aeruginosa* for the production of still unexplored enzymes. A number of lipolytic enzymes including phospholipases C (PlcA, PlcB, PlcH, PlcN), phospholipase D (PldA), phospholipase A2 (ExoU), phospholipase B (PlbF), lysophospholipase (TesA), lipases (LipA, LipC), and esterases (EstA, AchE, PA3859) were identified in *P. aeruginosa* (Table 1). Recently, we have newly demonstrated PLA2 and LysoPLA activities in *P. aeruginosa* PA01 associated with membrane and periplasmic localized enzymes [22].

Table 1
Lipolytic enzymes of *P. aeruginosa* PA01

PA code	Gene	Enzyme	Reference
PA0026	*plcB*	Zinc-dependent phospholipase C	[23]
PA0844	*plcH*	Phospholipase C	[24]
PA2856	*tesA*	Esterase and lysophospholipase	[9]
PA2862	*lipA*	Lipase	[25]
PA3319	*plcN*	Phospholipase C	[26]
PA3339	*plpD*	Lipase	[27]
PA3464	*plcA*	Zinc-dependent phospholipase C	[28]
PA3487	*pldA*	Phospholipase D	[29]
PA3859	–	Carboxylesterase	[30]
PA4813	*lipC*	Lipase	[31]
PA4921	*achE*	Acetylcholinesterase	[32]
PA5112	*estA*	Esterase	[33]

The industrial [18, 34, 35] and biomedical [36, 37] potential of lipolytic enzymes still provoke the discovery of novel (phospho) lipases with properties distinct from known enzymes. To this end, metagenomic libraries are constructed and directed evolution methods are applied to identify enzymes with novel properties. These methods require manipulation and screening of a large number of enzyme variants using practical and general activity assays applicable for a wide spectrum of lipolytic enzymes. In contrast, newly discovered enzymes need to be distinguished from already existing ones by using specific and sensitive activity assays with physiologically relevant substrates [38, 39].

A large number of methods have been developed to measure the catalytic activity of lipases [40] comprising mainly three types of assays: (a) quantification of released fatty acids by titrimetric or radiometric methods, (b) determination of released colored or fluorescent products by photometric methods, and (c) detection of changes in biophysical properties of substrates during lipolysis by conductometric, turbidometric, tensidometric, or microscopic methods. A very sensitive and sophisticated method is based on the measurement of changes in surface pressure at the interface of water and lipid substrates, which can be present as a lipid monolayer [40], bilayers [40] or drops [41, 42]. In short, the decrease in surface tension between water and for example a lipid monolayer upon hydrolysis by lipases can directly be measured by a tensiometer (e.g., EasyDyne Krüss, Germany). In general, methods directly detecting changes in biophysical properties of substrates are difficult to establish, require expensive equipment, and do not allow for a high throughput, and hence we do not discuss them in detail here.

Methods that rely on titrimetric measurements are convenient and widely used as long as inexpensive and physiological lipids are used as substrates. However, since large amounts of substrates (usually several hundreds of mg) are needed, the use of synthetic phospholipid substrates renders these assays expensive. Methods utilizing fluorogenic and chromogenic substrates are much more sensitive than titrimetric methods, but they cannot be applied for studying cellular functions of certain enzymes because they rely on the hydrolysis of artificial substrates.

A survey of the literature quickly reveals that even simple assays developed decades ago are used by different researches with slight modifications, thus hampering the comparison of biochemical properties of lipolytic enzymes. Here, we describe detailed protocols of commonly used lipolytic enzyme assays and we will discuss potential drawbacks and strategies how to overcome them.

2 Materials

Prepare all substrate solutions using ultrapure water with electrical resistivity of 18 M Ω cm and analytical grade reagents. All substrates and chemicals can be purchased from Sigma (St. Louis, MO, USA) or Avanti Polar Lipids (Alabaster, Alabama, USA) unless otherwise stated. For agar-plate assays, use distilled water and sterilize all solutions by autoclaving for 20 min at 120 °C or by filtration through filters with a pore size of 0.22 μm (unless indicated otherwise). We recommend to freshly prepare buffers needed for activity assays rather than to add toxic sodium azide for conservation. The accuracy of any activity assay strongly depends on the careful preparation of all solutions; therefore, pay attention when weighing small amounts of substrates and pipetting small volumes of solutions. Buffers can usually be stored for up to 2 weeks at room temperature or up to 1 month at 8 °C. Warm up the assay buffers to the temperature chosen for the assays and ensure that pH is adjusted at this same temperature (*see* **Note 1**).

2.1 Agar-Plate Assays

2.1.1 Tributyrin Emulsion Assay [43]

1. Luria–Bertani (LB) agar medium [44]: 10 g/L bacto-tryptone, 10 g/L sodium chloride, 5 g/L bacto-yeast extract, 15 g/L agar. Add distilled water, dissolve the components, adjust to pH 7 with NaOH, and autoclave.
2. Tributyrin emulsion: 50 % (v/v) tributyrin, sterilized by filtration through 0.22 μm membranes, 50 g/L sterile gum arabic dissolved in sterile distilled water. Mix the emulsion for 1 min using a homogenizer (e.g., Ultra Turrax, IKA Labortechnik, Germany) rinsed with 70 % (v/v) ethanol.

2.1.2 Rhodamine B Assay [45]

1. Luria–Bertani (LB) agar medium [44]: 10 g/L bacto-tryptone, 10 g/L sodium chloride, 5 g/L bacto-yeast extract, 15 g/L agar. Add distilled water, dissolve the components, adjust to pH 7 with NaOH, and autoclave.
2. Rhodamine B solution: 1 mg/mL dissolved in sterile distilled water; sterilize by filtration through 0.22 μm membranes.
3. Olive oil emulsion: mix 25 g/L olive oil with sterile distilled water, sterilize by filtration, and emulsify by mixing for 1 min with a homogenizer rinsed with 70 % (v/v) ethanol.

2.1.3 Egg Yolk Assay [46]

1. Luria–Bertani (LB) agar medium [44]: 10 g/L bacto-tryptone, 10 g/L sodium chloride, 5 g/L bacto-yeast extract, 15 g/L agar. Add distilled water, dissolve the components, adjust to pH 7 with NaOH, and autoclave.
2. Egg-yolk emulsion: 75 g/L egg yolk, 25 g/L taurocholic acid, 20 mM $CaCl_2$ dissolved in distilled water. Autoclave for 7 min at 120 °C. Emulsify by mixing for 1 min with a homogenizer rinsed with 70 % (v/v) ethanol.

2.2 Titrimetric Assays [47, 48]

1. Lipase and PLA substrates: tributyrine, triolein, and olive oil (consists mainly of triolein) are available as liquids; mono- and diacyl-phospholipids and egg yolk are available as solids (Sigma Aldrich, Fluka or Avanti Polar Lipids) (*see* **Note 2**).
2. Assay buffer: 1 mM Tris–HCl, pH 8; 15 mM NaCl; 2 mM $CaCl_2$; 4 mM taurodeoxycholate (*see* **Note 3**).
3. Titrant A: 1 M NaOH in water.
4. Titrant B: 50 mM NaOH in water.
5. Automatic titrator device (available for example from Mettler Toledo, Switzerland, or Radiometer Analytical, Denmark).
6. Homogenizer: bath sonicator, probe sonicator, or blender.

2.3 Colorimetric and Fluorometric Assays

2.3.1 Assays Using p-Nitrophenyl Acyl Ester Substrates

1. VIS spectrophotometer: for measurements in cuvettes (0.5–2 mL reaction mixture) or in microplates (20–200 μL reaction mixture) (e.g., microplate reader SpectraMax 250, Molecular Devices Corp.).
2. Cuvettes or microplates.
3. Plastic reaction tubes of 2 mL capacity.
4. Thermoconstant incubator, possibly with agitation.

p-Nitrophenyl-Palmitate Assay [49]

1. Substrate stock: 20 mM *p*-NP palmitate in isopropanol.
2. Assay buffer: 50 mM Na_2HPO_4; 50 mM KH_2PO_4; 5 mM sodium deoxycholate and 1 g/L arabic gum, adjust to pH 8 with NaOH.
3. *p*-NP standard stock solution: 20 mM *p*-nitrophenol in assay buffer.

p-Nitrophenyl-Butyrate Assay [50, 52]

1. Substrate stock: 20 mM *p*-NP butyrate in acetonitrile.
2. Assay buffer: 100 mM K_2HPO_4, pH 7.2.
3. *p*-NP standard stock solution: 20 mM *p*-nitrophenol in assay buffer.

2.3.2 Assays Using 4-Methylumbelliferyl Acyl Ester Substrates [53, 54]

1. Fluorescence spectrometer: for measurements in cuvettes (0.5–2 mL reaction mixture) or in microplates (20–200 μL reaction mixture).
2. Glass cuvettes (1 × 1 cm) or microplates suitable for fluorescence measurements (available from PerkinElmer) (*see* **Note 4**).
3. Plastic reaction tubes of 2 mL capacity.
4. Substrate stock: 1 mM 4-MU heptanoate in tetrahydrofuran (*see* **Note 5**).
5. Assay buffer: 20 mM Tris–HCl, pH 8; 1 mM EDTA; 300 μM taurodeoxycholate.
6. 4-MU standard stock solution: 1 mM sodium salt of 4-MU in assay buffer.

2.3.3 Assays Using Natural (Phospho)Lipid Substrates [55, 56]

1. VIS spectrophotometer: for measurements in cuvettes (0.5–2 mL reaction mixture) or in microplates (20–200 μL reaction mixture).
2. Cuvettes or microplates.
3. Plastic reaction tubes of 2 mL capacity.
4. Thermoconstant incubator allowing agitation.
5. Non-esterified fatty acid HR series (NEFA-HR) kit (Wako Chemicals, Richmond, USA) consists of two sets, reagent 1 set (R1) and reagent 2 set (R2). R1 set contains color reagent A and solvent A, and R2 set contains color reagent B and solvent B. The exact composition of all reagents is listed in the instruction manual provided by the manufacturer, http://www.wako-chemicals.de/DWD/_111327/upload/media_132965.pdf.
6. NEFA free fatty acid standards (Wako Chemicals).
7. Assay buffer: 40 mM Tris–HCl, pH 7.5; 1 % (v/v) Triton X-100.
8. Lipid substrates: 13.4 mM lipid (available from Sigma-Aldrich, Fluka, Avanti Polar Lipids) in assay buffer (*see* **Note 6**).

2.4 Adrenalin Fingerprinting Assay [57]

1. Substrates: 10 mM of compounds listed in Table 2 dissolved in acetonitrile.
2. Titrant 1: 10 mM $NaIO_4$ in water (*see* **Note 7**).
3. Titrant 2: 15 mM L-adrenaline hydrochloride in water.
4. Enzyme in 50 mM aqueous borate buffer, pH 8.0 (*see* **Note 8**).
5. Microplates.
6. Plate reader.

2.5 Quick E Enantioselectivity Assay [75]

1. Assay buffer: 50 mM Tris–HCl, pH 8; 4.5 g/L Triton X-100
2. Enantiomeric substrate solution S: 7.8 mM (*S*)-*p-nitrophenyl-2*-phenylpropanoate dissolved in acetonitrile.
3. Enantiomeric substrate solution R: 7.8 mM (*R*)-*p-nitrophenyl-2*-phenylpropanoate dissolved in acetonitrile.
4. Reference substrate solution: 1.6 mM resorufin tetradecanoate dissolved in acetonitrile.
5. Microplates (96-well plates).
6. Plate reader (*see* **Note 9**).

3 Methods

3.1 Agar-Plate Assays

LB-agar is the most widely used solid medium for the growth of bacteria which can be supplemented with antibiotics to maintain expression plasmids and with inducers of gene expression,

Table 2
Commercially available polyol acetate substrates used for lipase fingerprinting with the adrenalin assay [58]

Substrate name	CAS number	Supplier
Ethylene glycol diacetate	111-55-7	Sigma Aldrich
Diacetin, glycerol α,α′-diacetate	5395-31-7	TCI Europe
Propylene glycol diacetate	623-84-7	Sigma Aldrich
Triacetin	102-76-1	Sigma Aldrich
(*R*)-(+)-Dihydro-5-(hydroxymethyl)-2(3H)-furanone	52813-63-5	Sigma Aldrich
(*S*)-(+)-Dihydro-5-(hydroxymethyl)-2(3H)-furanone	32780-06-6	Sigma Aldrich
(1*R*,2*R*)-2-(Acetyloxy)cyclohexyl acetate	1759-71-3	Sigma Aldrich
β-D-Ribofuranose 1,2,3,5-tetraacetate	13035-61-5	AK Scientific
α-D-Mannose pentaacetate	4163-65-9	Sigma Aldrich
β-D-Ribopyranose 1,2,3,4-tetraacetate	4049-34-7	Sigma Aldrich
D-(+)-Sucrose octaacetate	126-14-7	Sigma Aldrich
4-O-(2,3,4,6-Tetra-O-acetyl-α-D-mannopyranosyl)-D-mannopyranose tetraacetate	123809-59-6	TRC
β-D-Galactose pentaacetate	4163-60-4	Sigma Aldrich
β-D-Ribopyranose 1,2,3,4-tetraacetate	4049-34-7	Sigma Aldrich
Lactose octaacetate	6291-42-5	TRC
D-(+)-Cellobiose octaacetate	3616-19-1	TRC
Mannitol hexaacetate	5346-76-9	MP Biomedicals
D-Sorbitol hexaacetate	7208-47-1	Sigma Aldrich
β-L-Glucose pentaacetate	66966-07-2	TRC
1,2,3,4-Tetra-O-acetyl-α-L-fucopyranoside	64913-16-2	TCI Europe
β-D-Ribopyranose 1,2,3,4-tetraacetate	4049-34-7	Sigma Aldrich

TRC Toronto research chemicals, *TCI Europe* Tokyo chemical industry Europe

e.g., isopropyl-β-D-galactopyranoside (IPTG). One key factor with regard to these assays is how the expression of the genes is regulated by the medium components, e.g., the expression of the *plcH* and *plcN* are strongly repressed by phosphate levels in the growth media. Substrates should be freshly emulsified before adding to the agar medium to obtain the emulsion with optimal properties. It is recommended to use fresh agar plates to increase assays sensitivity and reproducibility (*see* **Note 10**). The assays need to be adapted to the biophysical properties of the studied enzymes by varying the incubation time (from 1 up to 4 days)

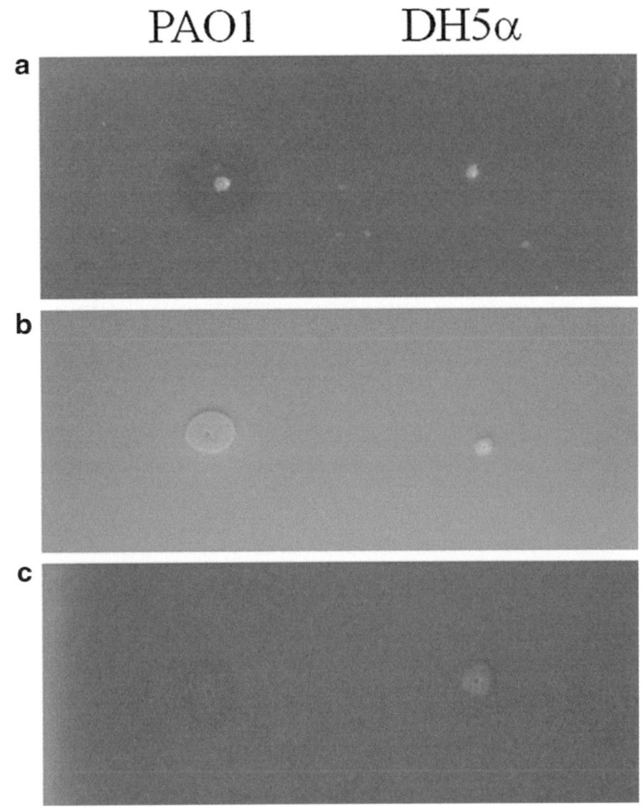

Fig. 2 Lipase producing strain *P. aeruginosa* PAO1 and *E. coli* DH5α, a negative control, grown overnight on (**a**) tributyrin agar plate (Subheading 2.1.1); (**b**) rhodamine B agar plate (Subheading 2.1.2); and (**c**) egg yolk agar plates (Subheading 2.1.3). Agar plates appear opaque due to the presence of lipid substrate drops in the solidified medium. Lipolytic enzymes produced by *P. aeruginosa* hydrolyze the substrates resulting in clearance of the medium around *P. aeruginosa* colonies grown on tributyrin and egg yolk agar plates. The rhodamine B agar plate was photographed under UV light and lipolytic activity around *P. aeruginosa* colony is visible as light blue fluorescence

and temperature (from 4 °C up to 50 °C) required for detection of an enzymatic activity. Not more than 500 clones per standard petri dish should be plated to achieve good resolution of clones and allow identification of single enzyme-producing clones.

3.1.1 Tributyrin Agar Plate Assay [43]

The most widespread plate assay to detect extracellular but also intracellular esterases and lipases uses tributyrin as the substrate. Clear zones appearing around the bacterial colonies indicate the production of a catalytic active enzyme (Fig. 2a).

1. Add 30 mL of tributyrin emulsion, the respective antibiotic and inducer (e.g., IPTG) to 1 L of melted LB agar medium cooled to a temperature of about 60 °C and mix thoroughly.

2. Pour 20 mL medium into appropriate petri plates and let it solidify for at least 20 min.
3. Plate bacterial clones and incubate at optimal growth temperature for at least 16 h.
4. Positive clones are identified after overnight growth or after prolonged (2–4 days) incubation in refrigerator for clones expressing low amounts of active enzymes (*see* **Note 11**).

3.1.2 Rhodamine B Agar Plate Assay [45]

This assay, contrary to the tributyrin assay, is specific for true lipases because it uses triolein as a substrate, which is a triglyceride ester containing long-chain (C-18:1) fatty acids. Fluorescent complexes are formed between the cationic rhodamine B and free fatty acids released from the substrate by a lipase. Around lipase-positive clones, a fluorescent halo can be observed after irradiation of the plate with UV light at 350 nm (Fig. 2b). Clones which do not produce lipases appear pink colored, but nonfluorescent (*see* **Note 12**).

1. Add 31.25 mL of olive oil emulsion, 10 mL RB solution, respective antibiotic and inducer (e.g., IPTG) to 1 L of melted LB agar medium at a temperature of about 60 °C and mix thoroughly.
2. Pour 20 mL medium into appropriate plates and let it solidify. Normally, plates are pink colored and have opaque appearance.
3. Plate the clones and incubate at optimal growth temperature for at least 16 h.
4. Positive clones are identified under UV light (e.g., hand lamp at 350 nm) by fluorescent halos around colonies (*see* **Note 13**).

3.1.3 Egg Yolk Agar Plate Assay [46]

This assay is generally used to detect phospholipase A and C activity. The egg yolk substrate consists of about 65 % triacylglycerols, about 30 % phospholipids, and 5 % cholesterol. Hence, the egg yolk assay is applicable for the analysis of phospholipases and lipases, too [48]. The assay is based on the formation of calcium complexes with free fatty acids released by phospholipases A and lipases, indicated as white precipitation halos around the colonies (Fig. 2c). Although assays for PLC are not covered here, it is worth to mention that the egg yolk LB agar assay is applicable for assaying PLC activities, too. Phospholipase C producing clones show precipitation halos as a result of complex formation between diacylglycerol, released by action of PLC, and vitellin, the major egg yolk protein [46].

1. Add 200 mL of egg-yolk emulsion to 800 mL melted LB medium at temperature of about 60 °C and mix thoroughly.
2. Pour 20 mL medium into appropriate plates and let it solidify for at least 20 min.

3. Plate the clones and incubate at optimal growth temperature for at least 16 h.
4. Positive clones are identified by appearance of clear halos around colonies.

3.2 Titrimetric Assays [47, 48]

These assays are based on the quantitative titration of carboxylic acids released from various types of ester substrates by the hydrolytic activity of lipases or phospholipases A. Titrimetric methods can be classified in three groups: titration of free fatty acids after extraction, end point titration, and continuous titration. We describe here the continuous automatic titration, also called pH-stat method.

This is a common and reliable method used for the quantification of lipase/PLA activities based on continuous monitoring the amount of NaOH required to neutralize the liberated free carboxylic acids. The advantages of this method include its sensitivity (1 μmol/min of free fatty acids released) and the possibility to use a wide spectrum of naturally occurring lipids (mono-, di-, and triacylglycerols for measuring lipase/esterase activities; and phospholipids for PLA activities). Expensive synthetic (phospho)lipids (provided by for example Avanti polar lipids, Alabama, USA; Sigma Aldrich, Missouri, USA), or inexpensive natural (phospho)lipids such as olive oil (for lipases) or egg yolk (for phospholipases) can be used in titrimetric assays.

1. Turn on pH-stat device, flush the burette and fill it with titrator. Set up constant temperature to 37 °C. Follow manufacturer's instructions regarding maintaining of temperature and preparing the device prior starting measurements (*see* **Note 14**).
2. Add 10 mM pure triacylglycerol or phospholipid or 50 g/L natural oil or phospholipid in assay buffer. 5 mL of emulsion is usually enough for one reaction, although the volume depends on the reaction vessels available.
3. Homogenize the emulsion by using (a) sample sonicator for 4 min (two times 2 min, with cooling in between) at maximum capacity, (b) blender for 10 min (two times 5 min with cooling in between) at a speed of 11,000 min^{-1}, or (c) bath sonicator for 10 min (two times 5 min with cooling in between) at maximal capacity. Emulsions are usually stable for 4–6 h (*see* **Note 15**).
4. Add substrate emulsion in the reaction vessel and place the electrode, magnetic stirrer and delivery tip into the emulsion. Stir the emulsion at constant speed and adjust the pH of the substrate emulsion to 8 with titrant A. Set up the end point of the titrator to pH 8.0.
5. Start the reaction by adding a defined amount of enzyme solution (up to 250 μL per 5 mL of emulsion).

6. Titrate the liberated fatty acids continuously with titrant B, usually for 5–10 min.
7. Calculate the lipase activity per mg or mL of enzyme using Eq. 1 [47]. For calculation, use only the data with linear response of the titrant spent for acid neutralization with time. The unit of lipase activity is defined as the amount of enzyme releasing 1.0 μmol of fatty acid per 1 min at defined temperature and pH.

$$U_{\text{lipase activity}} / \text{mg or ml} = \frac{c(\text{titrant B})\left[\frac{\text{mol}}{\text{L}}\right] \times V(\text{titrant B})[\text{mL}] \times 1{,}000}{t[\text{min}] \times \text{amount of enzyme}[\text{mg or ml}]} \quad (1)$$

c = concentration of titrant B, 50 mmol/L NaOH
V = volume of titrant B
t = time of enzymatic reaction
amount of enzyme = amount of mg or mL of enzyme added to the reaction

3.3 Colorimetric and Fluorometric Assays

These methods are based on the hydrolysis of ester substrates that, upon hydrolysis by lipolytic enzymes, release chromogenic or fluorogenic products which can be detected by UV/VIS or fluorescence spectrophotometry, respectively. The artificial profluorogenic or chromogenic substrate analogues are suitable for routine and accurate quantification of lipase activity using standard equipment usually available in biochemical laboratories. Mostly, chromogenic *p*-nitrophenyl (*p*-NP) or α/β-naphtyl and fluorogenic 4-methylumbelliferyl (4-MU) or resorufin esters of fatty acids are used as substrates. However, because these substrates naturally do not occur in living cells, the assays are not suitable to define cellular functions of lipolytic enzymes. On the other hand, chromogenic or fluorogenic substrates can be used for high throughput screening of a large number of samples with fluorogenic methods usually being much more sensitive than colorimetric methods.

3.3.1 Assay with p-Nitrophenyl Acyl Ester Substrates [49–52]

This method is based on the quantification of *p*-nitrophenolate (*p-NP*) released during the hydrolysis of *p*-nitrophenyl acyl esters by lipolytic enzymes. The *p*-NP is chromogenic with an absorption maximum at 410 nm and it can be quantified continuously using a spectrophotometer. As substrates can be used *p*-nitrophenyl esters with saturated or unsaturated fatty acids differing in chain length. Usually, *p*-NP butyrate (C4) or caproate (C6) are used for esterase assays and *p*-NP palmitate (C16) for lipases assays. Because substrates, *p*-NP acyl esters, are not water soluble they are dissolved in organic solvents and then diluted with assay buffer. The selection of suitable organic solvent is driven by the solubility and stability of short-chain and long-chain fatty acid *p*-NP esters. Acetonitrile is the most commonly used solvent for short-chain esters and isopropanol for long-chain esters. However, acetonitrile can also be used for assays with *p*-NP esters in the entire range of C2–C18 fatty acids. We present here two protocols, one for determination of esterase activity using *p*-NPC4 (also applicable for substrates with fatty acids up to 18 carbon atoms) and another for determination of lipase activity using *p*-NPC16. The major disadvantage of *p*-NP acyl ester substrates is their spontaneous hydrolysis under assay conditions. The activity of a purified enzyme should be assayed at the respective optimal temperature, pH, and buffer. Also 5 mM divalent cations (e.g., Ca^{2+} or Zn^{2+}) should be added in the buffers if it is required for activity or stability of an enzyme.

Similar to the assays that rely on hydrolysis of *p*-NP esters, organophosphonate esters of *p*-NP were used for determination of concentration of catalytically active lipases in solution by active-site titration methods. Standard methods for protein quantification with Bradford reagent or UV spectroscopic measurements do not provide data about concentration of enzymatically active enzyme required for kinetic and structural studies. In the active-site titration experiments, lipase is incubated with an irreversible inhibitor (e.g., methyl *p*-nitrophenyl *n*-hexylphosphonate [59]) that specifically binds to the active site serine yielding the chromophore *p*-nitrophenolate that is quantified spectrometrically. Optionally, 3,4-dichloroisocoumarin may be used as an inhibitor followed by measurement of residual lipase activity using *p*-NP assay [60].

1. Dilute substrate stock solution 20-fold with the assay buffer and vortex for 2–3 min.
2. Pipette enzyme sample into reaction tubes (5–50 μL) or in microplates (1–20 μL). Prepare the blank (non-enzymatic control) by adding buffer in which the enzyme is dissolved to the reaction tube or microplate well.
3. Set up the temperature of VIS spectrometer at 30 °C (*see* **Note 16**).
4. (A) Add substrate solution into reaction tubes (0.5–2 mL) to initiate the reactions. Start the timer, mix and transfer the solution from reaction tube into the cuvette and record absorbance at 410 nm (A_{410nm}) over the time course of 2–10 min.

   (B) Fill the microplate with the substrate (100–200 μL) (*see* **Note 17**) and place it into VIS microplate spectrophotometer (plate reader). Set up the 5 s mixing step prior measurements start and record A_{410nm} each 10–30 s over the time course of 10–15 min.
5. Prepare at least six standard solution of *p*-NP by diluting standard stock solution of *p*-NP in the assay buffer and by adding isopropanol (for *p*-NPC16 assay) or acetonitrile (for *p*-NPC4

assay) to yield final concentration of 5 % (v/v). Measure the A_{410nm} of *p*-NP dilutions in a range of 0.01–0.2 mM under the assay conditions. Prepare the standard curve by plotting the A_{410nm} versus the *p*-NP concentration.

6. Subtract the fluorescence of the blank from the measured values for the enzyme samples and convert the A_{410nm} into enzyme activity using the standard curve. Optionally, if the extinction coefficient is known, the absorbance measured may be used to calculate units of lipase activity (U) according to Eq. 2 (*see* **Note 18**).

$$U_{\text{lipase activity}} / \text{ml} = \frac{\left(A_{410\text{nm}}(t_1) - A_{410\text{nm}}(t_0)\right) \times V_{\text{total}}}{\varepsilon \times V_{\text{enzyme}} \times (t_1 - t_0) \times l} \tag{2}$$

$A_{410nm}(t_1)$ = absorbance at 410 nm measured after enzymatic reaction is finished

$A_{410nm}(t_0)$ = absorbance at 410 nm measured before enzymatic reaction is started

t_1 = time point when reaction is finished

t_0 = time point when reaction is started, usually 0 min

V_{total} = volume of reaction sample plus enzyme sample (V_{enzyme})

ε = molar extinction coefficient [$mmol^{-1}\ dm^3\ cm^{-1}$]

l = light path length [cm]

3.3.2 Assay with 4-Methylumbelliferyl Acyl Ester Substrates [53, 54]

Hydrolysis of 4-MU acyl esters by lipases can be followed continuously by monitoring the increase of fluorescence intensity due to the production of highly fluorescent 4-methylumbelliferone with an emission spectra maximum at 460 nm. In contrast to emission maxima, the fluorescence excitation maxima of 4-MU is pH dependent and varies from 330 nm at pH 4.6 to 385 nm at pH 10.4 [54]. Esters of 4-MU with different fatty acids (butyric, heptanoic, oleic acid) are commercially available. Short-chain 4-MU butyrate is used for assaying esterases, long-chain 4-MU-oleate is used for true lipases, and 4-MU heptanoate is suitable for both lipases and esterases. For activity calculations, it is necessary to correct for spontaneous hydrolysis of the substrates and for background fluorescence of the protein sample. Enzyme activity is defined as the amount of released 4-MU per minute, and can be calculated with a standard curve obtained by measuring the fluorescence of 4-MU under assay conditions.

1. Dilute substrate stock solution 100-fold into the assay buffer and vortex for 2–3 min.
2. Pipette enzyme sample into reaction tubes (5–50 μL) or in microplates (1–20 μL). Prepare the blank (non-enzymatic control) by adding the buffer in which enzyme is dissolved to the reaction tube or microplate. Prepare one reaction tube or

microplate well with the enzyme solution for measurement of intrinsic enzyme fluorescence.

3. Set up the temperature of spectrofluorometer at 30 °C.
4. (A) Add substrate solution into reaction tubes (0.5–2 mL) to initiate the reactions. Start the timer, mix and transfer the solution from the reaction tube into the cuvette and record fluorescence at 460 nm over a time course of 10–15 min.

   (B) Fill the microtiter plate with the substrate (100–200 μL) (*see* **Note 17**) and place it into fluorescence plate reader. Set up the 5 s mixing step prior measurements start and record the fluorescence at 460 nm each 10–30 s over a time course of 10–15 min.
5. Add the assay buffer only (without the substrate) to the reaction tube or microplate well with the enzyme sample and measure the intrinsic fluorescence of the enzyme.
6. Prepare at least six standard solution of 4-MU by diluting standard solution of 4-MU in the assay buffer to yield dilutions in a range 50–500 nM. Measure the fluorescence under the assay conditions. Prepare the standard curve by plotting the fluorescence versus the 4-MU concentration.
7. Subtract the fluorescence of the blank and intrinsic enzyme fluorescence from the sample fluorescence and convert the fluorescence into the enzyme activity using the standard curve.

3.3.3 Assay with Natural (Phospho)Lipid Substrates [55, 56]

Lipase, esterase, phospholipases A and lysophospholipase A activities can be determined with (phospho)lipids naturally occurring in living cells. Here, the ASC-ACOD-MEHA enzymatic-coupled method for the determination of released fatty acids can be used. Free fatty acids liberated upon enzymatic hydrolysis of natural (phospho)lipid substrates are determined colorimetrically by means of the NEFA-HR(2) kit according to the manufacturer's instructions. This method relies upon the synthesis of acyl-CoA from free fatty acid and coenzyme A in the presence of acyl-CoA synthetase (ACS). The acyl-CoA is subsequently oxidized by acyl-CoA oxidase (ACOD) to yield enoyl-CoA resulting in release of H_2O_2 as the side product. A peroxidase then uses this H_2O_2 to oxidize 4-aminoantipyrin (4-AA); which in its oxidized form reacts with 3-methyl-*N*-ethyl-*N*-(β-hydroxyethyl)-aniline (MEHA) to yield a quinoneimine which is purple colored and can be quantified spectrometrically at 550 nm. The concentration of non-esterified fatty acids (NEFA) can be calculated from the A_{550nm} by using the calibration curve. This enzymatic ASC-ACOD-MEHA method gains rapidly increasing interest for measurement of lipolytic activities because it is applicable for all ester substrate available, it is reliable even in the presence of numerous interfering compounds, it is simple, sensitive and requires only low amounts of substrates (in the μg range per reaction).

1. Dilute lipid substrate 20-fold with assay buffer by vortexing for 15 min at 37 °C and then subject to ultrasonication three times for 20 s.
2. Combine 25 μL of lipid substrate with 25 μL of enzyme sample in reaction tubes or microplates, tightly close reaction vessels and incubate at 37 °C in agitation mode (*see* **Note 19**). The incubation time, usually 30 min for very active samples and overnight for less active samples, is dependent on the activity of the analyzed samples and therefore it should be empirically estimated in preliminary experiments.
3. Prepare respective blank controls by treating them the same like enzyme samples.
4. Prepare at least six dilutions of oleic acid in the range of 0.1–1 mM in assay buffer (*see* **Note 20**). A standard curve may also be determined by using a commercially available calibrator kit (Wako Chemicals).
5. Prepare reagent solution R1 according to the supplier instructions by dissolving color reagent A in solvent A (*see* **Note 21**).
6. Prepare reagent solution R2 according to the supplier instructions by dissolving color reagent B in solvent B (*see* **Note 21**).
7. Pipette 2–10 μL of the reaction samples into new reaction tube or microplate and add reagent solution R1 (300 or 100 μL for assays in reaction tubes or microplates, respectively) followed by agitation for 5 min at 37 °C.
8. Add reagent B to the samples (150 or 50 μL for assays in reaction tubes or microplates, respectively), agitate at 30 °C for 5 min and measure absorbance at 550 nm.
9. Subtract the absorbance of the blank samples from the enzyme samples and convert the $A_{550\text{nm}}$ into the enzyme activity using the standard curve.

3.4 Adrenalin Fingerprinting Assay [57]

The simultaneous measurement of a single enzyme activity toward a range of different substrates provides a data set sufficient for a precise functional definition of a given enzyme, a so-called enzyme activity fingerprint. Because only small variations in substrate structure may significantly influence the enzyme activity, fingerprinting methods are confident for functional characterization of a single enzyme. The fingerprinting data are usually represented as images created by linking each measured data with one pixel which color intensity represents the activity (for example from white corresponding to 0 % to black corresponding to 100 % activity) (*see* Fig. 3). The assay requirements for such fingerprinting methods include applicability with usually tens of different substrates, high reproducibility and use of the same enzyme solution. These methods are valuable for definition of substrate promiscuity, identification of biological function, quality control, and medical diagnostics.

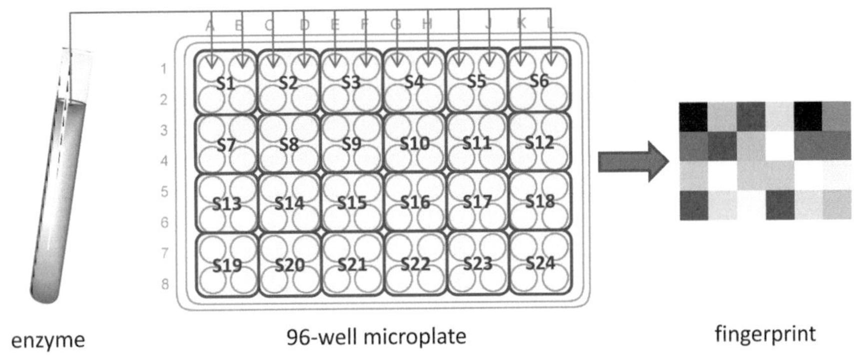

Fig. 3 The principle of the fingerprint method. The same enzyme sample is assayed under identical conditions with, in this case, 24 different substrates (S1–S24) resulting in an enzyme fingerprint

Although several fingerprinting methods for assaying lipase activities were developed with fluorogenic [61] and chromogenic [62] substrates, the limited range of commercially available tagged substrates hampers the creation of large substrate libraries for fingerprinting. To overcome the rather laborious synthesis of fluorogenic or chromogenic substrate libraries it is possible to use indirect assays with a range of commercially available esters substrates. In these assays, the products released by the hydrolytic activity of lipases a) may initiate a sequence of chemical reactions that result in generation of chromogenic or fluorogenic compounds (e.g., adrenalin assay) or b) may change the color of a pH indicator by changing the pH of the reaction. The amount of hydrolytic product may be indirectly quantified spectrophotometrically or fluorometrically.

Here, we describe the adrenalin assay [57], a fingerprinting method used for profiling substrate specificity of different lipases and esterases. This indirect fingerprinting assay uses 35 different commercially available polyol acetates substrates (Table 1) from which 1,2-diols are released upon enzymatic hydrolysis. The formed diols are oxidized with periodate and the remaining periodate is back-titrated with adrenaline to give a red colored adrenochrome product with an absorption maximum at 490 nm. These end point spectrometric measurements can be used for generating fingerprinting images.

1. Mix 10 μL substrate with 10 μL $NaIO_4$ and 80 μL of enzyme solution and incubate for 60 min at 37 °C.
2. Add 10 μL adrenalin solution and incubate for 5 min at 26 °C.
3. Measure the absorbance at 490 nm.

3.5 Quick E Enantioselectivity Assay [75]

The ability of many lipolytic enzymes to distinguish between two enantiomers of a given racemic substrate, designated as enantioselectivity, is a key feature for their industrial applications [13–15, 17–19].

The increasing demand for the synthesis of enantiomerically pure compounds by the pharmaceutical industries is driven by the fact that very often only one enantiomer has a desired biological activity, whereas the other enantiomer may cause undesirable side effects [63, 64]. Although several lipases often exhibit enantioselectivity, mostly, it is necessary to improve it for a specific substrate. For this purpose, molecular genetic methods have been developed and applied including rational protein design [65], random mutagenesis, saturation mutagenesis, iterative saturation mutagenesis [66], and DNA shuffling to increase the enantioselectivity of lipases [67]. Notably, the lipase LipA from *P. aeruginosa* represents the best studied enzyme with respect to directed evolution of enantioselectivity [68–74]. Initially, proof-of principle was demonstrated by increasing the enantioselectivity of LipA in the kinetic resolution of the substrate 2-methyldecanoic acid *nitrophenyl* ester to E=25,8 (the wild-type lipase is not selective at all showing an E=1,1) [72]. Later, diverse directed evolution methods were developed and optimized using this lipase. Finally, an enantioselectivity of E=106 was achieved for the hydrolysis of 2-phenylbutyric acid *p*-nitrophenyl esters (the wild-type lipase has E=2) [66].

The enantioselectivity is usually expressed as enantiomeric ratio (E) defined as the ratio of the initial rate of reaction with the preferred enantiomer (fast enantiomer) and the initial rate of reaction with the second enantiomer (slow enantiomer) (Eq. 3) [76]. The true enantiomeric ratio can be obtained by measuring the initial rates of reactions of each enantiomer in the presence of both enantiomers, under so-called competitive conditions. This limits the choice of detection methods to time consuming methods like for example HPLC or GC. Therefore, colorimetric methods were developed allowing screening of large libraries for enantioselective enzymes. Here, enzyme activities are determined with enantiomerically pure (S)- and (R)- substrates separately, e.g., (S)- and (R)-*p*-NP 2-methyl decanoate [77] or (S)- and (R)-*p-NP-2*-phenylpropanoate [78]. It should be noted that these assays reveal only estimated E-values that can significantly deviate from true E-values as shown for *Candida antarctica* lipase A (true E=1,4 and estimated E=4 for hydrolysis of (R)- and (S)-*p*-NP-2-phenylpropanoate [75]).

$$\text{Enantiomeric ratio} = E = \frac{\left(\frac{k_{\text{cat}}}{K_{\text{m}}}\right)\text{fast enantiomer}}{\left(\frac{k_{\text{cat}}}{K_{\text{m}}}\right)\text{slow enantiomer}} = \frac{\text{reaction rate (fast enantiomer)}}{\text{reaction rate (slow enantiomer)}} \quad (3)$$

Here, we describe a method for the fast determination of enantioselectivities, termed Quick E method. It was developed to overcome the lack of competition between two enantiomers.

It is based on the hydrolysis of each enantiomerically pure substrate, in the presence of a reference substrate that must not be enantiomerically pure [78] but both these substrates must be detectable simultaneously in one solution. The colorimetric assays with pure (*S*)- and (*R*)-*p*-NP-2-phenylpropanoate, and resorufin tetradecanoate as a reference substrate, were performed with several lipases [78]. The *p*-nitrophenolate and resorufin can be detected spectrophotometrically at 404 and 572 nm, respectively, thus providing initial rates of hydrolysis of each enantiomer. Principally, by performing two measurements, one with a mixture of (*S*)-*p*-NP-MD and RT and the second one with (*R*)-*p*-NP-MD and RT, the enantioselectivity for both reactions can be calculated using Eqs. 4 and 5 [75]. By dividing these two enantio selectivities, a Quick E-value can be calculated (Eq. 6 [75]). Such simulated competition between enantiomeric and reference substrates results in better agreement of Quick E-values with true E-values than estimated E-values with true E-values, as demonstrated e-.g. for lipases of *Pseudomonas cepacia* (true and Quick *E*-values = 29) and *Candida rugosa* (true and Quick *E*-values = 3,5) [75].

$$E = \frac{(R)\ \text{substrate}}{\text{reference}}\text{selectivity} = \frac{\dfrac{\text{reaction rate}(R\ \text{enantiomer})}{\text{concentration}(R\ \text{enantiomer})}}{\dfrac{\text{reaction rate}(\text{reference substrate})}{\text{concentration}(\text{reference substrate})}} \qquad (4)$$

$$E = \frac{(S)\ \text{substrate}}{\text{reference}}\text{selectivity} = \frac{\dfrac{\text{reaction rate}(S\ \text{enantiomer})}{\text{concentration}(S\ \text{enantiomer})}}{\dfrac{\text{reaction rate}(\text{reference substrate})}{\text{concentration}(\text{reference substrate})}} \qquad (5)$$

$$E_{\text{Quick}} = \frac{E1}{E2} = \frac{\dfrac{(R)-\text{substrate}}{\text{reference}}\text{selectivity}}{\dfrac{(S)-\text{substrate}}{\text{reference}}} = \frac{(R)-\text{substrate}}{(S)-\text{substrate}}\text{selectivity} \qquad (6)$$

1. Add 0.5 mL of (*S*)-substrate solution and 0.5 mL of reference substrate solution into 9 mL of assay buffer drop wise and vortex 2–3 min until emulsion is clear (*see* **Note 22**).
2. Add 0.5 mL of (*R*)-substrate solution and 0.5 mL of reference substrate solution into 9 mL of assay buffer drop wise and vortex 2–3 min until emulsion is clear.
3. Add 20 μL enzyme solution to the 180 μL substrate emulsion (containing (*S*)- and reference substrates) in microplate and measure the increase in absorbance at 404 and 572 nm at 25 °C for 5–30 min with steps of 15 s.

4. Add 20 μL enzyme solution to the 180 μL substrate emulsion (containing (*R*)- and reference substrates) in microplate and measure the increase in absorbance at 404 and 572 nm at 25 °C for 5–30 min with steps of 15 s.
5. Measure blank and substrate only samples the same as the enzyme samples (*see* **Note 23**).
6. Values are in absorbance per second × 1,000. Enantiomeric ratio for (*R*)- and (*S*)- enantiomer separately is calculated using Eqs. 4 and 5, respectively. Quick E enantiomeric ratio is calculated using Eq. 6.

4 Notes

1. pH of buffers may change with temperature; hence, for measuring temperature optima, use buffers having their pH adjusted accordingly.
2. The substrates with different acyl chain length, number of double bonds in the acyl chain and absence, or presence, of phosphoester moiety bound on the glycerol may be chosen for this assay dependent on the available information about the substrate specificity of tested enzyme.
3. pH 8 is required because most free fatty acids have a p*Ka* lower than 8. Also, optimal pH and stability of lipases is usually in the range of pH 7–9.
4. Black microplates have low background fluorescence and minimal light scatter, therefore they are recommended for fluorometric assays.
5. Optionally, 4-MU heptanoate may be dissolved in ethylene glycol monomethyl ether [54].
6. Mono-, di-, and tri-acylglycerol, phosphatidylglycerol, phosphatidylcholine, phosphatidyl-ethanolamine, lysophosphatidylcholine lysophosphatidylglycerol. and lysophosphatidyl-ethanolamine are usually used as (phospho)lipase substrates.
7. $NaIO_4$ solution in water should be prepared freshly.
8. Sodium phosphate buffer may be used depending on optimal stability and activity of studied enzyme. Do not use buffer that is oxidized with $NaIO_4$, e.g., Tris or triethanolamine buffers.
9. It is preferable to use a plate reader allowing the simultaneous measurement of absorbance at two different wavelengths.
10. The spontaneous hydrolysis of substrates can result in its decreased concentration upon long term storage of plates.
11. To increase the probability to find lipase-producing clones in gene libraries one can prepare several tributyrin agar plate replicas from one master plate after overnight incubation.

Each replica plate may then be incubated at a different temperature, for example 4, 37 and 50 °C in order to detect lipases with different temperature optima.

12. Bacteria accumulate intracellularly pink colored rhodamine B which is not fluorescent in absence of free fatty acids.
13. To prevent damaging effects of UV light apply UV radiation only for a short time (a few sec) if you want to propagate the clones from the same plate afterwards.
14. The temperature should be chosen based on the stability and optimal temperature of the enzyme.
15. The reproducibility of assays depends strongly on the emulsion properties, therefore, the same homogenization conditions should always be used.
16. The assay can be performed at temperatures up to 80–90 °C, although the spontaneous hydrolysis of the substrate at higher temperatures is significant.
17. Preferably use multichannel pipette to minimize time needed to fill microplate.
18. The molar extinction coefficient changes with temperature and pH.
19. Already 2 μL of reaction samples is sufficient to measure activities.
20. If you are using (phospho)lipid substrates with different fatty acid chain lengths than oleic acid, it is recommended to determine a calibration curve with the same fatty acid attached to the ester substrate. Fatty acids with 6 up to 18 carbon atoms can be detected with this method.
21. Preferably prepare and use the solution fresh or not older than 1 month when stored at 8 °C. Due to instability of the enzymes and ATP present in the solutions R1 and R2, older reagents may give false results.
22. So prepared emulsion remained stable for 2–3 h.
23. Substrate only sample should not show any increase of absorbance during 30 min.

References

1. Verger R (1997) Interfacial activation of lipases: facts and artifacts. Trends Biotechnol 15: 32–38
2. Ali YB, Verger R, Abousalham A (2012) Lipases or esterases: does it really matter? Toward a new bio-physico-chemical classification. Methods Mol Biol 861:31–51
3. Ferrato F, Carriere F, Sarda L, Verger R (1997) A critical reevaluation of the phenomenon of interfacial activation. Methods Enzymol 286: 327–347
4. Sarda L, Desnuelle P (1958) Actions of pancreatic lipase on esters in emulsions. Biochim Biophys Acta 30:513–521
5. Chahinian H, Nini L, Boitard E, Dubes JP, Comeau LC et al (2002) Distinction between esterases and lipases: a kinetic study with vinyl esters and TAG. Lipids 37:653–662

6. Chahinian H, Sarda L (2009) Distinction between esterases and lipases: comparative biochemical properties of sequence-related carboxylesterases. Protein Pept Lett 16:1149–1161
7. Fojan P, Jonson PH, Petersen MT, Petersen SB (2000) What distinguishes an esterase from a lipase: a novel structural approach. Biochimie 82:1033–1041
8. Lo YC, Lin SC, Shaw JF, Liaw YC (2003) Crystal structure of *Escherichia coli* thioesterase I/protease I/lysophospholipase L1: consensus sequence blocks constitute the catalytic center of SGNH-hydrolases through a conserved hydrogen bond network. J Mol Biol 330: 539–551
9. Lescic Asler I, Ivic N, Kovacic F, Schell S, Knorr J et al (2010) Probing enzyme promiscuity of SGNH hydrolases. Chembiochem 11:2158–2167
10. Schmid A, Dordick JS, Hauer B, Kiener A, Wubbolts M et al (2001) Industrial biocatalysis today and tomorrow. Nature 409:258–268
11. Kaewprapan K, Wongkongkatep J, Panbangred W, Phinyocheep P, Marie E et al (2011) Lipase-catalyzed synthesis of hydrophobically modified dextrans: activity and regioselectivity of lipase from *Candida rugosa*. J Biosci Bioeng 112:124–129
12. Zaidan UH, Abdul Rahman MB, Othman SS, Basri M, Abdulmalek E et al (2011) Kinetic behaviour of free lipase and mica-based immobilized lipase catalyzing the synthesis of sugar esters. Biosci Biotechnol Biochem 75:1446–1450
13. Bornscheuer UT, Kazlauskas RJ (2006) Phospholipases: sections 7.1–7.2. In: Bornscheuer UT, Kazlauskas RJ (eds) Hydrolases in organic synthesis. Wiley-VCH Verlag GmbH & Co. KGaA, Weinheim, pp 211–214
14. Bornscheuer UT, Kazlauskas RJ (2006) Lipases and esterases: sections 5.3–5.4. In: Bornscheuer UT, Kazlauskas RJ (eds) Hydrolases in organic synthesis. Wiley-VCH Verlag GmbH & Co KGaA, Weinheim, pp 141–183
15. Bornscheuer UT, Kazlauskas RJ (2006) Lipases and esterases: sections 5.1–5.2. In: Bornscheuer UT, Kazlauskas RJ (eds) Hydrolases in organic synthesis. Wiley-VCH Verlag GmbH & Co KGaA, Weinheim, pp 61–140
16. Schmidtke AJ, Hanson ND (2008) Role of ampD homologs in overproduction of AmpC in clinical isolates of *Pseudomonas aeruginosa*. Antimicrob Agents Chemother 52:3922–3927
17. Jaeger KE, Ransac S, Dijkstra BW, Colson C, van Heuvel M et al (1994) Bacterial lipases. FEMS Microbiol Rev 15:29–63
18. Jaeger KE, Reetz MT (1998) Microbial lipases form versatile tools for biotechnology. Trends Biotechnol 16:396–403
19. Jaeger KE, Dijkstra BW, Reetz MT (1999) Bacterial biocatalysts: molecular biology, three-dimensional structures, and biotechnological applications of lipases. Annu Rev Microbiol 53:315–351
20. Mahajan-Miklos S, Rahme LG, Ausubel FM (2000) Elucidating the molecular mechanisms of bacterial virulence using non-mammalian hosts. Mol Microbiol 37:981–988
21. Stover CK, Pham XQ, Erwin AL, Mizoguchi SD, Warrener P et al (2000) Complete genome sequence of *Pseudomonas aeruginosa* PA01, an opportunistic pathogen. Nature 406:959–964
22. Kovacic F (2010) Novel phospholipases A of *Pseudomonas aeruginosa*: biochemical characterisation and cellular localisation. PhD thesis, Heinrich-Heine-Universität Düsseldorf, Germany
23. Adam PB, Adriana IV, Alain F, Genevieve B, Paula JW et al (2004) A novel extracellular phospholipase C of *Pseudomonas aeruginosa* is required for phospholipid chemotaxis. Mol Microbiol 53:1089–1098
24. Luberto C, Stonehouse MJ, Collins EA, Marchesini N, El-Bawab S et al (2003) Purification, characterization, and identification of a sphingomyelin synthase from *Pseudomonas aeruginosa*. PlcH is a multifunctional enzyme. J Biol Chem 278:32733–32743
25. Wohlfarth S, Hoesche C, Strunk C, Winkler UK (1992) Molecular genetics of the extracellular lipase of *Pseudomonas aeruginosa* PAO1. J Gen Microbiol 138:1325–1335
26. Ostroff RM, Vasil AI, Vasil ML (1990) Molecular comparison of a nonhemolytic and a hemolytic phospholipase C from *Pseudomonas aeruginosa*. J Bacteriol 172:5915–5923
27. Salacha R, Kovacic F, Brochier-Armanet C, Wilhelm S, Tommassen J et al (2010) The *Pseudomonas aeruginosa* patatin-like protein PlpD is the archetype of a novel Type V secretion system. Environ Microbiol 12:1498–1512
28. Vasil M (2006) *Pseudomonas aeruginosa* phospholipases and phospholipids. In: Levesque R, Ramos J-L (eds) Pseudomonas. Springer, New York, pp 69–97
29. Wilderman PJ, Vasil AI, Johnson Z, Vasil ML (2001) Genetic and biochemical analyses of a eukaryotic-like phospholipase D of *Pseudomonas aeruginosa* suggest horizontal acquisition and a role for persistence in a chronic pulmonary infection model. Mol Microbiol 39:291–303
30. Pesaresi A, Lamba D (2005) Crystallization, X-ray diffraction analysis and phasing of carboxylesterase PA3859 from *Pseudomonas aeruginosa*. Biochim Biophys Acta 1752:197–201
31. Martinez A, Ostrovsky P, Nunn DN (1999) LipC, a second lipase of *Pseudomonas aeruginosa*,

is LipB and Xcp dependent and is transcriptionally regulated by pilus biogenesis components. Mol Microbiol 34:317–326

32. Sanchez DG, Otero LH, Hernandez CM, Serra AL, Encarnacion S et al (2012) A *Pseudomonas aeruginosa* PAO1 acetylcholinesterase is encoded by the PA4921 gene and belongs to the SGNH hydrolase family. Microbiol Res 167:317–325
33. Wilhelm S, Tommassen J, Jaeger KE (1999) A novel lipolytic enzyme located in the outer membrane of *Pseudomonas aeruginosa*. J Bacteriol 181:6977–6986
34. Jaeger KE, Eggert T (2002) Lipases for biotechnology. Curr Opin Biotechnol 13:390–397
35. Jaeger KE, Holliger P (2010) Chemical biotechnology—a marriage of convenience and necessity. Curr Opin Biotechnol 21:711–712
36. Sitkiewicz I, Stockbauer KE, Musser JM (2007) Secreted bacterial phospholipase A2 enzymes: better living through phospholipolysis. Trends Microbiol 15:63–69
37. Istivan TS, Coloe PJ (2006) Phospholipase A in Gram-negative bacteria and its role in pathogenesis. Microbiology 152:1263–1274
38. Arpigny JL, Jaeger KE (1999) Bacterial lipolytic enzymes: classification and properties. Biochem J 343(Pt 1):177–183
39. Hausmann S, Jaeger KE (2010) Lipolytic enzymes from bacteria. In: Timmis K (ed) Handbook of hydrocarbon and lipid microbiology. Springer, Berlin, pp 1099–1126
40. Beisson F, Tiss A, Rivière C, Verger R (2000) Methods for lipase detection and assay: a critical review. Eur J Lipid Sci Technol 102:133–153
41. Alberghina L, Schmid R, Verger R (1991) Lipases: structure, mechanism, and genetic engineering: contributions to the CEC-GBF international workshop, September 13 to 15, 1990. VCH, Braunschweig, Germany
42. Jaeger KE, Ransac S, Koch HB, Ferrato F, Dijkstra BW (1993) Topological characterization and modeling of the 3D structure of lipase from *Pseudomonas aeruginosa*. FEBS Lett 332: 143–149
43. Kugimiya W, Otani Y, Hashimoto Y, Takagi Y (1986) Molecular cloning and nucleotide sequence of the lipase gene from *Pseudomonas fragi*. Biochem Biophys Res Commun 141: 185–190
44. Bertani G (1951) Studies on lysogenesis. I. The mode of phage liberation by lysogenic *Escherichia coli*. J Bacteriol 62:293–300
45. Kouker G, Jaeger KE (1987) Specific and sensitive plate assay for bacterial lipases. Appl Environ Microbiol 53:211–213
46. Gubash SM (1991) Improved egg-yolk agar plate medium for the detection of clostridial phospholipase C activity. Res Microbiol 142: 87–93
47. Arzoglou P, Goudoula C, Tsantili P, Lessinger JM, Ferard G et al (1994) Transferability of lipase titrimetric assays: deductions from an interlaboratory study. Eur J Clin Chem Clin Biochem 32:773–777
48. Abousalham A, Verger R (2000) Egg yolk lipoproteins as substrates for lipases. Biochim Biophys Acta 1485:56–62
49. Winkler UK, Stuckmann M (1979) Glycogen, hyaluronate, and some other polysaccharides greatly enhance the formation of exolipase by *Serratia marcescens*. J Bacteriol 138:663–670
50. Kim HK, Park SY, Lee JK, Oh TK (1998) Gene cloning and characterization of thermostable lipase from *Bacillus stearothermophilus* L1. Biosci Biotechnol Biochem 62:66–71
51. Ma J, Zhang Z, Wang B, Kong X, Wang Y et al (2006) Overexpression and characterization of a lipase from *Bacillus subtilis*. Protein Expr Purif 45:22–29
52. Ahmad S, Kamal MZ, Sankaranarayanan R, Rao NM (2008) Thermostable *Bacillus subtilis* lipases: in vitro evolution and structural insight. J Mol Biol 381:324–340
53. Dolinsky VW, Douglas DN, Lehner R, Vance DE (2004) Regulation of the enzymes of hepatic microsomal triacylglycerol lipolysis and re-esterification by the glucocorticoid dexamethasone. Biochem J 378:967–974
54. Jacks TJ, Kircher HW (1967) Fluorometric assay for the hydrolytic activity of lipase using fatty acyl esters of 4-methylumbelliferone. Anal Biochem 21:279–285
55. Flieger A, Gong S, Faigle M, Deeg M, Bartmann P et al (2000) Novel phospholipase A activity secreted by *Legionella* species. J Bacteriol 182:1321–1327
56. Flieger A, Neumeister B, Cianciotto NP (2002) Characterization of the gene encoding the major secreted lysophospholipase A of *Legionella pneumophila* and its role in detoxification of lysophosphatidylcholine. Infect Immun 70:6094–6106
57. Fluxa VS, Wahler D, Reymond J-L (2008) Enzyme assay and activity fingerprinting of hydrolases with the red-chromogenic adrenaline test. Nat Protoc 3:1270–1277
58. Wahler D, Boujard O, Lefèvre F, Reymond J-L (2004) Adrenaline profiling of lipases and esterases with 1,2-diol and carbohydrate acetates. Tetrahedron 60:703–710
59. Rotticci D, Norin T, Hult K, Martinelle M (2000) An active-site titration method for lipases. Biochim Biophys Acta 1483:132–140
60. Asler IL, Kovacic F, Marchetti-Deschmann M, Allmaier G, Stefanic Z et al (2012) Inhibition

of extracellular lipase from *Streptomyces rimosus* with 3,4-dichloroisocoumarin. J Enzyme Inhib Med Chem. doi:10.3109/14756366.2012

61. Leroy E, Bensel N, Reymond J-L (2003) Fluorogenic cyanohydrin esters as chiral probes for esterase and lipase activity. Adv Synthesis Catal 345:859–865
62. Lagarde D, Nguyen H-K, Ravot G, Wahler D, Reymond J-L et al (2002) High-throughput screening of thermostable esterases for industrial bioconversions. Org Process Res Dev 6:441–445
63. Patel RN (2003) Microbial/enzymatic synthesis of chiral pharmaceutical intermediates. Curr Opin Drug Discov Dev 6:902–920
64. Straathof AJJ, Panke S, Schmid A (2002) The production of fine chemicals by biotransformations. Curr Opin Biotechnol 13:548–556
65. Rotticci D, Rotticci-Mulder JC, Denman S, Norin T, Hult K (2001) Improved enantioselectivity of a lipase by rational protein engineering. Chembiochem 2:766–770
66. Prasad S, Bocola M, Reetz MT (2011) Revisiting the lipase from *Pseudomonas aeruginosa*: directed evolution of substrate acceptance and enantioselectivity using iterative saturation mutagenesis. Chemphyschem 12:1550–1557
67. Bornscheuer UT (2002) Methods to increase enantioselectivity of lipases and esterases. Curr Opin Biotechnol 13:543–547
68. Reetz M, Jaeger K-E (1999) Superior biocatalysts by directed evolution. In: Fessner W-D, Archelas A, Demirjian DC, Furstoss R, Griengl H et al (eds) Biocatalysis—from discovery to application. Springer, Berlin, pp 31–57
69. Reetz MT, Jaeger KE (2000) Enantioselective enzymes for organic synthesis created by directed evolution. Chemistry 6:407–412
70. Jaeger KE, Reetz MT (2000) Directed evolution of enantioselective enzymes for organic chemistry. Curr Opin Chem Biol 4:68–73
71. Reetz MT, Wilensek S, Zha D, Jaeger KE (2001) Directed evolution of an enantioselective enzyme through combinatorial multiple-cassette mutagenesis. Angew Chem 40: 3589–3591
72. Liebeton K, Zonta A, Schimossek K, Nardini M, Lang D et al (2000) Directed evolution of an enantioselective lipase. Chem Biol 7:709–718
73. Reetz MT, Puls M, Carballeira JD, Vogel A, Jaeger KE et al (2007) Learning from directed evolution: further lessons from theoretical investigations into cooperative mutations in lipase enantioselectivity. Chembiochem 8:106–112
74. Bocola M, Otte N, Jaeger KE, Reetz MT, Thiel W (2004) Learning from directed evolution: theoretical investigations into cooperative mutations in lipase enantioselectivity. Chembiochem 5:214–223
75. Kazlauskas RJ (2006) Quantitative assay of hydrolases for activity and selectivity using color changes. In: Reymond JL (ed) Enzyme assays. Wiley-VCH Verlag GmbH & Co. KGaA, Weinhem, pp 15–39
76. Fersht A (1985) Enzyme structure and mechanism. CRC Press Inc, Boca Raton, FL
77. Reetz MT, Zonta A, Schimossek K, Jaeger K-E, Liebeton K (1997) Creation of enantioselective biocatalysts for organic chemistry by *in vitro* evolution. Angew Chem Int Ed Engl 36:2830–2832
78. Janes LE, Kazlauskas RJ, Quick E (1997) A fast spectrophotometric method to measure the enantioselectivity of hydrolases. J Org Chem 62:4560–4561

Chapter 13

Elastinolytic and Proteolytic Enzymes

Efrat Kessler and Mary Safrin

Abstract

Pseudomonas aeruginosa secretes into its environment at least seven extracellular proteases: pseudolysin (LasB protease; elastase), aeruginolysin (alkaline proteinase), staphylolysin (staphylolytic endopeptidase; LasA protease), lysyl endopeptidase (protease IV; PrpL), PASP (*P. aeruginosa* small protease), LepA (Large ExoProtease A), and an aminopeptidase. Their action on host proteins, both individually and synergistically, plays important roles in pathogenesis of *P. aeruginosa* infections. Methods to measure/detect their activities are fundamental for understanding their physiological functions, roles in pathogenesis, mechanisms of action, regulation, and secretion.

Most assays for determination/detection of proteolytic activity employ modified/non-modified casein or gelatin as substrates. In the quantitative assay, fragments generated from azocasein are separated from undigested substrate by trichloroacetic acid precipitation and their absorbance is measured. In non-quantitative assays, proteolytic activity is detected as clearing zones around bacterial growth or samples of culture supernatants on casein containing solid media formed due to local casein degradation. In zymography, individual proteases are detected as clear bands in gelatin/casein containing gels after SDS-PAGE separation, renaturation and protein staining. The elastinolytic capacity of *P. aeruginosa* is reflected by clearing zones on nutrient agar plates containing insoluble elastin instead of casein. Mueller-Hinton agar plates on which *S. aureus* cells are grown as a lawn are used to assess the susceptibility of *S. aureus* isolates to staphylolysin. A clear zone around a staphylolysin-containing sample indicates inhibition of *S. aureus* growth.

Methods for measuring the activity of individual proteases are based on their cleavage specificity. These include assays of elastinolytic activity of pseudolysin and/or staphylolysin using elastin-Congo red as a substrate, a method for determination of staphylolytic activity in which the rate of *S. aureus* cell lysis is determined spectrophotometrically, and methods for determination of peptidase activity of pseudolysin, staphylolysin, lysyl endopeptidase, and the aminopeptidase. The latter methods employ chromogenic or fluorogenic peptide derivatives comprising a short amino acid sequence matching the preferred cleavage site of the protease as substrates. As only one peptide bond is cleaved in each substrate, these assays permit kinetic studies.

Key words Elastase, LasB, Alkaline proteinase, Staphylolytic protease, LasA, Lysyl endopeptidase, Protease IV, Aminopeptidase, Proteolytic activity, Elastolytic activity

Alain Filloux and Juan-Luis Ramos (eds.), *Pseudomonas Methods and Protocols*, Methods in Molecular Biology, vol. 1149, DOI 10.1007/978-1-4939-0473-0_13, © Springer Science+Business Media New York 2014

Abbreviations

Abz	2-Aminobenzoyl
AMC	7-Amino-4-methylcoumarin (or 7-amido-4-methylcoumarin)
APS	Ammonium persulfate
Dabsyl	4-(Dimethylamino) azobenzene-4 sulfonyl chloride
DDW	Double distilled water (de-ionized water)
DFP	Diisopropyl fluorophosphate
DMF	Dimethylformamide
EDANS	5-(2-Aminoethylamino)-1-naphthalene sulfonic acid
FA	Furylacryloyl
FITC	Fluorescein isothiocyanate
LB	Luria broth
Nba	4-Nitrobenzylamide
PBS	Phosphate buffered saline
PMSF	Phenylmethylsulfonyl fluoride
pNA	*para*-Nitroanilide
RFU	Relative fluorescence units
SDS-PAGE	Sodium dodecyl sulfate polyacrylamide gel electrophoresis
SGAP	*Streptomyces griseus* aminopeptidase
Su	Succinyl
TCA	Trichloro acetic acid
TEMED	*N*,*N*,*N'*,*N'*-Tetramethylethylenediamine
TLCK	N^{α}-*p*-Tosyl-L-lysine chloromethyl ketone
Tris	2-Amino-2-(hydroxymethyl)-1,3-propanediol
TSB	Tryptic soy broth
Z	*N*-Carbobenzyloxy-

1 Introduction

1.1 Extracellular Proteases of *P. aeruginosa*

Pseudomonas aeruginosa is widely recognized for its medical importance as a leading opportunistic pathogen. It causes severe infections especially of the respiratory tract (cystic fibrosis, pneumonia), urinary tract, eye (keratitis), ear, wounds and burns [1]. The virulence of *P. aeruginosa* is mediated by multiple mechanisms, of which, production of extracellular proteases is a major contributor [2]. During infection, the secreted proteases degrade a wide array of host proteins, both independently and synergistically, thereby impairing host defenses and destroying physical barriers that normally prevent attachment and penetration of the bacteria.

At least seven extracellular proteases are known, of which, six are endopeptidases, i.e., enzymes cleaving internal peptide bonds within proteins and peptides while the seventh protease is an aminopeptidase, i.e., an exopeptidase that releases single amino acids from the N-terminus of proteins and peptides. Four of the six endopeptidases have been studied in great detail, including (1) pseudolysin, also called elastase because of its ability to degrade elastin or LasB protease because it is encoded by the *lasB* gene [3]; (2) aeruginolysin, also known as alkaline proteinase or Apr [4]; (3) staphylolysin, also known as staphylolytic endopeptidase because of its ability to lyse staphylococci or LasA protease because it is encoded by the *lasA* gene [5]; (4) lysyl endopeptidase [6], also known as protease IV [7–9] or PrpL [10]. The remaining endopeptidases, PASP (*P. aeruginosa* Small Protease) [11] and LepA (Large ExoProtease A) [12], are not yet fully characterized although PASP appears to mediate corneal erosions [11, 13] while LepA can activate NF-kB via protease-activated receptors [12]. Finally, the aminopeptidase, designated PAAP for *P. aeruginosa* aminopeptidase or leucine aminopeptidase because it acts preferentially on N-terminal leucine residues, has been well characterized biochemically but its role in pathogenesis remains to be elucidated [14, 15]. The aminopeptidase has been speculated, however, to complement the activity of the endopeptidases, releasing free amino acids/small peptides from protein fragments produced by the endopeptidases, thereby providing low molecular weight nutrients that can be taken up by the bacteria, which in turn, may promote bacterial proliferation [14].

Pseudolysin is the most abundant and most potent endopeptidase secreted by *P. aeruginosa* and a classic virulence factor. It is a zinc-dependent metalloprotease belonging to the M4 thermolysin-like peptidase family as classified in the MEROPS database [16]. Pseudolysin has a broad cleavage specificity although it favors hydrophobic or aromatic amino acid residues, with preference to Phe and Leu at the P1′ position. Cleavages occur on the amino side of these residues, a feature utilized in designing synthetic substrates and specific active site directed inhibitors for the enzyme [17–19]. Its activity on common protein substrates such as casein or azocasein is five- to tenfold higher than that of other *Pseudomonas* endopeptidases. This, along with its high abundance in most *P. aeruginosa* strains, implies that assays of general proteolytic activity in crude enzyme preparations from *P. aeruginosa* largely reflect pseudolysin activity [3].

Pseudolysin is unique in its ability to degrade elastin, a highly cross-linked connective tissue protein abundant in blood vessels and the lungs and practically resistant to most proteases [3]. In pulmonary infections, degradation of elastin by pseudolysin impairs lung elasticity and function [20]. As a highly potent protease, pseudolysin can degrade numerous host proteins in addition to elastin, including several collagens, proteoglycans, immunoglobulins, complement components, α1-proteinase inhibitor, fibrinogen, cytokines, surfactant proteins A and D (important components of the lung innate immune system), protease-activated receptor 2, and more (reviews 2, 3 and references therein). It is not surprising therefore that pseudolysin is a major contributor to pathogenesis of *P. aeruginosa* infections that has been the focus of countless studies since its discovery, almost six decades ago [21]. As would be expected, many methods have been used to determine pseudolysin activity, of which those based on insoluble elastin and defined peptide derivatives comprising its preferred cleavage site as substrates, are specific to pseudolysin. Its proteolytic activity, as that of other *P. aeruginosa* endopeptidases, is determined using casein, modified casein, or hide azure powder as substrates. Representative methods for determination of the elastinolytic, proteolytic, and peptidase activity of pseudolysin are described below.

Aeruginolysin is another well-studied protease of *P. aeruginosa* first described in 1963 [22]. It is a zinc-dependent metalloprotease, belonging to family M10B and a member of the metzincin superfamily [16]. Its proteolytic power is limited as compared to pseudolysin and its cleavage specificity remains quite unclear although it has been reported to cleave peptide bonds on the carboxyl side of arginine residues [4, 23, 24]. It was also proposed that the molecular size of the substrate is more important for aeruginolysin activity than the nature of the amino acid on both sides of the scissile bond [25]. To assay aeruginolysin activity, it is therefore recommended to use proteins such as casein and its chromogenic/fluorogenic derivatives as substrates. To minimize nonspecific proteolysis, assays of crude or semi-purified aeruginolysin preparations should be conducted in the presence of pseudolysin and lysyl endopeptidase inhibitors, phosphoramidon and TLCK, respectively. Despite its limited proteolytic efficiency, aeruginolysin can degrade a variety of biologically important proteins, including laminin (basement membranes), cytokines, immunoglobulins, protease inhibitors, etc. Its action on such substrates can inactivate biologically active molecules, alter signaling mechanisms, impair host defenses, cause hemorrhages, and promote bacterial cell spreading and septicemia, pointing at aeruginolysin as a virulence factor (see reviews 2, 4, 23 and references therein). As has been shown recently, aeruginolysin can also activate the epithelial sodium channel, suggesting a novel aeruginolysin-mediated virulence mechanism by which *P. aeruginosa* remodels its host cells to facilitate colonization and modulate virulence in the cystic fibrosis lung [26].

The *lasA* gene encoding staphylolysin was discovered in 1980 [27] and shown to be required for maximal expression of the elastinolytic phenotype of *P. aeruginosa*. It was therefore presumed to encode either pseudolysin or an enzyme involved in the maturation of pseudolysin. While the identification in 1988 of the *lasB* gene encoding pseudolysin [28] precluded the first possibility, the function of *lasA* remained unknown until 1993, when it was shown to be a staphylolytic endopeptidase [29]. By cleaving peptide bonds within the pentaglycine bridges that stabilize the cell wall peptidoglycan of *Staphylococci*, staphylolysin can lyse and kill *S. aureus* cells. In fact, staphylolysin cleavage requires at least two glycine residues in a raw, with preference to Gly, Ala or Phe in the P1′

(third) position [30]. This cleavage specificity accounts for both its staphylolytic activity and contribution to the elastinolytic potential of *P. aeruginosa*. A limited number of Gly-Gly-Ala and Gly-Gly-Phe sequences exist in elastin. Their cleavage by staphylolysin increases the susceptibility of elastin to pseudolysin and other proteases [30]. In addition to its role in elastin degradation, staphylolysin contributes to pathogenesis of *P. aeruginosa* infections by enhancing syndecan-1 shedding [31, 32]. The substrate involved remains to be identified. Since *P. aeruginosa* infections are often secondary to infections with *S. aureus*, staphylolysin may also play an important role at the colonization stage of the infection, when elimination of competing organisms such as *S. aureus* would be an advantage. By virtue of its staphylolytic activity, staphylolysin might also provide a therapeutic anti-staphylococcal treatment, as we have demonstrated in a rabbit model of keratitis [33] and a rat model of endophthalmitis [34].

Elucidation of the function and substrate specificity of staphylolysin was critical for the introduction of methods to determine its activity, first by measuring its staphylolytic activity using (killed) *S. aureus* cell suspension as a substrate [29], then by designing chromogenic/fluorogenic synthetic substrates containing at least two glycine residues in a raw, substrates that permit kinetic studies [17, 35]. The introduction of synthetic substrates for staphylolysin was important because whole *S. aureus* cells are poorly reproducible substrates and many complex variables may intervene between the action of the enzyme and the measured activity. Among these, synergistic action of pseudolysin and aeruginolysin with staphylolysin (which initiates the cell lysis reaction) that can increase the rate of cell lysis by up to 50 % as compared to staphylolysin alone [29, 36]. The ability of staphylolysin to cleave penta- and hexaglycine substrates is shared with lysostaphin, a highly potent staphylolytic endopeptidase produced by certain *S. aureus* strains [37]. A colorimetric microtiter plate assay for lysostaphin using N-blocked hexaglycine as a substrate was introduced [38] that might lend itself also to determination of staphylolysin activity. Protocols for determination of staphylolytic activity and peptidase activity of staphylolysin are described herein.

Lysyl endopeptidase (protease IV, PrpL) was first identified by Elliott and Cohen in 1986 [6]. They coined it Ps-1 and established its strict specificity to lysine, showing that cleavages occur exclusively on the carboxyl side of lysine residues in proteins and peptides. Lysyl endopeptidase is a serine protease, belonging to the chymotrypsin family S1 according to the MEROPS database [16]. It is very sensitive to the lysine specific serine protease inhibitor TLCK but is only partially inhibited by classical serine proteases inhibitors such as DFP and PMSF. Elliott and Cohen [6] showed that lysyl endopeptidase can cleave fibrinogen in vitro, suggesting it may interfere with clot formation/stabilization, leading to defective blood clotting and hemorrhages. Lysyl endopeptidase can cleave and inactivate surfactant proteins A, D, and B in bronchoalveolar lavage fluid, thereby abolishing their ability to aggregate bacteria and enhancing bacterial uptake by macrophages [39]. Thus, lysyl endopeptidase interferes with innate immune defense mechanisms. Lysyl endopeptidase cleaves β-casein on the carboxyl side of certain lysyl residues, producing a typical band pattern in SDS-PAGE [30]. This permits detection of even traces of lysyl endopeptidase in protease preparations from *P. aeruginosa* that are undetectable by protein staining. Lysyl endopeptidase acts as a virulence factor in keratitis [40].

Although lysyl endopeptidase can cleave various protein substrates such as casein/azocasein, ever since its strict specificity to lysine residues was established, lysyl endopeptidase activity is measured using Tosyl-Gly-Pro-Lys-*p*-nitroanilide as a substrate. In this (colorless) substrate, the carboxyl group of the lysine residue is coupled to the amino group of *p*-nitroaniline via an amide bond. Hydrolysis by lysyl endopeptidase of the amide bond releases the yellow chromophore *p*-nitroaniline whose absorbance is monitored at 405 nm. A protocol for this assay is provided below.

1.2 Assays of Proteolytic Activity

Proteolytic activity of *P. aeruginosa* endopeptidases can be measured using proteins impregnated with dyes (e.g., azo, azure) or coupled to a fluorophore such as FITC as substrates. Casein is often the substrate of choice because of its high sensitivity to proteolysis. Hydrolysis of internal peptide bonds liberates small fragments that in the colorimetric assays and some of the fluorimetric assays, their absorbance/fluorescence is measured after removal of undigested substrate by TCA precipitation and centrifugation [41, 42]. In more advanced fluorimetric assays, fluorescence of the fluorophore is quenched within the intact protein substrate while that of the protease-generated fragments increases due to "dequenching." The rate of increase in fluorescence following proteolysis is monitored continuously without prior separation of undigested substrate from the reaction products [26]. Hide azure powder and unmodified casein can also serve as substrates for *P. aeruginosa* proteases. However, hide azure powder is insoluble so that the reaction suspensions must be rotated throughout the reaction. Undigested substrate is removed at the end of the incubation by either filtration or centrifugation [43]. In assays using unmodified casein as the substrate, measurement of the amount of soluble fragments liberated by the protease(s) is based on tyrosine content [21, 44], which is determined by the method of Lowry et al. [45].

The overall proteolytic activity of *P. aeruginosa* strains can be evaluated by growing bacteria on skim milk (casein) agar plates, or by placing samples of cell free culture filtrates in wells made in the agar layer, and detecting clearing zones (halos) around the bacterial growth/wells [46, 47]. Identification of specific proteases secreted by *P. aeruginosa* is possible by zymography: separation of proteins by SDS-PAGE in gelatin/casein-containing acrylamide gels, renaturation, and protein staining [48]. Bands corresponding to proteases remain clear (unstained) due to local degradation of gelatin/casein. Gelatin/casein zymography is the method of choice for the detection and identification of unknown proteases (endopeptidases) as exemplified in the original studies on PASP [11] and LepA [12].

Of the many available methods for determination/detection of proteolytic activity, the most frequently used techniques, azocasein assay, skim milk agar plate assay, and gelatin zymography, are described below. The advanced fluorogenic protein substrates are commercially available (EnzCheck; Molecular Probes). They are provided as kits that include detailed protocols.

1.2.1 The Azocasein Assay

Azocasein consists of casein conjugated to an azo-dye and is used as a substrate for determination of proteolytic activity of endopeptidases. Degradation of casein liberates soluble dye that can be quantitatively analyzed after precipitation of undigested substrate by measuring the absorbance at 400 nm of the clear supernatant. The assay was introduced in 1947 [49] but due to its simplicity and low cost it remains one of the most frequently used assays of general proteolytic activity. With the exception of staphylolysin, azocasein is degraded to various degrees by all of the known *P. aeruginosa* endopeptidases. Thus, azocasein assays of crude culture supernatants of *P. aeruginosa* reflect the overall proteolytic capacity of a given strain under defined growth conditions.

1.2.2 Skim Milk Agar Plate Assay

This method is used for identification of protease producing *P. aeruginosa* strains. Samples of appropriately diluted bacterial cell suspensions are inoculated on TSB or LB agar plates containing skim milk to obtain single colonies. Proteolytic activity is revealed as clear halos around colonies of protease(s)-secreting bacteria due to degradation of the otherwise turbid casein protein. The same approach can be used to evaluate proteolytic activity in culture supernatants of *P. aeruginosa* by placing small filter paper disks on the plate, on which a few microliters of the culture supernatant were applied, or by pipetting the samples into small wells made before hand in the agar. Proteolytic activity is revealed as a clearing zone surrounding the filter paper/well [46, 47].

1.2.3 Gelatin Zymography

Zymography is the electrophoretic separation of proteins through a polyacrylamide gel containing a protein substrate such as casein or gelatin. After denaturing under non-reducing conditions and without boiling, proteins are separated by SDS-PAGE, renatured, and the gel is then incubated in an appropriate buffer to permit proteolysis of the protein substrate. Clear zones seen after Coomassie staining indicate active proteinases (*see* Fig. 1). In studies on *P. aeruginosa*

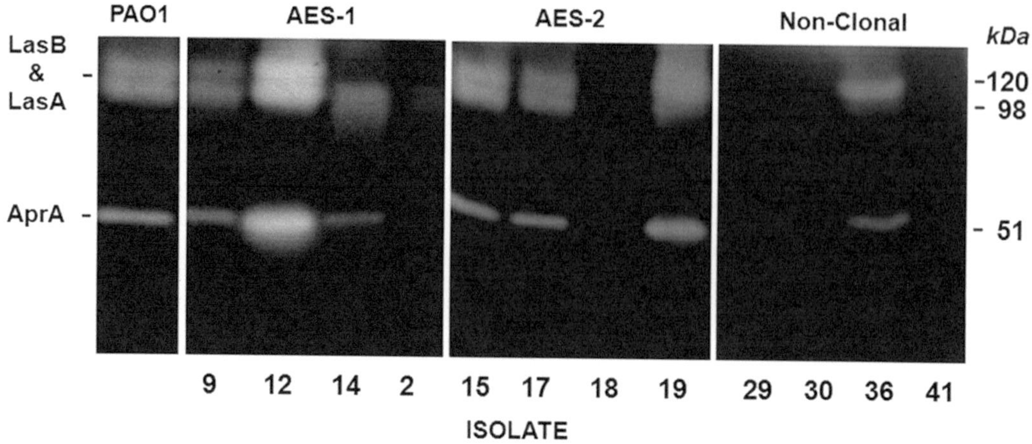

Fig. 1 Gelatin zymography to detect proteases secreted by clinical isolates of *P. aeruginosa* from lungs of adults with cystic fibrosis. Samples of unconcentrated culture supernatants were subjected to SDS-PAGE using 10 % gels. *PAO1* wild type control, *AES-1* and *AES-2* two groups of clonal isolates from the lungs of cystic fibrosis patients. Up to three bands were detected in most isolates, corresponding to molecular masses of ~120, 98 and 51 kDa, representing LasB (pseudolysin), LasA (staphylolysin), and alkaline proteinase (AprA), respectively. Isolates 18, 29, 30, and 41 were negative (secreted little or no proteolytic enzymes) and isolate 2 secreted a small amount of pseudolysin or staphylolysin. Reproduced from Tingpej et al. [61] with permission from the American Society for Microbiology

proteases, gelatin is preferred to casein because (1) unlike casein, gelatin is degraded by staphylolysin [7] and (2) pseudolysin and aeruginolysin require calcium for renaturation and stability; calcium ions, however, bind to and precipitate casein but do not interact with gelatin. The procedure described herein (Subheadings 2.3 and 3.3) is based on Twining et al. [48] and Caballero et al. [7].

1.3 Assays of Elastinolytic Activity

Pseudolysin and to a lesser extent staphylolysin, are unique among *P. aeruginosa* proteases in their ability to degrade elastin [3, 5]. The most widely used assay for determination of elastinolytic activity utilizes insoluble elastin impregnated with dye (e.g., Congo red, Orcein) as a substrate. Pseudolysin, like other elastases, liberates soluble dyed fragments and absorbance at an appropriate wave length of the soluble fraction provides a measure of elastinolytic activity (*see* Fig. 2). Staphylolysin has a basal elastinolytic activity in this assay due to cleavages at Gly-Gly-Phe or Gly-Gly-Ala sequences that occur in elastin [5, 30]. Although limited, this activity of staphylolysin enhances the elastinolytic activity of pseudolysin in this assay several fold, indicating that staphylolysin and pseudolysin action on elastin is synergistic [30]. To assess the elastinolytic potential of *P. aeruginosa* strains, bacteria are grown on nutrient agar plates containing insoluble elastin. A clear zone (halo) around and beneath the bacterial growth (*see* Fig. 3) indicates local degradation of insoluble elastin by pseudolysin and/or staphylolysin.

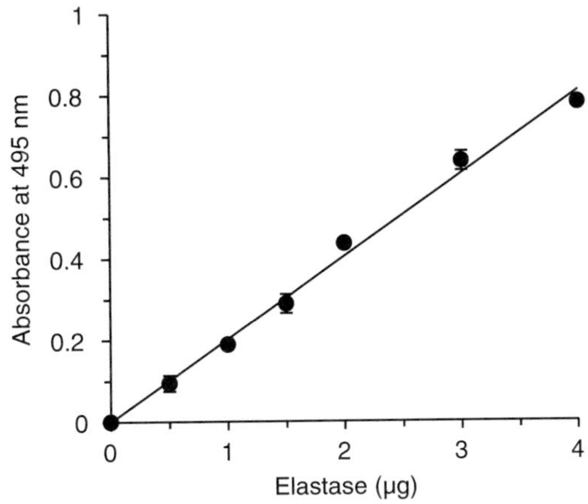

Fig. 2 Standard curve for elastinolytic activity of pseudolysin in the elastin-Congo red assay. The assay was performed as described in the text with 10 mg elastin-Congo red and increasing amounts of purified pseudolysin (elastase) per reaction. $n = 2$

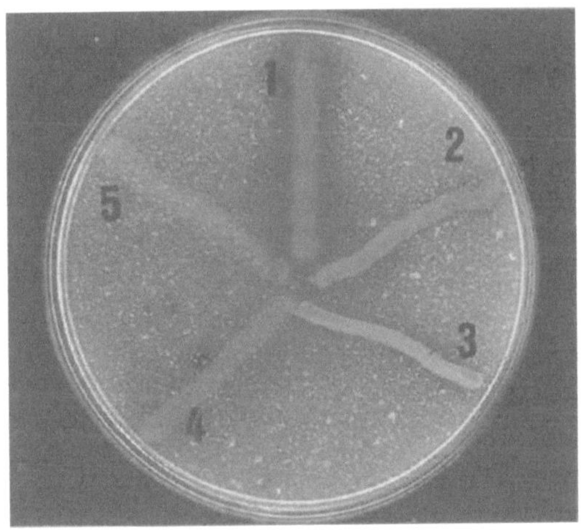

Fig. 3 Elastin nutrient agar plate assay revealing the elastinolytic phenotype of five *P. aeruginosa* strains. Growth was allowed for 24 h at 37 °C. The clearing zone seen under and around some of the strains is indicative of elastinolytic activity. Strains: *1*, PAO1 (wild type); *2*, PAO-E64 (*lasA* minus); *3*, PAO-B1 (*lasB* minus); *4*, PAO-E64-B1 (*lasA* and *lasB* minus); *5*, PAO-R1 (Δ*lasR*—a positive regulator of pseudolysin and staphylolysin expression). As seen, the presence of both staphylolysin and pseudolysin (PAO1) is required for extensive elastinolysis. Reproduced from Toder et al. [62] with permission from John Wiley and Sons

1.3.1 Elastin-Congo Red Assay

Originally developed by Naughton and Sanger for measuring the activity of pancreatic elastase [50], this assay is widely used for determination of the elastinolytic activity of pseudolysin and staphylolysin [7, 27, 30, 51, 52]. The assay utilizes insoluble elastin impregnated with Congo red dye as the substrate. When incubated with an elastinolytic endopeptidase, fragments containing the red dye are generated and become soluble. Remaining insoluble elastin is pelleted, the clear supernatant is collected, and elastinolytic activity is determined by measuring the absorbance of the clear supernatant at 495 nm (*see* Fig. 2). Elastin impregnated with orcein (Elastin-Orcein; Sigma E1500) can be used as an alternative substrate, in which case, absorbance of the supernatant is measured at 590 nm [19].

The protocol used in our laboratory [30] in which the reaction volume and amount of insoluble elastin are scaled down, is described below (Subheadings 2.4 and 3.4). Other protocols have been described [7, 27, 51, 52] and can also be used.

1.3.2 Elastin Nutrient Agar Plate Assay

This assay reveals the overall elastinolytic activity produced by specific *P. aeruginosa* strains during growth on elastin-containing agar plates consisting of a nutrient agar base and insoluble elastin-containing agar overlays. Bacterial strains are grown on the elastin-containing layer, and zones of elastin clearing around and below the bacterial growth indicate production of elastinolytic activity (52, 53; *see* Fig. 3).

1.4 Spectrophotometric and Fluorimetric Assays of the Peptidase Activity of Pseudolysin

The peptidase activity of pseudolysin can be determined using synthetic chromogenic or fluorogenic peptide derivatives with amino acid sequences comprising the preferred cleavage sites of pseudolysin (Gly-Leu, Gly-Phe, Ala-Phe) as substrates.

The spectrophotometric assays are based on chromogenic substrates such as FA-Gly-Leu-NH_2 [19] or FA-Gly-Leu-Ala [41]. Both are cleaved at the Gly-Leu peptide bond but cleavage of the tripeptide FA-Gly-Leu-Ala is 50- to 100-fold more efficient [19]. Hydrolysis leads to a decrease in absorbance at 340 nm, the rate of which can be monitored continuously using a spectrophotometer. Although both substrates are available commercially (Bachem), these chromogenic assays are rarely used nowadays because of their limited sensitivity (assays with FA-Gly-Leu-NH_2 require 20–50 μg pseudolysin). In addition, hydrolysis of these substrates is first order so the reaction cannot be followed for more than 1–3 min otherwise complex mathematical calculations are needed for determination of kinetic constants [41].

A highly sensitive and specific fluorogenic assay for pseudolysin relies on the fluorogenic substrate Dabsyl-Ala-Ala-Phe-Ala-Edans, an internally quenched compound with Edans serving as a fluorescent donor and Dabsyl as a quenching acceptor [17]. Fluorescence increases upon cleavage of the Ala-Phe bond. This substrate, however, is not available commercially, which makes the assay impractical at this time. In contrast, the fluorogenic substrate Abz-Ala-Gly-Leu-Ala-Nba, which was introduced by Nishino and Powers in 1980 [18], is commercially available. Thus, the assay relying on this substrate remains the assay of choice for determination of peptidase activity of pseudolysin, especially valuable in kinetic studies such as characterization of pseudolysin inhibitors [54]. In this assay, the basal fluorescence of the substrate is very low because its fluorogenic group (Abz) is in close vicinity to a quenching group (Nba). After cleavage, the fluorescent and quenching groups are separated by diffusion, resulting in increased fluorescence, allowing determination of the hydrolysis rate. Complete hydrolysis of the substrate leads to about 7 fold increase in fluorescence [18].

1.5 Assays of Staphylolysin Activity

1.5.1 Assay of Staphylolytic Activity

Staphylolysin/LasA protease can lyse *S. aureus* cells by cleaving the pentaglycine bridges within the peptidoglycan network of the cell wall [29]. This activity can be determined spectrophotometrically by monitoring the rate of decrease in absorbance at 595 nm of a *S. aureus* cell suspension due to enzymatic hydrolysis of the bacterial cell wall and cell lysis.

1.5.2 Disk Diffusion Susceptibility Assay

This assay is used to assess the susceptibility of *S. aureus* strains to staphylolysin. *S. aureus* cells are spread evenly on a Mueller-Hinton agar plate. Sterile filter paper disks impregnated with staphylolysin are placed on the agar surface and the plates are incubated overnight at 37 °C to permit bacterial growth. Clear zones around the disk(s) (*see* Fig. 4) indicate inhibition of growth, i.e., sensitivity to staphylolysin. The same procedure can be used to assess the staphylolytic potential of *P. aeruginosa* strains by applying cell free culture supernatants on the disks. The protocol described herein (Subheadings 2.8 and 3.8) is used in our laboratory to assess the

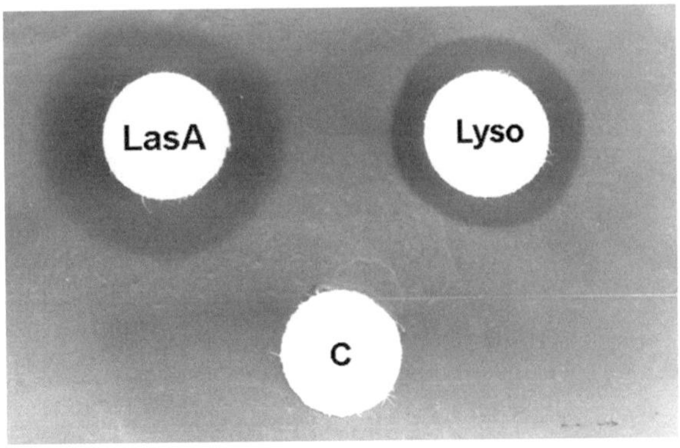

Fig. 4 Disk diffusion susceptibility assay of *S. aureus* growth inhibition by staphylolysin. *LasA*, 5 μg LasA protease (staphylolysin); *Lyso*, 5 μg lysostaphin (positive control); *C*, sterile buffer (negative control). A cell suspension of an ocular isolate of *S. aureus* was spread over the agar plate. Sterile disks containing the indicated samples were placed on the agar surface. The plate was incubated at room temperature for 24 h. Clear zones around the disks indicate inhibition of *S. aureus* growth by the staphylolytic enzymes, staphylolysin (LasA protease) and lysostaphin

susceptibility to staphylolysin of clinical *S. aureus* isolates and is a modification of a previously described method [55] designed to determine susceptibility of *S. aureus* strains to lysostaphin.

1.5.3 Assays of the Peptidase Activity of Staphylolysin

The peptidase activity of staphylolysin can be assayed spectrophotometrically or fluorimetrically using chromogenic/fluorogenic peptide derivatives containing at least two successive glycine residues followed by either another glycine residue or by a phenylalanine residue, the preferred cleavage sites of staphylolysin [30]. The spectrophotometric assay takes advantage of the combined action of staphylolysin and *Streptomyces griseus* aminopeptidase (SGAP) on the chromogenic substrate Su-Gly-Gly-Phe-pNA. Cleavage by staphylolysin of the Gly-Phe bond is followed by hydrolysis of the product Phe-pNA by SGAP and the rate of release of the chromophore *p*-nitroaniline is measured spectrophotometrically [35]. The Gly-Phe bond is cleaved slowly by pseudolysin. Assays of *P. aeruginosa* culture filtrates, which are likely to contain pseudolysin, should therefore be performed in the presence of a pseudolysin inhibitor such as phosphoramidon [35, 56]. In the fluorimetric assay of staphylolysin activity, the rate of hydrolysis of Dabsyl-Leu-Gly-Gly-Gly-Ala-Edans (an internally quenched substrate with Edans as fluorescent donor and Dabsyl as quenching acceptor) is followed continuously [17]. This assay has two advantages over the spectrophotometric assay: (1) The fluorogenic substrate is not cleaved by pseudolysin; (2) Fluorescence goes up upon cleavage independently of other components. However, because the fluorogenic substrate is not available commercially, the fluorimetric assay is not as practical at this time as the spectrophotometric assay. Therefore, only the protocol of the latter assay is described herein (Subheadings 2.9 and 3.9).

1.6 Assays of Lysyl Endopeptidase Activity

Lysyl endopeptidase (protease IV) cleaves specifically peptide bonds on the carboxyl side of lysine residues in proteins and peptides [6]. Its proteolytic activity can be determined with nonspecific protease substrates such as casein/azocasein. Such assays, however, are not very sensitive, requiring at least ~0.5–1 μg lysyl endopeptidase per reaction. Partial hydrolysis of β-casein into distinct fragments can be followed by SDS-PAGE [30, 57]. Although very sensitive (distinct fragments are generated by as little as 1 ng enzyme), this assay is not quantitative. The recommended assay of lysyl endopeptidase activity takes advantage of its strict specificity to lysine residues, measuring the rate of hydrolysis of the chromogenic substrate *N*-(*p*-Tosyl)-Gly-Pro-Lys-*p*-nitroanilide spectrophotometrically [6, 7]. Cleavage on the carboxyl side of the lysyl residue releases the yellow chromophore *p*-nitroaniline and the reaction rate is determined by monitoring the rate of increase of absorbance at 405 nm due to the release of *p*-nitroaniline. This assay is highly sensitive, specific (the substrate is not cleaved by any other *P. aeruginosa* protease), simple, and inexpensive.

1.7 Assays of Aminopeptidase Activity

The activity of *P. aeruginosa* aminopeptidase can be measured either spectrophotometrically using the chromogenic substrate leucine-*p*-nitroanilide (Leu-pNA) [14, 15, 58] or fluorimetrically using the fluorogenic substrate, leucine-7-amido-4-methylcoumarin (Leu-AMC) [59, 60]. In the spectrophotometric assay, hydrolysis of Leu-pNA leads to an increase in absorbance at 405 nm due to the release of the yellow chromophore *p*-nitroaniline, while in the fluorimetric assay, hydrolysis of Leu-AMC leads to an increase in fluorescence at 350 nm due to the release of the fluorophore 7-amino-4-methylcoumarin (AMC). The reaction rate is monitored using either a spectrophotometer or a fluorimeter, as appropriate. Protocols for both methods as used in our laboratory are described below (Subheadings 2.11, 2.12, 3.11 and 3.12).

2 Materials

2.1 Azocasein Assay of Proteolytic Activity

2.1.1 Equipment and Supplies

1. Double beam spectrophotometer.
2. Water bath (37 °C).
3. 10 ml round bottom centrifuge tubes.
4. Cuvettes.
5. Cotton.

2.1.2 Materials and Reagents

1. Azocasein (Sigma A2765; *see* **Note 1**).
2. Tris (2-Amino-2-(hydroxymethyl)-1,3-propanediol; Trizma base; Sigma T1503).
3. Reaction buffer: 0.05 M Tris–HCl, 0.5 mM $CaCl_2$, pH 7.5. Prepare a 10× stock buffer solution: 0.5 M Tris, 5 mM $CaCl_2$, pH 7.5. Place 6.05 g Trizma base and 73.5 mg $CaCl_2$ dihydrate form (mol. weight 147.02) in a 100 ml beaker. Add 80 ml double distilled water (DDW), dissolve by stirring while titrating to pH 7.5 with 6 M HCl. Adjust the volume to 100 ml with DDW. This buffer can be stored at 4 °C for up to a month. Before the assay, dilute 1:10 to obtain the reaction buffer by mixing 10 ml of the 10× stock buffer with 90 ml DDW. The reaction buffer can be stored at 4 °C for up to 2 weeks.
4. Azocasein solution: Dilute 5 ml of 10× stock reaction buffer 1:8 with DDW (final volume 40 ml). Add 200 mg azocasein powder to the diluted buffer and stir gently at room temperature for 30–60 min for maximal solubilization. Solubilization can be speeded up by mild heating (up to 40 °C maximum). Adjust the volume to 50 ml with DDW and mix well (final concentration 4 mg/ml in 1× reaction buffer). Remove insoluble material by centrifugation (8,000 × *g*, 15 min). Collect the clear supernatant, aliquot and store at −20 °C until use.

5. 10 % TCA: Dilute a 100 % solution of TCA (100 g TCA/100 ml H_2O; Sigma T0699) 1:10 in water (e.g., add 10 ml of 100 % TCA to 90 ml of DDW). This should be done slowly with gentle stirring to avoid damage to eyes or skin until a homogenous solution is obtained (*see* **Note 2**).
6. *Pseudomonas aeruginosa* culture supernatant, purified *P. aeruginosa* endopeptidase such as pseudolysin or aeruginolysin.

2.2 Skim Milk Agar Plate Assay

2.2.1 Equipment and Supplies

1. Autoclave.
2. Water bath (45 °C).
3. Incubator (37 °C).
4. Standard bacteriological plates (8.5–10 cm in diameter).
5. Sterile inoculation (Drigalski) rod.
6. Sterile tips.

2.2.2 Materials and Reagents

1. TSB without dextrose (Difco 286220) (*see* **Note 3**).
2. Bacto agar (Difco 0140-01-0).
3. Skim milk (Difco 232100).
4. *Pseudomonas aeruginosa* strain(s).

2.3 Gelatin Zymography

2.3.1 Equipment and Supplies

1. Mini gel apparatus (for example, Protean II Mini-cell [Bio-Rad] for gels 8 cm × 7.3 cm × 0.75 mm). A larger gel apparatus can be used if better resolution is required or a larger number of samples must be run on the same gel.
2. Power supply (200 V, 500 mA, max).
3. Water bath (55 °C) or microwave oven.
4. Incubator (37 °C).
5. Rocking platform or rotary shaker.
6. Glass or plastic containers with covers, for gel washing/incubation.

2.3.2 Materials and Reagents

1. Acrylamide: 40 % Acrylamide–Bis-acrylamide (37.5:1) (Bio-Rad 161-0148) (*see* **Note 4**).
2. High quality gelatin (e.g., Bio-Rad 170-6537).
3. 2.7× stock separating gel buffer: 1 M Tris–HCl, pH 8.8.
4. 8× stock stacking gel buffer: 1 M Tris–HCl, pH 6.8.
5. 20 % (w/v) sodium dodecyl sulfate (SDS; Bio-Rad 161-0302) in DDW.
6. TEMED (*N*,*N*,*N*′,*N*′-tetramethylethylenediamine). Use directly from the bottle. Store at 4 °C, protected from light.
7. Ammonium persulfate (APS): 10 % w/v in DDW. Make fresh.

8. 100× gelatin: 10 % (w/v; 100 mg/ml) gelatin in DDW. Warm in a water bath (~55 °C) to dissolve gelatin or heat gently in microwave oven. Store at −20 °C in 1 ml aliquots.
9. 5× (Laemmli) sample buffer: 0.31 M Tris–HCl pH 6.8, 10 % SDS, 50 % glycerol, 0.1 % bromophenol blue. Do not add reducing agent.
10. Molecular weight standards (*see* **Note 5**).
11. 10× electrophoresis running buffer: 0.25 M Tris (Trizma base, Sigma T1503), 1.92 M glycine (Sigma G7126), 1 % (w/v) SDS, pH 8.3. Store in a glass or plastic bottle at room temperature. Dilute to 1× with DDW before use.
12. Triton X-100: 2.5 % (v/v) in 0.05 M Tris–HCl, 5 mM $CaCl_2$, pH 8.0.
13. Development buffer: 0.05 M Tris–HCl, 0.1 M NaCl, 10 mM $CaCl_2$, 1 mM $ZnCl_2$, pH 8.0 (*see* **Note 6**).
14. Fixing/destaining solution: methanol–acetic acid–water (4.5:1:4.5, v:v:v). Store at room temperature.
15. Staining solution: 0.1 % Coomassie Brilliant Blue R-250 (w/v) in fixing/destaining solution. Filter using a Whatman number 1 paper to remove insoluble dye particles. Store in a dark bottle at room temperature.
16. Source of *Pseudomonas aeruginosa* proteases: cell free culture supernatant, purified protease(s): pseudolysin and aeruginolysin are available commercially (Elastin products Company, Inc. and United States Biochemicals, respectively).

2.4 Elastin-Congo Red Assay

2.4.1 Equipment and Supplies

1. Double beam spectrophotometer.
2. Thermo regulated shaking water bath or a roller drum tube rotator (e.g., New Brunswick-Eppendorf model TC-7) placed in a thermo regulated room.
3. Microfuge.
4. Microfuge tubes.
5. 25 ml glass scintillation vials or 10 ml round bottom tubes (*see* **Note 7**).

2.4.2 Materials and Reagents

1. Elastin-Congo red (Sigma E0502).
2. Enzyme: Cell free culture supernatant of *P. aeruginosa*, purified pseudolysin (Calbiochem; Elastin Products Company, Inc.) or purified staphylolysin.
3. Reaction buffer: 0.05 M Tris–HCl, 0.5 mM $CaCl_2$, pH 7.5. This buffer is prepared from a 10× buffer consisting of 0.5 M Tris–HCl, 5 mM $CaCl_2$, adjusted to pH 7.5 by titration with 6 M HCl. To prepare the reaction buffer, add 10 ml of 10× buffer to 90 ml of DDW, mix and store at 4 °C (the reaction

buffer can be stored for up to 2 weeks; the 10× buffer can be stored for a month).

4. Stopping solution: 120 mM EDTA, titrated to pH 7.5. Dissolve 4.47 g EDTA (di-sodium salt) in 80 ml DDW, and while stirring, titrate to pH 7.5 by adding slowly 0.5 M NaOH, switching to 0.1 M NaOH around pH 6.5. Adjust volume to 100 ml with DDW, mix and store at 4 °C (up to 4 weeks).

2.5 Elastin Nutrient Agar Plate Assay

2.5.1 Equipment and Supplies

1. Incubator (37 °C).
2. Shaking water bath (45 °C).
3. Sterile petri plates (15 × 100 mm).
4. Sterile inoculation loop or (Drigalski) inoculation rod.

2.5.2 Materials and Reagents

1. Nutrient broth (Difco laboratories).
2. Noble agar (Difco laboratories).
3. Elastin (Sigma E1625).
4. *P. aeruginosa* culture(s).

2.6 Fluorimetric Assay of Peptidase Activity of Pseudolysin. Hydrolysis of Abz-Ala Gly-Leu-Ala-Nba

2.6.1 Equipment and Supplies

1. Thermo regulated microplate fluorescence reader (e.g., Bio-Tek FLx 800).
2. 96 dark multi-well plates (Greiner 655076).

2.6.2 Materials and Reagents

1. Abz-Ala-Gly-Leu-Ala-Nba substrate (AAGLAN; Sigma SCP0185; Bachem H-6675).
2. DMF (solvent).
3. Purified pseudolysin (Elastin Products Company, MO) or *P. aeruginosa* culture filtrate (*see* **Note 8**).
4. Assay buffer: 0.05 M Tris–HCl, 2.5 mM $CaCl_2$, pH 7.2. Prepare 10× stock assay buffer: 0.5 M Tris–HCl, 25 mM $CaCl_2$, pH 7.2. Store at 4 °C for up to 4 weeks. To prepare the assay buffer, dilute 1:10 with DDW (e.g., mix 1 ml 10× buffer with 9 ml DDW). Store at 4 °C for up to 2 weeks.
5. 100× stock substrate solution: 11 mM AAGLAN in DMF. Dissolve 6.42 mg AAGLAN in 1 ml of DMF. Store in aliquots at −80 °C.
6. Pseudolysin stock solution: 100 μg/ml in assay buffer containing 1 % DMF. To prepare DMF containing assay buffer, mix 1 ml of 10× assay buffer with 8.9 ml DDW and 100 μl DMF. Store in aliquots at −80 °C.

2.7 Assay of Staphylolytic Activity

2.7.1 Equipment and Supplies

1. Thermo regulated double beam spectrophotometer.
2. Water bath (25 °C).
3. 1 ml cuvettes.

2.7.2 Materials and Reagents

1. Dried *Staphylococcus aureus* cells, strain Newman D_2C (Sigma S-0504).
2. *Pseudomonas aeruginosa* cell free culture filtrate or purified staphylolysin.
3. Reaction buffer: 0.02 M Tris–HCl, pH 8.5. Prepare a 10× stock buffer: 0.2 M Tris–HCl, pH 8.5. Place 2.42 g Trizma base (Sigma T1503) in a 100 ml beaker. Dissolve in 80 ml DDW and while stirring, titrate to pH 8.5 using 6 M HCl. Adjust the volume to 100 ml with DDW and mix. This buffer can be stored at 4 °C for 4 weeks. To prepare 1× reaction buffer, dilute 1:10 by mixing 10 ml of 10× stock buffer with 90 ml DDW. Store at 4 °C for up to 2 weeks.
4. Stock *S. aureus* cell suspension (substrate): Place 30 mg of lyophilized *S. aureus* powder in a 10 ml (heat resistant) tube. Add 3 ml reaction buffer, mix gently to obtain a homogenous suspension (final concentration 10 mg/ml). Cover with aluminum foil and heat to 100 °C for 10 min by placing the tube in boiling water to kill the bacteria. Cool to room temperature. Store in aliquots at −20 °C (frozen).
5. Working substrate suspension: Dilute the stock substrate suspension 1:30. For instance, add 400 μl stock substrate suspension to 11.6 ml reaction buffer. Mix well. The recommended absorbance at 595 nm should be 0.8–0.9. If the absorbance is above 0.9, adjust to ~0.9 by adding more reaction buffer as required.

2.8 Disk Diffusion Staphylolysin Susceptibility Assay

2.8.1 Equipment and Supplies

1. Double beam spectrophotometer.
2. Incubator (37 °C).
3. Water bath (45 °C).
4. Glass petri dishes.
5. Ten ml sterile tubes.
6. Filter paper disks (Whatman 3MM; chromatography grade).
7. Sterile swabs and inoculation (Drigalski) rods.
8. Mueller-Hinton agar plates.

2.8.2 Materials and Reagents

1. *Staphylococcus aureus* strains—frozen stocks.
2. Purified staphylolysin/concentrated cell-free *P. aeruginosa* culture filtrate.
3. Lysostaphin (e.g., Sigma L9043).

4. Enzyme dilution buffer: 0.02 M Tris–HCl, pH 8.5 (the same as the reaction buffer, Subheading 2.7.2, **item 3** above).
5. Sterile PBS.

2.9 Spectrophotometric Assay of Staphylolysin Peptidase Activity

2.9.1 Equipment and Supplies

1. Thermo regulated microplate reader (e.g., Bio-Tek ELx 808).
2. Water bath preheated to 30 or 69 °C as needed.
3. 96-well microplates (Costar 3596).

2.9.2 Materials and Reagents

1. Succinyl-Gly-Gly-Phe-*p*-nitroanilide (SuGGFpNA; Sigma S-1899).
2. *Streptomyces griseus* aminopeptidase (SGAP): Purified SGAP (Sigma A9934) is available; however, it is quite costly. It can be replaced by a cheaper product (Protease type XIV/pronase E; Sigma P5147) from which semi-purified but highly specific SGAP can be easily prepared (*see* Subheading 3.9, **step 1**).
3. Reaction buffer: 0.05 M Tris–HCl, 0.1 M NaCl, 1 mM $CaCl_2$, pH 8.0. Prepare a 10× stock buffer: 0.5 M Tris–HCl, 1 M NaCl, 10 mM $CaCl_2$, pH 8.0. Place 6.05 g Trizma base (Sigma T1503), 5.84 g NaCl and 0.147 g $CaCl_2$-dihydrate in a 100 ml beaker. Dissolve in 80 ml DDW, and while stirring, titrate to pH 8.0 with 6 M HCl. Adjust the volume to 100 ml with DDW and mix well. This buffer can be stored at 4 °C for 4 weeks. To prepare 1× reaction buffer, dilute 1:10 by mixing 10 ml of 10× stock buffer with 90 ml DDW. Store at 4 °C for up to 2 weeks.
4. SGAP buffer: 20 mM Tris–HCl, 60 mM NaCl, 10 mM $CaCl_2$, pH 7.7. Prepare a 10× stock buffer: 0.2 M Tris–HCl, 0.6 M NaCl, 0.1 M $CaCl_2$, pH 7.7. Place 2.42 g Trizma base, 3.5 g NaCl and 1.47 g $CaCl_2$ dihydrate in a 100 ml beaker. Add 80 ml DDW and while stirring, titrate to pH 7.7 using 6 M HCl. Adjust the volume to 100 ml with DDW. To obtain 1× SGAP buffer, add 10 ml of the 10× buffer to 90 ml DDW. Mix well and store at 4 °C for up to 2 weeks. The 10× SGAP buffer can be stored at 4 °C for up to 4 weeks.
5. SuGGFpNA (substrate) stock solution: Dissolve 10 mg SuGGFpNA in 5 ml reaction buffer; final concentration is 4 mM. Store in aliquots at −20 °C.
6. Cell free *P. aeruginosa* culture filtrate or purified staphylolysin (0.5–1 mg/ml).

2.10 Lysyl Endopeptidase Activity Assay

2.10.1 Equipment and Supplies

1. Thermo regulated microplate reader (e.g., Bio-Tek ELx 808) pre-equilibrated to 30 °C.
2. Water bath (30 °C).
3. 96-well microplates (Costar 3596).

2.10.2 Materials and Reagents

1. *N*-(*p*-Tosyl)-Gly-Pro-Lys-*p*-nitroanilide, acetate salt (also known as Chromozym PL; Sigma, T6140; mol. wt. 634.7).
2. Reaction buffer: 0.02 M Tris–HCl, pH 8.0. Prepare a 10× stock buffer: 0.2 M Tris–HCl, pH 8.0. Place 2.42 g Trizma base (Sigma T1503) in a 100–150 ml beaker. Add 80 ml DDW, dissolve by stirring while titrating to pH 8.0 with 6 M HCl. Adjust to 100 ml with DDW and mix well. This buffer can be stored at 4 °C for a month. Dilute 1:10 by adding 10 ml 10× stock buffer to 90 ml water. Mix well. Store at 4 °C for up to 4 weeks.
3. *N*-(*p*-Tosyl)-Gly-Pro-Lys-*p*-nitroanilide (substrate) stock solution: 7 mM *N*-(*p*-Tosyl)-Gly-Pro-Lys-*p*-nitroanilide in reaction buffer. Dissolve 5 mg substrate in 1.12 ml reaction buffer (*see* **Note 9**). Store in aliquots at −20 °C.
4. *Pseudomonas aeruginosa* culture supernatant or purified lysyl endopeptidase.

2.11 Spectrophotometric Assay of *P. aeruginosa* Aminopeptidase. Cleavage of Leu-pNA

2.11.1 Equipment and Supplies

1. Thermo regulated microplate reader (e.g., Bio-Tek ELx 808).
2. Water bath (30 °C).
3. 96-well microplates (Costar 3596).

2.11.2 Materials and Reagents

1. Leucine-*p*-nitroanilide hydrochloride (Leu-pNA) (Sigma L2504).
2. Reaction buffer: 0.05 M Tris–HCl, pH 8.5. Prepare a 10× stock buffer solution: 0.5 M Tris–HCl, pH 8.5. Place 6.05 g Trizma base (Sigma T1503) in a 100 ml beaker. Add 80 ml DDW. Dissolve by stirring while titrating to pH 8.5 using 6 M HCl. Adjust the volume to 100 ml with DDW and mix well. Store at 4 °C for up to 4 weeks. Dilute 1:10 by mixing 10 ml of 10× stock buffer with 90 ml DDW. This buffer can be stored at 4 °C for up to 2 weeks.
3. Leu-pNA (substrate) stock solution: 3 mM in reaction buffer. Dissolve 8.63 mg Leu-pNA in 10 ml reaction buffer. Store in aliquots at −20 °C. Before use, thaw and mix well.
4. Enzyme: *P. aeruginosa* culture supernatant or purified aminopeptidase [14, 15].

2.12 Fluorimetric Assay of *P. aeruginosa* Aminopeptidase. Cleavage of Leu-AMC

2.12.1 Equipment and Supplies

1. Thermo regulated microplate fluorescence reader (e.g., Bio-Tek FLx 800).
2. 96 dark multi-well plates (Greiner 655076).
3. Water bath (30 °C).

2.12.2 Materials and Reagents

1. Leucine-7-amido-4-methylcoumarin hydrochloride (Leu-AMC; Sigma L2145).
2. 7-Amino-4-methylcoumarin (AMC; Sigma A9891).
3. Dimethyl sulfoxide (DMSO).
4. Reaction buffer: 0.05 M Tris–HCl, pH 8.5. Prepare a 10× stock buffer: 0.5 M Tris–HCl, pH 8.5. Place 6.05 g Trizma base (Sigma T1503) in a 100 ml beaker. Add 80 ml DDW, dissolve, and titrate to pH 8.5 using 6 M HCl, with constant stirring. Adjust the volume to 100 ml with DDW and mix well. Store at 4 °C for up to 4 weeks. Dilute 1:10 by mixing 10 ml of 10× stock buffer with 90 ml DDW. This buffer can be stored at 4 °C for up to 2 weeks.
5. Leu-AMC (substrate) stock solution: Dissolve 4 mg Leu-AMC in 50 ml reaction buffer (0.08 mg/ml; 0.25 mM). Store in aliquots at −20 °C. Before the assay, thaw and equilibrate to 30 °C for 10 min, mixing well before use (*see* **Note 10**).
6. Fluorofore (AMC) stock solution: 1 mM AMC in DMSO. Dissolve 5.6 mg of 7-amino-4-methylcoumarin in 32 ml DMSO. Dilute 1:40 by mixing 100 μl of this solution with 3.9 ml reaction buffer (final concentration 0.025 mM) (*see* **Note 10**).
7. *P. aeruginosa* culture supernatant or purified aminopeptidase [14, 15].

3 Methods

3.1 Azocasein Assay of Proteolytic Activity

1. Pipette 0.75 ml of stock azocasein solution into 10 ml round bottom centrifuge tubes.
2. Add 250 μl of reaction buffer to blank tubes to be incubated without enzyme and 245 μl to reaction tubes, all in duplicates. Mix well. The final azocasein concentration is 3 mg/ml.
3. Equilibrate to 37 °C by placing the tubes in a preheated water bath for 10 min.
4. Add 5 μl of *P. aeruginosa* culture supernatant or purified protease to the reaction tubes. Vortex carefully and incubate at 37 °C for 15 min (*see* **Notes 11** and **12**).
5. Stop the reaction by adding 0.5 ml of 10 % TCA and mixing immediately (vortex) to bring the TCA concentration to 3.3 %. Leave at room temperature for at least 30 min to allow maximal precipitation of undigested/partially digested azocasein.
6. Centrifuge at 10,000 ×*g* for 20 min to pellet the insoluble material.
7. Draw the supernatants carefully with a Pasteur pipette and transfer to clean tubes (*see* **Note 13**). Using a double beam

spectrophotometer, measure the absorbance of the clear supernatants at 400 nm. Subtract the absorbance of the blanks (incubated without enzyme) from that of the enzyme reactions. Express the activity as ΔOD_{400}/min.

8. One unit of activity is defined as the amount of enzyme that produces an optical density increase of 1 U/min under the assay conditions.

3.2 Skim Milk Agar Plate Assay

1. Skim-milk-containing agar plates: Weigh 2.75 g TSB (w/o dextrose) powder, 1.5 g skim milk, and 1.5 g Bacto agar. Suspend in 100 ml DDW water in a 500 ml Erlenmeyer flask shaking well until the suspension is homogeneous.
2. Autoclave at 121 °C for 15 min.
3. Equilibrate to 45 °C in a preheated shaking water bath (*see* **Note 14**).
4. Pour aseptically (avoiding air bubbles formation) ~20 ml per plate and spread evenly by gently rotating the plate.
5. Let the agar solidify at room temperature. Leave on the bench overnight to obtain a firm layer of agar (*see* **Note 15**).
6. Dilute a fresh overnight culture of *P. aeruginosa* serially with sterile saline to obtain 20–40 colonies/200 μl. Pipette 200 μl of the diluted suspension on the agar surface and spread evenly using a sterile glass/plastic inoculation (Drigalski) rod.
7. Invert plates and incubate at 37 °C for 18 h.
8. Clearing zones around colonies identify protease producing bacteria. The width of the clearing zone is a semiquantitative measure of the overall proteolytic activity secreted by the bacteria.

3.3 Gelatin Zymography

1. Preparation of gelatin containing 10 % separation gel (*see* **Note 16**): Mix 3 ml (40 %) acrylamide stock solution, 4.4 ml 2.7× stock separating buffer, and 4.38 ml DDW. Degas by applying vacuum for 10 min (optional). Add 120 μl 100× gelatin solution and 60 μl 20 % SDS. Mix gently to avoid bubbles formation. To initiate polymerization, add 20 μl TEMED and 30 μl 10 % APS (final gelatin concentration 0.1 %; final volume 12 ml). Mix gently and cast two mini-gels using clean glass plates. Allow space for stacking gel and gently overlay with butanol or water to prevent contact with atmospheric oxygen and permit leveling the resolving gel solution.
2. Preparation of 4 % stacking gel (5 ml): When the separating gel has polymerized (at least 30 min), decant the liquid above the gel and rinse with 1× stacking gel buffer or H_2O. Combine the stacking gel components: 0.5 ml 40 % acrylamide solution, 0.625 ml 8× stock stacking gel buffer, and 3.82 ml DDW, mix

and degas (optional). Add 25 μl 20 % SDS stock solution, 5 μl TEMED, and 20 μl 10 % APS solution as described above. Pipet the stacking gel solution on top of the separating gel and insert the appropriate comb, taking care not to trap bubbles beneath the teeth. Allow the gel to polymerize for at least 30 min. Remove the comb, rinse the wells with 1× electrophoresis running buffer, assemble the apparatus, and fill upper and lower reservoir with 1× electrophoresis running buffer (*see* **Note 17**).

3. Preparation of samples and electrophoresis: Add 5 μl of 5× (Laemmli) sample buffer to 20 μl of protease sample, cell free conditioned medium of *P. aeruginosa* or purified protease (0.1–1 μg; as a standard) (*see* **Note 18**). Mix gently and leave on the bench for 10 min. Do not boil and do not add reducing agent. Load samples into the bottom of each well using gel loading pipet tips (new tip for each sample), taking care not to introduce bubbles or remove the pipet tip from the well until the entire sample has been loaded. Avoid cross-contamination of adjacent wells. Electrophorese samples at constant voltage or amperage (for two mini gels: ~125 V constant voltage or ~30 mA constant current) at 4 °C. The run is complete when the bromophenol blue tracking dye reaches the bottom of the gel.
4. Development of the gels: After electrophoresis, discard the stacking gels and transfer the separating gels to clean glass or plastic containers (one gel per container) and wash twice for 15–60 min in ~100 ml of Triton X-100 (2.5 % v/v) buffer by gentle agitation on a rocking platform or a rotary shaker at room temperature. After the second wash, remove all but 2–3 ml of the Triton X-100 solution and add development buffer (100 ml). Agitate for 15 min at room temperature and then transfer to a 37 °C incubator for 16–20 h. After incubation in development buffer, the gels can be transferred to small containers (e.g., the plastic boxes in which disposable pipet tips are packaged) to save on reagents.
5. Staining/destaining of the gels: Add Coomassie Blue solution to cover the gel and incubate with mild agitation for at least 2–4 h or overnight. Destain in fixing/destaining solution (replacing the solution several times) until the bands are clearly visible and contrast well with the background. At this stage, wash the gels twice with DDW and leave in DDW to stop further destaining.
6. Drying the gels: Gels can be dried between two sheets of cellophane (for example, cellophane membrane backing from Bio-Rad). Immerse the cellophane in DDW for a few minutes. Lay one piece of cellophane on a clean sheet of glass/plastic, arrange 1–2 gels on the cellophane, making sure that there are

no bubbles beneath them. Cover with a second piece of cellophane, smooth (using a glass rod or pipette) to eliminate bubbles, and clamp the edges of the cellophane to the glass/plastic with metal binder clips. Air dry overnight, preferably in a ventilated chemical hood. To avoid gel cracking, pre-equilibrate the gels and the cellophane in 15 % ethanol/5 % glycerol instead of DDW for 15 min before drying. Gels can be scanned wet or dry.

3.4 Elastin-Congo Red Assay

1. Weigh 10 mg elastin-Congo red and place in a vial or test tube (*see* **Note 19**).
2. Add 1.09 ml reaction buffer (or 1.1 ml for controls to be incubated without enzyme) to each vial/tube and mix/vortex immediately to prevent clumping of substrate.
3. Add 10 μl of *P. aeruginosa* cell free culture supernatant (*see* **Notes 20** and **21**), 0.5 μg purified pseudolysin, or 2 μg staphylolysin, each in 10 μl, to the reaction tubes/vials (total volume 1.1 ml) (*see* **Note 22**). For best results, all samples, including the blanks, should be tested in duplicates.
4. Incubate at 37 °C for 2 h with shaking (vials) or rotation (test tubes) to keep the substrate in suspension.
5. Stop the reaction by adding to each vial/tube 100 μl of 0.12 M EDTA (final EDTA concentration 10 mM) and mix well.
6. Transfer the suspensions to microfuge tubes and centrifuge (8,000 × *g*, 10 min) to pellet the insoluble material. From this step on, all procedures are performed at room temperature.
7. Using a Pasteur pipette, remove the supernatants into clean test tubes, taking care not to touch the (usually well packed) pellet.
8. Measure absorbance of the clear supernatants at 495 nm. Subtract the values of the blanks (incubated without enzyme) to obtain the net increase in absorbance resulting from enzyme digestion of the substrate.
9. One unit of activity is defined as the amount of enzyme that produces an OD_{495} increase of 1 U/h under the assay conditions. Conversion to μg pseudolysin is possible, based on a calibration curve generated simultaneously using known amounts of purified pseudolysin.

3.5 Elastin Nutrient Agar Plate Assay

This procedure is based on Rust et al. [52].

1. Preparation of the nutrient agar base: To 1 l water add 8 g nutrient broth powder and 20 g noble agar. Adjust to pH 7.5 and autoclave. Pour 10 ml per plate, allowing this layer to solidify at room temperature (*see* **Note 23**).

2. Preparation of the elastin agar overlay: To 300 ml water, add 2.4 g nutrient broth powder, 6 g noble agar and 1 g elastin. Adjust to pH 7.5 and autoclave. Cool to 45 °C using a 45 °C water bath with gentle shaking. While stirring, pour 5 ml (*see* **Note 24**) of the overlay suspension per plate and allow to solidify. Constant stirring of the suspension is needed to maintain an even distribution of elastin particles in the overlay.
3. Streak a fresh *P. aeruginosa* culture onto the elastin-nutrient agar plate. Alternatively, spread evenly (using a Drigalski rod) 100–200 μl aliquots of appropriately diluted cell suspension to assess elastinolytic activity of individual colonies.
4. Incubate for 24–48 h at 37 °C. Zones of elastin clearing in pseudolysin/staphylolysin producing strains will surround the bacterial growth (*see* Fig. 3 and **Note 25**).
5. Elastinolytic activity can be roughly estimated by measuring the width of the clearing zone and that of the streaking zone and calculating the ratio of culture width to the sum of zone of clearing plus culture width [52].

3.6 Fluorimetric Assay of Peptidase Activity of Pseudolysin. Hydrolysis of Abz-Ala-Gly-Leu-Ala-Nba

This protocol is based on the original assay of Nishino and Powers [18], scaled down to micro-wells and fluorescence plate reader as described by Catchart et al. [54].

1. Preparation of working Abz-Ala-Gly-Leu-Ala-Nba (AAGLAN; substrate) solution: Dilute the 100× stock AAGLAN substrate solution 1:100 with assay buffer. To prepare 5 ml, add 50 μl 100× substrate solution to a test tube containing 4.45 ml DDW premixed with 0.5 ml 10× assay buffer. The final concentrations will be 110 μM AAGLAN in assay buffer plus 1 % DMF.
2. Pipette 90 μl of the substrate working solution into each well. Add 10 μl reaction buffer to the first 3 wells to be incubated without enzyme (blank, in triplicates), then add 10 μl of appropriately diluted enzyme solution (1–5 ng of enzyme per reaction; *see* **Note 26**) into the reaction wells, in duplicates. Final reaction volume is 100 μl. Final substrate concentration is 100 μM. Pre-equilibrate the microplate fluorescence reader to 25 °C. Place the plate in the fluorescence plate reader and after brief equilibration, monitor the rate of increase in fluorescence at 415 ± 10 nm (excitation at 330 ± 10 nm), using the kinetic mode, reading fluorescence at 2–5 min intervals. Under these conditions the reaction is linear up to about 15 % hydrolysis [18, 54].
3. Calculate the rate of the reaction as ΔRFU/min. For conversion of RFU to nmoles product, perform the reaction with 10- to 20-fold more enzyme and follow the reaction to completion. At full hydrolysis, fluorescence in RFU corresponds to the theoretical concentration of the fluorescent product (100 μM) equivalent to 10 nmol/100 μl.

3.7 Assay of Staphylolytic Activity

1. Equilibrate 10 ml of the working substrate (killed *S. aureus* cells) suspension to 25 °C using a water bath preheated to 25 °C.
2. Pipette 0.9 ml of substrate suspension and 100 μl reaction buffer into two cuvettes. Seal with a small piece of parafilm and mix well by inverting the cuvettes upside down 2–3 times. Wipe the top edges of the cuvettes with soft tissue and place in the spectrophotometer, pre-equilibrated to 25 °C. Calibrate the absorbance at 595 nm to zero. Leave the reference cuvette in the chamber as a blank.
3. Discard the content of the second cuvette, pipette into it a new 0.9 ml sample of substrate suspension and 100 μl of enzyme sample (e.g., *P. aeruginosa* culture supernatant). Mix well using a piece of parafilm as above, place in the spectrophotometer chamber, and after about 2 min, start recording the rate of decrease in absorbance at 595 nm (relatively to the reference cuvette) for about 5 min, using the kinetic mode.
4. Express the rate of the reaction as ΔOD_{595}/min (*see* **Notes 27** and **28**).
5. One activity unit is defined as the amount of enzyme that produces a decrease in OD_{595} of 1 U/min under the assay conditions.

3.8 Disk Diffusion Staphylolysin Susceptibility Assay

This protocol is used in our laboratory to assess the susceptibility to staphylolysin of clinical *S. aureus* isolates and is a modification of a previously described method [55] designed to determine susceptibility of *S. aureus* strains to lysostaphin.

1. Preparation of Mueller-Hinton agar plates: Place 3.8 g of Mueller-Hinton medium (Difco 0252-01-4; containing 17 g/l Bacto agar) in a 250 ml Erlenmeyer flask containing 100 ml DDW. Heat in boiling water with agitation to completely dissolve the powder. Autoclave at 121 °C for 15 min. Cool to 45 °C. Pour 25 ml to sterile bacteriological plates and allow solidification at room temperature. Store at 4 °C until use.
2. Preparation of filter paper disks: Remove 6 mm round disks from a sheet of Whatman 3MM paper using a 6 mm puncher. Place the disks in a clean glass petri dish and autoclave at 121 °C for 15 min.
3. Fresh *S. aureus* cultures: Using a sterile swab, collect bacteria from the top of a frozen stock ampule, inoculate (for colonies isolation) a Mueller-Hinton agar plate and incubate at 37 °C for 18 h. On the following day, pick a single colony of *S. aureus*, inoculate a new Mueller-Hinton agar plate and incubate for 18 h at 37 °C as above.
4. *S. aureus* cell suspension: Using a new sterile swab, collect several colonies from the fresh plate and transfer to a sterile test tube

containing 3 ml of sterile PBS. Mix gently until a homogenous cell suspension is obtained. Measure absorbance of the suspension at 600 nm, adjusting it to ~0.6 OD units.

5. Pipette 300 μl of the bacterial cell suspension onto a new Mueller-Hinton agar plate and spread evenly using an inoculation (Drigalski) rod. Let the cells adsorb to the surface of the agar for 30 min at room temperature.
6. Apply onto each disk 5 μl of a staphylolysin solution containing 5 μg of staphylolysin and place carefully on the agar surface "face down" (the face on which the enzyme sample was applied should touch the agar). For a positive control, apply on a separate disk 5 μg of lysostaphin. Label the samples beneath the plate with a permanent pen.
7. Incubate overnight at either room temperature or at 37 °C (*see* **Note 29**).
8. Check for clearing zones around the disk(s) due to inhibition of *S. aureus* growth (*see* Fig. 4). The width of the clear zone provides a measure of the relative staphylolytic activity.

3.9 Spectrophotometric Assay of Staphylolysin Peptidase Activity

1. Preparation of semi-purified SGAP: Place 260 mg Pronase E (Sigma P5147) in a heat-resistant centrifuge tube. Dissolve in 5.2 ml SGAP buffer. Cover with aluminum foil and heat at 69 °C (in a water bath pre-equilibrated to 69 °C) for 5 h. Cool to room temperature and leave at 4 °C overnight for maximal precipitation of denatured nonrelevant heat-sensitive proteins (SGAP is heat stable). Centrifuge (10,000 × *g*, 10 min, 4 °C). Remove the clear supernatant carefully and dialyze against reaction buffer. Collect the dialyzed SGAP solution, centrifuge to remove additional insoluble material if forms during dialysis, and store in aliquots at −70 °C. The expected SGAP concentration is ~1 mg/ml.
2. Before the assay, in a test tube, prepare sufficient amount of the reaction solution, including the substrate SuGGFpNA, SGAP and buffer in the following ratios per reaction: 10 μl substrate (stock solution 4 mM), 10 μl semi-purified SGAP (~10 μg SGAP; *see* **Note 30**), and 70 μl reaction buffer. Mix and equilibrate to 30 °C in a preheated water bath. Pipette 90 μl of this mixture into each well.
3. Add 10 μl reaction buffer to blank wells to be incubated without staphylolysin in triplicates, 10 μl of *P. aeruginosa* culture filtrate or 10 μl purified staphylolysin (20–100 ng enzyme) to the reaction wells, in duplicates. The final substrate concentration and total reaction volume are 0.4 mM and 100 μl, respectively.
4. Place the multi-well plate containing the reaction solutions in the plate reader, pre-equilibrated to 30 °C.

5. Using the kinetic mode, monitor the increase in absorbance at 405 nm, reading absorbance at 1–2 min intervals for 8–10 min, shaking the plate automatically within the chamber before each reading.
6. Calculate the reaction rate as ΔmOD_{405}/min. One unit of activity is defined as the amount of enzyme that produces an increase in absorbance at 405 nm of 1 U/min under the assay conditions. The net ΔOD_{405}/min values can be converted to nmoles *p*-nitroaniline released per minute using the molar extinction coefficient of *p*-nitroaniline: $\varepsilon_{405} = 10{,}600\ M^{-1}\ cm^{-1}$ [58] (*see* **Note 31**).

3.10 Lysyl Endopeptidase Activity Assay

1. Prepare sufficient amount of reaction solution by mixing (per reaction) 3 μl *N*-(*p*-Tosyl)-Gly-Pro-Lys-*p*-nitroanilide (substrate) stock solution and 87 μl reaction buffer. Mix well and equilibrate to 30 °C. Pipette 90 μl into each well.
2. Add 10 μl reaction buffer to blank wells to be incubated without enzyme in duplicates, 10 μl of cell free culture filtrate or purified lysyl endopeptidase (10–20 ng/reaction) to the other wells, all in duplicates. The final substrate concentration and final reaction volume are 0.21 mM and 100 μl, respectively.
3. Place the micro-well plate in the plate reader pre-equilibrated to 30 °C, and follow the rate of increase in absorbance at 405 nm, using the kinetic mode. Read absorbance at 2–5 min intervals for 20–30 min, shaking the plate automatically before each reading.
4. Calculate the rate of the reaction as ΔmOD_{405}/min, converting it to μmole *p*-nitroaniline per minute, on the basis of its molar extinction coefficient $\varepsilon_{405} = 10{,}600\ M^{-1}\ cm^{-1}$ [6, 58] (*see* **Note 31**).
5. One activity unit is defined as the amount of enzyme that releases 1 μmol *p*-nitroaniline per minute under the assay conditions.

3.11 Spectrophotometric Assay of *P. aeruginosa* Aminopeptidase. Cleavage of Leu-pNA

1. Dilute the stock substrate (Leu-pNA) solution fourfold with reaction buffer (e.g., 0.25 ml stock substrate solution plus 0.75 ml reaction buffer), mix well and equilibrate to 30 °C using a preheated water bath. After equilibration, pipette 80 μl of the diluted substrate solution into each well.
2. Add 20 μl reaction buffer (pre-equilibrated to 30 °C as above) to blank wells to be incubated without enzyme in triplicates, 20 μl of purified/partially purified aminopeptidase in reaction buffer (0.2–0.3 μg aminopeptidase), or 20 μl of concentrated cell free culture supernatant to the remaining wells (*see* **Note 32**), each in duplicates. The final substrate concentration and total reaction volume are 0.6 mM and 100 μl, respectively.

3. Place the plate in the plate reader, equilibrate to 30 °C and follow the increase in absorbance at 405 nm using the kinetic mode. Read absorbance at 5 min intervals, shaking the plate before each reading. Record the increase in absorbance for 20–30 min, depending on the reaction rate (enzyme input). The rate of hydrolysis is linear up to an increase in absorbance of ~0.8 OD units.
4. Calculate the rate of the reaction, first in ΔmOD_{405}/min, then convert the absorbance values to μmole *p*-nitroaniline per minute, using its molar extinction coefficient $\varepsilon_{405} = 10{,}600\ M^{-1}\ cm^{-1}$ [58] (*see* **Note 31**).
5. One activity unit is the amount of enzyme that releases 1 μmol *p*-nitroaniline per minute under the assay conditions.

3.12 Fluorimetric assay of *P. aeruginosa* aminopeptidase. Cleavage of Leu-AMC

1. Pipette 90 μl substrate (Leu-AMC) solution (pre-equilibrated to 30 °C) into each well.
2. Add 10 μl reaction buffer to blank wells to be incubated without enzyme in triplicates and 10 μl of *P. aeruginosa* culture supernatant or purified aminopeptidase (*see* **Note 33**) to the remaining wells, all in duplicates. Final substrate concentration and reaction volume are 0.23 mM and 100 μl, respectively.
3. Place the plate in the fluorescence reader, re-equilibrate to 30 °C and, using the kinetic mode, with excitation at 340–360 and emission at 440–460 nm, follow the increase in fluorescence (emission) due to the release of free 7-amino-4-methylcoumarin at 5 min intervals, shaking the plate automatically before each reading. Start recording as soon as possible because the reaction is very rapid and the linear range is around 20 % hydrolysis.
4. Calculate the rate of the reaction in terms of Relative Fluorescence Units (RFU)/min.
5. Prepare a standard curve by reading the fluorescence of 7-amino-4-methylcoumarin (fluorophore) in the range 1–15 μM under the same conditions (volume, buffer, fluorimeter setting). Based on the fluorescence of a known amount of the fluorophore (also in RFU), convert the values RFU/min to nmol/min.
6. One activity unit is defined as the amount of enzyme that releases 1 nmol 7-amino-4-methylcoumarin per minute under the assay conditions.

4 Notes

1. The degree of substitution of casein by the azo-group varies from batch to batch. For long term studies, it is advisable to purchase the new lot of azocasein before the one in use is

exhausted so that two calibration curves can be generated for both in parallel using the same enzyme aliquots. If the slopes of the curves are significantly different, a factor can be calculated so that results obtained with the new batch of azocasein can be compared to those obtained with the previous one.

2. TCA is highly corrosive. Avoid contact with skin and eyes. Because of its short shelf life, a fresh solution of 10 % TCA should be prepared at least once a month and kept at room temperature in a tightly capped dark bottle as is the case for the original 100 % TCA stock solution.
3. Tryptic soy broth without dextrose is recommended because protease production by *P. aeruginosa* in the absence of glucose is higher than in its presence. LB can, however, be used instead with good results.
4. Acrylamide monomer is a neurotoxin and can be absorbed through unbroken skin. Wear gloves to avoid contact. Do not dispose of acrylamide solutions in the sewer system. Polymerize with an appropriate amount of 10 % ammonium persulfate and TEMED, and discard as solid waste.
5. Some commercial preparations of molecular weight standards contain reducing agents. It is advisable to leave an empty lane between the first unreduced protease sample and the standards. Unlike the protease samples, the molecular weight standard sample should be heated to 100 °C for 1 min prior to loading for complete denaturation.
6. Pseudolysin, aeruginolysin, and staphylolysin are zinc-dependent; the first two enzymes also require calcium for stability and activity. The development buffer is supplemented with Ca and Zn salts to permit correct refolding and restore enzyme activity.
7. The reaction can be performed in either 10 ml test tubes or 25 ml scintillation vials that are constantly rotated or shaken, respectively. Tubes are preferred if a roller drum tube rotator is available.
8. The culture filtrate of most wild type *P. aeruginosa* strains may contain 50–200 μg/ml of pseudolysin. For assays of pseudolysin activity in culture supernatants, the supernatants should be diluted so that enzyme aliquots will contain a few ng of pseudolysin and ΔRFU values will be within the linear range of the reaction.
9. The substrate *N*-(*p*-Tosyl)-Gly-Pro-Lys-*p*-nitroanilide comes in mg amounts per vial. It is recommended to dissolve the total amount of substrate in the original vial by adding the required volume of reaction buffer. In our hands, 1.12 ml reaction buffer are added to a 5 mg vial to obtain a 7 mM stock solution.

10. The substrate (Leu-AMC) and the fluorophore (AMC) are both light sensitive. Keep in the dark.
11. For precise timing, add the enzyme samples to the tubes at 30 s intervals using a stop watch. To terminate the reaction, add 10 % TCA at identical time intervals and the same order, starting exactly 15 min after initiation of the reaction in the first tube.
12. The recommended input of purified pseudolysin, lysyl endopeptidase or aeruginolysin (each alone) is 0.2, 1, or 3–4 μg, respectively, corresponding to their relative proteolytic efficiency under the assay conditions: pseudolysin > lysyl endopeptidase > aeruginolysin. Staphylolysin is practically inactive in this assay. To maximize the activity of aeruginolysin or lysyl endopeptidase, alone, the assay should be conducted at pH 8.5 or 8.0, respectively, using reaction buffers titrated to these pHs rather than pH 7.5. The specified incubation time and enzyme aliquots/input should be considered as suggestions. In each study, a calibration curve should be constructed using increasing amounts of enzyme (e.g., purified pseudolysin or aliquots of cell free culture filtrate) and values of unknown samples must fall within the linear range of the curve.
13. To avoid drawing small particles from the pellet while removing the supernatant, wrap the tip of the Pasteur pipette with a small piece of cotton so as to keep the particles out. After drawing the supernatant from each tube, replace the piece of cotton by a new one, using gloves to protect your fingers' skin.
14. The sterile agar medium can be stored at 4 °C for several weeks. For experiments, it should be melted by heating in a boiling water bath with occasional mixing, followed by equilibration to 45 °C using a 45 °C water bath with gentle shaking.
15. Unused sterile plates can be stored (upside down) in a tightly closed plastic container at 4 °C for up to 4 weeks.
16. Other acrylamide concentrations (in the range 8–12 %) can be used if needed. Change the amount of stock acrylamide solution and adjust the amount of DDW correspondingly to keep the total volume constant, 12 ml in the present protocol. If the total volume of the acrylamide solution differs from 12 ml, adjust the volume of each component proportionally.
17. If the gel is not going to be electrophoresed for a few hours it is better to overlay with 1× stacking gel buffer until you are ready to run the samples.
18. Of the *Pseudomonas* proteases that can be detected by gelatin zymography, only aeruginolysin migrates in the expected position, i.e., as a 50 kDa protein. Pseudolysin (mol. wt. 33 kDa), staphylolysin (mol. wt. ~20 kDa), lysyl endopeptidase/protease

IV (mol. wt. ~28 kDa) and PASP (mol. wt. ~18 kDa) migrate as larger proteins with approximate molecular weights (varying from one laboratory to another) of 160 kDa (pseudolysin and staphylolysin), 200 kDa (protease IV) and 80 kDa (PASP), respectively. LepA (110 kDa) migrates in casein zymography as expected [12]. Because of the unusual migration positions of some of the proteases in gelatin containing SDS-gels, it is important to run samples of purified proteases on the same gel as references.

19. The assay can be performed successfully with 5 mg of elastin-Congo red, which is recommended (for economic reasons) when assaying a large number of samples.
20. The total elastinolytic activity in crude enzyme preparations from most wild type *P. aeruginosa* strains represents the combined activity of pseudolysin and staphylolysin [30].
21. When assaying cell free culture supernatants, a control containing an equal volume of sterile unconditioned medium (negligible if a 10 μl aliquot is assayed) should be included and its absorbance should be subtracted from those obtained for the enzyme reactions. This control is essential when reaction volumes higher than 2 ml and culture filtrate aliquots higher than 50 μl are utilized (e.g., ref. 7, 27, 51, 52).
22. The enzyme input should be based on a calibration curve constructed using different amounts of enzyme (*see* Fig. 2). The amount of enzyme in unknown samples should yield a net OD_{495} value lying within the linear range of the calibration curve.
23. The nutrient agar plates with or without the elastin overlay can be stored at 4 °C until needed.
24. The volume of the overlay can be increased to 10 ml [53].
25. Due to its limited elastinolytic activity, prolonged (48 h) incubation may be required to detect zones of clearing produced by staphylolysin alone, i.e., in the absence of pseudolysin.
26. The stock enzyme solution can be used repeatedly by thawing and freezing several times. Diluted enzyme solutions (concentrations <100 μg/ml) should be prepared freshly before each assay and the remaining diluted solution should be discarded.
27. The linear range of the reaction is limited to a decrease in OD_{595} of ~40 %. Reaction rates should be derived from the linear portion of the curve.
28. Staphylolysin is critical for staphylolytic activity and it probably initiates the reaction. However, once hydrolysis by staphylolysin is initiated, other proteases in the culture filtrate of *P. aeruginosa*, especially pseudolysin, can increase the rate of lysis by up to 50 % [36]. The contribution of pseudolysin can be avoided by

including in the reaction buffer 1–10 μM phosphoramidon (Sigma R7385) – a specific inhibitor of pseudolysin [56].

29. Incubation at room temperature is recommended because in these conditions, the contrast between the turbidity of the bacterial lawn and the clearing zone due to growth inhibition is clearer.
30. The volume (10 μl) of SGAP added is based on the assumption that its concentration in the stock solution is 1 mg/ml.
31. To calculate the amount of *p*-nitroaniline released in the microwell, measure the absorbance at 405 nm of a *p-nitroaniline* solution of a known concentration (e.g., 0.1 mM) in both the ELISA reader and a standard spectrophotometer using a 1 cm path length cuvette. Calculate a correction factor by dividing the absorbance obtained for 1 cm path length by the respective absorbance derived from the ELISA plate reader. OD values obtained using the plate reader (path length is shorter than 1 cm) should be multiplied by this factor to determine the amount of *p*-nitroaniline released from its molar extinction coefficient. In our hands (for instance), the OD of a 0.1 mM *p*-nitroaniline solution is 0.195 in the plate reader and 0.79 in a standard spectrophotometer. The correction factor is 0.79:0.195 = 4.05. Thus, in our experiments, OD values obtained using the microplate reader are multiplied by 4 before calculating the concentrations of *p*-nitroaniline from its molar extinction coefficient.
32. The amount of aminopeptidase secreted into the culture medium of most wild type *P. aeruginosa* strains is relatively low (<1 μg/ml). It is recommended to concentrate the culture filtrate ~10-fold (e.g., by ammonium sulfate precipitation; *see* ref. 14) before assaying the activity or, alternatively, follow the reaction for a prolonged time (up to 120 min).
33. The fluorimetric method is more sensitive than the spectrophotometric method. Choosing the right sensitivity scale, aminopeptidase activity can be determined with as little as 10 ng enzyme. Thus, using the fluorimetric method, aminopeptidase activity in culture supernatants can be determined without prior concentration, as was recommended (in **Note 32**) for the spectrophotometric method.

References

1. Driscoll JA, Brody SL, Kollef MH (2007) The epidemiology, pathogenesis and treatment of *Pseudomonas aeruginosa* infections. Drugs 67:351–368
2. Hoge R, Pelzer A, Rosenau F, Wilhelm S (2010) Weapons of pathogen: proteases and their role in virulence of *Pseudomonas aeruginosa*. In: Mendez-Vilas A (ed) Current Research, technology and education topics in applied microbiology and microbial biotechnology, 2010th edn. Formatex Research Center, Badajoz, Spain, pp 383–395
3. Kessler E, Ohman DE (2013) Pseudolysin. In: Rawlings ND, Salversen GS (eds) Handbook

of proteolytic enzymes, 3rd edn. Academic, Oxford, pp 582–592

4. Honek JF (2013) Aeruginolysin. In: Rawlings ND, Salversen GS (eds) Handbook of proteolytic enzymes, 3rd edn. Academic, Oxford, pp 867–872
5. Kessler E, Ohman DE (2013) Staphylolysin. In: Rawlings ND, Salversen GS (eds) Handbook of proteolytic enzymes, 3rd edn. Academic, Oxford, pp 1553–1557
6. Elliott BW, Cohen C (1996) Isolation and characterization of a lysine-specific protease from *Pseudomonas aeruginosa*. J Biol Chem 261:11259–11265
7. Caballero AR, Moreau JM, Engel LS, Marquart ME, Hill JM, O'Callaghan RJ (2001) *Pseudomonas aeruginosa* protease IV enzyme assays and comparison to other *Pseudomonas* proteases. Anal Biochem 290:330–337
8. Engel LS, Hill JM, Caballero AR, Green LC, O'Callaghan RJ (1998) Protease IV, a unique extracellular protease and virulence factor from *Pseudomonas aeruginosa*. J Biol Chem 273:16792–16797
9. Thibodeaux BA, Caballero AR, Dajcs JJ, Marquart ME (2005) *Pseudomonas aeruginosa* protease IV: a corneal virulence factor of low immunogenicity. Ocul Immunol Inflamm 13: 169–182
10. Wilderman PJ, Vasil AI, Johnson Z, Wilson MJ, Cunliffe HE, Lamont IL, Vasil ML (2001) Characterization of an endoprotease (PrpL) encoded by a PvdS-regulated gene in *Pseudomonas aeruginosa*. Infect Immun 69:5385–5394
11. Marquart ME, Caballero AR, Chomnawang M, Thibodeaux BA, Twining SS, O'Callaghan RJ (2005) Identification of a novel secreted protease from *Pseudomonas aeruginosa* that causes corneal erosions. Invest Ophtalmol Vis Sci 46:3761–3768
12. Kida Y, Higashimoto Y, Inou H, Shimizu T, Kuwano K (2008) A novel secreted protease from *Pseudomonas aeruginosa* activates NF-kB through protease-activated receptors. Cell Microbiol 10:1491–1504
13. Tang A, Marquart ME, Fratkin JD, McCormick CC, Caballero AR, Gatlin HP, O'Callaghan RJ (2009) Properties of PASP: a *Pseudomonas* protease capable of mediating corneal erosions. Invest Ophtalmol Vis Sci 50:3794–3801
14. Cahan R, Axelrad I, Safrin M, Ohman DE, Kessler E (2001) A secreted aminopeptidase of *Pseudomonas aeruginosa*: identification, primary structure, and relationship to other aminopeptidases. J Biol Chem 276:43645–43652
15. Sarnovsky R, Rea J, Makowski M, Hertle R, Kelly C, Antignani A, Pastrana DV, FitzGerald DJ (2009) Proteolytic cleavage of a C-terminal prosequence, leading to autoprocessing at the N Terminus, activates leucine aminopeptidase from *Pseudomonas aeruginosa*. J Biol Chem 284:10243–10253
16. Rawlings ND, Barrett AJ, Bateman A (2010) MEROPS: the peptidase database. Nucleic Acids Res 38(Database issue):D227–D233
17. Elston C, Wallach J, Saulnier J (2007) New continuous and specific fluorometric assays for *Pseudomonas aeruginosa* elastase and LasA protease. Anal Biochem 368:87–94
18. Nishino N, Powers JC (1980) *Pseudomonas aeruginosa* elastase. Development of a new substrate, inhibitors, and an affinity ligand. J Biol Chem 255:3482–3486
19. Morihara K (1995) Pseudolysin and other pathogen endopeptidases of thermolysin family. Methods Enzymol 248:242–253
20. Starcher BC (1986) Elastin and the lung. Thorax 41:577–585
21. Morihara K, Tsuzuki H, Oka T, Inoue H, Ebata M (1965) *Pseudomonas aeruginosa* elastase. Isolation, crystallization and preliminary characterization. J Biol Chem 240:3295–3304
22. Morihara K (1963) *Pseudomonas aeruginosa* proteinase. I. Purification and general properties. Biochim Biophys Acta 73:113–124
23. Maeda H, Morihara K (1995) Serralysin and related bacterial proteinases. Methods Enzymol 248:395–413
24. Louis D, Kohlmann M, Wallach J (1997) Spectrophotometric assay for amidolytic activity of alkaline protease from *Pseudomonas aeruginosa*. Anal Chim Acta 345:219–225
25. Louis D, Bernillon J, Wallach J (1999) Use of 49-peptide library for a qualitative and quantitative determination of pseudomonal serralysin specificity. Int J Biochem Cell Biol 31:1435–1441
26. Butterworth MB, Zhang L, Heidrich EM, Myerburg MM, Thibodeau PH (2012) Activation of the epithelial sodium channel (ENaC) by the alkaline protease from *Pseudomonas aeruginosa*. J Biol Chem 287: 32556–32565
27. Ohman DE, Cryze SJ, Iglewski BH (1980) Isolation and characterization of a *Pseudomonas aeruginosa* PAO mutant that produces altered elastase. J Bacteriol 142:836–842
28. Bever RA, Iglewski BH (1988) Molecular characterization and nucleotide sequence of the *Pseudomonas aeruginosa* elastase structural gene. J Bacteriol 170:4309–4314
29. Kessler E, Safrin M, Olson JC, Ohman DE (1993) Secreted LasA of *Pseudomonas aeruginosa* is a staphylolytic protease. J Biol Chem 268:7503–7508

30. Kessler E, Safrin M, Abrams WR, Rosenbloom J, Ohman DE (1997) Inhibitors and specificity of *Pseudomonas aeruginosa* LasA. J Biol Chem 272:9884–9889
31. Park PW, Pier GB, Preston MJ, Goldberger O, Fitzgerald ML, Bernfield M (2000) Syndecan-1 shedding is enhanced by LasA, a secreted virulence factor of *Pseudomonas aeruginosa*. J Biol Chem 275:3057–3064
32. Park PW, Pier GB, Hinkes MT, Bernfield M (2001) Exploitation of syndecan-1 shedding by *Pseudomonas aeruginosa* enhances virulence. Nature 411:98–102
33. Barequet IS, Bourla N, Pessach YN, Safrin M, Yankovich D, Ohman DE, Rosner M, Kessler E (2012) Staphylolysin is an effective therapeutic agent for *Staphylococcus aureus* experimental keratitis. Graefes Arch Clin Exp Ophthalmol 250:223–229
34. Barequet IS, Habot-Wilner Z, Mann O, Safrin M, Ohman DE, Kessler E, Rosner M (2009) Evaluation of *Pseudomonas aeruginosa* staphylolysin (LasA protease) in the treatment of methicillin-resistant *Staphylococcus aureus* endophthalmitis in a rat model. Graefes Arch Clin Exp Ophthalmol 247:913–917
35. Kessler E, Safrin M, Blumberg S, Ohman DE (2004) A continuous spectrophotometric assay for *Pseudomonas aeruginosa* LasA protease (staphylolysin) using a two-stage enzymatic reaction. Anal Biochem 328:225–232
36. Gustin GK, Kessler E, Ohman DE (1996) A substitution at His-120 in the LasA protease of *Pseudomonas aeruginosa* blocks enzymatic activity without affecting propeptide processing or extracellular secretion. J Bacteriol 178:6608–6617
37. Kumar JK (2008) Lysostaphin: an antistaphylococcal agent. Appl Microbiol Biotechnol 80:555–561
38. Kline SA, de la Harpe J, Blackburn P (1994) A colorimetric microtiter plate assay for lysostaphin using a hexaglycine substrate. Anal Biochem 217:329–331
39. Malloy JL, Veldhuizen RAW, Thibodeux BA, O'Callaghan RJ, Wright JR (2005) *Pseudomonas aeruginosa* protease IV degrades surfactant proteins and inhibits surfactant host defense and biophysical functions. Am J Physiol Lung Cell Mol Physiol 288:L409–L418
40. Caballero A, Thibodeux BA, Marquart M, Traidej M, O'Callaghan RJ (2004) *Pseudomonas* keratitis: protease IV gene conservation, distribution, and production relative to virulence and other *Pseudomonas* proteases. Invest Ophthalmol Vis Sci 45:522–530
41. Kessler E, Israel M, Landshman N, Chechick A, Blumberg S (1982) In vitro inhibition of *Pseudomonas aeruginosa* elastase by metal chelating peptide derivatives. Infect Immun 38:716–723
42. Twining SS (1984) Fluorescein isothiocyanate-labeled casein assay for proteolytic enzymes. Anal Biochem 143:30–34
43. Poncz L, Gerken TA, Dearborn DG, Grobelny D, Galardy RE (1984) Inhibition of the elastase of *Pseudomonas aeruginosa* by Nα-phosphoryl dipeptides and kinetics of spontaneous hydrolysis of the inhibitors. Biochemistry 23:2766–2772
44. He H, Chen X, Li J, Zhang Y, Gao P (2004) Taste improvement of refrigerated meat treated with cold-adapted protease. Food Chem 84:307–311
45. Lowry OH, Rosenbrough NJ, Farr AL, Randall RJ (1951) Protein measurement with the Folin phenol reagent. J Biol Chem 193:265–275
46. Lin X, Xu W, Huang K, Mei X, Liang Z, Li Z, Guo J, Luo Y (2009) Cloning, expression and characterization of recombinant elastase from *Pseudomonas aeruginosa* in *Pichia pastoris*. Protein Expr Purif 63:69–74
47. Laux CD, Corson JM, Givskov M, Hentzer M, Møller A, Wosencroft KA, Olson JC, Krogfelt KA, Goldberg JB, Cohen PS (2002) Lysophosphatidic acid inhibition of the accumulation of *Pseudomonas aeruginosa* PAO1 alginate, pyoverdin, elastase and LasA. Microbiology 148:1709–1723
48. Twining SS, Kirschner SE, Mahnke LA, Frank DW (1993) Effect of *Pseudomonas aeruginosa* elastase, alkaline protease, and exotoxin A on corneal proteinases and proteins. Invest Ophthalmol Vis Sci 34:2699–2712
49. Charney J, Tomarelli RM (1947) A colorimetric method for the determination of the proteolytic activity of duodenal juice. J Biol Chem 171:501–505
50. Naughton MA, Sanger F (1961) Purification and specificity of pancreatic elastase. Biochem J 78:156–163
51. McIver K, Kessler E, Ohman DE (1991) Substitution of active-site His-223 in *Pseudomonas aeruginosa* elastase and expression of the mutated lasB alleles in *Escherichia coli* show evidence for autoproteolytic processing of proelastase. J Bacteriol 173:7781–7789
52. Rust L, Messing CR, Iglewski BH (1994) Elastase assays. Methods Enzymol 235:554–562
53. Hasan JAK, Macaluso P, Carnahan AM, Joseph SW (1992) Elastolytic activity among

Aeromonas spp. using a modified bilayer plate assay. Diagn Microbiol Infect Dis 15: 201–206

54. Catchart GRA, Quinn D, Greer B, Harriott P, Lynas JF, Gilmore BF, Walker B (2011) Novel inhibitors of the *Pseudomonas aeruginosa* virulence factor LasB: a potential therapeutic approach for attenuation of virulence mechanisms in Pseudomonal infection. Antimicrob Agents Chemother 55:2670–2678
55. Kusuma CM, Kokai-Kun JF (2005) Comparison of four methods for determining lysostaphin susceptibility of various strains of *Staphylococcus aureus*. Antimicrob Agents Chemotherapy 49:3256–3263
56. Morihara K, Tsuzuki H (1978) Phosphoramidon as an inhibitor of elastase from *Pseudomonas aeruginosa*. Jpn J Exp Med 48:81–84
57. Peters JE, Park JSJ, Darzins A, Freck LC, Saulnier JM, Wallach JM, Galloway D (1992) Further studies on *Pseudomonas aeruginosa* LasA: analysis of specificity. Mol Microbiol 6:1155–1162
58. Prescott JM, Wilkes SH (1976) *Aeromonas* aminopeptidase. Methods Enzymol 45:530–543
59. Zimmerman M, Yurewicz E, Patel G (1976) A new fluorogenic substrate for chymotrypsin. Anal Biochem 70:258–262
60. Matsumoto H, Rogi T, Yamashiro K, Kodama S, Tsuruoka N, Hattori A, Takio K, Mizutani S, Tsujimoto M (2000) Characterization of a recombinant soluble form of human placental leucine aminopeptidase/oxytocinase expressed in Chinese hamster ovary cells. Eur J Biochem 267:46–52
61. Tingpej P, Smith L, Rose B, Zhu H, Conibear T, Al Nassafi K, Manos J, Elkins M, Bye P, Willcox M, Bell S, Wainwright C, Harbour C (2007) Phenotypic characterization of clonal and nonclonal *Pseudomonas aeruginosa* strains isolated from lungs of adults with cystic fibrosis. J Clin Microbiol 45:1697–1704
62. Toder DS, Gambello MJ, Iglewski BH (1991) *Pseudomonas aeruginosa* LasA: a second elastase under transcriptional control of *lasR*. Mol Microbiol 5:2003–2010

Chapter 14

In vitro Assays to Monitor the Activity of *Pseudomonas aeruginosa* Type III Secreted Proteins

Stephanie L. Rolsma and Dara W. Frank

Abstract

Pseudomonas aeruginosa secretes numerous toxins and destructive enzymes that play distinct roles in pathogenesis. The Type III secretion system (T3SS) of *Pseudomonas* is a system that delivers a subset of toxins directly into the cytoplasm of eukaryotic cells. The secreted effectors include ExoS, ExoT, ExoU, and ExoY. In this chapter, we describe methods to induce T3S expression and measure the enzymatic activities of each effector in in vitro assays. ExoU is a phospholipase and its activity can be measured in a fluorescence-based assay monitoring the cleavage of the fluorogenic substrate, PED6. ExoS and ExoT both possess ADP-ribosyltransferase (ADPRT) and GTPase-activating protein (GAP) activity. ADPRT activity can be assessed by using radiolabeled nicotinamide adenine dinucleotide (NAD^+) and measuring the covalent incorporation of ADP-ribose into a target protein. GAP activity is measured by the release of radiolabeled phosphate from [γ-^{32}P]GTP-bound target proteins. In accordance with recent trends towards reducing the use of radioactivity in the laboratory, alternative assays using fluorescent or biotin-labeled reagents are described. ExoY is a nucleotidyl cyclase; cAMP production stimulated by ExoY can be monitored using reverse-phase HPLC or with commercially available immunological assays.

Key words *Pseudomonas aeruginosa*, Type III secretion, ExoU, ExoT, ExoS, ExoY, Bacterial toxins, ADP-ribosyltransferase, GTPase-activating protein, Phospholipase, Nucleotidyl cyclase

1 Introduction

The T3SS encodes a protein machinery that allows *P. aeruginosa* to directly inject effectors into the cytoplasm of mammalian host cells. Cells of the innate immune system are thought to be the central targets for T3SS intoxication. Intoxication of immune cells such as neutrophils, monocytes, and macrophages allows the bacteria to evade clearance mechanisms and establish an initial infection [1]. T3S is stimulated by specific environmental cues, including contact with mammalian cells [2, 3]. Interestingly each of the identified effectors possesses a domain whose enzymatic activity requires a eukaryotic cofactor, suggesting that these enzymes evolved to specifically target eukaryotic competitors. Intoxication of cells by the *P. aeruginosa* T3SS is thought to contribute to severe human

Alain Filloux and Juan-Luis Ramos (eds.), *Pseudomonas Methods and Protocols*, Methods in Molecular Biology, vol. 1149, DOI 10.1007/978-1-4939-0473-0_14, © Springer Science+Business Media New York 2014

disease [4–6] which is recapitulated in various animal models of acute infection [7–10]. The role of T3SS enzymes in the unique setting of *Pseudomonas* chronic infection remains controversial. While most environmental isolates secrete T3S proteins, the percentage of *Pseudomonas* isolates cultured from CF patients that are able to secrete T3S effectors decreases relative to increased duration of infection [11].

The *P. aeruginosa* T3SS is regulated by a member of the AraC/XylS family of transcriptional activators, ExsA, which is involved in controlling the ten promoters that drive expression of operons encoding the secretion machinery, chaperones and effectors [12, 13]. Inducing signals for ExsA expression include growth in a calcium-limited medium and bacterial contact with host cells [3, 14, 15]. To measure the extracellular accumulation of effector enzymes, simple mediums have been devised which contain chelators such as EGTA or nitrilotriacetic acid (NTA). The concentration of chelators varies between investigators but ranges between 2 and 10 mM [14–18]. A minimal medium has also been reported that supplements with a calcium chelator and magnesium salts [19, 20].

ExoU is an A_2 phospholipase [21–23] which targets cellular membranes, resulting in cytotoxicity in cellular models and pathology, such as lung damage, disseminated infection, and mortality in animal models [9, 20, 24–28]. In humans, infection with an ExoU-expressing strain is associated with fatal pneumonia [4, 5, 29]. ExoU activity can be measured by a fluorescence-based assay in which PED6, a lipid analog with a BODIPY dye-labeled *sn*-2 acyl chain and a dinitrophenol quencher-labeled head group [30], is utilized as a fluorogenic substrate for ExoU phospholipase activity. In intact PED6, the quencher group prevents fluorescence emission. Cleavage of the dye-labeled acyl chain of PED6 by an A_2 phospholipase eliminates the quenching effect and results in a fluorescent signal [31]. A eukaryotic cofactor is required for the in vitro activity of ExoU. This cofactor is present in yeast and mammalian cellular extracts [21, 22, 30, 32] and was initially identified as superoxide dismutase (SOD1) [30]. Later ubiquitin was shown to be the essential cofactor for ExoU activity. Ubiquitylated proteins such as SOD1 also activate ExoU [33]. Interestingly, ExoU itself can be ubiquitylated at a specific lysine residue [32]; however, ubiquitylation is not required for ExoU activity either in vitro [33] or in vivo [32].

ExoS and ExoT have 75 % amino acid identity [34] and are bifunctional enzymes with both GAP and ADPRT activity [34–39]. These activities target cellular signaling and the host cell cytoskeleton. Specifically, the GAP domain of ExoS and ExoT targets small GTPases such as Rho, Rac, and Cdc42 [36–38, 40]. GTPases cycle between an active GTP-bound state and an inactive GDP-bound state. Overexpression of a GAP, such as in a cell injected with ExoS or ExoT, results in the inactive GDP-bound form [36, 41]. GTPase activity can be measured by loading target proteins such as RhoA with [γ-^{32}P]GTP and determining the release of [^{32}Pi] from the target protein. *P. aeruginosa* inhibits phagocytosis by macrophages in vitro via the GAP activity of ExoS [42]. ExoS GAP activity disrupts the host cell's cytoskeleton and inhibits bacterial uptake [41, 43, 44]. ExoT GAP activity targets the cellular actin cytoskeleton and prevents bacterial internalization [45].

ADPRT activity involves transfer of an ADP-ribose moiety from NAD to a target protein. This covalent modification interferes with the normal function of the target protein. ADPRT activity can be measured by monitoring the incorporation of [^{32}P]adenylate phosphate-NAD$^+$ onto target protein arginine residues. ExoS and ExoT require a eukaryotic cofactor initially identified as the Factor-Activating Exoenzyme S (FAS) and later verified as members of the 14-3-3 protein family [39, 46–49]. Unlike most bacterial enzymes with ADPRT activity, ExoS has many targets, including small GTPases and members of the ezrin, radixin, and moesin family of proteins [35, 50–55]. ExoS ADPRT activity interferes with host cell endocytosis [54] and vesicular trafficking [44], is cytotoxic [56] and induces apoptosis in cellular models [57, 58]. ExoS ADPRT activity may also play a role in immune evasion through inhibiting secretion of the pro-inflammatory mediator IL-1β [59]. The targets of ExoT ADPRT activity, CrkI and CrkII adaptor proteins [39], are more limited than that of ExoS. ExoT ADPRT activity contributes to disruption of the host cell actin cytoskeleton, impairment of wound healing, and inhibits bacterial internalization [60]. Although not cytotoxic, ExoT has been shown to contribute to disease in animal models [27, 28, 45, 61].

ExoY was initially identified as an adenylyl cyclase. More recent analyses suggest its substrate specificity may be broader than initially thought, acting upon UTP, GTP, and CTP. In the future, these data may lead to the reclassification of ExoY as a nucleotidyl cyclase ([62], R. Seifert, personal communication). The full activity of ExoY also requires an unidentified eukaryotic cofactor, although basal adenylyl cyclase activity can be detected in the absence of a cofactor [63]. ExoY disrupts the actin cytoskeleton [63, 64], inhibits bacterial uptake [65] and causes junctional gaps in endothelial cells [66, 67]. The mechanism mediating endothelial gap formation appears to be associated with microtubule disruption as a result of Tau phosphorylation [66, 67]. Ultimately ExoY activity is postulated to increase lung permeability [66]. Recently, ExoY has also been shown to possess guanylyl cyclase activity that impacts endothelial gap formation and Tau phosphorylation, although not to as great an extent as adenylyl cyclase activity [62]. The specific role of ExoY in animal models is uncertain [28, 61]. Nucleotidyl cyclase activity can be measured by incubating toxin with its ATP substrate and monitoring cAMP formation by reverse-phase high-performance liquid chromatography (HPLC) [63], immunological assays measuring cAMP [62, 66], and mass spectrometry-based assays [68].

2 Materials

2.1 Induction of Type III Secretion and Harvest of Secreted Proteins from the Supernatant

1. 10× Vogel Bonner Minimal (VBM) medium: 900 ml dH_2O, 2 g $MgSO_4 \cdot 7H_2O$, 20 g citric acid (free acid), 100 g K_2HPO_4, 35 g $NaNH_4HPO_4 \cdot 4H_2O$. Adjust the pH to 7.0. Adjust final volume to 1 l. Autoclave. Store at room temperature (RT).
2. 1× VBM Agar medium: 450 ml water, 7.5 g Bacto Agar (BD). Autoclave. Place melted agar and 10× VBM in a 50 °C water bath and add 50 ml of sterile 10× VBM salts to the 450 ml agar/water. Swirl to mix. Pour into 10 cm bacteriological petri dishes.
3. 10× Trypticase Soy Broth Dialyzate (10× TSBD): Add 150 g of tryptic soy broth (Difco) to 450 ml of filtered water. Add 50 g Chelex 100 resin (Bio-Rad). Stir at RT for approximately 5.5 h or overnight (4 °C). To remove the Chelex resin subject the mixture to centrifugation at 10,000 × *g*, 4 °C for 20 min. Pour off the supernatant and filter it through Whatman #1 filter paper. Soak an Amicon PM30 76 mm membrane (30K MW cutoff) in water to remove glycine. Assemble a 400-ml capacity Amicon stirred ultrafiltration cell (EMD Millipore): place the filter shiny-side up on the bottom of the concentrator. Place the unit on the magnetic stir plate and start stirring. Connect the nitrogen gas line to the Amicon unit. Set the inlet pressure of the nitrogen cylinder to 40–70 psi. Rinse the membrane several times with filtered water. Pour the filtered medium into the concentrator. Collect the filtrate in a covered vessel. Filtration will take several hours at 4 °C. Continue filtering until ~75 ml medium is left in the Amicon unit. Save the filtered medium (filtrate) and freeze at −20 °C in 40 ml aliquots.
4. 1× TSBD: Thaw 10× filtered medium, dilute 1:10 with high-quality cation-deficient water (18 mΩ) and autoclave.
5. 1× TSBD++: Add glycerol to 1 % final concentration and monosodium glutamate (MSG) to 100 mM final concentration.
6. 2-10 mM NTA or EGTA.
7. Saturated ammonium sulfate.
8. SDS-PAGE loading buffer.

2.2 ExoU: Phospholipase Assay

1. *Pseudomonas* culture supernatant containing ExoU, unconcentrated.
2. Phospholipase Assay Buffer (pH 6.3): 50 mM MOPS, 50 mM NaCl, 250 mM MSG, 30 μM *N*-((6-(2,4-Dinitrophenyl)

amino)hexanoyl)-2-(4,4-Difluoro-5,7-Dimethyl-4-Bora-3a, 4a-Diaza-*s*-Indacene-3-Pentanoyl)-1-Hexadecanoyl-*sn*-Glycero-3-Phosphoethanolamine, Triethylammonium Salt (PED6) (Molecular Probes).

3. K48-linked polyubiquitin (stock solution in 10 mM KPO_4, pH 6.3) (Enzo Life Sciences).
4. 96-well black microplate.
5. Spectramax M5 microplate reader (Molecular Devices).

2.3 ExoS/T Bifunctional enzymes: GAP Assay

1. Purified ExoS [36, 69].
2. Recombinant ExoT [37].
3. Recombinant target protein RhoA [36].
4. [γ-^{32}P]GTP.
5. GAP Assay Buffer: 50 mM Tris–HCl, pH 7.5, 10 mM EDTA, 2 mM dithiothreitol (DTT).
6. $MgCl_2$.
7. GTP.
8. 0.4 μm nitrocellulose.
9. Scintillation counter.

2.4 ExoS/T Bifunctional Enzymes: ADPRT Assay

1. Secreted ExoS or ExoT: Supernatants from cultures grown as described in TSBD low-calcium medium [39].
2. Soybean trypsin inhibitor.
3. Purified Crk-I [39].
4. ADPRT Assay Buffer: 50 mM Tris–HCl (pH 7.4), 0.1 mM NAD^+ ([^{32}P]adenylate phosphate-NAD^+), 500 nM FAS [70], and 0.2 μg/μl bovine serum albumin.
5. SDS sample buffer.
6. Scintillation counter.

2.5 ExoY: Adenylyl Cyclase Assay

1. 1 μM recombinant ExoY [63].
2. ExoY Assay Buffer: 10 mM Tris–HCl (pH 8), 6 mM $MgCl_2$, 0.2 mM $CaCl_2$, 2 mM ATP, and 2 mM DTT.
3. Postnuclear extract (PNE) from CHO cells [71].
4. 100 % ethanol.
5. Reversed-phase HPLC C18 column (0.46 cm × 15 cm).
6. HPLC instrument.
7. Buffer A: 0.1 M potassium phosphate (pH 6.0).
8. Buffer B: 0.1 M potassium phosphate (pH 6.0) in 10 % CH_3OH.
9. Purified ATP, AMP, and cAMP for use as standards for HPLC.

3 Methods

3.1 Induction of Type III Secretion and Harvest of Secreted Proteins from the Supernatant

1. Culture *Pseudomonas* strain on VBM.
2. Incubate plates at 37 °C for 24 h, then store at RT for 1–2 days (do not store plates at 4 °C).
3. Emulsify a large loopful of bacteria in 1 ml of TSBD++ containing 0.01 M NTA. Obtain an optical density reading at 540 nm (OD_{540}).
4. Dilute the suspension to a final $OD_{540} = 0.02$ in 10 ml TSBD++, +/− antibiotic, +/− chelator: use either 2.5–10 mM NTA or 2–5 mM EGTA (*see* **Note 1**).
5. Grow 10–18 h at 32 °C with shaking to an $OD_{540} \approx 4.0$.
6. Harvest the supernatant in a microcentrifuge (13,500 rpm or 16,000 × *g*) for 10 min at 4 °C (or 8,000 × *g* in a preparative centrifuge).
7. For enzyme activity assays, unconcentrated supernatant material may be titrated into the assay immediately or frozen at −80 °C until the day the assay is to be performed (*see* **Notes 2–5**).
8. For SDS-PAGE analysis, concentrated supernatant material should be prepared. Continue from **step 6** as follows: Remove 0.64 ml of the supernatant to a microcentrifuge tube containing 0.8 ml saturated NH_4 sulfate (final concentration 55 %). Keep on ice 2 h or −20 °C overnight. Centrifuge for 10 min in the microcentrifuge (13,500 rpm or 16,000 × *g*) at 4 °C. Discard supernatant without disturbing the pellet. Repeat centrifugation. Remove remaining supernatant and suspend the pellet in 32 μl of 1× SDS-PAGE loading buffer. Final concentration of supernatant is 20× (*see* **Note 6**). *See* Fig. 1 for a sample gel showing concentrated supernatants from 4 strains of *P. aeruginosa*.

3.2 ExoU: Phospholipase Assay

Carry out all procedures at RT.

1. In a 50 μl total reaction volume, add 20 μl culture supernatant to Phospholipase Assay Buffer (*see* **Notes 7** and **8**). Add K48-linked polyubiquitin stock solution to a final amount of 0.4–3.0 μg (*see* **Notes 9** and **10**).
2. Vortex briefly.
3. Add to a 96-well black microplate.
4. Measure fluorescence intensity (relative fluorescence units, RFU) at 15-min increments for up to 120 min at RT at an excitation wavelength of 488 nm and an emission wavelength of 511 nm (495 nm cutoff filter) (*see* **Note 11**).
5. Normalize RFU values relative to growth of the bacterial culture by dividing RFU value by the OD_{540} value at time of harvest (*see* **Notes 12** and **13**).

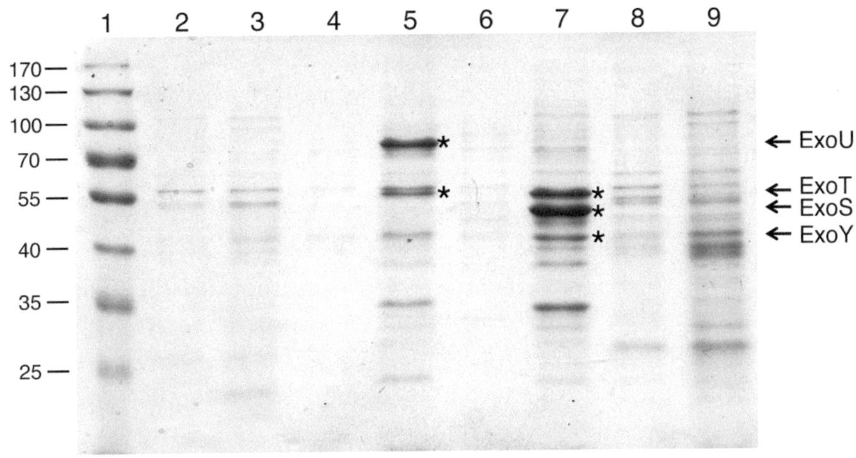

Fig. 1 Expression and secretion of the *Pseudomonas aeruginosa* Type III effectors. Four stains of *P. aeruginosa* were grown in Chelex-treated trypticase soy broth with monosodium glutamate, glycerol and in the absence (lanes *2*, *4*, *6*, and *8*) or in the presence of 10 mM nitrilotriacetic acid (lanes *3*, *5*, *7*, and *9*). Supernatant fractions were prepared and concentrated 20-fold as described in Subheading 3. Samples were normalized to the OD_{540} at the time of harvest and separated by SDS-PAGE (10 % acrylamide gel). The gel was stained with Coomassie blue. Lane *1* shows a prestained molecular weight standard. Lanes *2* and *3*, PA01; lanes *4* and *5*, PA103; lanes *6* and *7*, PAK and lanes *8* and *9*, the concentrated culture supernatants from strain PA14. Type III toxins secreted from PA103 and PAK are marked with an *asterisk* (*) in lanes *5* and *7*. Concentrated supernatants from strains with low protease expression (PA103, lanes *4* and *5*) and strains that overproduce type III effectors (PAK, lanes *6* and *7*) exhibit a clearly inducible pattern of protein bands in the extracellular supernatant fraction. In contrast, it can be difficult to detect the effectors in strains that produce proteases (PA01, lanes *2* and *3*) or strains that do not grow as well under these conditions (PA14, lanes *8* and *9*) (*see* **Note 3**). Methods that have a high level of sensitivity and specificity such as Western blot analysis with specific antibodies, silver stained gels, enzymatic activity, or mass spectrometry can be used as alternative approaches to identify the complement of secreted effectors

6. Alternatives to the described assay are described in Subheading 4 (*see* **Notes 14** and **15**).

3.3 ExoS/T Bifunctional enzymes: GAP Assay

1. Incubate recombinant Rho proteins at 2 μM final concentration (*see* **Note 16**) with 10 μM [γ-^{32}P]GTP for 5 min at 37 °C in GAP Assay Buffer.
2. For initiation of intrinsic GTPase activity, add 12 mM $MgCl_2$ and 2 mM unlabeled GTP to the reaction.
3. For GAP stimulation, add 100 nM ExoS or ExoT (*see* **Notes 17–19**) and incubate at 37 °C for 4 min.
4. Analyze GTPase activity by filter binding (*see* **Note 20**): Spot reaction mixtures on 0.4 μm nitrocellulose, wash filters with GAP Assay Buffer to remove free radiolabel, dry and then measure remaining radioactivity by scintillation counting.
5. Report results as percentage of GTP bound initially (100 %) (*see* **Note 21**).

3.4 ExoS/T Bifunctional Enzymes: ADPRT Activity Assay

1. Incubate 5 nM secreted ExoS or ExoT (*see* **Note 22**) with 3 μM soybean trypsin inhibitor (SBTI) (*see* **Note 23**) or purified Crk-I in ADPRT Assay Buffer (*see* **Notes 24–26**).
2. Stop reactions at 2, 4, 8, and 16 min by adding SDS sample buffer.
3. Fractionate samples by SDS-PAGE, followed by Coomassie staining.
4. Excise radioactive bands and subject to scintillation counting to measure incorporation of radiolabel (*see* **Note 27**).
5. Determine specific activity as moles of NAD^+ incorporation per mole of enzyme per minute (*see* **Note 28**).
6. Alternatives to the described assay are described in Subheading 4 (*see* **Notes 29** and **30**).

3.5 ExoY: Adenylyl Cyclase Assay (see Note 31)

1. In a 100 μl final reaction volume, incubate 1 μM ExoY (*see* **Note 32**) in ExoY Assay Buffer. 30 μg PNE from CHO cells may also be added to further stimulate ExoY activity (*see* **Note 33**).
2. Incubate at 30 °C for 4 h (30 min if +PNE).
3. Stop reaction with the addition of 186 μl of ethanol (65 %, final concentration).
4. Incubate at RT 5 min to precipitate protein.
5. Centrifuge at 14,000 × *g* for 5 min.
6. Harvest supernatants, lyophilize, and suspend in 50 μl distilled water.
7. Perform reverse-phase HPLC to resolve cAMP from ATP and AMP (*see* **Notes 34** and **35**): Inject samples into a C-18 column and resolve by washing for 5 min at 100 % Buffer A, followed by an 8-min gradient to 100 % Buffer B, hold for 5 min at 100 % Buffer B, and return to 100 % Buffer A for 5 min.
8. Monitor absorbance at 259 nm.
9. Calculate formation of cAMP as a percentage of original ATP substrate present in the reaction.
10. Calculate retention times using purified ATP, AMP, and cAMP as standards (*see* **Note 36**).

4 Notes

1. Alternate mediums for the expression of T3S are discussed in Subheading 1. The authors include the original medium developed by the Iglewski Laboratory [14] since it can be used in the absence of chelators to assay for extracellular exotoxin A activity [72].

2. *Pseudomonas* supernatant material has routinely been used as a source of toxin for enzymatic assays [39, 73].

3. The methods described for the induction of type III secretion have been optimized for strains PA103 and PAK. Depending on the Pseudomonas strain used, the expression of proteases may degrade secreted proteins and affect the yield of protein from secreted supernatants (as shown in Fig. 1). Importantly, some clinical isolates, particularly those from cystic fibrosis patients, are difficult to grow in the presence of a chelator. Therefore several induction parameters may need optimization depending on the strain used for a model system. These parameters include the type of medium, the concentration of chelator, the type of chelator, growth temperature, or length of time that the culture is induced. For strains that express and secrete alkaline protease or elastase, harvesting the culture at a lower optical density enhances the extracellular yield of ExoS and ExoT.

4. An alternate approach to using secreted supernatants as a source of Type III toxins is to express and purify recombinant toxins. This approach is particularly useful for assessing the activity of genetically modified toxins and determining structure–activity relationships.

5. Toxin activities can also be detected from infected host cells [31, 38, 39, 42, 62, 66, 74].

6. Total time required for the assay: 1 day is required for growth on plates followed by 10–18 h growth in liquid culture. Harvest of supernatant will take an additional 10 min. Total estimated time (minimum) of 2 days. Optional concentration of supernatant requires an additional 2–3 h.

7. This assay is sufficient to detect activity from unconcentrated *Pseudomonas* culture supernatants [73].

8. Recombinant ExoU may also be used successfully in this assay at a final concentration of approximately 34 nM [33].

9. Although other isoforms of ubiquitin, such as mono-, di-, tetra-, or octa-ubiquitin will activate ExoU, the affinity of ubiquitin for ExoU increases with increasing ubiquitin chain length. Polyubiquitin was chosen for this assay because of its high affinity for ExoU.

10. Other isoforms of ubiquitin may also be used in this assay with the concentration dependent upon the isoform used [33].

11. Determine normal background fluorescence intensity associated with PED6 by performing the assay in the absence of ubiquitin. Subtract this value from the sample value.

12. The described assay is highly sensitive and can provide kinetic data of ExoU activation. It is the first example in the literature

of an ExoU activity assay with enough sensitivity to detect catalytically active ExoU from an injected eukaryotic cell [31].

13. Total time required for the assay: This assay runs for 120 min, plus additional time to prepare reagents and the reaction mixture.
14. The activity of ExoU has also been measured with ^{14}C-labeled lipid substrates [23, 75]. Recombinant ExoU was incubated with labeled liposomes and eukaryotic cell extract (as a source for the cofactor). Hydrolysis of the substrate was measured by thin layer chromatography, or alternately by using a scintillation counter.
15. The activity of ExoU has also been indirectly measured with Ellman's reagent [76]. Briefly, ExoU-expressing yeast cells were lysed and incubated with a cleavable phospholipid substrate and then Ellman's reagent. Cleavage of the substrate yields a free thiol group that reacts with Ellman's reagent to form a colored product.
16. Recombinant Rac and Cdc42 may be substituted for Rho in this assay [36].
17. This assay has only been demonstrated for recombinant ExoS or ExoT expressed from *E. coli* [36, 37] and ExoS expressed from *P. aeruginosa* and subsequently purified by gel filtration and ion exchange chromatography [36, 69]. The methods described here apply to recombinant ExoS. Slight variations to the materials and methods (including concentration of RhoA, toxin, unlabeled GTP, and $MgCl_2$ and temperature of the assay) may be necessary for assaying activity of ExoT [37] or ExoS purified from *P. aeruginosa* [36].
18. Varying concentrations of ExoS or ExoT (5–100 nM) may be used in this assay. A linear rate of activity can be found at concentrations of ExoT/ExoS that stimulate hydrolysis of less than 25 % of the available Rho-GTP [37].
19. GAP activity of ExoT and ExoS does not require the presence of a eukaryotic cofactor.
20. An alternate approach to the filter-binding assay described would be to measure GDP:GTP:P(i) ratios by HPLC [77]. A similar reaction mixture could be used and radiolabeled nucleotides would not be necessary.
21. Total time required for the assay: This assay runs for approximately 10 min. Additional time for the filter-binding assay and scintillation counting (estimated less than one h) is required.
22. This assay is sufficient to detect activity from concentrated *Pseudomonas* culture supernatants.
23. For analysis of ExoS activity with a more physiologically relevant target than SBTI, a similar reaction can also be performed in the presence of 500 nM ezrin, radixin, moesin, or Ras [55].

24. ADPRT activity of ExoS and ExoT requires the presence of the cofactor FAS, which is present in the ADPRT Assay Buffer.
25. Control reactions should include a reaction without FAS. No activity should be detected in this reaction.
26. Rather than incubation with FAS and a specific target protein, enzyme and labeled NAD may be incubated with wheat germ extract or CHO/HeLa cell lysate, which are sources of both FAS and target proteins for ADP-ribosylation [39, 69].
27. As an alternative to band excision and scintillation counting, the reaction mixture can be fractionated by SDS-PAGE, subjected to autoradiography, and the incorporation of radiolabel determined by densitometry [78].
28. This assay runs for approximately 16 min. Additional time for SDS-PAGE and scintillation counting will be required.
29. A fluorescent assay for ExoS ADPRT activity has also been developed [79]. In this assay, ExoS is added to a reaction mixture of FAS, recombinant Ras, and fluorescent ε-NAD^+. Time-dependent change in fluorescence can then be monitored.
30. A biotin-based assay for ExoS ADPRT activity has also been developed [80]. Briefly, biotinylated-NAD is added to enzyme and target protein and incubated for 1 h. ADP-ribosylated proteins are then separated by SDS-PAGE, blotted onto nitrocellulose, detected with a streptavidin-HRP conjugate, and finally visualized by chemiluminescence. This assay avoids the use of radioactivity.
31. ExoY, initially identified as an adenylyl cyclase, has recently been demonstrated to have broader substrate specificity, acting upon UTP, GTP, and CTP, and may be reclassified as a nucleotidyl cyclase ([62], R. Seifert, unpublished).
32. Previously published ExoY activity assays have only been performed with recombinant ExoY or in ExoY-expressing eukaryotic cells.
33. Control reactions should include reactions without enzyme and with a post-nuclear extract of eukaryotic cells alone.
34. Formation of cAMP, cGMP, cCMP, and cUMP can also be detected via HPLC-MS/MS [68].
35. cAMP and cGMP formation can also be detected using commercially available immunological assays (Biomedical Technologies [62, 66]).
36. Total time required for the assay: The enzyme assay runs for approximately 16 min, with additional time required for SDS-PAGE and scintillation counting.

References

1. Diaz MH, Hauser AR (2010) *Pseudomonas aeruginosa* cytotoxin ExoU is injected into phagocytic cells during acute pneumonia. Infect Immun 78:1447–1456
2. Frank DW (1997) The exoenzyme S regulon of *Pseudomonas aeruginosa*. Mol Microbiol 26: 621–629
3. Vallis AJ et al (1999) Regulation of ExoS production and secretion by *Pseudomonas aeruginosa* in response to tissue culture conditions. Infect Immun 67:914–920
4. Roy-Burman A et al (2001) Type III protein secretion is associated with death in lower respiratory and systemic *Pseudomonas aeruginosa* infections. J Infect Dis 183:1767–1774
5. Hauser AR et al (2002) Type III protein secretion is associated with poor clinical outcomes in patients with ventilator-associated pneumonia caused by *Pseudomonas aeruginosa*. Crit Care Med 30:521–528
6. EL-Solh AA (2012) Clinical outcomes of type III *Pseudomonas aeruginosa* bacteremia. Crit Care Med 40:1157–1163
7. Sawa T et al (1999) Active and passive immunization with the *Pseudomonas* V antigen protects against type III intoxication and lung injury. Nat Med 5:392–398
8. Kurahashi K et al (1999) Pathogenesis of septic shock in *Pseudomonas aeruginosa* pneumonia. J Clin Invest 104:743–750
9. Allewelt M et al (2000) Acquisition of expression of the *Pseudomonas aeruginosa* ExoU cytotoxin leads to increased bacterial virulence in a murine model of acute pneumonia and systemic spread. Infect Immun 68:3998–4004
10. Holder IA, Neely AN, Frank DW (2001) Type III secretion/intoxication system important in virulence of *Pseudomonas aeruginosa* infections in burns. Burns 27:129–130
11. Jain M et al (2004) Type III secretion phenotypes of *Pseudomonas aeruginosa* strains change during infection of individuals with cystic fibrosis. J Clin Microbiol 42:5229–5237
12. Yahr TL, Wolfgang MC (2006) Transcriptional regulation of the *Pseudomonas aeruginosa* type III secretion system. Mol Microbiol 62:631–640
13. King JM et al (2012) Orientation of *Pseudomonas aeruginosa* ExsA monomers bound to promoter DNA and base-specific contacts with the P(exoT) promoter. J Bacteriol 194:2573–2585
14. Iglewski BH et al (1978) *Pseudomonas aeruginosa* exoenzyme S: an adenosine diphosphate ribosyltransferase distinct from toxin A. Proc Natl Acad Sci USA 75:3211–3215
15. Thompson MR et al (1980) Exoenzyme S: an ADP-ribosyltransferase produced by *Pseudomonas aeruginosa*. In: Sugimura T, Smulson M (eds) Novel ADP-ribosylation of regulatory enzymes and proteins. Elsevier/North-Holland, Amsterdam, pp 425–433
16. Dacheux D et al (1999) Cell death of human polymorphonuclear neutrophils induced by a *Pseudomonas aeruginosa* cystic fibrosis isolate requires a functional type III secretion system. Infect Immun 67:6164–6167
17. McCaw ML et al (2002) ExsD is a negative regulator of the *Pseudomonas aeruginosa* type III secretion regulon. Mol Microbiol 46:1123–1133
18. Kim J et al (2005) Factors triggering type III secretion in *Pseudomonas aeruginosa*. Microbiology 151:3575–3587
19. Nicas TI, Iglewski BH (1984) Isolation and characterization of transposon-induced mutants of *Pseudomonas aeruginosa* deficient in production of exoenzyme S. Infect Immun 45:470–474
20. Hauser AR, Kang PJ, Engel JN (1998) PepA, a secreted protein of *Pseudomonas aeruginosa*, is necessary for cytotoxicity and virulence. Mol Microbiol 27:807–818
21. Sato H et al (2003) The mechanism of action of the *Pseudomonas aeruginosa*-encoded type III cytotoxin, ExoU. EMBO J 22:2959–2969
22. Phillips RM et al (2003) *In vivo* phospholipase activity of the *Pseudomonas aeruginosa* cytotoxin ExoU and protection of mammalian cells with phospholipase A_2 inhibitors. J Biol Chem 278:41326–41332
23. Tamura M et al (2004) Lysophospholipase A activity of *Pseudomonas aeruginosa* type III secretory toxin ExoU. Biochem Biophys Res Commun 316:323–331
24. Finck-Barbançon V et al (1997) ExoU expression by *Pseudomonas aeruginosa* correlates with acute cytotoxicity and epithelial injury. Mol Microbiol 25:547–557
25. Finck-Barbançon V, Frank DW (2001) Multiple domains are required for the toxic activity of *Pseudomonas aeruginosa* ExoU. J Bacteriol 183:4330–4344
26. Pankhaniya RR et al (2004) *Pseudomonas aeruginosa* causes acute lung injury via the catalytic activity of the patatin-like phospholipase domain of ExoU. Crit Care Med 32: 2293–2299
27. Shaver CM, Hauser AR (2004) Relative contributions of *Pseudomonas aeruginosa* ExoU, ExoS, and ExoT to virulence in the lung. Infect Immun 72:6969–6977

28. Lee VT et al (2005) Activities of *Pseudomonas aeruginosa* effectors secreted by the Type III secretion system *in vitro* and during infection. Infect Immun 73:1695–1705
29. Schulert GS et al (2003) Secretion of the toxin ExoU is a marker for highly virulent *Pseudomonas aeruginosa* isolates obtained from patients with hospital-acquired pneumonia. J Infect Dis 188:1695–1706
30. Sato H, Feix JB, Frank DW (2006) Identification of superoxide dismutase as a cofactor for the pseudomonas type III toxin, ExoU. Biochemistry 45:10368–10375
31. Benson MA, Schmalzer KM, Frank DW (2010) A sensitive fluorescence-based assay for the detection of ExoU-mediated PLA_2 activity. Clin Chim Acta 411:190–197
32. Stirling FR et al (2006) Eukaryotic localization, activation and ubiquitinylation of a bacterial type III secreted toxin. Cell Microbiol 8: 1294–1309
33. Anderson DM et al (2011) Ubiquitin and ubiquitin-modified proteins activate the *Pseudomonas aeruginosa* T3SS cytotoxin, ExoU. Mol Microbiol 82:1454–1467
34. Yahr TL, Barbieri JT, Frank DW (1996) Genetic relationship between the 53- and 49-kilodalton forms of exoenzyme S from *Pseudomonas aeruginosa*. J Bacteriol 178: 1412–1419
35. Ganesan AK et al (1998) *Pseudomonas aeruginosa* exoenzyme S ADP-ribosylates Ras at multiple sites. J Biol Chem 273:7332–7337
36. Goehring UM et al (1999) The N-terminal domain of *Pseudomonas aeruginosa* exoenzyme S is a GTPase-activating protein for Rho GTPases. J Biol Chem 274:36369–36372
37. Krall R et al (2000) *Pseudomonas aeruginosa* ExoT is a Rho GTPase-activating protein. Infect Immun 68:6066–6068
38. Kazmierczak BI, Engel JN (2002) *Pseudomonas aeruginosa* ExoT acts *in vivo* as a GTPase-activating protein for RhoA, Rac1, and Cdc42. Infect Immun 70:2198–2205
39. Sun J, Barbieri JT (2003) *Pseudomonas aeruginosa* ExoT ADP-ribosylates CT10 regulator of kinase (Crk) proteins. J Biol Chem 278: 32794–32800
40. Henriksson ML et al (2002) Exoenzyme S shows selective ADP-ribosylation and GTPase-activating protein (GAP) activities towards small GTPases *in vivo*. Biochem J 367:617–628
41. Pederson KJ et al (1999) The amino-terminal domain of *Pseudomonas aeruginosa* ExoS disrupts actin filaments via small-molecular-weight GTP-binding proteins. Mol Microbiol 32: 393–401
42. Rocha CL et al (2003) Characterization of *Pseudomonas aeruginosa* exoenzyme S as a bifunctional enzyme in J774A.1 macrophages. Infect Immun 71:5296–5305
43. Krall R et al (2002) *In vivo* Rho GTPase-activating protein activity of *Pseudomonas aeruginosa* cytotoxin ExoS. Infect Immun 70: 360–367
44. Deng Q, Barbieri JT (2008) Modulation of host cell endocytosis by the type III cytotoxin, Pseudomonas ExoS. Traffic 9:1948–1957
45. Garrity-Ryan L et al (2000) The arginine finger domain of ExoT contributes to actin cytoskeleton disruption and inhibition of internalization of *Pseudomonas aeruginosa* by epithelial cells and macrophages. Infect Immun 68: 7100–7113
46. Coburn J et al (1991) *Pseudomonas aeruginosa* exoenzyme S requires a eukaryotic protein for ADP-ribosyltransferase activity. J Biol Chem 266:6438–6446
47. Fu H, Coburn J, Collier RJ (1993) The eukaryotic host factor that activates exoenzyme S of *Pseudomonas aeruginosa* is a member of the 14-3-3 protein family. Proc Natl Acad Sci USA 90:2320–2324
48. Liu S et al (1997) Biochemical relationships between the 53-kilodalton (Exo53) and 49-kilodalton (ExoS) forms of exoenzyme S of *Pseudomonas aeruginosa*. J Bacteriol 179: 1609–1613
49. Zhang L et al (1999) Residues of 14-3-3 zeta required for activation of exoenzyme S of *Pseudomonas aeruginosa*. Biochemistry 38: 12159–12164
50. Coburn J et al (1989) Exoenzyme S of *Pseudomonas aeruginosa* ADP-ribosylates the intermediate filament protein vimentin. Infect Immun 57:996–998
51. Coburn J et al (1989) Several GTP-binding proteins, including p21c-H-ras, are preferred substrates of *Pseudomonas aeruginosa* exoenzyme S. J Biol Chem 264:9004–9008
52. Coburn J, Gill DM (1991) ADP-ribosylation of p21ras and related proteins by *Pseudomonas aeruginosa* exoenzyme S. Infect Immun 59: 4259–4262
53. Riese MJ, Wittinghofer A, Barbieri JT (2001) ADP ribosylation of Arg41 of Rap by ExoS inhibits the ability of Rap to interact with its guanine nucleotide exchange factor, C3G. Biochemistry 40:3289–3294
54. Barbieri AM et al (2001) ADP-ribosylation of Rab5 by ExoS of *Pseudomonas aeruginosa* affects endocytosis. Infect Immun 69:5329–5334
55. Maresso AW, Baldwin MR, Barbieri JT (2004) Ezrin/Radixin/Moesin proteins are high affinity

targets for ADP-ribosylation by *Pseudomonas aeruginosa* ExoS. J Biol Chem 279: 38402–38408

56. Pederson KJ, Barbieri JT (1998) Intracellular expression of the ADP-ribosyltransferase domain of Pseudomonas exoenzyme S is cytotoxic to eukaryotic cells. Mol Microbiol 30:751–759
57. Jia J et al (2003) c-Jun NH2-terminal kinase-mediated signaling is essential for *Pseudomonas aeruginosa* ExoS-induced apoptosis. Infect Immun 71:3361–3370
58. Alaoui-El-Azher M et al (2006) ExoS of *Pseudomonas aeruginosa* induces apoptosis through a Fas receptor/caspase 8-independent pathway in HeLa cells. Cell Microbiol 8: 326–338
59. Galle M et al (2008) The *Pseudomonas aeruginosa* Type III secretion system plays a dual role in the regulation of caspase-1 mediated IL-1beta maturation. J Cell Mol Med 12:1767–1776
60. Garrity-Ryan L et al (2004) The ADP ribosyltransferase domain of *Pseudomonas aeruginosa* ExoT contributes to its biological activities. Infect Immun 72:546–558
61. Vance RE, Rietsch A, Mekalanos JJ (2005) Role of the type III secreted exoenzymes S, T, and Y in systemic spread of *Pseudomonas aeruginosa* PAO1 *in vivo*. Infect Immun 73:1706–1713
62. Ochoa CD et al (2012) *Pseudomonas aeruginosa* Exotoxin Y is a promiscuous cyclase that increases endothelial Tau phosphorylation and permeability. J Biol Chem 287:25407–25418
63. Yahr TL et al (1998) ExoY, an adenylate cyclase secreted by the *Pseudomonas aeruginosa* type III system. Proc Natl Acad Sci USA 95:13899–13904
64. Vallis AJ et al (1999) Biological effects of *Pseudomonas aeruginosa* type III-secreted proteins on CHO cells. Infect Immun 67: 2040–2044
65. Cowell BA, Evans DJ, Fleiszig SMJ (2005) Actin cytoskeleton disruption by ExoY and its effects on *Pseudomonas aeruginosa* invasion. FEMS Microbiol Lett 250:71–76
66. Sayner SL et al (2004) Paradoxical cAMP-induced lung endothelial hyperpermeability revealed by *Pseudomonas aeruginosa* ExoY. Circ Res 95:196–203
67. Sayner SL et al (2011) Filamin A is a phosphorylation target of membrane but not cytosolic adenylyl cyclase activity. AJP Lung Cell Mol Physiol 301:L117–L124
68. Beste KY et al (2012) Nucleotidyl cyclase activity of soluble guanylyl cyclase $\alpha_1\beta_1$. Biochemistry 51:194–204
69. Kulich SM, Frank DW, Barbieri JT (1993) Purification and characterization of exoenzyme S from *Pseudomonas aeruginosa* 388. Infect Immun 61:307–313
70. Masters SC et al (1999) Interaction of 14-3-3 with a nonphosphorylated protein ligand, exoenzyme S of *Pseudomonas aeruginosa*. Biochemistry 38:5216–5221
71. Xu Y, Barbieri JT (1995) Pertussis toxin-mediated ADP-ribosylation of target proteins in Chinese hamster ovary cells involves a vesicle trafficking mechanism. Infect Immun 63: 825–832
72. Liu PV (1973) Exotoxins of *Pseudomonas aeruginosa*. I. Factors that influence the production of exotoxin A. J Infect Dis 128:506–513
73. Schmalzer KM, Benson MA, Frank DW (2010) Activation of ExoU phospholipase activity requires specific C-terminal regions. J Bacteriol 192:1801–1812
74. Riese MJ et al (2002) Auto-ADP-ribosylation of *Pseudomonas aeruginosa* ExoS. J Biol Chem 277:12082–12088
75. Sato H et al (2005) Characterization of phospholipase activity of the *Pseudomonas aeruginosa* type III cytotoxin, ExoU. J Bacteriol 187:1192–1195
76. Rabin SDP, Hauser AR (2005) Functional regions of the *Pseudomonas aeruginosa* cytotoxin ExoU. Infect Immun 73:573–582
77. Hemsath L et al (2005) An electrostatic steering mechanism of Cdc42 recognition by Wiskott-Aldrich syndrome proteins. Mol Cell 20:313–324
78. Rocha CL et al (2005) Examination of the coordinate effects of *Pseudomonas aeruginosa* ExoS on Rac1. Infect Immun 73: 5458–5467
79. Arnoldo A et al (2008) Identification of small molecule inhibitors of *Pseudomonas aeruginosa* exoenzyme S using a yeast phenotypic screen. PLoS Genet 4:e1000005
80. Castagnini M et al (2012) Arginine-specific mono ADP-ribosylation *in vitro* of antimicrobial peptides by ADP-ribosylating toxins. PLoS One 7:e41417

Chapter 15

Cell Fractionation

Bérengère Ize, Véronique Viarre, and Romé Voulhoux

Abstract

Proteins within a cell are localized into specific cellular compartments, allowing compartmentalization of distinct tasks. If we consider lipid bilayers as compartments, then gram-negative bacteria such as *Pseudomonas aeruginosa* can target proteins to five distinct locations: the cytoplasm, the inner membrane, the periplasm, the outer membrane, and the extracellular environment. In this chapter, we describe how the different compartments can be selectively isolated by a combination of centrifugation and disruption techniques. Fractionation of the cells into subcellular compartments enables protein enrichment and is essential to accurately determine the localization of specific proteins, which is the first step towards understanding the function of a protein in the cell.

Key words *Pseudomonas aeruginosa*, Fractionation, Supernatant, Spheroplasts, Periplasm, Cytoplasm, Inner and outer membranes

1 Introduction

Gram-negative bacteria can be architecturally divided into four subcellular compartments. Moving from the inside to the outside of the cell, these compartments are the cytoplasm, the inner membrane, the periplasm, and the outer membrane. The extracellular milieu can be considered as a fifth compartment, especially in secreting bacteria such as *Pseudomonas aeruginosa*, which release dozens of exoproteins into the milieu. In this chapter, we describe detailed fractionation procedures that allow for the physical isolation of the five compartments defined above.

First, we present how to separate the extracellular milieu from cells by two consecutive centrifugations (Subheading 3.1). Next, we present how treatment of cells with $MgCl_2$ and a temperature shift allows for the release of the periplasmic content and its separation from the remainder of the cells containing the cytoplasm and both membranes [1, 2]. Sonication and ultracentrifugation of the remaining cells then allow for the separation of the cytoplasmic fraction from the total membranes (Subheading 3.2).

Alain Filloux and Juan-Luis Ramos (eds.), *Pseudomonas Methods and Protocols*, Methods in Molecular Biology, vol. 1149, DOI 10.1007/978-1-4939-0473-0_15, © Springer Science+Business Media New York 2014

Finally, we present how treatment of bacterial cells by French press fracturation and ultracentrifugation allows for the isolation of a soluble fraction containing cytoplasmic and periplasmic proteins and of an insoluble fraction containing total membranes. The total membrane fraction can be further separated by sucrose density gradient centrifugation to isolate the inner and outer membranes [3, 4] (Subheading 3.3).

2 Materials

2.1 Extracellular Medium Isolation

1. Trichloroacetic acid (TCA). Working solution at 75 % (w/v) (*see* **Note 1**) in ultrapure water made from pure TCA (powder).
2. Acetone 90 % (v/v) in ultrapure water.

2.2 Periplasm, Cytoplasm, and Total Membrane Isolation

1. 50 mM Tris–HCl pH 7.6.
2. 200 mM $MgCl_2$, 50 mM Tris–HCl pH 7.6.

2.3 Inner and Outer Membrane Separation by Sucrose Gradient Centrifugation

1. Buffer A: 10 mM Tris–HCl pH 7.4, 1 mM PMSF (phenylmethylsulfonyl fluoride), 10 μg/ml RNase, 10 μg/ml DNase, 20 % (w/w) sucrose.
2. French Press (Thermo) with a 20 K Mini French pressure cell (AMICON) for volumes <2 ml or a 40 K Maxi French pressure cell (Thermo) for volumes >2 ml.
3. 14 ml SW40 Ultra-Clear centrifuge tubes (BECKMAN).
4. Buffer B (sucrose-less working solution): 10 mM Tris–HCl pH 7.4, 5 mM EDTA (ethylenediaminetetraacetic acid), 1× freshly made inhibitor cocktail EDTA-free (ROCHE). Store at 4 °C.
5. Buffer C (55 % sucrose working solution): 55 % (w/w) sucrose in buffer B to be stored at 4 °C (*see* **Note 2**).
6. Intermediary sucrose working solutions were obtained by mixing buffers B and C in the following proportions (ml/ml): 0.9/9.1 (50 %); 1.8/8.2 (45 %); 2.7/7.3 (40 %); 3.6/6.4 (35 %); 4.5/5.5 (30 %); 6.4/3.6 (20 %).

3 Methods

3.1 Isolation of Extracellular Medium

1. Grow a bacterial culture under the required conditions.
2. Take a volume of culture equivalent to 10 OD_{600} (optical density at 600 nm) units and centrifuge for 10 min at 2,000 × g at room temperature (RT) (*see* **Notes 3** and **4**).
3. Pipette carefully 2/3 of the supernatant into a new tube.

4. Centrifuge for 5 min at 13,000 × *g* at RT.
5. Transfer 2/3 of the supernatant into a new tube (*see* **Notes 5** and **6**).
6. Precipitate proteins from the supernatant by adding 15 % (v/v) TCA (end concentration) and incubating for at least 1 h at 4 °C (*see* **Note** 7).
7. Centrifuge for 30 min at 13,000 × *g* at 4 °C (*see* **Note 8**).
8. Discard carefully the supernatant (*see* **Note 9**).
9. Wash the pellet with 0.5 ml of a solution of 90 % acetone.
10. Centrifuge for 10 min at 13,000 × *g* at 4 °C.
11. Discard carefully the supernatant (*see* **Note 9**).
12. Repeat **steps 9–11** once more and go to **step 13**.
13. Air-dry the pellet containing extracellular proteins at RT for 5–10 min (*see* **Note 10**).
14. Resuspend the isolated extracellular proteins in the desired buffer (*see* **Note 11**).

3.2 Isolation of Periplasm, Cytoplasm, and Total Membrane Fraction

1. Grow a bacterial culture under the required conditions.
2. Take a volume of culture equivalent to 5 OD_{600} units and centrifuge for 10 min at 2,000 × *g* (*see* **Note 4**).
3. Discard supernatant and wash cells once with 1 ml of 50 mM Tris–HCl (pH 7.6).
4. Resuspend cells in 0.5 ml of 200 mM $MgCl_2$, 50 mM Tris–HCl (pH 7.6) (*see* **Note 12**).
5. Incubate cells for 30 min at 30 °C with gentle shaking (*see* **Note 13**).
6. Cool cells on ice for 5 min.
7. Incubate cells for 15 min at RT.
8. Centrifuge for 10 min at 8,000 × *g* (4 °C) and keep supernatant as periplasmic fraction (*see* **Notes 14** and **15**).
9. Wash the cell pellet in 1 ml of 50 mM Tris–HCl (pH 7.6) (*see* **Note 16**).
10. Harvest the cells by centrifugation for 10 min at 8,000 × *g* (4 °C).
11. Resuspend in 0.5 ml of 50 mM Tris–HCl (pH 7.6).
12. Sonicate twice for 30 s, keeping the suspension cold during sonication (*see* **Note 17**).
13. Centrifuge at 2,000 × *g* for 15 min to remove intact cells and cell debris and take supernatant as cytoplasm and membrane fraction.

14. Transfer the supernatant into a tube compatible with ultracentrifugation.
15. Ultracentrifuge for 45 min at 120,000 × *g* at 4 °C.
16. Harvest the supernatant as the cytoplasmic fraction (*see* **Note 15**).
17. Resuspend the pellet corresponding to the total membrane fraction in the desired buffer.

3.3 Separation of Inner and Outer Membranes by Sucrose Density Gradient Centrifugation

1. Grow a bacterial culture under the required conditions (*see* **Note 18**).
2. Take a volume of culture equivalent to 250 OD_{600} units and centrifuge 15 min at 2,000 × *g* (*see* **Note 19**).
3. Resuspend the cells in 1.5 ml of buffer A (*see* **Note 20**).
4. Lyse the cells by French Press fracturation at 15,000 pressure units (PSI) (*see* **Notes 21** and **22**).
5. Remove cell debris by centrifugation for 15 min at 2,000 × *g* at 4 °C.
6. Discard the pellet and transfer the supernatant containing membrane vesicles into a tube compatible with ultracentrifugation.
7. Ultracentrifuge the supernatant for 45 min at 120,000 × *g* at 4 °C (*see* **Note 23**). The supernatant contains the soluble fraction (cytoplasm and periplasm) (*see* **Note 15**) and the pellet the insoluble fraction (total membranes).
8. Dissolve the membrane pellet in 0.5 ml of 20 % intermediary sucrose working solution (*see* **Note 24**).
9. Built the discontinuous sucrose gradient by filling centrifuge tubes with intermediary sucrose working solutions containing various concentrations of sucrose in the following order from the bottom to the top: 55 % (1.5 ml), 50 % (1.5 ml), 45 % (1.5 ml), 40 % (1.5 ml), 35 % (1.5 ml), and 30 % (1.5 ml) (*see* **Note 25**).
10. Apply 0.5 ml of the solubilized membrane fraction from **step 8** on the top of the precooled gradient.
11. Ultracentrifuge the tubes in a swing-out rotor for 36 h at 274,000 × *g* at 4 °C (*see* **Notes 26–28**).
12. Collect 0.5 ml fractions from the top to the bottom (*see* **Note 29**).
13. Identify inner and outer membrane fractions (*see* **Note 30**).

4 Notes

1. For every percentage (%) the nature of the ratio (weight (w) or volume (v)) will be mentioned all along the protocol. This is particularly important for preparation of sucrose fractions.
2. A 55 % (w/w) sucrose solution contains 55 g of sucrose (saccharose) and 45 g of buffer B. Solubilization of the sucrose

(from crystals) takes time and should be performed under constant stirring. It can be accelerated by heating the solution at 40 °C.

3. For example, 10 OD_{600} units of supernatant corresponds to 5 ml of a culture at an OD_{600} of 2.
4. The quantity of cells can be adapted according to the downstream application.
5. Leave 1/3 of the supernatant at the cell pellet interface to avoid collecting cells.
6. If a pellet of cells is still detected after the second centrifugation step, another centrifugation is recommended.
7. Precipitation can also be left overnight on ice in a cold room.
8. Since pellets are not always visible after centrifugation, tubes must be carefully positioned/marked in the centrifuge in order to guess the position of the pellet.
9. The supernatant can be removed using a vacuum pump siphon. The sucking tip/device should be positioned at the top of the tube and the tube slowly inverted to allow for the gentle removal of the supernatant without disturbing the pellet.
10. This is a critical step; the remaining acetone should be left to evaporate. This step can be performed by leaving the tube open in a flow hood. Never dry the pellet using a vacuum centrifuge. Never let the pellet air-dry completely, as this will drastically impede subsequent resuspension.
11. Solubilization of the pellet can be greatly improved by placing the tubes for 10 min on a shaker at maximum speed.
12. $MgCl_2$ treatment dissociates lipopolysaccharide from *P. aeruginosa*. Consequently, the quantity of lipopolysaccharide released is proportional to the release of periplasmic proteins [5]. This treatment avoids problems caused by EDTA treatment often used in fractionation procedures [6].
13. This step can be done in a thermomixer set at 500 rpm.
14. The **steps 5–7** can be repeated if after analysis, the periplasmic content has not been totally recovered.
15. The fraction can be concentrated by TCA precipitation following the protocol used for extracellular proteins (Subheading 3.1, **steps 6–12**).
16. This step is necessary to wash off any periplasmic proteins left.
17. Typically a Branson Microtip Sonifier 450 is used with a microtip probe and it is set at 90 % duty and an output level of 7.
18. Ideally logarithmically growing cells should be used. In any case, cells should not be grown too far in the stationary phase to avoid the formation of mixed vesicles which would compromise inner and outer membrane fractionation.

19. Cells can be frozen after this step to facilitate the fracturation procedure.
20. The cells should be resuspended in 1.5 ml of buffer A (final volume will be around 2 ml) in order to be lysed in one step by the Mini French pressure cell. Higher volumes can be lysed using a Maxi French pressure cell.
21. French Pressure fracturation is absolutely required for the formation of membrane vesicles compatible with sucrose gradient fractionation. French Pressure Cells must be precooled at 4 °C and broken cells collected into a bottle placed on ice.
22. If the sample is homogenous and the output flow low and constant, one passage through the French Press is sufficient to obtain full lysis of cells. Complete lysis can be visually checked by the clarity of the lysate as well as its pale orange color. If lysis is not complete, perform a second passage through the French press.
23. Use a Beckman coulter Optima TLX ultracentrifuge and a Beckman TLA55 fixed-angle rotor for small volume ultracentrifugation or similar.
24. The resuspension of the total membrane fraction is difficult and must be total. This step can be optimized by passing the sample through the needle of a syringe several times.
25. The gradient can be built while the total membrane fraction is collected by ultracentrifugation in order to be ready and precooled at 4 °C before the loading of the samples.
26. Use a Beckman coulter Optima L-90K ultracentrifuge and a Beckman SW40 swigging rotor or similar. In this rotor, the centrifugation speed is 39,000 RPM (rotation per minute). Program a slow acceleration and no brake for the deceleration.
27. Be sure that tubes, in their respective bucket, are perfectly equilibrated and hermetically sealed with SPINKOTE lubricant (Beckman Coulter) in order to avoid freezing of the sample during the centrifugation.
28. At the end of the centrifugation, tubes must be carefully removed from their bucket and securely placed on ice. Clear orange and white rings are usually visible at upper and lower density. They correspond to the inner and outer membrane fractions. Sucrose density can be measured with a refractometer.
29. Sucrose gradient fractions can be collected mechanically (with a fraction collector such as the Buchter auto density flow II apparatus connected to a peristaltic pump) or manually with a Pipetman. This step is crucial for the success of the fractionation. If it is done manually, pipetting must be very slow in order to avoid perturbation of the gradient.
30. Identification of proteins in the collected fractions can be done by SDS-PAGE (sodium dodecyl sulfate polyacrylamide gel

electrophoresis), western-blot analysis, or enzymatic assays with known outer and inner membrane markers. For example, *P. aeruginosa* inner membrane fractions can be localized using western blotting with antibodies directed against XcpY [7] or PilM [8] or using enzymatic activity such as NADH oxidase activity [3]. Outer membrane fractions can be localized using SDS-PAGE gels colored with Coomassie blue to detect porins [3] or western-blot analysis using antibodies directed against OprM [9] or EstA [10].

Acknowledgments

We would like to thank Pr. Jan Tommassen for careful reviewing and all members of our group for careful reading. R. Voulhoux laboratory is supported by the "Centre National de la Recherche Scientifique" (CNRS), the Agence Nationale de la Recherche (ANR-JC 07-183230 and ANR-08-PATH-004-01), the "Fondation pour le Recherche Médicale" (FRM), and "Vaincre la Mucoviscidose" (VLM).

References

1. Lu HM, Lory S (1996) A specific targeting domain in mature exotoxin A is required for its extracellular secretion from *Pseudomonas aeruginosa*. EMBO J 15:429–436
2. Voulhoux R, Taupiac MP, Czjzek M, Beaumelle B, Filloux A (2000) Influence of deletions within domain II of exotoxin A on its extracellular secretion from *Pseudomonas aeruginosa*. J Bacteriol 182:4051–4058
3. Viarre V, Cascales E, Ball G, Michel GP, Filloux A, Voulhoux R (2009) HxcQ liposecretin is self-piloted to the outer membrane by its N-terminal lipid anchor. J Biol Chem 284: 33815–33823
4. Casabona MG, Silverman JM, Sall KM, Boyer F, Couté Y, Poirel J, Grunwald D, Mougous JD, Elsen S, Attree I (2012) An ABC transporter and an outer membrane lipoprotein participate in posttranslational activation of type VI secretion in Pseudomonas aeruginosa. Environ Microbiol 15(2):471–486
5. Ingram JM, Cheng KJ, Costerton JW (1973) Alkaline phosphatase of *Pseudomonas aeruginosa*: the mechanism of secretion and release of the enzyme from whole cells. Can J Microbiol 19:1047–1455
6. Roberts NHG, Gray W, Wilkinson SG (1970) The bactericidal action of ethylenediaminetetraacetic acid on *Pseudomonas aeruginosa*. Microbios 2:189–208
7. Michel G, Bleves S, Ball G, Lazdunski A, Filloux A (1998) Mutual stabilization of the XcpZ and XcpY components of the secretory apparatus in *Pseudomonas aeruginosa*. Microbiology 144:3379–3386
8. Ayers M, Sampaleanu LM, Tammam S, Koo J, Harvey H, Howell PL, Burrows LL (2009) PilM/N/O/P proteins form an inner membrane complex that affects the stability of the *Pseudomonas aeruginosa* type IV pilus secretin. J Mol Biol 394:128–142
9. Ziha-Zarifi I, Llanes C, Köhler T, Pechère JC, Plésiat P (1999) *In vivo* emergence of multidrug-resistant mutants of *Pseudomonas aeruginosa* overexpressing the active efflux system MexA-MexB-OprM. Antimicrob Agents Chemother 43:287–291
10. Wilhelm S, Gdynia P, Tielen F, Rosenau F, Jaeger KE (2007) The autotransporter esterase EstA of *Pseudomonas aeruginosa* is required for rhamnolipid production, cell motility, and biofilm formation. J Bacteriol 189:6977–6986

Chapter 16

Characterization of Molecular Interactions Using Isothermal Titration Calorimetry

Tino Krell, Jesús Lacal, Cristina García-Fontana, Hortencia Silva-Jiménez, Miriam Rico-Jiménez, Andrés Corral Lugo, José Antonio Reyes Darias, and Juan-Luis Ramos

Abstract

Isothermal titration calorimetry (ITC) is based on a simple titration of one ligand with another and the small heat changes caused by the molecular interaction are detected. From one ITC experiment the complete set of thermodynamic parameters of binding including association and dissociation constants as well as changes in enthalpy, entropy, and free energy can be derived. Using this technique almost any type of molecular interaction can be analyzed. Both ligands are in solution, and there is no need for their chemical derivatization. There are no limits as to the choice of the analysis buffer, and the analysis temperature can be set between 4 and 80 °C. This technique has been primarily applied to study the interaction between various proteins of *Pseudomonas* with small molecule ligands. In addition, ITC has been used to study the binding of *Pseudomonas* proteins to target DNA fragments.

Key words Isothermal titration calorimetry, Thermodynamics, Molecular recognition, Ligand binding, Affinity, Enthalpy change, Entropy change, Ligand profile, Ligand specificity, Regulators, Promoters

1 Introduction

The specific interaction between two different molecules forms the very basis of life. Therefore, the characterization of the strength, nature, and specificity of these interactions is essential to understand the complexity of cellular processes. A wide range of different techniques exist to characterize binding and most of these techniques measure the equilibrium binding constant. A primary advantage of Isothermal Titration Calorimetry (ITC) over alternative techniques consists in the fact that it permits the direct measurement of two parameters, namely, the equilibrium association constant (K_A) and the enthalpy change (ΔH). From these parameters, the entropy change (ΔS), the change in Gibbs free

Alain Filloux and Juan-Luis Ramos (eds.), *Pseudomonas Methods and Protocols*, Methods in Molecular Biology, vol. 1149, DOI 10.1007/978-1-4939-0473-0_16, © Springer Science+Business Media New York 2014

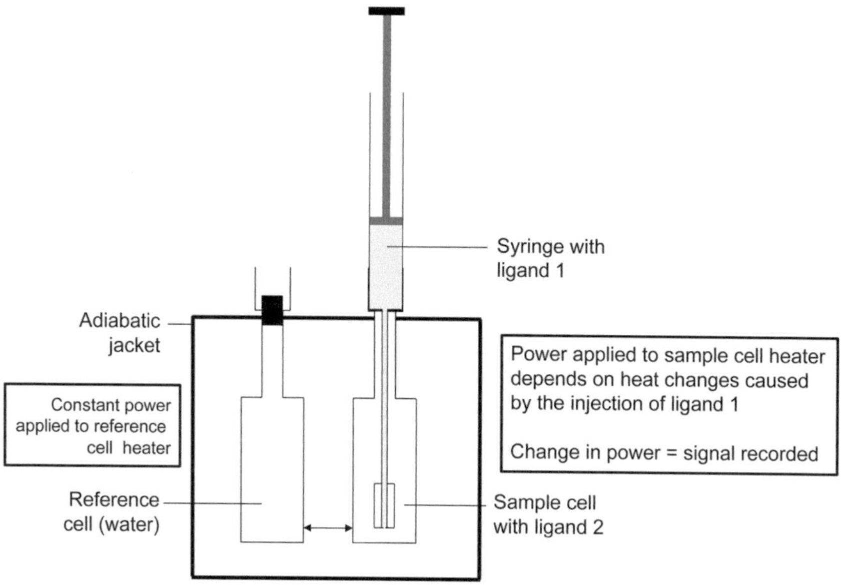

Fig. 1 Schematic representation of an adiabatic isothermal titration microcalorimeter

energy (ΔG), and the dissociation constant (K_D) can be calculated using the equations:

$$\Delta G = -\mathrm{RT} \ln K_A = \Delta H - T\Delta S \begin{pmatrix} \mathrm{R}: \text{gas constant;} \\ \mathrm{T}: \text{absolute temperature} \end{pmatrix} \text{and}\, K_D = 1/K_A$$

ITC permits thus the determination of the full set of thermodynamic parameters of ligand binding in a single experiment. The importance of the knowledge of the enthalpy and entropy changes, which drive a molecular interaction, has been reviewed in [1, 2] and has become of particular relevance in the improvement of lead compounds in drug development.

A schematic drawing of an adiabatic microcalorimeter is shown in Fig. 1 (Further information on the instrument is provided by Wiseman et al. [3]). The adiabatic chamber harbors two identical cells. The reference cell is typically filled with water, whereas one of the ligand solutions is placed into the sample cell. The computer controlled injector syringe, which contains the second ligand, is inserted into the sample cell. Below we will detail the criteria for the choice of which ligand goes into which compartment of the instrument. An ITC experiment consists in the titration of the ligand in the sample cell with aliquots of the injector syringe ligand. The heat changes resulting from this titration are measured. These heat changes are the sum of heats arising from molecular interaction and heats arising from other processes not related to molecular interaction as for example heats resulting from the simple

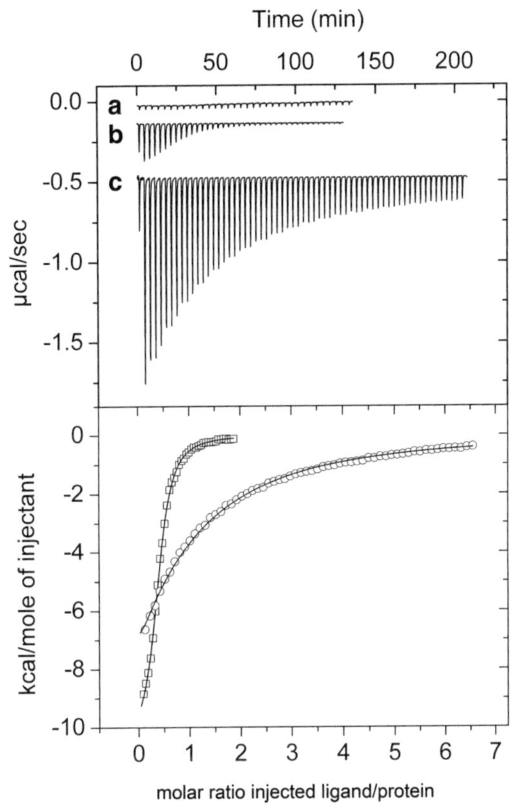

Fig. 2 Microcalorimetric analysis of the interaction of *Pseudomonas* proteins with its respective ligands. *Upper panel*: raw titration data. (**a**) Titration of buffer with 1 mM fumarate. (**b**) Titration of 12 μM of the sensor kinase TodS with 3.2 μl aliquots of 250 μM benzene. (**c**) Titration of 33 μM of the recombinant ligand binding domain of the McpS chemoreceptor (McpS-LBD) with 4.8 μl aliquots of 1 mM fumarate. *Lower panel*: Integrated, dilution corrected and heat normalized peak areas of the raw data. These data were fitted with the "One binding site" model of the ORIGIN software. □ TodS–benzene interaction. O McpS–fumarate interaction. The following thermodynamic parameters were obtained: TodS with benzene: $K_D = 760$ nM, $\Delta H = -11$ kcal/mol. McpS-LBD with fumarate: $K_D = 17$ μM, $\Delta H = -13.3$ kcal/mol

dilution of the syringe ligand, syringe ligand dissociation, or protonation/deprotonation events. In an optimal ITC experiment, the binding heats are largely superior to the nonspecific heats. To assess the magnitude of dilution heats, the syringe ligand is injected in a control experiment into buffer, which is shown in Fig. 2a. The resulting heats are small and uniform and largely inferior to the titration of the protein with the same ligand (Fig. 2b, c). To minimize dilution heats, it is essential that both ligands are in the same buffer. The concentrations of ligands as well as injection volumes have to be chosen in a manner that the ligand present in the sample cell is saturated with the syringe ligand in the course of the titration experiment. Titration raw data are then integrated, concentration normalized, and corrected for dilution effects prior to data fitting.

In addition to the above mentioned thermodynamic parameters, information can be obtained on the binding stoichiometry. In general, a simple noncooperative bimolecular interaction produces either sigmoid (typical for higher affinity binding) or hyperbolic titration curves (typically for lower affinity events, for further information refer to the section on *c*-values in Wiseman et al. [3]. For both curve shapes data analysis using the MicroCal version of ORIGIN will provide information on the binding stoichiometry. This information is highly precise for sigmoid curves where the stoichiometry of binding is very close to the ligand concentration at the point of inflection of the curve. In contrast, stoichiometry information derived from hyperbolic curves has to be considered with caution and is frequently associated with a significant error. The usefulness of such information in understanding binding processes is illustrated in [4].

Another major advantage of ITC as compared to alternative techniques consists in the possibility to analyze more complex binding events. Such interactions become evident by complex, multiphasic titration curves. One illustrative example is the titration of the transferrin receptor from *Neisseria meningitidis* with transferrin. This 300 kDa receptor complex is composed of two types of subunits, namely, TbpA and TbpB. A microcalorimetric titration of this receptor revealed two exothermic binding events. The analysis of these data in combination with the titrations of the individual receptor subunits permitted to conclude that transferrin binds independently to both subunits of the receptor but with much with higher affinity to TbpA as compared to TbpB [5]. The titration data of the transferrin receptor could be easily fitted with an algorithm for the binding of one ligand to two independent binding sites. In cases where an analysis of more complex titration data using models for the binding of molecules to independent binding sites is impossible, the binding process occurs frequently in a cooperative manner. One such example is the binding of the transcriptional regulator TtgR to a 40 bp DNA fragment comprising its operator site [6]. The resulting curve was biphasic and all attempts to analyze these data with model for the binding of a macromolecule to different independent binding sites failed. Analytical ultracentrifugation studies indicated that 2 TtgR dimers bind to this DNA fragment and a cooperative binding model was generated for the fitting of these ITC data [6]. Data analysis revealed that the two TtgR dimers bind with positive cooperativity to DNA and the binding of the initial dimer was found to increase binding affinity for the second dimer by a factor of around 20 [6].

Another advantage of ITC consists in the fact that there are no restrictions as to the composition of the analysis buffer and the choice of analysis temperature. Using the VP-ITC instrument (GE-healthcare, MicroCal, Northampton, USA) the analysis can be conducted at a temperature ranging from 4 to 80 °C. The generation of heat changes is a property of any molecular interaction, and therefore, literally all types of interactions can be monitored by ITC.

ITC has also a series of limitations. Firstly, relatively high amounts of ligands are necessary which are significantly superior to amounts needed for a Biacore experiment or protein intrinsic fluorescence titrations. For an VP-ITC instrument around 2 ml of ligand solution containing 10 μM of binding sites are necessary for an experiment. However, for the new generation, but more expensive auto-ITC instruments sample requirements have been reduced by a factor of around 7. Secondly, and as detailed below, an ITC experiment is based on the fact that the injection of the syringe ligand into buffer does not generate dilution heat effects. In some cases this cannot be achieved, mainly due to dissociation heats created for example for multimeric proteins and consequently no binding experiment can be performed. Thirdly, in our experience sometimes ITC thermograms are too complex to be analyzed in a satisfactory manner by mathematical models. This may be due to a complex nature of the molecular interaction and/or sample heterogeneity.

We have used this technique extensively to study various regulatory mechanisms in *Pseudomonas* and the universality of this technique is illustrated by the following examples. Small molecule-protein binding experiments have revealed for example the effector profiles of the TodS and TmoS sensor kinases [7–9], the TtgR, TtgV, HexR, PtxS, and Cra transcriptional regulators [10–14] or the McpS chemoreceptor [15]. The observation of an unusually high affinity between Crp and its effector cAMP, as measured by ITC, was fundamental in concluding that the function of Crp in *Pseudomonas* is different to that in *E. coli*. DNA–protein binding studies showed that the TodT and TtgR response regulator bind to their target promoters in a cooperative fashion [6, 16, 17]. The titration of different peptides with cations has led to the identification of a novel calcium binding motif in proteins [18]. Protein–protein binding studies were crucial in elucidating the mechanism of *toxA* expression in *P. aeruginosa*, which is based on a molecular interaction between PtxR and PtxS [19] and have also been used to characterize the binding of the TodT response regulator to the TodS sensor kinase [20]. In addition, ITC can be used to study the interaction between small molecules such as the binding of divalent cations to citrate [21]. A database on literature, classified according to the type of molecular interaction analyzed, can be found at http://microcal.com/reference-center/reference-list.asp.

2 Materials

1. Microcalorimeter.
2. Apparatus for sample degassing.
3. Two milliliter loading syringe with 20 cm needle.
4. Two and 10-ml plastic syringes and sterile filter units.
5. The analysis buffer of choice (50 ml).
6. Sample cell ligand (2 ml).
7. Syringe ligand (0.5 ml).

3 Methods

3.1 Design of an ITC Experiment

3.1.1 Choice of Buffer

Solubility and stability of both ligands in the same buffer: There are almost no limitations to the choice of buffer and the primary feature in selecting a buffer is its physiological relevance and its capacity to stabilize both molecules in its native state. In the case that evidence for a molecular interaction has been obtained with an alternative technique, the buffer used in these experiments would be a good choice for the ITC buffer. In the case that macromolecules require the presence of reducing agents please refer to *see* **Note 1**.

Minimization of heats caused buffer protonation/deprotonation: Some binding processes are accompanied by proton transfer, which implies that some molecular interactions are accompanied by a transfer of one or several protons from the ligands to the buffer or vice versa. Buffer protonation/deprotonation is not heat neutral and distorts the enthalpic contribution to binding but has no significant effect on the binding constant. To minimize an eventual contribution of proton transfer, a buffer system with a low ionization enthalpy should be chosen (phosphate, PIPES). Tris buffer has one of the highest ionization enthalpies and is therefore less suited.

3.1.2 Selection of Ligands for Sample Cell and Injector Syringe

An ITC experiment involves the titration of one ligand with another. The experiment can thus be conducted in two different setups which differ in whether a given ligand is present in the sample cell or injector syringe. The choice of where to place a given ligand depends on the following factors.

Sample availability: In an optimal experiment the ligand in the cell is saturated with an excess of syringe ligand, which implies that larger amounts of syringe ligand are necessary to conduct an experiment. If the supply of one ligand is restricted, it may be placed into the sample cell.

Solubility: The volume of the syringe is smaller than that of the sample cell. This, combined with the fact that excess syringe ligand is needed for saturation, implies that the syringe ligand concentration is significantly superior to that of the sample cell ligand. If one of the ligands has low solubility it should be placed into the sample cell to be titrated with the ligand with higher solubility. If solubility of ligands remains an issue, please *see* **Note 2**.

Avoiding dissociation events: As detailed in the Subheading 3.2.1 the injection of ligand into buffer must not create significant heat changes. Such undesired heat changes occur frequently when the ligand molecules dissociate upon injection. Therefore, compounds which dissociate upon dilution should not be placed into the syringe. Such compounds are for example protein multimers or micelle-forming compounds like lipopolysaccharides. ITC is typically used to analyze ligand solutions but can also be used to follow binding events to suspensions (*see* **Note 3**).

3.2 Conduct of an ITC Experiment

3.2.1 Sample Preparation

1. *Both ligands are in the same buffer*: To minimize dilution heats, both ligands have to be in the same buffer. In case both ligands are macromolecules, this can be easily achieved by dialysis. If a macromolecule-small molecule interaction is studied, it is advisable to dialyze the macromolecule and to make up a solution of the small molecule ligand in dialysis buffer.
2. *Filtering and degassing of samples*: Prior to the experiments the two ligand samples as well as 50 ml buffer are filtered using 0.22 μM cutoff filters and degassed (*see* **Note 4**) for around 5 min (note: sterile filtering removes bacteria, which create heats).
3. *Determine ligand concentration*: For solutions which have not been made up at a certain concentration, the exact ligand concentration has to be determined (Bradford assays etc. for proteins, spectrophotometrically for DNA).

3.2.2 The Dilution Control: Ligand into Buffer

1. Switch on instrument 1 h before use and set the analysis temperature.
2. Make sure that ligands are at temperature slightly below the analysis temperature. Using the loading syringe fill the sample cell with buffer, close, and leave equilibrating for around 10 min. Avoid the introduction of air bubbles during sample cell filling.
3. Fill the injector syringe with ligand. Close the loading port of the syringe, dry with a paper towel briefly the outside of the injector syringe needle and insert the injector syringe into the instrument. Leave equilibrating for approximately 5 min.
4. In the meanwhile set the following parameters in the VPviewer software (for Microcal instruments): reference power: 10 μcal/s; initial delay: 60 s (time from the beginning of experiment to first injection); stirring speed of injector: 300 rpm; feedback mode/gain: "high"; ITC equilibration options: activate

"No check temperature" and "Fast equilibration"; injection parameters: define volume of injections: 3–15 μl; speed of injection: 0.5 μl/s, spacing between two injections: 240 s; filter period: 2 s (every 2 s a data point is recorded).

5. Start the titration. Assess the data: In the case the peaks observed are small (below 0.1 μcal/s, ideally around 0.01–0.02 μcal/s, *see* Fig. 2a) proceed with Subheading 3.2.3. In the case peaks are large and not uniform, dilute the syringe ligand and repeat this control experiment. In the case that the signal measured is noisy, instrument cleaning (*see* **Note 5**) is recommended, which is then followed by another ligand-into-buffer control experiment.

3.2.3 The Ligand Titration Experiment

1. Rinse the sample cell excessively with buffer using the loading syringe. Make sure that there is no buffer left in the sample cell.
2. Load sample cell ligand with loading syringe, close instrument, and leave equilibrating for around 10 min.
3. Fill the injector syringe with ligand. Close the loading port of the syringe, dry with a paper towel briefly the outside of the injector syringe needle, and insert the injector syringe into the instrument. Leave equilibrating for 5 min.
4. Start experiment using the same experimental settings.

3.3 Data Analysis

Using the MicroCal version of ORIGIN proceed with the data analysis. The detailed steps of data analysis are provided in the manual. Briefly, data analysis consists in the following steps.

1. Open the raw data for the control and the titration experiment and create two individual project files.
2. Normalize data by the introduction of the exact concentrations for both ligands.
3. Manually optimize the automatically created baseline to assure optimal peak integration.
4. Correct the titration data by subtracting control data.
5. Curve fit to generate thermodynamic parameters.

3.4 Optimization of an ITC Experiment

The first titration of any two ligands gives rarely an optimal result. The inspection of the initial data can be used for an optimization of the experiments. The following points can be taken into consideration:

3.4.1 Sample Cell Ligand Not Saturated at the End of the Titration

To obtain high-quality data, the sample cell ligand needs to be saturated during an experiment. The fact whether the sample cell ligand is saturated with syringe ligand is determined by their affinity and sample concentrations. This is exemplified in Fig. 2b, c, which show interactions with K_D values of 0.76 and 17 μM, respectively.

In the first case a 1.5-fold molar excess of syringe ligand was sufficient to achieve complete saturation, whereas in the second case a sevenfold molar excess was necessary to approach ligand saturation. Therefore, in the case that ligand saturation has not been achieved, the experiment should be repeated with a higher syringe ligand concentration.

3.4.2 No or Very Weak Heat Signals

If no heat signals are observed please refer to *see* **Note 6**. In general the magnitudes of heat changes caused by a molecular interaction differ enormously. In some cases observed heat changes are very small which is due to the fact that enthalpic and entropic contributions to the total heats cancel out each other. However, the magnitude of heat observed is also a function of the analysis temperature. In general, an increase in the analysis temperature favors exothermic signals, whereas a reduction in temperature favors endothermic signals. In the case that very weak heat signals are obtained the experiment should be repeated at a different temperature.

3.4.3 Optimizing Signal-Noise Ratio Versus Number of Peaks

In an optimal ITC experiment there is a balance between the magnitude and the number of peaks. In general, some 20 data points are sufficient to precisely determine binding parameters. Suboptimal data are characterized by an excessive number of small peaks, of which its integration is subject to a significant error or by a reduced number of large peaks which will result in an increased error during the curve fitting process. The size and number of peaks depends, amongst other factors, on the concentrations of both ligands as well as the injection volumes. To optimize an ITC experiment it is frequently advisable to repeat the experiment with different ligand concentrations and/or injection volumes.

4 Notes

1. *Need of reducing agents*: Some proteins require the presence of reducing agents for activity. The air oxidation of reducing agents within the instrument causes a signal setoff into the exothermic range. To minimize this setoff, the use of β-mercaptoethanol should be avoided and not more than 1 mM DTT should be used.
2. *Compounds with reduced solubility*: Some compounds have a very low solubility in aqueous buffer systems, which does not permit an ITC analysis. To circumvent this problem a concentrated ligand solution in 100 % DMSO can be prepared which is then diluted tenfold with the analysis buffer. It is essential that 10 % of DMSO is added to the solution of the other ligand.
3. *Analysis of suspensions*: Binding events of molecules to ligand suspensions can also be monitored by ITC. The ligand suspension

is placed into the sample cell. It is recommended to increase the rotation speed of the injector syringe to 400 rpm.

4. *Volatility of compounds*: A large number of compounds, particularly hydrocarbons are volatile and tend to stick to plastic surfaces. To control their concentration the filter and degas steps are omitted in sample preparations. To prevent sticking to surfaces, samples are made up in glass vessels.
5. *Importance of instrument cleaning*: The reason for noisy data is in 99 % of the cases an insufficient cleaning of the instrument. The frequency of the cleaning depends, amongst other factors, on the propensity of ligands to precipitate or to attach to surfaces. However, it is recommended to clean the instrument at least once a week following the manufacturer's instructions. The sample compartments need to be rinsed extensively with water following cleaning. The best way to determine the state of cleanliness of an instrument is a titration of water with water.
6. *No heat changes measured: Does that mean there is no binding?* At a certain analysis temperature, it is possible that the exothermic and endothermic contributions to the final signal cancel out each other and as a result no heat changes are observed. The ratio of endothermic to exothermic contributions to the overall heat is a function of the analysis temperature. Therefore, to conclude from the absence of binding heats that there is no binding, the experiment needs to be repeated at a different temperature to avoid the possibility that, by chance, the experiment has been carried out at the temperature where endothermic and exothermic contributions are equal. If no heats are observed at a second analysis temperature, it can be concluded that there is no binding.

Acknowledgements

Work in Granada has been supported by Fondo Social Europeo and Fondos FEDER. We acknowledge financial support from the Andalusian regional government Junta de Andalucía (grants P09-RNM-4509, CVI-3010, CVI-7391) and the Spanish Ministry for Economy and Competitiveness (grant Bio2010-16937, BIO2010-17227, Consolider-Ingenio CSD2007-0005).

References

1. Kawasaki Y, Freire E (2011) Finding a better path to drug selectivity. Drug Discov Today 16:985–990
2. Schon A, Lam SY, Freire E (2011) Thermodynamics-based drug design: strategies for inhibiting protein–protein interactions. Future Med Chem 3:1129–1137
3. Wiseman T, Williston S, Brandts JF, Lin LN (1989) Rapid measurement of binding constants and heats of binding using a new titration calorimeter. Anal Biochem 179:131–137
4. Teran W, Krell T, Ramos J-L, Gallegos MT (2006) Effector-repressor interactions, binding of a single effector molecule to the operator-

bound TtgR homodimer mediates derepression. J Biol Chem 281:7102–7109

5. Krell T, Renauld-Mongenie G, Nicolai MC, Fraysse S, Chevalier M, Berard Y, Oakhill J, Evans RW, Gorringe A, Lissolo L (2003) Insight into the structure and function of the transferrin receptor from Neisseria meningitidis using microcalorimetric techniques. J Biol Chem 278:14712–14722
6. Krell T, Teran W, Mayorga OL, Rivas G, Jimenez M, Daniels C, Molina-Henares AJ, Martinez-Bueno M, Gallegos MT, Ramos J-L (2007) Optimization of the palindromic order of the TtgR operator enhances binding cooperativity. J Mol Biol 369:1188–1199
7. Busch A, Lacal J, Martos A, Ramos J-L, Krell T (2007) Bacterial sensor kinase TodS interacts with agonistic and antagonistic signals. Proc Natl Acad Sci USA 104:13774–13779
8. Silva-Jimenez H, Garcia-Fontana C, Cadirci BH, Ramos-Gonzalez MI, Ramos J-L, Krell T (2012) Study of the TmoS/TmoT two-component system: towards the functional characterization of the family of TodS/TodT like systems. Microb Biotechnol 5(4):489–500
9. Lacal J, Busch A, Guazzaroni ME, Krell T, Ramos J-L (2006) The TodS-TodT two-component regulatory system recognizes a wide range of effectors and works with DNA-bending proteins. Proc Natl Acad Sci USA 103:8191–8196
10. Guazzaroni ME, Gallegos MT, Ramos J-L, Krell T (2007) Different modes of binding of mono- and biaromatic effectors to the transcriptional regulator TTGV: role in differential derepression from its cognate operator. J Biol Chem 282:16308–16316
11. Guazzaroni ME, Krell T, Felipe A, Ruiz R, Meng C, Zhang X, Gallegos MT, Ramos J-L (2005) The multidrug efflux regulator TtgV recognizes a wide range of structurally different effectors in solution and complexed with target DNA: evidence from isothermal titration calorimetry. J Biol Chem 280:20887–20893
12. Daddaoua A, Krell T, Ramos J-L (2009) Regulation of glucose metabolism in Pseudomonas: the phosphorylative branch and entner-doudoroff enzymes are regulated by a repressor containing a sugar isomerase domain. J Biol Chem 284:21360–21368
13. Daddaoua A, Krell T, Alfonso C, Morel B, Ramos J-L (2010) Compartmentalized glucose metabolism in Pseudomonas putida is controlled by the PtxS repressor. J Bacteriol 192:4357–4366
14. Chavarria M, Santiago C, Platero R, Krell T, Casasnovas JM, de Lorenzo V (2011) Fructose 1-phosphate is the preferred effector of the metabolic regulator Cra of Pseudomonas putida. J Biol Chem 286:9351–9359
15. Lacal J, Alfonso C, Liu X, Parales RE, Morel B, Conejero-Lara F, Rivas G, Duque E, Ramos J-L, Krell T (2010) Identification of a chemoreceptor for tricarboxylic acid cycle intermediates: differential chemotactic response towards receptor ligands. J Biol Chem 285:23126–23136
16. Lacal J, Guazzaroni ME, Busch A, Krell T, Ramos J-L (2008) Hierarchical binding of the TodT response regulator to its multiple recognition sites at the tod pathway operon promoter. J Mol Biol 376:325–337
17. Lacal J, Guazzaroni ME, Gutierrez-del-Arroyo P, Busch A, Velez M, Krell T, Ramos J-L (2008) Two levels of cooperativeness in the binding of TodT to the tod operon promoter. J Mol Biol 384:1037–1047
18. Santamaria-Hernando S, Krell T, Ramos-Gonzalez MI (2012) Identification of a novel calcium binding motif based on the detection of sequence insertions in the animal peroxidase domain of bacterial proteins. PloS one 7:e40698
19. Daddaoua A, Fillet S, Fernandez M, Udaondo Z, Krell T, Ramos J-L (2012) Genes for carbon metabolism and the ToxA virulence factor in pseudomonas aeruginosa are regulated through molecular interactions of PtxR and PtxS. PloS one 7:e39390
20. Busch A, Guazzaroni ME, Lacal J, Ramos J-L, Krell T (2009) The sensor kinase TodS operates by a multiple step phosphorelay mechanism involving two autokinase domains. J Biol Chem 284:10353–10360
21. Lacal J, Garcia-Fontana C, Callejo-Garcia C, Ramos J-L, Krell T (2011) Physiologically relevant divalent cations modulate citrate recognition by the McpS chemoreceptor. J Mol Recognit 24:378–385

Chapter 17

Proteomic Analysis

Pascal Cosette and Thierry Jouenne

Abstract

Proteome provides highly valuable information on the amount, modifications, and subcellular localization of polypeptides. Accordingly, geneticists, molecular biologists, and biochemists have logically applied these new tools to respond to different lines of biological questions (inventory of proteins, impact of a mutation, dynamics of protein regulation under a given exposure, ...). However, even if the results obtained are very informative, this approach needs an excellent experimental design which ensures robustness and thus yields reproducibility. The present chapter gives appropriate methods for assessing the proteome of *Pseudomonas aeruginosa* by using a two-dimensional gel electrophoresis approach. Protocols for crude protein extraction, protein separation by using immobilized pH gradients, and protein identification by Liquid Chromatography coupled with tandem Mass Spectrometry (LC-MS/MS) are given.

Key words *P. aeruginosa*, Proteome, Two-dimensional gel electrophoresis, Mass spectrometry

1 Introduction

With the publication of the first complete genome sequence of a living organism in 1995, a new era has been opened in biology [1]. The exponential increase in genome sequence information that followed has prompted researchers to design new experiments dealing with functional genomics. Transcriptomics, relying on DNA arrays, provides information on the global gene expression pattern of a cell. Proteome—defined by Wilkins as all proteins expressed by a genome, cell, or tissue [2]—provides additional information on the amount, modification, and subcellular localization of polypeptides. The proteome and subproteomes of *Pseudomonas aeruginosa* have been extensively investigated [3–8]. Two-dimensional gel electrophoresis (2-DE) enables the separation of complex mixtures of proteins according to initially the isoelectric point and subsequently the molecular mass. An important development in 2-DE was the use of immobilized pH gradients (IPGs) in which the pH gradient is immobilized within an acrylamide matrix. Another innovation is the use of differential in-gel electrophoresis (DIGE)

Alain Filloux and Juan-Luis Ramos (eds.), *Pseudomonas Methods and Protocols*, Methods in Molecular Biology, vol. 1149, DOI 10.1007/978-1-4939-0473-0_17, © Springer Science+Business Media New York 2014

in which two pools of proteins are labelled with different fluorescent dyes [9]. Combined with protein identification by mass spectrometry (MS), the 2-DE is one of the two main techniques that can be routinely applied for parallel quantitative expression profiling of large sets of complex protein mixtures. The other one is the shotgun analysis which entails digestion of a mixture of intact proteins, followed by multidimensional separation of the resulting peptide mixture and mass spectrometry (MS) analysis. Though the shotgun approach shows some advantages over gel-based techniques in speed, sensitivity, scope of analysis, and dynamic range, 2-DE is still widely used. Thus, the size of bacterial genomes is relatively small and the number of proteins potentially expressed is quite close to the analytical possibilities of two-dimensional gel electrophoresis (up to 2,000 proteins). Another important criterion in favor of 2-DE lies in its ability to visualize directly the post-translational modifications borne by some proteins and that are responsible for essential signaling cascades for these organisms. However, these conclusions are not so easy to achieve and in order that these studies take place in the best conditions, this approach needs an excellent experimental design. The present chapter gives appropriate methods for assessing the whole proteome of *Pseudomonas aeruginosa* by using the 2-DE technique.

2 Materials

2.1 Bacterial Culture

1. Mueller-Hinton Broth (Difco).
2. Sterile flasks (*see* **Note 1**).

2.2 Preparation of Crude Protein Extracts

1. 0.1 M phosphate buffer, pH 7.
2. IEF buffer: 7 M urea, 2 M thiourea, 2 % (w/v) 3-[3-cholamidopropyl)dimethylammonio]-1-propanesulphonate (CHAPS), 2 mM tributyl phosphine (Caution, *see* **Note 2**), 20 mM dithiothreitol (DTT), 0.5 % 3-(4-Heptyl)phenyl-3-hydroxypropyl)dimethylammoniopropanesulfonate (C7BZ0).
3. Sonicator (Bioblock Scientific Vibrocell TM 75115).
4. Bio-Rad protein assay kit.
5. Gloves.

2.3 Rehydratation of IPG Strip and Isoelectric Focalisation

1. Eppendorf tubes.
2. Rehydratation/equilibration trays (Bio-Rad).
3. Carrier ampholytes (pH 3.5–10; Amersham).
4. IEF buffer.
5. Immobilized pH gradient (IPG) Strips (Immobiline Dry Strip 18 cm, pH 3–10, nonlinear).

6. Coomassie Brilliant Blue R-250 (Sigma).
7. Mineral oil (Sigma).
8. Protean IEF cell (Bio-Rad).
9. Mineral oil.
10. Micropipette.

2.4 Equilibration of IPG Strip

1. Equilibration buffer No. 1 containing 6 M Urea, 30 % (v/v) glycerol, 2 % (w/v) sodium dodecylsulfate (SDS) and 2 % (w/v) DTT in 50 mM Tris–HCl, pH 6.8.
2. Equilibration buffer No. 2 containing 6 M Urea, 30 % (v/v) glycerol, 2 % (w/v) SDS, 2.5 % (w/v) iodoacetamide, and 0.03 % (w/v) Coomassie Brilliant Blue R-250 (Sigma) in 50 mM Tris–HCl, pH 6.8.

2.5 Second Dimension: SDS-PAGE

1. Protean II xi cell electrophoresis apparatus (Bio-Rad).
2. Polyacrylamide solution (40 g polyacrylamide + 100 mL water).
3. 2 % (w/v) *bis*-acrylamide solution.
4. 20 % (w/v) SDS solution.
5. Tetramethyleneethylenediamine (TEMED) solution (Sigma).
6. 10 % (w/v) ammonium persulfate solution.
7. Stacking gel buffer composed of 1 M Tris–HCl, pH 6.8.
8. Resolving gel buffer composed of 1 M Tris–HCl, pH 8.8.
9. Electrode buffer solution containing 3 g/L Trizma base, 14 g/L glycine, 1 g/L SDS in deionized water.
10. Vacuum flask.
11. Spatula.

2.6 Gel Silver Staining

1. Fixation solution containing 10 % (v/v) acetic acid and 30 % (v/v) ethanol in deionized water.
2. 10 % (v/v) ethanol solution.
3. 0.02 % (w/v) sodium thiosulfate solution in deionized water.
4. 0.1 g/L silver nitrate solution in deionized water.
5. Development solution containing 24 g sodium carbonate, 200 μL of 37 % formaldehyde solution and 80 μL of 2 % sodium thiosulfate solution in 2 L deionized water.
6. 1 % (v/v) acetic acid solution.
7. 3 % (v/v) glycerol solution.

2.7 Gel Analysis

1. Gel scanning densitometer (e.g., ProXpressTM, PerkinElmer).
2. Gel analysis software (e.g., PDQuest (Bio-Rad) or SameSpots (Nonlinear Dynamics)).

2.8 Protein Identification

1. Sequencing grade modified trypsin (Promega, Madison, WI).
2. Sterile scalpel.
3. Acetonitrile (CH_3CN).
4. 10 mM ammonium bicarbonate solution.
5. Extraction solution containing water/CH_3CN/trifluoroacetic acid (TFA) (79/20/1).
6. 5 % (v/v) formic acid solution.
7. Mass spectrophotometer nanoLC-chip-MS/MS (Model 6370, Agilent).

3 Methods

3.1 Bacterial Culture

1. Fill sterile flasks with 100 mL of Mueller-Hinton broth.
2. Inoculate the flasks with a bacterial suspension (final bacteria concentration: 10^7 CFU/mL).
3. Put the flasks in a water bath maintained at 37 °C.
4. Incubate for 2 h.

3.2 Preparation of Crude Protein Extracts

1. Centrifuge the bacterial suspension at 1,500 × *g* for 15 min.
2. Wash the pellet three times in 0.1 M phosphate buffer pH 7.
3. Recover the pellet in 3 mL of IEF buffer.
4. Disrupt cells by 3 thermal shocks (from −24 °C to 20 °C) and then by ultrasonication (30 W; 15 pulses of 2 s separated by 2-s breaks). This last procedure must be performed at 4 °C.
5. Put the protein extract for 10 min at 4 °C and then centrifuge it at 10,000 × *g* for 10 min to eliminate cell debris.
6. Protein amount in the supernatant is evaluated using the Bio-Rad protein assay by measuring the absorbance at 595 nm (*see* **Note 3**).
7. Store the supernatant at −20 °C.

3.3 Rehydration of IPG Strips and Isoelectric Focalisation

1. Calculate the sample volume necessary to obtained 100 μg of proteins.
2. Transfer this sample to a 1.5-mL-Eppendorf tube and complete to 400 μL with IEF buffer containing 2 % (v/v) carrier ampholytes (pH 3.5–10, 1 % (m/v) and Coomassie Brilliant Blue R-250 (*see* **Note 4**).
3. Fill a Rehydradation/equilibration Tray channel (Bio-Rad) with the protein sample using a micropipette.
4. Pull off the protective cover of the IPG strips.
5. Gently put the strip gel side down onto the sample. The "+" and "pH 3–10" must be positioned at the left side of the tray.

Take care not to get the sample onto the plastic backing of the strip, and not to trap air bubbles beneath the strip.

6. Overlay the strip with 2 or 3 mL of mineral oil to prevent evaporation during the rehydratation process (*see* **Note 5**).
7. Cover the rehydratation/equilibration tray with the plastic lid provided.
8. The active rehydratation is programmed as the first step of the focalization process (50 V for 12 h).
9. After the rehydratation step, the program proceeds immediately to the first focusing step.
10. Use a linear voltage ramp to reach 10,000 V after 3 h.
11. Stop the focalization when 99,999 Volt-hours are reached. During this step the temperature is maintained at 15 °C by a Peltier platform.
12. After IEF, store the IPG strip between two sheets of plastic film at −24 °C.

3.4 Equilibration of IPG Gel Strips

1. Defrost the IPG strip.
2. Equilibrate successively the IPG strip in 15 mL of each equilibration buffer for 10 min each time, under slight agitation.
3. Blot the IPG strip to remove excess of equilibration buffer.

3.5 Second Dimension: SDS-PAGE (See Note 6)

1. Transfer in vacuum flask (for 1 gel) 15.70 mL of 40 % polyacrylamide solution, 8.39 mL of 2 % bis-acrylamide solution, 19.35 mL of Tris-buffer (pH 8.9), and 7.70 mL of deionized water (*see* **Note 7**).
2. Remove air under vacuum.
3. Add 258 μL of 20 % SDS solution, 40.50 μL of TEMED and 154.8 μL of 10 % ammonium persulfate solution.
4. Cast immediately the running gel by using a casting stand of the Protean II xi Cell apparatus [gel size $200 \times 250 \times 1$ mm^3]. Avoid introduction of any air bubble.
5. Carefully pipette 2 mL of overlay propan-2-ol 50 % in water onto the top of the gel.
6. Allow the gel to polymerize for 45–60 min at room temperature.
7. Remove propan-2-ol and add water from the top of the gel for the night.
8. Remove water.
9. Transfer in vacuum flask 923 μL of 40 % polyacrylamide solution, 491 μL of 2 % bis-acrylamide solution, 1.01 mL of Tris-buffer (pH 6.8), and 5.59 mL of water (*see* **Note 7**).
10. Remove air under vacuum.

11. Add 40.6 μL of 20 % SDS solution, 26.5 μL of TEMED, and 40.6 μL of 10 % ammonium persulfate solution.
12. Cast immediately the stacking gel onto the top of the running gel.
13. Add propan-2-ol 50 % on the top of the gel.
14. Allow the gel to polymerize for 45–60 min at room temperature.
15. Remove propan-2-ol and rinse with distilled water.
16. Place each IPG strip on the top of the stacking gel. Carefully press the IPG strip with a spatula onto the surface of the gel.
17. Fill the buffer tank of the electrophoresis apparatus with the electrode buffer.
18. Insert the gel cassette in the electrophoresis apparatus. Thermostate at 8 °C with polyethylene glycol.
19. Put the lid on the electrophoresis unit and connect cables.
20. Start SDS PAGE at 10 mA/gel for about 4 min with a limit of 150 V.
21. Put 20 mA/gel and a limit of 350 V.
22. When Coomassie Brilliant Blue tracking dye has migrated off the lower end of the gel, stop the run.
23. Remove the gel from the buffer tank.
24. Peel carefully the gel from the glass plate and remove the IPG strip and the stacking gel with a spatula.

3.6 Gel Silver Staining

1. Fix the gel overnight in the fixation solution.
2. Rinse twice the gel in the 10 % ethanol solution for 10 min.
3. Rinse 3 × 10 min in deionized water.
4. Soak gel for 1 min in 0.02 % sodium thiosulfate solution.
5. Rinse for 1 min in deionized water.
6. Impregnate for 30 min in 0.1 % silver nitrate solution.
7. Rinse in water for 30 s.
8. Develop image for 20–30 min in the development solution.
9. Stop development in 1 % acetic acid solution for 10 min (Fig. 1).
10. Soak in a 3 % glycerol solution.

3.7 Gel Analysis

1. Scan the gel using a densitometer.
2. Analyze the gel using a software (PDQuest, SameSpots, …).

3.8 Protein Identification

1. Excise the spots from the gel with a sterile scalpel and slice into small pieces.
2. Wash gel plugs twice for 15 min with 100 μL of deionized water.

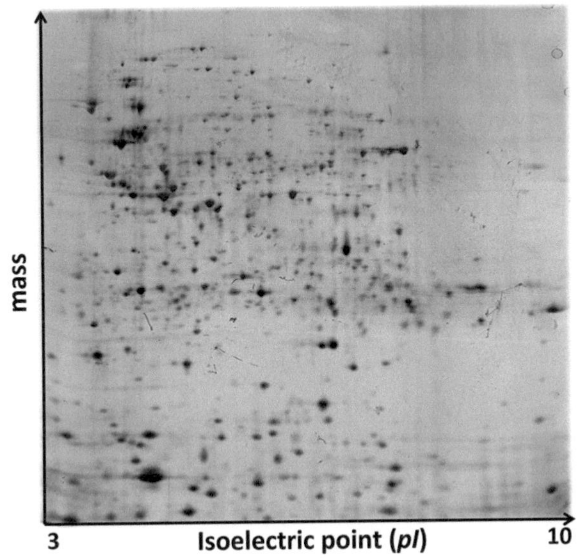

Fig. 1 2-DE gel obtained from whole proteins of *P. aeruginosa*; protein amount: 100 μg, silver nitrate staining

3. Wash plugs twice with 100 μL of H_2O/CH_3CN (1/1 v/v) for 15 min.
4. Place plugs in 20 μL of 100 % CH_3CN for 15 min.
5. Dry plugs using a SpeedVac centrifuge for a few minutes.
6. Add 15 μL of a 20 ng/μL trypsin solution.
7. After rehydratation (30 min), cover the plugs with 20 μL of 10 mM ammonium bicarbonate solution and after 3 h, add 35 μL of water.
8. Allow digestion overnight at 37 °C.
9. Collect the liquid phase containing peptides.
10. Add 20 μL of H_2O/CH_3CN/TFA mixture (79/20/1) to remove peptides remaining in the gel.
11. Dry in a vacuum centrifuge.
12. Redissolve in 10 μL of 5 % formic acid.
13. Peptides were enriched and separated using a lab-on-a-chip technology (Agilent, Massy, France) and fragmented using an online XCT mass spectrometer (Agilent) according to manufacturer.
14. The fragmentation data were interpreted using the Data Analysis software (version 3.4, Bruker Daltonic, Billerica, MA, USA). For protein identification, MS/MS peak lists were extracted and compared to the NCBInr protein database (restricted to *P. aeruginosa*), using the MASCOT Daemon (version 2.1.3) search engine. All searches were performed with no fixed modification and allowed for carbamidomethylation, oxidation, and a maximum of one missed trypsin cleavage.

4 Notes

1. All materials must be sterilized at 120 °C for 15 min before use.
2. Caution: explosive after contact with air, work in a chemical hood.
3. A calibration curve must be built by using increasing concentrations of bovine serum albumin (BSA, Sigma).
4. Better results are obtained by substituting Bromophenol Blue with Coomassie Brilliant Blue [10].
5. Add the mineral oil slowly.
6. The gel must be made the previous day.
7. The solutions must be precooled at 4 °C.

References

1. Fleischmann RD, Adams MD, White O, Clayton RA, Kirkness EF, Kerlavage AR et al (1995) Whole-genome random sequencing and assembly of *Haemophilus influenzae* Rd. Science 269:496–521
2. Wilkins MR, Sanchez JC, Gooley AA, Appel RD, Humphrey-Smith I, Hochstrasser DF et al (1995) Progress in proteome projects: why all proteins expressed by a genome should be identified and how to do it. Biotechnol Genet Eng Rev 13:19–50
3. Arevalo-Ferro C, Buschmann J, Reil G, Gorg A, Wiehlmann L, Tummler B et al (2004) Proteome analysis of intraclonal diversity of two *Pseudomonas aeruginosa* TB clone isolates. Proteomics 4:1241–1246
4. Choi DS, Kim DK, Choi SJ, Lee J, Choi JP, Rho S et al (2011) Proteomic analysis of outer membrane vesicles derived from *Pseudomonas aeruginosa*. Proteomics 11:3424–3429
5. Hare NJ, Solis N, Harmer G, Marzook NB, Rose B, Harbour C et al (2012) Proteomic profiling of Pseudomonas aeruginosa AES-1R, PAO1 and PA14 reveals potential virulence determinants associated with a transmissible cystic fibrosis-associated strain. BMC Microbiol 12:16. doi:10.1186/1471-2180-12-16
6. Imperi F, Ciccosanti F, Perdomo AB, Tiburzi F, Mancone C, Alonzi T et al (2009) Analysis of the periplasmic proteome of *Pseudomonas aeruginosa*, a metabolically versatile opportunistic pathogen. Proteomics 9:1901–1915
7. Nouwens AS, Cordwell SJ, Larsen MR, Molloy MP, Gillings M, Willcox MD et al (2000) Complementing genomics with proteomics: the membrane subproteome of *Pseudomonas aeruginosa* PAO1. Electrophoresis 21:3797–3809
8. Vilain S, Cosette P, Hubert M, Lange C, Junter GA, Jouenne T (2004) Comparative proteomic analysis of planktonic and immobilized *Pseudomonas aeruginosa* cells: a multivariate statistical approach. Anal Biochem 329:120–130
9. Patton WF (2002) Detection technologies in proteome analysis. J Chromatogr B771:3–31
10. Vilain S, Cosette P, Charlionet R, Hubert M, Lange C, Junter GA et al (2001) Substituting Coomassie brilliant blue for bromophenol blue in two dimensional electrophoresis buffers improves the resolution of 2-DE patterns. Electrophoresis 22:4368–4374

Chapter 18

Membrane Proteomics of *Pseudomonas aeruginosa*

Zofia Magnowska, Isabel Hartmann, Lothar Jänsch, and Dieter Jahn

Abstract

In recent years gel-free proteomics approaches have been increasingly used for global quantitative proteome analyses of multiple prokaryotic organisms, including *Pseudomonas aeruginosa*. A major advantage of this method is its suitability for the investigation of membrane proteomes. In this chapter, we present a protocol for preparation of proteins from the inner and outer membrane of *P. aeruginosa* PAO1 grown as a biofilm culture. Parameters for quantitative protein measurements by 2D-LC-MS/MS are described.

Key words *P. aeruginosa*, Membrane, Proteomics, Gel-free, 2D-LC-MS/MS

1 Introduction

For many years 2-dimensional gel electrophoresis (2-DE, *see* also Chapter 17) has been the method of choice for global comparative proteome studies of bacteria. A large body of knowledge providing valuable insights into several aspects of *P. aeruginosa* biology has been generated (e.g., 1–3). However, standard 2-DE has limitations with respect to the accessibility of certain groups of proteins, such as proteins with extreme p*I*s, low abundance, or high hydrophobicity [4]. Great technological advances have been seen in recent years in the fields of liquid chromatography (LC)-based protein purification, protein/peptide mass spectrometry (MS), and bioinformatics, leading to increased utilization of gel-free approaches for global proteome analyses. Currently, the most widely used method is 2D-LC-MS/MS, also referred to as MudPIT (multidimensional protein identification technology). In a typical 2D-LC-MS/MS experiment, proteins are first digested into peptides, yielding complex mixtures, which are fractionated in the first LC-dimension by strong cation exchange chromatography (SCX). These fractions are subsequently separated by a second LC step (typically C-18 reversed phase chromatography) and eluting peptides are analyzed online by tandem mass spectrometry (MS/MS). Finally, resulting MS/MS spectra are interpreted using appropriate

Alain Filloux and Juan-Luis Ramos (eds.), *Pseudomonas Methods and Protocols*, Methods in Molecular Biology, vol. 1149, DOI 10.1007/978-1-4939-0473-0_18, © Springer Science+Business Media New York 2014

protein identification software tools [5]. Major advantages of these methods include greater proteome coverage with routine identification of several thousands of proteins per experiment and an improved detection of less abundant as well as hydrophobic proteins (e.g., 6). The detection of hydrophobic proteins provides the basis for membrane proteomics, which is limited using gel-based approaches. Several powerful strategies for the gel-free comparative quantitative analysis of complex protein mixtures have been developed [7]. These include the isobaric tag for relative and absolute quantification (iTRAQ™)-technology used in the experiment described here. The iTRAQ™ approach is based on differential labeling of proteins or peptides from different samples to be compared. Those labels introduce tags of equal molecular weight, but differing in isotopic composition of their reporter groups. The samples are then combined and analyzed by 2D-LC-MS/MS. During the LC runs, the isobaric tags ensure identical separation behavior of identical peptides from different samples. The reporter groups released during MS/MS are distinguished in the MS spectra due to their different molecular weights. Based on the ratios of the signal intensities of the reporter groups, the relative quantity of any peptide in each of the original samples can be calculated [8]. Given the exciting and diverse new possibilities provided by the gel-free proteomics methods, these techniques have been increasingly applied to proteome analyses of *P. aeruginosa* in recent years (e.g., 9–11). Biofilm formation is a critical factor in the pathogenicity of *P. aeruginosa*. This organism is currently the best-studied model for biofilm formation amongst Gram-negative bacteria. It has become clear that the sessile mode of growth in a biofilm represents a physiological state distinct from the planktonic lifestyle [12–14].

The membranes of Gram-negative cell envelopes are an important cellular compartment providing many essential functions that are mediated by membrane proteins [15]. However, due to the methodological restrictions to the study of membrane proteins inherent to classical approaches like 2-DE, membrane proteomes remain insufficiently studied to date, leading to incomplete knowledge of the contributions of membrane proteins to processes such as biofilm formation. The 2D-LC-MS/MS approach was applied to the characterization of the biofilm membrane proteome in order to exploit its superior performance in identification of hydrophobic proteins.

In the workflow protocol presented in this chapter, membrane fractions from a *P. aeruginosa* PAO1 biofilm culture were extracted by isopycnic gradient centrifugation [16, 17]. Using this method, two membrane fractions were generated, one of which is enriched for cytoplasmic membrane proteins and the other for outer membrane proteins. The fractions were further purified by repeated carbonate extraction to remove proteins loosely attached to the membranes. As an example for comparative proteome analysis using iTRAQ™, the protein content of the two membrane fractions was compared [18, 19]. Equal amounts of protein from each of the fractions were separately trypsin-digested and resulting peptides differentially labeled with two isobaric tags. After combining the two samples at equal ratio, the pooled sample was subjected to 2D-LC-MS/MS using an Orbitrap Velos mass spectrometer. The raw data obtained were ready for comparative quantitative analysis using appropriate proteomics software.

2 Materials

2.1 Bacterial Culture

1. *Pseudomonas aeruginosa* PAO1 strain.
2. Luria–Bertani (LB) medium: 10 g/L Bacto tryptone, 5 g/L Bacto yeast extract, 7 g/L NaCl.
3. Sterile baffled glass flask (1 L).
4. Sterile nitrocellulose filters—100 mm diameter.
5. 100 mM phosphate buffered saline solution (PBS).

2.2 Bacterial Cell Disruption

1. Large plastic reaction tubes (15 mL).
2. 100 mM phosphate buffer, pH 7.
3. Cell disruption buffer: 100 mM CH_3CO_2K, 5 mM $Mg(CH_3COO)_2$, 50 mM 4-(2-hydroxyethyl)-1-piperazineethanesulfonic acid (HEPES), 0.05 % (v/v) 2-mercaptoethanol, pH 7.5 (*see* **Note 1**) containing 1× Complete™ protease inhibitor cocktail ethylenediaminetetraacetic acid (EDTA) free (Roche Diagnostics, Mannheim, Germany) as recommended by the manufacturer (*see* **Note 2**).
4. Benzonase (Merck, Darmstadt, Germany).
5. French press, 40 K cell (Thermo Electron, Waltham, MA, USA).
6. Biofuge™ Fresco centrifuge (Heraeus, Hanau, Germany).

2.3 Isopycnic Sucrose Gradient Centrifugation

1. 500 mM, 1.5 M and 2 M sucrose solutions in cell disruption buffer (*see* **Note 3**).
2. Ultra-Clear™ centrifuge tubes 16×100 mm (Beckman Coulter, Palo Alto, CA, USA).
3. J-30I High Performance Centrifuge Avanti (Beckman Coulter).
4. 10 mL pipettes.
5. Pasteur pipettes.
6. Complete™ protease inhibitor cocktail EDTA free (Roche Diagnostics).

2.4 Carbonate Extraction

1. 1 M tris(hydroxymethyl)aminomethane (TRIS), pH 7.5.
2. 100 mM sodium bicarbonate, pH 11.
3. Double-distilled water (ddH_2O).

4. Complete™ protease inhibitor cocktail (Roche Diagnostics).
5. Polycarbonate centrifuge tubes 13 × 51 mm (Beckman Coulter).
6. Sorvall Discovery™ MS120SE ultracentrifuge (Hitachi, Tokio, Japan).
7. Sonorex™ ultrasonic water bath DK 514 BP, 600 W (Bandelin, Berlin, Germany).

2.5 Wessel–Flügge Precipitation

1. Methanol.
2. Chloroform.
3. ddH_2O.
4. Large plastic reaction tubes (15 mL).
5. Vortex™ shaker (IKA-Werke, Staufen, Germany).
6. 2K15C centrifuge (Sigma Laborzentrifugen, Osterode am Harz, Germany).

2.6 Preparative SDS-PAGE

1. 1 M urea.
2. Sample buffer (4×): 240 mM TRIS, pH 6.8, 40 % (v/v) glycerol, 12 % (v/v) 2-mercaptoethanol, 270 mM sodium dodecyl sulfate (SDS), 0.004 % (w/v) bromophenol blue.
3. Collecting gel (4 %): 6.1 mL ddH_2O, 2.5 mL 500 mM TRIS, pH 6.8, 1.5 mL acrylamide–bisacrylamide (37.5:1), 100 μL 10 % (w/v) SDS, 50 μL TEMED, and 75 μL 10 % (w/v) ammonium persulfate (APS) (*see* **Note 4**).
4. Separating gel (12 %): 5.1 mL ddH_2O, 3.8 mL 1.5 M TRIS (pH 8.8), 6 mL acrylamide–bisacrylamide (37.5:1), 150 μL 10 % (w/v) SDS, 20 μL *N*,*N*,*N*′,*N*′-tetramethylethylenediamine (TEMED), 50 μL 10 % (w/v) APS.
5. Electrode buffer (10×), pH 8.9: 2 M glycine, 250 mM TRIS, 35 mM SDS.
6. Coomassie blue silver: 0.12 % (w/v) Coomassie BB-G-250, 10 % (v/v) H_3PO_4 (85 %), 20 % (v/v) methanol, 10 % (v/v) $(NH_4)_2SO_4$.
7. Gel fixing solution: 10 % (v/v) acetic acid, 40 % (v/v) ethanol, and 50 % (v/v) ddH_2O.
8. 50 mM triethyl ammonium bicarbonate (TEAB).
9. Destaining solution: 40 % (v/v) acetonitrile (ACN), 50 mM TEAB.
10. Low molecular weight calibration kit for SDS electrophoresis (GE Healthcare, Chalfont St Giles, UK).
11. Small plastic reaction tubes (1.5 mL).
12. Sterile scalpel.
13. Minigel system (Biometra, Goettingen, Germany).
14. Sonorex™ ultrasonic water bath DK 514 BP, 600 W (Bandelin).

2.7 Protein In-Gel Digestion and Peptide Extraction

1. Reducing solution, 20 mM 1,4 dithiothreitol (DTT), 50 mM TEAB.
2. Alkylating solution: 50 mM iodoacetamide, 50 mM TEAB.
3. Sequencing grade modified trypsin (Promega, Madison, WI, USA).
4. Digestion buffer: 50 mM TEAB, 10 % (v/v) ACN, pH 8.3.
5. 0.2 % (v/v) trifluoroacetic acid (TFA).
6. Small plastic reaction tubes (1.5 mL).
7. Large plastic reaction tubes (15 mL).
8. RP C_{18} chromatography μZipTip™ pipette tips (Millipore, Billerica, MA, USA).
9. SpeedVac™ RC1010 vacuum concentrator (Jouan, Winchester, UK).
10. Ultraflex MALDI TOF/TOF Mass Spectrometer (Bruker Daltonics, Bremen, Germany).

2.8 Quantitative iTRAQ™ Labeling of Peptides

1. RP wash buffer containing 0.2 % (v/v) TFA, pH 2.5.
2. RP elution buffer containing 0.2 % (v/v) TFA, 60 % (v/v) ACN, pH 2.5.
3. LiChroprep RP-18 SPE columns (Merck).
4. iTRAQ™ Reagents Multiplex Kit (Applied Biosystems, Foster City, CA, USA).
5. Small plastic reaction tubes (1.5 mL).

2.9 Strong Cation Exchange Chromatography (SCX)

1. SCX buffer A: 0.065 % (v/v) formic acid, 25 % (v/v) ACN.
2. SCX buffer B: 0.065 % (v/v) formic acid, 25 % (v/v) ACN, 0.5 M KCl.
3. Ultracentrifuge Sorvall Discovery™ MS120SE (Hitachi).
4. Ettan micro-LC system with MonoS PC1.6/5 column (GE Healthcare).
5. Fraction collector.
6. RP C_{18} chromatography μZipTip™ pipette tips (Millipore).

2.10 Protein Identification

1. UPLC buffer A: 0.1 % (v/v) formic acid.
2. UPLC buffer B: containing 0.1 % (v/v) formic acid and 99.9 % (v/v) ACN.
3. Polycarbonate centrifuge tubes 7 × 20 mm (Beckman Coulter).
4. Acquity ultraperformance LC system with analytical column 1.7-μm BEH130, 75 μm × 250 mm, controlled with AcquityUPLC software V1.30. (Waters, Milford, MA USA).
5. LTQ Orbitrap Velos mass spectrometer with Xcalibur 2.1.0.1139 software (Thermo Scientific, San Jose, CA, USA).
6. Mascot Daemon software (Matrix Science, Boston, MA, USA).

7. Mascot Server and Daemon software (Matrix Science, London, UK); free Mascot Server Web site:

   http://www.matrixscience.com/cgi/search_form.pl?FORMVER=2&SEARCH=SQ; Mascot Daemon: https://sysbio-mascot.wehi.edu.au/mascot/daemon_install.html.
8. Proteome Discoverer 1.3 software (Thermo Scientific).

3 Methods

3.1 Bacterial Culture

1. Fill sterile baffled glass flask with 100 mL of LB medium.
2. Inoculate the flask with *P. aeruginosa* PAO1 strain and cultivate at 37 °C overnight.
3. Spread 800 μL of the culture diluted 1:2,500 in PBS on the sterile nitrocellulose filter.
4. Place filter on LB agar plate and incubate at 37 °C.
5. Transfer filter after 24 h onto fresh LB agar plate and incubate for another 24 h.

3.2 Bacterial Cell Disruption

1. Harvest biofilm into 2 mL of ice-cold 100 mM phosphate buffer and resuspend by vigorous shaking.
2. Centrifuge the suspension at 4 °C for 5 min at 16,000 × *g* (*see* **Note 5**).
3. Remove supernatant and resuspend the pellet in cell disruption buffer.
4. Lyse cells by passing the culture three times through a French press at 1,000 psi (*see* **Note 6**).
5. Centrifuge suspension at 10,000 × *g* for 10 min at 4 °C to remove cell debris and unlysed cells.
6. Collect supernatant containing the membrane fractions.

3.3 Isopycnic Sucrose Gradient Centrifugation

1. Load isopycnic sucrose gradient into the centrifuge tube in the following order: 3 mL of 2 M sucrose solution, 3 mL of 1.5 M sucrose solution, 3 mL 0.5 M sucrose solution, and 6 mL membrane fraction (as presented in Fig. 1a).
2. Centrifuge gradient at 100,000 × *g* for 1 h at 4 °C.
3. Collect membrane fractions enriched on interphase between sucrose steps 0.5 M/1.5 M and 1.5 M/2 M of the gradient using a Pasteur pipette (as outlined in Fig. 1b). From this stage onwards, the membrane fractions should be processed separately.

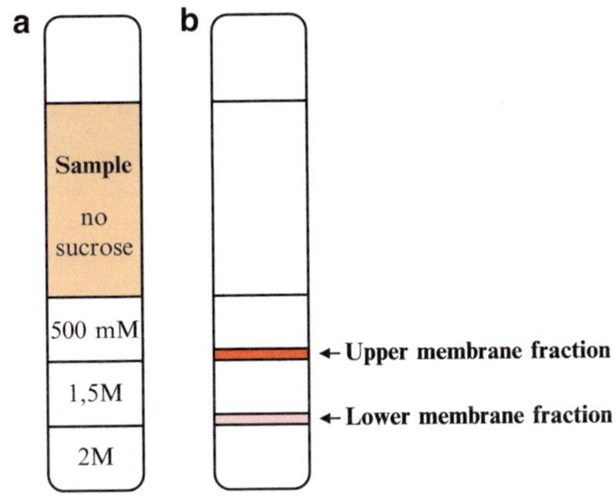

Fig. 1 Isopycnic sucrose gradient. Panel **a** presents the order of sucrose steps on the gradient prior to centrifugation. Panel **b** presents the localization of membrane bands after centrifugation

4. Wash membrane fractions with 14 mL ddH_2O water containing Complete™ EDTA free protease inhibitor cocktail and centrifuge again at 100,000 × *g* for 1 h at 4 °C.

3.4 Carbonate Extraction

1. Resuspend membrane pellets in 5 mL 1 M TRIS, pH 7.5 (two tubes per fraction), incubate on ice for 30 min, centrifuge for 30 min at 100,000 × *g*, and discard supernatant (*see* **Notes 7** and **8**).
2. Resuspend membrane pellets in 5 mL 100 mM $NaCO_3$, pH 11 (two tubes per fraction), sonicate for 5 min, incubate on ice for 30 min, centrifuge for 30 min at 100,000 × *g*, and discard supernatant. Repeat this step twice.
3. Resuspend membrane pellets in 5 mL ddH_2O (two tubes per fraction), sonicate for 5 min, incubate on ice for 30 min, centrifuge for 30 min at 100,000 × *g*, and discard supernatant. Repeat this step twice.

3.5 Wessel–Flügge Precipitation

1. Resuspend membrane pellets in 1 mL ddH_2O water.
2. Add these reagents in the following order: 4 mL methanol, 1 mL chloroform, and 3 mL water, and shake vigorously (*see* **Note 9**).
3. Incubate for 1 h at 4 °C.
4. Centrifuge at 5,000 × *g* for 1 h at 4 °C.
5. Collect proteins precipitated at the interface, resuspend in 3 mL of ice-cold methanol, and pellet by centrifugation at 5,000 × *g* for 1 h at 4 °C (*see* **Note 10**).

3.6 Preparative SDS-PAGE

1. Resuspend protein samples in 1 M urea and 4× SDS sample buffer (1:3 ratio), incubate for 5 min at 60 °C, and sonicate for 5 min (*see* **Note 11**).
2. Determine protein amount using Low Molecular Weight Calibration Kit according to manufacturer's protocol.
3. Calculate sample volume necessary to obtain 50 μg of proteins.
4. Load protein sample onto a preparative collecting/separating polyacrylamide gel (*see* **Note 12**).
5. Resolve proteins at 120 V until all samples have entered the separating gel but have not run more than 2 cm.
6. Incubate gel for 1 h in gel fixing solution.
7. Stain proteins with the Coomassie blue silver overnight (*see* **Note 13**).
8. Incubate gel in ddH_2O until unbound Coomassie dye is washed out.
9. Excise gel areas containing visible protein and disintegrate it by cutting into 1 mm^3 cubes (*see* **Note 14**).
10. Destain gel pieces in destaining solution overnight.

3.7 Protein In-Gel Digestion and Peptide Extraction

1. Wash gel pieces twice in 5 gel volumes of 50 mM TEAB.
2. Incubate gel pieces in 5 gel volumes of reducing solution for 30 min at 56 °C.
3. Incubate gel pieces in 5 gel volumes of alkylating solution for 1 h in the dark.
4. Dehydrate gel pieces in ACN and dry in a vacuum concentrator.
5. Re-swell the gel pieces in solution of trypsin in digestion buffer at 1:20 enzyme–protein ratio.
6. Digest proteins by incubation at 37 °C overnight with shaking (*see* **Note 15**).
7. Dilute 1 μL aliquots of digest solution in 15 μL 0.2 % TFA, purify by RP C_{18} chromatography (μZipTip™ pipette tips) and analyze with MALDI to check for peptide presence.
8. Collect digestion solution.
9. Elute peptides remaining in gel pieces with 5 gel volumes of ddH_2O, then sequentially 0.1 % TCA, 1 % TFA, 0.1 % TFA in 40 % ACN, and finally with pure ACN until the gel pieces are completely dehydrated.
10. Dry the digest solution and eluted peptides (in 500 μL portions) in one small reaction tube per protein sample in a vacuum concentrator (*see* **Note 16**).

3.8 Quantitative iTRAQ™ Labeling of Peptides

1. Desalt peptides on LiChroprep RP-18 columns according to instructions of manufacturer.
2. Dry eluted fraction to one small reaction tube per sample in a vacuum concentrator.
3. Dissolve dried peptide samples in 80 μL of iTRAQ dissolution buffer.
4. Add portions of labeling reagents (dissolved in 70 μL ethanol/portion) sufficient to label 50 μg of total protein:
   - Upper membrane fraction: label 114
   - Lower membrane fraction: label 116
5. Carry out labeling reaction for 80 min at room temperature in the dark. Consult iTRAQ™ kit manufacturer's protocol for the details of peptide labeling.
6. Purify labeled samples by LiChroprep RP-18 columns to remove free labels.
7. Perform approximate determination of peptide content by UPLC-MS/MS.
8. Combine labeled peptides from two membrane fractions in equal ratio.
9. Dry the sample in a vacuum concentrator.

3.9 Strong Cation Exchange Chromatography

1. Resuspend peptide mixture in 40 μL SCX buffer A and adjust the pH to <3 using 80 % formic acid.
2. Centrifuge samples for 20 min at 100,000 ×*g*.
3. Perform SCX chromatography on an Ettan micro-LC system with a linear gradient from 0 to 35 % SCX buffer B in 30 min at a flow rate of 150 μL/min.
4. Collect 150 μL fractions with a microfraction collector monitoring elution of peptides at 214 nm wavelength.
5. Dry fractions containing peptides, resuspend in 0.2 % TFA, desalted by RP C_{18} chromatography (μZipTip™ pipette tips) and again dry in a vacuum concentrator.

3.10 Protein Identification

1. Dissolve peptide samples in 13 μL buffer A and centrifuge at 100,000 ×*g* for 20 min
2. Separate the peptides on analytical column using linear 120 min gradient of UPLC buffer A and UPLC buffer B at a flow rate of 350 nL/min and analyze with connected mass spectrometer (electrospray ionization with voltage of ~1,900 kV, doubly and triply charged peptide ions automatically selected and fragmented).

3. Use the Mascot Daemon application to merge all peak lists from analyzed SCX fractions corresponding to one experiment for subsequent database searches and automatic protein identification using the Mascot Server. Appropriate search parameter settings are listed below:

Database	PAO1 NCBI (all predicted protein sequences of PAO1)
Taxonomy	*Pseudomonas aeruginosa*
Enzyme	Trypsin
Max missed cleavages	1
Fixed modification	iTRAQ (K), iTRAQ (N-terminus)
Variable modification	Carbamidomethyl (C), Oxidation (M)
Charge	2+ and 3+
Peptide tolerance	±10 ppm
MS/MS tolerance	±0.02 Da
Instrument type	ESI-FTICR

4. Accept proteins as identified only when at least two unique peptides show an individual score above 30.
5. Use the Proteome Discoverer 1.3 software for quantitative analysis.

4 Notes

1. Cell disruption buffer should be prepared freshly.
2. Until the carbonate extraction step, membranes should be processed in the presence of a protease inhibitor cocktail. For the sake of membrane integrity all solutions used until the step of sucrose gradient centrifugation should be devoid of EDTA.
3. Sucrose solutions should be prepared freshly.
4. Preparative polyacrylamide gel should be poured the previous day inside a fume hood.
5. Keeping the sample at low temperature is necessary for retaining membrane integrity and preventing cellular proteases release and protein degradation.
6. Efficient cell disruption is indicated by the cell solution turning opalescent.
7. All solutions for carbonate extraction should be ice-cold.

8. All steps of carbonate extraction should be performed in the presence of Complete™ protease inhibitor cocktail.
9. All solutions for Wessel–Flügge precipitation should be ice-cold.
10. Protein pellet should not be dried completely.
11. Membrane proteins are very difficult to solubilize, so increasing the volume of resuspension buffer may help. However, it is necessary to strike a balance between sample volume and sample concentration, keeping in mind that both need to be compatible with the following procedures.
12. In case of large volume, sample can be distributed to two wells on the gel.
13. I case of a problem with staining, repeat and extend the previous fixation step.
14. The distance of electrophoretic protein separation is short, therefore, after staining, proteins appear as a continuous area rather than separate protein bands. The whole of this stained area should be excised and disintegrated for further procedures.
15. Trypsin is maximally active in the range of pH 7–9.
16. Peptides adhere to reaction tube polymer; therefore, the surface of contact should be minimized.

Acknowledgement

We thank Uwe Kärst for critically reading the manuscript.

References

1. Nouwens AS, Walsh BJ, Cordwell SJ (2003) Application of proteomics to *Pseudomonas aeruginosa*. Adv Biochem Eng Biotechnol 83:117–140
2. Hare NJ, Cordwell SJ (2010) Proteomics of bacterial pathogens: *Pseudomonas aeruginosa* infections in cystic fibrosis—A case study. Proteomics Clin Appl 4:228–248
3. Eberl L, Riedel K (2011) Mining quorum sensing regulated proteins—role of bacterial cell-to-cell communication in global gene regulation as assessed by proteomics. Proteomics 11:3070–3085
4. Görg A, Drews O, Lück C, Weiland F, Weiss W (2009) 2-DE with IPGs. Electrophoresis 30:S122–S132
5. Yates JR, Ruse CI, Nakorchevsky A (2009) Proteomics by mass spectrometry: approaches, advances, and applications. Annu Rev Biomed Eng 11:49–79
6. Skipp P, Robinson J, O'Connor CD, Clarke IN (2005) Shotgun proteomic analysis of *Chlamydia trachomatis*. Proteomics 5:1558–1573
7. Wilm M (2009) Quantitative proteomics in biological research. Proteomics 9:4590–4605
8. Ross PL, Huang YN, Marchese JN, Williamson B, Parker K, Hattan S et al (2004) Multiplexed protein quantitation in *Saccharomyces cerevisiae* using amine-reactive isobaric tagging reagents. Mol Cell Proteomics 3:1154–1169
9. Toyofuku M, Roschitzki B, Riedel K, Eberl L (2012) Identification of proteins associated with the *Pseudomonas aeruginosa* biofilm matrix. J Proteome Res 11:4906–4915
10. Hare NJ, Solis N, Harmer C, Marzook NB, Rose B, Harbour C et al (2012) Proteomic profiling of

Pseudomonas aeruginosa AES-1R, PAO1 and PA14 reveals potential virulence determinants associated with a transmissible cystic fibrosis-associated strain. BMC Microbiol 12:16

11. Choi DS, Kim DK, Choi SJ, Lee J, Choi JP, Rho S et al (2011) Proteomic analysis of outer membrane vesicles derived from *Pseudomonas aeruginosa*. Proteomics 11:3424–3429
12. Häussler S (2010) Multicellular signalling and growth of *Pseudomonas aeruginosa*. Int J Med Microbiol 300:544–548
13. Harmsen M, Yang L, Pamp SJ, Tolker-Nielsen T (2010) An update on *Pseudomonas aeruginosa* biofilm formation, tolerance and dispersal. FEMS Immunol Med Microbiol 59: 253–268
14. Hall-Stoodley L, Costerton JW, Stoodley P (2004) Bacterial biofilms: from the natural environment to infectious diseases. Nat Rev Microbiol 2:95–108
15. Silhavy TJ, Kahne D, Walker S (2010) The bacterial cell envelope. Cold Spring Harb Perspect Biol 2(5):a000414
16. Fujiki Y, Fowler S, Shio H, Hubbard AL, Lazarow PB (1982) Polypeptide and phospholipid composition of the membrane of rat liver peroxisomes: comparison with endoplasmic reticulum and mitochondrial membranes. J Cell Biol 93:103–110
17. Wessel D, Flügge UI (1984) A method for the quantitative recovery of protein in dilute solution in the presence of detergents and lipids. Anal Biochem 138:141–143
18. Candiano G, Bruschi M, Musante L, Santucci L, Ghiggeri GM, Carnemolla B et al (2004) Blue silver: a very sensitive colloidal Coomassie G-250 staining for proteome analysis. Electrophoresis 25:1327–1333
19. Applied Biosystems iTRAQ™ Reagents—Chemistry Reference Guide.

Chapter 19

Construction of *Pseudomonas aeruginosa* Two-Hybrid Libraries for High-Throughput Assays

Sophie de Bentzmann and Christophe Bordi

Abstract

In *Pseudomonas aeruginosa*, identification of new partners of a protein of interest could give precious clues to decipher a biological process in which this protein is involved. However, genes encoding for partners of a protein of interest are unknown and frequently scattered throughout the genome. We describe herein the construction and the use of pan-genomic bacterial two-hybrid libraries to identify new partners of a protein of interest encoded by *P. aeruginosa*.

Key words Protein network, Bacterial two-hybrid, High-throughput screening, Protein partners

1 Introduction

Almost all biological processes in bacteria require (at their molecular level) interactions between different protein partners. Deciphering a biological process requires the combination of multiple and complementary approaches. Among techniques dedicated to the study of protein interactions involved in the studied process, is the bacterial two-hybrid system also called BACTH system developed by Karimova and collaborators [1]. This is a simple and rapid technique for detection of protein–protein interaction that occurs either in the cytosol or in the inner membrane. This technique is based on the reconstitution of the calmodulin-dependent adenylate cyclase activity of the toxin CyaA of *Bordetella pertussis* in an *Escherichia coli cya* deficient strain. The *B. pertussis* CyaA protein catalyzes the conversion of ATP into cAMP with the help of its catalytic domain made of two complementary sub-domains named T25 and T18 that are both required for the enzymatic activity [2] (Fig. 1a, b). Fused to proteins that can interact, these two sub-domains T25 and T18 can restore a functional adenylate cyclase activity involved in cAMP production [1] (Fig. 1c, d). This cAMP production can easily be monitored since its production triggers the transcription of several genes

Alain Filloux and Juan-Luis Ramos (eds.), *Pseudomonas Methods and Protocols*, Methods in Molecular Biology, vol. 1149, DOI 10.1007/978-1-4939-0473-0_19, © Springer Science+Business Media New York 2014

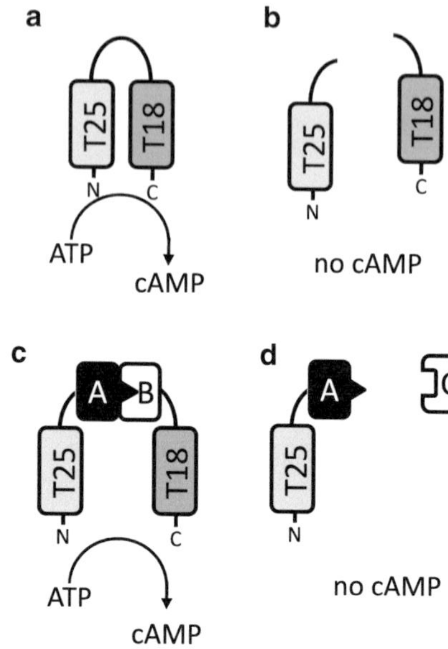

Fig. 1 Scheme of the principle of the bacterial two hybrid system and schematic representation of the CyaA adenylate cyclase activity of *B. pertussis* inspired from [1]. The T25 *box* and T18 *box* represent respectively the T25 (amino acids 1–224) and T18 (amino acids 225–399) subdomains. (**a**) Expression of the full-length catalytic domain (containing T25 and T18 subdomains) of the CyaA toxin generates cAMP from ATP hydrolysis in *E. coli.* (**b**) Co-production of the separated T25 and T18 subdomains in *E. coli* is unable to catalyze production of cAMP. (**c**) Co-production of chimera proteins made of the T25 or T18 subdomains fused with two putative interacting protein partners (A and B), respectively leads to reconstruction of the CyaA catalytic domain able to trigger cAMP production. (**d**) Co-production of chimera proteins made of the T25 or T18 subdomains fused with two non-interacting protein partners (A and C) results in absence of cAMP production

involved in the catabolism of carbohydrates such as lactose or maltose. Indeed, cAMP binds to the transcriptional activator CAP (*c*atabolite *a*ctivator *p*rotein) and this cAMP/CAP complex induces genes involved in the catabolism of lactose or maltose (Fig. 2). Thus, degradation of these carbohydrates can be detected on indicator medium like the MacConkey medium, containing maltose or lactose as the sole carbon source. With this bacterial two-hybrid system, a positive interaction is detected on MacConkey plates by the appearance of colonies with a red color (Fig. 2a), otherwise an absence of interaction leads to white color of colonies (Fig. 2b) (for review *see* [3, 4]).

As partners of regulation cascades in *Pseudomonas aeruginosa* are far from being all identified, and candidates being probably dispersed through the genome, we designed a pan-genomic strategy using the bacterial two-hybrid technique that is described herein. This approach has been applied to the identification of potential partners involved in the HptB/HsbR/HsbA pathway [5].

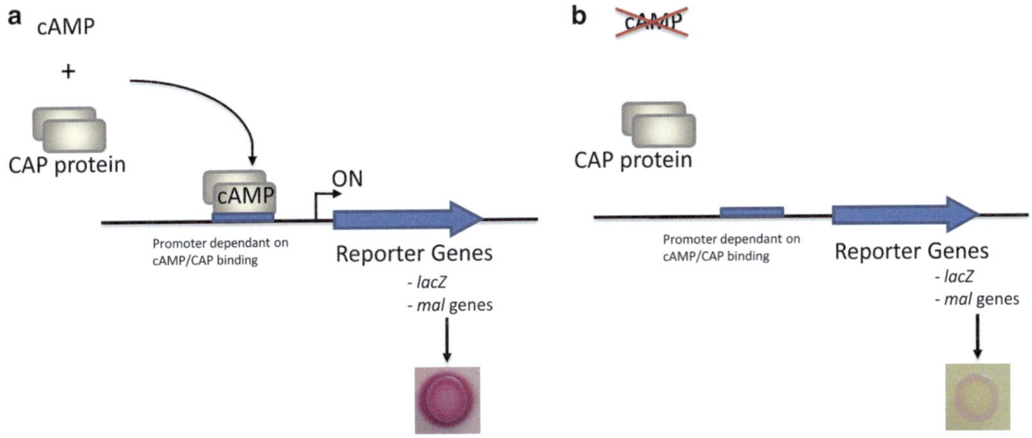

Fig. 2 Effect of cAMP production on the regulation of reporter genes exhibiting a cAMP/CAP dependent promoter. (**a**) Production of cAMP by the CyaA catalytic domain in an *E. coli cya* deficient strain leads to formation of a complex between CAP protein and cAMP (cAMP/CAP) which binds to cAMP/CAP dependent promoters (*blue box*) and induces expression of reporter genes like *lacZ* or *mal* genes. Activation of lactose or maltose catabolism can be detected in indicator medium like the MacConkey medium, containing maltose or lactose by the appearance of colonies with a *red color*. (**b**) Absence of cAMP production prevents formation of the cAMP/CAP complex leading to absence of activation of *lacZ* or *mal* genes. This absence of lactose or maltose catabolism is visualized on MacConkey medium, containing maltose or lactose by the appearance of white colonies

This powerful tool validated by the interaction previously described between HsbR and HsbA proteins, allowed us to demonstrate that the HsbR response regulator dimerizes through its PP2C/ATPase C-terminal effector domain, an observation further confirmed using pull down experiments. Among the other identified preys obtained in the libraries capable of interacting with HsbR and HsbA are interesting putative partners, which involvement in the HptB/HsbR/HsbA signaling pathway requires extensive studies that are in current investigation. This can be transposed easily to other physiological aspects of *P. aeruginosa*.

In this strategy, the DNA sequence encoding for a protein of interest that we call the "bait protein" is fused to the 3′- or 5′-terminal end of the DNA sequence encoding for T25 subdomain, This "bait" containing pKT25 vector is co-transformed in *E. coli cya* deficient strain (BTH101) with a DNA fragment library of *P. aeruginosa* fused to the 5′- or 3′-terminal part of the DNA sequence encoding the T18 sub-domain, thus forming a pUT18-derived prey vector bank. The co-transformants are screened on MacConkey media for identification of red colonies which may contain a pUT18-DNA fragment encoding a putative open reading frame (ORF) encoding a candidate able of interaction with the bait. This interaction remains to be checked with other techniques. This strategy has been used with success to decipher regulatory networks in *P. aeruginosa* [5].

2 Materials

2.1 Agarose Gel Components

1. TAE 50× Solution: dissolve 242.2 g of Tris powder in 100 mL water; add successively 57.1 mL of acetic acid and 18.5 g of EDTA (Ethylene Diamine Tetraacetic Acid). Complete to 1 L with water and adjust the pH to 8.4.
2. Loading dye 6× solution: dissolve 4 g sucrose, 25 mg bromophenol blue, and 25 mg xylene cyanol in 10 mL water. Store at 4 °C to avoid contaminants growing in the sucrose.
3. Weight 1.5 g of agarose powder and add 15 mL of 0.5× TAE solution and then warm the solution at 80 °C until complete dissolution of the agarose powder.
4. Pour the agarose solution in the agarose gel system and let polymerize the gel at least 30 min at room temperature.

2.2 Media, Solutions, and Plates

1. IPTG (Isopropyl-β-D-thiogalactoside) stock at 0.1 M: weight 0.239 g of IPTG in 10 mL of water, sterilize by filtration and keep at −20 °C.
2. Maltose 20 % solution: weight 20 g of maltose in 100 mL of water, sterilize by filtration and keep at +4 °C.
3. LB (Luria–Bertani) broth: dissolve in 1 L of water 10 g of NaCl, 10 g of tryptone, 5 g of yeast extract. Adjust pH to 7 with NaOH and autoclave at 121 °C for 20 min for sterilization. LB plates are prepared by adding 1.5 % Bacto agar (Difco™ ref 214050).
4. MacConkey plates: dissolve 40 g of MacConkey agar base powder (Difco™ ref 281810) in 1 L of water, mix thoroughly, heat at 80 °C with frequent agitation, and boil for 1 min to completely dissolve the powder. Autoclave at 121 °C for 15 min and after sterilization, add IPTG 100 μM, 1 % maltose, ampicillin 100 μg/mL and kanamycin 50 μg/mL. Pour this medium in petri dishes 145 mm × 20 mm for screening and in petri dishes 80 mm × 20 mm for re-streaking analysis.

2.3 Chemical Competent Bacteria

1. $CaCl_2$ 50 mM solution: dissolve 0.277 g of $CaCl_2$ powder in 100 mL water and keep the solution at +4 °C until use.
2. $CaCl_2$ 50 mM with 15 % glycerol: dissolve 0.277 g of $CaCl_2$ powder; add 18.75 mL of 80 % glycerol solution and complete to 100 mL with water and keep the solution at +4 °C until utilization.
3. The day before, inoculate in LB with the appropriate antibiotics (if necessary) the bacteria (DH5α or BTH101 *E.coli* strains) and incubate under agitation at 37 °C.
4. In a 500 mL flask, place 100 mL of fresh LB medium and add 2 mL of the overnight culture. Incubate the flask at 37 °C

under agitation until the $OD_{600\ nm}$ reaches a value between 0.6 and 0.8.

5. Pellet 100 mL of culture, remove the supernatant, and resuspend the pellet in 50 mL of 50 mM $CaCl_2$ solution.
6. Place the tube on ice for 20 min.
7. Pellet bacteria, remove the supernatant, and resuspend the pellet in 5 mL of 50 mM $CaCl_2$ with 15 % glycerol solution.
8. Aliquot by 100 μL the competent bacteria in sterile 1.5 mL eppendorf tubes and store at −80 °C until use.

2.4 Bacteria Transformation

1. Thaw the competent bacteria on ice.
2. Place 50 or 100 μL of competent bacteria in a sterile 1.5 mL eppendorf tube and add the indicated volume or quantity of DNA (ligation reaction or plasmid).
3. Incubate the eppendorf tube during 45 min on ice and successively place this tube at 42 °C in a water bath during 2 min and immediately on ice for 5 min.
4. Add to the reaction 900 μL of fresh LB medium and incubate at 37 °C under agitation (around 200 rpm) during 1 h.
5. Spread the transformed bacteria on the appropriate LB or MacConkey agar plates.

3 Methods

3.1 Generation of Four Independent P. aeruginosa Two-Hybrid Genome Fragment Libraries

The different steps of the workflow for the construction of the libraries are summarized in Fig. 3. Plasmids and strains used are listed in Table 1 and plasmids are presented in Fig. 4.

1. To create a representative and random DNA library, the *P. aeruginosa* chromosome has to be digested with *Sau*3AI. Pipette 25 μL of purified *P. aeruginosa* genomic DNA at the concentration of 100 ng/μL in a 1.5 mL eppendorf tube and add 0.3 U of *Sau*3AI restriction enzyme with the appropriate digestion buffer in a final volume of 30 μL. Incubate the eppendorf tube at 37 °C for 20 min and proceed immediately to the next steps in order to stop rapidly the digestion reaction (*see* **Note 1**).
2. Pour a six well TAE 0.8 % agarose gel and add in the eppendorf tube which contains the 30 μL of the digestion reaction, 5 μL of loading dye X6 and mix well. Load in one well the totality of the digestion reaction mixed with the loading dye and in a second well load a DNA ladder like the GeneRuler™ 1 kb DNA Ladder (Thermo Scientific-Fermentas ref SM0313). Submit to electrophoresis in TAE 0.5× buffer at 100 V during 25 min or until the dye front has reached the bottom of the gel.

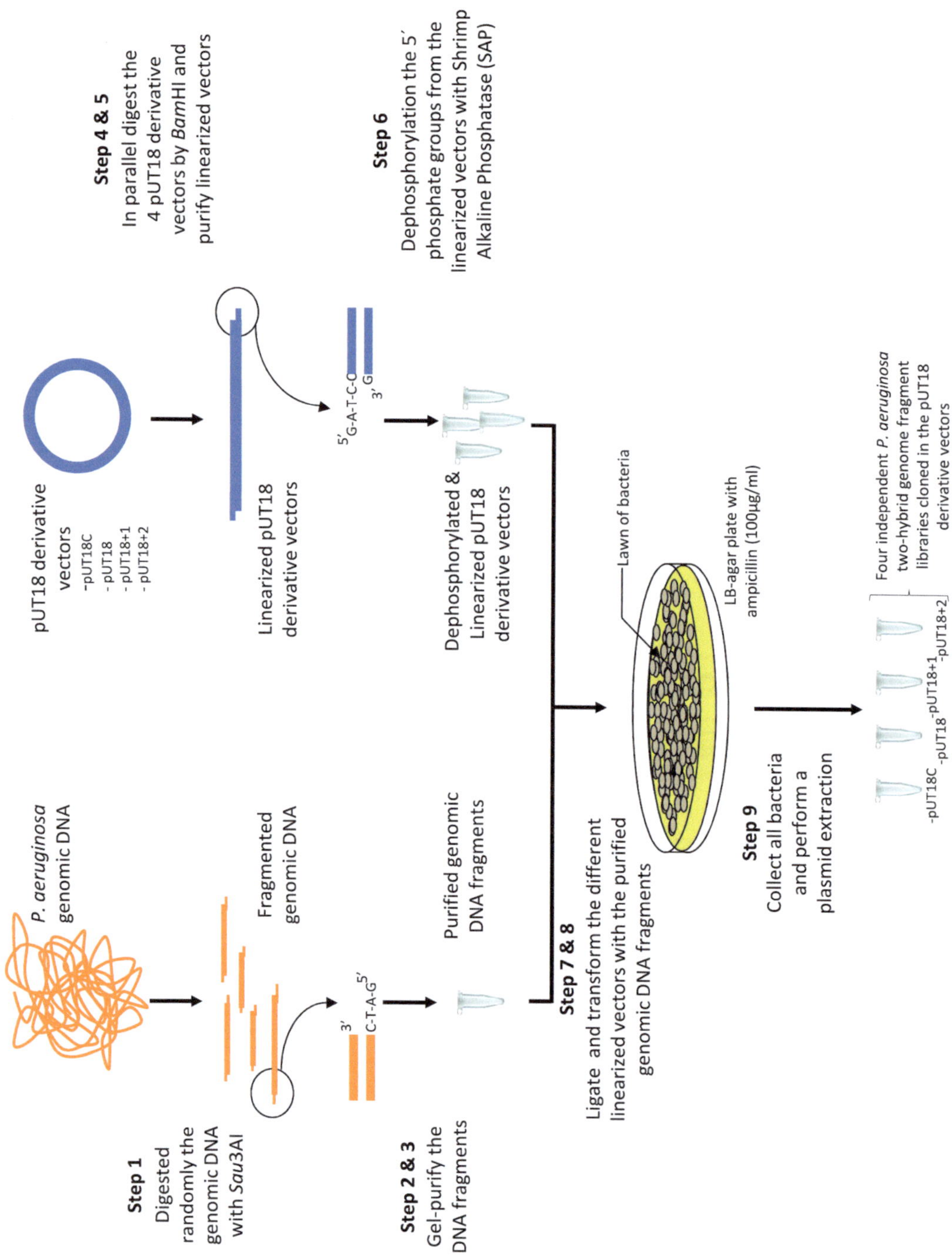

Fig. 3 Schematic representation of the different steps for construction of four independent *P. aeruginosa* two-hybrid genome fragment libraries

Table 1
Strains and plasmids used

Strains/ Plasmids	Relevant characteristics	References
Strains		
BTH101	F- *cya-99 araD139 galE15 galK16 rpsL1* (Str^r^) *hsdR2 mcrA1 mcrB1*	[7]
DH5α	*endA1 hsdR17 supE44 thi-1 recA1 gyrA relA1 Δ (lacZYA-argF) U169*	Invitrogen
	deoR (phi 80lacZ Δ M15)	
Plasmids		
pUT18	Vector encoding the T18 fragment of the CyaA toxin from *B. pertussis*; Amp^R^	[7]
pUT18+1	Modified version of pUT18 with reading frame shifted to +1	[8]
pUT18+2	Modified version of pUT18 with reading frame shifted to +2	[8]
pUT18C	Modified version of pUT18 with the polylinker located at the C-terminal end of T18	[7]
pKT25	Vector encoding the T25 fragment of the CyaA toxin from *B. pertussis cyaA*; Km^R^	[7]
pKNT25	Modified version of pKT25 with the polylinker located at the C-terminal end of T25	[7]

Stain the gel for 10–15 min with an ethidium bromide solution at the concentration of 0.5 μg/mL and visualize DNA fragments under UV illumination. SYBR green or Gel Red dye solutions could be used as an alternative to the solution ethidium bromide.

3. Under UV illumination, cut with a clean scalpel the loaded lane in order to collect the digested DNA fragments ranging in size from 150 to 2,000 bp (*see* **Note 2**). Gel-purify the DNA fragments with the QIAquick Gel Extraction Kit (QIAgen ref 28704) or equivalent and elute in a final volume of 40 μL of water in order to concentrate the digested DNA (*see* **Note 3**). At this step, the digested genomic DNA can be stored at −20 °C for at least 6 months.
4. Add in four independent 1.5 mL eppendorf tubes, 40 μL of the pUT18C, pUT18, pUT18+1, or pUT18+2 derivative vectors at the concentration of 200 ng/μL [7, 8] (*see* **Note 4**). The pUT18 derivative vectors bear the DNA sequence encoding the T18 sub-domain of CyaA with the position of the multiple cloning sites localized at the 3′ or 5′extremities of the T18 sub-domain, respectively (Fig. 4). In each reaction tube, add 1 U of *Bam*HI restriction enzyme with the appropriate

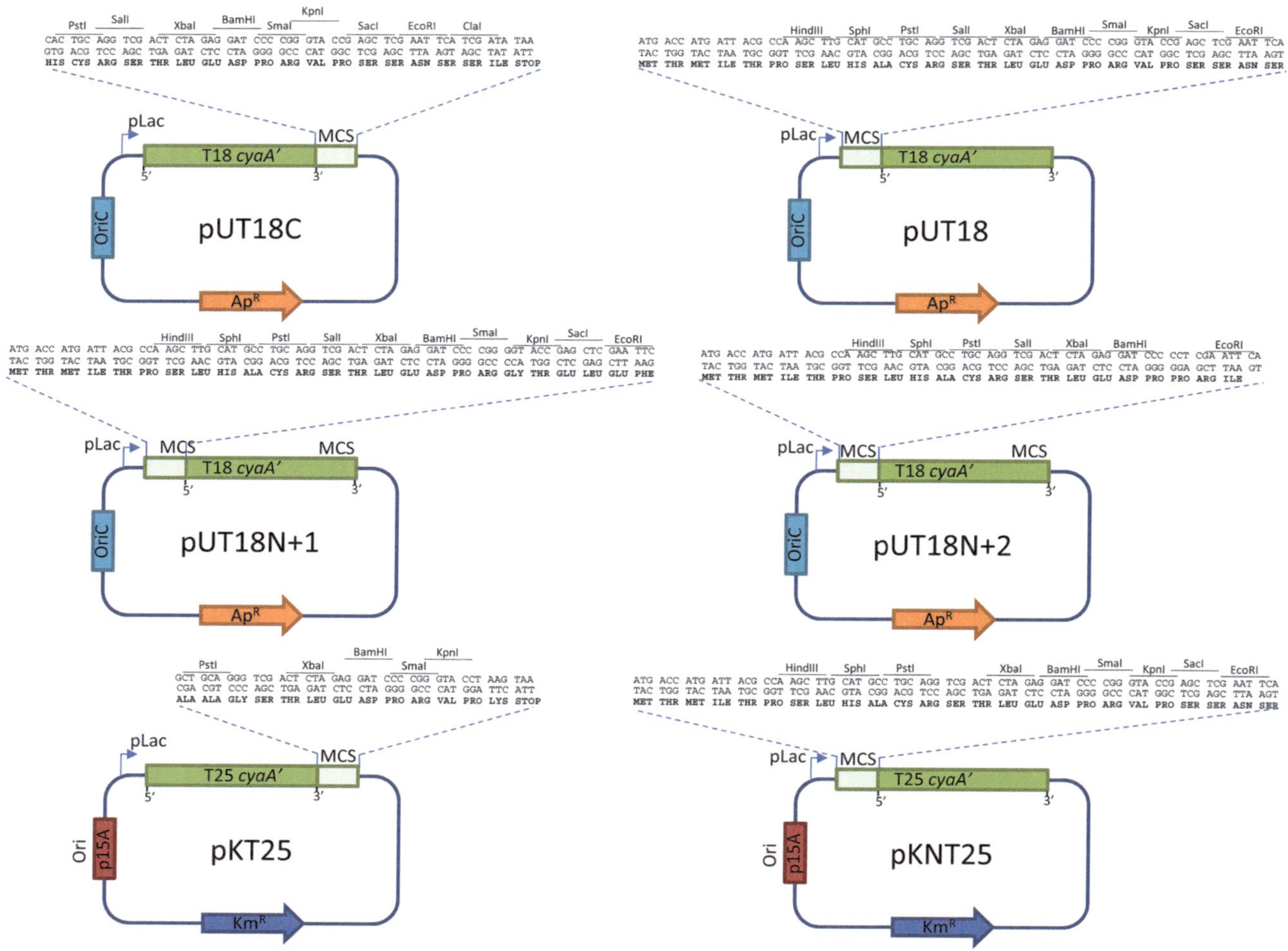

Fig. 4 Schematic representation of pKT25 and pUT18 derivative plasmids. The *open arrows* represent the ampicillin (*Ap*^{R}) or kanamycin (*Km*^{R}) resistance genes. The *green boxes* correspond to the open reading frame of T25-*cyaA′* or T18- *cyaA′* sub-domain with 5′ and 3′ extremities indicated below. The *open boxes* correspond to the multicloning site (MCS) sequences in which only unique sites are mentioned and the origin of replication of the plasmids is indicated by *blue* or *red boxes*

digestion buffer in a final volume of 50 μL. Incubate the reaction tubes at 37 °C for 2.5 h (*see* **Note 5**).

5. Purify the digested pUT18 derivative vectors in order to remove the *Bam*HI enzyme and the digestion buffer by using QIAquick PCR Purification Kit (QIAgen ref 28104) or equivalent and elute in a final volume of 40 μL of water in order to concentrate the linearized vectors. At this step, the linearized vectors can be stored at −20 °C for at least 6 months.
6. Add in four independent 1.5 mL eppendorf tubes, 30 μL of the pUT18C, pUT18, pUT18 + 1, or pUT18 + 2 linearized vectors at the concentration of 150 ng/μL and in each reaction tube, add 1 μL of Shrimp Alkaline Phosphatase (SAP) (Roche ref 11 758 250 00) or equivalent with 4 μL of 10× dephosphorylation buffer in a final volume of 40 μL. Incubate the reaction tubes at 37 °C for 1 h (*see* **Note 6**).

   In order to remove SAP, purify the dephosphorylated pUT18 derivative vectors by using the QIAquick PCR Purification Kit or equivalent.
7. Ligate the different linearized vectors with the resulting mixture of digested genomic DNA obtained at **step 3**. For this, in four independent 0.5 mL eppendorf tubes, put 4 μL of the pUT18C, pUT18, pUT18 + 1 or pUT18 + 2 linearized vectors at the concentration of 100 ng/μL, add 8 μL of the digested DNA at the concentration of 50 ng/μL, add 1.5 μL of T4 DNA Ligase (Roche® ref 11 635 379 001 or equivalent) with its appropriated T4 DNA Ligation Buffer in a final volume of 15 μL and incubate overnight at 16 °C (*see* **Note 7**).
8. Each resulting ligation mixture is independently transformed into *E. coli* DH5α competent cells. For each transformation reaction, spread the bacteria on two or three LB agar plates with ampicillin at 100 μg/mL. Incubate the plates overnight at 37 °C.
9. For each library, collect all bacteria present at the surface of the different plates with a glass spreader and resuspend them in 5 mL of LB medium. Perform a plasmid extraction procedure using The Wizard® *Plus* SV Minipreps DNA Purification System, (Promega ref A7100) or equivalent. At the end of this step, the plasmids collected represent the *Pseudomonas* libraries and can be stored at −20 °C (*see* **Note 8**).

3.2 Control of the Quality of the Bank and Determination of Its Size

Before using the different libraries for any high throughput assay, it is important to evaluate their quality. In particular, attention should be paid to the determination of their size, of the percentage of the genome coverage, and of the percentage of the plasmids of the library that have inserted a fragment of *P. aeruginosa* genomic DNA.

Table 2
Oligonucleotides used

Oligonucleotides	Sequence (5′ → 3′)
Sequencing of prey	
pUT18C	CGGATGTACTGGAAACGGTG
pUT18N	TGCGGAACGGGCGCCGGCGCGAGCG
Sequencing of bait	
pKT25	CGATTCGGTGACCGATTACC
pKNT25	TGTGGAATTGTGAGCGGATA
Library analysis	
Lib_pUT18NUp	TGTGGAATTGTGAGCGGATA
Lib_pUT18NDo	TTTCCACAACAAGTCGATGC
Lib_pUT18CUp	CGCCGGATGTACTGGAAAC
Lib_pUT18CDo	CGGCATCAGAGCAGATTGTA

1. For each library, transform 50 μL of DH5α *E. coli* chemical competent bacteria with 50 ng of each plasmid library. Perform a serial dilution from 10^0 up to 10^{-5}, plate 100 μL of each dilution on LB agar plates supplemented with ampicillin (100 μg/mL) and incubate at 37 °C.
2. To determine the library size, count the number of colonies on each plate for each dilution, multiple this number by the dilution factor and calculate the mean number of colonies present per plate. Multiply this value by 10 to obtain the library size.
3. To determine the percentage of plasmids that have inserted a fragment of *P. aeruginosa* genomic DNA, pick up randomly 100 clones from the dilution plates for each library and perform a colony PCR on each clone using the oligonuleotides Lib_pUT18CUp/Lib_pUT18CDo for the pUT18C derivative and Lib_pUT18NUp/Lib_pUT18NDo for the pUT18N derivatives (Table 2). This will give the percentage of empty plasmids and the mean size of the genomic fragments cloned in the different vectors (*see* **Note 9**).
4. Only one insert out of six will be in frame with the ORF encoding the T18 fragment. The probability of having a particular fragment cloned in the library is therefore calculated using the following formula:

$$P = 1 - \left(1 - \frac{i}{G}\right)^N$$

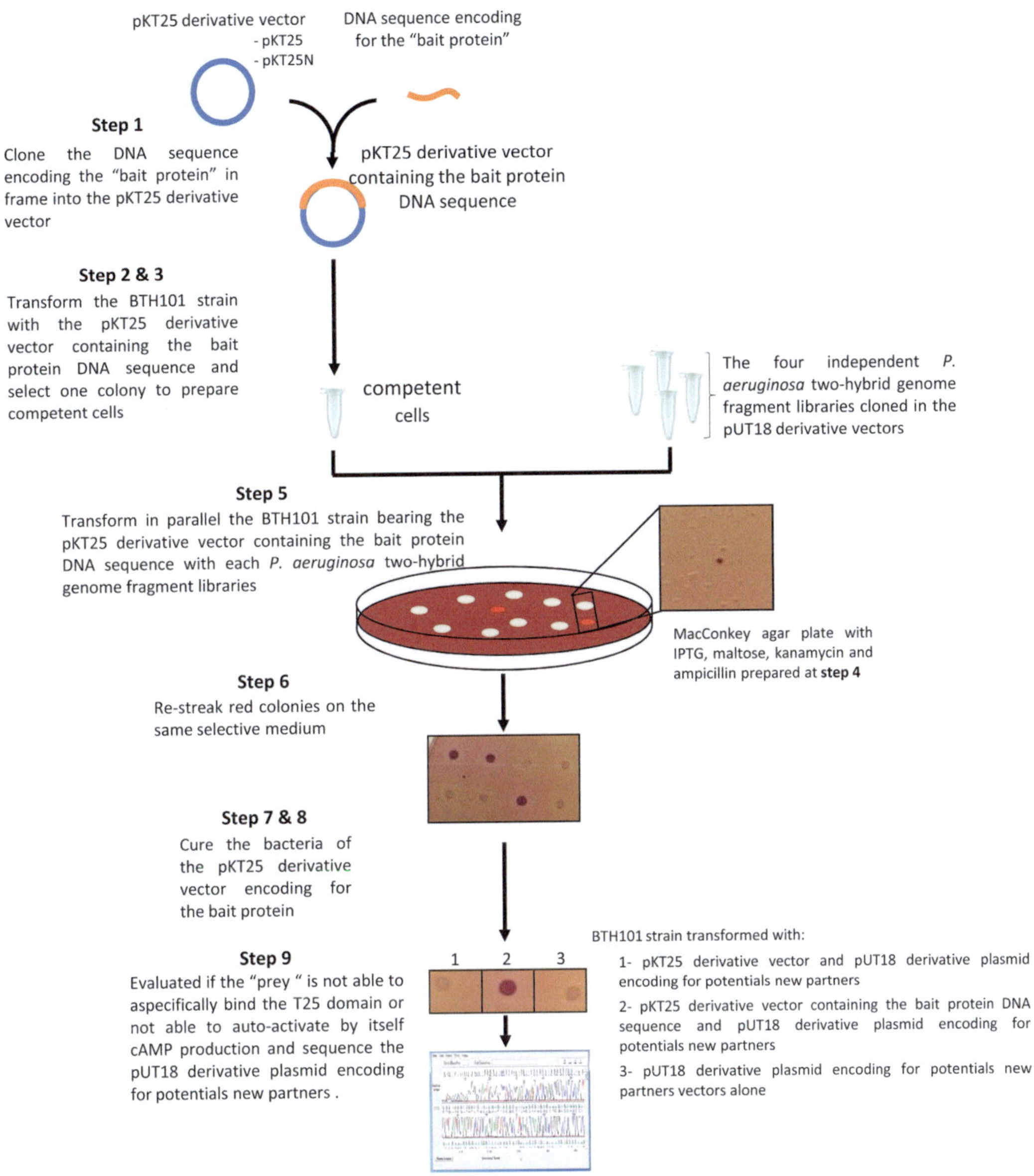

Fig. 5 Utilization of the *P. aeruginosa* two-hybrid libraries for high-throughput assays for identification of partners of a bait protein

Where i is the mean insert size of the genomic fragment cloned, G the genome size and N the number of clones obtained in the library [6]. Further calculation of the probability of having such a fragment inserted in the right orientation and in frame with the *cyaA* T18 fragment is performed by multiplying G by 6.

3.3 Libraries Screening

The different steps of the workflow using the *Pseudomonas* two-hybrid libraries for high-throughput assays are summarized in Fig. 5.

1. Before screening for new partners of your target protein that is called "bait protein" you need to clone the DNA sequence encoding the "bait protein" in frame into the pKT25 or pKNT25 vectors (*see* **Note 10**) and check for absence of mutation by DNA sequencing using respectively the pKT25 or pKNT25 oligonucleotides listed in Table 2.
2. Transform 50 μL of the BTH101 *E. coli* chemically competent bacterial cells (*see* **Note 11**) with 50 ng of the pKT25 derivative vector containing the bait protein DNA sequence obtained at the previous step and select on LB agar plates containing kanamycin at the concentration of 50 μg/mL (*see* **Note 12**).
3. From the LB agar plate where BTH101 strain clones with the pKT25 derivative vector containing DNA sequence of the "bait protein" have grown, select one colony to prepare new competent cells as described in Subheading 3.
4. Pour approximately 40–50 petri dishes 145 mm × 20 mm with the screening medium which contains MacConkey supplemented with 100 μM IPTG, 1 % maltose and the two antibiotics kanamycin (50 μg/mL) and ampicillin (100 μg/mL) (*see* Subheading 2 and *see* **Note 13**).
5. Perform in parallel four independent transformations with 100 μL of the BTH101 competent bacteria harboring the pKT25 derivative vector containing the bait protein sequence and 5–10 ng of each of library constructed in the Subheading 3.1. At the end the transformation reaction, measure the $OD_{600\text{ nm}}$ of each transformation reaction and plate around 10^6 bacteria per plate (145 mm × 20 mm). Incubate the plates for 24 h at 30 °C in the dark.
6. After 24 h, select the red colonies growing and re-streak them on the same selective medium (plates 80 mm × 20 mm). Place these new plates overnight at 30 °C in the dark to confirm the red color appearance (*see* **Note 14**). Leave the plates (plates 145 mm × 20 mm) at room temperature in the dark as well and check every 12 h for appearance of new red colonies during a maximal period of 5 days (*see* **Note 15**). Re-streak each new red colony on the same selective medium and place the plate overnight at 30 °C in the dark to confirm the red color of the colony.
7. If after re-streaking, the colonies are still red, cure the bacteria of the pKT25 derivative vector encoding the bait protein. For this, re-streak several times the red colonies on LB agar plate supplemented only with ampicillin (100 μg/mL) until the colonies growing on these plates are kanamycin sensitive and ampicillin resistant.
8. From each colony cured of the pKT25 derivative plasmid, collect the pUT18 derivative vector containing a DNA sequence

encoding the new partner of your "bait protein" using the QIAprep Spin Miniprep (QIAgen ref 27104) or equivalent.

9. Co-transform the BTH101 competent bacteria with the pKT25 derivative vector containing the DNA sequence encoding the bait protein and with the pUT18 derivative vector isolated at the previous step. In parallel co-transform the BTH101 competent bacteria with the empty pKT25 derivative vector and the pUT18 derivative vector isolated at the previous step (*see* **Note 16**). Plate the transformation reactions on MacConkey medium supplemented with 100 μM IPTG, 1 % maltose and the two antibiotic kanamycin (50 μg/mL) and ampicillin (100 μg/mL) and incubate overnight at 30 °C in the dark. If the colonies containing the pKT25 derivative vector containing the DNA sequence encoding for the bait protein and the pUT18 derivative vector isolated at the previous step are still red, and if the colonies containing the empty pKT25 derivative vector and the pUT18 derivative vector isolated at the previous step are white, perform the DNA sequencing of the pUT18 derivative vector which encodes a new partner of your bait protein using the pUT18C oligonuleotide for the pUT18C derivative and the pUT18N oligonuleotide for the pUT18N derivative (Table 2).

3.4 Validation of Interactions

When you have any positive hit, you need to validate the interaction observed by (1) cloning the sequences of the full-length prey protein and of the bait protein in the pKT25 and pUT18 derivative plasmids and retest the interaction (2) using another method like pull-down experiments or equivalent approaches.

4 Notes

1. It's one of the crucial steps for the construction of the library; the main goal of this step is to cut randomly the genomic DNA of *P. aeruginosa* in order to obtain a collection of DNA fragments with a range size between 150 and 2,000 kb. The quality and purity of the genomic DNA, but also the quality of the *Sau*3AI enzyme used could influence positively or negatively the efficiency of the digestion (that could lead to an insufficient or an excessive digestion of the genomic DNA). Optimization of this step can be achieved by adding or reducing the quantity of *Sau*3AI or DNA concentration in the digestion mix or also by increasing or reducing the duration of the digestion step at 37 °C.
2. Cut the gel as fast as possible, in order to prevent DNA deterioration due to thymine dimer formation. When DNA is damaged in such a way, it cannot be replicated or transcribed and this may reduce dramatically the coverage of the library.

3. At this step, monitor the DNA concentration by using a NanoDrop® system; the DNA fragment concentration after gel purification should not be below a concentration of 50 ng/μL. If the measured concentration is below this value: (1) redo the **steps 1–3** and pool with the previous fragment and/or concentrate the DNA fragment preparation, by using a vacuum concentrator like the SpeedVac® system.
4. It is important to kept in mind that a proportion of the DNA fragments inserted upstream the T18 DNA fragment (containing only an internal part of the ORF without start and stop codons) might never be in-frame with the downstream T18 coding region (statistically 2/3 of them) because only 1/3 of the fragment frame with plasmid translation signals in each pUT18, pUT18+1, or pUT18+2 derivative vectors (*see* Fig. 4). It is also important to consider that another proportion of the DNA fragments cloned upstream the T18 coding region containing the intact 5′ extremity of the ORF (including translation start codon ATG and Shine–Dalgarno) will have a chance to be in frame in one of the pUT18, pUT18+1, or pUT18+2 derivative vectors. Lastly, DNA fragments inserted upstream the T18 DNA and including a part of the ORF with stop codon, will never give a chimeric protein fused to T18 fragment. For these last cases, only 1/3 of the fragments will be on frame with the 3′ terminal part of the T18 DNA in the pUT18C plasmid.
5. Plasmid digestion by *Bam*HI will produce compatible cohesive ends (GATC) with *Sau*3AI, which will permit the correct cloning of the *P. aeruginosa* DNA fragments in the pUT18 derivative linearized vectors.
6. It is the second crucial step for the construction of the libraries; the main goal of this step is to catalyze the dephosphorylation of the 5′ phosphate groups from the linearized vectors with SAP with the aim to prevent self-religation of linearized plasmids. It is important to note that the dephosphorylation of the 5′ phosphate groups from the linearized vectors will influence the efficiency of the ligation step. A too short treatment with the SAP enzyme leads to an important percentage of self-religation of the plasmids. A too long treatment with the SAP enzyme dramatically reduces the insertion of DNA fragments into the linearized plasmids. Optimization of this step can be achieved by increasing or reducing the quantity of SAP enzyme or DNA concentration in the digestion mix or also by modifying the duration of the incubation at 37 °C.
7. During construction of libraries, pay attention to the origin of ligase. The best result has been obtained in our hands with freshly open tube of T4 DNA Ligase provided by Roche®.

8. To ensure a renewal of plasmid libraries transform 100 μL of DH5α competent bacteria with 250 ng of plasmid of each library (one transformation reaction by library) and redo the **steps** 7 and **8**.
9. Each library will generally and roughly contain 25,000 clones, will exhibit a percentage of clones with an insert around 70 % and 750 base pairs as the mean size of the genomic DNA fragment inserted. If a library displays a lower size or if the percentage of clones with an insert is low, the most probable cause is a too high or too low efficiency of SAP. Indeed a too high efficiency of the SAP enzyme generates a limited size of the library with a correct percentage of clones with an insert (≥65 %). At the opposite, a reduced efficiency of SAP results in a library with a large size but with a weak percentage of clones having an insert (*see* **Note 6** to solve this problem). Alternatively, if the library has only 10,000 clones but with a percentage of clones with an insert superior to 65 %, reconstruct a new library and pool the new library with this one.
10. The choice between pKT25 and pKNT25 vectors will condition the position of the T25 module in future bait proteins. In the pKT25 vector, the DNA sequence of the bait protein is fused at the 3′-terminal end of the DNA sequence corresponding to the T25 domain, while in the pKNT25 vector, the DNA sequences of the bait protein is fused at the 5′-terminal end of the DNA sequence corresponding to the T25 domain. For identification of interactions between inner membrane proteins cloning the DNA sequences of the protein bait in the pKNT25 vector is recommended.
11. The *E. coli* BTH101 strain is a non-reverting adenylate cyclase deficient *(cya)* strain and is frequently used in two-hybrid experiments because in this strain even weak interactions between proteins partners can be visualized. However, this strain exhibits some genetic instability due to its Rec^+ character. This instability leads to spontaneous Lac + and Mal + revertants (these revertants exhibit mutations in the *lac* of *mal* promoters leading to their independent cAMP/CAP activation). The frequencies of these spontaneous Lac^+ and Mal^+ revertants are respectively, 10^{-8} and 10^{-9} for BTH101. Before use of the BTH101 strain, it is recommended to re-streak the strain on a plate containing MacConkey supplemented with maltose 1 % and to grow it overnight at 37 °C under the dark. White colonies should be used to generate competent cells. Appearance of any red colony should be avoided, since they likely correspond to Lac^+ or Mal^+ revertants or are contaminant bacteria (note that the BTH101 strain is resistant to streptomycin (100 μg/mL)).
12. At this step, check the production of the chimera protein made of the T25 domain fused to the “bait protein” by western blot

analysis using the polyclonal anti-CyaA antibody from Santa Cruz Biotechnology (Note that this antibody can detect the T25 and T18 portions of CyaA adenylate cyclase). For liquid production of the chimera protein, induce with 100 μM IPTG.

13. The MacConkey medium is sensitive to light; a long exposition to light of the MacConkey plates leads to a bleaching of the medium. After pouring, protect the plates from light. For large scale screening, a M63 minimum medium supplemented with maltose can be used as an alternative to the MacConkey medium. Prepare the medium as follows: add 2 g $(NH_4)_2SO_4$, 13.6 g KH_2PO_4, 0.5 mg $FeSO_4$ and 15 g agar per liter of water (if its required adjust pH to 7.0 with KOH). After autoclaving, add 1 mL of $MgSO_4$, 10 mL of 20 % maltose, 2 mL of 0.05 % vitamin B1, add IPTG 100 μM, ampicillin 100 μg/mL and kanamycin 50 μg/mL.
14. Re-streaking each new red colony on the same selective medium (E.I MacConkey medium) is very important before further analysis. Indeed, some red colonies which are initially red on plate can sometimes turn white after re-streaking. These colonies are false positive and should be discarded.
15. The bacterial two-hydrid method is able to detect strong, weak as well as transient interactions. Protein partners displaying a strong interaction would correspond to clones appearing in the first days, while proteins partners displaying a weak or transient interaction will correspond to clones appearing after 4 or 5 days. However, after 6–7 days, most of the colonies display an irregular morphology and some of them possess red spot at their surface, all of these colonies are false positives. These red spots at the colony surface are probably due to apparition of spontaneous Lac^+ and Mal^+ revertants in the BTH101 strain.
16. The co-transformation of the BTH101 strain with the empty pKT25 derivative plasmid and the pUT18 derivative plasmids encoding potential new partner fused to the DNA sequence of the T18 sub-domain is an essential control to perform to ensure that the prey is not able to bind non specifically to the T25 domain or that the prey is not able to auto-activate *lac* or *mal* operons by self cAMP production.

Acknowledgements

The work of C. B. and S. de B. is supported by the French Cystic Fibrosis Foundation (VLM), and CNRS institutional and ANR grants: ERA-NET ADHRES 27481, PCV-ANR 27628, ANR Jeunes Chercheurs JC09-473544 and GDR3171, as well as Europathogenomics 2005–2010 REX LSHB-CT-2005512061-EPG.

References

1. Karimova G, Pidoux J, Ullmann A et al (1998) A bacterial two-hybrid system based on a reconstituted signal transduction pathway. Proc Natl Acad Sci U S A 95:5752–5756
2. Ladant D, Ullmann A (1999) Bordatella pertussis adenylate cyclase: a toxin with multiple talents. Trends Microbiol 7:172–176
3. Battesti A, Bouveret E (2012) The bacterial two-hybrid system based on adenylate cyclase reconstitution in Escherichia coli. Methods 58(4):325–334
4. Dautin N, Karimova G, Ladant D (2002) Bordetella pertussis adenylate cyclase toxin: a versatile screening tool. Toxicon 40:1383–1387
5. Houot L, Fanni A, de Bentzmann S et al (2012) A bacterial two-hybrid genome fragment library for deciphering regulatory networks of the opportunistic pathogen *Pseudomonas aeruginosa*. Microbiology 158:1964–1971
6. Clarke L, Carbon J (1976) A colony bank containing synthetic Col El hybrid plasmids representative of the entire *E. coli* genome. Cell 9: 91–99
7. Karimova G, Ullmann A, Ladant D (2001) Protein-protein interaction between *Bacillus stearothermophilus* tyrosyl-tRNA synthetase subdomains revealed by a bacterial two-hybrid system. J Mol Microbiol Biotechnol 3: 73–82
8. Handford J, Ize B, Buchanan G et al (2009) Conserved network of proteins essential for bacterial viability. J Bacteriol 191:4732–4749

Part III

Small Molecules and Polysaccharides

Chapter 20

Biosensors for Qualitative and Semiquantitative Analysis of Quorum Sensing Signal Molecules

Matthew Fletcher, Miguel Cámara, David A. Barrett, and Paul Williams

Abstract

Biosensors are biological tools that can be used to assay bacterial cultures for quorum sensing signal molecules (QSSMs) both qualitatively and semiquantitatively. QSSMs can be extracted from *Pseudomonas aeruginosa* cultures using organic solvents and tentatively identified via thin layer chromatography in combination with biosensor overlays. Alternatively, QSSMs can be quantified in spent culture supernatants or solvent extracts using biosensor-based spectrophotometric, luminescence, or fluorescence assays.

Key words *Pseudomonas aeruginosa*, Biosensor, Thin-layer chromatography (TLC), Quorum sensing, *N*-acylhomoserine lactones, *N*-butanoylhomoserine lactone, *N*-(3-oxo-dodecanoyl)homoserine lactone, *Pseudomonas* quinolone signal (PQS), 2-Heptyl-3-hydroxy-4(1*H*)-quinolone, 2-Heptyl-4-hydroxyquinoline (HHQ)

1 Introduction

Pseudomonas aeruginosa produces both *N*-acylhomoserine lactone (AHL) and 2-alkyl-4(1*H*)-quinolone (AQ) quorum sensing (QS) signal molecules (QSSMs) [1]. The primary AHLs are *N*-(3-oxododecanoyl)-L-homoserine lactone (3-oxo-C_{12}-HSL) and *N*-butanoyl-homoserine lactone (C_4-HSL) produced via the *las* and *rhl* systems respectively (Fig. 1a, b). Among AQs, the major QSSMs are 2-heptyl-3-hydroxy-4(1*H*)-quinolone (PQS) and its immediate biosynthetic precursor, 2-heptyl-4-hydroxyquinoline (HHQ) (Fig. 1c, d). While the AHL and AQ-dependent QS systems collectively control the expression of a large number of both common and distinct sets of target genes, they are also interlinked. The *las* system positively regulates both the *rhl* and *pqs* systems while *rhl* negatively controls AQ-dependent signaling. For a review of QS in *P. aeruginosa*, see Williams and Cámara [1].

Given the importance of the AHL and AQ-dependent QS systems to the biology of *P. aeruginosa* and as both systems are potential targets for anti-infective agents that attenuate virulence [1],

Alain Filloux and Juan-Luis Ramos (eds.), *Pseudomonas Methods and Protocols*, Methods in Molecular Biology, vol. 1149, DOI 10.1007/978-1-4939-0473-0_20, © Springer Science+Business Media New York 2014

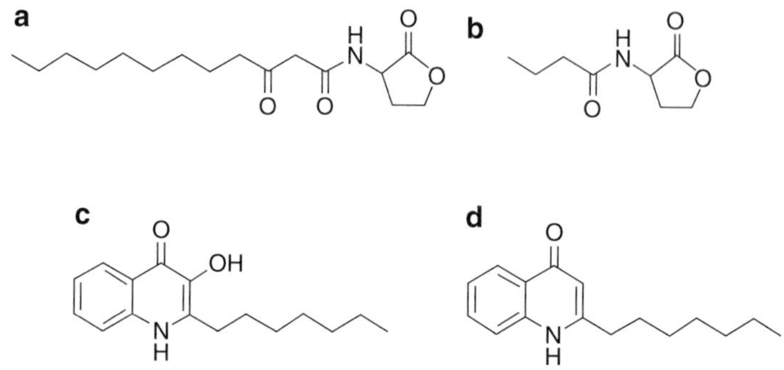

Fig. 1 Structures of the major QSSMs in *P. aeruginosa*. (**a**) *N*-(3-oxododecanoyl)-L-homoserine lactone (3-oxo-C_{12}-HSL); (**b**) *N*-butanoyl homoserine lactone (C_4-HSL); (**c**) 2-heptyl-3-hydroxy-4(1*H*)-quinolone (the *Pseudomonas* quinolone signal; PQS); (**d**) 2-heptyl-4-hydroxyquinoline (HHQ)

a number of methods for AHL and AQ detection and quantification have been developed. These usually employ mass spectrometry (*see* Chapter 20; [2]) or biosensor-based analytical techniques [3–5]. Biosensors developed for both AHL and AQ detection have exploited the QSSM-dependent activation of a response regulator protein to drive the expression of a target promoter fused to a suitable reporter gene(s) [3–5]. Ideally such constructs should (a) be sufficiently sensitive to detect to submicromolar concentrations of a specific QSSM, (b) be highly selective and unambiguous, and (c) possess low background reporter output in the absence of QSSM. The advantages of such biosensors are that that they can be cheap to use, provide rapid results, and require little technical knowledge or expensive equipment. They can usefully provide qualitative data on the presence or absence of a specific QSSM and can also be used to give indirect quantitative data. The disadvantages of biosensors are that they do not always discriminate between closely related QSSMs although this can sometimes usefully be exploited in conjunction with thin layer chromatography (TLC) to profile the range of QSSMs present [3, 5]. In addition they frequently have higher detection thresholds than methods such as mass spectrometry and cannot be used to unequivocally identify a given QSSM as they may be activated by unrelated compounds [6] giving rise to false positive results.

AHL biosensors usually consist of a genetically modified organism engineered to contain a functional LuxR-type protein plus a cognate target promoter (e.g., a *luxI* synthase promoter) which positively drives the transcription of a reporter gene(s) such as *luxCDABE*, *lacZ*, or *gfp* [3, 7]. The outputs from such reporters include bioluminescence, β-galactosidase enzyme activity, fluorescence via green-fluorescent protein, and pigment production. A number of AHL biosensors are available and for an extensive review of these, *see* Steindler and Venturi [7]. For example, for 3-oxo-C_{12}-HSL and C_4-HSL detection, Winson et al. [4] constructed the *Escherichia coli*-based bioreporter plasmids pSB1075 and pSB406 respectively in which LasR and RhlR drive the expression of *lasI'::luxCDABE* or *rhlI'::luxCDABE* in response to the cognate AHL. Similarly, biosensors for the AQs have also been constructed in which the functional *pqsR* (*mvfR*) gene product activates expression of the *pqsA* promoter coupled to *lux* or *lacZ* genes either in a *P. aeruginosa* AQ negative mutant or in *E. coli* [5, 8–10].

Since the same general biosensor methodology can be used for both AHL and AQ detection, their use will be illustrated in this chapter with reference to the AQs. The TLC and liquid assays outlined are essentially those originally described by Fletcher et al. [5, 9] and employ a *P. aeruginosa* biosensor strain rendered AQ negative through an in-frame, nonpolar, chromosomal deletion of *pqsA*, the first gene in the AQ biosynthetic operon and carrying a chromosomally integrated CTX::*pqsA'-luxCDABE* fusion. This biosensor strain produces light and pyocyanin in response to AQs that bind to PqsR and activate the *pqsA* promoter. These include PQS and HHQ as well as their C9 congeners, 2-nonyl-3-hydroxy-4(1*H*)-quinolone (C9-PQS) and 2-nonyl-4-hydroxyquinoline (NHQ) [5].

For qualitative analysis of both AHLs and AQs, cultures of interest are first extracted with an organic solvent such as ethyl acetate or dichloromethane. The QSSMs partition into the organic phase which can be removed and dried by evaporation to concentrate the extracted signal molecules. These can then be subjected to TLC using normal phase silica plates for AQs [5, 9], reversed-phase RP-18F254 plates for C_4-HSL [11], and RP-2F254 reversed-phase plates for 3-oxo-C_{12}-HSL [11] in conjunction with a mobile phase of dichloromethane-methanol (95:5 v/v) or methanol-water (60:40 v/v) or methanol-water (45:55 v/v) respectively. After chromatography the TLC plates are overlaid with a thin agar layer containing the relevant biosensor strain, incubated and examined for reporter output. Alternatively, for indirect semiquantitative analysis, culture supernatant extracts can be mixed with the biosensor strain and reporter output assayed using a spectrophotometer, luminometer, or fluorimeter depending on the nature of the readout. This works well for the AQs and 3-oxo-C_{12}-HSL although the latter may interfere with the quantification of C4-HSL using *E. coli*-based biosensors [12]. In addition it should also be noted that the AQ biosensor will only respond to AQs which function as QSSMs, i.e., those AQs which are capable of activating PqsR. Hence the assay will be an underestimate of the total concentration of all the AQs present since those such as the 2-alkyl-4-hydroxyquinoline-*N*-oxides which are produced at substantial levels are only weakly detected by the biosensor [2, 13].

2 Materials

2.1 Bacterial Cultures

1. PAO1. AQ-positive control.
2. PAO1Δ*pqsA*. AQ-negative control.
3. PAO1Δ*pqsH*. PQS-negative, HHQ-positive control.
4. PAO1Δ*pqsA* CTX::*pqsA′-luxCDABE*. AQ biosensor strain. Can be maintained in tetracycline 125 μg/ml.

All strains can be stored −80 °C in Luria-Bertani (LB) broth containing 25 % (v/v) glycerol, and revived by streaking onto LB agar containing the appropriate antibiotic and incubating at 37 °C overnight.

2.2 Growth Media

1. LB broth: tryptone (Bacto) 10 g/L, yeast extract (Oxoid) 5 g/L, sodium chloride 10 g/L in distilled water.
2. LB agar: as above but also containing technical agar No. 3 (Oxoid), 15 g/L.
3. Soft-top agar: agar technical no. 3 (Oxoid) 6.5 g/L, tryptone (Bacto) 10 g/L, sodium chloride 5 g/L in distilled water.

2.3 Reagents

1. Potassium dihydrogen phosphate solution: 5 % (w/v) KH_2PO_4 in distilled water.
2. Acidified ethyl acetate: 0.01 % (v/v) glacial acetic acid in ethyl acetate.
3. Normal phase 20×20 cm silica 60_{F254} TLC plates (Merck).
4. PQS and HHQ synthetic standards, 10 mM in methanol.
5. Dichloromethane-methanol mixture (95:5 v/v).

3 Methods

3.1 Growth of Bacterial Cultures for AQ Extraction

1. Streak out the test *P. aeruginosa* strain of interest and the control *P. aeruginosa* strains PAO1, PAO1Δ*pqsA*, and PAO1Δ*pqsH* onto fresh LB agar plates and incubate overnight at 37 °C.
2. Inoculate a single colony of each strain into 5 ml LB medium and grow overnight at 37 °C with shaking at 200 RPM.
3. Measure the optical density at 600 nm (OD_{600}) of each overnight culture and standardize the cultures to OD_{600} 1.0 by diluting with fresh LB (*see* **Note 1**).
4. Transfer 0.25 ml of the standardized cultures to 25 ml of LB medium in a 250 ml Erlenmeyer flask (*see* **Note 2**) and incubate at 37 °C with shaking at 200 RPM for 8 h (*see* **Note 3**).

3.2 Extraction of AQs from Bacterial Culture Supernatants

1. Transfer 10 ml of each culture to a 50 ml centrifuge tube and centrifuge at 10,000×*g* for 10 min (*see* **Note 4**).

2. Filter the supernatants through sterile 0.2 μm filters (Minisart) into new centrifuge tubes to remove any unpelleted cells from the extraction mixtures. To use this filtered supernatant for quantitative analysis of total AQ production using a combined spectrophotometer/luminometer, proceed to steps in Subheading 3.5.
3. Add an equal volume of acidified ethyl acetate to the supernatants (10 ml) and vortex vigorously for 30 s to mix the two phases.
4. Allow the two phases to settle, and the tubes can be centrifuged at 10,000 × *g* for 10 min to hasten this process if desired. Transfer the top organic layer to a clean centrifuge tube. Repeat **steps 3** and **4** twice and pool the collected ethyl acetate organic phases for each strain.
5. Evaporate the extraction mixtures to dryness, for example, by transferring each sample to appropriate glassware and using a rotary evaporator or a centrifugal evaporator. The samples can also be allowed to air-dry in a fume hood if time is not an issue.
6. Add 0.5 ml of methanol to the evaporated samples, vortex for 30 s, and transfer to an appropriate receptacle, e.g., 2 ml glass sample vials or Eppendorf tubes. Repeat this step with two further additions of 0.5 ml methanol and pool the collected methanol for each strain.
7. Evaporate the extraction mixtures to dryness, e.g., under a stream of nitrogen gas or in a centrifugal evaporator. The samples can be stored at −20 °C until required before resuspension in methanol. To use these concentrated extracts for quantitative analysis of total AQ production using a combined spectrophotometer/luminometer, proceed to steps in Subheading 3.5.

3.3 Analysis of AQ Production via TLC

1. Preparation of TLC plates. Soak normal phase silica 20 × 20 cm 60_{F254} TLC plates (Merck) in a 5 % (w/v) solution of KH_2PO_4 for 30 min. Bake the plates at high temperature (70–100 °C) for 1 h, e.g., in a hybridization oven. The TLC plates can be stored for several weeks if kept clean and dry.
2. Draw a faint pencil line around 4 cm from the bottom of the activated silica TLC plates as a guide for spotting sample extracts. Spot 2 μl of each of the 10 mM standard solutions of synthetic PQS and HHQ in methanol onto the TLC plate. Extracted samples can be reconstituted in 50 μl of methanol and 5 μl of each can be spotted onto the TLC plate (*see* **Note 5**). Space each spot sufficiently far apart along the pencil line, beginning at least 2 cm from the edge of the TLC plate. Allow the spots to dry.
3. Place the spotted TLC plate into a developing tank and run using a mixture of dichloromethane-methanol (95:5 v/v) as the mobile phase. Remove the TLC plate when the solvent

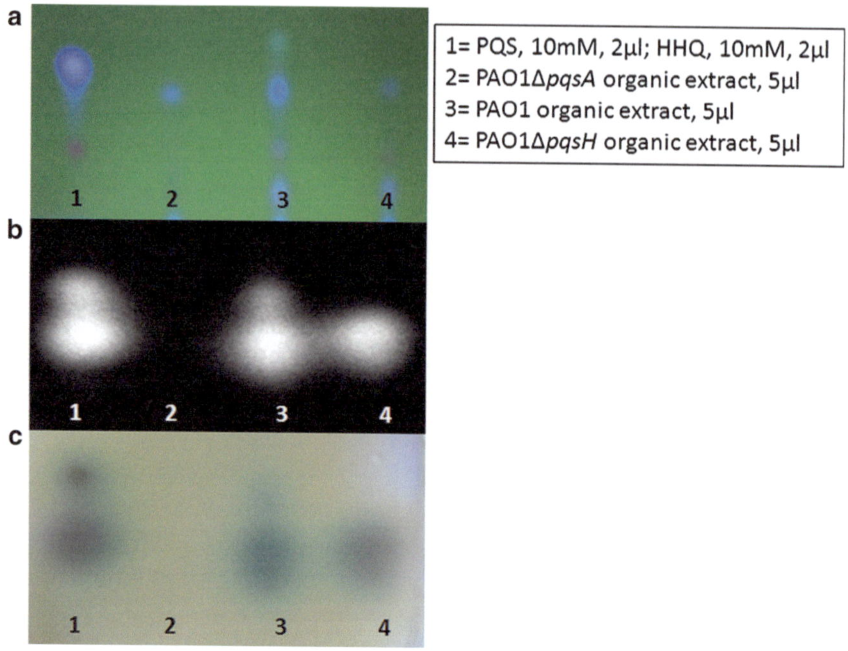

Fig. 2 TLC analysis of AQ production. (**a**) TLC plate showing PQS and HHQ standards together with organic solvent supernatant extracts of PAO1, PAO1Δ*pqsA* and PAO1Δ*pqsH*, visualized under UV light at 312 nm. (**b**) Overlay of the above TLC plate with AQ biosensor bacteria showing production of light in response to AQs, visualized using a luminograph photon camera; *upper spot*, PQS, *lower spot* HHQ. (**c**) Overlay of TLC plate with AQ biosensor bacteria showing production of the green pigment pyocyanin in response to AQs. *TLC lanes*: (1) PQS 10 mM, 2 µl and HHQ 10 mM, 2 µl, (2) PAO1Δ*pqsA* organic extract, 5 µl, (3) PAO1Δ*pqsH* organic extract, 5 µl, (4) PAO1 organic extract, 5 µl. The biosensor produces light and pyocyanin in response to PQS in the PAO1 wild type and to HHQ in the PAO1 wild type and the PAO1Δ*pqsH* mutant. The PAO1Δ*pqsA* mutant does not produce AQs so the biosensor is not activated

front approaches the top of the plate. The TLC plate can be photographed at this point, using a UV transilluminator at 312 nm to visualize the AQs (Fig. 2a).

4. Allow the TLC plate to dry and prepare for overlay of the biosensor by sticking autoclave tape to the underside edges of the TLC plate and then folding upward toward the silica surface so that the tape creates a well around the TLC plate. Place the TLC plate into an appropriate receptacle. This procedure is shown in Fig. 3 (*see* **Note 6**).

3.4 Overlay of TLC Plates with the Biosensor Strain

1. Streak out a 10 µl loop of the AQ biosensor PAO1Δ*pqsA* CTX::*pqsA′-luxCDABE* onto fresh LB agar plates containing 125 µg/ml tetracycline and grow overnight at 37 °C.
2. Inoculate a single colony of biosensor into 5 ml LB medium containing 125 µg/ml tetracycline and grow overnight at 37 °C with shaking at 200 RPM.
3. Melt 100 ml of soft top agar in a microwave and allow to cool to around 50 °C. Add 1 ml of the overnight culture of biosensor

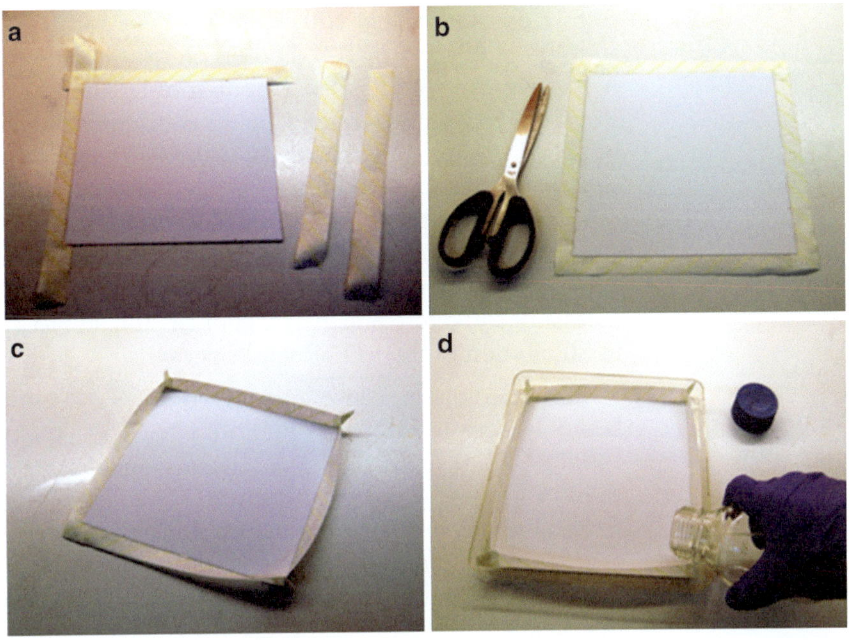

Fig. 3 Preparation of the TLC plate for biosensor overlay. (**a**) Attach autoclave tape to the underside edges of the TLC plate. (**b**) Trim the excess autoclave tape. (**c**) Pinch the autoclave tape at each corner of the TLC plate and then fold the tape upward toward the silica surface so that the tape creates a well around the TLC plate (**d**) Place the TLC plate into an appropriate container and slowly pour the molten soft top agar containing the biosensor bacteria onto the TLC plate

to the soft top agar and mix gently to avoid bubbles in the agar (*see* **Note 7**).

4. Pour the mixture slowly onto the TLC plate to minimize bubble formation. Make sure the agar is evenly spread (*see* **Note 8**).
5. Allow the agar to cool and set aseptically around a Bunsen flame before static incubation at 37 °C for 6–8 h.
6. Visualize the plates for light production using a luminograph photon camera (Fig. 2b) or overlay with X-ray film (e.g., Pierce CL-XPosure film). Alternatively, simply view production of the blue/green phenazine pigment pyocyanin by eye (Fig. 2c).

3.5 Analysis of AQ Production Using a Combined Spectrophotometer/Luminometer Assay

1. Grow the AQ biosensor overnight as described in Subheading 3.4, measure the OD_{600}, and adjust with fresh LB medium to OD_{600} 1.0.
2. Further dilute this standardized biosensor culture with LB medium to give both 1 in 50 and 1 in 100 dilutions.
3. Sterilize a black, transparent-bottomed, 96-well plate, e.g., under strong UV light for 15 min.
4. For each test well, mix 100 μl of the test bacterial supernatant with 100 μl of the 1 in 50 dilution of the biosensor to give a final

culture dilution of 1 in 100 in the well. For a negative control, add 200 μl of the 1 in 100 dilution of the AQ biosensor alone. A positive control of a 1 in 100 dilution of the biosensor plus PQS or HHQ synthetic standard at a concentration of 5 μM can be added to wells, or alternatively 100 μl of *P. aeruginosa* culture supernatant plus 100 μl of the 1 in 50 dilution of the biosensor can also be added to the assay (*see* **Note 9**). The organic solvent culture extracts described earlier in Subheading 3.2 may also be analyzed by this method. Reconstitute the samples in 50 μl of methanol and dilute 5 μl in 100 μl of LB and add to 100 μl of 1 in 50 dilution of the AQ biosensor per well.

5. Monitor bioluminescence and OD_{600} at 37 °C using a combined spectrophotometer/luminometer, for example, the Infinite F200 controlled by the i-Control 1.9 Software (TECAN). The program measures OD and bioluminescence from all wells every 30 min for 24 h. Luminescence is recorded as relative light units (RLU) per unit of OD (Fig. 4). If an automated combined spectrophotometer/luminometer is unavailable, readings can be taken manually at defined time points by growing the bacterial cultures under specific conditions and measuring the OD and bioluminescence of culture samples using a spectrophotometer and tube luminometer (e.g., EG & G Junior), respectively.

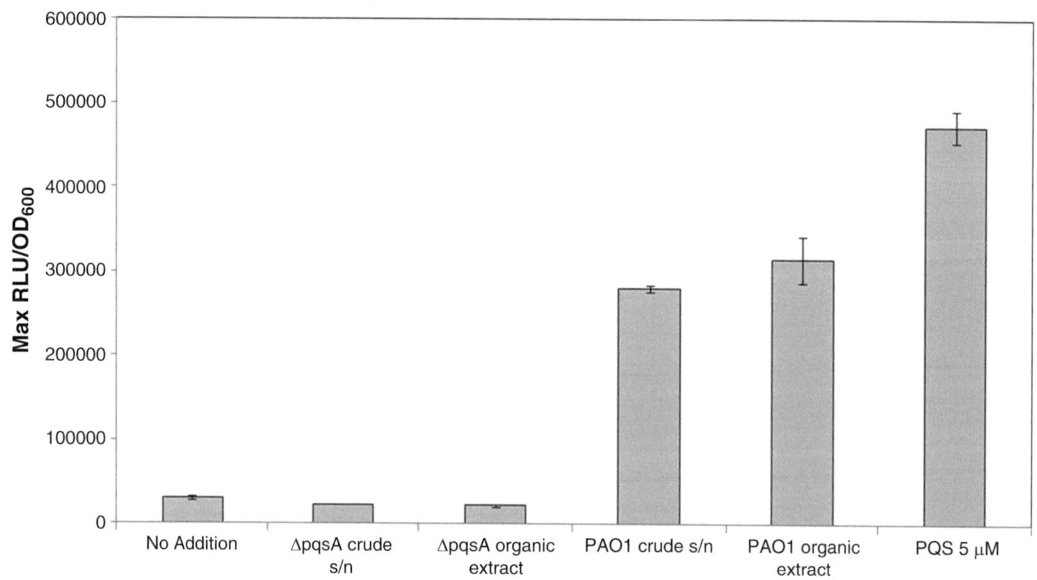

Fig. 4 Quantification of AQ production. Luminescence produced by the AQ biosensor, recorded as relative light units (RLU) per unit of OD by the Infinite F200 (TECAN), in response to the addition of crude supernatants and concentrated organic solvent extracts of PAO1 and PAO1Δ*pqsA*. In the absence of AQs, in the presence of crude supernatant or an organic solvent extract from the PAO1Δ*pqsA* mutant, the output of light is baseline. The AQ biosensor produces light in response to AQs present in the PAO1 wild-type supernatant and organic solvent extracts or to the PQS standard (5 μM)

4 Notes

1. This allows cultures to be compared with one another for AQ production by standardizing the initial inoculum. If direct comparison is not needed, this step can be omitted.
2. Growing the cultures in a small amount of medium in a large flask promotes good aeration, growth, and QSSM production.
3. The actual amount of time needed for growth depends on the growth characteristics of the strain but should be enough to provide the bacteria the opportunity to reach a high enough OD_{600} and produce AQs in sufficient quantity. A 1 in 100 dilution of OD_{600} 1.0 culture is inoculated here, but this can also be altered if necessary. For PAO1, a 10 ml extraction of culture after 8 h, grown to an approximate OD_{600} of 1.0–1.5, should produce enough AQs to be detected by the bioreporter. However, for other strains the exact amount of growth to gather enough AQs to be detected can vary and may have to be determined empirically.
4. The amount of culture to be extracted will depend on the purpose of the experiment and the AQ-producing characteristics of the strain. For those strains that do not produce a high quantity of AQs, more culture may have to be extracted. Accordingly, the specifics of the above method may have to be adjusted depending on the quantity of culture supernatant to be extracted.
5. The quantity of extract to be spotted can be altered as desired, depending on the quantity of AQs in the sample.
6. Make sure the autoclave tape is of decent quality and is pressed down firmly to form a tight seal around the bottom of the plate; otherwise leakage of agar may occur when the plate is overlaid with the biosensor.
7. Make sure that the agar has cooled sufficiently before adding the biosensor bacteria as too high a temperature will kill or harm the biosensor and attenuate growth. A temperature of 50 °C allows easy pouring of agar and gives sufficient time before the agar sets.
8. Do not delay too long or the agar will begin to solidify before pouring. If bubbles form, these can either be agitated to the edge of the plate before the agar sets or a Bunsen burner can be used to quickly and carefully flame the surface of the agar to remove the bubbles.
9. Test and control wells are usually performed in triplicate.

References

1. Williams P, Cámara M (2009) Quorum sensing and environmental adaptation in *Pseudomonas aeruginosa*: a tale of regulatory networks and multifunctional signal molecules. Curr Opin Microbiol 12:182–191
2. Ortori CA, Dubern JF, Chhabra SR, Cámara M, Hardie K, Williams P, Barrett DA (2011) Simultaneous quantitative profiling of *N-acyl-L*-homoserine lactone and 2-alkyl-4(1*H*)-quinolone families of quorum-sensing signalling molecules using LC-MS/MS. Anal Bioanal Chem 399:839–850
3. McClean KH, Winson MK, Fish L, Taylor A, Chhabra SR, Camara M, Daykin M, Lamb JH, Swift S, Bycroft BW, Stewart GS, Williams P (1997) Quorum sensing and *Chromobacterium violaceum*: exploitation of violacein production and inhibition for the detection of *N*-acylhomoserine lactones. Microbiology 143: 3703–3711
4. Winson MK, Swift S, Fish L, Throup JP, Jorgensen F, Chhabra SR, Bycroft BW, Williams P, Stewart GSAB (1998) Construction and analysis of *luxCDABE*-based plasmid sensors for investigating *N*-acyl homoserine lactone mediated quorum sensing. FEMS Microbiol Lett 163:185–192
5. Fletcher MP, Diggle SP, Crusz SA, Chhabra SR, Cámara M, Williams P (2007) A dual biosensor for 2-alkyl-4-quinolone quorum-sensing signal molecules. Environ Microbiol 9: 2683–2693
6. Holden MTG, Chhabra SR, de Nys R, Stead P, Bainton NJ, Hill PJ, Manefield M, Kumar N, Labatte M, England D, Rice S, Givskov M, Salmond GPC, Stewart GSAB, Bycroft BW, Kjelleberg S, Williams P (1999) Quorum sensing cross-talk: isolation and chemical characterization of cyclic dipeptides from *Pseudomonas aeruginosa* and other Gram-negative bacteria. Mol Microbiol 33:1254–1266
7. Steindler L, Venturi V (2007) Detection of quorum-sensing *N*-acyl homoserine lactone signal molecules by bacterial biosensors. FEMS Microbiol Lett 266:1–9
8. Müller C, Fetzner S (2012) A *Pseudomonas putida* bioreporter for the detection of enzymes active on 2-alkyl-4(1*H*)-quinolone signalling molecules. Appl Microbiol Biotechnol. doi:10.1007/s00253-012-4236-4
9. Fletcher MP, Diggle SP, Cámara M, Williams P (2007) Biosensor-based assays for PQS, HHQ and related 2-alkyl-4-quinolone quorum sensing signal molecules. Nat Protoc 2: 1254–1262
10. Hodgkinson J, Bowden SD, Galloway WRJD, Spring DR, Welch M (2010) Structure-activity analysis of the *Pseudomonas* quinolone signal molecule. J Bacteriol 192:3833–3837
11. Yates EA, Philipp B, Buckley C, Atkinson S, Chhabra SR, Sockett RE, Goldner M, Dessaux Y, Cámara M, Smith H, Williams P (2002) *N*-Acylhomoserine lactones undergo lactonolysis in a pH-, temperature-, and acyl chain length-dependent manner during growth of *Yersinia pseudotuberculosis* and *Pseudomonas aeruginosa*. Infect Immun 70:5635–5646
12. Pesci EC, Pearson JP, Seed PC, Iglewski BH (1997) Regulation of *las* and *rhl* quorum sensing in *Pseudomonas aeruginosa*. J Bacteriol 179:3127–3132
13. Déziel E, Lépine F, Milot S, He J, Mindrinos MN, Tompkins RG, Rahme LG (2004) Analysis of *Pseudomonas aeruginosa* 4-hydroxy-2-alkylquinolines (HAQs) reveals a role for 4-hydroxy-2-heptylquinoline in cell-to-cell communication. Proc Natl Acad Sci U S A 101: 1339–1344

Chapter 21

LC-MS/MS Quantitative Analysis of Quorum Sensing Signal Molecules

Catharine A. Ortori, Nigel Halliday, Miguel Cámara, Paul Williams, and David A. Barrett

Abstract

Extracts taken from spent growth media from *Pseudomonas aeruginosa* can be analyzed for *N*-acyl-L-homoserine lactones and 2-alkyl-4-(1*H*)-quinolones (AQs), including the known quorum sensing signalling molecules of *P. aeruginosa*, in a specific and sensitive manner by liquid chromatography coupled with tandem mass spectrometric detection. This analysis can be conducted in a quantitative manner by comparison with matrix-matched calibration samples.

Key words Quorum sensing, *N*-acylhomoserine lactones, 2-alkyl-4-quinolones, LC-MS/MS, Multiple reaction monitoring, Quantification, Bioanalytical methods, *Pseudomonas aeruginosa*

1 Introduction

In *Pseudomonas aeruginosa*, quorum sensing depends on two *N*-acyl-L-homoserine lactone (AHL) regulatory circuits (*las* and *rhl*) closely linked to a 2-alkyl-4-(1*H*)-quinolone (AQ) system. In the *las* system, the *lasI* gene product mainly directs the synthesis of *N*-(3-oxododecanoyl)-L-homoserine lactone (3-oxo-C_{12}-HSL) (Fig. 1), which interacts with the transcriptional regulator LasR to control target promoters [1, 2]. In the *rhl* system, RhlI directs the synthesis of *N*-butanoyl-L-homoserine lactone (C_4-HSL) (Fig. 1), which interacts with the cognate regulator RhlR and switches target gene expression. Apart from C_4-HSL and 3-oxo-C12-HSL, RhlI and LasI are also known to be responsible for the synthesis of additional AHLs, at much lower levels, including C_6-HSL [3] and 3-oxo-C_8-HSL, 3-oxo-C_{10}-HSL and 3-oxo-C_{14}-HSL, respectively [4]. Two major AQ signal molecules are released by *P. aeruginosa*, the

Catharine A. Ortori and Nigel Halliday are joint first authors.

Alain Filloux and Juan-Luis Ramos (eds.), *Pseudomonas Methods and Protocols*, Methods in Molecular Biology, vol. 1149, DOI 10.1007/978-1-4939-0473-0_21, © Springer Science+Business Media New York 2014

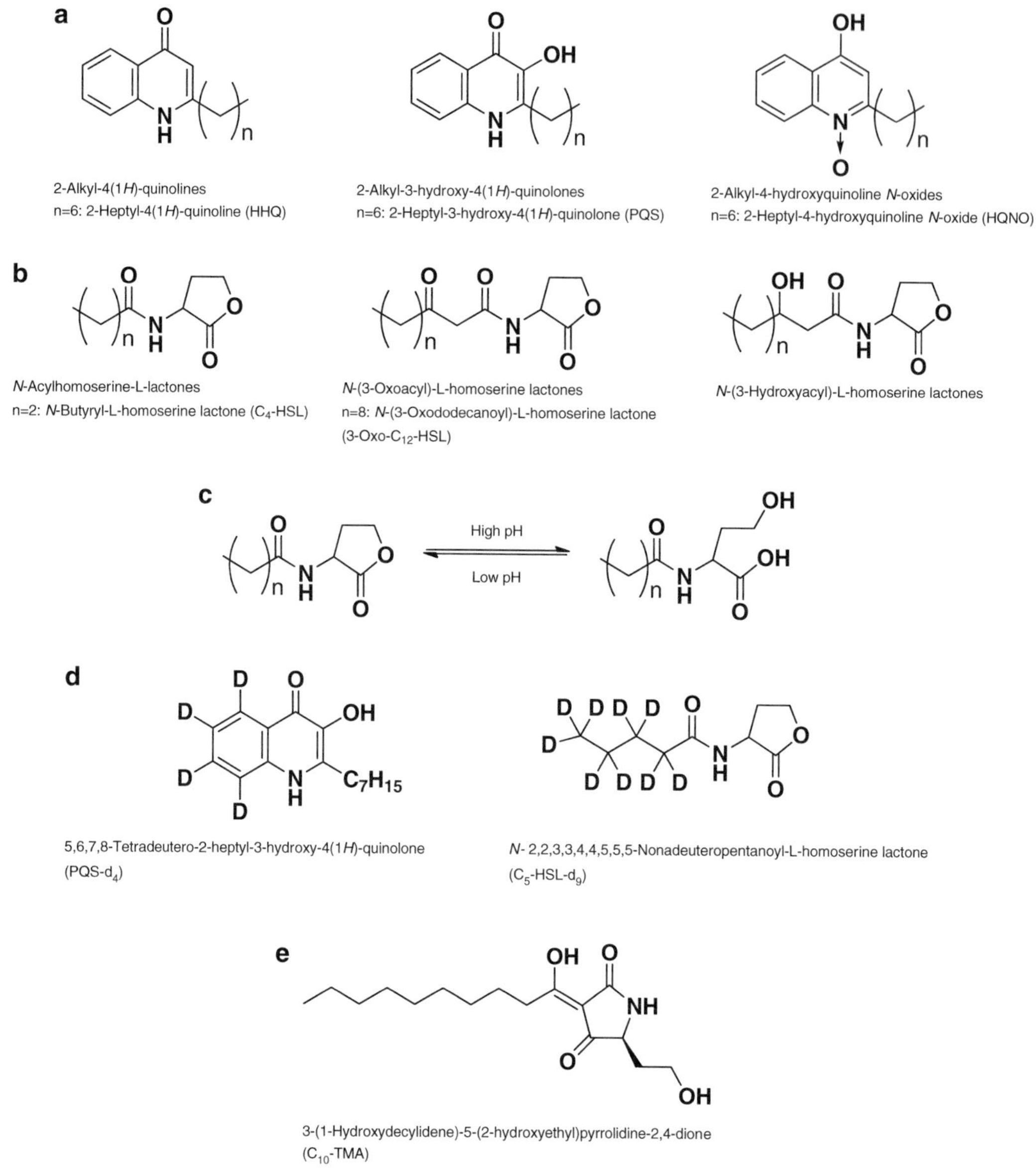

Fig. 1 Molecular structures: (**a**) 2-alkyl-4(1*H*)-quinolones, 2-alkyl-3-hydroxy-4(1*H*)-quinolones, and 2-alkyl-4-hydroxyquinoline *N*-oxides, highlighting the three main AQs produced by *P. aeruginosa* (HHQ, PQS, and HQNO). (**b**) Unsubstituted, 3-oxo and 3-hydroxy *N*-acylhomoserine lactones, highlighting the two main AHLs produced by *P. aeruginosa* (C4-HSL and 3-oxo-C12-HSL). (**c**) The lactone (cyclic ester) functionality of AHLs is hydrolyzed at high pH resulting in a ring-opened structure. (**d**) The deuterated internal standards spiked into all samples prior to extraction. (**e**) The tetramic acid that is produced by *P. aeruginosa* (C_{10}-TMA)

Pseudomonas quinolone signal (PQS; 2-heptyl-3-hydroxy-4(1*H*)-quinolone) and its immediate biosynthetic precursor, 2-heptyl-4-hydroxyquinoline (HHQ) (Fig. 1) [5]. In addition, *P. aeruginosa* releases a multitude of AQs belonging to structurally related families characterized by the presence of a hydrogen or a hydroxyl group at the 3 position, variations in length, and the degree of saturation of the 2-alkyl side chain as well as *N*-oxide derivatives of these variants [6, 7].

The analysis of AQs and AHLs in bacterial culture has traditionally used fast screening approaches by thin-layer chromatography (TLC) or by biological assay with a reporter (e.g., luminescent) bacteria biosensor system (*see* Chapter 19). However, these simple methods do not separate some individual quorum sensing signal molecules (QSSMs), can give rise to false-negative/positive results, and are often not sufficiently sensitive to detect low concentrations of QSSMs in growth media and other complex biological matrices such as biofluids or tissues. In addition, it is difficult to determine the exact chemical identity and concentrations of the different QSSMs produced by bacteria using a TLC or a bioassay. Quantification of QSSMs present in bacterial culture supernatants requires fast and selective separation of the structurally related AQ and AHL QSSMs using liquid chromatography linked directly with specific detection using tandem mass spectrometry (LC-MS/MS). Although several LC-MS/MS methods have been developed to measure either AHLs or AQs, few can simultaneously quantify the full range of both AHL and AQ classes of QSSMs in bacterial cultures.

For the purpose of acquiring quantitative data with high specificity and sensitivity it is necessary to use the tandem mass spectrometer in multiple reaction monitoring (MRM) mode. This mode allows the full power of the tandem mass spectrometer to select a specific ion representative of a specific AQ or AHL molecule, to break down this molecule under controlled conditions in a collision cell, and to detect and identify the resulting diagnostic fragment molecules. Modern LC-MS/MS systems allow for extremely fast scanning and can permit the simultaneous and quantitative monitoring of dozens of individual compounds. The choice of positive ionization electrospray (+ESI) MS detection mode was based upon a greater sensitivity for both AHL and AQ analytes and was relatively straightforward since all the analytes demonstrated protonated $[M+H]^+$ species as the dominant pseudo-molecular ions. The use of synthetic calibration standards is essential to provide accurate quantification using this methodology.

2 Materials

2.1 Bacterial Culture

1. *P. aeruginosa*: The method described here uses the wild-type PAO1 strain. The strain is stored as a cryo-preserved sample at −80 °C in LB broth containing 10 % (v/v) glycerol and revived by streaking onto LB agar (1.5 % w/v Oxoid technical agar No. 3 in LB) and incubating overnight at 37 °C.

2. Luria Bertani (LB) broth: Tryptone (Bacto) 10 g/L, yeast extract (Oxoid) 5 g/L, and sodium chloride 10 g/L in distilled water.
3. PAO1 growth is carried out with 50 mL volumes of LB broth in 250 mL conical flasks in a shaker-incubator at 37 °C (*see* **Note 1**).

2.2 Sample Preparation

1. Acidified ethyl acetate: 0.01 % (v/v) acetic acid (MS grade) in ethyl acetate (HPLC grade) (*see* **Note 2**).
2. Synthetic standards: A small selection of commonly used AHLs and AQs are commercially available (Sigma-Aldrich, Cayman Chemical). Otherwise, standards are synthesized in-house [8–11] (*see* **Note 3**).
3. Synthetic standard stock solutions: Prepare each individual AHL, AQ, and internal standard (IS) as a 1.0 mg/mL solution in acetonitrile and store at −20 °C (*see* **Note 4**).
4. IS mix: Combine and dilute synthetic standard stock solutions of C_5-HSL-d_9 and PQS-d_4 to give a 10 μM solution of both deuterium-labelled internal standards [12] in acetonitrile which is stored at −20 °C.

2.3 LC-MS/MS Analysis

1. Mobile phase A: 0.1 % (v/v) formic acid (MS grade) in ultra-pure water (> 18 MΩ/cm) from an Elga Maxima purification system (Elga, High Wycombe, UK), vacuum filtered through a 0.45 μm cellulose acetate filter membrane (Whatman, Maidstone, UK) (*see* **Notes 5** and **6**).
2. Mobile phase B: 0.1 % (v/v) formic acid (MS grade) in acetonitrile (HPLC grade), vacuum filtered through a 0.45 μm nylon filter membrane (Whatman, Maidstone, UK).
3. HPLC system: Shimadzu series 10 AD VP (Columbia, MD, USA) equipped with binary pumps, a vacuum degasser, a SIL-HTc autosampler, and column oven used with a post-column flow divert valve.
4. HPLC Column: Gemini reversed-phase C18, 150×2 mm, 5 μm particle size column (Phenomenex, Macclesfield, UK).
5. Mass spectrometer: 4000 QTRAP hybrid triple-quadrupole linear ion trap mass spectrometer (Applied Biosystem, Foster City, CA, USA), equipped with a Turbo-Ion source and used in positive ion electrospray mode. Nitrogen is used as both desolvation and collision gas.

3 Methods

3.1 Bacterial Cell Culture

1. Pick a single colony of PAO1 from a freshly streaked LB agar plate and use to inoculate 2 mL of LB which is then incubated overnight at 37 °C with shaking at 200 rpm.

2. Use 5 μL of the overnight culture to inoculate 50 mL of fresh LB medium which is then incubated at 37 °C, shaking at 200 rpm.
3. When samples need to be taken, remove approximately 1.5 mL of cell culture from the flasks and place in a 2 mL Eppendorf tube (*see* **Note 7**).
4. Remove the cells of PAO1 from the culture by first centrifuging at 13,000 rpm (16,000 × *g*) for 5 min and then by passing the supernatant through a 0.2 μm disposable syringe filter (Sartorius, Germany). The clarified samples can be stored frozen at −20 °C until a convenient time to proceed with the solvent extraction and sample preparation (*see* **Note 8**).

3.2 Solvent Extraction and Sample Preparation of Bacterial Supernatants

1. Spike 1.0 mL of each bacterial supernatant sample with 5.0 μL of the IS mix (*see* **Note 9**).
2. Add 1.0 mL of acidified ethyl acetate (*see* **Note 10**), and rapidly vortex-mix the samples for approximately 5 min.
3. Gentle centrifugation (1,000 rpm for 5 min) of the samples will facilitate separation of the organic and aqueous phases.
4. For each sample, remove the organic (upper) phase with a pipette and transfer to a clean tube.
5. Repeat **steps 2–4** twice more, pooling the solvent extracts for each sample into one 3 mL extract.
6. Evaporate the extracts to dryness in 1.5 mL Eppendorf tubes using a Jouan RC1022 vacuum concentrator attached to a RCT90 refrigerated cold trap (*see* **Note 11**).
7. Add 50 μL of 0.1 % (v/v) formic acid in MeOH to each dried sample and vortex-mix for approximately 5 min (*see* **Note 12**).
8. Remove any undissolved material by centrifugation (13,000 rpm, 16,000 × *g*, for 5 min), and transfer clarified samples to LC vials prior to LC-MS/MS analysis (*see* **Note 13**).

3.3 Preparation of Calibration and Quality Control Standards

1. Dilute synthetic standard stock solutions of individual AHLs and AQs to give a 1 mM acetonitrile solution for each analyte.
2. Combine and dilute the 1 mM stocks to prepare calibration standard mix samples in acetonitrile that are 300, 200, 100, 75, 50, 20, 10, and 2 μM with respect to C_4-HSL, 3-oxo-C_{12}-HSL, PQS, and HQNO (2-heptyl-4-hydroxyquinoline *N*-oxide) and tenfold less for HHQ.
3. Prepare quality control standard mix samples in a similar manner at concentrations of 200 and 10 μM with respect to C_4-HSL, 3-oxo-C_{12}-HSL, PQS, and HQNO and tenfold less for HHQ.
4. To prepare each calibration sample spike 1 mL of sterile LB with 50 μL of the calibration standard mix prepared in **step 2** (*see* **Note 14**), followed by 5.0 μL of the IS mix. These

samples are then extracted, dried, and reconstituted into LC samples in exactly the same manner as the bacterial supernatant sample described in Subheading 3.2, **steps 2–8**.

5. In an identical process, the quality control samples are prepared by spiking 1 mL aliquots of sterile LB with the quality control standard mixes prepared in **step 3** (*see* **Note 14**).

3.4 LC-MS/MS Analysis

The overall analytical process is summarized in the workflow diagram (Fig. 2). Samples are sequentially injected into the HPLC and undergo a chromatographic separation. The MS is set up in MRM mode to constantly screen the eluent from the LC column for all the AHL and AQ analytes of interest (Fig. 3) (*see* **Note 15**).

1. For analysis, 5.0 μL of each prepared sample is injected into the LC system flowing at a constant 450 μL/min. A mobile-phase gradient is applied. The initial 1 min after injection is an isocratic flow of 10 % B, followed by a linear gradient to 50 % B over the next 0.5 min. The linear gradient continues to 99 % B over 4 min and is maintained at this composition for 1.5 min. The column is then re-equilibrated for 2.9 min at the initial starting composition. The column oven is maintained throughout the run at 50 °C. The LC flow is diverted away from the MS at all times other than the period from 1 to 7 min (*see* **Note 16**).
2. Parameters that the MS is set to are curtain gas 20, nebulizer gas (GS1) 30, and turbo gas (GS2) 10. The ion source potential is held at 5 kV, and the temperature is set at 450 °C. The declustering potential and collision energies are analytical parameters that are unique to each analyte (*see* **Note 17**). A list of these, along with the precursor and product ions used to set up an MRM for each individual analyte, is contained in Table 1 (*see* **Notes 18** and **19**).
3. During data analysis Analyst software is used for automatic peak identification and integration. For each analyte a calibration line is constructed by calculating a mean ratio of analyte peak area/IS for the three replicate calibrants and plotting this value against concentration. The observed results for the quality control standards can be used to ensure that the overall quantitative process provides a result that is suitably accurate and precise (*see* **Note 20**). Quantitative results for PAO1 supernatant samples are determined from the calculated peak area/IS ratio for each observed analyte and the known slope and intercept of the corresponding straight line calibration graph.

   This process is illustrated for the analysis of C_4-HSL (Fig. 4). The calibration graph displays linearity over the range of 2–300 μM, with a slope of $y = 0.108x$ and intercept of 0.1113. The three replicate peak area/IS ratios for the low and high QC samples display a precision (8.7 and 1.6 %,

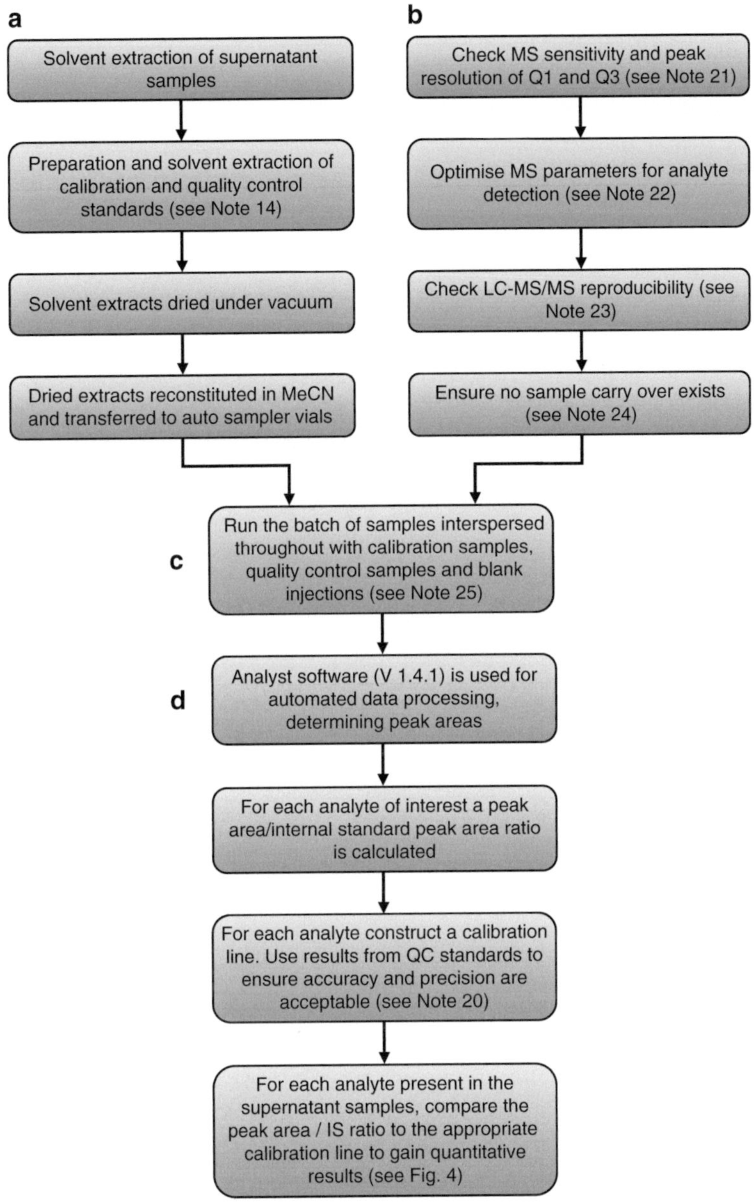

Fig. 2 Workflow diagram summarizing the procedure for the quantitative analysis of *P. aeruginosa* supernatant samples. (**a**) Preparation of samples and standards. (**b**) LC and MS setup. (**c**) Data acquisition. (**d**) Quantitative analysis of data

respectively) and accuracy (88.5 and 95.2 %, respectively) that are within acceptable values. The mean peak area/IS ratio for the PAO1 supernatant samples indicates a C_4-HSL concentration of 229 μM which corresponds to a concentration in the initial supernatant samples of 11.4 μM.

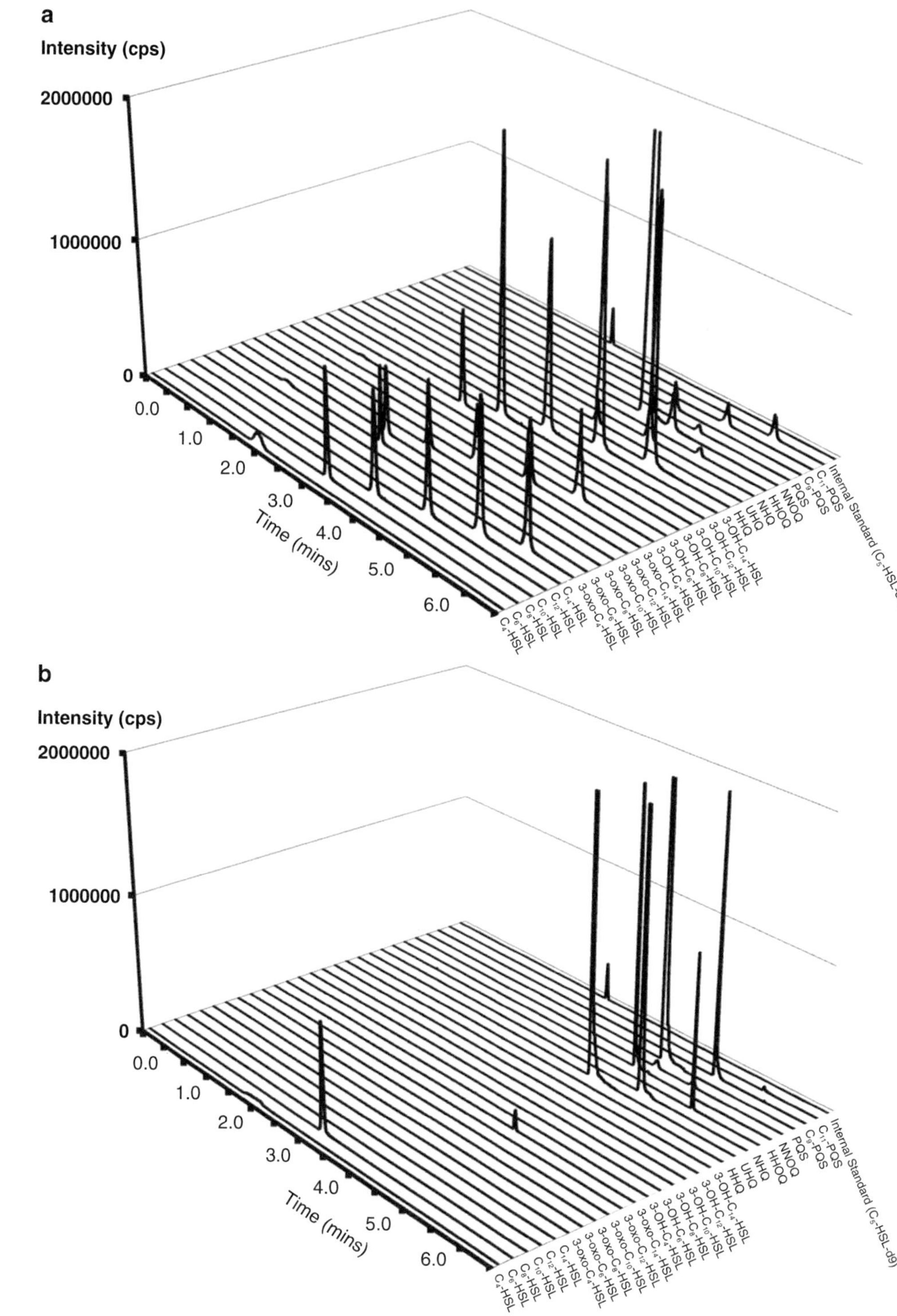

Fig. 3 MRM signals for (**a**) 1 μM of each synthetic standard, (**b**) analytes present in extracted sample of culture medium (LB broth) from *P. aeruginosa* PA01. Each analyte is represented as a chromatogram of a parent-product ion pair unique to that analyte. The *x*-axis represents analyte retention time (in minutes) on the HPLC column and the *y*-axis the mass spectrometer signal (counts per second, cps). As the analyte elutes from the column, the ion signal intensity increases and then returns to baseline. The chromatograms are tiered to allow ease of view

Table 1

A list of the analyte-specific variables used to set up the MS for MRM analysis of the LC eluent. A collision cell exit potential of 10 V was used for all analytes

Analyte	Precursor ion (*m/z*)	Product ion (*m/z*)	Declustering potential (V)	Collision energy (eV)
C_4-HSL	172.1	102.1	26	15
C_6-HSL	200.1	102.1	31	15
C_8-HSL	228.1	102.1	26	15
C_{10}-HSL	256.1	102.1	31	17
C_{12}-HSL	284.1	102.1	31	17
C_{14}-HSL	312.1	102.1	45	19
3-oxo-C_4-HSL	186.1	102.1	26	15
3-oxo-C_6-HSL	214.1	102.1	26	15
3-oxo-C_8-HSL	242.1	102.1	26	17
3-oxo-C_{10}-HSL	270.1	102.1	26	17
3-oxo-C_{12}-HSL	298.1	102.1	30	19
3-oxo-C_{14}-HSL	326.2	102.1	30	21
3-OH-C_4-HSL	188.1	102.1	26	15
3-OH-C_6-HSL	216.1	102.1	26	15
3-OH-C_8-HSL	244.1	102.1	31	17
3-OH-C_{10}-HSL	272.1	102.1	35	17
3-OH-C_{12}-HSL	300.1	102.1	35	19
3-OH-C_{14}-HSL	328.2	102.1	21	19
HHQ	244.1	159.1	31	20
NHQ	272.1	159.1	81	27
UHQ	300.1	159.1	81	27
HQNO	260.1	159.1	71	32
NQNO	288.1	159.1	96	20
PQS	260.1	175.1	71	22
C_9-PQS	288.1	175.1	111	25
C_{11}-PQS	316.1	175.1	116	35
C_5-HSL-d_9[a]	195.1	103.1	40	15
PQS-d_4[a]	264.1	179.1	80	42
C_{10}-TMA[b]	298.1	144.1	30	20
OR-C_4-HSL[c]	190.1	172.1	28	12
OR-3-oxo-C_{12}-HSL[c]	316.2	298.1	28	12

[a]Internal standards

[b]Tetramic acid product of 3-oxo-C_{12}-HSL

[c]Open-ring forms of C_4-HSL and 3-oxo-C_{12}-HSL

a

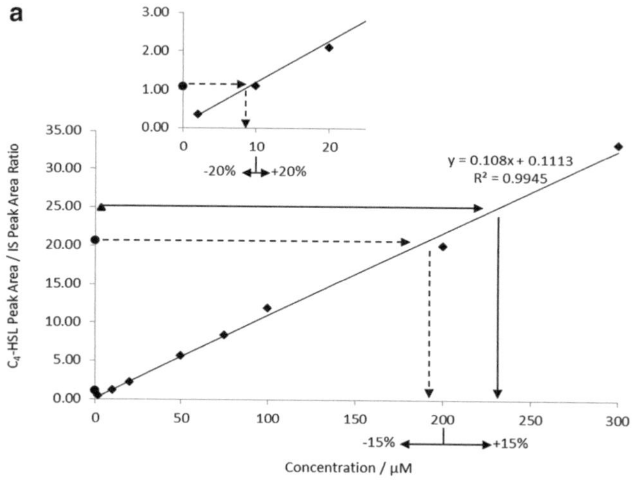

b

	Low QC	High QC
Replicate a	1.03	20.30
Replicate b	1.17	20.92
Replicate c	0.99	20.79
Mean	1.07	20.67
Standard Deviation	0.09	0.33
Precision	8.7%	1.6%
Calculated Concentration /μM	8.85	190.36
Accuracy	88.5%	95.2%

	PA01 WT
Replicate a	25.21
Replicate b	23.87
Replicate c	25.43
Mean	24.84
Calculated Concentration / μM	229.0
Supernatant Concentration / μM	11.4

Fig. 4 Quantitative analysis of data. (**a**) Eight-point calibration line for C_4-HSL (*diamond data points*), also indicating the mean high- and low-quality control standards (*circles*). The mean analyte peak area/IS peak area ratio (*triangle*) is then compared with the calibration line. (**b**) Worked example indicating that the quantification of C_4-HSL in a *P. aeruginosa* LB supernatant sample is suitably accurate and precise. Note that the numbers displayed for each replicate sample are calculated peak area/IS ratios

4 Notes

1. For growth experiments we typically set up cultures in triplicate so that our samples and quantitative data are based upon three independent cultures grown under identical conditions.
2. The integrity of the AHL ring structure is pH dependent [13]. At high pH the lactone ring is hydrolyzed resulting in a ring-opened structure (Fig. 1c). This process is reversible, and a reduction in pH to acidic conditions results in re-formation of an intact lactone. For this reason, acidic conditions are usually used for sample extraction and both the aqueous and organic LC mobile-phase components, maximizing AHL extraction and detection. However, for certain experiments it may be necessary to analyze samples for the open-ring forms, in which case acidic conditions should be avoided. The analyte-specific variables used to set up the MS for MRM of the open-ring forms of C_4-HSL and 3-oxo-C_{12}-HSL are included in Table 1.
3. The AHLs that were synthesized in-house are available at http://www.nottingham.ac.uk/quorum.
4. The solubility of AHLs in acetonitrile decreases as the length of acyl chains increases. As a consequence, stock solutions of the individual longer chain AHLs often contain precipitated material when removed from freezer storage which could potentially lead to inaccuracy making calibration standards.

Complete re-solvation can be ensured by rapid vortex mixing once the samples have warmed to room temperature.

5. PQS and other analogous compounds (including the PQS-d_4 internal standard) typically display distorted chromatographic peak shapes with considerable tailing often eluting over many minutes. This may result in an elevated base line, in severe cases running into subsequent sample injections, producing poorly reproducible results. In addition, automated peak integration software such as Analyst can struggle to distinguish and integrate peaks correctly, prompting time-consuming manual integration of these peaks for each and every sample. It is thought that the poor peak shape occurs due to the metal chelating properties of PQS and similar compounds [14], which interact with impurities in the chromatography column packing material. We have described how this can be overcome by the addition of EDTA, a known chelating compound, to the aqueous mobile phase [15]. However, the subsequent ion suppression has a detrimental effect on MS sensitivity, particularly for AHLs. This may be of little concern if analyzing samples exclusively for AQs or when analyzing samples where AHL concentrations are anticipated to be very high. However an alternative strategy is to precondition the chromatography column with a 1.0 mM EDTA solution and purge excess EDTA from the column by running repeated gradients of normal LC mobile phase prior to use in LC-MS/MS analysis.
6. If using EDTA in the mobile phase, add 18.6 mg of EDTA to 250 mL of mobile phase A (for a 0.2 μM solution) and dissolve by placing in an ultrasonic water bath for approximately 15 min. Vacuum filtration as previously described in Subheading 2 will remove any undissolved material prior to LC-MS/MS analysis.
7. PAO1 culture samples can be removed at any time during the growth of the bacteria. A frequently used protocol is to remove samples at regular time intervals from the lag phase through the exponential phase and into the stationary phase of growth. As well as determining overall concentrations of components, the quantitative data can be used to monitor the profile of AHLs and AQs with time.
8. The analytical methodology as described here has been used for studying supernatant samples from PAO1 wild-type and isogenic mutants grown in rich (LB broth) or chemically defined media (CDM), as either a batch or a continuous culture. The CDM base that we use for these purposes contains 20 mM D-glucose, 3 mM KCl, 3 mM NaCl, 12 mM $(NH_4)_2SO_4$, 3.2 mM $MgSO_4{\cdot}7H_2O$, 0.02 mM $FeSO_4{\cdot}7H_2O$, 1.2 mM K_2HPO_4, and 50 mM 3-(*N*-morpholino)-propanesulfonic acid (MOPS) buffered to a pH of 7.2.

9. The reason for adding internal standards during quantitative determinations is to correct the results for random and systematic errors that may occur during the extraction process and sample preparation. For this purpose we use a mix of the deuterated compounds PQS-d_4 and C_5-HSL-d_9 as stable isotope-labelled compounds, structurally similar to the analytes of interest (Fig. 1d). These are spiked at a set concentration into every supernatant, calibration standard, and quality control standard sample prior to extraction.
10. 15 mL polypropylene Falcon tubes are suitable for this.
11. Alternatively, the pooled solvent extracts can be dried under a stream of nitrogen gas using a Techne sample concentrator or other similar apparatus.
12. The extraction process results in a 20-fold sample concentration (i.e., 1.0 mL of bacterial culture supernatant is being extracted and made into a 50 μL LC sample), which needs to be remembered when calculating AQ and AHL concentrations in the initial supernatant sample.
13. LC vials with or without reduced volume inserts can be readily sourced. We recommend wide-neck vials (1.8 mL, 9 mm) with loose 0.25 mL inserts and short thread silicone/PTFE septa caps which offer flexibility across a wide range of uses and instrumentation.
14. Suppression or enhancement of instrument response to analytes in the presence of sample matrix is possible. For this reason, samples should be quantified using matrix-matched calibration standards. Aliquots of sterile LB broth (identical to the media used for growing *P. aeruginosa* in) are spiked with the various concentrations of calibration standard mix and the IS mix. These samples are then treated (solvent extraction, drying, and sample preparation) in an identical manner to the bacterial conditioned media. Although the useful concentration range for calibrants may vary for bacterial experiments involving different growth media, growth conditions, and bacterial strains, we typically prepare samples for an eight-point calibration line with the calibrant range up to 300 μM. Two sets of quality control samples are similarly prepared, at concentrations that are close to the top end and bottom end of the calibration line. Calibration and quality control samples are usually prepared in triplicate. Where resources and time permit, we recommend that an initial scouting experiment with a small sample subset is made to estimate levels and establish appropriate calibration points.
15. A tetramic acid displaying antibacterial activity (Fig. 1e), produced by the intramolecular rearrangement of 3-oxo-C_{12}-HSL, is also present in the conditioned media of *P. aeruginosa* [16]. Analyte-specific MRM parameters for this tetramic acid product are also listed in Table 1.

16. If using LC mobile phases containing EDTA, care should be taken to minimize contamination and buildup on the mass spectrometer's source with this nonvolatile salt. We recommend limiting the number of sample injections to no more than 120 before cleaning source components and diverting LC flow away from the mass spectrometer during time segments when data acquisition is not necessary.
17. Analyte-specific MRM parameters (precursor and product ion pairs, declustering potential, collision cell exit potential, and collision energy) were optimized with software automation, while a syringe pump infused a solution of 0.1–1.0 μM of each individual analyte in 50 % mobile phase B at a flow rate of 50 μL/min (Table 1).
18. The LC-MS/MS quantitative assay of AHLs and AQs can be conducted with any triple-quad MS instrument comparable to the 4000 QTRAP that we use. Other MS instrumentation will have equivalent instrument and analyte-specific settings to those that we have listed.
19. More sophisticated profiling of QSSMs can be achieved by using the advanced scanning features of a quadrupole linear ion trap (QTRAP) mass spectrometer. MRM, although a method targeted towards specific analytes, can on occasion result in unconnected chromatographic peaks. For example, odd chain length AHLs that are occasionally produced (e.g., 3-oxo-C_5-HSL) may be confused with commonly found even chain length AHLs (e.g., C_6-HSL), both having an identical precursor ion (m/z=200.1) and product ion (m/z=102.1). Although for this example the two analytes can easily be distinguished by their differing chromatographic retention times sometimes it is necessary to produce an unambiguous identification based upon a unique pattern of product ions. The 4000 QTRAP can be set up to acquire MS/MS spectra while simultaneously running MRM analysis. In this manner MRMs are used as a survey scan to trigger the acquisition of enhanced product ions in an information-dependant acquisition (MRM-IDA-EPI). The IDA parameters are set up such that if a precursor ion detection exceeds 2,000 counts per second, it is automatically selected for enhanced product ion formation resulting in the recording of an MS/MS spectra.

    Alternatively the QTRAP can be set up for precursor ion screening so that unusual or new compounds can be rapidly identified by monitoring for analytical structural identifiers unique to the families of AHLs and AQs. The detection of these unique product ions (i.e., m/z=102.1 for AHLs, m/z=175.1 for PQS-type compounds, and m/z=159.1 for HHQ-type compounds) triggers the MS to calculate the m/z of the precursor ion generating a specific product ion. This in turn also triggers the acquisition of a quality MS/MS spectrum.

These features of the MS system, using triggered combinations of triple-quadrupole and ion trap modes in the same LC-MS/MS run, offer comprehensive profiling of known and unknown QSSMs in bacterial extracts [17].

20. Calibration lines should demonstrate linearity over the full range of sample concentration anticipated for the quantitation. Precision (the standard deviation of the three QC replicates expressed as a percentage of the mean value) should typically be <15 % or <20 % for the QC samples at the bottom end of the calibration line. Accuracy (the calculated concentration of the QC samples expressed as a percentage of the nominal value) should typically be 100 ±15 % or ±20 % for the QC samples at the bottom end of the calibration line.
21. For MS calibration, a 2 μM solution of polypropylene glycol (PPG) standard mix, prepared according to the manufacturer's guideline, is infused directly into the MS, via a syringe pump, at a flow rate of 5.0 μL/min. Using positive electrospray ionisation mode, the sensitivity and peak resolution of both mass selection components (Q1 and Q3) in the MS are independently assessed. Each, after the summation of ten mass scans for all of the PPG components, should produce a peak for the PPG with a mass of 906.1 that is a minimum of 2×10^7 counts per second. If this minimum sensitivity is not reached, it is usually an indication that the curtain plate, orifice plate, and skimmer need cleaning. For optimal peak resolution the peak width for the PPG standards should be 0.7 ±0.1 amu full width at half maximum. If the observed peak width lies outside of this range then the resolution offset values can be adjusted to optimize the peak resolution. Note that these performance indicators apply to the particular 4000 QTRAP MS that we use but can be defined as appropriate for other triple-quad instruments.
22. A tuning solution of an analyte (0.1–1 μM prepared in the LC mobile phase) is infused, via a syringe pump, directly into the MS at a flow rate equal to that which is to be used for LC-MS/MS analysis. By constantly monitoring the MS response for this analyte the electrospray source position as well as various temperature, gas flow, and voltage settings can be adjusted to optimize source-dependant parameters, maximizing peak size.
23. To check LC and MS function, several repeat injections of a system suitability sample (containing all analytes of interest at a concentration of approximately 1 μM) should be made to ensure the consistency of chromatographic retention time, peak shape, and peak area. Peak areas should display a CV ≤5 %.
24. Blank injections of acetonitrile can be made to gauge if carryover (contamination from one sample run into the next) is an issue. Autosampler wash parameters are adjusted to limit carryover to less than 0.01 % of peak area of the previous sample.

25. Within the overall batch of samples to be analyzed, the three sets of replicate calibration standards (run from the lowest concentration to highest) and the quality control standards should be included at the beginning, middle, and end of the sequence. This is to assess if the response of the mass spectrometer is deteriorating over the batch runtime. It is also useful to inject negative control samples prepared by extracting fresh sterile media, and a reagent blank sample prepared by extracting distilled water, to help determine the origin of any sample contamination, should it occur.

Acknowledgements

The contribution of Mavis Daykin to the development of this methodology is gratefully acknowledged as is the work of Ram Chhabra and Alex Truman in synthesizing AHLs and AQs. The work was supported by grants from the Biotechnology and Biological Sciences Research Council (BBSRC grant number BBS/B/09740, BB/C/509082, and P18575) and the Medical Research Council (MRC grant number G0801558).

References

1. Williams P, Winzer K, Chan W, Cámara M (2007) Look who's talking: communication and quorum sensing in the bacterial world. Philos Trans R Soc Lond B Biol Sci 362:1119–1134
2. Diggle SP, Cornelis P, Williams P, Cámara M (2006) 4-Quinolone signalling in *Pseudomonas aeruginosa*: old molecules, new perspectives. Int J Med Microbiol 296:83–91
3. Winson MK, Cámara M, Latifi A, Foglino M, Chhabra SR, Daykin M, Chapon V, Bycroft BW, Salmond GPC, Lazdunski A, Stewart GSAB, Williams P (1995) Multiple quorum sensing modulons interactively regulate virulence and secondary metabolism in *Pseudomonas aeruginosa*: identification of the signal molecules *N*-butanoyl-L-homoserine lactone and *N*-hexanoyl-L-homoserine lactone. Proc Natl Acad Sci U S A 92:9427–9431
4. Charlton TS, de Nys R, Netting A, Kumar N, Hentzer M, Givskov M, Kjelleberg S (2000) A novel and sensitive method for the quantification of *N*-3-oxoacyl homoserine lactones using gas chromatography-mass spectrometry: application to a model bacterial biofilm. Environ Microbiol 2:530–541
5. Williams P, Cámara M (2009) Quorum sensing and environmental adaptation in *Pseudomonas aeruginosa*: a tale of regulatory networks and multifunctional signal molecules. Curr Opin Microbiol 12:182–191
6. Lépine F, Milot S, Déziel E, He J, Rahme LG (2004) Electrospray/mass spectrometric identification and analysis of 4-hydroxy-2-alkylquinolines (HAQs) produced by *Pseudomonas aeruginosa*. J Am Soc Mass Spectrom 15:862–869
7. Heeb S, Fletcher MP, Chhabra SR, Diggle SP, Williams P, Cámara M (2011) Quinolones: from antibiotics to autoinducers. FEMS Microbiol Rev 35:247–274
8. Chhabra SR, Stead P, Bainton NJ, Salmond GPC, Stewart GSAB, Williams P, Bycroft BW (1993) Autoregulation of carbapenem biosynthesis in *Erwinia carotovora* by analogs of *N*-(3-oxohexanoyl)-L-homoserine lactone. J Antibiot 46:441–454
9. Chhabra SR, Harty C, Hooi DSW, Daykin M, Williams P, Telford G, Pritchard DI, Bycroft BW (2003) Synthetic analogues of the bacterial signal (quorum sensing) molecule N-(3-oxododecanoyl)-L-homoserine lactone as immune modulators. J Med Chem 46:97–104
10. Somanathan R, Smith KM (1981) Synthesis of some 2-alkyl-4-quinolone and 2-alkyl-4-methoxyquinolone alkaloids. J Heterocycl Chem 18:1077–1079

11. Pesci EC, Jared BJ, Milbank JBJ, Pearson JP, McKnight S, Kende AS, Greenberg PE, Iglewski BH (1999) Quinolone signalling in the cell-to-cell communication of Pseudomonas aeruginosa. Proc Natl Acad Sci U S A 96:11229–11234
12. Lepine F, Deziel E, Milot S, Rahme LG (2003) A stable isotope dilution assay for the quantification of the Pseudomonas quinolone signal in *Pseudomonas aeruginosa* cultures. Biochim Biophys Acta 1622:36–41
13. Yates EA, Philipp B, Buckley C, Atkinson S, Chhabra SR, Sockett RE, Goldner M, Dessaux Y, Cámara M, Smith H, Williams P (2002) *N*-Acylhomoserine lactones undergo lactonolysis in a pH-, temperature-, and acyl chain length-dependent manner during growth of *Yersinia pseudotuberculosis* and *Pseudomonas aeruginosa*. Infect Immun 70:5635–5646
14. Bredenbruch F, Geffers R, Nimtz M, Buer J, Häussler S (2006) The *Pseudomonas aeruginosa* quinolone signal (PQS) has an iron-chelating activity. Environ Microbiol 8:1318–1329
15. Ortori CA, Dubern JF, Chhabra SR, Cámara M, Hardie K, Williams P, Barrett DA (2011) Simultaneous quantitative profiling of *N*-acyl-L-homoserine lactone and 2-alkyl-4(1*H*)-quinolone families of quorum-sensing signalling molecules using LC-MS/MS. Anal Bioanal Chem 399:839–850
16. Lowery CA, Park J, Gloeckner C, Meijer MM, Mueller RS, Boshoff HI, Ulrich RL, Barry CE, Bartlett DH, Kravchenko VV, Kaufmann GF, Janda KD (2009) Defining the mode of action of tetramic acid antibacterials derived from *Pseudomonas aeruginosa* quorum sensing signals. J Am Chem Soc 131: 14473–14479
17. Ortori CA, Atkinson S, Chhabra SR, Cámara M, Williams P, Barrett DA (2007) Comprehensive profiling of *N*-acylhomoserine lactones produced by *Yersinia pseudotuberculosis* using liquid chromatography coupled to hybrid quadrupole-linear ion trap mass spectrometry. Anal Bioanal Chem 387: 497–511

Chapter 22

LC/MS/MS-Based Quantitative Assay for the Secondary Messenger Molecule, c-di-GMP

Yasuhiko Irie and Matthew R. Parsek

Abstract

The secondary messenger molecule, 3′,5′-cyclic diguanosine monophosphate (c-di-GMP), controls various cellular processes in bacteria. Direct measurement of intracellular concentration of c-di-GMP is fast becoming an important tool for studying prokaryotic biology. Here, we describe a comprehensive extraction protocol from live bacteria and quantitative analysis using LC/MS/MS.

Key words c-di-GMP, LC/MS/MS, Quantitative analysis, *Pseudomonas aeruginosa*, Nucleotide extraction

1 Introduction

Bacteria can sense dynamic changes in the environment and modify gene expression or protein function. One of the regulatory elements that control such adaptation is the secondary intracellular messenger molecule, 3′,5′-cyclic diguanosine monophosphate (c-di-GMP). This signaling molecule is synthesized by diguanylate cyclase proteins harboring the GGDEF amino acid motif and degraded by phosphodiesterases containing EAL or HD-GYP domains. By controlling the expression and/or activities of these enzymes, bacteria manipulate their c-di-GMP content, leading to changes in a variety of cellular processes.

c-di-GMP was first discovered to directly regulate cellulose production in *Gluconacetobacter xylinus* [1]. Since then, c-di-GMP has been implicated in the control of physiological processes such as biofilm formation, motility, cell cycle, and virulence in various Gram-negative and Gram-positive species [2–4]. In *Pseudomonas aeruginosa*, elevated levels of intracellular c-di-GMP have been associated with a switch from a motile to a sessile lifestyle by reducing flagellar motility and increasing expression levels of biofilm-associated factors [5–11]. The *P. aeruginosa* PAO1 strain encodes

Alain Filloux and Juan-Luis Ramos (eds.), *Pseudomonas Methods and Protocols*, Methods in Molecular Biology, vol. 1149, DOI 10.1007/978-1-4939-0473-0_22, © Springer Science+Business Media New York 2014

for 17 GGDEF proteins, 5 EAL proteins, 16 proteins with both GGDEF and EAL domains, and 3 HD-GYP proteins [12], suggesting a complex regulatory system capable of responding to multiple environmental cues.

Mass spectrometry-based quantitative analyses of intracellular c-di-GMP have been described by several groups. Simm et al. described a method using MALDI-TOF [13]. However, MALDI instruments are not typically ideal for quantitative measurements, necessitating the development of LC/MS/MS-based analytical methodologies [7, 14, 15]. In addition, several of these studies involved complex multi-step protocols for nucleotide extraction and isolation prior to MS analyses, which included multiple centrifugation steps to concentrate bacterial cells prior to lysis. A previous study described rapid changes in bacterial nucleotide pools upon brief centrifugation [16]. This implied that centrifuging cells can be detrimental to acquiring accurate physiological concentrations of c-di-GMP.

In this chapter, we describe a simplified method for nucleotide extraction that does not involve centrifugation of the cells until after they have been lysed. This also has the advantage of shortening the protocol to avoid possible problems such as the chemical instability of c-di-GMP as previously suggested [17]. We therefore adapted the extraction protocol described by Hickman et al. [7] and, in addition, introduced the use of an internal standard to improve data resolution (Fig. 1). We developed a modified protocol [7] for quantitative analysis of intracellular c-di-GMP that used 2-chloro-adenosine-5′-O-monophosphate rather than c-di-AMP [13] as the internal standard. This is because c-di-AMP has recently been found in bacteria to be a biologically active molecule [18, 19], disqualifying it to be an appropriate internal standard compound. We found no traces of 2-chloro-adenosine-5′-O-monophosphate in *P. aeruginosa* total nucleotide extracts (Fig. 1). Unlike previous reports [14, 15], we did not find interfering biological signals using our protocol.

2 Materials

2.1 Culture Medium

Volger–Bonner minimal medium (VBMM) 10× concentration: 2 g/L $MgSO_4{\cdot}7H_2O$ (0.976 g/L for anhydrous), 20 g/L citric acid (30.62 g/L for sodium citrate) (*see* **Note 1**), 35 g/L $NaNH_4HPO_4{\cdot}4H_2O$, 100 g/L K_2HPO_4, add into solution in the following order: $MgSO_4{\cdot}7H_2O$, citric acid, $NaNH_4HPO_4{\cdot}4H_2O$, and K_2HPO_4, adjust pH to 7.0, bring to 1 L volume, then filter sterilize.

2.2 Nucleotide Extraction Reagents

1. LC/MS/MS-grade water.
2. 70 % vol/vol perchloric acid.

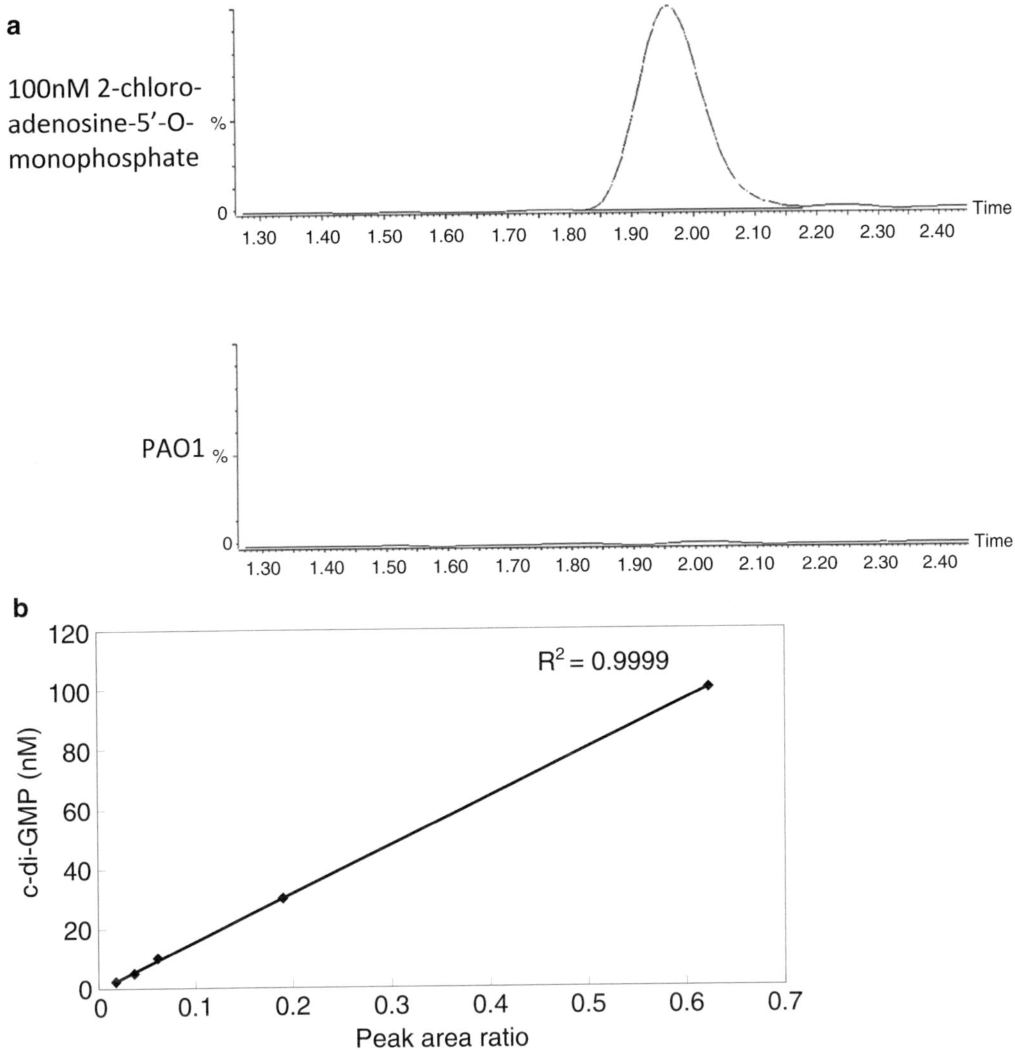

Fig. 1 (**a**) 2-chloro-adenosine-5′-O-monophosphate is not present in biological samples as the corresponding peak is absent in *P. aeruginosa* PAO1 extract, validating 2-chloro-adenosine-5′-O-monophosphate as an appropriate internal standard compound for c-di-GMP analyses. (**b**) Standard curve demonstrating that the introduction of 2-chloro-adenosine-5′-O-monophosphate into the assay allows for a highly accurate measurement of c-di-GMP as low as 2.5 nM

3. 250.3 g/L $KHCO_3$.
4. c-di-GMP (*see* **Note 2**).
5. 2-chloro-adenosine 5′-O-monophosphate (*see* **Note 2**).

2.3 Nucleotide Extraction Equipment and Materials

1. Ice.
2. 1.5 ml microfuge tubes.
3. 15 ml conical tubes.
4. Bench top centrifuge (either refrigerated or refrigeration capable).

5. 4 °C room.
6. –80 °C freezer.

2.4 Protein Quantitation Reagents (Example)

1. 480.48 g/L urea (up to 6 M final concentration of sample solution is tolerated for many Bradford reagents).
2. Bradford assay reagent.
3. Protein of choice for performing standard curve measurements (e.g., bovine serum albumin solutions of known concentrations).

2.5 Protein Quantitation Equipment and Materials

1. Microfuge tubes.
2. Desktop centrifuge.
3. 100 °C heat block.
4. Absorbance reader (if using colorimetric assay).
5. Cuvette or plate (depending on the recommended protocol by the assay reagent's manufacturer).

2.6 LC/MS/MS Reagents

1. LC/MS/MS-grade formic acid.
2. LC/MS/MS-grade water.
3. LC/MS/MS-grade acetonitrile.

2.7 LC/MS/MS Equipment and Materials

1. Acuity UPLC with Synergi 4μ Hydro RP 80A column (Phenomenex).
2. C18 Guard Cartridge (Phenomenex).
3. Instrument capable of performing reverse-phase liquid chromatography attached to a triple-quadrupole charged needle electrospray tandem mass spectrometer capable of detecting low nanomolar quantities of c-di-GMP (the authors have used Premier XL manufactured by Waters Corporation).

3 Methods

3.1 Nucleotide Extraction

The basic outline of the c-di-GMP extraction protocol is shown in Fig. 2. The compositions of each sample are shown in Fig. 3. Samples for standard curve construction are subjected to the same nucleotide extraction procedure as the biological samples and are aliquoted prior to storing at –80 °C to avoid freeze–thaw cycles. For each extraction batch, quality control samples are included to ensure consistency of extraction efficiency. All bacterial cultures are grown in independent biological quadruplicates on separate days. All solutions were prepared using Optima LC/MS grade water (Fisher).

1. Add 70 % vol/vol perchloric acid to a final concentration of 0.6 M (~94.2 μl) along with 10 μl of a 10 μM solution of 2-chloro-adenosine-5′-O-monophosphate to 990 μl of a *P. aeruginosa*

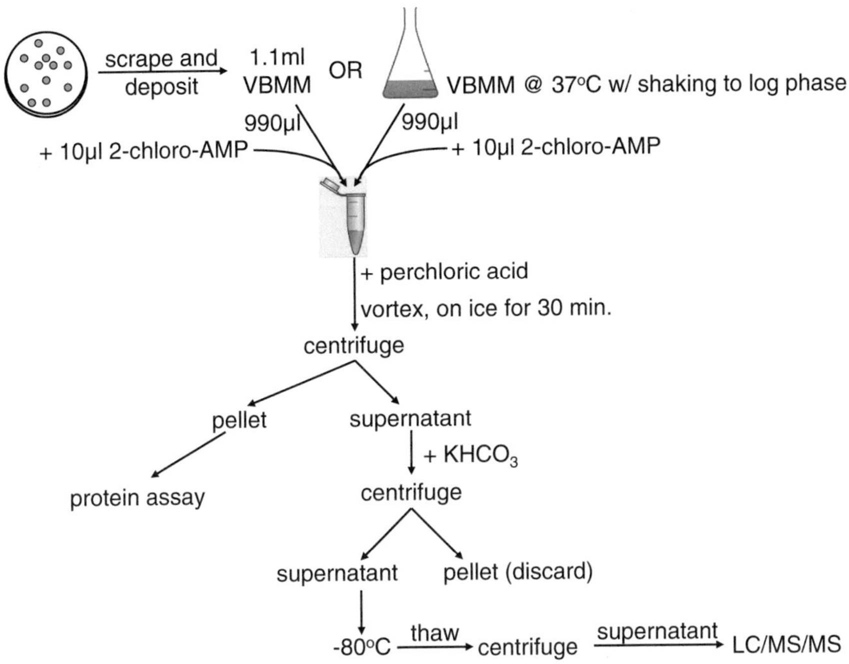

Fig. 2 Schematic representation of c-di-GMP extraction

Standard curve

	0nM	2.5nM-100nM
c-di-GMP	0µl	100µl 25nM-1µM c-di-GMP
10µM 2-chloro-AMP	10µl	10µl
10× VBMM	100µl	100µl
LC/MS H_2O	890µl	790µl
total	1ml	1ml

Biological samples

culture/sample	990µl
10µM 2-chloro-AMP	10µl
total	1ml

(dilution factor = 1000/990)

Quality controls (per extraction batch)

	10nM	25nM
c-di-GMP	100µl 100nM c-di-GMP	100µl 250nM c-di-GMP
10µM 2-chloro-AMP	10µl	10µl
10× VBMM	100µl	100µl
LC/MS H_2O	790µl	790µl
total	1ml	1ml

Fig. 3 Compositions of standard curve samples, biological samples, and quality control samples prior to perchloric acid treatment

culture in the mid-logarithmic phase of growth and vortex (*see* **Note 3**). For these experiments cells are grown in VBMM broth (*see* **Note 4**).

2. Alternatively, surface-grown cells (for example, on agar plates) can be scraped off and immediately resuspended in VBMM prior to aspirating 990 μl as described in **step 1** (*see* **Note 5**).
3. Incubate samples initially on ice for 30 min with all subsequent steps performed at 4 °C.
4. Centrifuge samples at maximum speed for 10 min.
5. Transfer supernatant to a 15 ml conical tube. Keep precipitated fractions for subsequent protein quantitation (*see* **Note 6**).
6. Add 1/5 volume 2.5 M $KHCO_3$ (~219 μl) to the nucleotide extracts to neutralize the pH (*see* **Note 7**).
7. Briefly centrifuge the tubes to collect the samples, and transfer the entire contents to fresh 1.5 ml microfuge tubes.
8. Centrifuge for 10 min to remove the perchlorate salt precipitates.
9. Aliquot the samples and store at −80 °C (*see* **Note 8**).

3.2 LC/MS/MS

LC/MS/MS analysis was undertaken using an Acuity UPLC with Synergi 4μ Hydro RP 80A column and C18 Guard Cartridge attached to a Premier XL triple-quadrupole charged needle electrospray mass spectrometer (*see* **Note 9**) essentially as previously described [7] but with several modifications. A gradient mobile phase consisting of 10 mM formic acid and acetonitrile in water was used as described in Table 1. The m/z 691 > 152 transition is used for c-di-GMP and 382 > 170 for 2-chloro-adenosine-5′-O-monophosphate. The cone voltages and collision energies are 40 V/30 eV and 35 V/20 eV, respectively. Peak area ratio of c-di-GMP:2-chloro-adenosine-5′-O-monophosphate is used for all samples.

Table 1
Liquid chromatography gradient system

Start time	Formic acid	Acetonitrile	Flow rate
0 min	98 %	2 %	0.25 ml/min
2 min	70 %	30 %	0.25 ml/min
2.5 min	20 %	80 %	0.35 ml/min
3 min	100 %	0 %	0.35 ml/min
5 min	98 %	2 %	0.35 ml/min
6.9 min	98 %	2 %	0.25 ml/min
Total time: 7.5 min			

Absolute molarities of c-di-GMP in each sample are calculated using the standard curves, measured in technical triplicates or quadruplicates during the mass spectrometry run. Extraction batches can be validated by comparing the actual and measured concentrations of the quality control samples (within ±10 % error). The c-di-GMP concentration is divided by 0.99 to compensate for the dilution caused by adding 2-chloro-adenosine-5′-O-monophosphate at the beginning of the nucleotide extraction. The values are then divided by the protein concentrations of the respective biological samples to normalize to total proteins of the cultures. The resulting concentration unit is expressed as pmol c-di-GMP per mg total protein (*see* **Note 10**).

4 Notes

1. Carbon sources can be altered depending on the experimental design.
2. Pure c-di-GMP and 2-chloro-adenosine-5′-O-monophosphate compounds were purchased from BIOLOG.
3. According to a previous study, nucleotide pools in bacteria dropped significantly upon removal of cultures from a shaker [16]. Immediate acid treatment is more important than the 30-min incubation time. Each culture needs to be acid treated individually, and a longer extraction time than 30 min is allowed. Unless specifically studying dynamic changes in c-di-GMP levels during growth phases, we find that c-di-GMP levels are most robust and the differences among strains/mutants the greatest during logarithmic growth [15].
4. We have verified that VBMM does not produce interfering signals in c-di-GMP/2-chloro-adenosine-5′-O-monophosphate channels. We have determined that both tryptone and yeast extract interfere with the assay (data not shown). We therefore recommend using chemically defined growth media and to perform nucleotide extraction and MS measurements on uninoculated growth media. When changing culture media, alter the compositions shown in Fig. 3 accordingly.
5. Unlike liquid cultures, cells grown on undefined rich solid media (such as LB) do not produce interfering signals during MS analysis measurements.
6. Acid-precipitated materials largely contain proteins. Precipitates can be resuspended and boiled in a denaturant and quantitated for proteins using standard protein assays.
7. Addition of $KHCO_3$ will produce large quantity of CO_2 gas and insoluble perchlorate salts. The gas production is the reason for a temporary transfer of the extracts to a larger volume 15 ml conical tube to prevent explosion and sample loss.

8. Contrary to a previous report [20], in our hands, repeated freeze–thaw cycles are detrimental to c-di-GMP (data not shown). Aliquoting ensures that all samples undergo exactly one freeze–thaw cycle. As such, stock solutions of c-di-GMP and 2-chloro-adenosine-5′-O-monophosphate for standard curves and quality controls need to be aliquoted.
9. It is important for users to test their LC/MS/MS's capabilities of detecting c-di-GMP and 2-chloro-adenosine-5′-O-monophosphate prior to running any assays. In our experience, LC/MS/MS incapable of detecting low to sub-nanomolar quantity of pure c-di-GMP solution within a linear standard curve is insufficient to reliably measure intracellular c-di-GMP of many wild-type strains and species. As each LC/MS/MS differs in their technologies, lowest possible detection limit must be determined for each instrument used. We have encountered several instruments that are fully capable of sensitive detection of other molecules except c-di-GMP for reasons unknown. Many modern ion trap instruments are sensitive enough but are impractical due to their per sample run time.
10. According to our estimates, we calculate a wild-type *P. aeruginosa* cell to contain 15 zeptomoles or ~9,000 c-di-GMP molecules. Average c-di-GMP concentrations in wild-type and Δ*wspF* strains are approximately 11 and 34.4 μM, respectively, which are consistent with the observed activity levels of c-di-GMP-binding protein FleQ and its published K_d for c-di-GMP of 15–20 μM [7], validating that our measurements are biologically accurate. This protocol is applicable to other Gram-negative organisms (we have so far successfully tested *Vibrio vulnificus* and *Klebsiella pneumoniae*, data not shown), provided the measurement values fall within standard curve measurements.

Acknowledgments

The authors would like to thank Dale Whittington, Ross Lawrence, and the late Thomas Kalhorn for their expertise in LC/MS/MS. This work was supported in part by NIH grant MCB0822405 and AI061396-06 to MRP. YI was a UW CFF RDP Fellow.

References

1. Ross P, Mayer R, Weinhouse H, Amikam D, Huggirat Y, Benziman M, de Vroom E, Fidder A, de Paus P, Sliedregt LA (1990) The cyclic diguanylic acid regulatory system of cellulose synthesis in *Acetobacter xylinum*. Chemical synthesis and biological activity of cyclic nucleotide dimer, trimer, and phosphothioate derivatives. J Biol Chem 265:18933–18943
2. Cotter PA, Stibitz S (2007) c-di-GMP-mediated regulation of virulence and biofilm formation. Curr Opin Microbiol 10:17–23
3. Hengge R (2009) Principles of c-di-GMP signalling in bacteria. Nat Rev Microbiol 7:263–273
4. Sondermann H, Shikuma NJ, Yildiz FH (2012) You've come a long way: c-di-GMP signaling. Curr Opin Microbiol 15:140–146

5. Boehm A, Kaiser M, Li H, Spangler C, Kasper CA, Ackermann M, Kaever V, Sourjik V, Roth V, Jenal U (2010) Second messenger-mediated adjustment of bacterial swimming velocity. Cell 141:107–116
6. Borlee BR, Goldman AD, Murakami K, Samudrala R, Wozniak DJ, Parsek MR (2010) *Pseudomonas aeruginosa* uses a cyclic-di-GMP-regulated adhesin to reinforce the biofilm extracellular matrix. Mol Microbiol 75:827–842
7. Hickman JW, Harwood CS (2008) Identification of FleQ from *Pseudomonas aeruginosa* as a c-di-GMP-responsive transcription factor. Mol Microbiol 69:376–389
8. Irie Y, Borlee BR, O'Connor JR, Hill PJ, Harwood CS, Wozniak DJ, Parsek MR (2012) Self-produced exopolysaccharide is a signal that stimulates biofilm formation in *Pseudomonas aeruginosa*. Proc Natl Acad Sci U S A 109: 20632–20636
9. Malone JG, Jaeger T, Spangler C, Ritz D, Spang A, Arrieumerlou C, Kaever V, Landmann R, Jenal U (2010) YfiBNR mediates cyclic di-GMP dependent small colony variant formation and persistence in *Pseudomonas aeruginosa*. PLoS Pathog 6:e1000804
10. Starkey M, Hickman JH, Ma L, Zhang N, De Long S, Hinz A, Palacios S, Manoil C, Kirisits MJ, Starner TD, Wozniak DJ, Harwood CS, Parsek MR (2009) *Pseudomonas aeruginosa* rugose small-colony variants have adaptations that likely promote persistence in the cystic fibrosis lung. J Bacteriol 191:3492–3503
11. Ueda A, Wood TK (2009) Connecting quorum sensing, c-di-GMP, pel polysaccharide, and biofilm formation in *Pseudomonas aeruginosa* through tyrosine phosphatase TpbA (PA3885). PLoS Pathog 5:e1000483
12. Kulasakara H, Lee V, Brencic A, Liberati N, Urbach J, Miyata S, Lee DG, Neely AN, Hyodo M, Hayakawa Y, Ausubel FM, Lory S (2006) Analysis of *Pseudomonas aeruginosa* diguanylate cyclases and phosphodiesterases reveals a role for bis-(3′-5′)-cyclic-GMP in virulence. Proc Natl Acad Sci U S A 103:2839–2844
13. Simm R, Morr M, Remminghorst U, Andersson M, Römling U (2009) Quantitative determination of cyclic diguanosine monophosphate concentrations in nucleotide extracts of bacteria by matrix-assisted laser desorption/ionization-time-of-flight mass spectrometry. Anal Biochem 386:53–58
14. Spangler C, Böhm A, Jenal U, Seifert R, Kaever V (2010) A liquid chromatography-coupled tandem mass spectrometry method for quantitation of cyclic di-guanosine monophosphate. J Microbiol Methods 81:226–231
15. Waters CM, Lu W, Rabinowitz JD, Bassler BL (2008) Quorum sensing controls biofilm formation in *Vibrio cholerae* through modulation of cyclic di-GMP levels and repression of *vpsT*. J Bacteriol 190:2527–2536
16. Buckstein MH, He J, Rubin H (2008) Characterization of nucleotide pools as a function of physiological state in *Escherichia coli*. J Bacteriol 190:718–726
17. Wagner VE, Bushnell D, Passador L, Brooks AI, Iglewski BH (2003) Microarray analysis of *Pseudomonas aeruginosa* quorum-sensing regulons: effects of growth phase and environment. J Bacteriol 185:2080–2095
18. Witte G, Hartung S, Büttner K, Hopfner KP (2008) Structural biochemistry of a bacterial checkpoint protein reveals diadenylate cyclase activity regulated by DNA recombination intermediates. Mol Cell 30:167–178
19. Woodward JJ, Iavarone AT, Portnoy DA (2010) c-di-AMP secreted by intracellular *Listeria monocytogenes* activates a host type I interferon response. Science 328:1703–1705
20. Lee ER, Baker JL, Weinberg Z, Sudarsan N, Breaker RR (2010) An allosteric self-splicing ribozyme triggered by a bacterial second messenger. Science 329:845–848

Chapter 23

Metabolic Footprinting: Extracellular Metabolomic Analysis

Volker Behrends, Huw D. Williams, and Jacob G. Bundy

Abstract

Uptake and excretion of nutrients is an integral part of a cell's physiology. Using analytical chemistry techniques, metabolite uptake and excretion from the culture medium can be quantified. As cellular metabolism changes throughout growth, additional information is available if transient and growth phase-dependent changes are monitored. Here, we describe time-resolved metabolic footprinting (TReF), a technique which employs nuclear magnetic resonance spectroscopy and nonlinear curve fitting to understand and visualize metabolite utilization of *P. aeruginosa*.

Key words NMR, Metabolomics, Nonlinear fitting, Metabolite uptake, Excretion, Utilization

1 Introduction

Metabolic profiling or metabolomics has proven to be a powerful approach for the characterization of the physiology of organisms at a functional level [1]. The approach employs analytical chemistry techniques—most commonly nuclear magnetic resonance (NMR) spectroscopy, or mass spectrometry (MS), generally with a prior chromatographic separation step—to measure the low-molecular-weight components of a cell or a biological system (*see* **Note 1**). It offers information complementary to other -omics techniques like transcriptomics and proteomics and is useful in the context of functional genomics and hypothesis generation [2, 3].

Compared to eukaryotic or multicellular organisms where each subcellular compartment has its own metabolome, and there may be multiple cell and tissue types, bacterial systems are comparatively simple. The metabolic entities are the exo-metabolome (culture media) and the endo-metabolome (the inside of the cell). The two are by no means independent: cellular metabolism is to a great degree influenced by extracellular conditions, such as osmolarity or nutrient availability, while the culture medium becomes a cumulative picture of intracellular activity, as it is changed by the cells' efforts to take up nutrients and maintain a working homeostasis [4].

Alain Filloux and Juan-Luis Ramos (eds.), *Pseudomonas Methods and Protocols*, Methods in Molecular Biology, vol. 1149, DOI 10.1007/978-1-4939-0473-0_23, © Springer Science+Business Media New York 2014

Experimentally, quantification of the endo-metabolome (metabolic fingerprinting) is more complex than for the exo-metabolome (metabolic footprinting). Sampling cultures for footprinting analysis only requires separating media from cells, but sampling the intracellular metabolome typically requires "quenching," i.e., rapid halting of metabolism; the efficient separation of cells and media; and a reliable nonselective extraction of metabolites. Therefore, studies of the intracellular metabolome are technically demanding and results are at greater risk of being biased than studies of the culture media [5, 6].

The data gathered on culture media composition can be utilized in a variety of ways. By quantifying the utilization differences of a large number of clinical isolates from CF patients, we were able to describe metabolic adaptations of *P. aeruginosa* to the CF lung [7] and identify an osmotic tolerance phenotype of the *mucA* mutation [8]. Additionally, vertebrate and bacterial studies have shown that explicitly considering "through-time" responses (metabolic trajectories) adds considerably to the description and understanding of biological events [9, 10]. Here, we show a general protocol for sampling and quantifying the exo-metabolome by NMR spectroscopy and several examples as to how to exploit data gathered through time.

2 Materials

If not otherwise stated, all chemicals are available from Sigma Aldrich, Gillingham, UK. Always use ultrapure (>18 MΩ) water.

2.1 Culture Media

We have successfully used metabolic footprinting for a variety of defined and rich media. We recommend using one of the following:

1. LB: Dissolve tryptone (10 g) (Oxoid, Basingstoke, UK), 5 g of yeast extract (Oxoid, Basingstoke, UK), and 5 g of NaCl (BDH Chemicals, Poole, UK) in 1 L deionized water and autoclave at 121 °C, 15 psi, for 20 min.
2. Synthetic cystic fibrosis medium, SCFM [11]:
   (a) Stock solutions: 1 M NaH_2PO_4: dissolve 120 g NaH_2PO_4 in 1 L of water; 1 M Na_2HPO_4: dissolve 142 g in 1 L of water; 1 M KNO_3: dissolve 101 g in 1 L of water; 0.2 M $CaCl_2$: dissolve 22 g in 1 L of water; 2 M $MgCl_2$: dissolve 190.4 g in 1 L of water; 0.02 M $FeSO_4$: dissolve 56 mg in 10 ml of water.

   (b) Base medium: Add 6.25 ml 1 M Na_2HPO_4, 6.25 ml 1 M NaH_2PO_4 (*see* **Note 2**), 0.348 ml 1 M KNO_3, 8.77 ml 0.2 M $CaCl_2$, 0.303 ml 2 M $MgCl_2$, 1.114 g KCl, 0.122 g NH_4Cl, and 3.03 g NaCl to ca. 700 ml water; adjust pH to 7; and autoclave.

(c) Amino acids: Prepare 100 mM stocks of the amino acids mentioned below, and filter sterilize the following amounts into the autoclaved base medium: 14.46 ml L-serine, 10.72 ml L-threonine, 17.8 ml L-alanine, 12.03 ml L-glycine, 16.61 ml L-proline, 11.2 ml L-isoleucine, 16.09 ml L-leucine, 11.17 ml L-valine, 8.27 ml L-aspartate, 15.49 ml L-glutamate, 5.3 ml L-phenylalanine, 8.02 ml L-tyrosine, 0.13 ml L-tryptophan, 21.28 ml L-lysine, 5.19 ml L-histidine, 3.06 ml L-arginine, 6.76 ml L-ornithine, 1.6 ml L-cysteine, 6.33 ml L-methionine (*see* **Note 3**).

(d) Carbon sources and iron: Filter sterilize 2.7 ml D-glucose, 10 ml L-lactate, and 0.18 ml $FeSO_4$ into the medium from 20 %, 10 %, and 0.02 M stocks, respectively.

(e) Also *see* original publication, ref. 11.

3. Artificial sputum medium, ASM [12]:

(a) Stock solution: 100× diethylene triamine pentaacetic acid (DTPA): Dissolve 0.059 g in 100 ml water.

(b) Base medium: Add 10 ml 1 M Na_2HPO_4, 10 ml 1 M NaH_2PO_4 (for preparation *see* SCFM), 10 ml DTPA stock solution, 5 g NaCl, and 2.2 g KCl to ca. 300 ml water; adjust to pH 6.9; and autoclave.

(c) Mucin: Dissolve 5 g of mucin (mucin from porcine stomach, type II) in 200 ml of H_2O using a stirrer hot plate at 80 °C for 1.5 h.

(d) DNA: Dissolve 4 g DNA (DNA from fish sperm) in 200 ml of H_2O using a stirrer hot plate at 50 °C overnight.

(e) Amino acids: Weigh out 0.25 g of all 20 proteinogenic amino acids and dissolve in 10 ml water each (*see* **Note 3**).

(f) Mix all components, add 5 ml sterile egg yolk emulsion, adjust the pH to 6.9, and make up to 1 L (*see* **Note 4**).

(g) Also *see* original publication, ref. 12.

2.2 NMR Reagents

1. NMR buffer: 100 % ^2H_2O (D_2O) containing 5 mM 2,2-dimethyl-2-silapentane-d_6-5-sulfonic acid (DSS) and 25 mM NaN_3. Weight out 49.1 mg DSS and 81.3 mg NaN3 and dissolve in 50 ml D_2O (*see* **Note 5**).
2. NMR tubes (5 mm NMR Sample Tubes, 178 mm, VWR).
3. NMR tube caps (VWR).

2.3 Software

1. Various platforms are available for spectral processing. We use iNMR (Nucleomatica, Molfetta, Italy, http://www.inmr.net).
2. For assignment of spectral resonances we use BioMagResBank (http://www.bmrb.wisc.edu, [13]) and HMDB (http://www.hmdb.ca, [14]) and in-house data.

3. We have developed in-house code for quantification of spectral resonances in Matlab (Mathworks, Oxford, UK). Alternatively, commercial packages like iNMR or NMR Suite (Chenomx, Edmonton, Canada, [15]) or freeware packages like BATMAN [16] can be used.
4. Microsoft Excel or any comparable spreadsheet software.

2.4 Miscellaneous Equipment

1. Reaction tubes, 1.5 ml.
2. Pipettes (20–200 μl and 100–1000 μl) and tips.
3. Spectrophotometer to read OD_{600}.
4. Benchtop centrifuge for 1.5 ml reaction tubes.

2.5 Specialist Equipment

1. An NMR spectrometer (*see* **Note 6**): We routinely measure on a Bruker Avance DRX600 (Bruker BioSpin, Rheinstetten, Germany), equipped with an inverse 5 mm probe.

3 Methods

3.1 Sampling

Strains should be inoculated from pre-cultures to a starting OD_{600} of <0.05 and can be grown aerobically as well as anaerobically using standard microbiological techniques (*see* **Notes 7** and **8**). We advise to use at least three biological replicates but recommend $n = 5$.

1. Label two 1.5 ml reaction tubes per sample, as one is used for sample collection and centrifugation and one for mixing with the NMR buffer.
2. Sample 750 μl of culture at appropriate time points along the growth curve (we use 1- or 2-h intervals). It is important to include a $t=0$ sample of the medium immediately following inoculation but before incubation. Use 100 μl (+900 μl media) to record the OD_{600}, and centrifuge the remainder of the sample (13k rpm, $16{,}100 \times g$, RT, 5 min). Transfer 480 μl of the supernatant to the second reaction tube and mix with 120 μl NMR buffer.
3. Transfer 550 μl of the mixture to an NMR tube.

3.2 NMR

Acquisition of spectra on an NMR spectrometer—in our case with a magnetic field strength of 14.1T and resulting ^{1}H resonance frequency of 600 MHz—follows an approach outlined in [17]. Using this approach, spectra are recorded on the ^{1}H channel using a one-dimensional pre-saturation pulse sequence (noesypr1d Bruker sequence) for water suppression. Data are acquired into 64 K data points over a spectral width of 12 kHz, with 8 dummy scans and 64 scans per sample and an additional longitudinal relaxation recovery delay of 3.5 s per scan, giving a total recycle time of about 5 s (*see* **Note 9**).

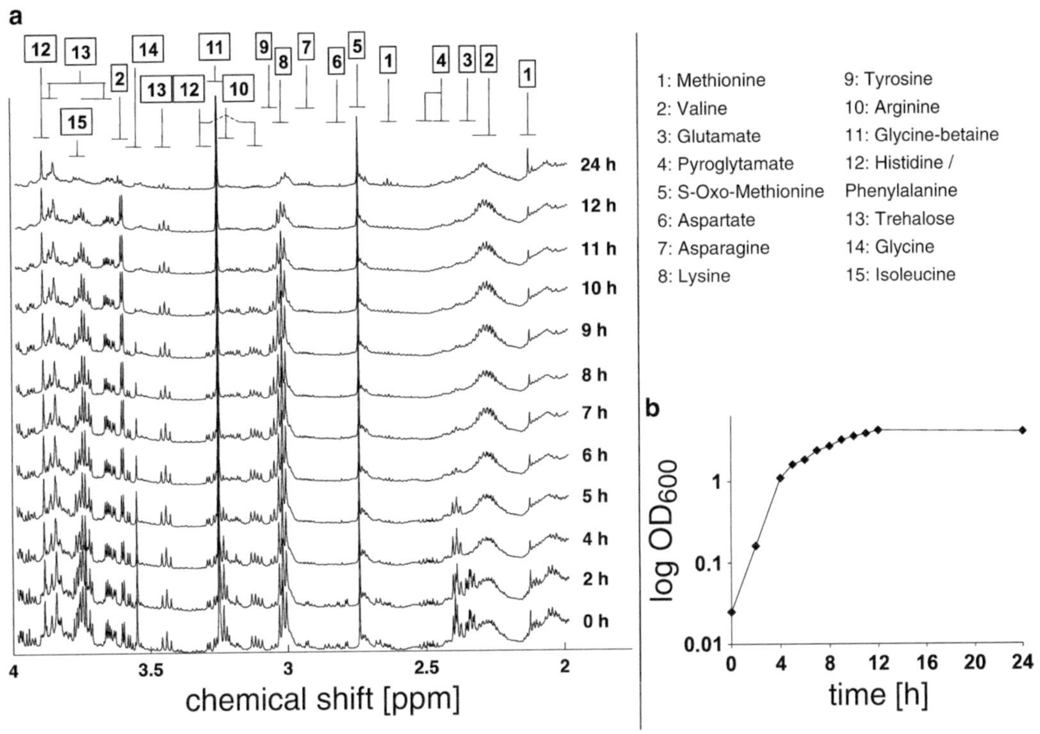

Fig. 1 Differential utilization of compounds from rich medium LB. (**a**) Section of a 600 MHz ^{1}H NMR spectra (4–2 ppm) for a single *P. aeruginosa* PAO1 culture supernatant sampled over the course of growth. Cultures were sampled at the times indicated, and spectra were normalized to DSS. (**b**) Growth curve

3.3 Spectral Processing and Integration (See Note 10)

1. Generate an Excel spreadsheet, which contains the metadata (strain, sampling time, OD_{600}, replicate, etc.) for each sample. Samples should be in rows and metadata in columns.
2. To preprocess, open spectra in iNMR and apply a Fourier transformation introducing an exponential apodization function equivalent to 0.5 Hz line broadening. For phasing, baseline correction and referencing iNMR offers automatic algorithms for batch-processing sets of spectra (*see* **Note 11**).
3. Sort spectra according to experiment number, and display with appropriate vertical offsets to allow integration (Fig. 1).
4. Using iNMR's integration tool (Tools → integrator), integrate the relevant resonances. For a list of resonances useful for quantification of LB and SCFM, *see* [10]. A preliminary assignment of the resonances is very useful for this step.
5. Export the data (Edit → Copy… → Integrals/Report the midpoints) to the Excel sheet, and align it with the metadata.
6. Normalize all integrals to DSS (the internal standard, present at 1 mM in all samples).

3.4 Data Analysis

The resulting integrals can be analyzed in different ways. As can be seen from Fig. 1, metabolites are taken up from the medium at different points of growth. As a first step, the integrals of metabolites can be plotted and the relative decrease or increase in signal can be monitored in Excel (Fig. 2).

To describe the decrease in a mathematically more robust manner, the metabolic changes can be described by a nonlinear function (*see* **Note 12**). For metabolic changes over growth, a sigmoid curve (Fig. 3) is best suited for most metabolites (*see* **Note 13**). In addition, the parameters of the fitted curves can be interpreted as intrinsically biologically meaningful data that can also be employed for visualization.

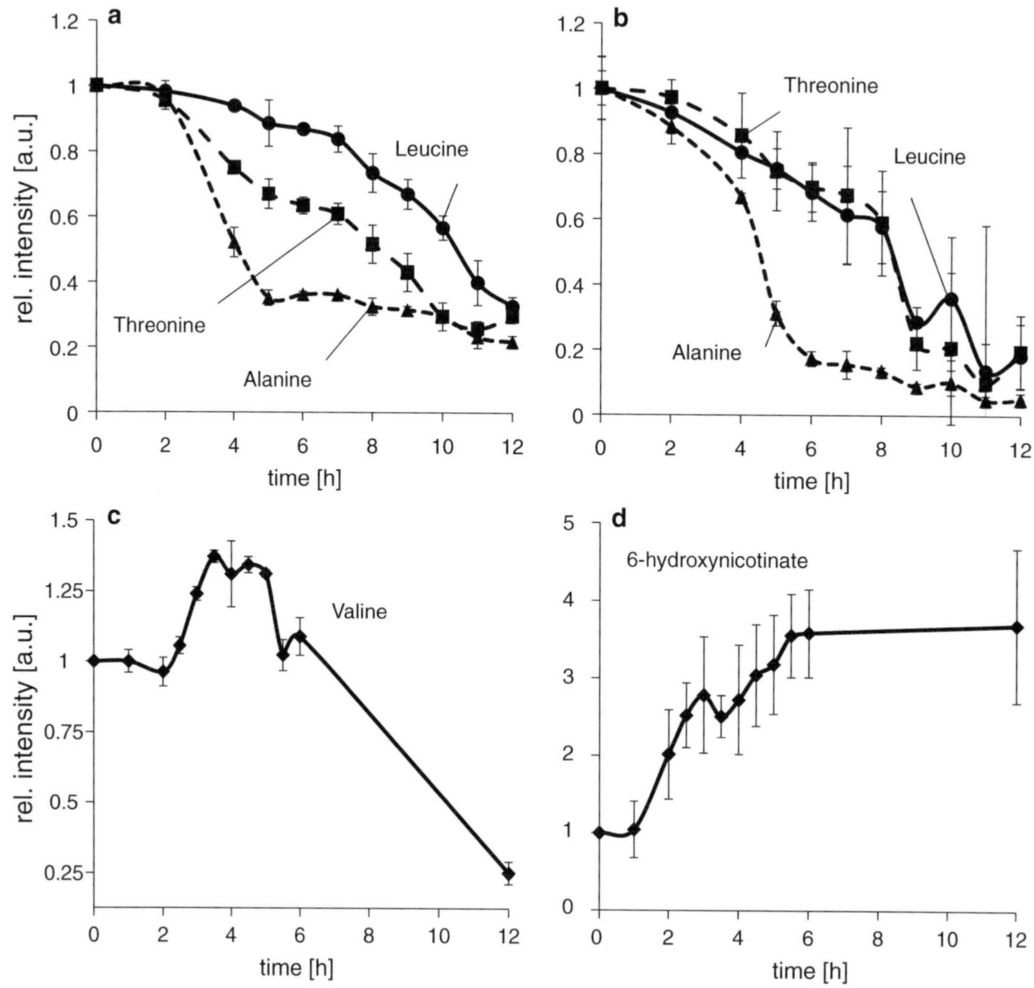

Fig. 2 Utilization of amino acids alanine (*triangles*), threonine (*squares*), and leucine (*circles*) for LB (**a**)- and SCFM (**b**)-grown PA01 cultures. Examples of transient (**c**) and continuous (**d**) excretion are shown for valine and 6-hydroxynicotinate, respectively. Figure 1a has been reprinted with permission from ref. 10. Error bars represent SD of four independent biological replicates

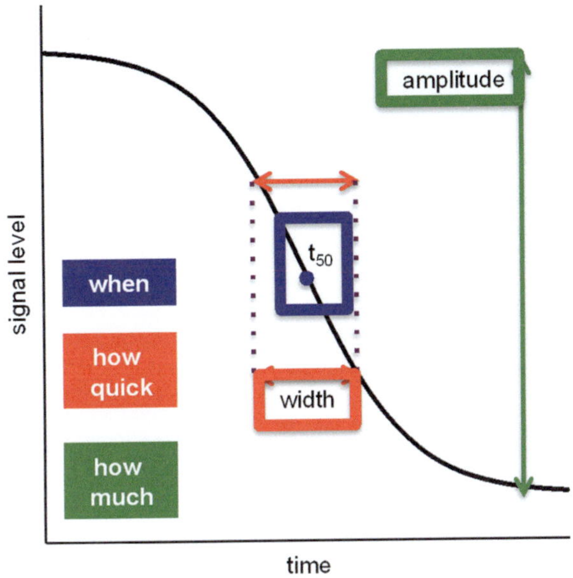

Fig. 3 Graphical representation of the sigmoid parameters and their biological meaning

$$y = \frac{a}{1+e^{b-cx}} + d = \frac{amplitude}{1+e^{-\frac{x-t_{50}}{width}}} + offset$$

Fig. 4 Generic logistic sigmoid equation and sigmoid equation with the parameters used here by substituting *a* with *amplitude*, *b* with t_{50}/*width*, *c* with *1*/*width*, and *d* with *offset*. The meaning of the parameters is explained in the text. *x* corresponds to the time the sample was taken, and *y* corresponds to the intensity of the integrated resonance

Nonlinear fitting:

The relative concentrations of many compounds increase or decrease in a sigmoid fashion. To fit, use a general sigmoid equation (Fig. 4).

Fitting the data to a sigmoid curve compresses the time dimension of the data set into four parameters, of which three can be readily related to biologically meaningful information. The *amplitude* is the change in relative concentration over the fitted time; the *offset* is the relative concentration at the starting time point. The t_{50} value is the time at which the concentration has changed by half the amplitude. The *width* is the time during which the relative concentration decreases from 75 % of the starting value to 25 % of the starting value (Fig. 3).

Sigmoid parameters can be fitted to all integrals, but not all integrals are appropriate for fitting with this model. To ensure better data quality, certain cutoff criteria should be applied. We devised a relative error (rmspe) by dividing the sum of residuals by the difference of highest and lowest values (Fig. 5). This way, rmspe takes

$$rmspe = \frac{\sqrt{\sum (y_i - y_i^{pred})^2}}{(y_{i_{max}} - y_{i_{min}})}$$

Fig. 5 Calculation of the relative error term *rmspe*. y_i corresponds to the intensity of the integrated resonance, and y_i^{pred} is the intensity predicted by the fitted sigmoid equation

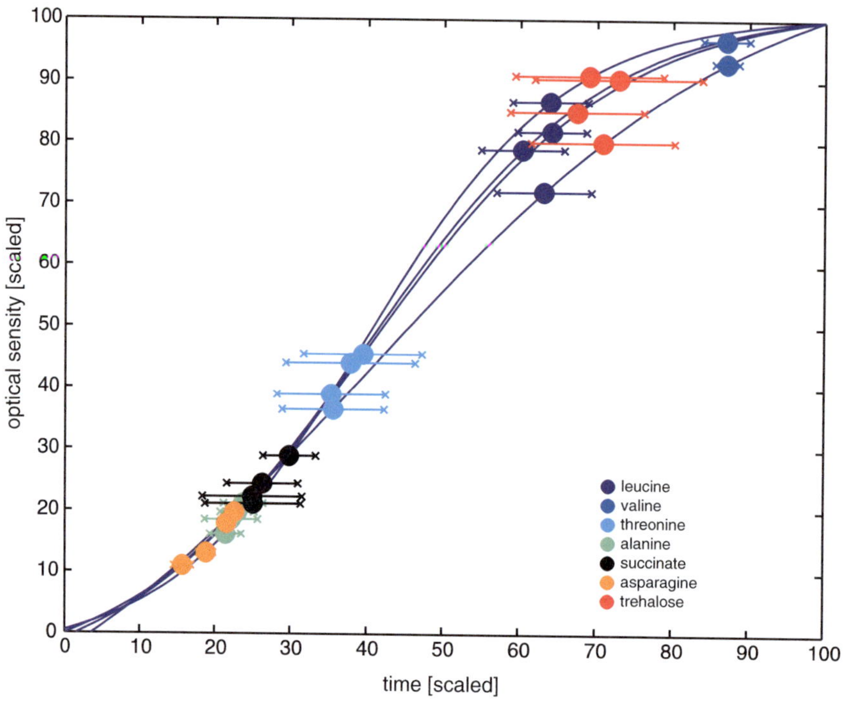

Fig. 6 Uptake plot of seven compounds visualizes differential compound uptake for LB-grown PAO1. Each compound's t_{50} is back-projected upon the growth curve of the replicate. The "error bars" represent calculated width (*see* Fig. 3 for illustration of t_{50} and width). Note that both the time as well as the optical density axes have been scaled to 100 % of the maximum value. The figure has been reprinted from ref. 10

into account the fact that variables with large intensities have greater intrinsic variability. We further recommend the following cutoff criteria for meaningful sigmoid fitting:

1. t_{50}: 0 h < t_{50} < end of experiment
2. Width: 0 h < width < 3 h
3. rmspe < 0.35

Uptake plots:

As both the t_{50} and width are in units of time, they define a utilization time span or "uptake window" for each fitted metabolite. By projecting these back onto the growth curve, three things are achieved (Fig. 6). Firstly, differences and hierarchies of compound uptake can be directly visualized. Secondly, comparing different replicates allows visual assessment of the quality of the fitting for various compounds. Thirdly, uptake plots can be used to compare compound uptake between different physiological growth conditions or between different species or strains (*see* **Note 14**).

4 Notes

1. Metabolic profiling is not a complete-coverage technique like transcriptomics, and it is not possible to measure the true "metabolome" as originally defined by Oliver [18], which includes all low-molecular-weight components of the cell or the cellular compartment due to the variety of chemical natures of metabolites and concentration ranges at which they are present. For culture media profiling, NMR offers the advantage of universal detection, while its disadvantage of having a relatively high detection threshold is alleviated by the relatively high concentrations of compounds in the media. However, it would be perfectly feasible to adopt the approach that we describe in this chapter with a different analytical method such as GC-MS.
2. The original recipe for SCFM published by Palmer et al. contains 10 mM 3-(N-morpholino)-propanesulfonic acid (MOPS) and 2.5 mM phosphate buffer. MOPS has pH-sensitive NMR-visible resonances that obscure large regions of the NMR spectrum, and so we omit MOPS from our version of SCFM and instead increase the phosphate buffer concentration to 12.5 mM.
3. All amino acids for SCFM and ASM can be dissolved in water at room temperature, except the following: L-methionine, L-phenylalanine, L-tyrosine, L-cysteine, L-asparagine, and L-aspartic acid, which are dissolved by dropwise adding conc. HCl, and tryptophan and histidine, which are dissolved in hot H_2O.
4. After mixing, ASM can be used directly but is cloudy, and bacterial growth cannot therefore be monitored by changes in optical density. For ASM preparation, we therefore adopted the filtration approach shown in [19]. To clear and sterilize the medium by filtration we employ an integrated filtration system (bottle top filter unit, 1 L collection flasks, 0.22 μm PES membrane, VWR collection). To shorten filtration times, we recommend "pre-filtering" the medium with a rough-textured filter to avoid clogging of the 0.22 μm filter by the medium's components.
5. The ^2H_2O provides a field frequency lock for the spectrometer, and the DSS serves as an internal chemical shift reference. Depending on the buffering capacity of the medium, the NMR sample can also be buffered by adding phosphate buffer (typically to a final concentration of 50–250 mM, with

100 mM being a good compromise between buffering ability and compatibility with NMR), pH 7 in D_2O to the NMR buffer to avoid pH-induced resonance shifting.

6. NMR spectrometers are specialized pieces of equipment, and it is generally unrealistic to expect a microbiology lab to be able to obtain one. However most large institutions will have an NMR spectrometer or spectrometers, and they are frequently operated as a service for multiple users.
7. Each sampling step removes 750 μl from the cultures. If aeration status is crucial, the volume should be adjusted so that no more than 20 % of the total volume is removed from the culture.
8. We routinely grow strains in conical flasks or universals in batch culture. Sampling from fermenters or flow-through chamber is also possible but is unlikely to be compatible with sigmoidal curve fitting due to media replenishment.
9. The number of increments can be increased for better signal-to-noise, and the relaxation delay (D1) should not be shorter than 2 s.
10. We carry out the preprocessing steps in iNMR, then export the resulting data as an ASCII matrix, and employ in-house code written in the Matlab environment for integration. Though the code is available on request, we have decided to describe the principal operations necessary to replicate the integration as operations that can be carried out in iNMR and Excel. Metabolite quantitation can also be carried out by fitting spectra to spectral templates using BATMAN or NMR Suite. The fitting approach is potentially more accurate than taking simple integrals because it allows for deconvolution of overlapping signals, but it is also more complex. For greatest ease of use and software support, we recommend NMR Suite; for highest throughput, we recommend BATMAN, as it runs in automated fashion. However in our experience it is likely to take considerable skilled user input to optimize the input parameters for BATMAN to work reliably, and attention should be paid to monitoring the quality of output of the automated fitting process.
11. In iNMR, we use the "Process folder" option in the "Process" menu, selecting "Metabolomic phase-correction," "Automatic first order baseline correction," "Locate reference at: 0 ± 0.1 ppm," and "External documents as windowless overlays."
12. As before, we use in-house functions to fit the metabolite data to a sigmoid function. The main function used from the statistics toolbox in Matlab is "nlinfit," supplying time, variable

$$\Delta t_{50_i} = \frac{t_{50_i} - t_{50_{OD}}}{width_{OD}}$$

Fig. 7 Modification to the t_{50} parameter of sigmoid-fitted variables to account for growth rate differences between samples

intensity, the sigmoid function given in Fig. 4, and a rough estimate of the sigmoid parameters as arguments. The function returns the four parameters of the sigmoid. We then used the parameters to calculate y^{pred} and rmspe.

13. Not all changes in the medium can be described as sigmoid functions. Metabolites can also show transient excretion, followed by depletion or transient depletion, followed by excretion (Fig. 2). In these cases, sigmoid fitting will not result in sensible values.
14. If strains with different growth rates are compared, this will result in higher t_{50} values for slower growing strains. One way to circumvent this is to correct the sigmoid parameters for growth curve bias by expressing them as relative values. To do this, the OD values are fitted to a sigmoid curve, giving a set of four parameters (amplitude, t_{50}, width, and offset) for the OD data. To express the t_{50} of each metabolite relative to the t_{50} of the growth curve, the t_{50OD} is subtracted from the t_{50} of each fitted metabolite with the result being divided by the $width_{OD}$ (Fig. 7).

5 Conclusion

Metabolic footprinting allows the comparison of bacterial physiology in different nutritional environments. The method has been successfully applied in studies investigating physiology [8], phylogenetic classification [9], or long-term adaptation [7, 20] and is generally applicable across the bacteria and not limited to use with *Pseudomonas*.

Acknowledgements

The authors thank Dr Olaf Beckonert and Dr Anthony Maher for technical assistance with NMR spectroscopy and Prof. Jeremy Nicholson and Prof. Elaine Holmes for support and access to facilities.

References

1. Ryan D, Robards K (2006) Metabolomics: the greatest omics of them all? Anal Chem 78:7954–7958
2. Nicholson JK, Lindon JC (2008) Systems biology: metabonomics. Nature 455:1054–1056
3. Raamsdonk LM et al (2001) A functional genomics strategy that uses metabolome data to reveal the phenotype of silent mutations. Nat Biotechnol 19:45–50
4. Kell DB (2004) Metabolomics and systems biology: making sense of the soup. Curr Opin Microbiol 7:296–307
5. Mashego MR, Rumbold K, De Mey M, Vandamme E, Soetaert W, Heijnen JJ (2007) Microbial metabolomics: past, present and future methodologies. Biotechnol Lett 29:1–16
6. van Gulik WM (2010) Fast sampling for quantitative microbial metabolomics. Curr Opin Biotechnol 21:27–34
7. Behrends V, Ryall B, Zlosnik JEA, Speert DP, Bundy JG, Williams HD (2012) Metabolic adaptations of *Pseudomonas aeruginosa* during cystic fibrosis chronic lung infections. Environ Microbiol 15:398–408. doi:10.1111/j.1462-2920.2012.02840.x
8. Behrends V, Ryall B, Wang X, Bundy JG, Williams HD (2010) Metabolic profiling of *Pseudomonas aeruginosa* demonstrates that the anti-sigma factor MucA modulates osmotic stress tolerance. Mol Biosyst 6:562–569
9. Allen J, Davey HM, Broadhurst D, Heald JK, Rowland JJ, Oliver SG, Kell DB (2003) High-throughput classification of yeast mutants for functional genomics using metabolic footprinting. Nat Biotechnol 21:692–696
10. Behrends V, Ebbels TMD, Williams HD, Bundy JG (2009) Time-resolved metabolic footprinting for nonlinear modeling of bacterial substrate utilization. Appl Environ Microbiol 75:2453–2463
11. Palmer KL, Aye LM, Whiteley M (2007) Nutritional cues control *Pseudomonas aeruginosa* multicellular behavior in cystic fibrosis sputum. J Bacteriol 189:8079–8087
12. Sriramulu DD, Lünsdorf H, Lam JS, Römling U (2005) Microcolony formation: a novel biofilm model of *Pseudomonas aeruginosa* for the cystic fibrosis lung. J Med Microbiol 54: 667–676
13. Markley JL, Ulrich EL, Berman HM, Henrick K, Nakamura H, Akutsu H (2008) BioMagResBank (BMRB) as a partner in the Worldwide Protein Data Bank (wwPDB): new policies affecting biomolecular NMR depositions. J Biomol NMR 40:153–155
14. Wishart DS et al (2007) HMDB: the Human Metabolome Database. Nucleic Acids Res 35:D521–D526
15. Weljie AM, Newton J, Mercier P, Carlson E, Slupsky CM (2006) Targeted profiling: quantitative analysis of 1H NMR metabolomics data. Anal Chem 78:4430–4442
16. Hao J, Astle W, De Iorio M, Ebbels TMD (2012) BATMAN—an R package for the automated quantification of metabolites from nuclear magnetic resonance spectra using a Bayesian model. Bioinformatics 28:2088–2090
17. Beckonert O, Keun HC, Ebbels TMD, Bundy J, Holmes E, Lindon JC, Nicholson JK (2007) Metabolic profiling, metabolomic and metabonomic procedures for NMR spectroscopy of urine, plasma, serum and tissue extracts. Nat Protoc 2:2692–2703
18. Oliver SG, Winson MK, Kell DB, Baganz F (1998) Systematic functional analysis of the yeast genome. Trends Biotechnol 16: 373–378
19. Kirchner S, Fothergill JL, Wright EA, James CE, Mowat E, Winstanley C (2012) Use of artificial sputum medium to test antibiotic efficacy against *Pseudomonas aeruginosa* in conditions more relevant to the cystic fibrosis lung. J Vis Exp. doi:10.3791/3857
20. Lawrence D, Fiegna F, Behrends V, Bundy JG, Phillimore AB, Bell T, Barraclough TG (2012) Species interactions alter evolutionary responses to a novel environment. PLoS Biol 10:e1001330. doi:10.1371/journal.pbio.1001330

Chapter 24

Pyoverdine and Pyochelin Measurements

Françoise Hoegy, Gaetan L.A. Mislin, and Isabelle J. Schalk

Abstract

Siderophores are small organic chelators (of molecular weight between 200 and 2,000 Da), having a very high affinity for iron (10^{17}–10^{43} M^{-1}). They are synthesized by bacteria and secreted into their environment in order to get access to iron, an essential element for bacterial growth. Pyoverdine (also called fluorescins or pseudobactins) and pyochelin are the two major siderophores produced by *Pseudomonas aeruginosa* in iron-limited media. Methods to specifically detect and measure the amount of pyoverdine and pyochelin in a bacterial culture are provided here. These methods are based on the spectral properties of these two siderophores.

Key words Siderophores, Pyoverdine, Pyochelin, Iron uptake, *Pseudomonas*

1 Introduction

Siderophores are organic chelators produced by microorganisms to make iron accessible [1]. They all have a low molecular weight (molecular weight between 200 and 2,000 Da) and present a high and specific affinity for iron(III) (10^{17}–10^{43} M^{-1}). *Pseudomonas aeruginosa* produces two siderophores in iron-limited media: pyoverdine (PVD), which gives also this typical green color to the *Pseudomonas* culture, and pyochelin (PCH). Pyochelin structure was established as (4′*R*, 2″*R*, 4″*R*)-2′-(2-hydroxyphenyl)-3″-methyl-4′,5′,2″,3″,4″,5″-hexahydro-[4′,2″]bisthiazolyl-4″-carboxylic acid (Fig. 1) and chelates iron with an affinity of 10^{17} M^{-1} and a stoichiometry of 2:1 for PCH:Fe [2, 3]. PVD is a more complex siderophore than PCH, since each species/strain of fluorescent pseudomonads produces its own PVD. So far more than 60 pyoverdines from different strains and species of *Pseudomonas* have been chemically identified [4–8]. All these PVDs are composed of three parts: a dihydroquinoline-type chromophore responsible for their spectral characteristics, a strain-specific peptide comprising 6–12 amino acids, and a side chain bound to the nitrogen atom at position

Alain Filloux and Juan-Luis Ramos (eds.), *Pseudomonas Methods and Protocols*, Methods in Molecular Biology, vol. 1149, DOI 10.1007/978-1-4939-0473-0_24, © Springer Science+Business Media New York 2014

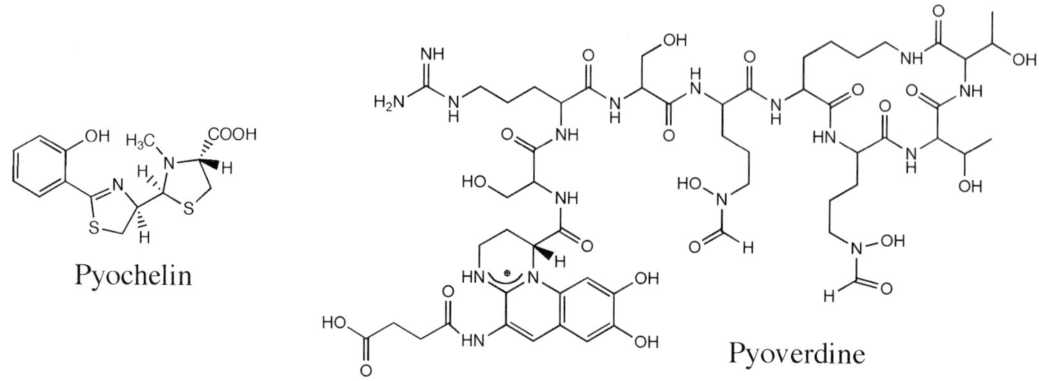

Fig. 1 Pyochelin and pyoverdine (PVD), the two siderophores produced by *P. aeruginosa* PAO1

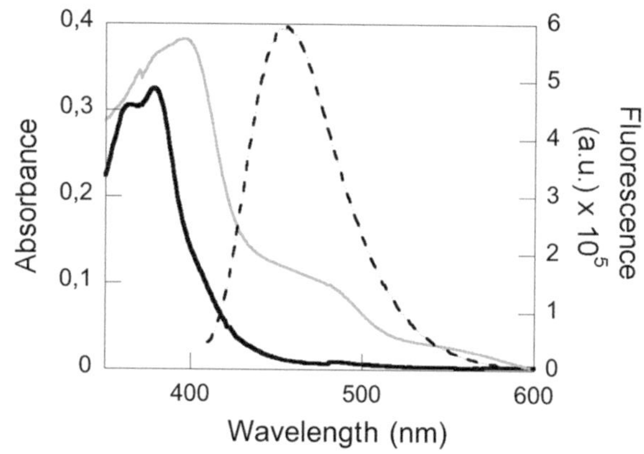

Fig. 2 UV and fluorescence spectra of PVD and PVD-Fe. UV spectra of 20 μM PVD (*black line*) and 20 μM PVD-Fe (*grey line*) in 50 mM Pyr/AcOH pH 5.0 buffer and fluorescent spectrum of 500 nM PVD (*dashed line*) in 50 mM Tris–HCl pH 8.0 buffer

C-3 of the chromophore [4–8]. PVD produced by *P. aeruginosa* PAO1 (Fig. 1) chelates iron with a 1:1 (PVD:Fe) stoichiometry and an affinity of 10^{32} M^{-1} [9]. Since the chromophore is common to each of these PVDs, the spectral method presented here to detect and quantify PVDs in Pseudomonads cultures can be used whatever is the PVD. PVDs have an absorbance at 380 or 400 nm depending on the pH (pH 5.0 and 8.0, respectively), is fluorescent at 447 nm [10] (Fig. 2), and can be directly detected and quantified in bacterial cultures. The ferric form has an absorbance at 400 and 450 nm and is not fluorescent [10] (Fig. 2). PCH is more difficult to detect in bacterial cultures and has to be extracted from the culture media. This siderophore is characterized by an absorbance at 313 nm and a fluorescence at 430 nm [2] (Fig. 3). For reviews describing PVD and PCH biosynthesis and iron uptake by these siderophores see the corresponding articles [11–13].

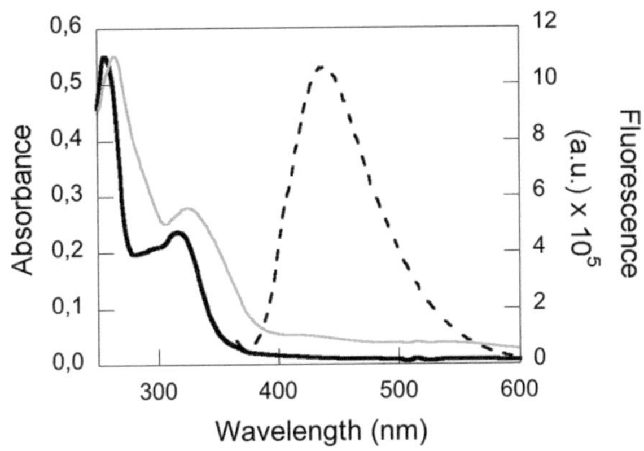

Fig. 3 UV and fluorescence spectra of PCH and PCH-Fe. UV spectra of 50 μM PCH (*black line*) and 50 μM PCH-Fe (*grey line*) in MeOH/H_2O 50/50 and fluorescent spectrum of 50 μM PCH (*dashed line*) 50 mM Tris–HCl pH 8.0 buffer

2 Materials

Prepare all solutions using bidistillated water and analytical grade reagents. Prepare and store all reagents at room temperature (unless indicated otherwise). Diligently follow all waste disposal regulations when disposing waste materials.

2.1 Bacterial Strains and Growth Media

1. *P. aeruginosa* strains are stored in LB (L-Broth or Luria Bertani, AthenaSE) with 20 % glycerol in 1.5 mL microtubes at −80 °C.
2. LB medium: Weigh 20 g of LB, transfer into a 1 L bottle, and complete to 1 L with water. Sterilize immediately after preparation, and store the solution at room temperature.
3. Succinate medium is an iron-restricted medium: Weigh 6 g of K_2HPO_4, 3 g of KH_2PO_4, 1 g of $(NH_4)_2PO_4$, 0.2 g of $MgSO_4$, 4 g of succinic acid, and 1.1 g of NaOH. Add water to a volume of 500 mL. Mix and adjust pH at 7.0 with NaOH. Make up to 1 L with water. Sterilize immediately after preparation, and store at room temperature.
4. Centrifugation tubes, Falcon 50 mL (Becton Dickinson, Franklin Lakes, USA).
5. Centrifuge 5804 (Eppendorf, Hamburg, Germany).
6. Semi-micro disposal cuvettes for spectroscopy, 1.5 mL, optical PS (Kartell, Noviglio, Italy).
7. Spectrophotometer UV-Visible, BioPhotometer (Eppendorf).

2.2 Pyochelin Extraction

1. Citric acid (Sigma-Aldrich).
2. Sintered glass funnel (porosity 2).
3. Filter paper Durieux N°111 green.

4. Chemist vacuum flask (5 L).
5. 250 mL separating funnel (Schott, Duran).
6. CH_2Cl_2 stabilized with amylene (Carlo Erba).
7. Desiccant: $MgSO_4$ anhydrous (Carlo Erba).
8. Evaporation of solvents: Rotavapor Büchi R-200 coupled to a vacuubrand PC 101 NT pump.

2.3 UV Spectrophotometry

1. Prepare 50 mM pyridine–acetic acid buffer pH 5.0 by mixing 4.05 mL pyridine and 2.86 mL acetic acid in 993.09 mL of water (*see* **Note 1**).
2. Prepare diluted MeOH (MeOH/H_2O, 80/20 by weight).
3. Spectrophotometer UV-Visible, Specord 205 (Analytikjena, Wembley, UK).
4. Quartz cuvette, 10 mm (catalog number 104B-QS 10 mm) (Hellma) for UV spectroscopy (*see* **Note 2**).

2.4 Spectrofluorimetry

1. Tris buffer: 50 mM Tris–HCl pH 8.
2. QuantaMaster PTI Spectrofluorimeter (Photon Technology International, Birmingham, NJ).
3. Quartz cuvettes (catalog number 109.004F-QS, 10×4 mm) (Hellma) for fluorescence spectroscopy.

2.5 TLC

1. Thin-layer chromatography (TLC) Geduran Kieselgel Si 60 (40–63 μm), 20×20 cm on alumina sheets (Merck).
2. 1-Butanol (Riedel-de Haën puriss).
3. Acetic acid for analysis 96 % (Merck).
4. Iron trichloride solution in water (27.5–29 %) (Prolabo).
5. UV lamp Vilber-Lourmat VL-4LC, with 2 tubes (4 W, 365 nm and 4 W, 254 nm).
6. Preparation of methanolic solution of ion trichloride: Dilute 10 mL of commercial iron trichloride solution in 100 mL of MeOH.

3 Methods

Carry out all procedures at room temperature unless otherwise specified.

3.1 Bacterial Culture and Preparation

1. Bacterial growth: Transfer 10 μL of glycerol bacterial solution stored at −80 °C into 10 mL of LB medium, and grow the cells overnight at 37 °C under agitation (220 rpm). The next day, dilute 10 mL of this culture in 100 mL of succinate medium,

and run a 24-h culture at 30 °C under agitation (220 rpm). The following day, dilute again the culture in succinate (10 mL in 100 mL medium) and allow them to grow for 24 h at 30 °C under agitation (*see* **Note 3**).

3.2 PVD Detection by UV-Visible Spectrophotometry

1. Before starting measurements, set the spectrophotometer wavelength range from 350 to 600 nm and realize a baseline measurement.
2. Prepare a quartz cuvette containing 1 mL of Pyr/AcOH 50 mM buffer pH 5.0, and place it in the control position of the spectrophotometer.
3. Prepare a quartz cuvette containing 100 μL of pseudomonads culture, complete to 1 mL with Pyr/AcOH 50 mM buffer pH 5.0, mix the solution, place the cuvette in the sample position of the machine, and run a measurement.
4. If PVD is present in the sample, a specific band with a maximum of absorption at 380 nm should appear, preceded by a substantial shoulder at 360 nm (Fig. 2). PVD can be quantified since absorbance is linearly correlated with concentration following the Beer–Lambert law, in the range of 0.1–1 of absorbance. At 380 nm, the molar extinction coefficient (ε) of PVD is 16,500 $M^{-1} \cdot cm^{-1}$ in Pyr/AcOH 50 mM buffer pH 5.0 [9] (*see* **Note 4**).
5. If PVD is in the presence of iron, the formation of PVD-Fe will occur to form a 1:1 complex (1 molecule of PVD chelating 1 iron ion). The formation of this compound causes modifications of the absorption spectra, mainly the disappearance of the 360 nm shoulder, the shift of the 380 nm band to 400 nm ($\varepsilon = 19{,}000\ M^{-1}\ cm^{-1}$), and appearance of a low band at 450 nm (Fig. 2) (*see* **Note 5**).

3.3 PVD Detection by UV Spectrofluorimetry

1. Only apo PVD is fluorescent; PVD in complex with iron is not [10].
2. Before starting measurements, set the spectrofluorimeter excitation wavelength at 400 nm and the wavelength range from 410 to 600 nm. The excitation and emission slits should be at 1.25 nm, the power of the lamp set at 60 W, and the photomultiplier at 900 V.
3. Prepare a quartz cuvette (for fluorescence spectroscopy) containing 100 μL of pseudomonads culture, complete to 1 mL with Tris–HCl pH 8.0 buffer, mix the solution, place the cuvette in the sample position of the machine, and run a measurement. Apo PVD, when excited at 400 nm, emits fluorescence at 447 nm (Fig. 2).

3.4 Pyochelin Extraction

1. Pellet 25 mL of pseudomonads culture in a 50 mL centrifugation tube at 4,000 × *g* for 8 min.
2. Collect the supernatant.
3. Filter 100 mL of the supernatant under vacuum on a sintered glass funnel fitted with an adapted filter paper disk (sintered glass funnel, vacuum flask) (*see* **Notes 6** and **7**).
4. Add solid citric acid to the filtrate until a pH of 3.0 is reached.
5. Extract twice the acidified supernatant with 50 mL CH_2Cl_2 in a separating funnel (*see* **Note 8**).
6. Collect and dry the organic phases over $MgSO_4$.
7. Filtrate the desiccant with a sintered glass funnel over a vacuum flask.
8. Evaporate the solvent under reduced pressure (rotavapor). PCH is isolated either as an oil (when produced in small amounts) or as a yellow solide (when produced in large amounts).
9. Store PCH at −20 °C (*see* **Note 9**).

3.5 Pyochelin Detection by UV-Visible Spectrophotometry

1. Before starting measurements, set the spectrophotometer wavelength range from 230 to 600 nm and realize a baseline measurement.
2. Prepare a solution of PCH in MeOH/H_2O (*see* **Note 10**).
3. Prepare a quartz cuvette containing 1 mL of MeOH/H_2O mixture, and place it in the control position of the spectrophotometer.
4. Prepare a cuvette containing the PCH solution, place it in the sample position of the machine, and run a measurement.
5. If PCH is present in the sample, one specific band with a maximum of absorption at 313 nm should appear (Fig. 3). PCH can be quantified since absorbance is linearly correlated with concentration following the Beer–Lambert law, in the range of 0.1–1 of absorbance. At 313 nm, the molar extinction coefficient (ε) of PCH is 4,900 M^{-1} cm^{-1} in MeOH/H_2O (*see* **Note 11**).
6. If PCH is in the presence of iron, formation of PCH-Fe will occur by chelation of iron by the siderophore with a 2:1 stoichiometry (2 molecules of PCH for 1 iron ion), but it does not affect the UV spectrum (Fig. 3) (*see* **Note 12**).

3.6 Pyochelin Detection by Spectrofluorimetry

1. Before starting measurements, set the spectrofluorimeter excitation wavelength at 355 nm and the wavelength range from 365 to 600 nm (*see* **Note 13**). The excitation and emission slits should be at 1.25 nm, the power of the lamp set at 60 W, and the photomultiplier at 900 V.
2. Prepare 1 mL of PCH at 100 μM in Tris–HCl buffer 50 mM pH 8 in a quartz cuvette, and run a measurement. PCH,

when excited at 355 nm, emits fluorescence at 430 nm (Fig. 3). At acidic pH, fluorescence signal of PCH reaches baseline (*see* **Note 14**).

3.7 Pyochelin Detection by Thin-Layer Chromatography

1. Eluent preparation: Mix 1-butanol, water, and acetic acid in the proportions 4/1/1.
2. TLC preparation: Cut commercial TLC sheets to the desired size (*see* **Note 15**).
3. Dissolve PCH in a small amount of MeOH, and spot it on the TLC.
4. Elute TLC sheets in the previously described eluent.
5. The natural diastereoisomers of PCH can be visualized on the TLC plate under UV light at 365 nm and, at a lesser extent, at 254 nm. Migration reference of PCH diastereoisomers produced by *P. aeruginosa* on TLC is as follows: R_f (major diastereoisomer) = 0.40 cm; R_f (minor diastereoisomer) = 0.66 cm; and R_f for retardation factor (*see* **Note 16**).
6. PCH diastereoisomers can also be revealed by spraying the TLC plate with a methanolic solution of ion trichloride (*see* **Note 17**).

4 Notes

1. Prepare this buffer under a hood to avoid bad smell and toxicity of pyridine and acetic acid.
2. Quartz is optically clear from 190 to 700 nm; it is useful in UV wavelength range.
3. Cells are rather grown at 30 °C at this stage because of the higher production of siderophores, compared to 37 °C cultures. Besides, one 24-h culture in iron-limited succinate medium is not enough to get the bacteria iron starved and to get a high production of siderophores.
4. If the maximum of absorption is more than 1, dilute the sample with Pyr/AcOH 50 mM buffer pH 5.0 to get an absorbance value between 0.1 and 1. If the absorption value obtained is less than 0.1, run a measurement of undiluted culture; in this case, the control cuvette should contain succinate medium. At pH 7.0, maximum of absorption of PVD will occur at 400 nm, with $\varepsilon = 16{,}500\ M^{-1}\cdot cm^{-1}$.
5. PVDI has a yellow-green color, and PVD in complex with iron is dark brown.
6. PCH being not highly stable, the culture supernatant should be either immediately filtered and PCH extracted or stored at −20 °C. Do not store the supernatant at pH 7.0 at 4 °C or at room temperature.

7. The filter paper protects the sintered glass funnel from clogging. Depending on the density of bacterial debris in the supernatant and the volume of supernatant to be filtered, the filter paper might be changed frequently.
8. The first extraction sometimes gives a thick emulsion. This can be avoided by saturating the aqueous phase with sodium chloride. The filtration of emulsion on a coton pad in a funnel can in many cases separate the organic and the aqueous phase.
9. Do not store PCH at 4 °C but always at −20 °C as a powder. When in solution, PCH is only stable for a few days even at −20 °C.
10. Due to its toxicity, use standard individual protection equipment when using methanol.
11. If the maximum of absorbance is more than 1, dilute the sample in the same solvent to get an absorbance value between 0.1 and 1. If the absorption value obtained is less than 0.1, run a measurement of a more concentrated PCH solution.
12. In concentrated MeOH solution, PCH is yellow. When complexed with iron, it becomes deep dark brown, nearly black.
13. The emission wavelength range should not include the excitation wavelength to avoid excessive lightening of the photomultiplier, which could cause long-term damage.
14. To improve fluorescence of the sample, you may have to modify different parameters of the apparatus (opening of the slits for example, or increasing lamp power, in the limits of the apparatus use) or concentrate the samples.
15. Usually 3 cm width for 8 cm length is enough for routine TLC analysis of PCH samples.
16. PCH has three chiral centers, and therefore different diastereoisomers can be produced by pseudomonads [14, 15].
17. The revelation of TLC with the lamps should be performed before the treatment with the ferric solution.

Acknowledgement

This work was supported by CNRS and ANR grants (ANR-08-BLAN-0315-01 and 05-JCJC-0181-01).

References

1. Hider RC, Kong X (2011) Chemistry and biology of siderophores. Nat Prod Rep 27: 637–657
2. Brandel J, Humbert N, Elhabiri M, Schalk IJ, Mislin GLA, Albrecht-Garry A-M (2012) Pyochelin, a siderophore of *Pseudomonas aeruginosa:* physicochemical characterization of the iron(III), copper(II) and zinc(II) complexes. Dalton Trans 41:2820–2834
3. Tseng CF, Burger A, Mislin GLA, Schalk IJ, Yu SS-F, Chan SI, Abdallah MA (2006) Bacterial siderophores: the solution

stoichiometry and coordination of the Fe(III) complexes of pyochelin and related compounds. J Biol Inorg Chem 11:419–432

4. Budzikiewicz H (1997) Siderophores of fluorescent pseudomonads. Z Naturforsch C 52:713–720
5. Budzikiewicz H (2004) Siderophores of the *Pseudomonadaceae sensu stricto* (fluorescent and non-fluorescent Pseudomonas spp.). Fortschr Chem Org Naturst 87:81–237
6. Budzikiewicz H, Schafer M, Fernandez DU, Matthijs S, Cornelis P (2007) Characterization of the chromophores of pyoverdins and related siderophores by electrospray tandem mass spectrometry. Biometals 20:135–144
7. Demange P, Wendenbaum S, Linget C, Mertz C, Cung MT, Dell A, Abdallah MA (1990) Bacterial siderophores: structure and NMR assignment of pyoverdins PaA, siderophores of Pseudomonas aeruginosa ATCC 15692. Biol Metals 3:155–170
8. Fuchs R, Budzikiewicz H (2001) Structural studies of pyoverdins by mass spectrometry. Curr Org Chem 5:265–288
9. Albrecht-Gary AM, Blanc S, Rochel N, Ocacktan AZ, Abdallah MA (1994) Bacterial iron transport: coordination properties of pyoverdin PaA, a peptidic siderophore of *Pseudomonas aeruginosa*. Inorg Chem 33: 6391–6402
10. Folschweiller N, Gallay J, Vincent M, Abdallah MA, Pattus F, Schalk IJ (2002) The interaction between pyoverdin and its outer membrane receptor in *Pseudomonas aeruginosa* leads to different conformers: a time-resolved fluorescence study. Biochemistry 41:14591–14601
11. Schalk IJ (2008) Metal trafficking via siderophores in Gram-negative bacteria: specificities and characteristics of the pyoverdine pathway. J Inorg Biochem 102:1159–1169
12. Schalk IJ, Guillon L (2013) Pyoverdine biosynthesis and secretion in *Pseudomonas aeruginosa*: implications for metal homeostasis. Environ Microbiol 15(6):1661–1673
13. Youard ZA, Wenner N, Reimmann C (2011) Iron acquisition with the natural siderophore enantiomers pyochelin and enantio-pyochelin in *Pseudomonas species*. Biometals 24:513–522
14. Hoegy F, Lee X, Noël S, Mislin GL, Rognan D, Reimmann C, Schalk IJ (2009) Stereospecificity of the siderophore pyochelin outer membrane transporters in fluorescent pseudomonads. J Biol Chem 284: 14949–14957
15. Youard ZA, Mislin GL, Majcherczyk PA, Schalk IJ, Reimmann C (2007) Pseudomonas fluorescens CHA0 produces enantio-pyochelin, the optical antipode of the Pseudomonas aeruginosa siderophore pyochelin. J Biol Chem 282:35546–35553

Chapter 25

Measurement of Phenazines in Bacterial Cultures

Suzanne E. Kern and Dianne K. Newman

Abstract

Certain pseudomonads are capable of producing phenazines—pigmented, reversibly redox-active metabolites that induce a variety of physiological effects on the producing organism as well as others in their vicinity. Environmental conditions and the specific physiological state of cells can dramatically affect the absolute amounts and relative proportions of the various phenazines produced. The method detailed here—high-performance liquid chromatography coupled to detection by UV–Vis absorption—can be used to separate and quantify the amount of phenazines in a *Pseudomonas* culture. Simple spectrophotometric measurements of filtered culture supernatants can be used to quantify certain oxidized phenazines, such as pyocyanin, in cultures. For cases where the conditions under study are not planktonic cultures (e.g., soil or biofilms) extracting the phenazines may be a necessary first step.

Key words Phenazine, Pyocyanin, Phenazine-1-carboxylic acid, 1-Hydroxyphenazine, PCA, PYO, HPLC, UV–Vis spectroscopy

1 Introduction

Phenazines are brightly colored small molecules naturally produced by pseudomonads and other microorganisms (Fig. 1). For example, pyocyanin (PYO) is a well-studied blue phenazine produced by the opportunistic human pathogen *Pseudomonas aeruginosa*. In pseudomonads, biosynthesis of the precursor phenazine molecule phenazine-1-carboxylic acid (PCA) from chorismate is encoded by the *phzABCDEFG* operon, with additional phz genes responsible for adding specific functional groups [1, 2]. By virtue of their redox activity, phenazines can generate reactive oxygen species upon reducing molecular oxygen [3, 4] and enhance electricity generation in microbial fuel cells [5, 6]. In ecological contexts, phenazines are important for the virulence of *P. aeruginosa* [7] as well as the protection afforded by certain pseudomonads against fungal pathogens of plants [1, 8]. In addition, they can directly benefit producing cells by making iron more bioavailable [9, 10],

Alain Filloux and Juan-Luis Ramos (eds.), *Pseudomonas Methods and Protocols*, Methods in Molecular Biology, vol. 1149, DOI 10.1007/978-1-4939-0473-0_25, © Springer Science+Business Media New York 2014

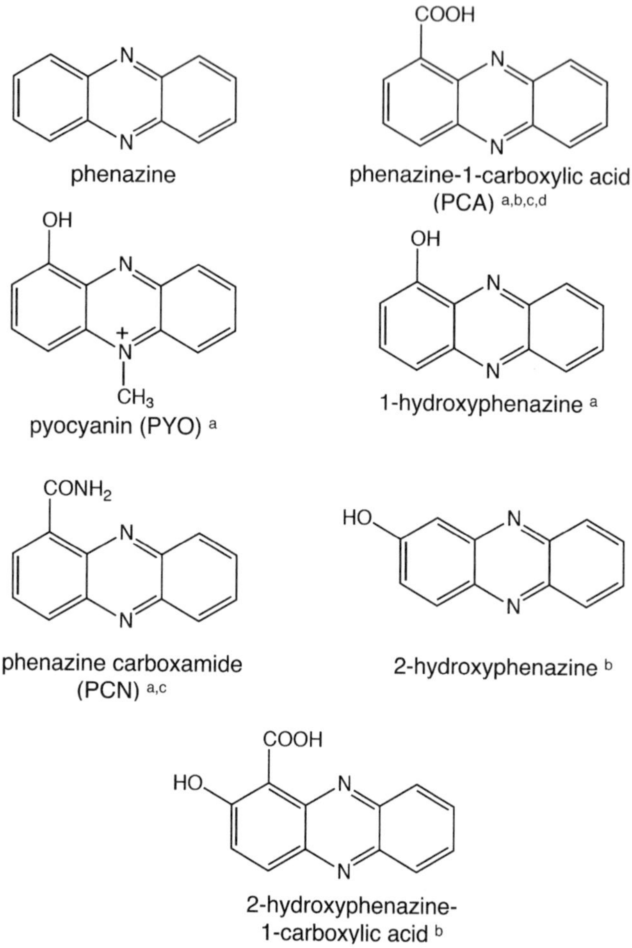

Fig. 1 Structures of the core phenazine molecule and selected phenazines produced by pseudomonads. Note that the structures indicate the oxidized, protonated forms. Examples of species that may produce each phenazine: (**a**) *P. aeruginosa,* (**b**) *P. aureofaciens,* (**c**) *P. chlororaphis,* (**d**) *P. fluorescens* [1]

enhancing biofilm development [11–13], oxidizing the intracellular NAD(P)(H) pool [14], enabling anaerobic survival [15], and acting as inter- and intracellular signals [16].

In recent years, several methods have been developed to probe phenazine levels in specific environments and conditions. These exploit the various properties of phenazines, for instance, (1) their redox activity, via cyclic voltammetry studies of biofilms and surrounding medium [17]; (2) their particular fluorescent properties, using multiphoton microscopy for in situ detection [18]; and (3) their light absorption properties, as measured by ultraviolet and visible spectrum spectroscopy [19]. Here we elaborate on the method of high-performance liquid chromatography (HPLC) separation and UV absorption for quantifying phenazines, using PYO, PCA, and 1-hydroxyphenazine (1-OH-PHZ) as examples. This technique is very sensitive (capable of detecting concentrations as low as 2 μM for PYO and PCA and 20 μM for 1-OH-PHZ) and involves equipment that can be found in many biology and chemistry laboratories. Figure 1 shows examples of phenazines that may *be analyzed using reverse-phase chromatography as outlined here. For other phenazines, including* 5-methyl-phenazinium-1-carboxylate, alternative separation methods may be necessary [20]. Although this method only describes sample processing for planktonic cultures, HPLC analysis can also be used on other liquid samples, for instance, the sputum from individuals with cystic fibrosis [21], and solutions of phenazines extracted from more complex conditions, such as biofilms (*see* **Note 1**). We also include instructions for determining the concentration of PYO by direct spectroscopy of filtered culture supernatants (*see* **Note 2**).

2 Materials

2.1 Reagents

All solutions, including water, should be of HPLC grade.

1. Acidified water solution: 0.1 % trifluoroacetic acid (TFA) in water.
2. Acidified acetonitrile solution: 0.1 % TFA in acetonitrile.
3. Growth medium or buffer to serve as blank samples and for background subtraction.
4. Mixtures of growth medium or buffer with known concentrations of phenazines.

2.2 Equipment and Supplies

1. Pipettes and tips.
2. Microcentrifuge tubes.
3. 0.2 μm pore size filters, either syringe filters or filter tubes (e.g., Costar Spin-X HPLC Micro Centrifuge Filter Tubes with a nylon membrane).
4. Microcentrifuge.
5. HPLC system with a diode array UV–Visible light detector and software for analyzing peak area.
6. C_{18} reverse-phase column (e.g., Waters Symmetry® 5 μm particle size, 4.6 mm by 250 mm).
7. Glass sample vials with septum caps, as appropriate for use with an autosampler.

3 Methods

Diligently follow all regulations when disposing waste materials.

1. Prepare 0.2 μm filtered solutions of all phenazines of interest at known concentrations (in the same medium as any cultures to be analyzed) for the purpose of generating a standard curve (*see* **Note 3**). In most cases, a standard curve including five or more concentrations ranging from 1 to 100 μM will be sufficient for quantification of phenazines in samples.
2. Obtain a 0.2 μm filtered culture supernatant by loading 200–700 μL of a culture into the top of a filter-column microcentrifuge tube and then centrifuging for 1 min at maximum speed (16,000 rcf) (*see* **Notes 3** and **4**).
3. Set up HPLC solvents as follows (*see* **Note 5**):

   Solvent A = 0.1 % TFA in water.

   Solvent B = 0.1 % TFA in acetonitrile.
4. Set up the separation profile as follows:

   Flow rate = 1 mL/min.

   0–2 min: linear gradient from 0 to 15 % solvent B.

   2–14 min: linear gradient from 15 to 83 % solvent B.

   14–16 min: linear gradient from 83 to 0 % solvent B.

   16–20 min: at 0 % solvent B.
5. Follow all procedures for preparing the HPLC instrument for use, including priming pumps, turning on detectors and lamps, and ensuring that volumes of solvents are adequate for the number of samples to be analyzed.
6. Inject 100 μL of the filtrate onto the HPLC column, either manually or using an autosampler, as appropriate for your HPLC setup (*see* **Notes 6** and **7**).
7. Use the same separation protocol to analyze solutions of known concentrations of phenazines dissolved in the medium used for culturing. Also run a blank of the medium alone.
8. Make a standard curve using the peak areas at 365 nm for each phenazine (*see* **Note 8** and Fig. 2a).

Fig. 2 (continued) the shift in retention times caused by the complex LB medium. A comparison of wild-type PA14 grown in a complex versus minimal medium demonstrates that growth conditions affect the production of different types of phenazines. One of the peaks between PYO and PCA in the LB-grown culture supernatant (with a retention time near 15.5 min) is likely due to PCN, which is commonly produced by *P. aeruginosa*. MOPS-buffered minimal medium: 100 mM MOPS, 40 mM sodium succinate, 9.3 mM NH_4Cl, 43 mM NaCl, 2.2 mM KH_2PO_4, 1 mM $MgSO_4$, 3.6 μM $FeSO_4$. LB medium: 10 g/L NaCl, 10 g/L tryptone, 5 g/L yeast extract. *PYO* pyocyanin, *PCA* phenazine-1-carboxylic acid, *1-OH-PHZ* 1-hydroxyphenazine

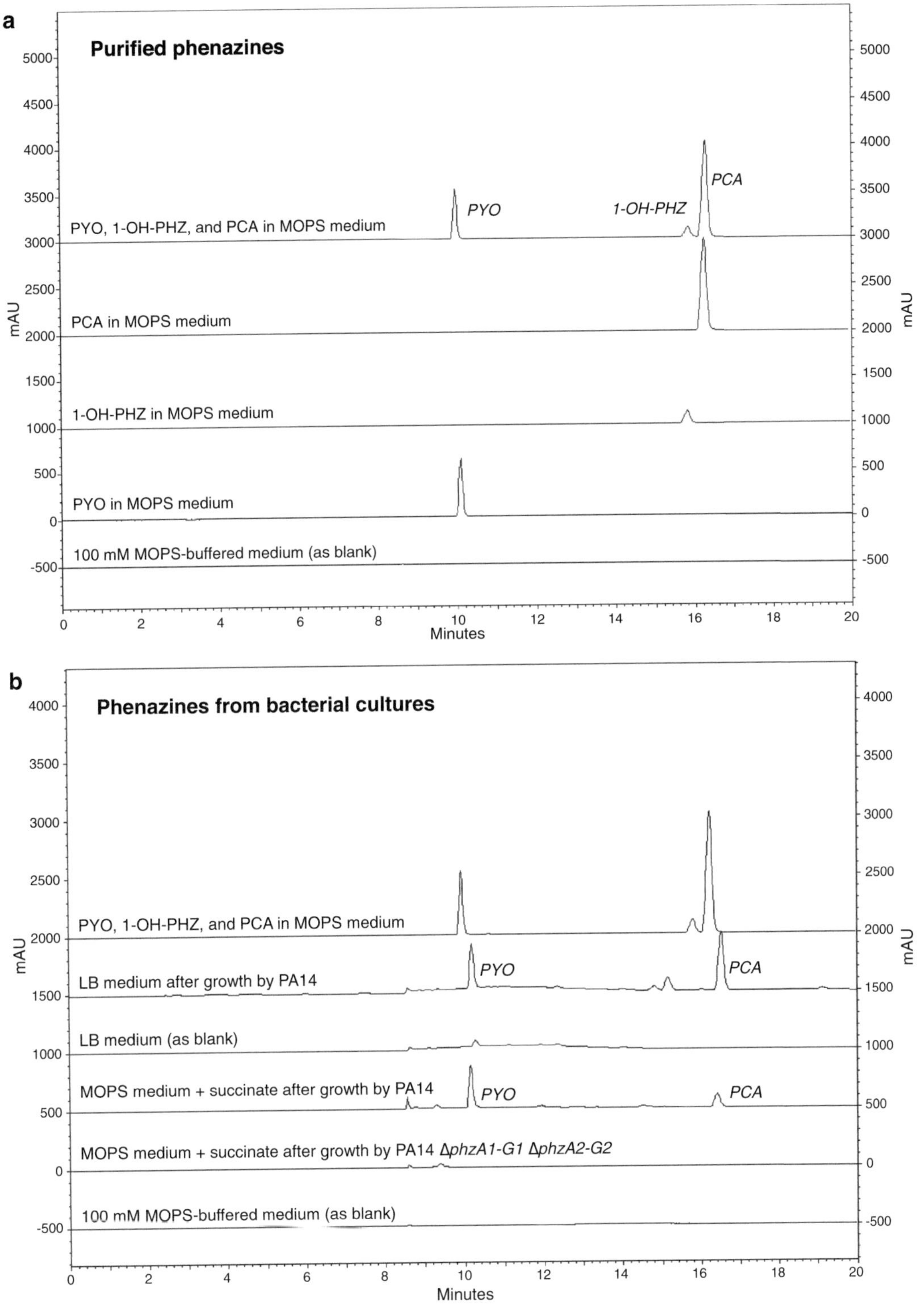

Fig. 2 (**a**) HPLC traces at 365 nm (4 nm width) of approximately 100 μM PYO, PCA, and 1-OH-PHZ separately and combined in a minimal MOPS-buffered medium. (**b**) HPLC traces of wild-type *P. aeruginosa* PA14 cultures grown in LB and MOPS-buffered minimal medium with succinate and the phenazine-null mutant (PA14 Δ*phzA1-G1* Δ*phzA2-G2* [16]) also grown in MOPS + succinate. Notice the signal interference at 365 nm and

9. Using the standard curves, determine the concentration of phenazine in each sample based on the corresponding peak area (*see* **Note 8**).

4 Notes

1. This protocol is intended for measuring phenazines in planktonic cultures. For other sample types (e.g., soil, biofilm), phenazines can be extracted and concentrated using chloroform and other solvents, with the particular procedure depending on the phenazine of interest. For more information, see Supplemental Materials in [5]. The extract can be dissolved in a small volume of acetonitrile and run according to the HPLC protocol outlined in Subheading 3. In the case of complex solutions or gel-like samples, extensive centrifugation steps may be used to separate the clear liquid containing phenazines from the solid phase of the sample [21].
2. Solutions containing oxidized PYO can be analyzed very simply in cuvettes using a spectrophotometer. At circumneutral pH, PYO absorbs maximally at 691 nm, with a molar absorptivity of 4.31 mM^{-1} cm^{-1} (also known as an extinction coefficient, ε_{691}) [16]. Pellet cells, pass the supernatant through a 0.2 μm filter (such as a filter-column microcentrifuge tube), and then vortex the solution for 30 s to ensure complete oxidation of any PYO by ambient oxygen. Measure the absorbance at 691 nm (A_{691}) of the culture filtrate and the relevant blank solution in 1-cm-pathlength cuvettes. Subtract A_{691} of the blank solution from the A_{691} of the filtered culture supernatant to give the absorbance of PYO (A_{PYO}). Calculate the concentration of PYO according to the equation [PYO] (mM) = A_{PYO} cm^{-1}/4.31 mM^{-1} cm^{-1}.
3. Because the medium used for culturing can cause differences in retention time due to matrix effects it is necessary to run known amounts of phenazines dissolved in the same medium that serves as the background for any biological samples. In addition, a complex medium can interfere with the efforts to determine the peak area for specific phenazines (for example, cultures grown in lysogeny broth (LB-Miller) have more background at 365 nm than those in a 3-morpholinopropane-1-sulfonic acid (MOPS)-buffered minimal medium, see Fig. 2b). When possible, use a simple, defined medium as the background for phenazine quantification.
4. If the culture of interest has a moderate or high cell density (OD_{600} > 0.3), avoid clogging the spin-column filter by first pelleting cells in a microcentrifuge tube and then passing the cleared supernatant through the 0.2 μm pore size filter column.

5. If a large number of samples will be analyzed by this method, make up solutions of TFA in new bottles of HPLC-grade water and HPLC-grade acetonitrile and label the amended bottles accordingly. For example, to a 4-L bottle of HPLC-grade water, add 4 mL TFA, cap the bottle, and swirl or invert to mix.
6. When using an autosampler, place >200 μL of filtrate into a clean HPLC sample vial (use an insert for small volumes if sample size is very low).
7. To avoid potential carryover of phenazine from one sample to the next, run a water blank between every sample.
8. Make standard curves based on the peak area at the retention time determined by running individual standards. Confirm that the peaks have maximal light absorbance at the following characteristic wavelengths: PYO 280 nm, PCA 250 nm, and 1-hydroxyphenazine 262 nm [22]. Using the traces at 365 nm, set the boundaries for peak area integration where the peak of interest meets the baseline. If the medium is too complex and there are overlapping peaks (that is, the peak of interest does not reach all the way to the baseline), try (1) repeating the analysis for cells cultured in a less complex medium, (2) extending the duration of the relevant segment of the separation protocol (**step 4**), or (3) noting that the calculated areas are only approximate.

Acknowledgments

S.E.K. was supported by the National Science Foundation Graduate Research Fellowship Program, and D.K.N. is a Howard Hughes Medical Institute (HHMI) Investigator. We thank the HHMI for supporting our work. R. Hunter, H. Sakhtah, C. Okegbe, and L. Dietrich provided constructive criticism of earlier drafts of this chapter.

References

1. Mavrodi DV, Blankenfeldt W, Thomashow LS (2006) Phenazine compounds in fluorescent *Pseudomonas* spp. biosynthesis and regulation. Ann Rev Phytopath 44:417–445
2. Mavrodi DV, Bonsall RF, Delaney SM, Soule MJ, Phillips G, Thomashow LS (2001) Functional analysis of genes for biosynthesis of pyocyanin and phenazine-1-carboxamide from Pseudomonas aeruginosa PAO1. J Bacteriol 183:6454–6465
3. Davis G, Thornalley PJ (1983) Free radical production from the aerobic oxidation of reduced pyridine nucleotides catalysed by phenazine derivatives. Biochim Biophys Acta 724:456–464
4. Hassan HM, Fridovich I (1980) Mechanism of the antibiotic action of pyocyanine. J Bacteriol 141:156–163
5. Rabaey K, Boon N, Hofte M, Verstraete W (2005) Microbial phenazine production enhances electron transfer in biofuel cells. Environ Sci Technol 39:3401–3408
6. Yong Y-C, Yu Y-Y, Li C-M, Zhong J-J, Song H (2011) Bioelectricity enhancement via overexpression of quorum sensing system in *Pseudomonas aeruginosa*-inoculated microbial fuel cells. Biosens Bioelectron 30:87–92
7. Lau GW, Hassett DJ, Ran H, Kong F (2004) The role of pyocyanin in *Pseudomonas aeruginosa* infection. Trends Mol Med 10:599–606

8. Chin-A-Woeng T, Bloemberg G, Lugtenberg B (2003) Phenazines and their role in biocontrol by *Pseudomonas* bacteria. New Phytologist 157:503–523
9. Wang Y, Newman DK (2008) Redox reactions of phenazine antibiotics with ferric (hydr) oxides and molecular oxygen. Environ Sci Technol 42:2380–2386
10. Hernandez ME, Kappler A, Newman DK (2004) Phenazines and other redox-active antibiotics promote microbial mineral reduction. Appl Environ Microb 70:921–928
11. Ramos I, Dietrich LEP, Price-Whelan A, Newman DK (2010) Phenazines affect biofilm formation by *Pseudomonas aeruginosa* in similar ways at various scales. Res Microbiol 161:187–191
12. Wang Y, Wilks JC, Danhorn T, Ramos I, Croal L, Newman DK (2011) Phenazine-1-carboxylic acid promotes bacterial biofilm development via ferrous iron acquisition. J Bacteriol 193: 3606–3617
13. Dietrich LEP, Okegbe C, Price-Whelan A, Sakhtah H, Hunter RC, Newman DK (2013) Bacterial community morphogenesis is intimately linked to the intracellular redox state. J Bacteriol 195(7):1371–1380
14. Price-Whelan A, Dietrich LEP, Newman D (2007) Pyocyanin alters redox homeostasis and carbon flux through central metabolic pathways in *Pseudomonas aeruginosa* PA14. J Bacteriol 189:6372–6381
15. Wang Y, Kern SE, Newman DK (2010) Endogenous phenazine antibiotics promote anaerobic survival of *Pseudomonas aeruginosa* via extracellular electron transfer. J Bacteriol 192:365–369
16. Dietrich LEP, Price-Whelan A, Petersen A, Whiteley M, Newman DK (2006) The phenazine pyocyanin is a terminal signalling factor in the quorum sensing network of *Pseudomonas aeruginosa*. Mol Microbiol 61:1308–1321
17. Koley D, Ramsey MM, Bard AJ, Whiteley M (2011) Discovery of a biofilm electrocline using real-time 3D metabolite analysis. Proc Natl Acad Sci U S A 108:19996–20001
18. Sullivan NL, Tzeranis DS, Wang Y, So PTC, Newman D (2011) Quantifying the dynamics of bacterial secondary metabolites by spectral multiphoton microscopy. ACS Chem Biol 6:893–899
19. Reszka KJ, O'Malley Y, McCormick ML, Denning GM, Britigan BE (2004) Oxidation of pyocyanin, a cytotoxic product from *Pseudomonas aeruginosa*, by microperoxidase 11 and hydrogen peroxide. Free Radic Biol Med 36:1448–1459
20. Morales DK, Jacobs NJ, Rajamani S, Krishnamurthy M, Cubillos-Ruiz JR, Hogan DA (2010) Antifungal mechanisms by which a novel Pseudomonas aeruginosa phenazine toxin kills Candida albicans in biofilms. Mol Microbiol 78:1379–1392
21. Hunter RC, Klepac-Ceraj V, Lorenzi MM, Grotzinger H, Martin TR, Newman DK (2012) Phenazine content in the cystic fibrosis respiratory tract negatively correlates with lung function and microbial complexity. Am J Respir Cell Mol Biol 47:738–745
22. Fernandez RO, Pizarro RA (1997) High-performance liquid chromatographic analysis of *Pseudomonas aeruginosa* phenazines. J Chromat A 771:99–104

Chapter 26

Extraction and Measurement of NAD(P)$^+$ and NAD(P)H

Suzanne E. Kern, Alexa Price-Whelan, and Dianne K. Newman

Abstract

Nicotinamide adenine dinucleotides are critical redox-active substrates for countless catabolic and anabolic reactions. Ratios of NAD$^+$ to NADH and NADP$^+$ to NADPH are therefore considered key indicators of the overall intracellular redox potential and metabolic state. These ratios can be measured in bulk conditions using a highly sensitive enzyme cycling-based colorimetric assay (detection limit at or below 0.05 μM or 1 pmol) following a simple extraction procedure involving solutions of acid and base. Special considerations are necessary to avoid measurement artifacts caused by the presence of endogenous redox-active metabolites, such as phenazines made by diverse *Pseudomonas* species (*see* Chapter 25).

Key words NAD$^+$, NADH, NADP$^+$, NADPH, Phenazines, Metabolism, Redox state, *Pseudomonas aeruginosa*

1 Introduction

Oxidized nicotinamide adenine dinucleotide (NAD$^+$) is an enzyme cofactor that participates in redox transformations involved in the oxidative conversion of substrates to metabolic end products, such as CO_2 and H_2O in the case of sugar metabolism. In the process, NAD$^+$ accepts two electrons and a proton. In its reduced form (NADH) the cofactor can donate electrons to the respiratory chain through NADH dehydrogenase, regenerating NAD$^+$. The related nicotinamide adenine dinucleotide phosphate (NADP(H)) is commonly involved in anabolic reactions, such as the synthesis of fatty acids, which require the input of reducing equivalents. The membrane-bound transhydrogenase enzyme interconverts NAD(H) and NADP(H) by coupling NADP$^+$ reduction to NADH oxidation at the expense of moving one proton from the periplasm to the cytoplasm [1, 2]. This reaction is reversible under physiological conditions and permits the maintenance of appropriate cellular NAD(H) and NADP(H) redox levels. Overall, the relative

Alain Filloux and Juan-Luis Ramos (eds.), *Pseudomonas Methods and Protocols*, Methods in Molecular Biology, vol. 1149, DOI 10.1007/978-1-4939-0473-0_26, © Springer Science+Business Media New York 2014

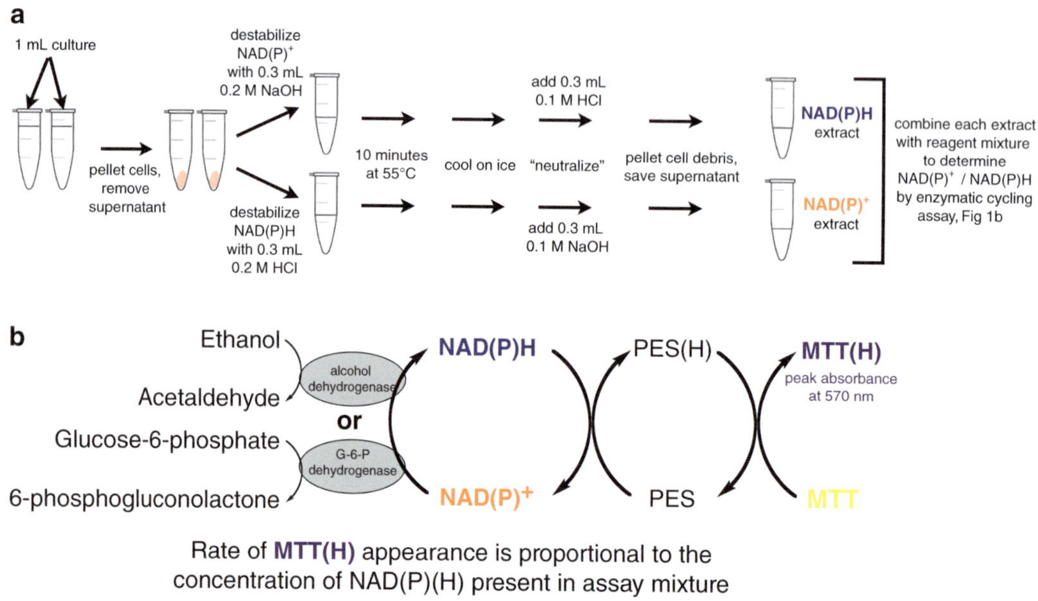

Fig. 1 (**a**) Schematic overview of the extraction of NAD(P)(H) from cell pellets by destabilization of the reduced cofactor in acid and the oxidized cofactor in base. (**b**) Representation of the enzyme-cycling assay used to detect the level of NAD(H) or NADP(H) in a sample. *MTT* methylthiazolyldiphenyl-tetrazolium bromide, *PES* phenazine ethosulfate

levels of reduced and oxidized nicotinamide cofactors within living cells determine which metabolic reactions can occur [1, 3, 4].

These compounds can be detected by several methods:

1. Direct measurement of NADH fluorescence of cells in culture [3–5] or of cell extracts [5, 6].
2. Measurement of NAD(H)-binding GFP fluorescence [6].
3. In vitro colorimetric enzyme-mediated assays of cell extracts [2, 7, 8].
4. Time-resolved fluorescence profiling of cells in situ using multiphoton microscopy [2, 9].

It is important to be aware that directly measuring the levels of NAD(P)(H) in cell cultures using only spectral properties may be compromised by other endogenous compounds with similar spectral properties, e.g., 1-hydroxyphenazine and the iron chelator pyoverdine [1, 9]. Because of its simplicity, sensitivity, and specificity, a time-resolved enzymatic colorimetric assay based on the procedures developed by San et al. [3, 4, 10] and Bernofsky and Swan [5, 7] is our preferred approach [6, 8] and the focus of this chapter. With an exchange of the enzyme and substrate, this assay is specific to either NAD(H) or NADP(H) and a single set of cell extracts can be used for both (Fig. 1). NAD(P)(H) levels as low as 0.05 μM (<1 pmol in the assay mixture, equivalent to ~5 nmol/10^9 cells) can be measured by this method. One drawback to the extraction procedure is the length of time that passes before quenching metabolism, which occurs upon the addition of acid or base to the cell pellet following a 1-min centrifugation step to concentrate the sample and exclude any extracellular NAD(P)(H) [7, 8, 11]. For time-sensitive measurements, this is an important caveat [4].

This method has been applied widely to samples grown in well-mixed bacterial cultures [8, 10, 12] and has been adapted for use in measuring the NADH/NAD$^+$ of bacterial colony biofilms [13]. Extractions from phenazine-producing organisms, such as pseudomonads, require special consideration: phenazines, another class of redox-active molecules, can oxidize NAD(P)H and participate in cycling reactions to reduce extracellular oxidants, such as oxygen and ferric iron [14–16]. Because the nature of this extraction procedure is to destabilize the reduced or the oxidized nicotinamide cofactor [17, 18] in paired samples, subsequent reactions with phenazines in the cell-free extract can result in the degradation of NAD(P)H, which can artificially skew the measured ratio. In this method, we include notes and precautions for successfully circumventing this problem.

2 Materials

Prepare all reagents using distilled, deionized (18.2 MΩ resistivity) water. Diligently follow all waste disposal regulations when disposing waste materials.

2.1 Extraction Reagents

1. Sodium hydroxide: 0.2 and 0.1 M.
2. Hydrochloric acid: 0.2 and 0.1 M.

2.2 Extraction Equipment and Supplies

Some extractions may need to be performed in the absence of oxygen (*see* **Notes 1** and **2**):

1. Water or sand bath at 50 °C.
2. Ice.
3. Vortex.
4. Pipettes and tips.
5. Microcentrifuge tubes.
6. Microcentrifuge.

2.3 NAD(H) Measurement Reagents

1. Bicine buffer solution: 1 M at pH 8.0. Can be prepared in advance and stored at room temperature.
2. Ethylenediaminetetraacetic acid (EDTA) solution: 40 mM at pH 8.0. Can be prepared in advance and stored at room temperature.

3. Methylthiazolyldiphenyl-tetrazolium bromide (MTT, also known as thiazolyl blue tetrazolium bromide) solution: 4.2 mM. Can be prepared in advance and stored at room temperature or frozen at −20 °C in aliquots.
4. Phenazine ethosulfate (PES) solution: 16.6 mM. Can be prepared in advance and must be protected from light. Aliquots can be stored in the dark and frozen at −20 °C for at least 6 months (*see* **Note 3**).
5. Ethanol: 100 %.
6. Alcohol dehydrogenase II (ADH II) enzyme solution: 1 mg/mL in 0.1 M bicine buffer, pH 8.0. Use Sigma A-3263, which has very low background levels of NAD bound. Make this just before use, in Measurement **step** 7 of Subheading 3.2.
7. ß-nicotinamide adenine dinucleotide (NAD^+) and/or reduced ß-nicotinamide adenine dinucleotide (NADH) solution(s): 0.01–100 μM, according to concentrations needed for a standard curve (*see* **Note 4**). If preparing in advance, dissolve NADH in slightly basic solution (0.01 M NaOH) and dissolve NAD^+ in slightly acidic solution (0.01 M HCl). Single-use aliquots of NAD^+ and NADH can be stored at −80 °C for at least 6 months.

2.4 NADP(H) Measurement Reagents

Materials are identical to those for NAD(H) measurement (Sect. 2.3), with only the following changes:

1. Glucose-6-phosphate solution: 25 mM. Prepare aliquots and freeze at −20 °C for up to 2 weeks.
2. Glucose-6-phosphate dehydrogenase (G6PD) enzyme solution: 0.1 mg/mL in 0.1 M bicine buffer, pH 8.0. This can be prepared in advance and frozen in single-use aliquots at −80 °C.
3. ß-nicotinamide adenine dinucleotide phosphate ($NADP^+$) and/or reduced ß-nicotinamide adenine dinucleotide phosphate (NADPH) solution(s): 0.05–50 μM, according to concentrations needed for a standard curve (*see* **Note 4**). If preparing in advance, mix NADPH with anoxic water under an oxygen-free atmosphere. Aliquots of $NADP^+$ and NADPH can be stored at −80 °C for at least 6 months.

2.5 NAD(P)(H) Measurement Equipment and Supplies

1. Plate reader capable of maintaining a constant internal temperature (at about 30 °C) and recording absorbance at 570 nm at regular time intervals (for example, Biotek Synergy 4).
2. Optically clear flat-bottom styrene 96-well microtiter plate (for example, Thermo Scientific #9205).
3. Single- and multichannel pipettes (for 90 and 5 μL volumes) and tips.
4. Reagent reservoir for multichannel pipetting (for <25 mL volumes).

3 Methods

3.1 Extraction

Unless otherwise specified, perform all steps at room temperature.

1. Remove a sample of the culture for measuring viable cell counts, cell mass, or optical density in order to calculate NAD(P)(H) concentrations relative to the amount of cells, if desired.
2. Per sample, pipette 1 mL of culture into each of the two labeled microfuge tubes, one for NAD(P)H extraction and the other for $NAD(P)^+$ extraction (*see* **Notes 5** and **6**; if cell density is low, *see* **Notes** 7 and **8**).
3. Pellet cells: 1 min at maximum speed (16,000 rcf).
4. Remove supernatant (*see* **Note 6**).
5. Resuspend pellet in 0.2 M acid or base as follows (*see* **Note 5**):
   - 300 μL 0.2 M sodium hydroxide (for NAD(P)H extraction).
   - 300 μL 0.2 M hydrochloric acid (for $NAD(P)^+$ extraction).
6. Incubate tubes at 50 °C (using heat block or water bath) for 10 min. Then cool on ice for about 5 min.
7. With the cap open, add 0.1 M acid or base dropwise to each tube while vortexing at low speed (*see* **Note 9**) as follows:
   - 300 μL 0.1 M hydrochloric acid (for NAD(P)H extraction).
   - 300 μL 0.1 M sodium hydroxide (for $NAD(P)^+$ extraction).
8. Pellet debris: 5 min at maximum speed (16,000 rcf).
9. Transfer 150–300 μL of supernatant to a clean, labeled microfuge tube (*see* **Note 10**).
10. Either store extracts at −80 °C for up to a week or proceed right away to Measurement **step 1** of Subheading 3.2(*see* **Note 1**).

3.2 NAD(H) Measurement

1. Turn on the plate reader, and set it to preheat to 30 °C.
2. Prepare a mix of the assay reagents according to the number of samples and expected concentrations of NAD(H), and keep the mix protected from light (*see* **Notes 3**, 7, and **8**).
   - 1 vol 1 M bicine buffer, pH 8.0.
   - 1 vol 100 % ethanol.
   - 1 vol 40 mM EDTA.
   - 1 vol 4.2 mM MTT.
   - 2 vol 16.6 mM PES.
   - 3 vol water.

   For a full 96-well plate, one volume = 1 mL.

3. Prepare a range of concentrations of NAD^+ and/or NADH in water for a standard curve (*see* **Note 4**).
4. Pipette 5 μL of the standards, water blanks, and sample supernatants into separate wells of the 96-well microtiter plate (*see* **Note 11**; for higher sensitivity *see* **Note 8**).
5. In dim light, use a multichannel pipette to add 90 μL of the reagent mix to each well. (For higher sensitivity *see* **Note 8**.)
6. Keeping the plate protected from light, incubate the plate for approximately 10 min at 30 °C (*see* **Note 12**).
7. Prepare a fresh solution of alcohol dehydrogenase enzyme:
   - Weigh out 0.7–2.0 mg of ADH II.
   - Gently dissolve the enzyme powder to 1 mg/mL in 0.1 M bicine buffer, pH 8.0.
   - Pipette the enzyme solution into a reagent reservoir for multichannel pipetting.
8. In dim light, using a multichannel pipette, add 5 μL of the enzyme solution to each well containing the reagent mix with sample or standard.
9. Return the plate to the plate reader, and begin recording absorbance at 570 nm according to the following program:
   - Hold temperature at 30 °C.
   - Start kinetic read, repeating each of the steps below for at least 20 min. For maximum resolution, use the software to calculate the minimum interval required between each round. (For higher sensitivity, *see* **Note 8**.)
     - Shake the plate for 3 s at medium speed.
     - Read the absorbance at 570 nm (A_{570}) for each well.

3.3 NADP(H) Measurement

Follow the procedure outlined for NAD(H) Measurement **step 2** of Subheading 3.2, with the following changes:

1. Prepare a mix of the assay reagents according to the number of samples and expected concentrations of NADP(H), and keep the mix protected from light (*see* **Notes 3**, **7**, and **8**).
   - 1 vol 1 M bicine buffer, pH 8.0.
   - 1 vol 25 mM glucose-6-phosphate.
   - 1 vol 40 mM EDTA.
   - 1 vol 4.2 mM MTT.
   - 2 vol 16.6 mM PES.
   - 3 vol water.

   For a full 96-well plate, one volume = 1 mL.
2. Prepare a range of concentrations of $NADP^+$ and/or NADPH in water for a standard curve (*see* **Note 4**).

3. Thaw a frozen aliquot or prepare a fresh solution of G6PD:
   - Weigh out 0.07–1.0 mg of G6PDH.
   - Gently dissolve the enzyme powder to 0.1 mg/mL in 0.1 M bicine buffer, pH 8.0.
   - Pipette the enzyme solution into a reagent reservoir for multichannel pipetting.

3.4 Data Work-Up

1. Obtain a data table of absorbance at 570 nm (A_{570}) for each well over time.
2. To determine absolute concentrations of NAD(P)(H) from samples, start by creating a standard curve (*see* Fig. 2a):
   (a) Make a plot of A_{570} versus time for the standards and water blanks.
   (b) Define a time frame for which all standards show a linear increase in A_{570}/time.
   (c) Calculate the slope for each standard during the same time frame (e.g., from 1 to 8 min).
   (d) Plot or use a linear regression to obtain a standard curve and an equation relating the known concentrations of the standards versus A_{570}/time (*see* **Note 13**).
3. Use the standard curve to calculate the concentration of each sample based on its slope (A_{570}/time, *see* Fig. 2b).
4. Calculate the ratio of NAD(P)+ to NAD(P)H for each sample and/or report absolute concentrations of each relative to the amount of cells in the original sample (*see* **Note 14**).

4 Notes

1. The reduced form of pyocyanin tends to associate with the cell pellet, and this can lead to carryover of redox-active compounds that can interfere with the stability of NAD(P)H extracts. For samples with phenazines present, particularly under low-aeration conditions, it is advisable to perform extraction **step 2** of Subheading 3.1 (for anaerobic growth) or extraction **step 5** of Subheading 3.1 (for aerobic growth) through measurement **step 5** of Subheading 3.2 in an oxygen-free glove box using reagents and supplies that have equilibrated to the anoxic atmosphere. To determine if this precaution is necessary, pellet a sample of cells and remove the supernatant. If, upon resuspending the pellet in base, the extract turns faintly blue (evidence of oxidized pyocyanin in *P. aeruginosa* cultures), then anaerobic extraction is essential. With oxidized phenazines present, NADP(H) reacts to form NAD(P)+, which then degrades in the basic extraction solution

a

	A570	A570	A570	A570	A570
Kinetic read	**1 μM**	**0.5 μM**	**0.1 μM**	**0.05 μM**	**0 μM**
0:00:05	0.059	0.058	0.059	0.06	0.055
0:00:26	0.065	0.061	0.059	0.061	0.055
0:00:47	0.072	0.065	0.06	0.062	0.056
0:01:08	0.078	0.068	0.061	0.063	0.056
0:01:29	0.086	0.072	0.062	0.068	0.056
0:01:50	0.092	0.075	0.062	0.064	0.056
0:02:11	0.099	0.079	0.063	0.064	0.056
0:02:32	0.106	0.083	0.064	0.065	0.057
0:02:53	0.113	0.086	0.065	0.065	0.057
0:03:14	0.12	0.09	0.065	0.066	0.057
0:03:35	0.127	0.094	0.066	0.066	0.057
0:03:56	0.133	0.098	0.067	0.067	0.057
0:04:17	0.14	0.101	0.068	0.068	0.057
0:04:38	0.147	0.103	0.069	0.068	0.059
0:04:59	0.154	0.107	0.069	0.069	0.058
0:05:20	0.16	0.11	0.07	0.069	0.058
0:05:41	0.168	0.113	0.071	0.07	0.058
0:06:02	0.174	0.117	0.071	0.07	0.058
0:06:23	0.181	0.12	0.072	0.07	0.058
0:06:44	0.188	0.123	0.073	0.071	0.058
0:07:05	0.194	0.127	0.074	0.072	0.059
0:07:26	0.201	0.13	0.074	0.072	0.058
0:07:47	0.207	0.134	0.075	0.072	0.059
0:08:08	0.214	0.137	0.076	0.072	0.058
0:08:29	0.22	0.14	0.076	0.073	0.058
0:08:50	0.227	0.144	0.077	0.074	0.058
0:09:11	0.233	0.147	0.078	0.073	0.059
0:09:32	0.24	0.15	0.079	0.074	0.058
0:09:53	0.246	0.154	0.08	0.075	0.058
0:10:14	0.253	0.158	0.08	0.075	0.06
0:10:35	0.259	0.16	0.081	0.075	0.058
0:10:56	0.265	0.163	0.082	0.075	0.059
0:11:17	0.272	0.166	0.083	0.076	0.059
0:11:38	0.279	0.169	0.083	0.076	0.059
0:11:59	0.284	0.173	0.084	0.077	0.059
0:12:20	0.291	0.176	0.085	0.077	0.059
0:12:41	0.297	0.179	0.086	0.078	0.059
0:13:02	0.304	0.183	0.086	0.078	0.06
0:13:23	0.309	0.186	0.087	0.079	0.059
0:13:44	0.316	0.189	0.088	0.079	0.06
0:14:05	0.322	0.192	0.088	0.079	0.059
0:14:26	0.328	0.195	0.089	0.079	0.059
0:14:47	0.334	0.198	0.09	0.08	0.059
0:15:08	0.34	0.201	0.09	0.081	0.06
0:15:29	0.347	0.204	0.091	0.081	0.06
0:15:50	0.352	0.207	0.091	0.081	0.059
0:16:11	0.359	0.21	0.092	0.082	0.06
0:16:32	0.364	0.213	0.093	0.082	0.06
0:16:53	0.371	0.216	0.093	0.082	0.06
0:17:14	0.377	0.219	0.095	0.082	0.06
0:17:35	0.383	0.222	0.095	0.085	0.06
0:17:56	0.389	0.226	0.096	0.083	0.06
0:18:17	0.395	0.229	0.096	0.083	0.06
0:18:38	0.401	0.232	0.097	0.083	0.06
0:18:59	0.406	0.235	0.098	0.084	0.06
0:19:20	0.412	0.239	0.098	0.084	0.06
0:19:41	0.418	0.24	0.099	0.085	0.06
0:20:02	0.424	0.243	0.099	0.085	0.061
A570/time:	27.11	13.69	3.075	1.793	0.334
known [NAD+] (μM):	1	0.5	0.1	0.05	0

← *"known_x's"* *=SLOPE(known_y's, known_x's)* *=INTERCEPT(known_y's, known_*

← *"known_y's"* **slope: 0.0374** **intercept: -0.0145**

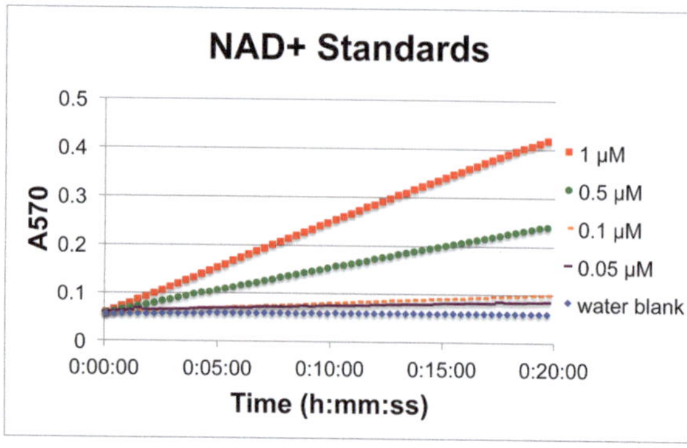

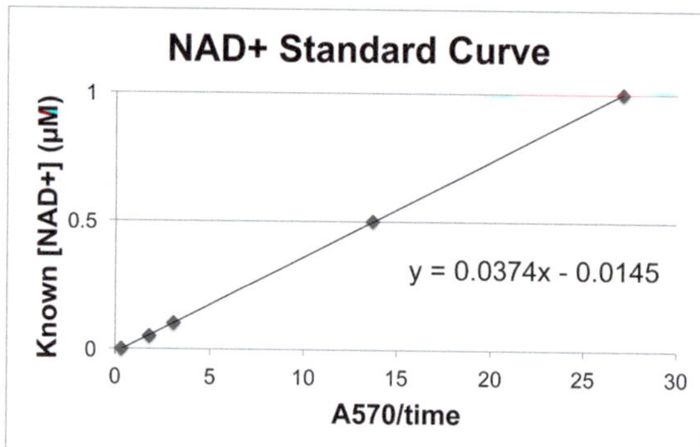

Fig. 2 (**a**) Example of a standard curve calculated from A_{570}/time of solutions of known NAD(P)(H) concentration using the "high-sensitivity" reagent mixture and sampling procedure (*see* **Note 8**). The range of data points used for A_{570}/time calculation is highlighted in *white* and *bracketed*. Using Microsoft Excel, the SLOPE and INTERCEPT functions give the same equation components as shown by the linear regression overlaid on the data plot. (**b**) Example of extraction sample data work-up: A_{570}/time and concentrations were calculated using the equation determined by the standard curve. The final ratio of NADH/NAD^+ is represented on both a linear and log-scale plot

b

Kinetic read	A570 NADH	A570 NAD+
0:00:05	0.077	0.065
0:00:26	0.082	0.067
0:00:47	0.088	0.068
0:01:08	0.093	0.07
0:01:29	0.098	0.072
0:01:50	0.103	0.074
0:02:11	0.109	0.076
0:02:32	0.114	0.077
0:02:53	0.119	0.079
0:03:14	0.125	0.081
0:03:35	0.13	0.083
0:03:56	0.136	0.084
0:04:17	0.141	0.086
0:04:38	0.145	0.088
0:04:59	0.151	0.09
0:05:20	0.156	0.091
0:05:41	0.161	0.093
0:06:02	0.166	0.094
0:06:23	0.172	0.096
0:06:44	0.177	0.098
0:07:05	0.181	0.099
0:07:26	0.187	0.101
0:07:47	0.192	0.102
0:08:08	0.197	0.104
0:08:29	0.202	0.106
0:08:50	0.207	0.107
0:09:11	0.211	0.109
0:09:32	0.216	0.11
0:09:53	0.222	0.112
0:10:14	0.227	0.114
0:10:35	0.232	0.115
0:10:56	0.236	0.116
0:11:17	0.241	0.118
0:11:38	0.246	0.12
0:11:59	0.251	0.121
0:12:20	0.256	0.123
0:12:41	0.261	0.124
0:13:02	0.265	0.126
0:13:23	0.27	0.127
0:13:44	0.274	0.128
0:14:05	0.279	0.13
0:14:26	0.284	0.131
0:14:47	0.289	0.133
0:15:08	0.293	0.135
0:15:29	0.298	0.136
0:15:50	0.302	0.137
0:16:11	0.307	0.139
0:16:32	0.312	0.14
0:16:53	0.317	0.141
0:17:14	0.322	0.143
0:17:35	0.326	0.144
0:17:56	0.331	0.146
0:18:17	0.336	0.147
0:18:38	0.34	0.148
0:18:59	0.344	0.15
0:19:20	0.349	0.151
0:19:41	0.353	0.153
0:20:02	0.358	0.154
A570/time:	20.77	6.621
Calculated [NAD(H)] (μM):	0.763	0.233
NADH / NAD+:	**3.27**	

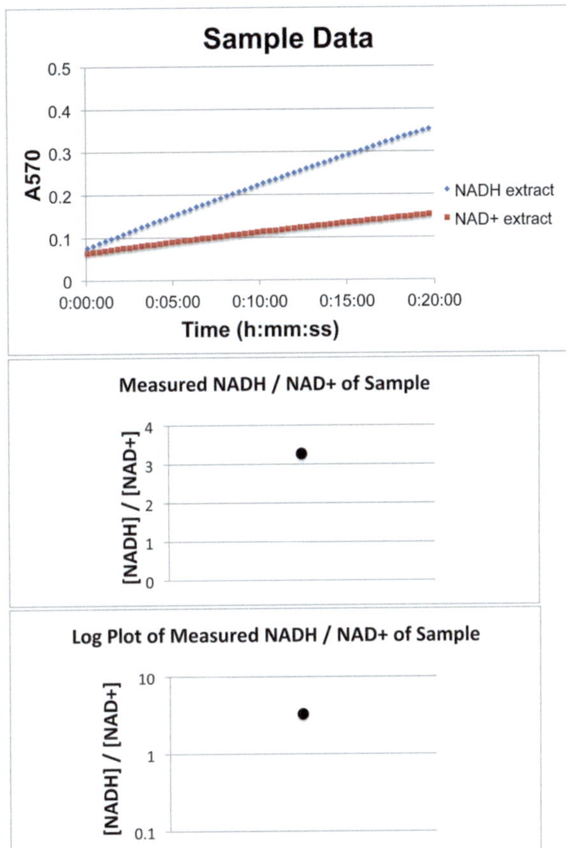

slope: 0.0374 **intercept: -0.0145**

← calculated [NAD(H)] = (A570/time) * 0.0374 - 0.0145

Fig. 2 (continued)

and leads to artificially low NAD(P)H readings. Anaerobically prepared extracts can be stored for up to 24 h in an oxygen-free chamber in the dark at room temperature. To be absolutely sure that no oxygen interferes, switch the order of Measurement **steps 4** and **5** of Subheading 3.2, first filling wells of the assay plate with the reagent mix and then adding each sample individually, all in an oxygen-free environment. Residual oxygen associated with tubes and microtiter plates can be sufficient to oxidize phenazines and, therefore, destroy NAD(P)H. For strains that produce little or no pyocyanin, it is worthwhile to perform parallel NAD(P)H extractions under ambient atmospheric conditions and in an oxygen-free context and compare the measured levels.

2. Equipment and supplies for anaerobic extractions: Oxygen-free glove box with all equipment inside—heat block at 50 °C,

ice or −20 °C freezer block for microfuge tubes, vortex, pipettes and tips, microfuge tubes (equilibrated under oxygen-free conditions for at least 24 h), and microcentrifuge.

3. PES is very sensitive to light and degrades to form a compound that is dark in color, which interferes with the A_{570} readings of reduced MTT. When pipetting PES or the reagent mix, turn off bright lights and perform these steps in dim lighting, away from windows. Cover tubes of PES aliquots (may be stored at −20 °C) and the reagent mix in foil. Protect the final mix of samples and reagents from light by covering the 96-well plate with foil until it is moved inside of the plate reader.

4. Make standard solutions based on concentrations calculated using the following extinction coefficients: NAD(P)H: $\varepsilon_{340} = 6.22\ mM^{-1}\ cm^{-1}$, $NAD(P)^+$: $\varepsilon_{260} = 18.0\ mM^{-1}\ cm^{-1}$ (Sigma product information sheets). Because the measurement depends on a cycling reaction, both the reduced and oxidized substrate should result in the same calibration curve.

5. There are many ways to conserve pipette tips during this procedure: For extraction **step 2** of Subheading 3.1, pipette up and down once in the culture to wet the pipette tip before removing the paired 1-mL samples to get equal volumes. In extraction **step 5** of Subheading 3.1 the 0.2 M acid or base can be added using the same tip and the pellet resuspended by vortexing. In measurement **steps 4** and **5** of Subheading 3.2, the reagent mix can be added by multichannel pipette first, followed by the individual standards and samples.

6. Work quickly through Extraction **step 5** of Subheading 3.1 to rapidly quench cellular metabolism. Because the cell pellet starts to resuspend back into the supernatant, it may be best to process only two or three paired extraction samples at a time. To remove the supernatant (extraction **step 4** of Subheading 3.1) determine what works best: pouring and tapping out (faster, may leave varying amounts of residual medium) or pipetting (slower, more accurate unless a long time elapses before later samples get processed). For aerobically cultured cells, it is particularly important to sample as quickly as possible upon removing the culture from its incubation conditions and to proceed rapidly through Extraction **step 5**.

7. Determine whether the higher sensitivity adaptations are necessary for a given sample:

   (a) Is the cell density below OD_{500} 0.2 (for actively growing cells) or 0.8 (for dying/late stationary-phase cells)?

   (b) Will the redox state be strongly skewed, such that the abundance of either $NAD(P)^+$ or NAD(P)H is comparatively low?

In either case, it may help to increase the sensitivity by following the adaptations outlined in **Note 8**, below.

8. For very low levels of NAD(P)(H), use the following adaptations. Use 2-mL microcentrifuge tubes and pellet 1.8 mL of culture. Add only 200 μL acid and base for extraction **steps 5** and 7 of Subheading 3.1 (instead of 300 μL). Use these proportions for a higher sensitivity reagent mix (keep mix protected from light):
   - 2 vol 1 M bicine buffer, pH 8.0.
   - 1 vol 100 % ethanol or 25 mM glucose-6-phosphate.
   - 1 vol 40 mM EDTA.
   - 1 vol 4.2 mM MTT.
   - 2 vol 16.6 mM PES.
   - 1 vol water.

   For a full 96-well plate, one volume = 1 mL.

   For each reaction, combine 80 μL reagent mix, 15 μL sample extract or standard, and 5 μL enzyme solution (total volume remains at 100 μL per well). The larger volume of sample requires stronger pH buffering. The kinetic read in measurement **step 9** of Subheading 3.2 can be extended for up to 40 min in order to discern very slight increases in A_{570}/time. Include several water blanks so that very low levels can be accurately distinguished from zero. Levels as low as 10 nM NAD(H) and 50 nM NADP(H) can be detected by following these procedural changes.
9. Use a vortex speed that swirls the total volume (600 μL) to somewhere below the cap. A high speed will spin the liquid into the cap, but with the proper setting, the cap can be left open and the acid or the base can be safely added dropwise while the solution is mixed.
10. The final pellet in the NAD(P)H extract is very fluffy and viscous. To avoid disturbing the pellet, carefully pipette from the top portion of the liquid and consider taking the supernatants from all NAD(P)H extracts as soon as centrifugation finishes, followed by the $NAD(P)^+$ extracts.
11. In order to minimize measurement error due to pipetting, include technical replicates of the samples and standards in the 96-well plate and average the resulting A_{570}/time values calculated in Subheading 3.4, **steps 2** and **3**. Since it may be difficult to see the wells in dim lighting (applicable when the reagent mix is added to the 96-well plate before the samples), consider using a guide, for instance a piece of label tape with numbers 1–12 placed directly below the row of wells into which you add samples or standards. After pipetting the last sample of a row, move the tape down to just below the next row.

12. Use aluminum foil to completely cover the microtiter plate, and protect the reagent mix from light. Foil is not necessary if the plate reader, which is dark inside, is used to incubate the plate at 30 °C.
13. If using Excel to process the plate reader data, make use of predefined functions. Use the SLOPE function: select the A_{570} values (known *ys*) and time stamps (known *xs*) over the linear time range to obtain A_{570}/time. To generate a standard curve, use INTERCEPT and SLOPE, inputting the known standard concentrations for the *y* values and the corresponding A_{570}/time for the *x* values (*see* Fig. 2a). Then the concentration of the samples can be easily determined by multiplying the A_{570}/time of each sample by the calculated slope and adding the intercept as provided by the standard curve.
14. Representing NAD(P)H/NAD(P)$^+$ (or the inverse) on a log scale conveys the biologically relevant ratio, yet allows for straightforward interpretation of ratios both greater and less than one. By comparison, a linear scale makes fractional ratios difficult to discriminate, while reporting percentages of the total pool does not directly indicate the ratio, which is what ultimately defines the equilibrium state of reactions in the cell.

Acknowledgments

S.E.K. was supported by the National Science Foundation Graduate Research Fellowship Program, and D.K.N. is a Howard Hughes Medical Institute (HHMI) Investigator. We thank the HHMI for supporting our work. N. Glasser provided constructive criticism of earlier drafts of this chapter.

References

1. de Graef MR, Alexeeva S, Snoep JL, Teixeira de Mattos MJ (1999) The steady-state internal redox state (NADH/NAD) reflects the external redox state and is correlated with catabolic adaptation in *Escherichia coli*. J Bacteriol 181:2351–2357
2. Hoek JB, Rydström J (1988) Physiological roles of nicotinamide nucleotide transhydrogenase. Biochem J 254:1–10
3. Harrison D, Chance B (1970) Fluorimetric technique for monitoring changes in the level of reduced nicotinamide nucleotides in continuous cultures of microorganisms. Appl Microbiol 19:446–450
4. Chen F, Xia Q, Ju L-K (2003) Aerobic denitrification of *Pseudomonas aeruginosa* monitored by online NAD(P)H fluorescence. Appl Environ Microbiol 69:6715–6722
5. London J, Knight M (1966) Concentrations of nicotinamide nucleotide coenzymes in micro-organisms. J Gen Microbiol 44:241–254
6. Hung YP, Albeck JG, Tantama M, Yellen G (2011) Imaging cytosolic NADH-NAD(+) redox state with a genetically encoded fluorescent biosensor. Cell Metab 14:545–554
7. Bernofsky C, Swan M (1973) An improved cycling assay for nicotinamide adenine dinucleotide. Anal Biochem 53:452–458
8. Price-Whelan A, Dietrich LEP, Newman DK (2007) Pyocyanin alters redox homeostasis and carbon flux through central metabolic

pathways in *Pseudomonas aeruginosa* PA14. J Bacteriol 189:6372–6381

9. Sullivan NL, Tzeranis DS, Wang Y, So PTC, Newman DK (2011) Quantifying the dynamics of bacterial secondary metabolites by spectral multiphoton microscopy. ACS Chem Biol 6:893–899
10. San K-Y, Bennett GN, Berríos-Rivera SJ, Vadali RV, Yang Y-T, Horton E, Rudolph FB, Sariyar B, Blackwood K (2002) Metabolic engineering through cofactor manipulation and its effects on metabolic flux redistribution in *Escherichia coli*. Metab Eng 4:182–192
11. Wos M, Pollard P (2006) Sensitive and meaningful measures of bacterial metabolic activity using NADH fluorescence. Water Res 40: 2084–2092
12. Watanabe S, Zimmermann M, Goodwin MB, Sauer U, Barry CE, Boshoff HI (2011) Fumarate reductase activity maintains an energized membrane in anaerobic *Mycobacterium tuberculosis*. PLoS Pathog 7:e1002287
13. Dietrich LEP, Okegbe C, Price-Whelan A, Sakhtah H, Hunter RC, Newman DK (2013) Bacterial community morphogenesis is intimately linked to the intracellular redox state. J Bacteriol 195(7):1371–1380
14. Denning GM, Iyer SS, Reszka KJ, O'Malley Y, Rasmussen GT, Britigan BE (2003) Phenazine-1-carboxylic acid, a secondary metabolite of *Pseudomonas aeruginosa*, alters expression of immunomodulatory proteins by human airway epithelial cells. Am J Physiol Lung Cell Mol Physiol 285:L584–L592
15. Kito N, Ohnishi Y, Kagami M, Ohno A (1974) Reduction by a model of NAD(P)H. Construction of electron bridges. Chem Lett 3:353–356
16. Wang Y, Newman DK (2008) Redox reactions of phenazine antibiotics with ferric (hydr) oxides and molecular oxygen. Environ Sci Technol 42:2380–2386
17. Burton RM, Kaplan NO (1963) The reaction of reduced pyridine nucleotides with acid. Arch Biochem Biophys 101:150–159
18. Burton RM, Kaplan NO (1963) The reaction of diphosphopyridine nucleotide and related pyridinium salts with alkali. Arch Biochem Biophys 101:139–149

Chapter 27

Cyanide Measurements in Bacterial Culture and Sputum

Chandrika Goh Nair, Ben Ryall, and Huw D. Williams

Abstract

Cyanide is produced by a few bacterial species, including *Pseudomonas aeruginosa*, and it has a role in the opportunistic infections of this bacterium including in cystic fibrosis lung infections. We describe two methods for determining cyanide in culture and patient sputum samples. One uses an ion-selective electrode to provide a convenient, rapid method of cyanide quantitation in culture or sputum, and the second is a semiquantitative method using Feigl–Anger paper that is useful for screening large numbers of bacterial strains for cyanide production.

Key words Ion-selective electrode, Hydrogen cyanide, HCN, Quorum sensing, Cyanogenesis

1 Introduction

Bacteria produce a wide spectrum of secondary metabolites including antibiotics, enzymes, siderophores, and toxins. One such toxin, produced by a number of mainly proteobacteria and some cyanobacteria, is hydrogen cyanide [1–4]. Cyanide is a potent respiratory inhibitor due to its reactivity with cytochrome *c* oxidases and other metalloenzymes that leads to the inhibition of aerobic respiration. Bacterial HCN production is best understood in cyanogenic *Pseudomonas* species, e.g., *P. aeruginosa*, where it is made by enzymatic oxidation of the amino acid glycine. The enzyme system responsible for HCN production, the HCN synthase, is encoded by the *hcnABC* gene cluster and is subject to regulation by a number of factors including quorum sensing [1]. The biological function of bacterial cyanogenesis is not clear, but it probably confers a selective advantage over competitors and it plays a role in the biocontrol properties of *P. fluorescens* CHA0 [5]. However, there is accumulating evidence for a role of cyanide in pathogenicity. It is the mediating factor in the paralytic killing of *Caenorhabditis elegans* by *P. aeruginosa* [6], and cyanide production plays a role in the pathogenicity of *P. aeruginosa* to *Drosophila melanogaster* [7]. Cyanide has been detected in burn wounds caused by *P. aeruginosa*

Alain Filloux and Juan-Luis Ramos (eds.), *Pseudomonas Methods and Protocols*, Methods in Molecular Biology, vol. 1149, DOI 10.1007/978-1-4939-0473-0_27, © Springer Science+Business Media New York 2014

[8], is detectable in sputum from cystic fibrosis (CF) and non-CF bronchiectasis patients infected with *P. aeruginosa*, and is associated with impaired lung function [9, 10]. Given that cyanide is highly toxic and is potentially of clinical relevance, it is important to have methods for quickly and easily detecting cyanide in human samples as well as methods for screening bacterial cyanide production for research purposes.

In this chapter we describe two protocols for determining cyanide levels. One is a convenient, straightforward, rapid, and quantitative method for use in culture or to detect bacterial production of cyanide in infected sputum using an ion-selective electrode (ISE) and the second a semiquantitative method based on Feigl–Anger paper—a cyanide detection filter paper. This latter method provides a convenient, semiquantitative way of screening large numbers of strains for cyanide production, for example, mutant libraries and sets of clinical isolates. It is based around methods originally described by [11] and [12] and recently applied in a different context by [13]. The Feigl–Anger paper is impregnated with two chemical compounds: copper ethylacetoacetate and tetra base (4,4′-tetramethyldiaminodiphenylmethane). The paper reacts with HCN gas, and an oxidation product of the tetra base gives a blue color.

2 Materials

2.1 Measuring Bacterial Cyanide Production Using a Cyanide Ion-Selective Electrode During Growth in Liquid Culture, on Agar Plates, and in Infected Sputum

2.1.1 Culture Media

We have successfully measured cyanide production by *P. aeruginosa* and other bacteria in a variety of defined and rich media, although usually we use LB medium.

LB: Dissolve tryptone (10 g) (Oxoid, Basingstoke, UK), 5 g of yeast extract (Oxoid, Basingstoke, UK), and 5 g of NaCl (BDH chemicals, Poole, UK) in 1 l deionized water and autoclave at 121 °C, 15 psi, for 20 min. Add agar at 15 g/l when solid medium is required.

2.1.2 Reagents

If not otherwise stated, all chemicals are available from Sigma Aldrich, Gillingham, UK.

1. Electrolyte solution: Provided by electrode supplier.
2. 10 M NaOH (ionic strength adjustment buffer).
3. 4 M NaOH.
4. Deionized water (dH_2O).
5. Cyanide stock solution: 0.5 M KCN in 2 g/l KOH (KCN is highly toxic and appropriate safety procedures should be taken when handling, *see* **Note 1**).

6. Cyanide standards for ISE measurements: 10 ml of 10 mM KCN standard solution is made in growth medium (e.g., LB) from the 0.5 M cyanide stock solution. This is then used to make standard solutions as follows: 10 mM KCN (0.2 ml 0.5 M KCN stock, 9.7 ml LB medium, 100 μl 10 M NaOH), 1 mM KCN (1 ml 10 mM KCN, 8.91 ml LB medium, 90 μl 10 M NaOH), 500 μM KCN (5 ml 1 mM KCN, 4.95 ml LB medium, 50 μl10 M NaOH), 250 μM KCN (5 ml 500 μM KCN, 4.95 ml LB medium, 50 μl 10 M NaOH), 100 μM KCN (1 ml 1 mM KCN, 8.91 ml LB medium, 90 μl 10 M NaOH), 50 μM KCN (5 ml 100 μM KCN, 4.95 ml LB medium, 50 μl 10 M NaOH).

2.1.3 Software

Graphpad Prism, Canada.

2.1.4 Miscellaneous Equipment

1. Sterile 250 ml flasks containing 50 ml LB.
2. Sterile 25 ml Universal tubes.
3. Microfuge tubes, 1.5 ml.
4. Pipettes (20–200 and 100–100 μl) and tips.
5. Spectrophotometer to read OD_{600}.
6. Disposable 1 ml cuvettes.
7. Microfuge centrifuge for 1.5 ml tubes.
8. Sterile, disposable plastic spreaders.
9. Agar plates.
10. Plastic weighing boats.
11. Parafilm.
12. 140 mm large plastic Petri dishes.

2.1.5 Specialist Equipment

1. Cyanide ion-selective microelectrode ISM-146CN (Lazar Research Laboratories, Los Angeles, CA, USA).
2. Voltmeter or pH meter set to read in mV.
3. Class II biosafety cabinet for handling sputum samples.

2.2 Use of Cyanide-Sensitive Paper for High-Throughput Detection of Bacterial Cyanide Production

2.2.1 Culture Media

We use LB medium for the growth of bacteria for this protocol (*see* Subheading 2.1.1), but any medium in which cyanogenesis takes place will be suitable.

2.2.2 Cyanide-Sensitive Paper Detection Reagents

If not otherwise stated, all chemicals are available from Sigma Aldrich, Gillingham, UK.

1. Cyanide detection solution (per sheet): 10 mg copper(II) ethylacetoacetate (Alfa Aesar, Zurich, Switzerland) and 10 mg

4,4′-methylenebis (*N,N*-dimethylaniline) are dissolved in 2 ml chloroform.

2. 3 MM Whatman filter paper (Whatman, Maidstone, UK): Cut the filter paper into 8 × 12 cm sheets and wet with 2 ml cyanide detection solution by pipetting the solution onto the paper in a fume hood. Allow the sheets to dry in the hood for 15 min, and then they can be stored for up to 7 days wrapped in aluminum foil at 4 °C.

2.2.3 Software

1. Microsoft Excel or any comparable spreadsheet software.
2. Plate reader software (we use Magellan by Tecan, Crailsheim, Germany).

2.2.4 Miscellaneous Equipment

1. 96-Well round-bottom microtiter plates.
2. 12-Prong multichannel pipette (10–100 μl), pipettes (20–200 μl and 100–1,000 μl), and tips.
3. Breathe-Easy sealing membranes (Sigma Aldrich, Gillingham, UK).
4. Plate roller (Sigma-Aldrich, Gillingham, UK).

2.2.5 Specialist Equipment

1. Microtiter plate reader to read OD_{600}.
2. Fume hood.

3 Methods

3.1 Measuring Bacterial Cyanide Production During Growth in Liquid Culture Using a Cyanide Ion-Selective Electrode

3.1.1 General Points

The method is quick and straightforward as long as basic precautions are adopted in the use of the electrode and preparation of samples [9, 14, 15]. A new calibration curve should be produced each day the electrode is used. Since the electrode needs preconditioning (see below) before each use, it is best to perform the calibration and then move directly on to measuring samples. Do not store the electrode in air between readings, but place it in the 100 μM KCN standard. At the end of use rinse the electrode tip in dH_2O and then cover with the cap (provided with electrode). The electrode plugs into a pH meter, which needs to be switched to mV reading.

3.1.2 Before You Start

1. Check buffer level in electrode (should be just below fill-hole level) and refill if needed with electrolyte provided by the supplier.
2. Check for air bubbles in stem of the electrode, which if present need to be removed. This can be done by holding the electrode firmly by the wire that connects to mV meter and then rotating the electrode rapidly to force the air bubbles out of the end of the electrode (this is the procedure suggested in the manufacturer's instructions).

3. Each time the electrode is used it needs preconditioning first by placing it in 1 mM KCN solution for 10 min.

The electrode suffers from a hysteresis effect (lagging or carryover effect) in that it takes time to adjust from one reading to the next; this is especially evident when going between two very different cyanide concentrations. We usually put the electrode in a 100 μM KCN solution for 10 min after preconditioning as this is around the concentration usually measured in culture or sputum (*see* **Notes 2–4**).

3.1.3 Preparation of a Calibration Curve

Standard solutions (*see* Subheading 2.1.3) are made up in the medium employed for bacterial growth. The ISE is subject to hysteresis effects, so always carry out calibration in the order outlined below measuring the lowest concentration first and working up the standard concentration range.

1. Make up standard solutions (*see* Subheading 2.1.3).
2. Precondition electrode in 1 mM KCN solution for 10 min.
3. Remove electrode from 1 mM KCN solution, rinse, and place in 100 μM KCN solution for a further 10 min.
4. Remove and rinse electrode, place in 50 μM KCN solution for exactly 2 min, and then record mV reading. Repeat until consecutive readings are within a few mV of each other before moving on to measure the remaining standards.
5. Rinse electrode, place in 100 μM KCN solution for 2 min, and record mV reading.
6. Repeat with 250 μM, then 500 μM, and then 1,000 μM KCN solutions remembering to rinse the electrode each time.
7. After calibration is complete place electrode in 100 μM solution and leave until ready to measure samples.
8. Generate a standard curve by plotting mV reading on the x-axis and Log_{10} cyanide concentration on the y-axis. This should be a straight line, and you should get a good fit for a linear equation with r^2 of >0.95. Obtain equation for curve and use to convert mV readings for samples into Log_{10} CN concentration, and then take inverse log to get CN concentration (Fig. 1).

3.1.4 Measurement of Cyanide in Liquid Culture Samples

1. At appropriate time point remove 1 ml of LB-grown culture from culture flask and put into a 1.5 ml microfuge tube. If a medium other than LB is used be aware of potential interferences (*see* **Note 5**).
2. Spin in microfuge at 13,000 × g for 5 min.
3. While spinning prepare a new 1.5 ml microfuge tube for each sample containing 8 μl of 10 M NaOH (ionic strength adjustment buffer) (*see* **Note 6**).

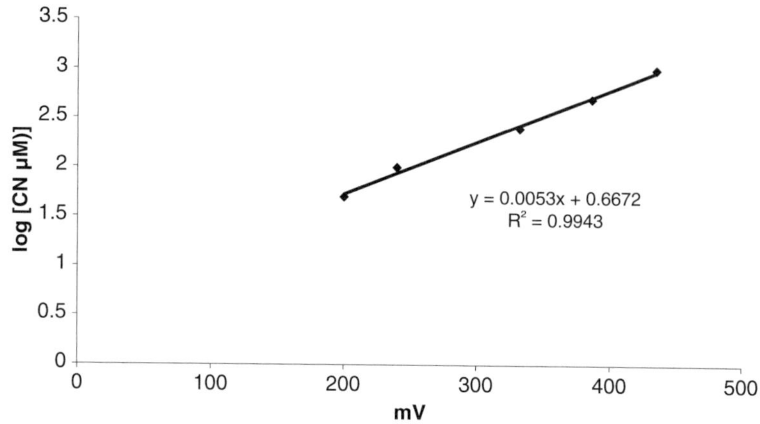

Fig. 1 Calibration curve for ion-selective electrode. A cyanide concentration series made up in LB as described in Subheading 2.1.2 was assayed as in Subheading 3.1.3

4. Add 800 µl of the culture supernatant from spun-down microfuge tubes to the 8 µl of 10 M NaOH in the new microfuge tubes (the NaOH adjusts the ionic strength and raises the pH to 12 so that the electrode works optimally and so that any cyanide present is not lost as HCN gas).
5. Measure cyanide in the tubes by placing electrode in sample tubes for 2 min and recording mV reading. This is done following preconditioning of the electrode in exactly the same way as for calibration measurements (*see* Subheading 3.1.2 above), remembering to rinse the electrode between samples.

3.2 Measuring Bacterial Cyanide Production During Growth on Agar Plates Using Cyanide Ion: Selective Electrode

This method is adapted from ref. 6.

1. Liquid cultures of *P. aeruginosa* are grown in LB broth to an OD_{600} of 0.5.
2. Pipette 100 µl of culture onto an LB agar plate, and spread it over the plate using a disposable sterile plastic spreader. Incubate the plates overnight at 37 °C.
3. Place plates, with now confluent growth, with lids removed, inside a sterile 140 mm large Petri dish, alongside a plastic weighing boat containing 1 ml 4 M NaOH.
4. Seal the 140 mm Petri dish with Parafilm and incubate for 4 h at 37 °C. During this time any HCN gas leaving the plated culture is trapped in the adjacent sodium hydroxide solution.
5. Dilute the 4 M NaOH solution 1:10 in dH20, and the concentration of cyanide trapped in the NaOH solution is measured using the cyanide electrode in the same way as described above in Subheading 3.1.3, except that the KCN standards are

made up in 0.4 M NaOH (note: it is not necessary to add the 10 M NaOH ionic strength adjustment buffer). If cyanide concentration in the NaOH is too low for measurement (below calibrated range) then the undiluted NaOH solution can be used for measurement of the trapped cyanide but alongside a calibration curve carried out in 4 M NaOH.

3.3 Methods: Measurement of Cyanide Levels in Infected Sputum

This procedure is based on that of [9].

1. Handling sputum: A class II safety cabinet is used for all handling of sputum samples, including all stages of the sputum cyanide measurement, with the cyanide electrode and mV recorder being moved into the cabinet for measurements.
2. Calibration of the electrode for use in sputum using KCN-spiked sputum is not possible routinely due to the difficulty in obtaining sputum that is cyanide free for this purpose. Therefore, routine calibration of the electrode is carried out using the KCN solutions dissolved in sodium hydroxide as described above (Subheading 3.1.3). We have tested the reliability of this method using KCN-spiked sputum samples (*see* **Note 7**).
3. Collection of sputum: Patients are given a 50 ml polypropylene, conical bottomed, and screw top tube in which to cough into to produce a sputum sample.
4. Immediately seal the tubes with a SubaSeal rubber stopper (Scientific Laboratory Supplies Nottingham, UK) and place on ice. Cyanide concentration measurements are then made on the sputum sample a maximum of 1 h after collection. Immediately before measurement inject 2× volume/weight sputum 0.1 M NaOH through the SubaSeal, and vortex the sample to ensure that any HCN gas that has escaped from the sputum redissolves and to adjust the pH and ionic strength of the sample ready for cyanide measurement.
5. Cyanide concentration is then determined using the cyanide ISE by placing the conditioned electrode (Subheading 3.1.2) into the sample and recording the mV reading after 2 min.

3.4 Methods: Use of Cyanide-Sensitive Paper for High-Throughput Detection of Bacterial Cyanide Production

This method is most conveniently applied when mutants or clinical strains are arranged in a 96-well microtiter plate format. Bacterial strains of interest are grown overnight aerobically at 37 °C to an $OD_{600} > 1$. Overnight cultures are then diluted to a starting OD_{600} of <0.05 and grown to mid-log phase to an OD_{600} 0.5–0.6. We recommend the use of at least three biological replicates ($n = 3$) as well as suitable control strains (*see* **Note 8**).

3.4.1 Assay Setup

1. Inoculate 10 μl culture at OD_{600} 0.5–0.6 into 190 μl LB in 96-well microtiter plates (1:20 dilution). If starter cultures have been grown up in 96-well format then this is done using a multichannel pipette.

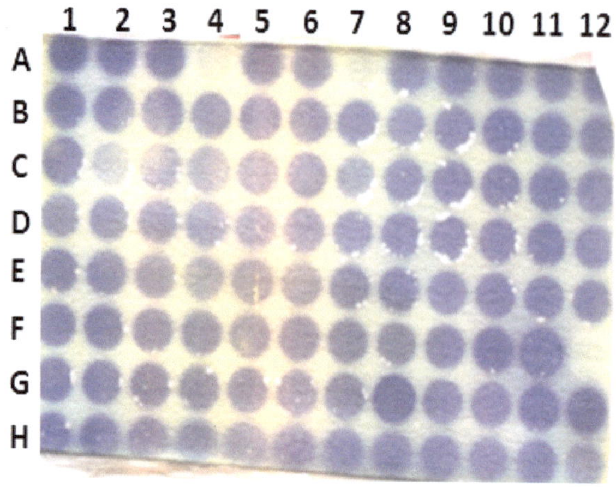

Fig. 2 Microtiter plates containing *P. aeruginosa* strains can be screened for production of volatile hydrogen cyanide, turning the cyanide detection paper blue. Fainter blue spots in wells identify cyanide-deficient mutants, for example in wells A4, A7, C2, and F12

2. Remove the microtiter plate lid, and cover the plate with a cyanide detection paper sheet. Discretely label with a pencil the top left corner of the sheet "A1."
3. Seal detection paper tightly in place using a Breathe-Easy membrane and a plate roller. Discard microtiter plate lid (*see* **Note 9**).
4. Incubate plate at 35–37 °C without shaking for 24 h.
5. After 24 h, remove cyanide detection paper and measure OD_{600} for each well of the plate using a plate reader (*see* **Note 10**).

3.4.2 Data Analysis

Generate an Excel spreadsheet containing OD_{600} values for each strain. Make note of wells with OD_{600} values <0.8 and discard these from analysis. Perform a visual assessment of cyanide detection papers. The absence of a blue spot above a well is indicative of an acyanogenic strain. The intensity of the color of the spot correlates to the amount of volatile HCN produced and allows for the identification of cyanide-deficient mutants and cyanide overproducers (Fig. 2), and this is most usefully done in comparison to wild-type and mutant strains affected in HCN production, such as Δ*hcnA*, Δ*gacS*, and Δ*rsmA* (overproducer) (Fig. 3).

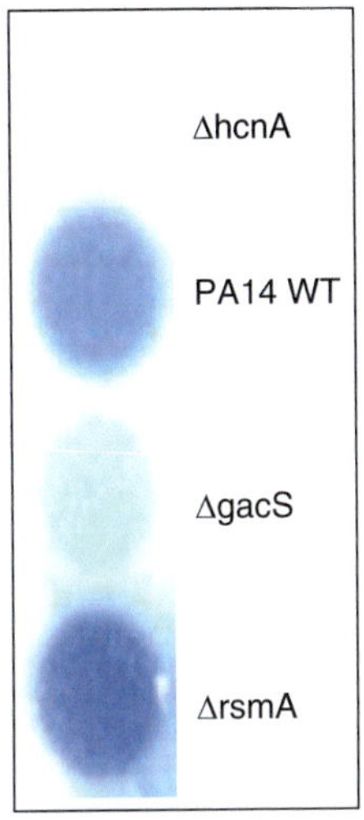

Fig. 3 Cyanide production detected by cyanide-sensitive paper for different *P. aeruginosa* strains—PA14, wild-type strain; Δ*hcnA*, no cyanide produced; Δ*gacS*, impaired cyanogenesis; Δ*rsmA*, cyanide overproducer

4 Notes

1. We use solid KCN for making our cyanide stock solution as it is crystalline and so less prone to dispersing in air when handling. The following precautions were taken for storing and handling solid KCN and the 0.5 M KCN stock solution: (a) The solid KCN is kept in a locked cupboard. (b) Gloves and a dust mask are worn when handling. (c) Solid KCN is added directly to 1 ml of 2 g/l KOH in a 20 ml Sterilin tube. N.B. Highly toxic HCN gas is released from KCN on contact with acid; dissolving the KCN in alkali ensures against this. (d) To minimize handling a single amount of solid KCN of approximately the correct weight to make 5 ml of 0.5 M KCN solution is added to the 1 ml KOH and the increase in weight recorded; further KOH is then added to make a 0.5 M KCN solution. (e) Gloves are worn when handling the 0.5 M KCN stock, and the tube is clearly labeled as containing a toxic substance.
2. *Electrode performance.* The electrode performance will deteriorate over time; signs that it is starting to fail include the following: loss of ability to accurately measure at the lower limit of its range and readings start to cycle between high and low mV outputs. The definitive sign that the electrode has reached the end of its life is if it *cannot* produce a standard curve with an r^2 of above 0.95.

3. *Normal electrode readings.* The electrode reading will not be completely stable; when you first put it in the sample it usually starts high and then decreases, rapidly at first, and then stabilizes over a couple of minutes but will continue to fall slowly. We have found that measuring values after a 2 min leads to reproducible values at a given cyanide concentration.
4. *Zero cyanide readings.* The relationship between mV reading and cyanide concentration is not linear. Within the working range of the electrode (~50 μM—several 1,000 μM) the relationship is logarithmic, i.e., mV reading is proportional to LOG (cyanide) (this relationship does not hold outside the working range). The main consequence of this is that you still get a mV reading even if no cyanide is present.
5. *Interference.* Other ions can interfere with the electrode, in particular sulphide and silver, so be aware of this if measuring in media other than LB, perform blank (0) cyanide readings, and also make sure that you get a good standard curve if you use any other media.
6. *Sample and standard preparation.* 10 M NaOH (1 μl per 100 μl) needs to be added to samples/standards. This is referred to as ionic strength buffer; it adjusts the ionic strength to a relatively high value to limit differences in total ionic strength of samples (total ionic strength affects readings). It is also important to make standards for calibration in the same medium that your measurements will be made in. The standard and sample preparation described is for LB, which as we have shown does not suffer too much from interference and in which we get good reproducibility. Other media can be used, but be aware that sulphide and silver ions are major interfering ions and will have a large effect on readings.
7. It is not practical to routinely calibrate the electorate in sputum, and we have found that using a NaOH calibration curve for sputum measurement (instead of KCN-spiked sputum) overestimates cyanide concentration because of the different ionic matrix properties between sodium hydroxide and sputum. However, we have found the overestimation to be identical in separate sputum-spiking experiments using sputum from three different patients. This data indicated that the relationship between actual and measured cyanide is linear which allows an equation describing the relationship with actual and measured cyanide to be generated. In our hands this equation is

$$\text{Actual}\ [\text{cyanide}] = 0.675\ (\text{apparent}\ [\text{cyanide}] - 52.64) \quad (1)$$

We use this equation to convert sputum cyanide measurements using a calibration curve made up in NaOH (Subheading 3.1.3) to actual sputum cyanide concentrations.

However we recommend each laboratory determines the relationship between actual and measured cyanide concentration by carrying out a cyanide-spiking experiment in which cyanide is added to the sputum from a patient known not to be infected with the cyanogenic bacteria *P. aeruginosa* and *Burkholderia cepacia* complex, then determining whether the cyanide levels measured in the spiked sample agree with the actual (or known) amount added, and determining cyanide concentrations and comparing these actual cyanide concentrations [9]. A linear plot of actual against measured cyanide is then generated from which the relationship between measured and actual cyanide concentration can be derived.

8. We recommend including known cyanogenesis controls in the screen. These include a wild-type *P. aeruginosa* strain (we use PA14), a known acyanogenic mutant (we use Δ*hcnA*), an underproducer such as a quorum sensing deficient mutant (we use Δ*gacS*), and a known overproducer (we use Δ*rsmA*) (*see* Fig. 3).
9. A tight fit of the cyanide detection paper to the microtiter plate is essential for obtaining discrete spots. Care should be taken not to allow the culture to come into contact with the cyanide detection paper as this will interfere with the cyanide detection reagents.
10. As cyanide production occurs in stationary phase, strains that have not grown to an $OD_{600} > 0.8$ after 24 h cannot be assessed for cyanide production. Incubation time can be extended to allow these to reach a higher OD_{600}. If an $OD_{600} > 0.8$ cannot be achieved, the strains should be considered as growth-deficient rather than cyanide-deficient mutants.

References

1. Blumer C, Haas D (2000) Mechanism, regulation, and ecological role of bacterial cyanide biosynthesis. Arch Microbiol 173:170–177
2. Ryall B, Lee X, Zlosnik JE, Hoshino S, Williams HD (2008) Bacteria of the *Burkholderia cepacia* complex are cyanogenic under biofilm and colonial growth conditions. BMC Microbiol 8:108
3. Ryall B, Mitchell H, Mossialos D, Williams HD (2009) Cyanogenesis by the entomopathogenic bacterium Pseudomonas entomophila. Lett Appl Microbiol 49:131–135
4. Williams HD, Ziosnik JEA, Ryall B (2007) Oxygen, cyanide and energy generation in the cystic fibrosis pathogen *Pseudomonas aeruginosa*. Adv Microb Physiol 52:1–71
5. Haas D, Defago G (2005) Biological control of soil-borne pathogens by fluorescent pseudomonads. Nat Rev Microbiol 3:307–319
6. Gallagher LA, Manoil C (2001) *Pseudomonas aeruginosa* PAO1 kills *Caenorhabditis elegans* by cyanide poisoning. J Bacteriol 183: 6207–6214
7. Broderick KE, Chan A, Balasubramanian M, Feala J, Reed SL et al (2008) Cyanide produced by human isolates of *Pseudomonas aeruginosa* contributes to lethality in *Drosophila melanogaster*. J Infect Dis 197:457–464
8. Goldfarb WB, Margraf H (1967) Cyanide production by *Pseudomonas aeruginosa*. Ann Surg 165:104–110
9. Ryall B, Davies JC, Wilson R, Shoemark A, Williams HD (2008) *Pseudomonas aeruginosa*, cyanide accumulation and lung function in CF and non-CF bronchiectasis patients. Eur Respir J 32:740–747
10. Sanderson K, Wescombe L, Kirov SM, Champion A, Reid DW (2008) Bacterial

cyanogenesis occurs in the cystic fibrosis lung. Eur Respir J 32:329–333

11. Feigl F, Anger V (1966) Replacement of benzidine by copper ethylacetoacetate and tetra base as spot-test reagent for hydrogen cyanide and cyanogen. Analyst 91:282–284
12. Castric PA (1983) Hydrogen cyanide production by *Pseudomonas aeruginosa* at reduced oxygen levels. Can J Microbiol 29:1344–1349
13. Takos A, Lai D, Mikkelsen L, Abou Hachem M, Shelton D et al (2010) Genetic screening identifies cyanogenesis-deficient mutants of *Lotus japonicus* and reveals enzymatic specificity in hydroxynitrile glucoside metabolism. Plant Cell 22:1605–1619
14. Zlosnik JE, Williams HD (2004) Methods for assaying cyanide in bacterial culture supernatant. Lett Appl Microbiol 38:360–365
15. Zlosnik JE, Tavankar GR, Bundy JG, Mossialos D, O'Toole R et al (2006) Investigation of the physiological relationship between the cyanide-insensitive oxidase and cyanide production in *Pseudomonas aeruginosa*. Microbiology 152: 1407–1415

Chapter 28

Monitoring Iron Uptake by Siderophores

Françoise Hoegy and Isabelle J. Schalk

Abstract

Iron is an important element for almost all forms of life. In order to get access to this essential nutriment, *Pseudomonads* produce two major siderophores, pyoverdine PVD and pyochelin (PCH). Uptake of iron in bacterial cells can be monitored accurately using ^{55}Fe. Bacteria cells are incubated in the presence of either PVD or PCH loaded with ^{55}Fe. After incubation, extracellular iron ions are separated from those accumulated in the bacteria cells by either centrifugation or filtration on glass microfiber filters, for the PCH and PVD assays, respectively. ^{55}Fe contained in the harvested cells on the filter or in the cell pellet is counted in scintillation cocktail. The number of moles of ^{55}Fe transported can be determined using the specific activity of the radionuclide.

Key words Siderophore, Pyoverdine, Pyochelin, Iron uptake, *Pseudomonas aeruginosa*, Iron homeostasis

1 Introduction

Iron is an essential nutriment for the growth and development of almost all living organisms. It acts as a catalyst in some of the most fundamental enzymatic processes, including oxygen metabolism, electron transfer, and DNA and RNA synthesis. Analyses of *Pseudomonas aeruginosa* genome indicated the presence of at least 11 iron acquisition pathways [1], illustrating the importance of iron for this microorganism. Iron depletion leads to the increased expression of these iron acquisition systems and decreased expression of biological systems that rely on relatively large amounts of iron. On the opposite, the presence of an excess of iron ion becomes toxic for bacterial cells, because of oxidative stress and interactions of iron with nonspecific targets. Consequently, iron homeostasis needs to be finely and tightly regulated. The cytoplasmic ferric uptake regulator (Fur) plays a key role in this process. In *P. aeruginosa*, Fur directly or indirectly controls the expression of a large number of genes and operons involved in iron uptake and virulence factors [2, 3]. Under iron-replete conditions, the Fur protein

Alain Filloux and Juan-Luis Ramos (eds.), *Pseudomonas Methods and Protocols*, Methods in Molecular Biology, vol. 1149, DOI 10.1007/978-1-4939-0473-0_28,

© Springer Science+Business Media New York 2014

becomes ferrated and binds to the Fur box, in the promoters of genes required for iron uptake, thereby preventing their transcription [4, 5]. Fur can also contribute to the increased expression of genes via the repression of two small regulatory RNAs, PrrF1 and PrrF2 [6]. These small RNAs contribute to iron homeostasis by causing the degradation of mRNAs encoding iron-containing proteins [6, 7].

Despite its importance in biology, iron is found in insoluble oxide hydrate complexes in most aerobic microbial habitats. During infections, iron is similarly not freely accessible for the pathogen, as it is tightly sequestered by proteins or hemes in the host. Therefore, the free iron concentrations are always extremely low in the environment of microorganisms (10^{-7} to 10^{-24} M) [8]. To get access to iron, *P. aeruginosa* produces two major siderophores: pyoverdine (PVD) and pyochelin (PCH). The structure of PCH was established as (4′*R*, 2″*R*, 4″*R*)-2′-(2-hydroxyphenyl)-3″-methyl-4′,5′,2″,3″,4″,5″-hexahydro-[4′,2″]bithiazolyl-4″-carboxylic acid as described in Chapter 23. In the extracellular medium, PCH chelates Fe^{3+} with a 2:1 (PCH:Fe(III)) stoichiometry, with one molecule of PCH tetradentately coordinated to Fe^{3+} and the second molecule bound bidentately to complete the hexacoordinate octahedral geometry [9, 10]. The affinity of PCH for iron is 10^{18} M^{-1} [9]. Uptake of iron by PCH in *P. aeruginosa* cells involves a TonB-dependent outer membrane transporter, FptA [11], and probably the permease FptX for translocation across the inner membrane [12].

PVD is a chromopeptide produced by all fluorescent pseudomonads as described in Chapter 23. All these PVD are composed of three parts: a dihydroquinoline-type chromophore responsible for their fluorescence, a strain-specific peptide comprising 6–12 amino acids and a side chain bound to the nitrogen atom at position C-3 of the chromophore [13]. The sequence of the peptide moiety is highly variable among species and even between strains of the same species [14]. This involves that each strain of fluorescent pseudomonads is only able to use its proper PVD and not the one produced by another strain. It is important for the protocol described below to take into account this high strain specificity of the PVD pathways. PVD produced by *P. aeruginosa* chelates Fe^{3+} with a 1:1 (PVD:Fe(III)) stoichiometry and an affinity of 10^{32} M^{-1} [15].

2 Materials

Prepare all solutions using bidistillated water and analytical grade reagents. Prepare and store all reagents at room temperature (unless indicated otherwise). Diligently follow all waste disposal regulations when disposing waste materials, especially for radioactive wastes.

2.1 Bacterial Strains and Growth Media

1. *P. aeruginosa* strains are stored in LB (L-Broth or Luria Bertani) with 20 % glycerol in 1.5 mL micro tubes at −80 °C.
2. LB medium: Weigh 20 g of LB, transfer into a 1 L bottle, and complete to 1 L with water. Sterilize immediately after preparation, and store the solution at room temperature.
3. Succinate medium is an iron-restricted medium: Weigh 6 g of K_2HPO_4, 3 g of KH_2PO_4, 1 g of $(NH_4)_2PO_4$, 0.2 g of $MgSO_4$, 4 g of succinic acid, and 1.1 g of NaOH. Add water to a volume of 500 mL. Mix and adjust pH at 7.0 with NaOH. Make up to 1 L with water. Sterilize immediately after preparation, and store at room temperature.
4. Centrifugation tubes, Falcon 50 mL (Becton Dickinson, Franklin Lakes, USA).
5. Centrifuge 5804 (Eppendorf, Hamburg, Germany).
6. Semi-micro disposal cuvettes for spectroscopy, 1.5 mL, optical PS (Kartell, Noviglio, Italy).
7. Spectrophotometer UV–Visible, BioPhotometer (Eppendorf).

2.2 Siderophore–^{55}Fe Complex

1. Purified PVD: PVD is purified from *P. aeruginosa* cultures as described in [13] and prepared at 10 mM in water using its molar extinction coefficient at 380 nm at pH 5 ($\varepsilon = 16{,}500\ M^{-1}\ cm^{-1}$) and its molecular mass (MM = 1,334 g/mol) (*see* **Notes 1** and **2**).
2. Purified PCH: PCH is purified as described in [16] and prepared at 10 mM in MeOH (molecular mass of PCH being 324 g/mol and molar extinction coefficient 4,900 $M^{-1}\ cm^{-1}$) (*see* **Notes 3** and **4**).
3. 2 mCi of $^{55}FeCl_3$ in HCl 0.1 N at 3.3 mM (Perkin Elmer, Waltham, USA).
4. Nonradioactive $FeCl_3$: 27.5 % solution in water (Prolabo, Fontenay-sous-Bois, France).

2.3 Uptake Assay Components

1. Tris–HCl buffer: 50 mM Tris–HCl, pH 8.0.
2. Carbonyl cyanide *m*-chlorophenylhydrazone (CCCP): 20 mM solution in EtOH (Sigma Chemical Company, St. Louis, MO, USA).
3. GFB filters, 25 mm circles (Whatman, Maidstone, UK).
4. Minisorp 70 × 11 Nunc immuno tubes (Thermo Fischer Scientific, Roskilde, Denmark) (*see* **Note 5**).
5. Micro tubes PP 1.5 mL (Trefflab, Degersheim, Switzerland).
6. Heating block, 60 positions for 1.5 mL tubes (Thermolyne, Dubuque, USA).
7. Vortex Genie 2 shaker (Scientific Industries, NY, USA).
8. Multipette plus (Eppendorf).

9. Diaphragm pumping unit (Vacuubrand, Wertheim, Germany).
10. Filtration manifold, 10 wells 25 mm, ref FH 225V (Hoefer, Inc., Holliston, USA).
11. MiniSpin Centrifuge, 12 positions for 1.5 mL tubes (Eppendorf France, Le Pecq, France).
12. Scintillation cocktail, Ultima Gold MV (Perkin Elmer).
13. Scintillation vials, Snaptwist 6.5 mL, 16 mm × 57 mm, Polyethylene (Simport, Beloeil, Canada).
14. Liquid scintillation counter, Tricarb 2100 TR (Perkin Elmer).
15. Chronometer (Dutscher, Brumath, France).

3 Methods

Carry out all procedures at room temperature unless otherwise specified.

3.1 Bacterial Culture and Preparation

1. Bacterial growth: Transfer 10 μL of glycerol bacterial solution stored at −80 °C into 10 mL of LB medium, and grow the cells overnight at 37 °C under agitation. The next day, dilute 10 mL of this culture in 100 mL of succinate medium and run a 24-h culture at 30 °C under agitation. The following day, dilute again the culture in succinate (10 mL in 100 mL medium) and allow them to grow for 24 h at 30 °C under agitation (*see* **Note 6**).
2. Pellet 25 mL of the cell culture in succinate medium in a 50 mL centrifugation tube at 4,000 × *g* for 8 min (*see* **Note 7**).
3. Remove the supernatant, suspend the cells with 50 mL of 50 mM Tris–HCl buffer pH 8.0, and centrifuge them at 4,000 × *g* for 8 min. Repeat this cell washing step (*see* **Note 7**).
4. After removal of the supernatant, suspend the cells in 5 mL of Tris–HCl buffer pH 8.0.
5. Dilute 100 μL of the cell suspension in 1 mL of Tris–HCl buffer pH 8.0, in a 1 mL cuvette, and measure the turbidity of the cells on the spectrophotometer settled at 600 nm.
6. Adjust the buffer volume of the cell sample to obtain an OD at 600 nm of 1.

3.2 Siderophore–^{55}Fe Complex Preparation (See Note 8)

3.2.1 Solution of PVD–^{55}Fe Complex

1. Prepare a solution of ^{55}Fe at 250 μM in HCl 0.1 N. Prepare the same solution with nonradioactive Fe. Mix 1 volume of the ^{55}Fe solution at 250 μM with 9 volumes of the nonradioactive Fe solution (*see* **Notes 9** and **10**).
2. Prepare a stock solution of PVD in water at 1 mM as described above.
3. Mix 4 μL of isotopically diluted ^{55}Fe at 250 μM with 4 μL of purified PVD at 1 mM in water. Incubate for 15 min to allow

the metal chelation by the siderophore to occur (*see* **Notes 11** and **12**).

4. Complete the volume to 100 μL with 92 μL of 50 mM Tris–HCl buffer pH 8.0 to obtain a 10 μM PVD–^{55}Fe complex (iron being the limiting factor) (*see* **Note 12**).

3.2.2 Solution of PCH–^{55}Fe Complex

1. Prepare a solution of ^{55}Fe at 250 μM in HCl 0.1 N. Prepare the same solution with nonradioactive Fe. Mix 1 volume of ^{55}Fe with 9 volumes of nonradioactive Fe (*see* **Notes 9** and **10**).
2. Prepare a stock solution of PCH in MeOH at 1 mM as described above.
3. Mix 4 μL of isotopically diluted ^{55}Fe at 250 μM with 10 μL of purified PCH at 1 mM in MeOH. Incubate for 15 min to enable the metal chelation by the siderophore to occur (*see* **Notes 11** and **12**).
4. Complete the volume to 100 μL with 86 μL of 50 mM Tris–HCl buffer pH 8.0 to obtain a 10 μ M PCH–^{55}Fe complex (iron being the limiting factor) (*see* **Note 12**).

3.3 Iron Uptake with the Siderophore PVD

1. Pour 1 mL of bacterial cells prepared in 50 mM Tris–HCl pH 8.0 at $OD_{600\ nm}$ of 1 in one Minisorp tube preheated in the heating block set at 37 °C. One tube will be used per kinetic.
2. In parallel, as a control, pour 1 mL of bacterial cells prepared in 50 mM Tris–HCl pH 8.0 at $OD_{600\ nm}$ of 1 in one Minisorp tubes refrigerated on ice and containing 10 μL of 20 mM CCCP. Mix the sample.
3. Incubate the cells for 15 min at the corresponding temperatures (*see* **Note 13**).
4. Prepare 18 GFB filters in 50 mM Tris–HCl buffer pH 8.0, with incubation at room temperature.
5. Place 18 scintillation vials in an appropriate rack (*see* **Note 14**).
6. Switch on the pump that equips the filtration manifold, and place the filters on it. Suck up each filter by turning on the tap of the corresponding well (*see* **Note 15**).
7. Add 10 μL of the solution of PVD–^{55}Fe complex to the 1 mL bacterial cell solutions incubated at 37 °C and on ice. Mix efficiently on a vortex shaker.
8. Start the kinetic with an instantaneous 100 μL aliquot withdrawals made immediately after PVD–^{55}Fe addition to the bacterial cells incubated at 37 °C and on ice (*see* **Notes 16** and **17**).
9. Wash the filters with 4 mL of 50 mM Tris–HCl buffer pH 8.0 (*see* **Notes 18** and **19**).
10. Remove the filters from the manifold, and place them in the appropriate scintillation vials. Replace them by new filters for the following kinetic points.

11. Repeat the 100 μL withdrawals at 3, 6, 9, 12, 15, 20, 25, and 30 min after addition of PVD–^{55}Fe to the bacterial cells at 37 °C and on ice (**steps 8–10**).
12. Add 2.5 mL of scintillation cocktail to each vial.
13. Close the scintillation vials with their adapted caps.
14. Agitate each scintillation vial energetically, and incubate them in the darkness for an hour (*see* **Notes 20** and **21**).
15. Place the tubes in the liquid scintillation counter, and choose an appropriate counting program (*see* **Note 22**).
16. To determine the specific activity (dpm or cpm per mol of PVD–^{55}Fe) of the 10 μM PVD–^{55}Fe solution, count 5 μL of this solution in a scintillation vial in the presence of 1 mL scintillation cocktail.

3.4 Iron Uptake with the Siderophore PCH

1. Pour 1 mL of bacterial cells prepared in 50 mM Tris–HCl pH 8.0 at $OD_{600\ nm}$ of 1 in one Minisorp tube preheated in the heating block set at 37 °C. One tube will be used per kinetic.
2. In parallel, as a control, pour 1 mL of bacterial cells prepared in 50 mM Tris–HCl pH 8.0 at OD_{600nm} of 1 in one Minisorp tubes kept at room temperature and containing 10 μL of 20 mM CCCP (*see* **Note 23**).
3. Incubate the cells for 15 min at the corresponding temperatures.
4. Prepare 18 1.5 mL micro tubes.
5. Place 18 scintillation vials in an appropriate rack.
6. Add 10 μL of the solution of PCH–^{55}Fe complex to the 1 mL bacterial cell solutions incubated at 37 °C and at room temperature. Mix efficiently on a vortex shaker.
7. Start the kinetic with an instantaneous 100 μL aliquot withdrawals made immediately after PCH–^{55}Fe addition to the bacterial cells incubated at 37 °C and at room temperature (*see* **Note 16**), and place them in a micro tube. Add 500 μL of ice-cold Tris–HCl buffer pH 8.0 (*see* **Notes 17** and **24**).
8. Repeat **step** 7 at 3, 6, 9, 12, 15, 20, 25, and 30 min of incubation of the bacterial cells incubated at 37 °C and at room temperature.
9. Pellet the samples in the micro tube in a minispin centrifuge for 2 min at 3,000 ×*g*.
10. Remove the supernatants (*see* **Note 25**).
11. Cut the caps of the micro tubes, and place them in a vial (*see* **Note 26**).
12. Add 0.5 mL of scintillation cocktail to each micro tube.
13. Close the vials with their adapted caps.
14. Agitate each tube energetically, and place them in the darkness for an hour (*see* **Note 21**).

15. Place the tubes in the liquid scintillation counter, and choose an appropriate counting program (*see* **Note 22**).
16. To determine the specific activity (dpm or cpm per mol of PCH–^{55}Fe) of the 10 μM PCH–^{55}Fe solution, count 5 μL of this solution in a scintillation vial in the presence of 1 mL scintillation cocktail.

4 Notes

1. Because of the high strain specificity of the PVD pathways, it is important to isolate the PVD from the strain used in the uptake assay.
2. Purified PVD from *P. aeruginosa* PAO1 and from other fluorescent pseudomonads can be obtained from our group. Contact isabelle.schalk@unistra.fr to have information.
3. PCH can be prepared as described in Chapter 23, or highly purified PCH from *P. aeruginosa* and *P. fluorescens* can be obtained from our group. Contact isabelle.schalk@unistra.fr to have information.
4. Due to its toxicity, use standard individual protection equipment when using MeOH. Since PCH is poorly soluble in water, MeOH is used to solubilize this siderophore. The dilutions are done in water to avoid any cell death due to high concentrations of MeOH.
5. Pay attention to the size of the tubes used: they must fit into the heating block, and they should be high enough to avoid pouring of the sample while shacking it.
6. Cells are rather grown at 30 °C at this stage because of the higher production of proteins involved in the iron uptake pathways, compared to 37 °C cultures. Besides, one 24-h culture in iron-limited succinate medium is not enough to get the bacteria iron starved and a high expression level of the proteins involved in iron uptake.
7. The centrifuge and its rotor have to be at room temperature. If they are at low temperature, they must be heated before pelleting the cells. Cells at low temperature do not transport iron efficiently.
8. PVD–^{55}Fe and PCH–^{55}Fe have to be prepared freshly before use and cannot be stored at 4 °C or −20 °C.
9. Iron is poorly soluble at neutral pH; therefore, iron solutions have to be prepared exclusively in HCl 0.1 N to avoid iron precipitation.
10. According to the ALARA principle (as low as reasonably achievable) about the use of radioactivity, the quantities of ^{55}Fe

used should be optimized to the lowest level possible. In this purpose, an isotopic dilution allows to divide the quantities of radioactivity used without compromising the monitoring of iron accumulation in the bacterial cells.

11. To prevent ^{55}Fe precipitation, which would increase the noise of the radioactive signal, a ratio of 4 equivalent of PVD for 1 equivalent of ^{55}Fe ion has to be used (in the case of the uptake assay with PCH, this ratio has to be increased to 10 equivalent PCH for 1 equivalent ^{55}Fe ion due to a lower affinity of this siderophore for iron).
12. Incubation times were optimized to allow ^{55}Fe chelation to occur without compromising the stability of PVD or PCH at low pH. For the same reason, it is important to increase the pH of the solution by addition of 50 mM Tris–HCl buffer pH 8.0.
13. Since temperature is an important factor for cells to be able to incorporate ferri–siderophore complexes, the ^{55}Fe uptake kinetics are driven at 37 °C.
14. At this point, and before beginning the time-based part of the experiment, it is important to pay attention to the layout of the material. The operator will have to be fast, especially for the first kinetic points. Therefore it is necessary to have all the needed materials ready and accessible.
15. Before addition of the PVD–^{55}Fe complex to the cells, it can be useful to test the capacity of the pump to suck up the filters. Leave the taps turned on when beginning the experiment; that way, as soon as the cells are dropped on the filter, PVD–^{55}Fe complexes not incorporated into the bacteria will be immediately separated from the cells by filtration.
16. It is essential to efficiency shake the samples to have a homogenous bacterial solution before each withdrawal.
17. It is important to have a first kinetic point immediately after addition of PVD–^{55}Fe to the bacterial cells. The operator will thus pay attention to make this first withdrawal as quickly as possible.
18. The use of a multipette will allow constant washing volumes.
19. If problems of iron precipitation are observed (high radioactive background), it can be useful to replace washing of the filters with Tris–HCl buffer pH 8.0 by washing with HCl 0.1 N to lower the radioactive background.
20. The purpose of shaking the vials is to increase the contact of the cells with the scintillation cocktail; it will improve the signal of scintillation and the quality of the experiment.
21. The vials are stored for 1 h in the darkness to allow the scintillation phenomenon to occur without any natural lightning disturbing it. Cover of the liquid scintillation counters is therefore dark stained too.

22. We chose to use a counting program that accumulates decays during 10 min maximum. Shorter time counting programs can also be used, but accuracy of the measurements can thus be affected.
23. With the siderophore PCH, problems of increasing radioactive noise were observed when the cells were incubated on ice, probably due to the lower affinity of PCH for iron ion compared to PVD. Therefore, it is important to work at room temperature with the siderophore PCH to avoid this problem. This incubation at room temperature seems to have no major impact on the quality of the experiment.
24. The addition of ice-cold Tris–HCl buffer pH 8.0 immediately stops the uptake mechanism, since *P. aeruginosa* cells are unable to incorporate iron via siderophores when incubated at temperature lower than 15 °C.
25. Removing of the supernatant can be carried out either with a pipet or with a pumping system.
26. If the caps of the micro tubes are properly removed, they just fit in the scintillation vials and the 1.5 mL micro tubes containing the bacterial cell pellets can directly be placed into them for radioactive counting.

Acknowledgement

This work was supported by CNRS and ANR grants (ANR-08-BLAN-0315-01 and 05-JCJC-0181-01).

References

1. Llamas MA, Sparrius M, Kloet R, Jimenez CR, Vandenbroucke-Grauls C, Bitter W (2006) The heterologous siderophores ferrioxamine B and ferrichrome activate signaling pathways in *Pseudomonas aeruginosa*. J Bacteriol 188: 1882–1891
2. Leoni L, Orsi N, de Lorenzo V, Visca P (2000) Functional analysis of PvdS, an iron starvation sigma factor of *Pseudomonas aeruginosa*. J Bacteriol 182:1481–1491
3. Vasil ML, Ochsner UA (1999) The response of *Pseudomonas aeruginosa* to iron: genetics, biochemistry and virulence. Mol Microbiol 34:399–413
4. Griggs DW, Konisky J (1989) Mechanism for iron-regulated transcription of the *Escherichia coli cir* gene: metal-dependent binding of *fur* protein to the promoters. J Bacteriol 171: 1048–1054
5. Prince RW, Cox CD, Vasil ML (1993) Coordinate regulation of siderophore and exotoxin A production: molecular cloning and sequencing of the *Pseudomonas aeruginosa fur* gene. J Bacteriol 175:2589–2598
6. Wilderman PJ, Sowa NA, Fitzgerald DJ, Fitzgerald PC, Gottesman S, Ochsner UA, Vasil ML (2004) Identification of tandem duplicate regulatory small RNAs in *Pseudomonas aeruginosa* involved in iron homeostasis. Proc Natl Acad Sci U S A 101:9792–9797
7. Vasil ML (2007) How we learnt about iron acquisition in *Pseudomonas aeruginosa*: a series of very fortunate events. Biometals 20: 587–601
8. Raymond KN, Dertz EA, Kim SS (2003) Enterobactin: an archetype for microbial iron transport. Proc Natl Acad Sci U S A 100: 3584–3588

9. Brandel J, Humbert N, Elhabiri M, Schalk IJ, Mislin GLA, Albrecht-Garry A-M (2012) Pyochelin, a siderophore of *Pseudomonas aeruginosa*: Physicochemical characterization of the iron(III), copper(II) and zinc(II) complexes. Dalton Trans 41:2820–2834
10. Tseng CF, Burger A, Mislin GLA, Schalk IJ, Yu SS-F, Chan SI, Abdallah MA (2006) Bacterial siderophores: the solution stoichiometry and coordination of the Fe(III) complexes of pyochelin and related compounds. J Biol Inorg Chem 11:419–432
11. Cobessi D, Celia H, Pattus F (2004) Crystallization and X-ray diffraction analyses of the outer membrane pyochelin receptor FptA from *Pseudomonas aeruginosa*. Acta Crystallogr D Biol Crystallogr 60:1919–1921
12. Michel L, Bachelard A, Reimmann C (2007) Ferripyochelin uptake genes are involved in pyochelin-mediated signalling in *Pseudomonas aeruginosa*. Microbiology 153:1508–1518
13. Demange P, Wendenbaum S, Linget C, Mertz C, Cung MT, Dell A, Abdallah MA (1990) Bacterial siderophores: structure and NMR assignment of pyoverdins PaA, siderophores of Pseudomonas aeruginosa ATCC 15692. Biol Metals 3:155–170
14. Meyer JM, Geoffroy VA, Baida N, Gardan L, Izard D, Lemanceau P, Achouak W, Palleroni NJ (2002) Siderophore typing, a powerful tool for the identification of fluorescent and non-fluorescent pseudomonads. Appl Environ Microbiol 68:2745–2753
15. Albrecht-Gary AM, Blanc S, Rochel N, Ocacktan AZ, Abdallah MA (1994) Bacterial iron transport: coordination properties of pyoverdin PaA, a peptidic siderophore of *Pseudomonas aeruginosa*. Inorg Chem 33:6391–6402
16. Braud A, Hannauer M, Mislin GLA, Schalk IJ (2009) The *Pseudomonas aeruginosa* pyochelin-iron uptake pathway and its metal specificity. J Bacteriol 191:5317–5325

Chapter 29

Exopolysaccharide Quantification

Irina Sadovskaya

Abstract

The extracellular (EC) matrix is a key feature of mature *P. aeruginosa* biofilms. Exopolysaccharides are considered as major components of this biofilm matrix. They include alginate, LPS, glucans, and *psl*- and *pel*-dependent products. Here, we describe a method of quantification of the *psl*-dependent mannose-rich exopolysaccharide, based on the quantification of mannose in carbohydrate-enriched cell-associated extracts and growth media. Mannose is quantified by GC or GC-MS with an internal standard, after acid hydrolysis and conversion into volatile alditol acetates.

Key words Extracellular matrix, Exopolysaccharide, Psl, Monosaccharide analysis, Alditol acetates, GC

1 Introduction

The presence of an extracellular (EC) matrix is a key feature of mature *P. aeruginosa* biofilms. This matrix, composed of extracellular polymeric substances (often referred to as EPS), holds the biofilm cells together, protects them from shear forces in fluid environment [1], and confers increased resistance to antimicrobial agents and host defenses. The matrix of *P. aeruginosa* biofilms is a complex mixture of several types of macromolecules: extracellular DNA (eDNA), proteins, and polysaccharides. Exopolysaccharides (also referred to as EPS, which may lead to confusion) are considered as major biofilm matrix components [2, 3].

Alginate is an essential polysaccharide constituent of biofilms of mucoid *P. aeruginosa* strains [4]. Other EC carbohydrate polymers identified in the matrix and the growth media include LPS, cyclic phosphorylated β-(1, 3), and linear β-(1,2) and β-(1,6) glucans [5–8].

Pel and *psl* gene clusters are reported to encode proteins involved in polysaccharide biosynthesis [2]. Friedman and Kolter [9] suggested that Pel polysaccharide was a glucose-rich polymer, but not cellulose. To date, the chemical nature of Pel remains unknown.

Alain Filloux and Juan-Luis Ramos (eds.), *Pseudomonas Methods and Protocols*, Methods in Molecular Biology, vol. 1149, DOI 10.1007/978-1-4939-0473-0_29, © Springer Science+Business Media New York 2014

[→3)-β-D-Man*p*-(1→3) -β-D-Man*p*-(1→3) -α-L-Rha*p*-(1→3)-β-D-Glc*p*-(1→]$_n$
2
↑
1
α-D-Man*p*

Fig. 1 Structure of the pentasaccharide repeating unit of the Psl exopolysaccharide

The chemical structure of the *psl*-dependent polysaccharide (Psl), released in the growth media, was recently established. It was shown to be composed of pentasaccharide repeating units, containing three D-mannose, one D-glucose, and one L-rhamnose residue ([10], Fig. 1). However, there may be structural modifications between this extracellular and cell surface-associated Psl.

Taking into account the diversity of extracellular carbohydrates of *P. aeruginosa*, it is difficult to suggest a general technique of exopolysaccharide quantification. The response of phenol-sulfuric assay, a general method for quantitation of carbohydrates, depends on the monosaccharide composition of the EPS. In addition, nucleic acids, often present in extracellular extracts, and residual simple sugars, used as a carbon source in the growth media, can falsify the results.

Congo red (CR) binding assay has been used for quantification of EPS in *P. aeruginosa* [11, 12]. This assay was developed by Spiers et al. [13] as a quantitative assay for "cellulistic products" in the extracellular matrix of *P. fluorescens*. Indeed, the CR dye is known to bind to "amyloid fibers" and cellulose. However, CR binds primarily to proteins of different classes of secondary structure (α, α + β, β, and parallel β-helical proteins). The dye molecule is sandwiched between two protein molecules, causing protein oligomerization [14]. In this context, CR cannot be considered as a specific reagent for the quantification of exopolysaccharides.

Thus, it seems more reliable to quantify separately different polysaccharides, making use of the characteristic features of their chemical structure.

Alginate is an HMW linear copolymer of β-(1,4)-linked manuronic and guluronic acid [2]. The carbazole colorimetric assay for uronic acids [15, 16], which was also adapted as a 96-well assay for rapid processing of a number of samples [17], is commonly used for its quantification [18, 19]. It should be taken into account, however, that the method suffers from interference from neutral sugars [20, 21]. Alternatively, a more specific colorimetric test for uronic acids with meta-hydroxy-diphenyl can be used [20].

Psl in surface-bound polysaccharide extracts and culture supernatants is usually quantified by immunoblotting and ELISA, using the Psl-specific antiserum [10, 19]. However, this method does not allow detection of the free *psl*-dependent polysaccharide of MW 3–6 kDa, released in the growth medium.

Here, we describe a method of quantification of the Psl polysaccharide based on the quantification of mannose, which is a monosaccharide characteristic for Psl structure (Fig. 2, [10]). This general method, based on quantification of an EPS of a known structure using a monosaccharide characteristic for its structure and absent in other contaminating carbohydrates, can be applied to a variety of complex mixtures. Another example of this approach is a widely used quantification of LPS using a Kdo-specific colorimetric assay [6, 22].

To quantify Psl in the growth medium, the latter is first partially purified from abundant eDNA and proteins by precipitation with TCA, followed by dialysis. Mannose is quantified by gas chromatography (GC) in the carbohydrate-enriched extract after acid hydrolysis and conversion into volatile alditol acetates by conventional methods [23] in the presence of myo-inositol as an internal standard.

In this traditional method of quantification in GC, a peak area of a known quantity of a standard (Ino) is used to determine the quantity of a target compound (Man). A known amount of Ino is added to the extracts after the dialysis step, and thus sample losses during freeze-drying, transfer, and alditol acetate preparation do not have an effect on the quantification. The ratio of the peak areas (or highs) Man/Ino is proportional to quantities of Psl. In an example shown in Fig. 3, identical amount of Ino is added to 0.5 mg (A) and 1 mg (B) of a Psl preparation [10]. Relative integration ratios Ino/Man (1:1.3 and 1:2.6, respectively) are in agreement with the quantities of Psl in preparations A and B (1:2). This allows to assess the amount of Psl in crude extracts (Fig. 3).

2 Materials

2.1 Growth Media (See Note 1)

M63 minimal medium: 2 g/L $(NH_4)_2SO_4$, 13.6 g/L KH_2PO_4, 5 g/L casamino acids (Difco). pH is adjusted to 7.0. After autoclaving, the medium is supplemented with 0.5 mg/L $FeSO_4$, 1 mM (120 mg/L) $MgSO_4$, and a carbon source (glucose or arabinose).

2.2 Sample Preparation

1. Centrifuge and adapted centrifuge tubes. Teflon centrifuge tubes with screw cups.
2. Sonicator (IKA Labortechnik, Staufen, Germany).
3. Dialysis tubing (Visking) 22 mm diameter, dialysis reservoirs, closures.
4. Lyophilizer and adapted glassware.
5. 50 % (w/w) Trichloroacetic acid (TCA) solution: In a glass beaker, weight 100 g of TCA, and add 100 mL of water. Stir in a magnetic stirrer till the TCA is completely dissolved, and transfer into glass bottle (*see* **Note 2**).
6. Solution of myo-inositol (Ino), 1.8 mg/mL (10 mM).

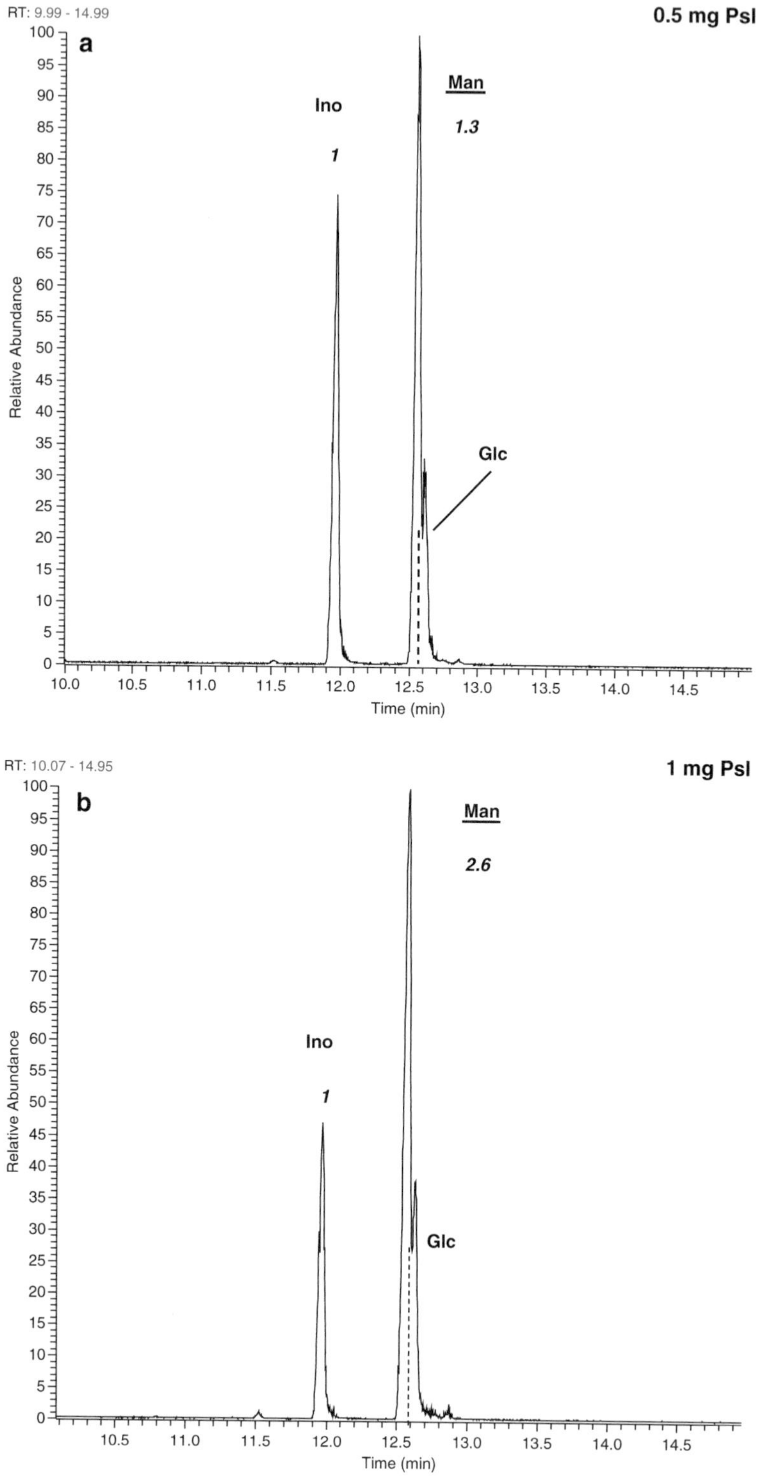

Fig. 2 Fragments of GC-MS profiles of alditol acetate preparations of 0.5 mg (**a**) and 1 mg (**b**) of a purified Psl preparation [10] with 72 μg of Ino. Numbers in italic correspond to relative integration ratios of mannose and inositol. GC-MS was performed on Trace GC Ultra apparatus, coupled with DSQ II Mass spectrometer (Thermo Scientific), and equipped with TR-5MS capillary column (30 m × 0.25 mm ID × 0.25 μm film), with helium carrier gas using a temperature gradient 170 °C (3 min) and 250 °C at 5 °C·min^{-1}

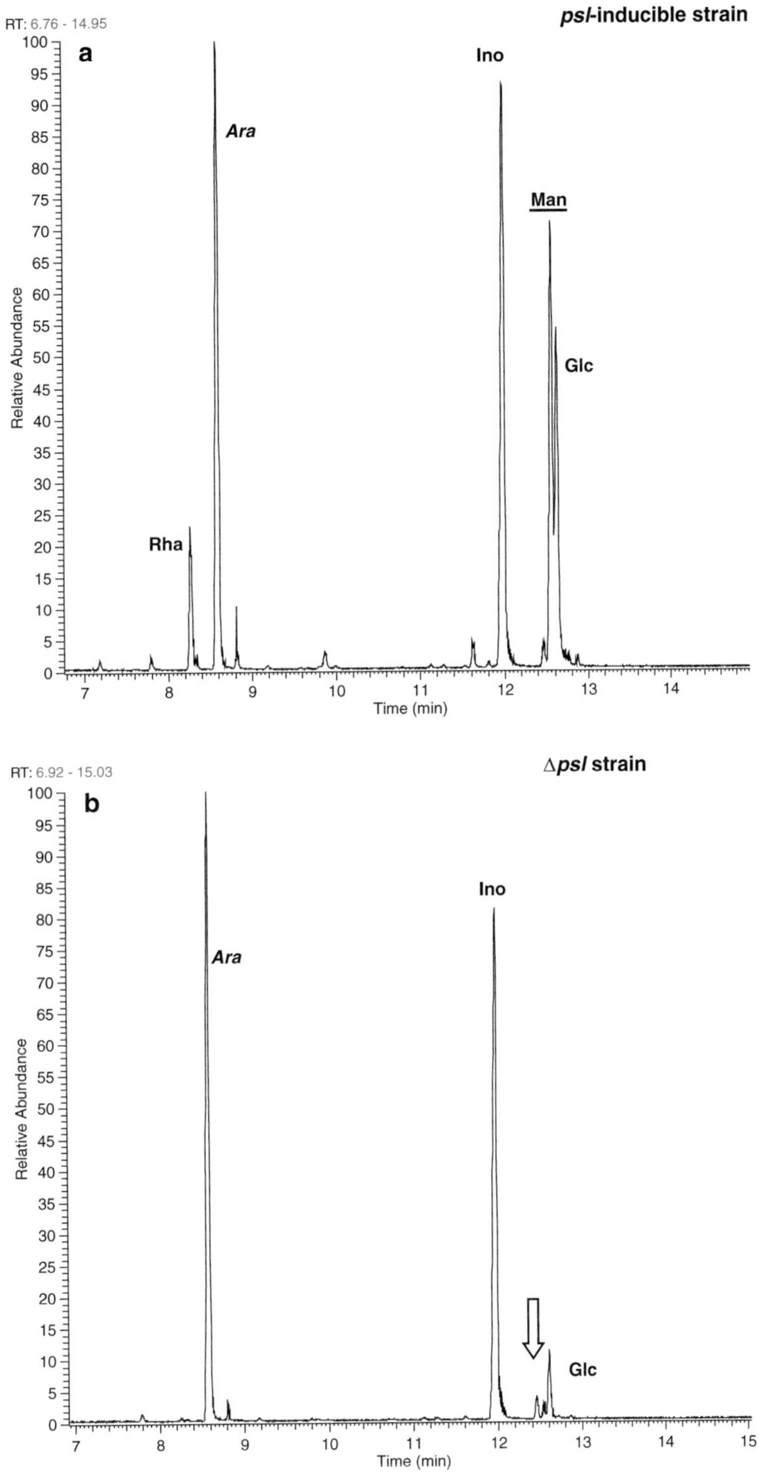

Fig. 3 GC-MS profiles of the alditol acetate preparations, corresponding to the growth media (20 mL) of the *P. aeruginosa psl*-inducible strain WFPA801 (**a**) and the *psl* promoter deletion strain WFPA800 (**b**). 72 μL of Ino internal standard was added to each preparation. The abundant peak of arabinose derives from the growth media. Note the quasi absence of mannose, characteristic for the Psl polysaccharide, in the WFPA800 preparation (indicated with an *arrow*). Preparations contain approximately 4 (**a**) and 0.1 (**b**) μg/mL Psl

2.3 Alditol Acetate Preparation and Analysis

1. Reacti-Vap Evaporator with the Reacti-Therm Heating Modules (*see* **Note 3**).
2. Screw-cap 4-mL glass vials, adapted for the Reacti-block aluminium block, with PTFE-lined caps.
3. 4 M trifluoroacetic acid (TFA): Put 50 mL of water in a glass cylinder. Very carefully, add 30.6 mL of pure TFA (HPLC grade). Adjust the volume to 100 mL, transfer into glass bottle, and mix carefully (*see* **Note 2**).
4. 1 M NH_4OH ammonium hydroxide: Under the fume hood, add 6 mL of concentrated ammonia to 94 mL of water, mix, and transfer into a bottle.
5. $NaBH_4$.
6. 10 % AcOH in MeOH: Under the fume hood, mix 10 mL of glacial acetic acid and 90 mL of methanol.
7. Methanol.
8. Pyridine (HPLC grade).
9. Acetic anhydride (HPLC grade).
10. Toluene (HPLC grade).
11. CH_2Cl_2 (HPLC grade).
12. GC with an FID detector or GC-MS, equipped with the capillary column suitable for separation of alditol acetates of monosaccharides.

3 Methods

3.1 Bacterial Growth

Grow the cells for 24 or 48 h in M63 medium, supplemented with 0.5 % casamino acids and a carbon source (0.2 % Glc or L-arabinose, if *psl*-inducible strain is used).

3.2 Sample Preparation

3.2.1 Crude Media and Cell-Associated Extracts

1. Transfer the cell culture (20 mL, *see* **Note 4**) into a centrifuge tube, and collect the cells with the cell-bound EC material by centrifugation (6,000 × *g*, 4 °C, 15 min). The clear supernatant (growth media) is treated separately.
2. Suspend the pellet in distilled water, and sonicate in ice, in order to detach the cell surface-associated polymers (50 % cycle, intensity 0.5; 3 × 30 s). Centrifuge (6,000 × *g*, 4 °C, 15 min). Add a fresh portion of water, and sonicate in the same way. Joined supernatants correspond to the crude cell-associated extract.

3.2.2 Crude Polysaccharide Extracts

Culture media and crude cell-associated extracts are treated in the same way.

1. Add 1/10 volume of the 50 % TCA in order to precipitate eDNA and proteins (*see* **Note 5**).

2. Centrifuge (10,000 ×*g*, 10 min). The precipitate can be either discarded or used to assess the quantities of polysaccharide associated with proteins and eDNA. Collect the supernatant.
3. Prepare dialysis tubing by cutting an appropriate length, soaking and closing one edge with a closure or by making a knot. Fill approx. two-thirds of the volume of the tubing with the solution, and close the tubing at the top.
4. Dialyze for at least 48 h against deionized water (*see* **Note 6**).
5. After dialysis, transfer the content of the tubing into a lyophilization container. Add a known amount of internal standard (Ino). It is important that after conversion into alditol acetates, peaks corresponding to Ino and Man are of the same range (ratio of the peak areas does not exceed 1:10). In an example given in Fig. 3, 40 μL of 10 mM Ino (72 μg) is added to the preparation corresponding to a 20 mL culture extract.
6. Freeze-dry to a small volume, transfer into a 20-mL bottle, and freeze-dry to dryness. The sample contains the exopolysaccharides and the internal standard. It is hydrolyzed to monosaccharides which are converted into alditol acetates and analyzed by GC equipped with an FID detector or by GC-MS.

3.3 Preparation of Alditol Acetates (See Note 7)

3.3.1 Hydrolysis

1. Dissolve the freeze-dried residue in 2–3 mL of 4 M TFA and transfer into a 4-mL screw-cup glass vial.
2. Close well with the screw cup, and heat for 3 h at 110 °C in the heating block. Cool and evaporate TFA to dryness with the stream of nitrogen at 60–70 °C.

3.3.2 Reduction

Take the residue into 0.5 mL of water, and add 3–4 drops of 1 M NH_4OH (to adjust pH of the solution to ~9) and 2–3 small crystals of $NaBH_4$ (~5 mg). Leave the reduction for 1 h or overnight.

3.3.3 Elimination of Boric Acid (See **Note 8**)

1. Add 2 mL of 10 % acetic acid in methanol to the reaction mixture, close well, and vortex. Wait until the release of hydrogen from the excess of $NaBH_4$ is complete, and then evaporate to dryness.
2. Dissolve the residue in 2 mL of 10 % AcOH in MeOH and evaporate. Repeat once more with 10 % AcOH in MeOH and twice with pure MeOH.

3.3.4 Acetylation

1. Add 0.4 mL of acetic anhydride and 0.4 mL of pyridine, close, and vortex well. Heat at 100 °C for 1 h.
2. Cool, add 1 mL of toluene which forms an azeotrope with acetic anhydride, and evaporate to dryness. If needed add another portion of toluene to complete the evaporation.
3. Take the residue into 1 mL of $CHCl_3$ or CH_2Cl_2, add 1–2 mL of water, and vortex. Allow the separation of two phases.

Take the lower organic phase and filter it through a filtering pipette (*see* **Note 9**). The solution can be further concentrated and analyzed by GC.

3.4 Analysis of Alditol Acetates

Monosaccharides are identified by GC or GC-MS. Results presented in Figs. 2 and 3 were obtained using GC-MS performed on Trace GC Ultra apparatus, coupled with DSQ II Mass spectrometer (Thermo Scientific), and equipped with TR-5MS capillary column (30 m × 0.25 mm ID × 0.25 μm film), with helium carrier gas using a temperature gradient 170 °C (3 min) and 250 °C at 5 °C·min^{-1}. Use manual integration to separate peaks, if resolution is not sufficient (Fig. 2).

3.5 Psl Quantification

In the example shown in Fig. 3, the GC-MS profile corresponds to 20 mL of growth media of preparations of the *P. aeruginosa psl-inducible* strain WFPA801 (A) and the *psl* promoter deletion strain WFPA800 (B). 72 μL of Ino internal standard was added to each preparation. Integration of peaks of Ino and Man shows the ratio of 1:0.7 (A) and 1:0.02 (B), corresponding to approx. 50 and 1.5 μg Man, respectively. Taking into account the MW of the Psl repeating unit ($Man_3RhaGlc$), Man corresponds to ~61 % (w/w) of Psl (486/794). Consequently, in these experimental conditions strain WFPA801 produces ~82 μg Psl per 20 mL (4 μg/mL) and strain WFPA800—about 0.1 μg/mL of Psl.

4 Notes

1. This quantification is only possible when cells are grown in a chemically defined medium, not containing polysaccharides, such as M63 supplemented with casamino acids. Media containing yeast extracts, rich in yeast mannans (such as LB, LBNS, T, or TNS media) should be excluded. If necessary, a control experiment can be performed with the equivalent amount of growth media.
   This method is based on the quantification of monosaccharides and can be adapted for other polysaccharides of bacterial origin, containing a characteristic monosaccharide residue.
2. TCA and TFA are very aggressive agents causing burns. Always wear gloves and work in the fume hood. Take special care to always close well the bottle containing TFA solutions, since TFA is volatile.
3. Reacti-Vap Evaporator (Thermo Scientific, previously PIERCE, ex. Ref. TS-18825) is a gassing manifold designed as accessories to Reacti-Therm Heating Modules for attachment to a supply of inert gas (nitrogen) to enable needle-directed dry-down of liquid reaction samples. The manifolds split the

incoming flow of gas into 9 or 27 needle ports that point downward into open sample vials placed in compatible Reacti-Block Aluminum Blocks. All 9 or 27 ports can be used at once or selective ports can be capped off with the supplied plugs. The module is placed in the fume hood and attached to a nitrogen tank.

4. Cell-associated extract often contain less Psl than the growth media. For the quantification of Psl in cell-associated extract the volume of culture can be adapted, according to the relative yield of Psl and sensitivity of the GC.
5. Cultures of several *P. aeruginosa* strains, including PAO1, become blue-green due to secreted pigments. The color changes to deep reddish brown upon the addition of acid.
6. Water should be changed several times during the first hours of dialysis and then twice a day. The efficiency of dialysis can be checked with the simple pH paper placed on the bottom of the dialysis reservoir, detecting the accumulation of acid.

   The volume of dialysate increases significantly during dialysis. Neutral pH of the solution indicates that dialysis is complete; however, Glc or Ara from the growth media could still be present in variable amounts.
7. All the reaction steps of the preparation of alditol acetates, except the reduction, are performed in the heating block; Reacti-Vap Evaporator is used for evaporation steps.
8. A simplified method of elimination of boric acid (E. Vinogradov, personal communication): Add 1 mL of 2 M HCl, dry; add 2 mL of methanol, dry. Repeat the evaporation with methanol once more.
9. This is a way of quick drying of the organic phase. Filtering pipette is made by inserting a small piece of cotton into the top of a disposable Pasteur pipette and pushing it down close to the tip. Once filtered, the cloudy organic phase becomes transparent, since the cotton absorbs the residual water still present in the organic phase.

Acknowledgements

I am thankful to Prof. D. J. Wozniak (Center for Microbial Interface biology, the Ohio State University, Columbus, OH, USA) for providing *P. aeruginosa* strains WFPA800 and WFPA801. I like to acknowledge my former supervisors and older colleagues who taught me the theory and practice of the carbohydrate chemistry: Prof. A.I. Usov (Zelinskii Institute of Organic Chemistry, Moscow) and Drs. Eleonora Altman, Malcolm Perry, and E. Vinogradov (IBS, National Research

Council, Ottawa, Canada). I am thankful to Dr. E. Vinogradov for fruitful discussions, Dr. Ph Lencel for help with GC-MS and a critical reading of the manuscript, and Dr. T. Grard (BPA, ULCO) for his support. This work was supported by Nord-Pas de Calais Region (ARCIR 54R01).

References

1. Ma L, Conover M, Lu H, Parsek MR, Bayles K, Wozniak DJ (2009) Assembly and development of the *Pseudomonas aeruginosa* biofilm matrix. PLoS Pathog 5:e1000354
2. Ryder C, Byrd M, Wozniak DJ (2007) Role of polysaccharides in *Pseudomonas aeruginosa* biofilm development. Curr Opin Microbiol 10: 644–648
3. Ma L, Wang S, Wang D, Parsek MR, Wozniak DJ (2012) The roles of biofilm matrix polysaccharide Psl in mucoid *Pseudomonas aeruginosa* biofilms. FEMS Immunol Med Microbiol 65: 377–80
4. Wozniak DJ, Wyckoff TJ, Starkey M, Keyser R, Azadi P, O'Toole GA, Parsek MR (2003) Alginate is not a significant component of the extracellular polysaccharide matrix of PA14 and PAO1 *Pseudomonas aeruginosa* biofilms. Proc Natl Acad Sci U S A 100:7907–7912
5. Mah TF, Pitts B, Pellock B, Walker GC, Stewart PS, O'Toole GA (2003) A genetic basis for *Pseudomonas aeruginosa* biofilm antibiotic resistance. Nature 426:306–310
6. Coulon C, Vinogradov E, Filloux A, Sadovskaya I (2010) Chemical analysis of cellular and extracellular carbohydrates of a biofilm-forming strain *Pseudomonas aeruginosa* PA14. PLoS One 5:e14220
7. Lequette Y, Rollet E, Delangle A, Greenberg EP, Bohin JP (2007) Linear osmoregulated periplasmic glucans are encoded by the *opgGH* locus of *Pseudomonas aeruginosa*. Microbiology 153:3255–3263
8. Sadovskaya I, Vinogradov E, Li J, Hachani A, Kowalska K, Filloux A (2010) High-level antibiotic resistance in *Pseudomonas aeruginosa* biofilm: the *ndvB* gene is involved in the production of highly glycerol-phosphorylated beta-(1->3)-glucans, which bind aminoglycosides. Glycobiology 20:895–904
9. Friedman L, Kolter R (2004) Genes involved in matrix formation in *Pseudomonas aeruginosa* PA14 biofilms. Mol Microbiol 51:675–690
10. Byrd MS, Sadovskaya I, Vinogradov E, Lu H, Sprinkle AB, Richardson SH, Ma L, Ralston B, Parsek MR, Anderson EM, Lam JS, Wozniak DJ (2009) Genetic and biochemical analyses of the *Pseudomonas aeruginosa* Psl exopolysaccharide reveal overlapping roles for polysaccharide synthesis enzymes in Psl and LPS production. Mol Microbiol 73:622–638
11. Lee YW, Jeong SY, In YH, Kim KY, So JS, Chang WS (2010) Lack of O-polysaccharide enhances biofilm formation by *Bradyrhizobium japonicum*. Lett Appl Microbiol 50:452–456
12. Ueda A, Wood TK (2009) Connecting quorum sensing, c-di-GMP, pel polysaccharide, and biofilm formation in *Pseudomonas aeruginosa* through tyrosine phosphatase TpbA (PA3885). PLoS Pathog 5:e1000483
13. Spiers AJ, Bohannon J, Gehrig SM, Rainey PB (2003) Biofilm formation at the air–liquid interface by the *Pseudomonas fluorescens* SBW25 wrinkly spreader requires an acetylated form of cellulose. Mol Microbiol 50:15–27
14. Khurana R, Uversky VN, Nielsen L, Fink AL (2001) Is Congo red an amyloid-specific dye? J Biol Chem 276:22715–22721
15. Bitter T, Muir HM (1962) A modified uronic acid carbazole reaction. Anal Biochem 4: 330–334
16. Knutson CA, Jeanes A (1968) A new modification of the carbazole analysis: application to heteropolysaccharides. Anal Biochem 24: 470–481
17. Cesaretti M, Luppi E, Maccari F, Volpi N (2003) A 96-well assay for urinic acids carbazole reaction. Carbohydr Polymers 54:59–61
18. Ma S, Selvaraj U, Ohman DE, Quarless R, Hassett DJ, Wozniak DJ (1998) Phosphorylation-independent activity of the response regulators AlgB and AlgR in promoting alginate biosynthesis in mucoid *Pseudomonas aeruginosa*. J Bacteriol 180: 956–968
19. Ma L, Wang J, Wang S, Anderson EM, Lam JS, Parsek MR, Wozniak DJ (2012) Synthesis of multiple *Pseudomonas aeruginosa* biofilm matrix exopolysaccharides is post-transcriptionally regulated. Environ Microbiol 14:1995–2005
20. Blumenkrantz N, Asboe-Hansen G (1973) New method for quantitative determination of uronic acids. Anal Biochem 54:484–489

21. Taylor KA, Buchanan-Smith JG (1992) A colorimetric method for the quantitation of uronic acids and a specific assay for galacturonic acid. Anal Biochem 201:190–196
22. Cadieux JE, Kuzio J, Milazzo FH, Kropinski AM (1983) Spontaneous release of lipopolysaccharide by *Pseudomonas aeruginosa*. J Bacteriol 155:817–825
23. Lindberg B, Lonngren J (1978) Methylation analysis of complex carbohydrates: general procedure and application for sequence analysis. Methods Enzymol 50:3–33

Chapter 30

Liquid Chromatography/Mass Spectrometry for the Identification and Quantification of Rhamnolipids

Ahmad Mohammad Abdel-Mawgoud, François Lépine, and Eric Déziel

Abstract

Rhamnolipids (RL) are surface-active glycolipids produced by *Pseudomonas aeruginosa*. They are always produced by this bacterium as a complex mixture of congeners, each composed of one or two rhamnose molecules linked to a dimer of 3-hydroxyfatty acids with a chain length of 8–12 carbons. Increasing interest for RL drives the need for efficient analytical methods to characterize these mixtures of molecules.

High-performance liquid chromatography (HPLC) coupled with tandem mass spectrometry (MS/MS) is a very precise and relatively high-throughput method for the identification of each congener and their quantification in bacterial cultures. Using ^{13}C-labeled RL as internal standards can further enhance the precision of the quantification. Collision-induced dissociation (CID) experiments by MS/MS is a powerful tool for the detection and identification of structural variations in RL produced by various *Pseudomonas* strains or by a specific strain under different culture conditions. CID even allows the discrimination between isomers with subtle structural variations, like Rha-C_8-C_{10} and Rha-C_{10}-C_8, which are almost inseparable chromatographically. We are presenting here the detailed protocols for HPLC/MS and HPLC/MS/MS analysis of RL and their lipid precursors, the 3-(3-hydroxyalkanoyloxy)alkanoic acids (HAA), directly in bacterial culture supernatants.

Key words HPLC, Reversed phase, Tandem mass spectrometry, Electrospray negative ionization, Rhamnolipids, Congeners, HAA, ^{13}C-labeled RL, Identification, Quantification

1 Introduction

While the structures and the biosynthetic pathway of rhamnolipids (RLs) produced by *Pseudomonas aeruginosa* were initially described in considerable details [1–3], their exact biological role remained poorly understood for a long time. Over the years, many functions were attributed to these unique glycolipids with surface-active properties [4]. They were at first described as a heat-stable hemolysin [5, 6]. Then, their role in the assimilation of hydrocarbons was investigated extensively [7–10]. Their involvement in infections as virulence factors and as immunomodulators was described [11–14] and further defined over recent years [15–17]. In the mid-1990s, the genes required for the biosynthesis of RLs were

Alain Filloux and Juan-Luis Ramos (eds.), *Pseudomonas Methods and Protocols*, Methods in Molecular Biology, vol. 1149, DOI 10.1007/978-1-4939-0473-0_30, © Springer Science+Business Media New York 2014

identified [18], including those responsible for their regulation, the RhlR/RhlI quorum sensing system [19, 20]. In the following years, the role of RLs and of their biosynthetic precursors 3-(3-hydroxyalkanoyloxy)alkanoic acids (HAAs) in bacterial surface motility, especially swarming [21–24], and also twitching [25, 26], emerged as an important function that is increasingly linked to biofilm development [27–30].

Nowadays, the primary interests for RLs among scientists investigating *P. aeruginosa* stem from their involvement in surface motility/biofilm development and their role as virulence factors. Concurrently, these biosurfactants are still attracting a lot of attention as alternatives to synthetic surfactants and a great deal of literature is devoted to the development of production processes for their industrial applications [31, 32]. Since we know that the level of rhamnosylation and the length of the side chains impact the function and activity of these tensioactive compounds [22, 24], it is important to know which molecules are produced out of the more than 50 congeners known [33] and in which quantity.

A number of methods have been developed and used over the years to characterize the presence of RLs [32, 34]. These include (a) the drop-collapsing test, which indirectly indicates the presence of a surfactant in a liquid as revealed by a reduction in surface tension [35]; (b) the methylene blue agar plate, which allows the semiquantitative on-plate detection of the production of anionic amphiphilic molecules by growing bacteria [36]; (c) the colorimetric quantification of rhamnose after hydrolysis of RL, using orcinol, as an indirect measurement [10, 37]; (d) chromatographic separation of the more abundant congeners by thin-layer [38] or liquid chromatography (LC) [39] using various detection methods; and, finally, (e) mass spectrometry (MS), as a very precise and efficient technique [33, 40]. We describe here how to use LC/MS to directly detect and quantify RLs/HAAs in bacterial cultures.

2 Materials

The chromatography eluents consist of high-performance liquid chromatography (HPLC)-grade solvents. The aqueous solutions are prepared using Milli-Q water with a resistivity of 18 MΩ cm at 25 °C. Otherwise, reagents are of analytical or ACS grade. Prepare and store all reagents at room temperature (unless otherwise indicated). Diligently follow all waste disposal regulations.

2.1 Bacterial Strain, Culture Conditions, and Sample Preparation

1. *Pseudomonas aeruginosa*, strain PA14 [41], is the RL-producing bacterium used for the presented example. Nonetheless, any RL-producing *Pseudomonas* strain would be applicable with this protocol. PA14 strain is preserved in glycerol stock (15 % v/v) and stored at −80 °C.

2. Tryptic soy broth (TSB) medium (Difco) or any other rich medium for overnight culture of frozen cells.
3. In principle, any typical culture medium is appropriate to perform the cultures for RL production. An optimized mineral salts medium (MSM) with glycerol as the carbon source is used for the presented protocol. MSM has the following composition: Na_2HPO_4 (0.9 g/L), KH_2PO_4 (0.7 g/L), $CaCl_2 \cdot 2H_2O$ (0.1 g/L), $MgSO_4 \cdot 7H_2O$ (0.4 g/L), $NaNO_3$ (2 g/L), tryptone (1 g/L, Fisher Scientific), trace element solution (TES, 2 mL/L), pH 7±0.1. Composition of TES is as follows: $FeSO_4 \cdot 7H_2O$ (2 g/L), $MnSO_4 \cdot H_2O$ (1.5 g/L), $(NH_4)_6Mo_7O_{24} \cdot 4H_2O$ (0.6 g/L). Store TES at 4 °C. Glycerol is added as the carbon source at 15 g/L (*see* **Note 1**).
4. Sterile borosilicate glass test tubes (18×150 mm).
5. Incubator adjusted at 34 °C and equipped with a TC-7 roller drum (New Brunswick Scientific).
6. Polypropylene 1.5 mL microtubes and pipette tips.
7. Mini centrifuge to achieve 10,000×*g*.
8. Freezer at −20 °C.

2.2 HPLC/MS Solutions and Standards

1. Concentrated ammonium hydroxide, ACS grade (28–30 %).
2. Ammonium acetate solution: 100 mM ammonium acetate solution in Milli-Q water, pH 8.55. Weigh 1.9271 g of ammonium acetate in 200 mL Milli-Q water, adjust the pH to 8.55 by the addition of concentrated ammonium hydroxide (28–30 %) (*see* **Note 2**), and complete the volume to 250 mL using a volumetric flask. Store at 4 °C.
3. Mobile phase, solvent A: 2 mM ammonium acetate in Milli-Q water, pH 8.0–8.5. In a 500 mL volumetric flask, add 10 mL of 100 mM ammonium acetate solution, then add 400 mL of Milli-Q water, adjust the pH with concentrated ammonium hydroxide (28–30 %) to 8.0–8.5, and then complete the volume to 500 mL with Milli-Q water (*see* **Note 3**).
4. Mobile phase, solvent B: 2 mM ammonium acetate in acetonitrile, pH 8.0–8.5. In a 500 mL volumetric flask, add 10 mL of 100 mM ammonium acetate solution, then add 400 mL acetonitrile, adjust the pH with ammonium hydroxide (28–30 %) to 8.0–8.5, and then complete the volume to 500 mL with acetonitrile (*see* **Note 4**).
5. Internal standard (IS) stock solutions: ^{13}C-labeled mono-RL (Rha-C_{10}-C_{10}; $^{13}C_{26}H_{48}O_9$, with a mass shift of 26 Da) and di-RL (Rha-Rha-C_{10}-C_{10}, $^{13}C_{32}H_{58}O_{13}$ with a mass shift of 32 Da) (*see* **Note 5**). These ^{13}C-labeled RLs are produced by PA14 fed with dextrose-$^{13}C_6$ as sole carbon source, extracted from the cultures with ethyl acetate, and purified using 1 mm thin-layer chromatography (TLC) plates (Partisil PK6F,

Whatman) developed with methanol:acetic acid:ethyl acetate (10:1:89). TLC-purified HAA-$^{13}C_{20}$ with a mass shift of 20 Da is prepared from an ethyl acetate extract of HAAs produced by a *rhlB* mutant strain of PA14 fed with dextrose-$^{13}C_6$ as sole carbon source. All internal standard solutions are prepared in ACS-grade methanol at 5,000 mg/L for labeled RL and 2,500 mg/L for labeled HAAs. Store all internal standard solutions at −20 °C.

6. HPLC vials: Screw capped, borosilicate glass vials 11.6 × 32 mm.

2.3 HPLC/MS Instrumentation

1. HPLC instrument (Alliance HT, 2795 Separation Module, Waters) equipped with a C8 reversed phase column (4.6 × 150 mm Eclipse XDB-C8, Agilent, USA, particle size 5 μm) with a C8 guard column (*see* **Note 6**).
2. The HPLC is coupled to the MS (Quattro Premier XE, triple quadrupole, Micromass) through a Tee splitter with a fixed 10:1 split ratio (Agilent). This is achieved by fitting the third outlet of the splitter with a tube of internal diameter and length such that 10 % of the initial flow goes to the electrospray ionization interface of the MS operated in negative ionization mode.
3. Nitrogen is used as the cone and desolvation gas. Argon is used as collision gas in collision-induced dissociation (CID) experiments or in cases quantification is performed in multiple reaction monitoring (MRM) mode.
4. The MassLynx™ software with its QuanLynx™ function (version 4.1, Waters) is used for automatic peak integration.

3 Methods

RLs are rhamnosylated β-hydroxy fatty acids initially isolated from *P. aeruginosa* cultures. Currently, nearly 60 RL homologues have been identified in a few bacteria with β-hydroxy fatty acid of various lengths and unsaturation level [4]. The di-RL (Rha-Rha-C_{10}-C_{10}) and mono-RL (Rha-C_{10}-C_{10}) are the most abundant congeners produced by *P. aeruginosa* (Fig. 1). This bacterium also releases the di-lipidic precursors of RL (HAAs) in the supernatant, also produced as a series of homologues with carbon chains varying from C_8 to C_{12} [42]. The most abundant of these HAA congener is 3-(3-hydroxydecanoyloxy)-decanoic acid (C_{10}-C_{10}) (Fig. 1).

HPLC coupled with MS is a very powerful tool for the separation, identification, and quantification of RL congeners directly in bacterial culture supernatants. Moreover, tandem MS/MS even allows the discrimination of HPLC-inseparable structural isomers such as Rha-C_{10}-C_8 from Rha-C_8-C_{10}, according to their fragmentation patterns.

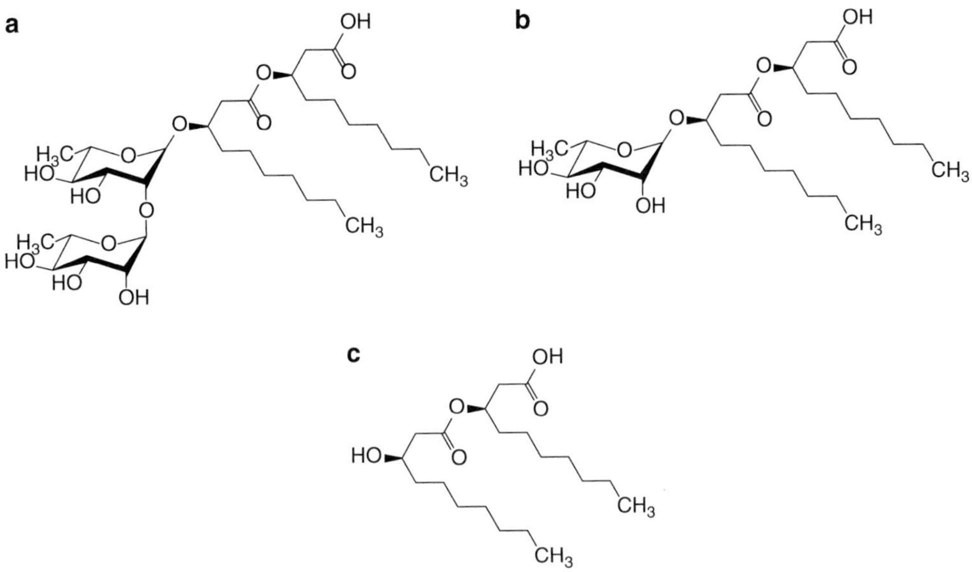

Fig. 1 Chemical structure of the two rhamnolipids produced by *P. aeruginosa*: (**a**) di-RL (Rha-Rha-C_{10}-C_{10}), (**b**) mono-RL (Rha-C_{10}-C_{10}), and their di-lipid precursor (**c**) 3-(3-hydroxydecanoyloxy)decanoic acid (C_{10}-C_{10}) (HAA)

HPLC separation is performed using a mobile-phase gradient of water to acetonitrile, each buffered with 2 mM ammonium acetate using a reversed-phase column. Mass spectrometric analyses of RL and HAA are performed under negative electrospray ionization (ESI) mode, which generates $[M–H]^-$ pseudomolecular ions corresponding to the mass of the neutral molecule minus 1 Da as a result of the loss of one proton. Mass spectrometers analyze ions according to their mass divided by their charge (m/z), but since only one charge is applied under the conditions used, the m/z ratio is equal to the mass of the pseudomolecular ion. The usefulness of ESI is that it is considered as a soft method of ionization with little fragmentation of the primary pseudomolecular ion $[M–H]^-$.

The HPLC/MS method described here is for RL produced by *P. aeruginosa*, with particular emphasis on the two most abundant RL congeners Rha-Rha-C_{10}-C_{10} (m/z 649) and Rha-C_{10}-C_{10} (m/z 503). This protocol is also valid for other RL homologues produced by *Pseudomonas* species with β-hydroxy fatty acid chain lengths ranging from C_8 to C_{12}, taking into consideration that 14 Da must be added to their masses for every additional methylene unit ($–CH_2–$) and subtracted by 2 Da for each additional unsaturation. In addition to their pseudomolecular ions, RL can be identified using tandem MS/MS in which the pseudomolecular ion is fragmented by colliding with argon, generating fragments of specific masses (Fig. 2).

Depending on their structure, RL congeners will produce fragment ions with different m/z values according to their number of rhamnose (one or two) or β-hydroxy fatty acid moiety (one or two), the length of their carbon chains, or the presence of

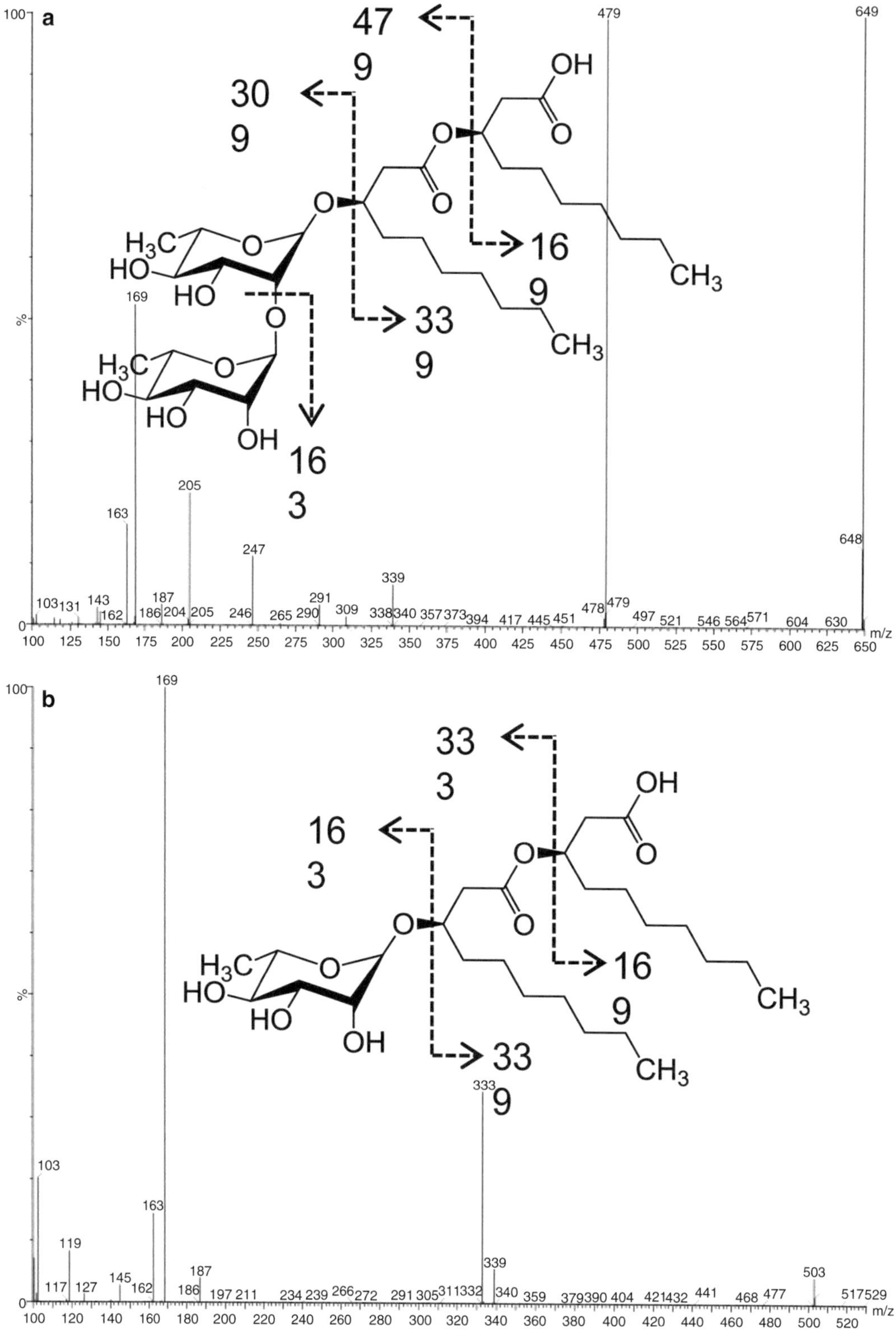

Fig. 2 Common fragment ions produced upon fragmentation of (**a**) di-RL (Rha-Rha-C_{10}-C_{10}) and (**b**) mono-RL (Rha-C_{10}-C_{10}) using tandem MS/MS

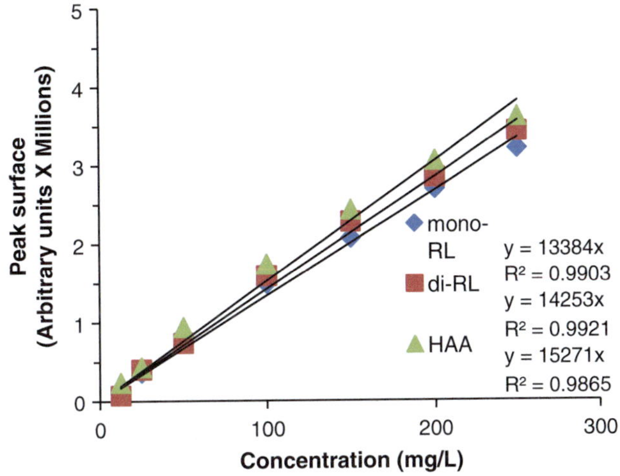

Fig. 3 Range of concentrations showing linearity of MS detection for di-RL, mono-RL, and HAAs

unsaturations or substitutions in the β-hydroxy fatty acid moiety. The masses of these fragment ions can be used to tentatively identify structural changes in new RL congeners.

In addition to the structural information it provides, HPLC/MS also allows precise quantification of these congeners by comparing the peak area of the pseudomolecular ion relative to that of an internal standard added at a known concentration during sample preparation. In this approach, RL can be quantified in the culture supernatants after undergoing chromatographic separation, although quantification can even be performed by direct infusion without prior chromatographic separation [40]. However, because in direct infusion mode other components of the matrix will also compete for the charges to be applied on RL molecules, the sensitivity of MS detection and hence precision of quantification are considerably improved by chromatographic separation.

Under the conditions described in the methods presented here, the linear detection range of mono- and di-RL lies between 10 and 250 mg/L in the full scan mode and that of HAAs lies between 10 and 150 mg/L (Fig. 3). Accordingly, to achieve precise quantification, the supernatants should be diluted when RL concentrations exceed this range.

Detection limits below 10 mg/L can be achieved using an alternative scanning mode called "MRM". In MRM, tandem MS/MS is used to fragment selected pseudomolecular ions of RL congeners and monitor the intensity of a selected fragment ion, from which the concentration of the parent ion can be deduced. MRM provides a much better signal-to-noise ratio than the full scan mode. However, MRM requires the prior determination of the ions to fragment and the fragment to monitor. Thus, novel RLs with unexpected structures will not be detected resulting in a potential loss of information that would have been provided in full scanning mode. Therefore, MRM is advised only when maximal sensitivity is needed for detection and quantification of a limited number of predetermined RL congeners. Otherwise, full scanning mode will provide sufficient sensitivity for detection and quantification of a wider range of molecules that can be detected in negative ionization mode.

An important advantage of the method described here is that it analyzes RLs and HAAs directly in culture supernatants without any prior extraction steps (typically with organic solvents), thus avoiding the inevitable losses encountered during purification. Nevertheless, if an even greater sensitivity is required or the culture medium is not amenable to direct injection (e.g., high concentrations of interfering exoproducts), RLs can be extracted from the culture medium with ethyl acetate, the solvent evaporated to dryness, and the residue dissolved in a minimum amount of a 30:70 water:acetonitrile solution.

For quantification purposes, it is imperative that a fixed concentration of an internal standard be added in the sample to analyze in order to compensate for the frequently encountered day-to-day or run-to-run variations in the sensitivity of the MS detector and hence the peak area-based calculations of RL concentrations. The internal standards used in the presented method are fully ^{13}C-labeled Rha-Rha-C_{10}-C_{10} and Rha-C_{10}-C_{10}, which are directly added in the culture sample prior to centrifugation to act as reference for all di- and mono-RL congeners, respectively. Similarly, the fully ^{13}C-labeled HAA, $^{13}C_{10}$-$^{13}C_{10}$, is used as internal standard for quantification of all HAA homologues. These internal standards are ideal for precise RL quantification as they possess essentially identical physicochemical, chromatographic, and ionization properties to natural RL while remaining discriminable by the mass analyzer. With an elution program of 20 min, this HPLC/MS method allows the quantification of RL and HAAs at a rate of three samples per hour. Higher throughput can be achieved using isocratic gradients. The preparation of the samples for RL and HAA analysis and details of HPLC/MS method setup are described below.

3.1 Rhamnolipid Production and Sample Preparation for HPLC/MS Analysis

1. Transfer 1 mL of a PA14 preculture (*see* **Note 7**) grown overnight in TSB (200 rpm, OD_{600} of 5.0–6.0) into a 1.5 mL microtube and centrifuge at 10,000 ×g for 2 min. Discard the supernatant, add 1 mL of sterile MSM–glycerol, vortex, and centrifuge at 10,000 ×g for 2 min. Repeat the washing with MSM–glycerol twice (*see* **Note 8**). Resuspend the cell pellet in 1 mL of MSM–glycerol, and measure the OD_{600}. Transfer a calculated volume (~100 μL) of this washed cell suspension into a borosilicate test tube containing 5 mL of MSM–glycerol to achieve an initial OD_{600} of 0.1.
2. Incubate the inoculated MSM–glycerol test tube under rotation (200 rpm) at 34 °C for 3–4 days (*see* **Note 9**).

3. Transfer about 500 μL of culture broth into a 1.5 mL microtube, and record the OD_{600} reading. Centrifuge at 10,000 × *g* for 10–15 min to remove cells and debris. Carefully transfer 250 μL of the supernatant into another 1.5 mL microtube. This supernatant can be stored at −20 °C until the day of analysis as RLs remain stable for unlimited time at this temperature.
4. Prepare the sample diluent solution containing 52 mg/L of each of the two IS, di-RL-$^{13}C_{32}$ and mono-RL-$^{13}C_{26}$, in 30:70 v/v water:acetonitrile. Add 480 μL of this IS-containing solution into an HPLC borosilicate vial, add 20 μL of sample supernatant (*see* **Note 10**), and vortex for 10 s. It is recommended to prepare a stock solution of IS (52 mg/L) to be distributed to each vial as this takes less time and avoids sample-to-sample variations in IS concentration. The final concentration of each IS. in the vial will be 50 mg/L, and the supernatant will have been diluted 25-fold.

3.2 Quantitative Analysis of RL Using HPLC/MS

1. Ensure the stability of the HPLC conditions (tubing, injector, and separation column) as well as the MS sensitivity (*see* **Note 11**).
2. Place the HPLC borosilicate vials in the HPLC tray chamber, and set the injection volume at 20 μL.
3. Set the mobile-phase flow rate to 400 μL/min with a 20-min gradient of water (solvent A) and acetonitrile (solvent B). Set the gradient as follows: 0 min, 30 % A + 70 % B; 16 min, 10 % A + 90 % B; 17 min, 30 % A + 70 % B; and 20 min, 30 % A + 70 % B.
4. Verify that the split ratio at the Tee splitter is fixed at 10:1, so that the flow at the mass spectrometer is at 40 μL/min.
5. Configure the mass spectrometer tuning parameters according to the following: Negative ESI mode; capillary voltage 3.0 kV; cone voltage 20 V; source and desolvation temperatures 120 and 150 °C, respectively; and cone and desolvation gas (N_2) flow 15 and 100 L/h, respectively.
6. Set the MS method to scan over a *m/z* range of 130–800 every 1.7 s (for an instrument calibration scan speed of 400 Da/s). A characteristic total ion chromatogram (TIC) and selected ion chromatograms are generated using the MassLynx 4.1 software (Fig. 4).
7. For quantitation of a specific RL congener, display the single-ion chromatogram corresponding to the pseudomolecular ions $[M{-}H]^-$, such as the *m/z* ions 649 and 503, which correspond to the pseudomolecular ions of natural di-RL (Rha-Rha-C_{10}-C_{10}) and mono-RL (Rha-C_{10}-C_{10}), respectively. Similarly, display the ion chromatogram of the masses of ^{13}C-labeled RL acting as internal standards, having *m/z* of 681 and 529 for di-RL-^{13} C32 and mono-RL-$^{13}C_{26}$, respectively (Fig. 4) (*see* **Note 12**).

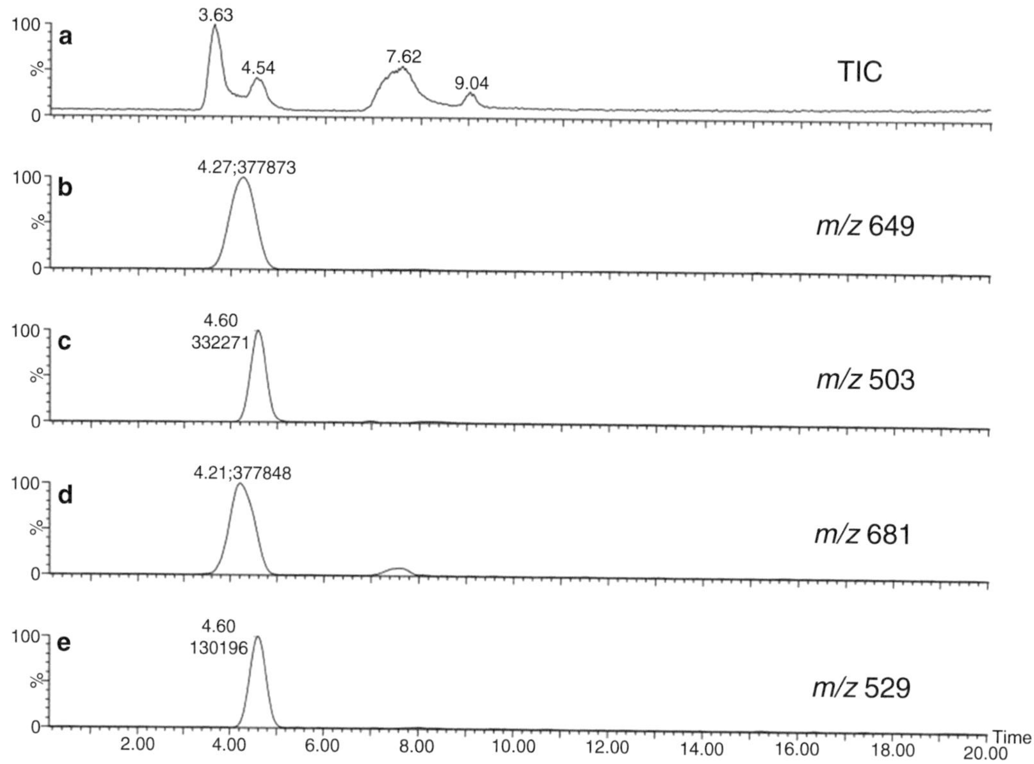

Fig. 4 Chromatograms of the supernatant of *P. aeruginosa* PA14 after 3 days of cultivation in MSM–glycerol at 34 °C. (**a**) Total ion chromatogram (TIC). (**b**) The *m/z* 649 ion chromatogram corresponding to $[M–H]^-$ of di-RL (Rha-Rha-C_{10}-C_{10}) with retention time (R_t) at 4.27 min. (**c**) The *m/z* 503 ion chromatogram corresponding to $[M–H]^-$ of mono-RL (Rha-C_{10}-C_{10}) with R_t at 4.6 min. (**d**) The *m/z* 681 ion chromatogram corresponding to $[M–H]^-$ of ^{13}C-labeled di-RL-$^{13}C_{32}$ which acts as internal standard for di-RL. (**e**) The *m/z* 528 ion chromatogram corresponding to $[M–H]^-$ of ^{13}C-labeled mono-RL-$^{13}C_{26}$ which acts as internal standard for mono-RL

8. Integrate the analyte and internal standard peaks using MassLynx 4.1 integration function, and use the peak areas to calculate the analyte concentration using the following formula:

$$C_T = \mathrm{DF} \times (C_S \times A_T / A_S)$$

where C_T is the concentration of the RL congener in mg/L, and DF is the dilution factor which corrects for the 25-fold dilution of the culture supernatant as described in **step 4** under Subheading 2.1. C_S is the concentration of corresponding internal standard (^{13}C-labeled RL), 50 mg/L in this case. A_T is the peak area of the RL congener to quantify. A_S is the peak area of the corresponding internal standard (*see* **Note 13**).

3.3 Identification of New RL Homologues Using HPLC–Tandem MS/MS

Tandem MS/MS is useful to identify a new RL congener observed in different strains of *P. aeruginosa* cultures or under different culture conditions [4]. Tandem MS/MS is also very useful for isotope tracking for the study of the biosynthesis of RL. The following protocol illustrates the identification of the homologues of the two major RL congeners produced by *P. aeruginosa*.

1. Prepare samples as described in Subheading 3.1, and then follow the HPLC/MS **steps 1–6** in Subheading 3.2.
2. Identify the pseudomolecular ions $[M–H]^-$ of the RL congener of interest.
3. Conduct a CID experiment to fragment the ion of interest. Use argon as the collision gas at a pressure of 3.28×10^{-3} mBar. Scan for fragment ions at a mass range of *m/z* 100–530 for mono-RL homologues and of *m/z* 100–650 for di-RL. Set collision energy at 20 eV for both molecules (*see* **Note 14**).
4. Display the spectrum of the fragment ions, identify the fragment ions characteristic to RL, and tentatively identify the structural changes according to the fragmentation pathways presented in Fig. 2.

3.4 Quantitative Analysis of HAAs Using HPLC/MS

1. Prepare samples as described in Subheading 3.1, but add labeled HAA ($^{13}C_{10}$-$^{13}C_{10}$) as internal standard at a final concentration of 25 mg/L. The sample supernatant should be diluted to get the concentration within the linear range of detection.
2. Follow **steps 1–6** under Subheading 3.2.
3. For quantitation of a given HAA congener, display the ion chromatogram of the pseudomolecular ions $[M–H]^-$ of interest. For example, the *m/z* 357 ion corresponds the $[M–H]^-$ of 3-(3-hydroxydecanoyloxy)-decanoic acid, C_{10}-C_{10}. Display also the ion chromatogram of the internal standard, $^{13}C_{10}$-$^{13}C_{10}$, at *m/z* 377.
4. Integrate the analyte and internal standard peak areas, and calculate the analyte concentration using the following formula:

$$C_T = \mathrm{DF} \times \left(C_S \times A_T \,/\, A_S\right)$$

where C_T is the concentration of the HAA congener in mg/L, and DF is the dilution factor which corrects for the 25-fold dilution of the culture supernatant as described in **step 4** under Subheading 2.1. C_S is the concentration of corresponding internal standard ($^{13}C_{10}$-$^{13}C_{10}$) at 25 mg/L. A_T is the peak area of the HAA congener. A_S is the peak area of corresponding internal standard.

4 Notes

1. Dissolve the MSM's ingredients in the listed order using a magnetic stirrer. Sterilize by autoclaving at 121 °C for 20 min. Take the medium out of the autoclave immediately after sterilization.

If the medium appears cloudy while hot, shake well in order to avoid any precipitation. The medium turns clear when it cools down to room temperature. The composition of MSM might have to be changed if a different source of carbon is used. For instance, if dextrose is provided instead of glycerol, higher buffering capacity is required (e.g., MSM–100 mM phosphate), as assimilation of dextrose by *P. aeruginosa* results in significant medium acidification to a pH as low as 3, which inhibits RL production.

2. About 425 μL of ammonium hydroxide (28–30 %) solution is required to achieve pH 8.55 per 250 mL of 100 mM ammonium acetate solution.
3. A few drops of concentrated ammonium hydroxide (28–30 %) will suffice. We find that it is better to use a freshly prepared solvent A as, upon exposure to air, the pH of the solution tends to decrease to below 7.0 over time, probably due to dissolution of atmospheric CO_2. Microbial contamination can also be a problem upon long-term storage at room temperature. Solution A should not be used when older than 24 h. It is important to ensure a pH above 8.0 throughout all the HPLC/MS runs as this can affect the RL retention times and the efficiency of the electrospray negative ionization.
4. Solvent B can be used for up to 1 week after preparation as its pH does not decrease much upon storage. Yet, the pH should be checked and readjusted to 8.0–8.5 before use, if required.
5. If ^{13}C-labeled RLs as internal standards are not available, 5,6,7,8-tetradeutero-4-hydroxy-2-heptylquinoline (HHQ-d_4) or 16-hydroxyhexadecanoic acid has been used [43]. However, in this case a specific response factor must be calculated, as done previously [40].
6. A C18 reversed-phase column with the corresponding guard column can be used alternatively [40].
7. Besides strain PA14, other *P. aeruginosa* strains also produce RLs. Some other *Pseudomonas* and a few *Burkholderia* species produce RLs as well [4].
8. Inoculation is performed using washed cells instead of directly inoculating from the overnight TSB culture broth in order to avoid carryover from the rich nutritive ingredients of TSB into the nitrogen-limited MSM medium.
9. For *P. aeruginosa*, the optimal temperature for RL production is close to 30 °C and for growth at 37 °C. We typically select 34 °C as a compromise. At this temperature, in MSM–glycerol at 200 rpm, PA14 reaches the stationary phase at the third day of incubation and RL production peak at the fourth day; under these conditions, RL start to be degraded by the bacteria afterwards.

10. It is important that the supernatant sample and the internal standard solutions be at room temperature before mixing. Culture supernatants are typically diluted to bring the RL concentration within the linear range of detection (Fig. 3). When *P. aeruginosa* PA14 is cultivated under optimized conditions, for instance in the MSM–glycerol described here, it can achieve RL yields higher than 2,000 mg/L. However, cultivation in commonly used rich media (e.g., LB, TSB) will not require dilution, since, in these media, concentrations in the range of 100 mg/L are typically expected.
11. Attaining appropriate MS detector sensitivity is critically important, especially when switching from the positive ionization in previous analyses to the negative ionization mode. This can be assessed by first injecting 2–3 blank samples containing only the internal standard and monitoring the increase in the peak area of the IS that usually attains its maximum after three runs.
12. The response factor of mono-RL-$^{13}C_{26}$ can be directly used for the quantification of all the mono-RL congeners [4]. Similarly, the response factor of di-RL-$^{13}C_{32}$ is directly used for quantification of all di-RL congeners [4].
13. Make sure that the peak area of the analyte does not exceed the linear range for the estimation of concentration to be valid.
14. The mass range of fragment ions should be adjusted based on the predicted structural changes, e.g., in case of isotopically enriched RL. Collision energy should also be optimized if incomplete or no fragmentation is obtained; this can be carried out by varying the value upward or downward by 5 eV increments.

References

1. Jarvis FG, Johnson MJ (1949) A glycolipid produced by *Pseudomonas aeruginosa*. J Am Chem Soc 71:4124–4126
2. Hauser G, Karnovsky ML (1954) Studies on the production of glycolipid by *Pseudomonas aeruginosa*. J Bacteriol 68:645–654
3. Burger MM, Glaser L, Burton RM (1963) The enzymatic synthesis of a rhamnose-containing glycolipid by extracts of *Pseudomonas aeruginosa*. J Biol Chem 238:2595–2602
4. Abdel-Mawgoud AM, Lépine F, Déziel E (2010) Rhamnolipids: diversity of structures, microbial origins and roles. Appl Microbiol Biotechnol 86:1323–1336
5. Kurioka S, Liu PV (1967) Effect of the hemolysin of *Pseudomonas aeruginosa* on phosphatides and on phospholipase c activity. J Bacteriol 93:670–674
6. Sierra G (1960) Hemolytic effect of a glycolipid produced by *Pseudomonas aeruginosa*. Antonie Van Leeuwenhoek 26:189–192
7. Hisatsuka K-I, Nakahara T, Sano N, Yamada K (1971) Formation of rhamnolipid by *Pseudomonas aeruginosa* and its function in hydrocarbon fermentation. Agr Biol Chem 35: 686–692
8. Itoh S, Suzuki T (1972) Effect of rhamnolipids on growth of *Pseudomonas aeruginosa* mutant deficient in *n*-paraffin-utilizing ability. Agr Biol Chem 36:2233–2235
9. Syldatk C, Lang S, Wagner F, Wray V, Witte L (1985) Chemical and physical characterization of four interfacial-active rhamnolipids from *Pseudomonas* spec. DSM 2874 grown on n-alkanes. Z Naturforsch C 40:51–60
10. Koch AK, Käppeli O, Fiechter A, Reiser J (1991) Hydrocarbon assimilation and biosurfactant production in *Pseudomonas aeruginosa* mutants. J Bacteriol 173:4212–4219
11. Read RC, Roberts P, Munro N, Rutman A, Hastie A, Shryock T, Hall R, McDonald-Gibson W, Lund V, Taylor G (1992) Effect of

Pseudomonas aeruginosa rhamnolipids on mucociliary transport and ciliary beating. J Appl Physiol 72:2271–2277

12. Kownatzki R, Tummler B, Doring G (1987) Rhamnolipid of *Pseudomonas aeruginosa* in sputum of cystic fibrosis patients. Lancet 1: 1026–1027
13. McClure CD, Schiller NL (1992) Effects of *Pseudomonas aeruginosa* rhamnolipids on human monocyte-derived macrophages. J Leukoc Biol 51:97–102
14. Kharazmi A, Bibi Z, Nielsen H, Hoiby N, Döring G (1989) Effect of *Pseudomonas aeruginosa* rhamnolipid on human neutrophil and monocyte function. APMIS 97: 1068–1072
15. Jensen PO, Bjarnsholt T, Phipps R, Rasmussen TB, Calum H, Christoffersen L, Moser C, Williams P, Pressler T, Givskov M, Hoiby N (2007) Rapid necrotic killing of polymorphonuclear leukocytes is caused by quorum-sensing-controlled production of rhamnolipid by *Pseudomonas aeruginosa*. Microbiology 153:1329–1338
16. Alhede M, Bjarnsholt T, Jensen PO, Phipps RK, Moser C, Christophersen L, Christensen LD, van Gennip M, Parsek M, Hoiby N, Rasmussen TB, Givskov M (2009) *Pseudomonas aeruginosa* recognizes and responds aggressively to the presence of polymorphonuclear leukocytes. Microbiology 155:3500–3508
17. Van Gennip M, Christensen LD, Alhede M, Phipps R, Jensen PO, Christophersen L, Pamp SJ, Moser C, Mikkelsen PJ, Koh AY, Tolker-Nielsen T, Pier GB, Hoiby N, Givskov M, Bjarnsholt T (2009) Inactivation of the *rhlA* gene in *Pseudomonas aeruginosa* prevents rhamnolipid production, disabling the protection against polymorphonuclear leukocytes. APMIS 117:537–546
18. Ochsner UA, Fiechter A, Reiser J (1994) Isolation, characterization, and expression in *Escherichia coli* of the *Pseudomonas aeruginosa rhlAB* genes encoding a rhamnosyl transferase involved in rhamnolipid biosurfactant synthesis. J Biol Chem 269:19787–19795
19. Ochsner UA, Koch AK, Fiechter A, Reiser J (1994) Isolation and characterization of a regulatory gene affecting rhamnolipid biosurfactant synthesis in *Pseudomonas aeruginosa*. J Bacteriol 176:2044–2054
20. Ochsner UA, Reiser J (1995) Autoinducer-mediated regulation of rhamnolipid biosurfactant synthesis in *Pseudomonas aeruginosa*. Proc Natl Acad Sci U S A 92:6424–6428
21. Caiazza NC, Shanks RM, O'Toole GA (2005) Rhamnolipids modulate swarming motility patterns of *Pseudomonas aeruginosa*. J Bacteriol 187:7351–7361
22. Déziel E, Lépine F, Milot S, Villemur R (2003) *rhlA* is required for the production of a novel biosurfactant promoting swarming motility in *Pseudomonas aeruginosa*: 3-(3-hydroxyalkanoyloxy)alkanoic acids (HAAs), the precursors of rhamnolipids. Microbiology 149:2005–2013
23. Köhler T, Curty LK, Barja F, Van Delden C, Pechère J-C (2000) Swarming of *Pseudomonas aeruginosa* is dependent on cell-to-cell signaling and requires flagella and pili. J Bacteriol 182:5990–5996
24. Tremblay J, Richardson AP, Lepine F, Deziel E (2007) Self-produced extracellular stimuli modulate the *Pseudomonas aeruginosa* swarming motility behaviour. Environ Microbiol 9: 2622–2630
25. Glick R, Gilmour C, Tremblay J, Satanower S, Avidan O, Deziel E, Greenberg EP, Poole K, Banin E (2010) Increase in rhamnolipid synthesis under iron-limiting conditions influences surface motility and biofilm formation in *Pseudomonas aeruginosa*. J Bacteriol 192: 2973–2980
26. Pamp SJ, Tolker-Nielsen T (2007) Multiple roles of biosurfactants in structural biofilm development by *Pseudomonas aeruginosa*. J Bacteriol 189:2531–2539
27. Davey ME, Caiazza NC, O'Toole GA (2003) Rhamnolipid surfactant production affects biofilm architecture in *Pseudomonas aeruginosa* PAO1. J Bacteriol 185:1027–1036
28. Schooling SR, Charaf UK, Allison DG, Gilbert P (2004) A role for rhamnolipid in biofilm dispersion. Biofilms 1:91–99
29. Boles BR, Thoendel M, Singh PK (2005) Rhamnolipids mediate detachment of *Pseudomonas aeruginosa* from biofilms. Mol Microbiol 57:1210–1223
30. Lequette Y, Greenberg EP (2005) Timing and localization of rhamnolipid synthesis gene expression in *Pseudomonas aeruginosa* biofilms. J Bacteriol 187:37–44
31. Nitschke M, Costa SGVAO, Contiero J (2011) Rhamnolipids and PHAs: Recent reports on Pseudomonas-derived molecules of increasing industrial interest. Process Biochem 46: 621–630
32. Abdel-Mawgoud AM, Hausmann R, Lépine F, Müller MM, Déziel E (2011) Rhamnolipids: detection, characterization, biosynthesis, genetic regulation and bioengineering of production. In: Soberón-Chávez G (ed) Biosurfactants: from genes to applications, 1st edn. Springer, Berlin, pp 13–55

33. Déziel E, Lépine F, Dennie D, Boismenu D, Mamer OA, Villemur R (1999) Liquid chromatography/mass spectrometry analysis of mixtures of rhamnolipids produced by *Pseudomonas aeruginosa* strain 57RP grown on mannitol or naphthalene. Biochim Biophys Acta 1440:244–252
34. Heyd M, Kohnert A, Tan TH, Nusser M, Kirschhofer F, Brenner-Weiss G, Franzreb M, Berensmeier S (2008) Development and trends of biosurfactant analysis and purification using rhamnolipids as an example. Anal Bioanal Chem 391:1579–1590
35. Jain DK, Collins-Thompson DL, Lee H, Trevors JT (1991) A drop-collapsing test for screening surfactant-producing microorganisms. J Microbiol Meth 13:271–279
36. Siegmund I, Wagner F (1991) New method for detecting rhamnolipids excreted by *Pseudomonas* species during growth on mineral agar. Biotechnol Tech 5:265–268
37. Chandrasekaran EV, BeMiller JN (1980) Constituent analysis of glucosaminoglycans. In: Whistler RL (ed) Methods in carbohydrate chemistry. Academic Press, Inc., New York, pp 89–96
38. Syldatk C, Lang S, Matulovic U, Wagner F (1985) Production of four interfacial active rhamnolipids from n-alkanes or glycerol by resting cells of Pseudomonas species DSM 2874. Z Naturforsch C 40:61–67
39. Schenk T, Schuphan I, Schmidt B (1995) High-performance liquid chromatographic determination of the rhamnolipids produced by *Pseudomonas aeruginosa*. J Chromatogr A 693:7–13
40. Déziel E, Lépine F, Milot S, Villemur R (2000) Mass spectrometry monitoring of rhamnolipids from a growing culture of *Pseudomonas aeruginosa* strain 57RP. Biochim Biophys Acta 1485:145–152
41. Rahme LG, Stevens EJ, Wolfort SF, Shao J, Tompkins RG, Ausubel FM (1995) Common virulence factors for bacterial pathogenicity in plants and animals. Science 268:1899–1902
42. Lépine F, Déziel E, Milot S, Villemur R (2002) Liquid chromatographic/mass spectrometric detection of the 3-(3-hydroxyalkanoyloxy) alkanoic acid precursors of rhamnolipids in *Pseudomonas aeruginosa* cultures. J Mass Spectr 37:41–46
43. Dubeau D, Déziel E, Woods DE, Lépine F (2009) *Burkholderia thailandensis* harbors two identical *rhl* gene clusters responsible for the biosynthesis of rhamnolipids. BMC Microbiol 9:263

Chapter 31

LPS Quantitation Procedures

Joseph S. Lam, Erin M. Anderson, and Youai Hao

Abstract

Lipopolysaccharide is the predominant component of the Gram-negative cell wall occupying the outer leaflet of the outer membrane of *Pseudomonas aeruginosa*. Wild-type bacteria produce smooth LPS composed of lipid A, core oligosaccharide, and long O-antigen polysaccharide. In contrast, mutant bacteria defective in LPS biosynthesis produce rough LPS lacking the long O-antigen side chains. LPS is also a major virulence factor and proven to be crucial for full elaboration of other virulence factors and for a range of cellular functions. In order to determine the relationship between LPS and other cellular functions, a means to measure changes in the quantities of LPS being produced under certain growth/environmental conditions is important. Hence, the objective of this chapter is to provide readers with the methodologies for analyzing LPS of *P. aeruginosa* both qualitatively and quantitatively. As a prerequisite to quantifying LPS, one must be able to isolate LPS from the cell envelope; therefore, Subheading 2.1 is devoted to describing several standard LPS preparation methods. This is followed by Subheading 2.2, which deals with a number of practical methods for analyzing and/or quantifying whole-molecule LPS or assays for quantifying specific sugar constituents that are present within *P. aeruginosa* LPS. The methods described herein should be broadly applicable to the studying of LPS of other pseudomonads as well as *Burkholderia* species.

Key words LPS, Phenol–hot-water method, LAL assay, Kdo, Western blots, Densitometry

1 Introduction

A considerable number of protocols have been developed for the extraction of LPS from *P. aeruginosa* cells. When choosing an LPS preparation method one must take into account the originating species, the degree of purity of LPS required, and whether downstream applications require accurate quantitation or the LPS will simply be used to screen for presence or absence. By far, the most commonly used and historically recognized as a standard procedure for LPS isolation is the hot-aqueous-phenol method originally developed by Westphal and Jann in 1965 [1]. This was developed to preferentially isolate smooth-type LPS (S-LPS) (i.e., containing lipid A, core, and O-antigen [2, 3]). For extraction of rough-type LPS (R-LPS) (i.e., containing lipid A and core, minus

Alain Filloux and Juan-Luis Ramos (eds.), *Pseudomonas Methods and Protocols*, Methods in Molecular Biology, vol. 1149, DOI 10.1007/978-1-4939-0473-0_31, © Springer Science+Business Media New York 2014

O-antigen), the phenol–chloroform–petroleum (PCP) ether extraction method that favors partitioning of the more hydrophobic form of LPS was developed [4]. Alternatively, the Darveau and Hancock (D&H) method [5] can be used to efficiently extract both S-LPS and R-LPS. The Hitchcock and Brown preparation (H&B prep) method [6], on the other hand, is a rapid technique for quick visualization of the phenotype of LPS, i.e., S-LPS or R-LPS, produced by a certain bacterial isolate. After extraction of LPS from *P. aeruginosa* cells, further purification can be achieved by using polymyxin-affinity chromatography.

LPS can be detected and quantified using several methods outlined in Subheading 2.2. Limulus amoebocyte lysate (LAL) is derived from the amoebocytes (similar to white blood cells in humans) from the Limulidae family (horseshoe crab), which includes *Limulus polyphemus* (Atlantic) and *Tachypleus tridentatus* (East Asia). LAL has the natural property of reacting to form a gel-like substance when exposed to Gram-negative bacteria. Hence, assays based on amoebocyte lysate are highly sensitive at detecting the whole-molecule LPS, can be utilized either in whole-cell preparations or with LPS extractions, and have become standard methods for detecting ppm quantities of the lipid A moiety of LPS (also known as endotoxin) to ensure the sterility of fluids or drug formulations prepared for medical and clinical usage. Another way to determine the quantity of LPS is to assay for the presence of certain sugar constituents, for instance, uronic acid in *P. aeruginosa* LPS, or 2-keto-3-deoxyoctonate (Kdo), an 8-carbon sugar that is uniquely found in the core oligosaccharide region of the LPS. Changes in LPS production often reflect that *P. aeruginosa* has acquired mutation in LPS biosynthesis genes or that the LPS biosynthesis pathways are downregulated while the bacteria have adopted a biofilm lifestyle. To quantitatively examine such changes, we have outlined a method [7] whereby Western blots of the LPS can be probed with LPS-specific monoclonal antibodies, and then densitometry scans of the blots were created in order to quantitate the resolved LPS bands. This way LPS produced by clinical variants can be compared to that of standard wild-type bacterial strains.

2 Materials

2.1 LPS Preparation

2.1.1 Hitchcock and Brown LPS Preparation

Equipment and Supplies

1. Shaker incubator for maintaining cultures at 37 °C.
2. Microcentrifuge containing rotor for 1.5–2.0 ml microcentrifuge tubes (e.g., Eppendorf Model 5415).
3. Spectrophotometer capable of reading at 600 nm. Boiling water bath or steamer and 55 °C water bath.

Materials and Reagents

1. Media for culturing *P. aeruginosa* include Pseudomonas Isolation Agar (PIA) (Difco) and lysogeny broth (LB, also called Luria-Bertani medium) containing suitable antibiotic or no antibiotic.
2. By addition of 1.5 % Bacto-agar, LB can be made as an LB agar medium.
3. Standard phosphate-buffered saline (PBS) (137 mM NaCl, 2 mM KH_2PO_4, 10 mM Na_2HPO_4, 2.7 mM KCl, pH 7.4).
4. Lysis buffer (1.0 M Tris pH 6.8, 10 % v/v glycerol, 2 % w/v SDS) for lysing bacterial cells and releasing LPS from the cell envelope.
5. Proteinase K, at 30 μg, is added to each sample for degradation of all the protein from the lysing mixture.

2.1.2 Darveau and Hancock LPS Preparation

Equipment and Supplies

1. Shaker incubator for growth of cultures at 37 °C.
2. Centrifuge and a rotor for 15 ml centrifuge tubes (e.g., Sorvall Legend RT Plus) or rotor capable of handling the volume required.
3. Ultracentrifuge, with rotor capable of handling tubes with capacity >15 ml.
4. French Pressure cell for cell breakage.
5. Sonicator (e.g., Model 500 Ultrasonic Dismembrator) with microtip.
6. Vortex mixer.
7. Incubating water bath set at 85 °C.
8. Lyophilizer (e.g., Labconco FreeZone freeze dry system).

Materials and Reagents

1. Culturing of bacteria will be identical as described in the previous method.
2. Tris buffer: 10 mM Tris (pH 8.0), 2 mM $MgCl_2$, 2 % SDS, 4 N NaOH, EDTA at 0.1 and 0.5 M.
3. Reagents: 95 % ethanol, 200 μg/ml DNase, 50 μg/ml RNase, and protease (or pronase, Sigma cat. No. P5147).

2.1.3 Hot-Phenol–Water LPS Preparation

Equipment and Supplies

1. Water baths set at 37, 55, and 68 °C.
2. Hot plate with magnetic stir capabilities.
3. Fumehood.
4. Lyophilizer (e.g., Labconco FreeZone freeze dry system).
5. Centrifuge and rotor for handling 500 ml phenol-resistant centrifuge bottles (e.g., Beckman Avanti J series).
6. Ultracentrifuge, ultracentrifuge tubes.

Materials and Reagents

1. Proteose peptone No. 2 medium.
2. 100 mM NaCl.
3. Liquid phenol.
4. Dialysis tubing (with molecular cutoff of 3,500 Da).
5. $MgCl_2$.
6. DNase.
7. RNase.
8. Proteinase K.
9. 20 mM sodium acetate (pH 7.5).
10. Ultrapure H_2O (obtained from MilliQ or SuperQ (a higher output model) of water purification system, Millipore).

2.1.4 PCP Ether LPS Extraction

Equipment and Supplies

1. Centrifuge containing rotor for 500 ml phenol-resistant centrifuge bottles (e.g., Beckman Avanti J series).
2. Rotary evaporator with round-bottomed flask.
3. Pyrex/Corning glass centrifuge bottles.
4. Lyophilizer.

Materials and Reagents

1. Proteose peptone No. 2 culture medium.
2. The PCP mix for extracting R-LPS is prepared by adding liquid PCP ether in a 2:5:8 ratio (v/v/v); while stirring, slowly add solid phenol until the solution turns clear.
3. 80 % Phenol (prepared by adding 7 ml of ddH_2O to 53 ml of 90 % liquid phenol).
4. Ice.
5. Acetone.
6. Ultrapure water (using MilliQ or SuperQ as described before, Millipore).
7. Whatman No. 3 paper.

2.1.5 Polymyxin-B Affinity Column Chromatography

Equipment and Supplies

1. BIO-RAD disposable columns that will handle 10 ml or more of bed volume.
2. Nutator mixer.

Materials and Reagents

1. 50 mM NH_4HCO_3 (pH 8.0), polymyxin B-agarose (Pharmacia).
2. Buffer: 300 mM triethylamine acetate (pH 6.4) containing 10 % ethylene glycol, 2.0 M urea in 0.1 M NH_4HCO_3 (pH 8.0), 1 % sodium deoxycholate (DOC) in 0.1 M NH_4HCO_3 (pH 8.0), and 8.0 M guanidine HCl.

2.2 LPS Quantitation

2.2.1 Limulus Amoebocyte Lysate Assay

Equipment and Supplies

1. Multi-well plate reading spectrophotometer.
2. Multichannel pipettor.
3. Vortex mixer.

Materials and Reagents

1. LAL assays reagent kit from Associates of Cape Cod Inc. and other companies.
2. Pyrogen-free water.
3. Lyophilized endotoxin for use as standards to plot standard curves.
4. Recombinant factor C (rFC) [18].
5. Chromatic [19] or bioluminescent substrate [20].

2.2.2 LPS Sugar Constituent Assays

Kdo Assay

Equipment and Supplies

For Procedure 1

1. Heating block set at 100 °C.
2. Vortex mixer.
3. Spectrophotometer capable of measurement at 552 and 509 nm.
4. Microcentrifuge containing a rotor for 1.5–2.0 ml microcentrifuge tubes (e.g., Eppendorf Model 5415).

For Procedure 2

1. Heating block set at 100 °C.
2. Vortex mixer.
3. Multi-well plate reading spectrophotometer capable of reading at 550 nm.

Materials and Reagents

1. Kdo (Sigma cat. #K2755) to be used to prepare the standard curve.
2. H_2SO_4.
3. Periodic acid ($HIO_4 \cdot 2H_2O$).
4. Sodium arsenite ($NaAsO_2$) in 0.5 N HCl.
5. Thiobarbituric acid.
6. Butanol reagent (5.0 % v/v concentrated 0.5 N HCl; 95 % v/v *n*-butanol).
7. DMSO.

Purpald Assay

Equipment and Supplies

1. Multi-well plate reading spectrophotometer capable of measurement at 550 nm.

Materials and Reagents

1. Kdo.
2. $NaIO_4$ (32 and 64 mM).
3. Purpald reagent (136 mM in 2 N NaOH).
4. 2-Propanol.

2.2.3 LPS Western Blotting and Densitometry

Equipment and Supplies

Polyacrylamide vertical gel electrophoresis system (BIO-RAD, Mini-Protean Tetra, or equivalent). Power supply capable of loads up to 200 V and 180 mA (BIO-RAD, PowerPac Universal Power supply, or equivalent). Trans-blot electrophoretic transfer (such as BIO-RAD, Mini Trans-blot cell). Nitrocellulose membrane (such as Pall Life Sciences, BioTrace™ NT Nitrocellulose Transfer Membranes). Platform shaker for the washing and developing of the blots with antibody probes and substrate development by enzyme-conjugated antibodies. Densitometer with CCD camera (e.g., BIO-RAD GS-800 Calibrated Densitometer). Computer software for image analysis (e.g., ImageJ, MCID Analysis, or Quantity One), and statistical software for *t*-test (e.g., Microsoft Excel or GraphPad (www.graphpad.com/quickcalcs) and ANOVA (e.g., SPSS, SAS, R, Microsoft Excel)).

Materials and Reagents

Reagents to be used include 12 % discontinuous polyacrylamide running gel (Stacking: 4 % (39:1) acrylamide:bis, 126 mM Tris pH 6.8, 0.1 % w/v SDS, 1.5 % v/v TEMED, 0.05 % v/v APS. Resolving: 12 % (39:1) acrylamide:bis, 375 mM Tris pH 8.8, 0.1 % w/v SDS, 1.5 % v/v TEMED, 0.05 % v/v APS.), SDS-running buffer (0.25 M Tris, 1.92 M glycine, 1 % w/v SDS), Western transfer buffer (192 mM glycine, 25 mM Tris, 20 % v/v methanol, 0.01 % w/v SDS), PBS (137 mM NaCl, 2 mM KH_2PO_4, 10 mM Na_2HPO_4, 2.7 mM KCl, pH 7.4), buffer A (100 mM Tris pH 9.5, 100 mM NaCl, 5 mM $MgCl_2$), monoclonal antibodies against various regions of *P. aeruginosa* LPS (available either from commercial sources or by request to Lam laboratory at the University of Guelph), goat-anti-mouse antibodies conjugated to alkaline phosphatase to be used at dilutions recommended by the manufacturer, and substrates for developing the blots made up of a mixture of 5 % w/v 5-bromo-4-chloro-3-indoyl phosphate (BCIP) in PBS and 5 % w/v nitro blue tetrazolium (NBT) in PBS.

3 Methods

3.1 Hitchcock and Brown LPS Preparation Method

3.1.1 Concept

Hitchcock and Brown preparation [6] has been widely accepted and is by far the most popular LPS isolation method, albeit the quality of LPS is not considered to be pure. It is utilized as a quick protocol to prepare LPS from one to many bacterial strains that can be screened by SDS-polyacrylamide gel electrophoresis (SDS-PAGE) followed by silver staining [8, 9]. However, other gel staining protocols have been proposed and/or modified for LPS with varying sensitivities [10–14]. For increased selectivity, LPS banding pattern resolved by the SDS-PAGE gels can be subsequently transferred onto nitrocellulose membrane by Western blotting and then the membrane can be probed with LPS-specific antibodies, as discussed in Subheading 2.2.3. No cell fractionation is required, as protein and other moieties from the cells are simply destroyed by the action of detergent, enzymatic hydrolysis, and heating.

3.1.2 Procedures

1. Inoculate bacteria on PIA, or a suitable medium containing appropriate antibiotic, and grow overnight at 37 °C.
2. Select a single colony to inoculate into 5 ml LB medium, or a suitable medium containing appropriate antibiotic, and grow overnight at 37 °C with shaking at 200 rpm (*see* **Note 1**).
3. Take 1 ml of the culture, and centrifuge using a microcentrifuge at 15,600 × *g* for 1 min.
4. Suspend pellet in 1 ml of standard PBS.
5. Wash the pellet three times using the same centrifugation speed and PBS buffer as in **steps 3** and **4**. Washing will remove any pyocyanin production which can interfere with spectroscopy.
6. Dilute the cell suspension in PBS, and adjust the optical density at 600 nm (OD_{600}) to 0.45. This OD_{600} adjustment allows standardization of the cell density before LPS is extracted from each sample.
7. Take 1 ml of OD_{600}-standardized culture, centrifuge at 15,600 × *g* for 1 min, and then discard the supernatant without disturbing the pellet.
8. Suspend pellet in 250 μL lysis buffer (*see* **Note 2**).
9. Heat samples at 100 °C (boil or steam) for a minimum of 10 min or up to 30 min.
10. Add 30 μg proteinase K.
11. Incubate at 55 °C for a minimum of 4 h or at 25 °C overnight.
12. Cool to room temperature, and load onto PAGE gel. *See* Subheading 2.2.3.
13. Samples can be stored at −20 °C (*see* **Note 3**).

3.2 Darveau and Hancock LPS Preparation Method

3.2.1 Concept

The cold-ethanol–magnesium chloride precipitation protocol proposed by Darveau and Hancock in 1983 produces a relatively cleaner LPS preparation of both smooth and rough LPS that can be useful for a much broader application range. However, this method requires more time and has many more steps in comparison to the H&B method. DNA, RNA, protein, lipid, and other contaminants are degraded and removed through sequential enzymatic incubation, detergent solubilization, and centrifugation, while LPS is retained due to precipitation with high concentrations of $MgCl_2$ and cold ethanol. The final product from this protocol is lyophilized (white flake appearance) containing predominantly LPS which can be solubilized into water or a suitable buffer as required for use in downstream applications. D&H LPS preparations retain small amounts of contaminants, including less than 0.1 % protein, 1.0 % nucleic acid, and 5.0 % phospholipids per μg purified LPS [5].

3.2.2 Procedures

1. Inoculate *P. aeruginosa* bacteria into 1 l of LB medium containing appropriate antibiotic, from an overnight 5 ml culture, and grow at 37 °C with shaking at 200 rpm till an OD_{600} at 0.6–0.8 is attained.
2. Centrifuge at 9,200 × *g* for 10 min. Cell pellet can be lyophilized for quantitation purposes.
3. Suspend cell pellet in 15 ml of 10 mM Tris (pH 8.0) and 2 mM $MgCl_2$ (*see* **Note 4**).
4. Add 1,500 μg DNase and 375 μg RNase.
5. Break cells by passing through the French Pressure cell at 15,000 psi. Repeat once.
6. Sonicate for 2 × 30-s bursts at 75 % probe intensity. By this point there should be <1 % intact cells.
7. Add additional DNase to a final concentration of 200 μg/ml and RNase to a final concentration of 50 μg/ml.
8. Incubate at 37 °C for 2 h.
9. Add 5 ml of 0.5 M EDTA in 10 mM Tris (pH 8.0), 2.5 ml of 20 % SDS in 10 mM Tris (pH 8.0), and 2.5 ml of 10 mM Tris (pH 8.0) to give a final volume of 25 ml. Thoroughly agitate this mixture by using a vortex.
10. Centrifuge the above mixture at 50,000 × *g* for 30 min at 20 °C. This removes peptidoglycan from the mixture.
11. Add protease to the supernatant to a final concentration of 200 μg/ml.
12. Incubate at 37 °C overnight with shaking. Occasionally a precipitate may form; if so, centrifuge at 200 × *g* for 10 min.
13. Add two volumes of 0.375 M $MgCl_2$ in 95 % ethanol that was stored at −20 °C.
14. Centrifuge at 12,000 × *g* for 15 min at 0–4 °C.
15. Suspend pellet in 25 ml of 2 % SDS, 0.1 M EDTA, and 10 mM Tris (pH 8.0).
16. Sonicate, as above, 2 × 30-s burst at 75 % probe intensity.
17. Incubate at 85 °C for 10–30 min.
18. Ensure that pH is ~9.5. Adjust pH by dropwise addition of 4N NaOH, if necessary.

19. Add protease to a final concentration of 25 μg/ml.
20. Incubate overnight at 37 °C with shaking.
21. Precipitate LPS with 2–3 volumes of 0.375 M $MgCl_2$ in 95 % ethanol stored at −20 °C. The supernatant should be clear.
22. Centrifuge at 12,000 × *g* for 15 min at 4 °C.
23. Resuspend pellet in 15 ml of 10 mM Tris (pH 8.0).
24. Sonicate, as above, 2 × 30-s burst at 75 % probe intensity.
25. Centrifuge at 200 × *g* for 5 min. This step helps to remove insoluble Mg^{2+}–EDTA crystals. Occasionally, R-LPS may pellet at this step, which can be recovered by resuspending in fresh 10 mM Tris (pH 8.0) or water and repeating centrifugation.
26. Add $MgCl_2$ to a final concentration of 25 mM.
27. Centrifuge at 200,000 × *g* for 2 h at 15 °C. For mucoid strains, $MgCl_2$ can be omitted from this final centrifuge step to prevent contamination from polysaccharide Mg^{2+} complexes [15].
28. Pellet containing LPS can be suspended in water or lyophilized for indefinite storage.

3.3 Hot-Phenol–Water LPS Preparation Method

3.3.1 Concept

The principle of the hot-phenol–water method is that when a 1:1 ratio of phenol and water is mixed with a cell suspension and heated to 65–68 °C, the resulting mixture will become one homogenous phase, which when cooled can be separated into three phases. LPS and nucleic acids will remain in the upper water phase, while proteins remain in the lower phenol phase, and the insoluble cell remnants are at the interface between the upper and lower phase or at the bottom of the centrifuge tube. However, on occasion, trace amounts of R-LPS may partition into the phenol phase; therefore, all phases should be analyzed for LPS content. The LPS-containing phase is then dialyzed against water until all residual phenol is removed. DNA, RNA, and protein contaminants are further degraded by treating the LPS-containing sample with DNase, RNase, and protease. LPS can then be sedimented by ultracentrifugation. Below is the protocol from Westphal and Jann [1] with only minor modifications.

3.3.2 Procedures

1. Grow 10 l of bacteria culture in proteose peptone No. 2 or suitable medium containing appropriate antibiotics to stationary phase.
2. Centrifuge at 8,000 × *g* for 10 min.
3. Resuspend the cell pellet in 1 l of 100 mM NaCl, and repeat **step 2**.
4. Freeze-dry the cell pellet.
5. To 5 g of cell pellet or less, suspend cells in 50 ml of 100 mM NaCl. Ensure that no cell clumping remains.

6. Heat cell suspension to 68 °C in a water bath.
7. Add 50 ml of hot phenol (68 °C).
8. Stir vigorously for 1 h in a fume hood, maintaining at 68 °C, with the container covered to prevent excessive evaporation of the water.
9. Transfer to a glass centrifuge bottle, and chill on ice for 10 min (*see* **Note 5**).
10. Centrifuge at 12,000 ×*g* for 15 min at 4 °C to clearly separate the phases. Transfer the upper (water) layer to another container. Be careful to avoid removing the lower (phenol) phase and interphase material.
11. To the lower phase, add 50 ml of 100 mM NaCl, and stir vigorously for another 30–60 min at 68 °C.
12. Repeat **step 10**, and combine the extracts from the two upper phases.
13. Dialyze the extracts (molecular weight cutoff: 3,500 Da) against water. Continue dialysis until you can no longer smell any phenol. Change the water a few times during the process.
14. To the dialyzed extract, add $MgCl_2$ to a final concentration of 5 mM (*see* **Note 6**).
15. Add 20 μg/ml DNase, and incubate for 2 h at 37 °C.
16. Add 20 μg/ml RNase, and incubate for 2 h at 37 °C.
17. Add 30 μg/ml proteinase K, and incubate for 2 h at 55 °C.
18. Freeze the sample at −80 °C and lyophilize.
19. Resuspend crude LPS in 25 ml of 20 mM NaOAc, pH 7.5 (*see* **Note 7**).
20. Centrifuge at 100,000 ×*g*, overnight, at 4 °C to sediment LPS micelles. Discard supernatant.
21. Resuspend LPS pellets in a minimal volume of ultrapure water (Millipore).
22. Lyophilize and calculate yield.

3.4 PCP Ether LPS Extraction Method

3.4.1 Concept

R-LPS is highly hydrophobic; hence, it naturally partitions into the phenol phase when using the standard hot-phenol–water method of Westphal and Jann. To address this problem, Galanos et al. developed the PCP ether method for preparing R-LPS of high purity [4]. During the process, lyophilized bacteria pellets are resuspended and homogenized with an extraction mixture of liquid PCP ether (2:5:8). The cell suspension is centrifuged, the supernatant containing LPS is filtered, and the organic solvents, chloroform and petroleum ether, are removed by rotary evaporation. The R-LPS remaining in the phenol-containing supernatant can then be precipitated by adding water. Below is a protocol from Galanos et al. [4] with slight modifications.

3.4.2 Procedures

1. Grow 10 l bacteria culture in proteose peptone No. 2, or a suitable medium, containing appropriate antibiotics.
2. Centrifuge at 8,000 × *g* for 10 min.
3. Suspend cell pellet in ultrapure H_2O or ddH_2O, and repeat **step 2**.
4. Lyophilize the cell pellet.
5. To 5 g of cell pellet or less, add 50 ml PCP mixture to the dry cell pellet. Cover, and stir for 60 min at room temperature (*see* **Note 8**).
6. Centrifuge at 4,000 × *g* for 10 min at room temperature.
7. Filter supernatant through Whatman No. 3 paper into a flask.
8. Resuspend pellet in 70 ml PCP mixture, cover, and stir for 60 min at room temperature.
9. Repeat **step 6**. Combine the two extractions.
10. Fill rotary evaporator cold trap with ice water or dry ice–ethanol mixture. Heat rotary evaporator (water or oil) bath to 37 °C.
11. Place the PCP-extracted LPS sample in a rotary evaporator round-bottomed flask. The petroleum ether and chloroform from the extracts will be evaporated. When the volume of the mixture is reduced to <50 ml, transfer it to a 50 ml glass centrifuge tube (*see* **Note 9**).
12. Using a Pasteur pipet, add three drops of water and swirl the centrifuge tube. The sample should begin to turn cloudy. Repeat this step 3–5 times. Centrifuge at 3,000 × *g* for 10 min to obtain precipitated LPS (*see* **Note 10**).
13. Resuspend the pellet in 10 ml of 80 % phenol. Centrifuge at 3,000 × *g* for 10 min, and remove supernatant with a Pasteur pipet.
14. Repeat **step 13** three times.
15. Resuspend pellet in 10 ml 100 % acetone. Centrifuge at 3,000 × *g* for 10 min, and remove the supernatant with a Pasteur pipet.
16. Repeat **step 15** for a total of four times.
17. Lightly cover the tube and leave in fume hood overnight to dry the pellet.
18. Dissolve the LPS precipitate in 5 ml of distilled water. Centrifuge the solution at 100,000 × *g* for 4 h.
19. Lyophilize overnight, and measure the dry weight of the R-LPS.

3.5 Polymyxin-B Affinity Column Chromatography Method

3.5.1 Concept

LPS prepared by the above methods can be further purified using gel filtration, anion-exchange, or affinity chromatography methods to further remove any contaminants. Gel filtration chromatography isolates heterogeneous LPS based on size differences and is often performed in the presence of detergent to disperse LPS, which has a tendency to form micelle aggregates. Columns composed of various Sephadex/Superose resins in different buffer systems can be used, reviewed by Wang et al. [16]. Anion-exchange chromatography utilizes the DEAE-coated resins to bind the overall negative charge of the *P. aeruginosa* LPS molecule, which can be eluted by a salt gradient, whereas affinity chromatography using polymyxin B-agarose, which specifically binds lipid A of LPS with high affinity [17], is a quick and efficient method for purifying LPS.

3.5.2 Procedures According to Forsberg and Carlson [17]

1. Suspend crude LPS extracts in 50 mM NH_4HCO_3, pH 8.0.
2. Apply up to 30 ml of the extract to a 10 ml bed volume column of polymyxin B-agarose equilibrated in 50 mM NH_4HCO_3, pH 8.0.
3. Incubate resin with extract, rocking overnight on a Nutator mixer at room temperature.
4. Wash the column with 300 mM triethylamine acetate buffer (pH 6.4) containing 10 % ethylene glycol.
5. Elute weakly bound non-LPS components with 2.0 M urea in 0.1 M NH_4HCO_3 (pH 8.0).
6. Elute the LPS with 1 % sodium DOC in 0.1 M NH_4HCO_3, pH 8.0, or 8.0 M guanidine HCl (*see* **Note 11**).

3.6 Limulus Amoebocyte Lysate Assay Method

3.6.1 Concept

Due to its simplicity, sensitivity, and ability to quantify LPS, the LAL assay has been a highly popular method for the quantitation of ppm amounts of endotoxin (lipid A). When exposed to endotoxin, LAL undergoes a serine protease cascade reaction and forms a gel-like substance as a means of host defense response within the Limulidae family (horseshoe crab) [21]. Excellent in-depth reviews of LAL gelling mechanisms have been written by Iwanaga et al. [22] and Williams [23]. Briefly, bacterial phospholipids, particularly acidic phospholipids including LPS (lipid A), interact and activate factor C, which in turn activates factor B, whose reaction produces the clotting enzyme, coagulase (Fig. 1). Coagulase functions to cleave two bonds (Arg-Thr and Arg-Gly) in coagulogen, releasing peptide C and coagulin. The loss of peptide C from coagulogen exposes a hydrophobic region that can interact with the hydrophobic tail of another coagulin or coagulogen, forming the insoluble homopolymer gel [24].

For the LAL assays, any of the three commonly used protocols including the gel-clot assay (qualitative), the turbidimetric method

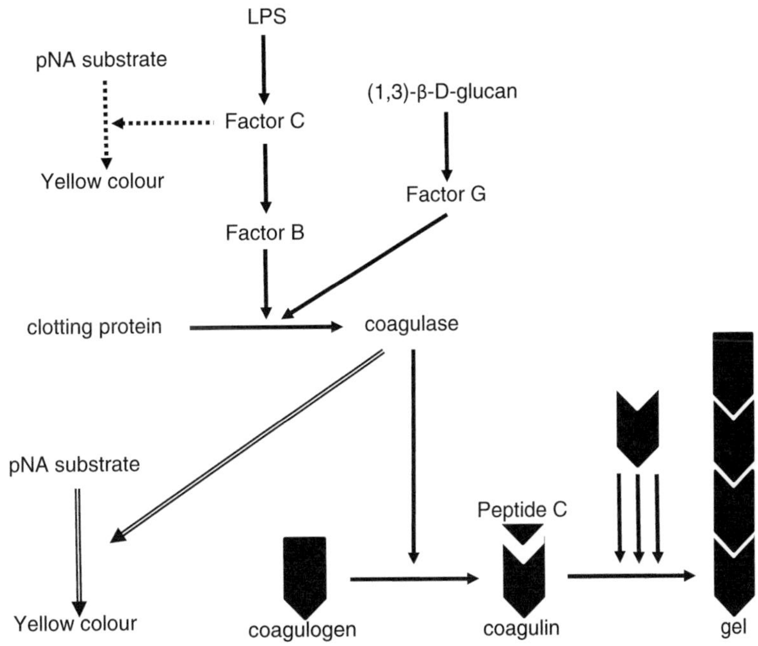

Fig. 1 The serine protease cascade reactions of the immune response pathway from the Limulidae family of horseshoe crabs. These reactions are the basis of the LAL assays, whereby coagulin is produced from coagulogen, and ultimately a homopolymeric gel is formed. Environmental toxins, LPS (Gram-negative bacteria) and (1,3)-β-D-glucan (fungi), are recognized by factor C and factor G, respectively. The activated factor B/G in turn activates the clotting protein producing the clotting enzyme, coagulase. Coagulase cleaves two bonds in coagulogen (depicted as an irregular *pentagon*) to produce coagulin (depicted as an *arrowhead*) and peptide C (depicted as a *triangle*). Multiple coagulin and/or coagulogen interact to produce the visible gel. *Double lines* indicate the adaptation for the chromatic kinetic assay; *dotted line* indicates the adaption for the rFC chromatic kinetic assay (adapted from Osaki and Kawabata [61])

(quantitative), and the highly sensitive chromatic method (quantitative) can be chosen. The following protocol can be used for either the turbidimetric or the chromatic method depending on the substrate used. The turbidimetric assay is the LAL–endotoxin reaction measured as an increase in turbidity over time at OD_{430} [25]. The chromatic assay is accomplished by coagulase cleavage of a synthetic peptide from either a chromogenic substrate, e.g., *para*-nitroaniline (pNA) [19], or a bioluminescent substrate, e.g., luciferin [20] (Fig. 1). The rate of development of the colored pigment is proportional to the concentration of LPS, and the absorbance/fluorescence is measured at A_{405} (pNA) or at $A_{550-620}$ (luciferin).

3.6.2 Procedures

1. Preheat a 96-well microtiter plate to 37 °C either in an incubator or in a plate reader with temperature control.
2. To prepare endotoxin standards dilute lyophilized endotoxin in pyrogen-free water (generally supplied by the manufacturer) and agitate vigorously with a vortex for 15 min. Prepare a serial dilution series of test tubes containing 0.01, 0.1, and 1.0 EU/ml of the standard endotoxin; mix with a vortex for 1 min between dilutions to ensure proper distribution of endotoxin (*see* **Note 12**).
3. Add endotoxin samples to the wells of the microtiter plate and incubate at 37 °C for 10 min.
4. Mix substrate (LAL or chromatic substrate) in pyrogen-free water or buffer in the ratio specified by the manufacturer. Incubate at 37 °C for 10 min.
5. To the microtiter plate containing the endotoxin samples in the wells, add 1:1 (v/v) endotoxin sample to substrate, and mix gently. This step can be achieved quickly by using a multichannel pipettor.
6. Be careful not to produce bubbles. Ensure that all wells contain the same volume.
7. Place the microplate into a multiplate reader, and take initial absorbance reading (at wavelength appropriate for the substrate chosen, i.e., A_{405} for pNA or OD_{430} for turbidimetric) and a reading every 1 min for 1 h (*see* **Note 13**).
8. Correct sample reading by subtraction against control blanks (i.e., wells with pyrogen-free water only and not endotoxin sample or LAL).
9. Plot a log graph with the "reading" against the endotoxin concentrations from the standard endotoxin concentrations (standard curve) (*see* **Note 14**).
10. Read the unknown sample concentrations directly from the linear standard curve.

3.7 Kdo Assay Method

3.7.1 Concept

The 8-carbon sugar, Kdo, is a unique structural component of the inner core of LPS. It is attached to lipid A by an acid-labile ketosidic linkage and can be released by mild acid hydrolysis in most cases [26]; however, in some cases, strong acid hydrolysis will be needed [27, 28]. Since Kdo is exclusive to core oligosaccharide of LPS, it is an ideal candidate for detection and quantitation of LPS. In 1959, Weissbach and Hurwitch [29] developed the first method for the quantitative analysis of LPS using Kdo, which was later modified by Osborne in 1963 [30]. In this procedure, LPS is hydrolyzed under mild acidic conditions (such as 1 % acetic acid) to release the Kdo sugar and then oxidized with HIO_4 and $NaAsO_2$, thereby forming formylpyruvic acid which can react with thiobarbituric acid (TBA) to give a chromophore with an absorption maximum at A_{549} (*see* **Note 15**). A striking disadvantage is the development of turbidity in the reaction mixture at room temperature which can interfere with spectroscopy. Turbidity issues can be overcome by either extraction of the chromophore with cyclohexanone or acidified butanol [31] or by addition of dimethyl sulfoxide (DMSO) [32, 33]. Two protocols are currently used by most researchers. Protocol 1 is posted on the website of Dr. Robert E. Hancock's lab (http://cmdr.ubc.ca/bobh/methods/KDOASSAY.html) and utilizes acidified butanol extraction. Protocol 2 was published by Lee and Tsai [34], who reported a procedure that is a modification of the method described by Karkhanis et al. [32] using DMSO as an additive to extract the chromophore.

Procedure 1: Kdo Assay Utilizing Butanol Extraction (*See* ***Note 16***)

1. Add 1:1 LPS sample or Kdo standard (0, 40, 80, 120, 160 μg Kdo) to 0.5 N H_2SO_4 for 100 μl total volume, and agitate the mixture by using a vortex (*see* **Note 16**).
2. Place in 100 °C heating block for 8 min to hydrolyze. Cool to room temperature (*see* **Note 17**).
3. Add 50 μl periodic acid ($HIO_4{\cdot}2H_2O$) (22.8 mg/ml), and mix using a vortex.
4. Incubate for 10 min at room temperature.
5. Add 200 μl arsenite reagent (40 mg/ml sodium arsenite ($NaAsO_2$) in 0.5 N HCl), and mix using a vortex.
6. Add 800 μl TBA reagent (6 mg/ml TBA), and mix using a vortex (*see* **Note 18**).
7. Place in 100 °C heating block for 10 min.
8. Cool to room temperature, add 1.5 ml butanol reagent (5.0 % v/v concentrated HCl; 95 % v/v *n*-butanol), and mix using a vortex.
9. Centrifuge at 400 × *g* for 5 min.
10. Collect the upper butanol layer (~800 μl).
11. Measure absorbance at 552 and 509 nm, respectively. An absorbance difference(A_{552}–A_{509}) of 19 = 1 μM Kdo.
12. To calculate the concentration of LPS, the measured Kdo concentration is divided by the number of Kdo residues in the LPS.

Procedure 2: Kdo Assay Utilizing DMSO [34]

1. To 50 μl of LPS samples or Kdo standards (0, 0.05, 0.1, 0.2, 0.3, 0.4, and 0.5 mM Kdo), add 60 μl of 0.018 N H_2SO_4, and mix using a vortex.
2. Incubate at 100 °C for 20 min to hydrolyze.
3. Cool to room temperature, add 25 μl of periodic acid (9.1 mg/ml in 0.125 N H_2SO_4), and incubate in the dark for 20 min to oxidize.

4. Add 50 μl of arsenite reagent (2.6 % $NaAsO_2$ in 0.5 N HCl). Mix until the yellowish color disappears.
5. Add 250 μl of 0.3 % TBA, and boil at 100 °C for 10 min.
6. While still hot, add 125 μl of DMSO and mix thoroughly.
7. Cool to room temperature, transfer 200 μl to duplicate wells in a 96-well plate, and measure absorbance at A_{550}.
8. Prepare a standard curve; the Kdo concentration can be determined using linear regression.
9. The LPS concentration = The calculated Kdo concentration/ the number of reactive Kdo per LPS (*see* **Note 19**).

3.8 Purpald Assay Method

3.8.1 Concept

Purpald assay was developed for the measurement of formaldehyde and has been used widely in environmental pollution surveillance [35], industry [36], and carbohydrate research [37, 38]. Quesenberry and Lee [39] first established a rapid method using the Purpald reagent (4-amino-3-hydrazino-5-mercapto-1,2,4-triazole) under periodation conditions [39] for quantification of formaldehyde; this was later adapted by Lee and Tsai for quantification of bacterial LPS [34] (*see* **Note 20**). Through oxidation by periodate, the unsubstituted terminal vicinal glycol (UTVG) of the Kdo and heptose residues in LPS produce formaldehyde, which reacts with the Purpald reagent to yield a colorless intermediate. This intermediate is further oxidized by periodate to form a purple product with maximum absorption at 550 nm [34, 39]. The detail of the chemical reaction that leads to the purple product has been illustrated in [34].

3.8.2 Procedure

1. In a 96-well microtiter plate, add 50 μl of LPS samples or standards (0, 0.05, 0.1, 0.2, 0.3, 0.4, and 0.5 mM Kdo) in duplicate wells.
2. Add 50 μl of 32 mM $NaIO_4$, and incubate at room temperature for 25 min.
3. Add 50 μl of 136 mM Purpald reagent in 2 N NaOH, and incubate for 20 min at room temperature.
4. Add 50 μl of 64 mM $NaIO_4$, and incubate for another 20 min.
5. Add 20 μl of 2-propanol which helps to eliminate any foam in the well.
6. Measure the absorbance at A_{550}.
7. Construct a standard curve of the molarity of UTVG against absorbance at A_{550}. The UTVG concentration in the LPS samples can be determined using linear regression from the standard curve.
8. The LPS concentration = The calculated UTVG concentration/the number of UTVG per LPS (*see* **Note 21**).

3.9 LPS Western Blotting and Densitometry Method

3.9.1 Concept

Western blotting has become a standard method for protein analysis starting in the early 1980s [40, 41]. Western blotting of LPS has arisen as one of the most useful analytical techniques, whereby utilization of primary antibodies with exquisite epitope specificity allows both visualization and verification of the LPS components (*see* **Note 22**). Monoclonal antibodies with specificities against different regions of the *P. aeruginosa* LPS have been developed, each with specificity against lipid A, inner core, outer core, common polysaccharide antigen, and the O-specific antigen for all 20 *P. aeruginosa* serotypes [42–45] (Table 1).

Quantification of macromolecules resolved after gel electrophoresis or blotting methods by densitometry has been utilized for protocols such as Southern, Northern, Western, PAGE, and TLC, regardless of whether the detection procedure is via colorimetric, chemiluminescent, radioactive, or fluorescent detection. The main drawback of the quantification of LPS via Western blot densitometry is that there are no commercially available LPS standards that are recognized by the same primary antibody. Therefore, linear regression analysis is very time consuming, as standard curves must be produced using stringently purified and quantified LPS through one or the other methods within this chapter. The densitometry of Western blot technique, however, offers a quick and easy method for the quantification of the change in relative amounts of LPS in comparison to a wild-type LPS control (Fig. 2). This technique was recently used for the comparison between various point mutations of Wzx (O-antigen flippase) against the unmodified Wzx to detect changes in functionality of the protein due to site-specific mutagenesis [7]. Substitutions of any particular amino acid residues in the LPS assembly protein Wzx which caused a partial or a complete loss of O-specific LPS could be analyzed, and the information was useful for understanding the contribution of a certain region of the protein to its structure and function. This technique could be similarly applied to changes in how LPS production is affected by changes in growth conditions and whether a particular strain of *P. aeruginosa* is growing as biofilms or as planktonic cultures.

3.9.2 Procedure

1. Prepare H&B LPS samples, as described in Subheading 2.1.1, in triplicate, to facilitate statistical analysis of the densitometry scans in later steps. Each test will require six H&B preparation samples: three for the standard (e.g., wild type) and three for the sample (e.g., mutant) to be compared to standard (*see* **Note 23**).
2. Load 3 μl of H&B LPS samples into the wells of a 12 % discontinuous PAGE mini-slab gel in a vertical gel electrophoresis system and run at 200 V for 50 min (*see* **Note 24**).

Table 1

Hybridoma-derived monoclonal antibodies available for specific detection of distinct serotypes and specific regions of LPS of *Pseudomonas aeruginosa*

mAb	Epitope specificity	Isotype	Serotype of *P. aeruginosa* that mAb is reactive with	Cross-reactivity	Reference
MF25-1[a]	O Ag	IgM	IATS O1		[47]
MF71-2[a]	O Ag	IgM	IATS O2	IATS O5, O16, O18, O20	[46]
MF57-9[a]	O Ag	IgM	IATS O3		[47]
MF60.5[a]	O Ag	IgM	IATS O4		[46]
MF15-4[a]	O Ag	IgM	IATS O5	IATS O2, O16, O18, O20	[17, 47]
MF23-2[a]	O Ag	IgG3	IATS O6		[47]
MF29-2[a]	O Ag	IgM	IATS O7	IATS O8	[47]
MF30-3[a]	O Ag	IgM	IATS O8	IATS O7	[47]
MF43-3[a]	O Ag	IgM	IATS O9		[47]
MF76-2[a]	O Ag	IgM	IATS O10	IATS O19	[46]
MF55-1[a]	O Ag	IgM	IATS O11		[47]
MF35-4[a]	O Ag	IgM	IATS O12		[47]
MF57-2[a]	O Ag	IgM	IATS O13	IATS O14	[46]
MF45-3[a]	O Ag	IgM	IATS O14	IATS O13	[47]
MF58-2[a]	O Ag	IgG3	IATS O15		[46]
MF47-4[a]	O Ag	IgM	IATS O16	IATS O2, O5, O18, O20	[46]
MF59-2[a]	O Ag	IgM	IATS O17		[46]
18-19[b]	Core + one O-repeat	IgM	IATS O5	IATS O2, O16, O18, O20	[17, 21]
N1F10[b]	Common polysaccharide antigen	IgM	All serotypes except IATSO7, O12, O13, O14, O16		[17, 48]
5c-101[b]	Outer core—(terminal glucose residue)	IgG3	IATS O2, O5, O7, O8, O10, O16, O18, O19, O20		[17, 21]
5c-7-4[b]	Inner core (phosphorylated)	IgG2b	All serotypes	Reactive with other *Pseudomonas* species	[17, 21]

[a]These mAb against each of serotypes O1–O17 can be obtained from a commercial company called MediMabs Inc. (https://www.medimabs.com/)

[b]Note that these antibodies are proprietary from the laboratory of the authors. They are not for sale at present but will be made available to researchers upon request. A small cost-recovery charge may apply

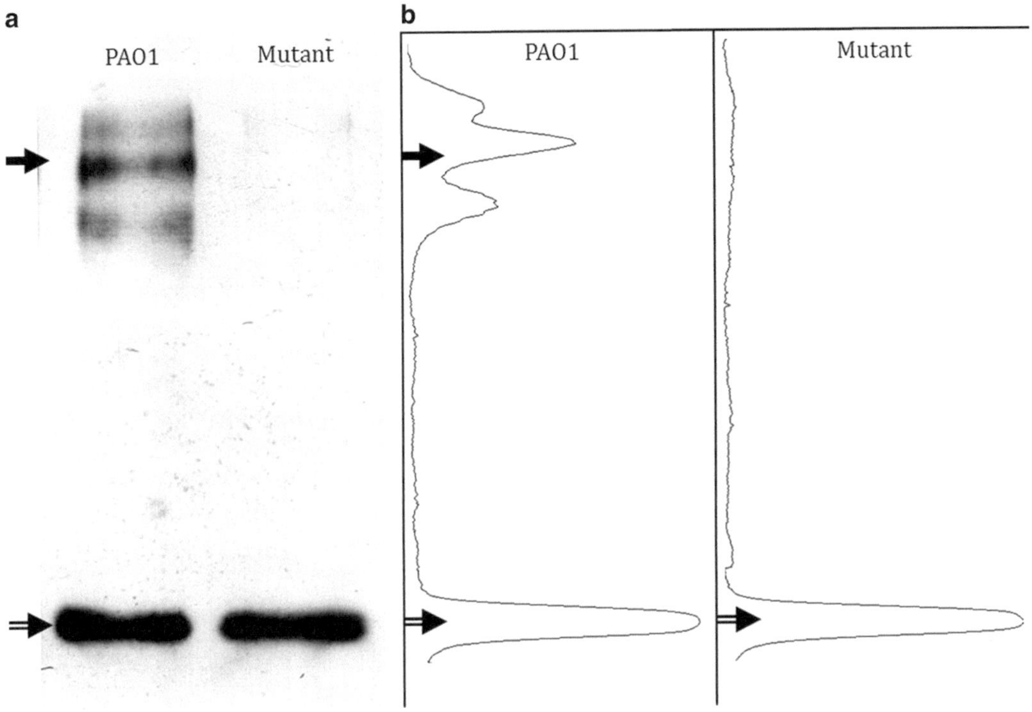

Fig. 2 Western blot densitometry of LPS from *P. aeruginosa* PAO1 wild type and a mutant defective in O-specific antigen biosynthesis. *Panel A*—the Western blot was probed with a mixture of mAb MF15-4 (O-specific) (reactive to high-molecular-weight O–Ag bands, *solid arrow*) and mAb 5c-101 (outer core) (reactive to fast-migrating core oligosaccharide band, *double arrow*), followed by anti-mouse Fab-conjugated alkaline phosphatase secondary antibody, and developed with BCIP/NBT. The mutant shows a complete loss of O-specific antigen in comparison to the wild-type PAO1 LPS profile. Core density remains equivalent in the mutant and PAO1 and functions as a loading control. *Panel B*—the associated histogram of the densitometry scans for each lane of the Western blot, illustrating the regions for densitometry analysis

3. Electrophoretically transfer LPS from polyacrylamide gel onto nitrocellulose membrane at 180 mA for 60 min. Make sure that the apparatus does not become warm; one may need to pack ice in the external compartment of the blotting device.
4. Block membrane with 5 % skim milk in PBS, for 20 min at room temperature, shaking gently on a platform shaker. Be sure to always keep membrane completely submerged as drying will produce mottled and higher background staining.
5. Wash three times in PBS for 10 min at room temperature, shaking gently.
6. Add primary antibodies, if needed, and dilute in PBS (*see* **Note 25**).
7. Incubate overnight, shaking at room temperature (*see* **Note 26**).
8. Wash three times in PBS for 10 min at room temperature, shaking gently.

9. Add secondary antibodies (e.g., goat-anti-mouse–alkaline phosphatase conjugate) diluted in PBS according to the manufacturer's instructions.
10. Incubate for 1 h, shaking gently at room temperature. Use a longer incubation time if required.
11. Wash two times in PBS, for 10 min at room temperature, shaking gently.
12. Wash once in buffer A, for 10 min at room temperature, shaking gently.
13. Develop using 5 % BCIP and 5 % NBT in buffer A at room temperature, shaking gently. Allow enough time to develop a sufficiently dark staining of the appropriate banding, anywhere from 30 s to 5 min. Extended development time can result in increased background staining.
14. Wash with water to stop development.
15. Scan each Western blot using a densitometer with CCD camera (*see* **Note 27**).
16. Using computer software, such as ImageJ, MCID Analysis, or Quantity One, highlight regions containing similar LPS banding patterns, such as outer core or O-specific regions. Care must be taken to isolate individual regions to ensure proper reading of the sample and background. Always utilize the same box for highlighting similar areas.
17. Calculate the density profile, peak height, peak intensity (average OD of the band, INT), or peak volume (average OD of the band times its area, INT*mm^2) for each LPS region (core or O-specific region). Choice of calculated value depends on program choice. Quantity One only allows calculation of INT or INT*mm^2. INT is preferable over INT*mm^2 [46].
18. Subtract background density to obtain the adjusted density (AD). Each program handles background subtraction differently.
19. For each lane, take the ratio of the AD of the O-specific region compared to the core region. Utilizing the O-specific to core ratio for further calculations equilibrates for differences in loading:

$$\textit{Adjusted density ratio}\left(ADR\right) = AD\ of\ O\quad specific\ /\ AD\ of\ core$$

20. Calculate the within-blot average-adjusted density ratio (AADR) of the standard using the following formula:

$$\begin{aligned}&\textit{Average adjusted density ratio}\left(AADR\right)\\&= Average\left(AD\ of\ O\ specific\ /\ AD\ of\ core\right)\ of\ triplicate\end{aligned}$$

For each sample ADR, calculate the percent change (%C) from the standard AADR:

$$\%C = (Sample\ ADR\ /\ standard\ AADR) \times 100$$

21. Repeat **steps 4–10** for each of the triplicate Western blots.
22. Calculate the average %C for the triplicate (H&B preparations) of triplicate (Western blot) reactions.
23. Calculate the standard deviation (s) and/or standard error (SE_x) for the triplicate of triplicate reactions:

$$s = \sqrt{\frac{1}{N-1}\sum_{i=1}^{N}(x_i - \bar{x})^2} \quad SE_{\bar{x}} = \frac{s}{\sqrt{n}}$$

where x_i = ADR, $\bar{x}$ = AADR , N = number of samples, and n = number of replicates.

24. Calculate the significance of the %C using statistical analysis, ANOVA, or *t*-test (*see* **Notes 28–30**).

4 Notes

1. Alternative: Samples can be taken directly from agar medium culture and suspended in PBS or LB, and proceed with **step 3**.
2. When bacteria are suspended in the lysis buffer, the ionic detergent SDS disrupts cellular membranes to facilitate the release of LPS from the cell envelope; protein contaminants are degraded by the addition of 30 μg proteinase K. Peptides and other contaminants, such as DNA/RNA and various sugars moieties, remain in the LPS preparation. By virtue of the silver staining of the SDS-PAGE gels that has been optimized to visualize the LPS bands, resolved LPS bands are easily visible (*see* Subheading 2.2.3). Hence this protocol is generally regarded as a "quick and dirty" LPS extraction method. As a result, this method may not be suitable for precise structural or quantitative determination of the LPS.
3. Concerning storage of H&B-prepared LPS samples, the frozen samples may lose clarity after prolonged storage; if so, reheat samples at 100 °C for 10 min before applying them to the SDS-PAGE gels.
4. This protocol is designed for working with ca. 500 mg of dry cell pellet or the equivalent of approximately 1 l culture with a cell density OD_{600} at 0.6–0.8. The protocol can be modified for different dry pellet weights by adjusting the volume of buffer added in each step.
5. Chill the centrifuge bottle to <15 °C; as the temperature drops, the mixture partitions into two phases. The pre-chilling will also help prevent collapse of the centrifuge bottle (if phenol-resistant polycarbonate bottles are used) in the centrifuge due to cooling-air contraction.

6. $MgCl_2$ is required for DNase to be active. The optimum concentration for the best enzyme activity is 5–10 mM $MgCl_2$.
7. If it is difficult to resuspend the crude LPS, it could be due to Mg^{2+}-induced changes to LPS micelle size. To aid LPS solubility, add EDTA to match the molarity of Mg^{2+}, chill sample on ice for 1 h, and then centrifuge at 10,000 × *g* for 10 min to remove insoluble EDTA–Mg^{2+} complexes.
8. The PCP extraction mix should be prepared fresh and used immediately each time.
9. After the rotary evaporation step, PCP-extracted LPS should be transferred to a Pyrex/Corning glass centrifuge bottle because plastic centrifuge bottles may not be resistant to phenol.
10. The objective of this step is to add enough water to start precipitating the LPS without forming two distinct phases. If one is concerned about adding too much water, a total of 8–10 drops of water could be added, and then centrifuge the mixture to collect the first LPS pellet. The supernatant can be transferred to a new tube, and more water can be added without worrying about losing the first sample.
11. The addition of DOC in the NH_4HCO_3 buffer will help the purified LPS molecules to be monodispersed instead of aggregating into clumps.
12. Detergents, salts, chemicals, antibiotics, antibodies, and proteins can significantly diminish or enhance LAL activity [47–50]. In addition, (1-3)-β-D-glucan, an integral component of fungal cell walls, can activate factor G contained in LAL which can stimulate the protease cascade reaction causing a false-positive reaction [51]. Therefore, precautions must be taken to ensure proper handling, dilution, and controls. The manufacturer, Associates of Cape Code Inc. (East Falmouth, MA, USA), has produced specific additives that can decrease interference and increase sensitivity down to 0.001 endotoxin units (EU)/ml. Conversion from EU to ng can vary depending on the LAL source. All materials should be depyrogenated (by heating at 250 °C for 30 min) prior to use, and great care must be taken to eliminate endotoxin contamination throughout the protocol.
13. The chromatic and turbidimetric assays can be analyzed through endpoint or kinetic methods. The kinetic assay monitors activity over time and the rate of the change (slope) or the time to reach a specific OD. Endpoint analysis allows the reaction to occur for a set time period; the reaction can be quenched by the addition of 10 % acetic acid or SDS and the substrate conversion measured.

14. For kinetic assay "reading" will be the rate of enzyme activity measured as the slope of the change in OD/fluorescence reading over time. Make sure to exclude the lag and plateau phase of the sigmoidal curve as part of the calculation of the slope. For kinetic time assay: "reading" will be the time to change a preset amount, e.g., 0.05–0.20 OD above background. Ensure that this reading is within the exponential region of the sigmoidal curve for all standard concentrations. This is the typical analysis performed by purchased LAL-specific kinetic programs. For endpoint assay "reading" will be the OD or the fluorescence reading at the final time point. The correlation coefficient $(r) \geq 0.98$.
15. Kdo functional group composition-, incomplete hydrolysis-, or acid hydrolysis-induced modifications have been noted to interfere with the TBA reaction [52, 53]. If one intends to use this method to quantitate LPS extracted from species other than *Pseudomonas*, one should note that Kdo residues in some species (e.g., *Haemophilus influenzae* [54–56], *Bordetella pertussis* [55, 57, 58], and *Vibrio cholerae* [27, 57]) contain substitutions at the C4 or the C5 position which are not susceptible to acid digestion, thus limiting or completely abolishing detection by the use of the Kdo assay.
16. The concentration and volume of the reagents used in this protocol are suitable for using either test tubes or cuvettes.
17. This step is omitted for Kdo standards. For LPS samples, hydrolysis time can be changed to obtain maximal release of Kdo. For *Pseudomonas* species and *E. coli*, 8 min is usually sufficient. For more refractory LPS samples (e.g., *Burkholderia cepacia*), 12–15 min is required.
18. For preparation of the TBA reagent, which should be made up fresh daily, warming should help to more efficiently dissolve the chemical.
19. Due to substitution at the C4, C5, C6, and C7 positions of Kdo, the number of reactive Kdo (free, unsubstituted Kdo after acid hydrolysis) is not the same as the total number of Kdo in LPS molecule. So the structure of the LPS investigated needs to be known to decide the number of reactive Kdo.
20. The Purpald assay was reported to have similar sensitivity for quantification of LPS as the Kdo assay [34]. This assay has a number of advantages. First, unlike the Kdo assay which requires heating of the sample under acidic conditions, the Purpald assay is performed at room temperature and does not require acid hydrolysis. Second, since the Purpald assay reacts with the UTVG in both Kdo and heptose residues in LPS, it can be used for quantification of LPS from *Haemophilus influenza* [54–56], *Bordetella pertussis* [55, 57, 58], and *Vibrio*

cholerae [27, 57] that contain substitution or modifications which are undetectable by Kdo assay. However, there is a caveat: since the UTVG group is not exclusive to the LPS molecule in bacteria, the Purpald assay is less specific for detection of LPS compared to the Kdo assay.

21. The number of UTVG per LPS molecule is calculated from the Kdo and heptose residues within the LPS. Theoretically, each free unsubstituted Kdo or heptose produces two UTVG. However, due to partial substitution at the C6 and C7 positions of heptose, and at the C4, C5, C6, and C7 positions of Kdo, the theoretical number of UTVG per LPS molecule is usually not simply twice of the number of heptose plus Kdo in the LPS.
22. The main difference in the protocols for transfer of proteins and LPS using the Western blotting technique is the slight alteration in buffers and the specificity of the antibodies.
23. To obtain accurate densitometry data for statistical analysis, it is important to load each H&B preparation in triplicate within a polyacrylamide gel and generate triplicate blots from three separate polyacrylamide gels blotted. Also, it is important to ensure that the LPS banding patterns are well resolved with clearly visible bands.
24. In choosing the type of polyacrylamide gel for LPS separation, Tris-glycine-polyacrylamide gels are commonly used and are ideal for separating O-specific regions. However, Tris-tricine-polyacrylamide gels are ideal for separating core regions because they provide better resolution of low-molecular-weight LPS bands [59].
25. For densitometry, use primary antibodies against both O-specific region and core region. This will allow equilibration of loading by comparison of the core region densities, which should remain constant across samples. Of course, care must be taken if mutations or effects in the core region are expected.
26. For stronger binding antibodies, incubation from one–several hours at room temperature is sufficient. To reduce higher background staining, 0.2 % Triton X-100 or 0.05 % Tween 20 can be added, but only for the strongest binding antibodies. Addition of detergents may affect the ability of weaker binding antibodies to bind to respective antigens and could result in no signal development in later stages.
27. Use a scanner dedicated for collecting densitometry data. Do not use typical flat-bed scanners purchased from common computer outlets as such instruments may not have the resolution required and they modify images to obtain an ideal density range [46].

28. Statistical analysis: Western blot detection of LPS is highly variable, and each blot will vary in signal intensity. Therefore, statistical analysis as well as within- and between-blot controls for variation are highly important. For comparison between two samples, the student's *t*-test is sufficient. Comparison of three or more samples will require an ANOVA and post hoc analysis.
29. The student's *t*-test assumes a normal–(*t*)-distribution (bell shaped) around the mean and requires equal sample sizes. The *t*-test indicates significant differences between the sample means when the *t*-value <0.05, i.e., there is less than 5 % chance the two populations are the same. The *t*-test is highly susceptible to type 1 statistical errors and should not be utilized for comparison of greater than two populations due to increasing probabilities of obtaining a false result.
30. For Western quantitation, a one-way ANOVA F-test should be performed. A significant difference between populations is observed if the calculated F_{value} (between-group variability) is above the $F_{critical}$ (within-group variability). To determine which of the populations are producing any observed significant differences, a post hoc test must be performed such as Scheffe and Tukey. The ANOVA-type method is a contemporary statistical method that when used in circumstances of high variance, multiple outliers in the data, or populations that are not normally distributed will limit type 1 errors better than conventional ANOVA [60].

Acknowledgements

Research in the laboratory of J.S.L. is funded by operating grants from the Canadian Institutes of Health Research (CIHR) (MOP-14687) and Cystic Fibrosis Canada (CFC). Y.H. is the recipient of a CFC postdoctoral fellowship. J.S.L. holds a Canada Research Chair in Cystic Fibrosis and Microbial Glycobiology.

References

1. Westphal O, Jann K (1965) Bacterial lipopolysaccharides. Extraction with phenol-water and further applications of the procedure. In: Whistler RL (ed) Methods in carbohydrate chemistry. Academic Press, New York, pp 83–91
2. Hickman J, Ashwell G (1966) Isolation of a bacterial lipopolysaccharide from *Xanthomonas campestris* containing 3-acetamido-3,6-dideoxy-D-galactose and D-rhamnose. J Biol Chem 241:1424–1428
3. Kasai N, Nowotny A (1967) Endotoxic glycolipid from a heptoseless mutant of *Salmonella minnesota*. J Bacteriol 94:1824–1836
4. Galanos C, Luderitz O, Westphal O (1969) A new method for the extraction of R lipopolysaccharides. Eur J Biochem 9:245–249
5. Darveau RP, Hancock RE (1983) Procedure for isolation of bacterial lipopolysaccharides from both smooth and rough *Pseudomonas aeruginosa* and *Salmonella typhimurium* strains. J Bacteriol 155:831–838

6. Hitchcock PJ, Brown TM (1983) Morphological heterogeneity among *Salmonella* lipopolysaccharide chemotypes in silver-stained polyacrylamide gels. J Bacteriol 154:269–277
7. Islam ST, Fieldhouse RJ, Anderson EM, Taylor VL, Keates RA, Ford RC, Lam JS (2012) A cationic lumen in the Wzx flippase mediates anionic O-antigen subunit translocation in *Pseudomonas aeruginosa* PAO1. Mol Microbiol 84:1165–1176
8. Fomsgaard A, Freudenberg MA, Galanos C (1990) Modification of the silver staining technique to detect lipopolysaccharide in polyacrylamide gels. J Clin Microbiol 28: 2627–2631
9. Tsai CM, Frasch CE (1982) A sensitive silver stain for detecting lipopolysaccharides in polyacrylamide gels. Anal Biochem 119:115–119
10. Hardy E, Pupo E, Castellanos-Serra L, Reyes J, Fernandez-Patron C (1997) Sensitive reverse staining of bacterial lipopolysaccharides on polyacrylamide gels by using zinc and imidazole salts. Anal Biochem 244:28–32
11. Kido N, Ohta M, Kato N (1990) Detection of lipopolysaccharides by ethidium bromide staining after sodium dodecyl sulfate-polyacrylamide gel electrophoresis. J Bacteriol 172:1145–1147
12. Kittelberger R, Hilbink F (1993) Sensitive silver-staining detection of bacterial lipopolysaccharides in polyacrylamide gels. J Biochem Biophy Meth 26:81–86
13. Redwan EM (2012) Simple, sensitive, and quick protocol to detect less than 1 ng of bacterial lipopolysaccharide. Prep Biochem Biotech 42:171–182
14. Zhu ZX, Cong WT, Zhou X, Chen M, Choi JK, Jin LT, Li XK (2012) Negative staining of lipopolysaccharides on polyacrylamide gels by using eosin B. Anal Biochem 426:1–3
15. Hancock RE, Mutharia LM, Chan L, Darveau RP, Speert DP, Pier GB (1983) *Pseudomonas aeruginosa* isolates from patients with cystic fibrosis: a class of serum-sensitive, nontypable strains deficient in lipopolysaccharide O side chains. Infect Immun 42:170–177
16. Wang X, Zhang C, Shi F, Hu X (2010) Purification and characterization of lipopolysaccharides. Subcell Biochem 53:27–51
17. Forsberg LS, Carlson RW (1998) The structures of the lipopolysaccharides from *Rhizobium etli* strains CE358 and CE359. The complete structure of the core region of R. etli lipopolysaccharides. J Biol Chem 273: 2747–2757
18. Ding JL, Navas MA 3rd, Ho B (1995) Molecular cloning and sequence analysis of factor C cDNA from the Singapore horseshoe crab, *Carcinoscorpius rotundicauda*. Mol Mar Biol Biotechnol 4:90–103
19. Lindsay GK, Roslansky PF, Novitsky TJ (1989) Single-step, chromogenic *Limulus* amebocyte lysate assay for endotoxin. J Clin Microbiol 27:947–951
20. Noda K, Goto H, Murakami Y, Ahmed AB, Kuroda A (2010) Endotoxin assay by bioluminescence using mutant firefly luciferase. Anal Biochem 397:152–155
21. Bang FB (1956) A bacterial disease of *Limulus polyphemus*. Bull Johns Hopkins Hosp 98:325–351
22. Iwanaga S, Miyata T, Tokunaga F, Muta T (1992) Molecular mechanism of hemolymph clotting system in *Limulus*. Throm Res 68:1–32
23. Williams KL (2001) LAL discovery, mechanism and application. In: Williams KL (ed) Endotoxins. Pyrogens, LAL testing, and depyrogenation, 2nd edn. Revised and expanded, Marcel Dekker Inc., New York, pp 178–203
24. Kawasaki H, Nose T, Muta T, Iwanaga S, Shimohigashi Y, Kawabata S (2000) Head-to-tail polymerization of coagulin, a clottable protein of the horseshoe crab. J Biol Chem 275:35297–35301
25. Teller JD, Kelly KM (1979) A turbidimetric *Limulus* amebocyte assay for the quantitative determination of Gram negative bacterial endotoxin. Prog Clin Biol Res 29:423–433
26. Unger FM (1981) The chemistry and biological significance of 3-deoxy-D-manno-2-octulosonic acid (KDO). Adv Carbohydr Chem Biochem 38:323–388
27. Brade H (1985) Occurrence of 2-keto-deoxyoctonic acid 5-phosphate in lipopolysaccharides of *Vibrio cholerae* Ogawa and Inaba. J Bacteriol 161:795–798
28. Hurlbert RE, Weckesser J, Mayer H, Fromme I (1976) Isolation and characterization of the lipopolysaccharide of *Chromatium vinosum*. Eur J Biochem 68:365–371
29. Weissbach A, Hurwitz J (1959) The formation of 2-keto-3-deoxyheptonic acid in extracts of *Escherichia coli* B. I. Identification. J Biol Chem 234:705–709
30. Osborn MJ (1963) Studies on the Gram-negative cell wall. I. Evidence for the role of 2-keto-3-deoxyoctonate in the lipopolysaccharide of *Salmonella Typhimurium*. Proc Nat Acad Sci U S A 50:499–506
31. Warren L (1959) The thiobarbituric acid assay of sialic acids. J Biol Chem 234:1971–1975
32. Karkhanis YD, Zeltner JY, Jackson JJ, Carlo DJ (1978) A new and improved microassay to determine 2-keto-3-deoxyoctonate in lipo-

polysaccharide of Gram-negative bacteria. Anal Biochem 85:595–601

33. Skoza L, Mohos S (1976) Stable thiobarbituric acid chromophore with dimethyl sulphoxide. Application to sialic acid assay in analytical de-O-acetylation. Biochem J 159:457–462

34. Lee CH, Tsai CM (1999) Quantification of bacterial lipopolysaccharides by the purpald assay: measuring formaldehyde generated from 2-keto-3-deoxyoctonate and heptose at the inner core by periodate oxidation. Anal Biochem 267:161–168

35. Nondek L, Rodler DR, Birks JW (1992) Measurement of sub-ppbv concentrations of aldehydes in a forest atmosphere using a new HPLC technique. Environ Sci Tech 26:1174–1178

36. Jacobson NW, Dickinson RG (1974) Spectrometric assay of aldehydes as 6-mercapto-3-substituted-S-triazolo(4,3-B)-S-tetrazines. Anal Chem 46:298–299

37. Guthrie RD (1962) Periodate oxidation. In: Whistler RL, Wolfrom ML (eds) Methods in carbohydrate chemistry. Academic Press, New York, pp 432–435

38. Speck JCJ (1962) Advances in carbohydrate chemistry. In: Whistler RL (ed) Methods in carbohydrate chemistry. Academic Press, New York, pp 441–445

39. Quesenberry MS, Lee YC (1996) A rapid formaldehyde assay using purpald reagent: application under periodation conditions. Anal Biochem 234:50–55

40. Burnette WN (1981) "Western blotting": electrophoretic transfer of proteins from sodium dodecyl sulfate–polyacrylamide gels to unmodified nitrocellulose and radiographic detection with antibody and radioiodinated protein A. Anal Biochem 112:195–203

41. Towbin H, Staehelin T, Gordon J (1979) Electrophoretic transfer of proteins from polyacrylamide gels to nitrocellulose sheets: procedure and some applications. Proc Nat Acad Sci U S A 76:4350–4354

42. de Kievit TR, Lam JS (1994) Monoclonal antibodies that distinguish inner core, outer core, and lipid A regions of *Pseudomonas aeruginosa* lipopolysaccharide. J Bacteriol 176: 7129–7139

43. Lam JS, MacDonald LA, Lam MY (1987) Production of monoclonal antibodies against serotype strains of *Pseudomonas aeruginosa*. Infect Immun 55:2854–2856

44. Lam JS, MacDonald LA, Lam MY, Duchesne LG, Southam GG (1987) Production and characterization of monoclonal antibodies against serotype strains of *Pseudomonas aeruginosa*. Infect Immun 55:1051–1057

45. Lam MY, McGroarty EJ, Kropinski AM, MacDonald LA, Pedersen SS, Hoiby N, Lam JS (1989) Occurrence of a common lipopolysaccharide antigen in standard and clinical strains of *Pseudomonas aeruginosa*. J Clin Microbiol 27:962–967

46. Gassmann M, Grenacher B, Rohde B, Vogel J (2009) Quantifying Western blots: pitfalls of densitometry. Electrophoresis 30:1845–1855

47. Huszar G, Jenei B, Szabo G, Medgyesi GA (2002) Detection of pyrogens in intravenous IgG preparations. Biologicals 30:77–83

48. Ochiai M, Yamamoto A, Kataoka M, Toyoizumi H, Horiuchi Y (2001) Interfering effect of diphtheria-tetanus-acellular pertussis combined (DTaP) vaccines on the bacterial endotoxin test. Biologicals 29:55–58

49. Pfeiffer M, Weiss AR (1987) Removal of LAL-test interfering low molecular weight substances by ultrafiltration. Prog Clin Biol Res 231:251–262

50. Tsai CM, Frasch CE, Rivera E, Hochstein HD (1989) Measurements of lipopolysaccharide (endotoxin) in meningococcal protein and polysaccharide preparations for vaccine usage. J Biol Stand 17:249–258

51. Roslansky PF, Novitsky TJ (1991) Sensitivity of *Limulus* amebocyte lysate (LAL) to LAL-reactive glucans. J Clin Microbiol 29:2477–2483

52. Brade H, Galanos C, Luderitz O (1983) Differential determination of the 3-deoxy-D-mannooctulosonic acid residues in lipopolysaccharides of *Salmonella minnesota* rough mutants. Eur J Biochem 131:195–200

53. Greenberg ID, McNicholas P (1987) Positional changes in a hemangioma demonstrated by red blood cell imaging. Clin Nuc Med 12:668–670

54. Helander IM, Lindner B, Brade H, Altmann K, Lindberg AA, Rietschel ET, Zahringer U (1988) Chemical structure of the lipopolysaccharide of *Haemophilus influenzae* strain I-69 Rd-/b+. Description of a novel deep-rough chemotype. Eur J Biochem 177:483–492

55. Wu LH, Tsai CM, Frasch CE (1987) A method for purification of bacterial R-type lipopolysaccharides (lipooligosaccharides). Anal Biochem 160:281–289

56. Zamze SE, Ferguson MA, Moxon ER, Dwek RA, Rademacher TW (1987) Identification of phosphorylated 3-deoxy-manno-octulosonic acid as a component of *Haemophilus influenzae* lipopolysaccharide. Biochem J 245: 583–587

57. Carof M, Lebbar S, Szabo L (1987) Do endotoxins devoid of 3-deoxy-D-manno-2-octulosonic acid exist? Biochem Biophy Res Commun 143:845–847

58. Carof M, Lebbar S, Szabo L (1987) Detection of a 3-deoxy-2-octulosonic acid in thiobarbiturate-negative endotoxins. Carbohydr Res 161:C4–C7

59. Lesse AJ, Campagnari AA, Bittner WE, Apicella MA (1990) Increased resolution of lipopolysaccharides and lipooligosaccharides utilizing tricine-sodium dodecyl sulfate-polyacrylamide gel electrophoresis. J Immunol Meth 126:109–117

60. Wilcox RR (2003) Applying contemporary statistical techniques. Academic, San Diego, CA

61. Osaki T, Kawabata S (2004) Structure and function of coagulogen, a clottable protein in horseshoe crabs. Cell Mol Life Sci 61: 1257–1265

Chapter 32

Monitoring Lectin Interactions with Carbohydrates

Sophie de Bentzmann, Annabelle Varrot, and Anne Imberty

Abstract

Protein–carbohydrate interactions are often involved in the first step of infection and *Pseudomonas aeruginosa* produces several proteins that are able to bind specifically to glycan epitopes present on host epithelia. The experimental approaches for studying protein–carbohydrate interaction have been inspired, with some adaptations, from those commonly used for protein–protein or protein–ligand interactions. A range of methods are described herein for detecting lectin activity, screening for monosaccharide or oligosaccharide specificity, determining the affinity of binding together with thermodynamics and kinetics parameters, and producing crystal of lectin–carbohydrate complexes for further structural studies.

Key words Lectins, Adhesins, Glycans, Interactions

1 Introduction

All organisms are covered with an array of free or covalently attached glycans that interact with receptors, cells, and microorganisms [1]. These glycans that present a wide range of structural diversity are often involved in the first step of pathogen infection since they can mediate specific host recognition and tissue adhesion [2–4]. The role of human glycosylation in infection has been well characterized in infections caused by viruses, for example Norwalk virus which specifically recognizes histo-blood group antigens (HBGAs) [5], bacteria such as uropathogenic *Escherichia coli* (ETEC) [6], or parasites such as *Toxoplasma gondii* [7]. Among molecules able to recognize and bind to glycans are lectins. In bacteria, the lectins may belong to different families gathering soluble lectins, proteins like toxins in which subdomains are able to bind sugars [8] or insoluble proteins part of multiprotein architecture appendages such as pili and flagella. In that way, *Pseudomonas aeruginosa* is a good example of a human pathogen, exhibiting a wide range of those types of proteins (Fig. 1), including two soluble lectins LecA and LecB [9, 10], five lectin-like proteins named adhesins assembled in Type 1 pili [11], one in Type IVa pili [12]),

Alain Filloux and Juan-Luis Ramos (eds.), *Pseudomonas Methods and Protocols*, Methods in Molecular Biology, vol. 1149, DOI 10.1007/978-1-4939-0473-0_32, © Springer Science+Business Media New York 2014

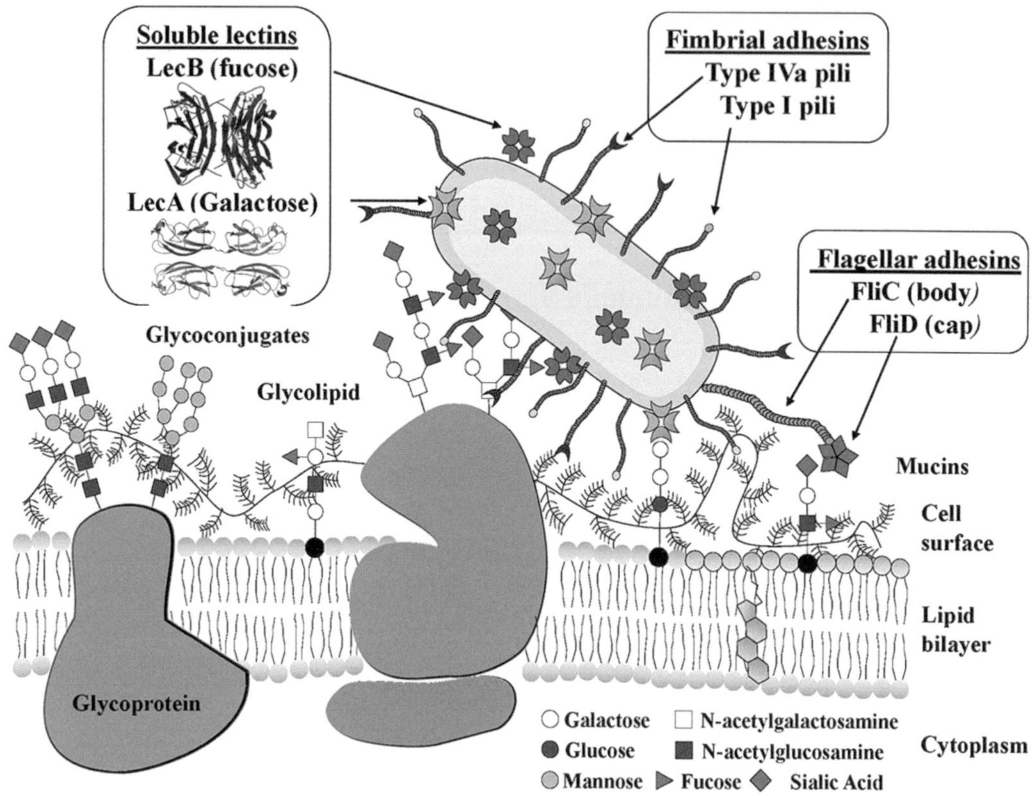

Fig. 1 Schematic representation of lectins and adhesins of *Pseudomonas aeruginosa* interacting with glycoconjugates on host surface

and the two flagellar proteins, the flagellin FliC and the flagellar cap protein FliD, from the unique polar flagellum [13, 14].

LecA and LecB are two soluble lectins specific for galactose and fucose, respectively. They are mainly intracellular but can be found on the outer membrane. They have been proposed to play a role in the virulence and in the formation or maturation of the biofilm [15, 16]. LecA appeared to be most toxic for airway cells and to cause lung tissue damages in murine model of infection [17]. These tetrameric proteins bind monosaccharides through one (LecA) or two calcium ions (LecB) [18–20]. Their interactions with human oligosaccharide epitopes have been studied at a molecular level [21–23]. These studies served as the basis for designing and testing glycomimetics with high affinity towards LecA [24–26] and LecB [27, 28]. Adhesins are more challenging to study at a molecular level since they are part of organelles and are difficult to produce in soluble form or to obtain in monodisperse solution.

Here are described the methodologies that can be used to produce LecA and to characterize its interactions with carbohydrate ligands. The methodologies used to test biological properties (natural carbohydrate binding or competitors such as glycomimetics) of this soluble lectin include hemagglutination assays, Enzyme-Linked Lectin assays (ELLA), Glycan-array assays, Isothermal Titration Microcalorimetry (ITC), Surface Plasmon Resonance (SPR), and Thermal Shift Assays. The protocols described herein can be easily adapted to characterize the other carbohydrate-binding lectins from *P. aeruginosa*.

2 Materials

2.1 Production of Recombinant Lectin

1. The plasmid *pa1l* coding the LecA protein is obtained according to previously published protocol [21]. *Escherichia coli* BL21 (DE3) cells (from pre-culture dish) harboring the pET25-*pa1l* plasmid are grown in 2×1 L of Luria broth medium (LB) (supplemented with 100 μg/mL of ampicillin) at 37 °C. When the culture reaches an optical density of 0.6 at 600 nm, isopropyl-β-D-thiogalactopyranoside (IPTG) is added to a final concentration of 0.5 mM. The cells are harvested (5,000×*g*, 20 min) after further 3 h incubation at 30 °C, washed, and resuspended in 30 mL of the loading buffer (20 mM Tris–HCl and 100 μM $CaCl_2$, pH 7.5). The cells are broken by cell disruption at 1.9 kbars (Constant Cell Disruption System, UK).
2. After centrifugation at 50,000×*g* for 30 min, the supernatant is purified by affinity chromatography on Sepharose 4B (GE Healthcare) equilibrated with Tris–HCl 20 mM pH 7.5. LecA is then eluted with 1 M NaCl in loading buffer.
3. The purified protein (about 20–30 mg/2 L) is intensively dialyzed against distilled water for 4 days, lyophilized, and kept at −20 °C.

2.2 Hemagglutination Assays

1. U-shaped 96-well plates (Nunc, Rochester NY).
2. Rabbit erythrocytes (Biomérieux, Lyon).
3. Tris 20 mM pH 7.5 100 mM NaCl, 100 μM $CaCl_2$. (This is the Tris buffer that is used in most experiments) (*see* **Note 1**).

2.3 Enzyme-Linked Lectin Assays

1. 96-Well plates Maxisorb (Nunc, Rochester, NY).
2. β-PAA-galactose (Lectinity, Moscow), phosphate buffer tablets, bovine serum albumin (BSA), sodium carbonate, and Tween 20 for plate preparation.
3. Biotinamidohexanoyl-6-aminohexanoic acid *N*-hydroxysuccinimide ester (Sigma-Aldrich) for lectin biotinylation.
4. Streptavidin peroxidase labelled (Sigma) and o-phenylenediamine (OPD) (Sigma) for detection of biotinylated lectin.
5. ELISA plate-reader Bio-Rad 680.

2.4 Glycan-Array Assays

1. Alexa-fluor 488 (Fisher) and buffer indicated in the manufacturer sheet.
2. Polyacrylamide desalting column 6 K (Fisher).

2.5 Isothermal Titration Microcalorimetry

Microcalorimeter VP ITC Microcal (GE Healthcare).

2.6 Surface Plasmon Resonance

1. Biacore CM5 chips (GE Healthcare).
2. Biotinylated β-PAA-sugars (Lectinity), streptavidin, EDC/NHS solution, Acetate buffer (pH 5) for functionalization of the chips and ethanolamine for blocking the sensor surface.
3. HEPES 10 mM, NaCl 150 mM, $CaCl_2$ 10 mM, Tween P20 0.005 %, pH 7.4 as running buffer.
4. D-galactose, 100 mM in running buffer for chips regeneration.
5. Biacore 3000 (GE Healthcare).

2.7 Thermal Shift Assays

1. Low-profile 48 well PCR Multiplates and Optical Flat 8-cap strips (Bio-Rad).
2. SYPRO Orange (Sigma).
3. MiniOpticon MJ Mini Personal thermal cycler (Bio-Rad).

2.8 Crystallization

1. 24 or 48 well VDX plates plus sealant (Hampton research).
2. Siliconized coverslips.
3. Commercial screening kits: Clear Strategy Screen I/II, Morpheus, Midas (Molecular Dimension Ltd), crystal screen I and II (Hampton Research) in first intention.
4. Stock solutions of ultrapure compounds in MilliQ water and filter sterilized on 0.2 μm.
5. Mounted cryoloops.
6. X-taLight™ 100—Microscope UV (Molecular Dimension Ltd).
7. Microscope SZX-ILLK200 (Olympus).
8. Liquid nitrogen and cryogenic dewars.

3 Methods

3.1 Determination of Lectin Activity

3.1.1 Hemagglutination Assays

Lectins are often referred to as hemagglutinins because of their capacity to bind to red cell glycans and to reticulate them through multivalent interactions. Hemagglutination is therefore a simple, sensitive, and quasi-quantitative assay for detecting lectin activity (*see* **Note 2**).

1. Dilute rabbit erythrocytes as 4 % solution in NaCl 150 mM.
2. Deposit 25 μL of lectin solution (2 mg/mL in Tris buffer) in a 96 wells of a U-shape microtiter plate with sequential (twofold) lectin dilutions.

3. Deposit 25 μL of the 4 % erythrocyte solution in each well and incubate 1 h at 25 °C.
4. Determine hemagglutination unit (HU) as the minimum concentration of lectin required to observe hemagglutination.

3.1.2 Thermal Shift Assays

1. Thermal shift assays are based on the monitoring of protein thermal unfolding with use of environmentally sensitive fluorescent dye. The ligand-binding affinity can be evaluated from the shift of the unfolding temperature obtained by the stabilization in presence of ligand (*see* **Note 3**).
2. Deposit 23 μL of lectin–glycan complex solution (2 mg/mL lectin and 10 mM glycan in Tris buffer) in 48 well PCR microtiter plates.
3. Add 2 μL of SYPRO Orange 50× and seal the plate with the optical flat cap strip (*see* **Note 4**).
4. Heat the plate from 20 to 100 °C in increments of 1 °C per min using the MiniOpticon Real-Time PCR Detection System that records the change in fluorescence.
5. Determine the denaturation temperature (T_m) of the complex by analyzing the variation of fluorescence as a function of temperature with the CFX Manager™ software.

3.2 Determination of Lectin Specificity

Inhibition of hemagglutination is a simple and rapid way to screen for carbohydrate with specificity for a given lectin.

3.2.1 Inhibition of Hemagglutination

1. Prepare a lectin solution in Tris buffer at a concentration corresponding to four times the HU determined in Subheading 3.1.1. Incubate 12.5 μL of this lectin solution with 12.5 μL of sequential dilutions of glycans for 2 h at 25 °C.
2. Add 25 μL of 4 % erythrocyte solution and incubate at 25 °C for 30 min. The minimum inhibitory concentration (MIC) for each glycan is the minimum concentration of glycan required to disrupt hemagglutination.

3.2.2 Enzyme-Linked Lectin Assays

The enzyme-linked lectin assay (ELLA) approach is based on modifications of the enzyme-linked immunosorbent (ELISA) test. This method permits to study the lectin specificity by determining the ability of a soluble compound to inhibit interaction between a lectin and a reference ligand.

1. Prepare 96-well microtiter plates by incubating 100 μL of 5 μg/mL solution of β-PAA-Gal in carbonate buffer, pH 9.6 for 1 h at 37 °C then block at 37 °C for 1 h with 100 μL of 3 % (w/v) BSA in PBS per well.
2. Biotinylate the lectin through dissolution in amine-free buffer like PBS (pH 7.2–8) at a concentration of 2–10 mg/mL. Add a volume corresponding to 20 molar excess of biotin from a

100 mM solution of Biotinamidohexanoyl-6-aminohexanoic acid *N*-hydroxysuccinimide ester in DMSO or DMF. Stir gently during 30 min at 20 °C. Remove the excess of non-reacted biotin by desalting on columns or dialysis.

3. Determine appropriate lectin concentration by adding in the wells, 100 μL of lectin solution (100 μL) with twofold dilution starting from 30 μg/mL. After 1 h incubation at 37 °C and 3 washes with T-PBS (PBS containing 0.05 % Tween 20), add 100 μL of horseradish streptavidin–peroxidase (HRP) conjugate (dilute 1:5,000 from 2 mg/mL solution). After 1 h incubation at 37 °C, add 100 μL of phosphate/citrate buffer 0.05 M containing OPD (0.4 mg/mL) and urea hydrogen peroxide (0.4 mg/mL), incubate for 15 min and stop the coloration with 50 μL of 30 % sulfuric acid. Read the absorbance at 490 nM using a microtiter plate reader and plot the relative absorbance versus lectin concentration. Select the concentration yielding 70 % maximum response in the linear part of the curve for subsequent inhibition experiments (*see* **Note 5**).
4. Using plates functionalized as in **step 1**, add 50 μL of carbohydrate ligand with threefold serial dilutions (1/3) with PBS-BSA 0.3 % (w/v). Add 50 μL of biotinylated lectin solution at appropriate concentration (*see* **step 3**). Perform incubation, washing, reaction and reading as in **step 3**. Plot the inhibition percentage versus inhibitor concentration and sigmoidal fitting to determine IC_{50} (*see* **Notes 6** and 7).

3.2.3 Glycan Array Assays

1. Submit a request for using the glycan array resource at the Consortium for Functional Glycomics (www.functionalglycomics.org) with details about the lectin to be tested and aim of the work. Wait for approval before preparing the protein.
2. Dissolve the lectin in carbonate buffer 100 mM pH 9.3 at a concentration of 2–10 mg/mL. Dissolve 1 mg of Alexa fluor 488 in 100 μL of DMF. Add 10 μL of Alexa solution for each 1 mg of lectin to be labelled. Stir gently during 30 min at 20 °C.
3. Use desalting column to separate labelled lectins from non-labelled lectin and non-reacted Alexa probes. Measure Absorbance of each fraction at 280 and 495 nm to determine labelled lectins containing fractions.
4. Send your labelled lectin according to the website (http://www.functionalglycomics.org/) instructions.

3.3 Determination of Lectin Affinity for Carbohydrate Ligand

3.3.1 Isothermal Titration Microcalorimetry

Isothermal titration calorimetry (ITC) directly quantifies the heat effects accompanying association between molecular entities, and yields thermodynamic of the interaction in a single experiment (Fig. 2). Lectin–carbohydrate interactions are strongly exothermic, due to large number of hydrogen bonds and are very well suited for this approach.

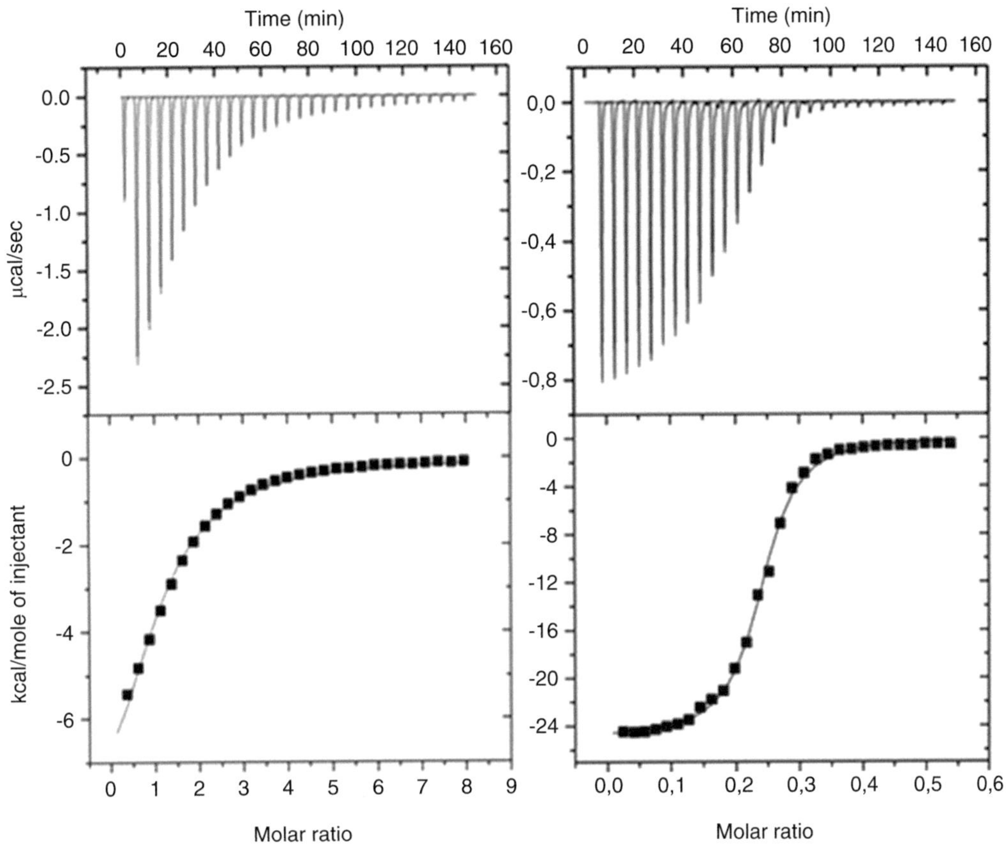

Fig. 2 Typical ITC measurements representing the raw ITC data (*top*) and integrated titration curves (*bottom*) for the titration of glycans in LecA. *Left*: digalactoside (Galα1-2GalαOMe from [22], 1.7 mM) in LecA 0.048 mM. *Right*: tetravalent galactosylated glycocluster (TetraGalEG3-Calix[4]-1,3-alt, compound **29** from [24], 0.12 mM) in LecA 0.05 mM

1. Dissolve the lectin in 2 mL Tris buffer (50 and 120 μM depending on ligand affinities) and degas the solution, check the protein concentration by measurement of optical density using a theoretical molar extinction coefficient of 28,000, and charge it in the 1.4478 mL sample cell (*see* **Note 8**).
2. Dissolve carbohydrate ligands in the same buffer (400 μL) at appropriate concentration (*see* **Note 9**) and charge it in the injection 300 μL syringe.
3. Perform the titration at 25 °C by 30 injections of 10 μL carbohydrate ligand every 300 s (after a first injection of 22 μL). Fit the data with MicroCal Origin 7 software, according to standard procedures.
4. Perform blank titration by injecting the carbohydrate ligand in buffer at same concentration as in the experiment. If the signal is significant, subtract it from the experimental data according to software instructions.

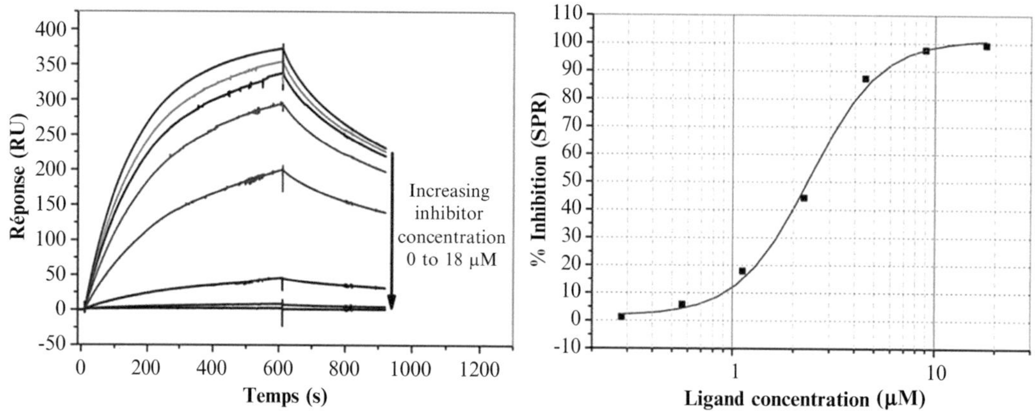

Fig. 3 Typical SPR measurements for the interaction of tetravalent galactosylated glycocluster (TetraGal EG3-Calix[4]-cone, compound **28** from [24] at various concentrations), with LecA (10 mM) injected on a CM5 chip coated with streptavidin/Biotin–PAA–α-D galactose. *Left*: sensorgram, *Right*: corresponding inhibition curve

3.3.2 Surface Plasmon Resonance

SPR biosensor experiments involve immobilizing of one reaction partner on a biochip surface and monitoring its interactions with a second component passing over the surface. The main advantage of this method is its ability to characterize binding reactions in real-time without labeling and to provide the kinetics of the interaction (Fig. 3).

1. Prepare two channels by coating with α-L-fucose (channel 1: control channel) and α-D-galactose (channel 2: binding channel) as follow (*see* **Note 10**). Activate both channels by injecting a fresh mixture of EDC/NHS (35 μL, 420 s), and then a solution of streptavidin (100 μg/mL in sodium acetate pH 5 buffer). Remaining reacting species are inactivated by injecting ethanolamine (1 M, 35 μL, 420 s). Immobilization of biotinylated PAA-sugars (200 μg/mL) is performed at 25 °C using running buffer at 5 μL/min for 10 min.
2. Inject LecA at 5 μM (150 μL, 10 μL/min, dissociation: 120 s) to observe the adhesion of the lectin onto the galactose-coated surface (0 % inhibition). Inhibition studies consist in the injection (150 μL, 10 μL/min, dissociation: 120 s) of incubated (>1 h at room temperature) mixtures of LecA (5 μM) and various concentrations of glycans (twofold cascade dilutions). After each injection, the CM5 chip is fully regenerated by injections of D-Galactose (2 × 30 μL, 100 mM in running buffer). Binding curves are analyzed by subtracting channel 1 signal from channel 2 signal. Plot the inhibition percentage versus inhibitor concentration and sigmoidal fitting to determine IC_{50}.

3.4 Structure of Lectin–Carbohydrate Complex

In order to obtain the structure of lectins in complex with carbohydrate, two methods can be used: soaking if the solvent channels and crystal contacts allow the diffusion and binding of the carbohydrate or co-crystallization where the sugar can participate to the crystallization process. Binding of oligosaccharides can modify crystal contacts and to obtain the complex structure, the screening of new crystallizing conditions is often required.

3.4.1 Co-crystallization Experiments

1. Prepare a concentrated solution of sugar (10–100 mM depending on the sugar solubility) in water if possible (*see* **Note 11**).
2. Dilute 1/10 the sugar solution in LecA solution at 0.8 μM (*see* **Note 12**).
3. Incubate for at least 1 h at room temperature.
4. Make 2 μL drops containing a 50:50 (v/v) mix of protein and reservoir solutions.
5. Incubate the crystallization plates at 19 °C.
6. Observe the crystal growth under the microscope after 1 night, 3 days and 1 week.
7. Transfer the crystal in a cryoprotectant solution with a cryoloop.
8. Flash freeze the crystal after a few second in a cryoloop in liquid nitrogen.
9. Store in a cryogenic dewar.
10. Check crystal diffraction.

3.4.2 Soaking Experiments

1. Prepare a concentrated solution of sugar (10–100 mM depending on the sugar solubility) in water if possible.
2. Dilute 1/10 the sugar solution in the reservoir solution containing the appropriate cryoprotectant if necessary.
3. Transfer a crystal in the above solution using a cryoloop.
4. Soak for time between 30 s to several min (*see* **Note 13**).
5. Flash freeze the crystal in a cryoloop in liquid nitrogen.
6. Store in a cryogenic dewar.

3.4.3 Determination of Crystal Structure

Structure determination requires access to X-rays (via home sources or synchrotrons) and knowledge in protein crystallography. Crystal freezing and data collection are also critical steps. We advise you at this point to enter in contact with a crystallographer.

4 Notes

1. It is recommended to maintain the presence of $CaCl_2$ in all buffers for calcium-dependent lectins such as LecA and LecB. This is not necessary for other lectins.

2. Hemagglutination can also be performed on crude soluble protein extract obtained from *P. aeruginosa* cultures.
3. Thermal shift assays can also be used to evaluate the stability of the lectin as a function of buffer composition, pH, additives and these data are useful for setting up crystallization conditions.
4. Reference (no test compound added) and negative control (no protein added) samples should also be included.
5. The appropriate concentration for LecA is in the range of 5 μg/mL.
6. Do not forget to include negative and positive controls.
7. Be aware that IC_{50} can be calculated using negative and positive controls as references, or by using minimum and maximum of each plot (IC_{50rel}).
8. Volumes are indicated for a VP-ITC. When using microcalorimeter with smaller cells, such as ITC-200, the volumes needed are smaller but one should consider that higher concentrations are usually requested in order to get strong enough signal.
9. Choice of LecA and ligand concentrations depends on the affinity. If the affinity is high enough (micromolar), a protein concentration [M] yielding to c-value between 10 and 100 ($c = K_a \times [M]$ for single site system) is recommended in order to obtain sigmoidal integration curve [29]. In this case, a ligand concentration 10 times higher than [M] will result in titration occurring in the middle of the curve. This has to be adapted for multivalent ligands. If the affinity is expected to be low, it is recommended to work with excess of ligand (500–1,000 higher than [M]) [30].
10. Channel 1 should contain a non-bound carbohydrate.
11. For insoluble compounds, DMSO or organic solvent concentration should not be superior to 50 % (5 % in the final mix) in order not to interfere with the crystallization process.
12. Excess of sugar to be tested with ratio protein/sugar of 1/5–1/10 is recommended for co-crystallization experiment.
13. If the crystal dissolves or breaks itself, lower concentration of sugar or shorter soaking time should be tested as well as soaking first in a solution with the sugar and without the cryoprotectant and then transfer in the cryoprotecting solution prior to flash-freezing of the crystal.

Acknowledgments

The authors acknowledge funding from ANR grants PA-Antiadh (ANR-09-JCJC-0047) and Glycoasterix (ANR-08-PCVI-0028). Support from GDR Pseudomonas and Association Vaincre la Mucoviscidose is also acknowledged.

References

1. Gabius HJ, Siebert HC, Andre S, Jimenez-Barbero J, Rudiger H (2004) Chemical biology of the sugar code. Chem Bio Chem 5: 740–764
2. Imberty A, Varrot A (2008) Microbial recognition of human cell surface glycoconjugates. Curr Opin Struct Biol 18:567–576
3. Karlsson KA (2001) Pathogen-host protein–carbohydrate interactions as the basis of important infections. Adv Exp Med Biol 491: 431–443
4. Sharon N (1996) Carbohydrate–lectin interactions in infectious disease. Adv Exp Med Biol 408:1–8
5. Lindesmith L, Moe C, Marionneau S, Ruvoen N, Jiang X, Lindblad L, Stewart P, LePendu J, Baric R (2003) Human susceptibility and resistance to Norwalk virus infection. Nat Med 9:548–553
6. De Greve H, Wyns L, Bouckaert J (2007) Combining sites of bacterial fimbriae. Curr Opin Struct Biol 17:506–512
7. Blumenschein TM, Friedrich N, Childs RA, Saouros S, Carpenter EP, Campanero-Rhodes MA, Simpson P, Chai W, Koutroukides T, Blackman MJ, Feizi T, Soldati-Favre D, Matthews S (2007) Atomic resolution insight into host cell recognition by *Toxoplasma gondii*. EMBO J 26:2808–2820
8. Merritt EA, Hol WGJ (1995) AB5 toxins. Curr Opin Struct Biol 5:165–171
9. Gilboa-Garber N (1982) *Pseudomonas aeruginosa* lectins. Meth Enzymol 83:378–385
10. Imberty A, Wimmerova M, Mitchell EP, Gilboa-Garber N (2004) Structures of the lectins from *Pseudomonas aeruginosa*: insights into molecular basis for host glycan recognition. Microb Infect 6:222–229
11. Giraud C, Bernard CS, Ruer S, de Bentzmann S (2010) Biological "glue" and "Velcro": molecular tools for adhesion and biofilm formation in the hairy and gluey bug Pseudomonas aeruginosa. Environ Microbiol Rep 2:343–358
12. Hazes B, Sastry PA, Hayakawa K, Read RJ, Irvin RT (2000) Crystal structure of Pseudomonas aeruginosa PAK pilin suggests a main-chain-dominated mode of receptor binding. J Mol Biol 299:1005–1017
13. Arora SK, Ritchings BW, Almira EC, Lory S, Ramphal R (1998) The Pseudomonas aeruginosa flagellar cap protein, FliD, is responsible for mucin adhesion. Infect Immun 66: 1000–1007
14. Scharfman A, Arora SK, Delmotte P, Van Brussel E, Mazurier J, Ramphal R, Roussel P (2001) Recognition of Lewis x derivatives present on mucins by flagellar components of Pseudomonas aeruginosa. Infect Immun 69: 5243–5248
15. Diggle SP, Stacey RE, Dodd C, Camara M, Williams P, Winzer K (2006) The galactophilic lectin, LecA, contributes to biofilm development in Pseudomonas aeruginosa. Environ Microbiol 8:1095–1104
16. Tielker D, Hacker S, Loris R, Strathmann M, Wingender J, Wilhelm S, Rosenau F, Jaeger K-E (2005) *Pseudomonas aeruginosa* lectin LecB is located in the outer membrane and is involved in biofilm formation. Microbiology 151:1313–1323
17. Chemani C, Imberty A, de Bentzman S, Pierre P, Wimmerová M, Guery BP, Faure K (2009) Role of LecA and LecB lectins in *Pseudomonas aeruginosa* induced lung injury and effect of carbohydrates ligands. Infect Immun 77: 2065–2075
18. Cioci G, Mitchell EP, Gautier C, Wimmerova M, Sudakevitz D, Pérez S, Gilboa-Garber N, Imberty A (2003) Structural basis of calcium and galactose recognition by the lectin PA-IL of *Pseudomonas aeruginosa*. FEBS Lett 555: 297–301
19. Mitchell E, Houles C, Sudakevitz D, Wimmerova M, Gautier C, Pérez S, Wu AM, Gilboa-Garber N, Imberty A (2002) Structural basis for oligosaccharide-mediated adhesion of *Pseudomonas aeruginosa* in the lungs of cystic fibrosis patients. Nat Struct Biol 9:918–921
20. Sabin C, Mitchell EP, Pokorná M, Gautier C, Utille J-P, Wimmerová M, Imberty A (2006) Binding of different monosaccharides by lectin PA-IIL from *Pseudomonas aeruginosa:* Thermodynamics data correlated with X-ray structures. FEBS Lett 580:982–987
21. Blanchard B, Nurisso A, Hollville E, Tétaud C, Wiels J, Pokorná M, Wimmerová M, Varrot A, Imberty A (2008) Structural basis of the preferential binding for globo-series glycosphingolipids displayed by *Pseudomonas aeruginosa* lectin I (PA-IL). J Mol Biol 383:837–853
22. Nurisso A, Blanchard B, Audfray A, Rydner L, Oscarson S, Varrot A, Imberty A (2010) Role of water molecules in structure and energetics of *Pseudomonas aeruginosa* PA-IL lectin interacting with disaccharides. J Biol Chem 285: 20316–20327
23. Perret S, Sabin C, Dumon C, Pokorná M, Gautier C, Galanina O, Ilia S, Bovin N, Nicaise M, Desmadril M, Gilboa-Garber N, Wimmerova M, Mitchell EP, Imberty A (2005) Structural basis for the interaction between human milk oligosaccharides and the bacterial lectin PA-IIL of *Pseudomonas aeruginosa*. Biochem J 389:325–332

24. Cecioni S, Lalor R, Blanchard B, Praly JP, Imberty A, Matthews SE, Vidal S (2009) Achieving high affinity towards a bacterial lectin through multivalent topological isomers of calix[4]arene glycoconjugate. Chem Eur J 15: 13232–13240
25. Imberty A, Chabre YM, Roy R (2008) Glycomimetics and glycodendrimers as high affinity microbial antiadhesins. Chem Eur J 14:7490–7499
26. Kadam RU, Bergmann M, Hurley M, Garg D, Cacciarini M, Swiderska MA, Nativi C, Sattler M, Smyth AR, Williams P, Cámara M, Stocker A, Darbre T, Reymond J-L (2011) A glycopeptide dendrimer inhibitor of the galactose-specific lectin LecA and of *Pseudomonas aeruginosa* biofilms. Angew Chem Int Ed Engl 50:10631–10635
27. Marotte K, Preville C, Sabin C, Moume-Pymbock M, Imberty A, Roy R (2007) Synthesis and binding properties of divalent and trivalent clusters of the Lewis a disaccharide moiety to Pseudomonas aeruginosa lectin PA-IIL. Org Biomol Chem 5:2953–2961
28. Marotte K, Sabin C, Preville C, Moume-Pymbock M, Wimmerova M, Mitchell EP, Imberty A, Roy R (2007) X-ray structures and thermodynamics of the interaction of PA-IIL from *Pseudomonas aeruginosa* with disaccharide derivatives. ChemMedChem 2: 1328–1338
29. Wiseman T, Williston S, Brandts JF, Lin LN (1989) Rapid measurement of binding constants and heats of binding using a new titration calorimeter. Anal Biochem 179: 131–137
30. Turnbull WB, Daranas AH (2003) On the value of c: can low affinity systems be studied by isothermal titration calorimetry? J Am Chem Soc 125:14859–14866

Part IV

Pseudomonas Genes and Gene Expression

Chapter 33

Mining the *Pseudomonas* Genome

Geoffrey L. Winsor and Fiona S.L. Brinkman

Abstract

Pseudomonas species were targeted early for genomic studies since they were noted for their diverse metabolic capacity, ability to inhabit a wide range of environments and hosts, and include notable human and agriculturally relevant pathogens. As more genomes are sequenced, the power of genome-scale analyses are increasing and a wide range of analyses are now possible. The *Pseudomonas* Genome database has contributed to this effort by providing peer-reviewed, continually updated annotations of the *Pseudomonas aeruginosa* PAO1 reference strain genome plus integrated data and analyses of related *Pseudomonas* species. Analyses are now available via multiple resources to facilitate identification and characterization of drug targets, virulence factors, regulatory elements, genomic islands, genome rearrangements, orthologs, single nucleotide polymorphisms, and multiple other gene/protein-based analyses from gene expression to protein structure. We describe here how the *Pseudomonas* Genome Database and other bioinformatics resources can be leveraged to help *Pseudomonas* researchers "mine" *Pseudomonas* genomes, and associated genome-scale data, to facilitate new discovery.

Key words *Pseudomonas aeruginosa*, Genome annotation, Comparative genomics, Bioinformatics, Genomic islands, Genome alignment, Orthologs, Database, Sequence

1 Introduction

At the completion of sequencing in 2000, the 6.3 Mbp genome of *Pseudomonas aeruginosa* PAO1 was the largest sequenced genome to date [1]. A hallmark of its genome is the high number of regulatory genes working to modulate a complex system of adaptations for exploiting diverse ecological niches. *P. aeruginosa* is capable of thriving in soil, freshwater, marine and animal habitats, while exhibiting a high degree of intrinsic resistance to antibiotics and other antimicrobial agents, and causing serious opportunistic infections-making it a high priority microorganism to study. Reflecting this, a search of PubMed reveals that it is still the fourth-most cited bacterium. Using the wealth of research already available, the *Pseudomonas aeruginosa* Community Annotation Project (PseudoCAP) was formed to meet the need for conservative, peer-reviewed annotation of the

Alain Filloux and Juan-Luis Ramos (eds.), *Pseudomonas Methods and Protocols*, Methods in Molecular Biology, vol. 1149, DOI 10.1007/978-1-4939-0473-0_33, © Springer Science+Business Media New York 2014

P. aeruginosa PAO1 genome sequence, using an Internet-based approach to community-assisted genome annotation (essentially early "crowdsourcing"). This project resulted in the development of the first version of the *Pseudomonas* Genome Database [2, 3]. The scope of the database was soon expanded to host the *Pseudomonas syringae* DC3000 genome sequence [4] managed by the *Pseudomonas*–Plant Interaction (PPI) Web site. A sharp rise in the number of *Pseudomonas* genomes being sequenced cast light on the importance of comparative genome analysis and the integration of completely sequenced *Pseudomonas* genomes from NCBI began [5, 6]. As of fall 2012, the database currently hosts data for 32 completely sequenced and four permanent draft *Pseudomonas* genomes. With the increased speed and quality by which genomes are being sequenced, a challenge has arisen to genomics researchers on how best to interpret this data and capitalize on it. This chapter focuses on the how the *Pseudomonas* Genome Database and other computational resources can play a role in assisting researchers who wish to analyze or "mine" *Pseudomonas* genomes, highlighting selected examples of the wealth of possible analyses that may be performed.

2 Materials

This tutorial is primarily focused on resources available at the *Pseudomonas* Genome Database (http://www.pseudomonas.com) Web site. The database integrates genome annotations from a number of external resources which we will introduce in Subheading 3. We also make reference to the following third-party software packages.

2.1 GBrowse: For Visualization of Alternate Genome Annotations

We currently use Generic Genome Browser version 2.48 (downloaded from the GMOD Web site at http://www.gmod.org/wiki/GBrowse). It is installed on a dedicated Debian GNU/Linux server containing 4 GB RAM and utilizing four CPUs. GBrowse can be accessed from the top navigation bar at the *Pseudomonas* Genome Database Web site.

2.2 Formatting RNA-Seq Data for Upload to GBrowse

All file manipulations were performed in a Debian GNU/Linux environment. We used SAMTools version 0.1.19 (http://sourceforge.net/projects/samtools/) to manipulate alignment files resulting from an RNA-Seq analysis of *P. aeruginosa* PAO1 total RNA. For finding and replacing text in large SAM files, we used the "sed" (stream editor, version 4.2.1) command.

2.3 Mauve and Mauve Contig Mover

In order to describe how to perform whole-genome alignments of *P. aeruginosa* genome sequences, our example utilizes the features available in Mauve 2.3.1 downloaded from http://gel.ahabs.wisc.edu/mauve/.

Table 1
Summary of the product name classification system at the *Pseudomonas* Genome Database

Class 1	Function experimentally demonstrated in *P. aeruginosa*
Class 2	Function of highly similar gene experimentally demonstrated in another organism (and gene context consistent in terms of pathways it is involved in, if known)
Class 3	Function proposed based on presence of conserved amino acid motif, structural feature or limited sequence similarity to an experimentally studied gene
Class 4	Homologs of previously reported genes of unknown function, or no similarity to any previously reported sequences

Table 2
Summary of the subcellular localization confidence system at the *Pseudomonas* Genome Database

Class 1	Subcellular localization experimentally demonstrated in strain of the same species
Class 2	Subcellular localization of highly similar gene experimentally demonstrated in another organism. BLAST expect value of 10e-10 for query within 80–120 % of subject length
Class 3	Subcellular localization computationally predicted by PSORTb V3.0

3 Methods

3.1 *Pseudomonas* Genome Database: A Source for High-Quality Genome Annotations

The *Pseudomonas* Genome Database was initially developed to be a source of critical, conservative genome annotations obtained from a combination of in-house literature curation and participation by members of the *Pseudomonas* research community. In the latter case, suggested annotations can be entered using a submission form on the Web site or by sending an e-mail to the mailing list. Upon submission of an annotation, the content is examined for clarity, and functional evidence supporting a name or other annotation change is assigned a code to indicate what type of evidence is being used to support it. The system at www.pseudomonas.com is based on a "Product Name Confidence" score from 1 to 4 (Table 1), although other well established biological databases including the Gene Ontology (GO) consortium [7] and MetaCyc [8] make use of evidence classification systems. As of early 2013, over 800 of the more than 5,600 genes in the *P. aeruginosa* PAO1 genome have been given a high-quality Class 1 product name classification. A similar classification system in the database is also applied to protein subcellular localization annotations (Table 2).

In addition to high-quality gene name, protein function, and subcellular localization annotations historically available through our Web site, we have recently started to curate/annotate other relevant categories including *Pseudomonas* virulence factors, operons, regulatory motif data and manually mapped GO terms.

An emphasis will continue to be placed on providing classification systems and references for clarifying the degree of confidence in a particular annotation.

3.2 Enhanced User Interface for Searching Annotations and Sequences: From Drug Targets to Virulence Associations to Motifs

The search interface is continually being updated as we identify new features that are of value to *Pseudomonas* researchers. Annotations may be searched using simple or advanced Boolean-based search forms utilizing a wide range of searchable fields and filters. In most cases, a simple keyword search using a name or locus tag will suffice when looking for a gene or protein of interest. However, it is important to note that the quality of annotated genomes can vary considerably (*see* **Note 1**). Complex queries requiring more stringent criteria for whole genome analyses are best addressed using the advanced search form. From here, you can specify a wider range of search fields or restrict results to those with exact matches to your query while limiting the results to one or more strains or genes with a specified gene size cut-off.

3.3 Application of Search Filters: Limiting to Subcellular Localization, Essential Genes, Some Known Drug Targets

Filters at the bottom of the advanced search form can be applied to help obtain whole-genome data relevant for studies including those identifying putative drug target candidates or diagnostic markers. For example, it may be desirable to obtain a list of all genes coding for easily accessible protein targets including secreted or outer membrane proteins; this can be accomplished by selecting the appropriate localization under "Specify subcellular localization." Obtaining a list of proteins with known three-dimensional structures is often cited as being important for selecting drug target candidates; however, such a list is usually expanded to include those with high similarity to proteins of known 3D structure. By selecting the "Filter proteins with 3D structures" option underneath the advanced search form, you can select for proteins with greater than 90 % sequence similarity to those in the Protein Data Bank [9]. It may also be desirable to limit your list to essential genes by selecting "Filter Transposon Mutants" option, which restricts your search to genes not containing a transposon mutant under saturating conditions. This feature is currently only available for *P. aeruginosa* strains PAO1 and PA14 based on three different studies [10–12]. Other filters can be used to select for proteins with or without similarity to human proteins downloaded from the Ensembl database [13] or view genes that are currently annotated as being drug targets [14].

3.4 Browse Virulence Factors, Virulence-Associated Genes and Functional Gene Groupings

In addition to the above approaches, there is a growing consensus that targeting virulence factor genes or specific metabolic pathways is a useful step in developing anti-infective drugs, plus new potential virulence genes may be uncovered through simply their association with strain/species virulence phenotypes. By going to the "Browse Annotations" page, you will be presented with a number of options for browsing gene annotations by their biological role or function. In the case of *P. aeruginosa*, several genes have also been manually curated as virulence factors and a list can be retrieved by using the "Browse Virulence Factors" form on the above page. Please note that since this list is currently limited to two *P. aeruginosa* strains, a more comprehensive list of "pathogen-associated" genes identified from a wider range of *Pseudomonas* genomes [15] may be of interest, and can be obtained from a form on the same page. Furthermore, if you are interested in obtaining a list of genes based on a functional or classification, the browse page offers variety of lists containing functional or phylogenetic classifications including TIGR Roles and SubRoles [16], Clusters of Orthologous Genes (COGs) [17] and PseudoCAP functional classifications. The latter grouping was developed by the original annotators of the *P. aeruginosa* PAO1 sequencing project and continue to constitute a very well curated list of functional groupings of genes suitable for genome-level analyses. GO mappings though will become increasingly valuable.

3.5 Searching for Sequence Motifs in a Pseudomonas Genome

When searching for novel occurrences of motifs such as transcription factor binding sites, bioinformaticians can generate a profile or frequency matrix from an alignment of experimentally validated sequences and use it to search a genome of interest. Any matches are then overlaid with other genomic features to help come to a conclusion about their biological relevance. However, frequently a non-bioinformatician may want to more simply assess the presence of a particular motif in their genome of interest. Therefore, the *Pseudomonas* Genome Database developed a DNA motif search tool that allows input of a consensus sequence formatted to reflect an unknown number of nucleotides or variable nucleotides at certain positions. For example, the formatted consensus sequence "AAGS{3,8}TTN{3,20}TTGAC" represents a sequence where "S{3,8}" means a C or G appears between 3 and 8 times (starting at position 4) and "N{3,20}" means any nucleotide appearing 3–20 times. The result of the search is a downloadable list of sequences and their coordinates meeting the criterion and details highlighting where it is found, including distance from adjacent upstream and downstream genes.

3.6 Visualization of Alternate Genome Annotations and Genome-Scale Data Using GBrowse

Interpretation of the increasing volume and diversity of genome annotations from multiple sources is a nontrivial task often best handled by visualization of the data in a genome browser such as GBrowse [18]. GBrowse allows you to view third-party and PseudoCAP genome annotations as separate horizontal tracks in context of a coordinate-based system, making it ideal for interpreting a growing list of curated and computer-predicted regulatory motifs, transcription factor binding sites and operons hosted on the *Pseudomonas* Genome Database. The ability to easily customize your view, upload your own annotation data, and download other tracks adds to its value. We are also in the early stages of incorporating sequence "read" data from RNA-Seq projects based on total RNA or mapped transcription start sites, which has potential for greatly expanding the detail in which regulatory analyses can be performed.

You can access GBrowse through a number of different links on the pseudomonas.com Web site, or by visiting gbrowse.pseudomonas.com. The default view is based on the *P. aeruginosa* PAO1 genomic sequence; however, you can easily switch to another strain by selecting from the drop-down box under "Data Source." After your sequence of interest is displayed, you can easily add or remove data tracks by going to the "Select Tracks" tab and highlighting the appropriate boxes. It is often very useful to reorder the tracks relative to each other by clicking and dragging the titles located directly above the individual tracks. If you would like to download data from any of the tracks, select "Download Track Data" from the drop-down list to the right side above the tracks and any data associated with the current view will be downloaded in GFF3 format. More advanced options, including setting the download to include features across the entire genome or embed the sequence into the download file, can be set by clicking on "Configure" to the right of the list.

3.7 Uploading RNA-Seq Data to GBrowse

The number of labs performing RNA-Seq analysis is increasing rapidly; however, the number of Web-accessible tools capable of displaying the data in an intuitive manner is lacking. Fortunately, GBrowse does an excellent job at displaying XY plots, density plots, and individual reads (Fig. 1) based on the standard SAM/BAM file format output by most alignment tools and manipulated using the SAMtools software package [19]. In order to upload your own BAM sequence alignment files into GBrowse, go to "Custom Tracks" and click on "From a file" and select a sorted BAM file stored on your computer. The most important considerations when uploading this kind of data are:

1. The recommended maximum file size is 200 MB. While many BAM files are much larger than this, we strongly recommend reducing the number of aligned reads in the BAM file (*see* **Note 2**), which should still provide an adequate representation for visualization purposes sorting. If not, errors can occur, including time-out errors while loading data for visualization. If time out errors do occur during your analysis, you may also consider zooming in on a smaller area.
2. Ensure that the sequence name that reads are mapped to is matching the GBrowse RefSeq accession found in the "Landmark or Region" box (*see* **Note 3**). Provided they are the same, the uploaded BAM file will be automatically indexed when the upload completes.

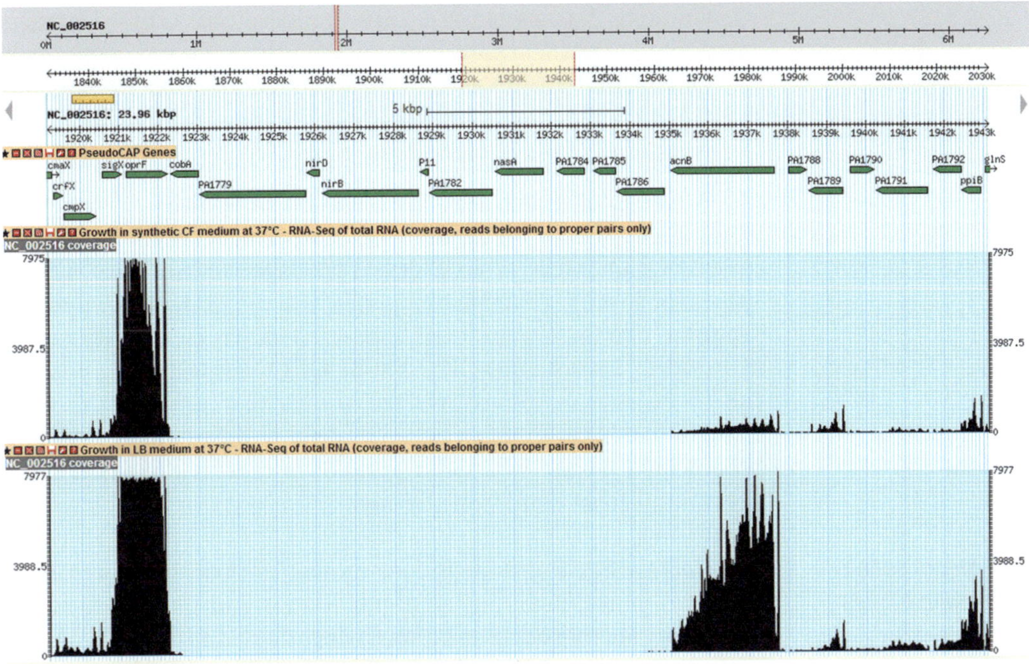

Fig. 1 GBrowse view comparing *P. aeruginosa* PA01 gene annotations (*upper track*) to RNA-Seq coverage represented as *XY* plots based for PA01 grown in synthetic cystic fibrosis medium (*middle track*) and LB medium (*lower track*)

If the file is successfully uploaded, a colored box containing the file name, date and number of bytes will appear under "Source files." The track should appear by default and have a name matching the name of the uploaded file. You can also edit an automatically generated configuration file in order to specify how your track appears by clicking on the "[edit]" link while further help with customizing the basic appearance such as height and color of the figure is provided on a help page linked to from the top of "Custom Tracks" box.

If you wish to share any current screenshot with others, or export an image of it, simply go the "File" link at the top left of the page where you can select "Bookmark this" to get a permanent link or select "Export as…" for a list of suggested ways to export your image.

3.8 Viewing Gene Expression Data

While RNA-Seq approaches to measuring gene expression are growing in popularity, and the above analyses allow one to capitalize on this, a wealth of *Pseudomonas* species' expression data based on microarray analysis also exists. One of the most comprehensive repositories for hosting microarray data, along with expression data from next-generation sequencing studies is the Gene Expression Omnibus (GEO) at NCBI [20]. GEO is organized as:

1. A large primary database of submitter-supplied meta-data organized into Platform, Sample, and Series records linked to raw data files uploaded by the submitter.

2. Curated data organized into standardized GEO DataSets. A single DataSet represents consistently processed, experimentally related groups of GEO Sample records that are linked to GEO Profiles containing expression values for individual genes measured across all samples in a DataSet.

A variety of search options are available at http://www.ncbi.nlm.nih.gov/geo/ which allow you to perform specific queries or browse GEO DataSets. In order to facilitate viewing of expression values for specific *Pseudomonas* genes, you can go to a *Pseudomonas* Genome Database gene card and scroll down to the Microarray Expression Data section. The NCBI GEO link will bring you to a list of expression values associated with that gene in the GEO Profiles database. This can be very useful if, for example, you want to quickly evaluate whether a gene of interest is expressed under certain infection conditions, or not.

3.9 Identification of Genomic Islands

Pseudomonas genomes include a mosaic of probable horizontally acquired genes, including genomic islands and phage which play an important role adaptation to environmental niches. To identify genomic islands, our lab developed IslandViewer [21] a computational tool integrating the two most accurate sequence composition-based island prediction methods, SIGI-HMM [22] and Island Path-DIMOB [23], with the most precise comparative genomics-based prediction method, IslandPick [24]. Views of the data in IslandViewer have been integrated into the *Pseudomonas* Genome database from a "browse genomic island" page and from individual strain summary pages. These sections link to IslandViewer Web site pages containing circular chromosome images overlaid with details of genomic islands identified by the various methods with links to download the results for further analysis. You can also customize the IslandPick analysis, selecting what genomes you want to compare to, to for example identify only species-specific islands or islands associated with a particular strain phenotype. Pathogen-associated genes, virulence genes (from the Virulence Factor Database), and antimicrobial resistance genes (from the Antimicrobial Resistance gene Database) can be viewed as an overlay on the islands, and more island annotation information for the genes can be viewed and downloaded. If your genome of interest is not currently available in the precomputed IslandViewer data, you also have the option of going to the IslandViewer "Genome Upload" page and submitting a GenBank or EMBL-formatted file of your genome sequence. Once the analysis is completed, a Web site link to the results can be e-mailed to you. These results are not accessible to anyone else (the Web site URL for this results page essentially acts as a password for access). IslandViewer currently accepts only complete genome sequences but we are in the process of expanding the ability to upload files containing multiple contigs such as those associated with draft genomes.

3.10 Identification and Analysis of Orthologs

The ability to confidently identify orthologs in different strains/species is an essential component of comparative genomics analyses and is typically performed using the reciprocal best-BLAST hit (RBBH) based method. However, RBBH sometimes returns false positives when an ortholog is missing from one isolate, due to gene loss or incomplete genome sequence data [25]. Therefore, high-throughput RBBH-based ortholog predictions in our database have undergone further assessment by a high-precision method called Ortholuge which examines phylogenetic distance ratios between two comparison species and an outgroup species [25]. The Ortholuge algorithm examines putative ortholog pairs, such as those arising from RBBH-based analysis, and assigns them to one of three groups whose phylogenetic ratios are supportive of species divergence and are (a) probable orthologs, (b) those that resemble probable paralogs, and (c) those that cannot be classified as true orthologs or true paralogs. Each gene details page on the *Pseudomonas* Genome Database Web site contains links to Ortholuge-validated ortholog predictions for the specific gene. If you go to this page, you will see that each ortholog is given with an image of the local gene context of neighboring genes. Since gene synteny is usually conserved between true orthologs, this provides another degree of evidence for establishing whether two genes are orthologs.

If you are interested in expanding your view of orthologs outside the *Pseudomonas* genus or are interested in other types of ortholog comparisons, OrthologeDB (http://www.pathogenomics.sfu.ca/orthologedb) was also recently developed as a comprehensive database of ortholog predictions for bacterial and archaeal genomes [26]. OrthologeDB provides more in-depth analysis including grouping of orthologs into five different hierarchical levels where enhanced viewing of connections between different members of the group can be viewed as a graph, with genes being represented as nodes and ortholog and in-paralog relationships being represented as differentially colored edges between genes. In addition, orthologs in multiple genomes can be viewed in a phyletic matrix that quickly shows which genes have orthologs in a list of user-specified comparison genomes. Using a phyletic matrix search, one can quickly formulate complex queries that determine which orthologs are common to one set of species or isolates with a certain phenotype, but are absent in another set of species [26]. The phyletic range of a given gene can also easily be queried through a gene-specific search.

3.11 Clustering of Orthologous Genes into Pseudomonas Orthologous Groups

Note that a caveat to the Ortholuge approach is that it is difficult to compare the phylogenetic distance ratios of highly similar genomes (e.g., isolates of the same strain or species). We therefore developed a complementary method for clustering genes into orthologous groups called *Pseudomonas* Orthologous Groups (POGs) containing genes that are very likely orthologs or in-paralogs, as judged by BLASTp analyses and filtering criteria [5]. By going to an individual gene details page and scrolling down to the "Orthologs/Comparative Genomics" section, you can see if there is a POG entry for the current gene.

The clustering approach to constructing POGs has some differences versus to the method used by Ortholuge to compare two candidate RBBH-predicted orthologs to an outgroup species. There are often instances where more than one RBBH candidate is returned so the similarity between the query's and hit's flanking genes are examined in order to narrow down the list of probable orthologs. In cases where candidate genes are directly adjacent, they are both labelled as RBBHs to the query gene and are given a flag to indicate putative paralogy. We also perform intra-genome BLASTp analysis to acquire in-paralog information, indicative of gene duplication occurring after species divergence. Therefore, if two genes from the same genome are reciprocally more similar to each other than genes in the other genomes, they are flagged as being putative in-paralogs. Ortholog groups are built by taking a seed gene and finding all genes that are RBBHs or in-paralogs. Every time a new gene is added to an ortholog group, it is treated as the seed gene and the process is repeated until all qualifying genes have been assigned a group. The POG approach to clustering is an effective tool that *Pseudomonas* researchers studying particularly large paralogous gene families (e.g., porins and outer membrane efflux proteins) can use to identify an inclusive set of homologous proteins. Due to the approach we use to identify in-paralogs, it is important to note that this method works best when clustering orthologous genes from closely related genomes, such as those in the genus *Pseudomonas* and that it is not a replacement for definitively inferring orthology as occurs when using Ortholuge. Also note that each group contains only genes with the highest level of sequence similarity to the other members of the group, so not all genes in the database are members of an orthologous group. POGs are an example of a genera-specific or clade-specific analysis that is not available through larger genome databases, complementing such resources.

3.12 Application of Ortholog Predictions and Comparative Genomics to Specialized Searches: From Genome Rearrangements, to Genes, to SNPs

The predicted orthologs can be used for a variety of genome level analyses, many of which can be performed on the *Pseudomonas* Genome Database Web site under the "Comparative Search" section. From here, you can find groups of genes that meet a wide range of search criteria. Below are some suggested uses. Note that individual mappings of genes/proteins in one strain to orthologs in all other strains can be downloaded from files under the "Orthologs" tab of the http://www.pseudomonas.com/download.jsp page.

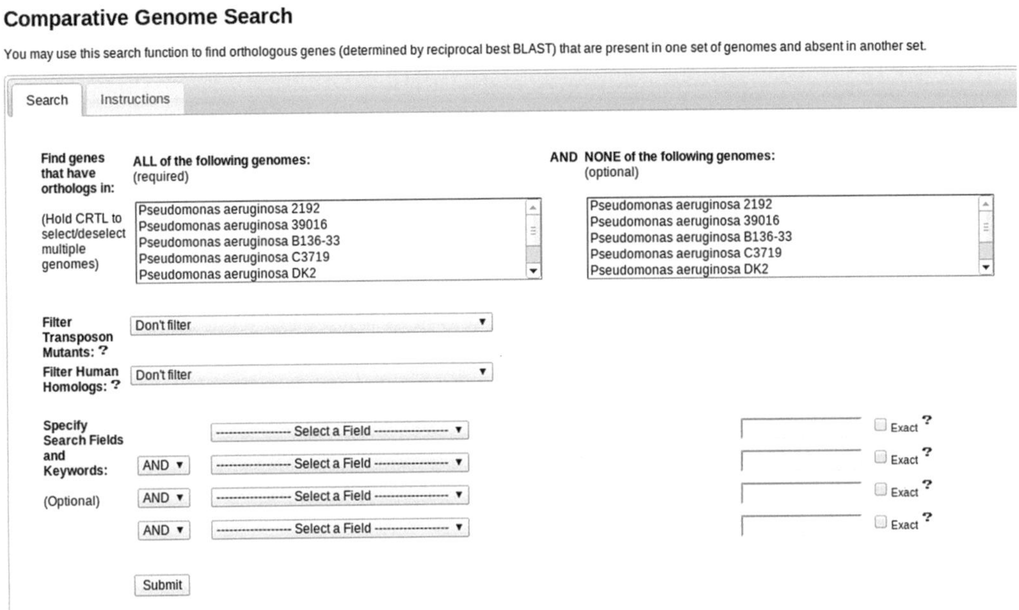

Fig. 2 Search form for finding sets of genes based on orthologs/no orthologs in separate genomes

3.13 Find a List of Genes that Have Orthologs in One Set of Genomes but Not in Another Set of Genomes

You can use the search form (Fig. 2) to find a list of genes that have orthologs in *P. aeruginosa* PAO1 and *P. aeruginosa* LESB58 but do not have orthologs in a third genome such as *P. aeruginosa* PA7. In order to do this analysis you would select *P. aeruginosa* PAO1 and *P. aeruginosa* LES from the box on the left while selecting *P. aeruginosa* PA7 from the right box and clicking on submit. A list of genes meeting the user-specified criteria (one per row) will be returned; however, it will be up to the user to map orthologs in this list to each other if that is what they are looking for. If you wish to perform such an analysis but include non-*Pseudomonas* genomes (for example, identify genes common to respiratory pathogens, including *P. aeruginosa*, but not found in other pathogens), you can also perform such a query in OrthologeDB.

3.14 Find Genes in One Genome That Have No Orthologs in All the Other Genomes

If you are interested in finding a list of genes that have no predicted orthologs in the other strains in the database, you can simply select its name under the left box and select all other strains in the right box.

3.15 Obtain a List of Putative Core Genes Based on the Strains Currently Stored in the Database

If you wish to obtain a list of putative core genes as defined by a list of genes that have orthologs in all the *Pseudomonas* genomes, this can easily be done by selecting all strains from the box on the left while leaving the right box empty and clicking submit. You can also examine the core genome through expanded searches that include other genera through OrthologeDB.

3.16 Whole Genome Rearrangement Analysis-Alignment Using Mauve

While the identification of genomic islands is one application for comparative genomics analysis of *Pseudomonas* genomes, other genomic rearrangements of interest include large scale inversions, duplications and losses, plus smaller segmental duplications and local nucleotide conservation of narrow genomic regions. A number of tools have been developed for identifying and visualizing these rearrangements and one of the most widely used is the Mauve aligner [27]. Mauve is a software package first developed to align multiple closely related genome sequences using a method that identifies locally collinear blocks (LCBs) of local alignments occurring in the same order and orientation in a pair of genomes. These LCBs are considered to be homologous sequences containing no internal rearrangements within them. An improvement in the algorithm called progressiveMauve was introduced in 2009 and overcame a limitation preventing identification of large conserved regions in subsets of genomes while also improving on alignment of more distantly related sequences [28]. The Mauve viewer displays horizontal panel views of the final alignments with a reference sequence at top serving the function of displaying a reference orientation for all of the multiple colored LCBs. As for the comparison genomes underneath the reference genome, the LCBs may appear above or below a horizontal black line whereby those appearing below have an inverse orientation in relation to the reference sequence. Any regions appearing outside colored LCBs are considered to have no homology to the other input genomes. Mauve will also draw vertical similarity profile bars inside the LCBs that indicate average level of conservation within that region of the genome sequence. This level of view is very useful for identifying higher level evolutionary events including horizontal transfer, inversion and duplication.

In order to study nucleotide level variation and smaller segmental variation, an alternative "backbone" view displays regions conserved amongst all of the genomes in the color mauve (the backbone) while coloring regions conserved in subsets of genomes different colors. If the input genomes were provided in GenBank format containing gene or protein feature annotations, zooming in on this or the previous LCB view will display them alongside the sequence similarity profiles. Visualization of the backbone and differential coloring of nucleotide segment subsets, combined with these annotated sequence features can help gain important insight into events that alter regulation of gene expression or otherwise affect gene function (*see* **Note 4**).

3.17 Contig Viewing Using Mauve

The large number of genome sequences being currently released as draft assemblies contain hundreds or even thousands of contigs that must be correctly ordered and oriented relative to a reference sequence in order for comparative genomic analysis to be most effective. The Mauve package contains a tool called the Mauve Contig

Mover (MCM) that performs an iterative series of alignments and successive reordering of the contigs in order to reduce the number of LCBs being produced and thus providing a more likely representation of the contig order [29]. The output contains several numbered folders containing Mauve alignments, although the final iteration (with the largest number) is considered to contain the optimal ordering of contigs.

The MCM is relatively easy to use for novices, and takes about 20–30 min to run on a current desktop computer. Unlike progressiveMauve, the MCM takes as input a single reference sequence in FASTA or GenBank format and a second multi-fasta or multi-GenBank formatted file containing the contigs that you would like to have reordered (*see* **Note 5**). In addition to enabling comparative analysis of draft *Pseudomonas* genomes, using the MCM will help identify regions in the draft genome that are misassembled, help evaluate annotations that span contig boundaries and also serve as an aid in closing gaps. MCM will output an assembly file and multi-fasta file containing the new order and orientation of contigs as well as a text file listing the ordered contigs and genomics coordinates (*see* **Note 6**). In order to obtain sequences and annotations of draft genomes to use with the MCB, go to the Whole Genome Shotgun (WGS) sequencing projects page at NCBI (http://www.ncbi.nlm.nih.gov/Traces/wgs/) and search for your strain of interest where a list of matches with information about the contigs and links to the WGS project data will be returned. For more details on how to install the Mauve software package, perform alignments in progressiveMauve, reorder contigs and visualize the results, visit the Mauve Web site at: http://asap.ahabs.wisc.edu/mauve/.

3.18 Comparing SNPs and Indels Using the Pseudomonas Genome Database

In addition to the backbone view of Mauve, the *Pseudomonas* Genome Database provides access to a view of SNPs and Indels computed from whole genome alignments by MUMmer [30] of *Pseudomonas* genomes against the *P. aeruginosa* PAO1 reference genome. Such structural variation is listed on the details pages of single genes or intergenic regions and contains information on the type of mutation (transition or transversion) and evidence for its validation. A SNP map found below this list contains the sites of any annotated SNPs and provides hyperlinks to their details. We are continuing to evaluate additional methods for optimally viewing SNPs in a growing list of comparison genomes.

3.19 Moving Forward in the Genome Era

As genome sequencing costs become lower and the barrier to labs performing their own whole-genome sequencing analyses falls, databases that host genomic data—large and small—face mounting challenges. They include the management of an increasing volume of data while identifying and reducing annotation errors and standardizing annotation practices to better facilitate transfer of existing annotations or incorporation of new data from researchers [31]. Another issue being currently faced is the rapid shift in the number of permanent draft genomes versus complete genome sequences being released, which have the potential of introducing annotation errors, missed genomic features and incorrect assemblies. We expect that this will be more of a short-term problem that can be overcome with continuing advances in sequencing technology. In time, the more pressing need being anticipated is the thousands of genome sequences and associated genome-scale data that will become available. Many genome sequences will be highly similar and require methods to effectively cluster related strains around better-curated reference genomes. New ways to visualize and analyze the data in interactive ways, that facilitate more population-based genomics analysis, will become increasingly important. This will need to occur while not losing site of the importance of linking high-quality, curated experimental data to genome sequence information. Model organism databases like the *Pseudomonas* Genome Database will confront these challenges. The *Pseudomonas* Genome Database aims to continue to play an important role, complementing larger genomic databases by contributing clade-specific analyses and higher quality annotation updates customized for the *Pseudomonas* research community.

4 Notes

1. We recommend complementing an annotation search with a BLAST search to identify homologous genes/proteins of interest that do not show up in your original search (e.g., many sequencing projects provide no 4-letter gene names). From the results of a BLASTP search you can select features of interest to store in a clipboard utility for further study using a side-by-side comparison.
2. The easiest way to generate a smaller BAM file suitable for uploading is to use the "view -s" option in SAMtools [19]. A new feature (as of version 0.1.19) allows you to randomly sample a small number of reads from the large BAM file and output them to a smaller file. If we want to reduce a 2 GB BAM file to a manageable size of around 200 MB, the correct syntax to use is:

   ```
   samtools -b -s 0.1 largefile.bam > smallfile. bam
   ```

   The "-b" indicates that output should be in BAM format and the lower-case "-s 0.1" specifies that 10 % of the reads should be sampled into the output file. Before uploading the BAM file, please ensure that the reads within it have been sorted by their leftmost coordinates using:

   ```
   samtools sort smallfile.bam smallfile_sorted
   ```

   The .bam extension is not required for the final argument and the final output will be a BAM file called smallfile_sorted.bam.

3. If you need to check that the correct sequence name is given in the BAM file, take the "smallfile.bam" mentioned in **Note 2** above and convert it to the human-readable SAM format:

```
samtools view smallfile.bam > smallfile.sam
```

The third column in "smallfile.sam" is the sequence name. If it does not match the sequence name given in the GBrowse Landmark or Region box (e.g., NC_002516 for *P. aeruginosa* PAO1), you can quickly edit it in Linux using the "sed" command. For example, to change all occurrences of the name "Chromosome" to "NC_002516":

```
sed -e "s/Chromosome/NC_002516/g" < smallfile.
sam > renamed.sam
```

Convert the "renamed.sam" file back into BAM format for upload. You will need to have the genomic DNA fasta file (e.g., NC_002516.fa) containing the correct sequence name ">NC_002516" in the definition line and run the following commands:

```
samtools faidx NC_002516.fa
samtools view -bt NC_002516.fai.fai renamed.
sam > renamed.bam
```

Please note that the file `NC_002516.fa.fai` is automatically generated in the first of the two commands. Also ensure that the BAM file is correctly sorted prior to upload (as described in **Note 2**).

4. The *Pseudomonas* Genome Database provides pre-computed Mauve alignments for many completely sequenced genomes that may be downloaded and viewed using your own local installation of Mauve (available for Windows, Linux, and Mac).
5. It is important for the reference sequence to be provided first in the form. If you wish to compare feature annotations between the two genomes, you need to include GenBank files with the annotated sequence features already in place before running the program.
6. Please be aware that the final FASTA output of the ordered contigs will not infer distance between contigs or fill the gaps with Ns due to how unreliable it is to predict this.

Acknowledgment

We wish to acknowledge the efforts of the many genome sequencing projects and *Pseudomonas* researchers that have made our analysis possible. FSLB is a Michael Smith Foundation for Health Research (MSFHR) Senior Scholar. Critical funding was provided by Cystic Fibrosis Foundation Therapeutics.

References

1. Stover CK et al (2000) Complete genome sequence of *Pseudomonas aeruginosa* PAO1, an opportunistic pathogen. Nature 406:959–964
2. Brinkman FS et al (2000) Sequencing solution: use volunteer annotators organized via internet. Nature 406:933
3. Winsor GL et al (2005) Pseudomonas aeruginosa genome database and PseudoCAP: facilitating community-based, continually updated, genome annotation. Nucleic Acids Res 33 (Database Issue):D338–D343
4. Joardar V et al (2005) Whole-genome sequence analysis of *Pseudomonas syringae* pv. phaseolicola 1448A reveals divergence among pathovars in genes involved in virulence and transposition. J Bacteriol 187:6488–6498
5. Winsor GL et al (2009) Pseudomonas genome database: facilitating user-friendly, comprehensive comparisons of microbial genomes. Nucleic Acids Res 37:D483–D488
6. Winsor GL et al (2011) Pseudomonas genome database: improved comparative analysis and population genomics capability for Pseudomonas genomes. Nucleic Acids Res 39:D596–D600
7. Ashburner M et al (2000) Gene ontology: tool for the unification of biology. The gene ontology consortium. Nat Genet 25:25–29
8. Caspi R et al (2012) The MetaCyc database of metabolic pathways and enzymes and the BioCyc collection of pathway/genome databases. Nucleic Acids Res 40:D742–D753
9. Kouranov A et al (2006) The RCSB PDB information portal for structural genomics. Nucleic Acids Res 34:D302–D305
10. Liberati NT et al (2006) An ordered, nonredundant library of *Pseudomonas aeruginosa* strain PA14 transposon insertion mutants. Proc Natl Acad Sci U S A 103:2833–2838
11. Lewenza S et al (2005) Construction of a mini-Tn5-luxCDABE mutant library in Pseudomonas aeruginosa PAO1: a tool for identifying differentially regulated genes. Genome Res 15: 583–589
12. Jacobs MA et al (2003) Comprehensive transposon mutant library of *Pseudomonas aeruginosa*. Proc Natl Acad Sci U S A 100:14339–14344
13. Flicek P et al (2012) Ensembl 2012. Nucleic Acids Res 40:D84–D90
14. Knox C et al (2011) DrugBank 3.0: a comprehensive resource for 'omics' research on drugs. Nucleic Acids Res 39:D1035–D1041
15. Ho Sui SJ et al (2009) The association of virulence factors with genomic islands. PLoS One 4:e8094
16. Haft DH et al (2003) The TIGRFAMs database of protein families. Nucleic Acids Res 31:371–373
17. Tatusov RL et al (2003) The COG database: an updated version includes eukaryotes. BMC Bioinformatics 4:41
18. Stein LD (2013) Using GBrowse 2.0 to visualize and share next-generation sequence data. Brief Bioinform 14:162–171
19. Li H et al (2009) The sequence alignment/map format and SAMtools. Bioinformatics 25:2078–2079
20. Barrett T et al (2011) NCBI GEO: archive for functional genomics data sets—10 years on. Nucleic Acids Res 39:D1005–D1010
21. Langille MG et al (2009) IslandViewer: an integrated interface for computational identification and visualization of genomic islands. Bioinformatics 25:664–665
22. Waack S et al (2006) Score-based prediction of genomic islands in prokaryotic genomes using hidden Markov models. BMC Bioinformatics 7:142
23. Hsiao W et al (2003) IslandPath: aiding detection of genomic islands in prokaryotes. Bioinformatics 19:418–420
24. Langille MG et al (2008) Evaluation of genomic island predictors using a comparative genomics approach. BMC Bioinformatics 9: 329
25. Fulton DL et al (2006) Improving the specificity of high-throughput ortholog prediction. BMC Bioinformatics 7:270
26. Whiteside MD et al (2013) OrtholugeDB: a bacterial and archaeal orthology resource for improved comparative genomic analysis. Nucleic Acids Res 41:D366–D376
27. Darling AC et al (2004) Mauve: multiple alignment of conserved genomic sequence with rearrangements. Genome Res 14:1394–1403
28. Darling AE et al (2010) progressive Mauve: multiple genome alignment with gene gain, loss and rearrangement. PLoS One 5:e11147
29. Rissman AI et al (2009) Reordering contigs of draft genomes using the mauve aligner. Bioinformatics 25:2071–2073
30. Kurtz S et al (2004) Versatile and open software for comparing large genomes. Genome Biol 5:R12
31. Klimke W et al (2011) Solving the problem: genome annotation standards before the data deluge. Stand Genomic Sci 5:168–193

Chapter 34

Identification of Bacterial Small RNAs by RNA Sequencing

María Gómez-Lozano, Rasmus Lykke Marvig, Søren Molin, and Katherine S. Long

Abstract

Small regulatory RNAs (sRNAs) in bacteria are known to modulate gene expression and control a variety of processes including metabolic reactions, stress responses, and pathogenesis in response to environmental signals. A method to identify bacterial sRNAs on a genome-wide scale based on RNA sequencing (RNA-seq) is described that involves the preparation and analysis of three different sequencing libraries. As a significant number of unique sRNAs are identified in each library, the libraries can be used either alone or in combination to increase the number of sRNAs identified. The approach may be applied to identify sRNAs in any bacterium under different growth and stress conditions.

Key words Small RNA, sRNAs, Sequencing, rRNA depletion, RNA-seq, Illumina, Transcriptome, Data analysis

1 Introduction

Small RNAs (sRNAs) in bacteria play important regulatory roles in controlling various physiological processes such as carbon metabolism, iron homeostasis, and virulence in response to environmental cues. Bacteria contain hundreds of sRNAs that exist as a heterogeneous group of transcripts with a typical size range of 70–500 nucleotides (nt) [1, 2]. Many sRNAs function by affecting the expression of mRNA targets via base-pairing, while others act by binding to proteins and altering their activity. The base-pairing sRNAs fall into two broad classes: *cis*-encoded antisense sRNAs and *trans*-encoded sRNAs. The former are encoded on the DNA strand opposite the target RNA and therefore have perfect complementarity with their targets. *Trans*-encoded sRNAs are encoded in intergenic regions and often have multiple targets with which there is only limited complementarity. Despite the fact that some bacterial sRNAs have been known for some time, earlier studies to identify these RNAs have often depended on serendipity, direct detection due to high abundance, protein copurification, and

Alain Filloux and Juan-Luis Ramos (eds.), *Pseudomonas Methods and Protocols*, Methods in Molecular Biology, vol. 1149, DOI 10.1007/978-1-4939-0473-0_34, © Springer Science+Business Media New York 2014

computational predictions based on sequence conservation [3]. Recent efforts to identify sRNAs on a genome-wide scale have been based mainly on the extremely sensitive, probe-independent RNA sequencing approach (RNA-seq) that has revolutionized the field and led to the discovery of hundreds of novel sRNAs in diverse bacteria [4, 5].

We have recently developed a robust RNA-seq method and used it to identify over 500 novel sRNAs in *Pseudomonas aeruginosa* [6]. The approach is based on three different sequencing libraries prepared from different RNA populations that can either be used alone or in combination to increase the number of identified sRNAs. In order to enrich the RNA samples for sRNA detection, ribosomal RNAs (23S, 16S, and 5S rRNAs) are depleted from the total RNA. One of the libraries (LIB > 100) is prepared using a standard RNA-seq protocol for full transcriptomes and contains information on all mRNAs transcribed by the bacteria. The other two libraries (LIB < 500 and LIB < 200) are prepared from size-selected RNA containing transcripts shorter than 500 nt and 200 nt, respectively. A significant number of unique sRNAs are detected in each library, suggesting that the set of identified RNAs obtained depends strongly on the specific library preparation strategy used [6]. The detailed experimental procedures to prepare the three sequencing libraries are described in the first sections of the chapter.

Using RNA-seq allows counting the number of reads that align to specific parts of a genome, producing results similar to those of gene expression microarrays. While the end results are similar, the informatic challenges of analyzing RNA-seq data are fundamentally different than those of analyzing microarrays. Although there are several ways in which RNA-seq data can be analyzed, our analysis pipeline to find novel *P. aeruginosa* intergenic and antisense sRNAs is detailed in the last sections of this chapter.

2 Materials (*See* Note 1)

2.1 General Materials and Equipment

1. Disposable, nuclease-free pipette tips with filter, 1.5-mL microcentrifuge tubes, 15-mL conical tubes.
2. RNase-ZAP (Ambion) to remove ribonuclease (RNase) contamination from glass and plastic (*see* **Note 1**).
3. Glycogen (5 mg/mL) (Ambion).
4. Nuclease-free water (not DEPC-treated water).
5. TE buffer (10 mM Tris–HCl, pH 7, 1 mM EDTA).
6. 100 and 70 % Ethanol.
7. Phase-Lock Gel tubes, Heavy, 2 mL (5 PRIME).
8. 3 M sodium acetate (NaOAc) solution, pH 5.5 (Ambion).

9. PCI (Phenol–Chloroform–Isoamyl Alcohol) solution (25:24:1, v/v).
10. RiboLock RNase inhibitor (40 U/μL) (Thermo Scientific).
11. Microcentrifuges (room temperature and 4 °C), centrifuge (4 °C).
12. Vortexer.
13. Heat block.
14. Bioanalyzer (Agilent).
15. RNA 6000 Nano Kit (Agilent).
16. Qubit® Fluorometer (Invitrogen).
17. Qubit® dsDNA HS or BR Assay Kit.
18. Qubit® RNA Assay Kit.
19. Low-volume spectrophotometer, for example, the Nanodrop (Thermo Fisher Scientific).

2.2 Preparation of Total RNA

1. Isopropanol.
2. Trizol (Invitrogen).
3. Chloroform.
4. Phenol.
5. Deoxyribonuclease (DNase) I, nuclease-free (1 U/μL) (Thermo Scientific).

2.3 Depletion of Ribosomal RNAs (rRNAs)

1. MICROB*Express*™ Bacterial mRNA Enrichment Kit (Ambion).
2. DynaMag-2 Magnet (Invitrogen).
3. HPLC-purified 5S oligo: 5′-AAAAAAAAAAAAAAAAAAGCG TTTCACTTCTGAGTTCGGCA-3′.

2.4 Size Selection of Transcripts Shorter than 500 nt (for LIB<500 Library)

1. Mini-PROTEAN Electrophoresis System for vertical gel electrophoresis (Bio-Rad).
2. 10 % Mini-PROTEAN TBE-Urea Gel (Bio-Rad, cat no. 456-6033). This is a precast 10 % polyacrylamide gel (10-well, 30 μL well capacity) for use with the Mini-PROTEAN system (*see* **Note 2**).
3. Gel Loading Buffer II: 95 % Formamide, 18 mM EDTA, 0.025 % SDS, 0.025 % xylene cyanol, 0.025 % bromophenol blue (Ambion).
4. Nuclease-free, sterile 10× TBE solution: 890 mM Tris base, 890 mM boric acid, 20 mM EDTA, pH 8.0. From this stock prepare 1× TBE solution.
5. 0.4 M sodium chloride (NaCl) solution.
6. Low range ssRNA ladder (New England Biolabs).
7. 5 μm filter tube (IST Engineering).

8. Gel Breaker tube (IST Engineering).
9. Sterile scalpel blade.
10. SYBR Gold Nucleic Acid Gel Stain (Invitrogen).
11. Tube shaker or rotator.
12. Safe Imager 2.0 Blue Light Transilluminator (Invitrogen) or similar.

2.5 Library-Specific RNA Preparation Steps

1. *mir*Vana™ miRNA Isolation Kit (Ambion) (for LIB < 200 library).
2. Tobacco Acid Pyrophosphatase (10 U/μL) (Epicentre) (for LIB < 500 and LIB < 200 libraries).
3. RNase III (1 U/μL) (Invitrogen) (for LIB < 500 library).

2.6 Library Preparation

1. Magnetic Stand-96 (Ambion).
2. 96-well thermal cycler.
3. 10-, 200-, and 1,000-μL multichannel pipettes.
4. Agencourt AMPure XP 60 mL kit (Beckman Coulter Genomics).
5. 96-well 250-μL PCR plates and caps for PCR strips compatible with the plates.
6. 8-well PCR strip tubes and caps.
7. SuperScript II Reverse Transcriptase (Invitrogen).
8. 10 mM Tris–Cl solution, pH 8.5, containing 0.1 % Tween 20.
9. Microplate centrifuge.

2.6.1 Library Preparation (for LIB > 100 Library)

1. TruSeq™ RNA Sample Preparation Kit v2 (Illumina) (*see* **Notes 3** and **4**).
2. DNA 1000 kit (Agilent).

2.6.2 Library Preparation (for LIB < 500 and LIB < 200 Libraries)

1. TruSeq™ Small RNA Sample Preparation Kit (Illumina) (*see* **Notes 3** and **5**).
2. T4 RNA Ligase 2, truncated (New England Biolabs).
3. High Sensitivity DNA kit (Agilent).

2.7 List of Programs Used

The programs referred to in the data analysis pipeline are listed below with name, version, and URL:

1. Flexbar v2.2, http://sourceforge.net/projects/theflexibleadap/.
2. Bowtie 2, http://bowtie-bio.sourceforge.net/bowtie2/index.shtml.
3. SAMtools v1.4, http://samtools.sourceforge.net/.
4. Tablet v1.12.08.29, http://bioinf.scri.ac.uk/tablet/.

3 Methods

A general scheme of the method is illustrated in Fig. 1, and the library characteristics are summarized in Table 1. The libraries prepared in this protocol are compatible with the Illumina technology (*see* **Note 6**). The sequencing libraries are called LIB > 100, LIB < 500,

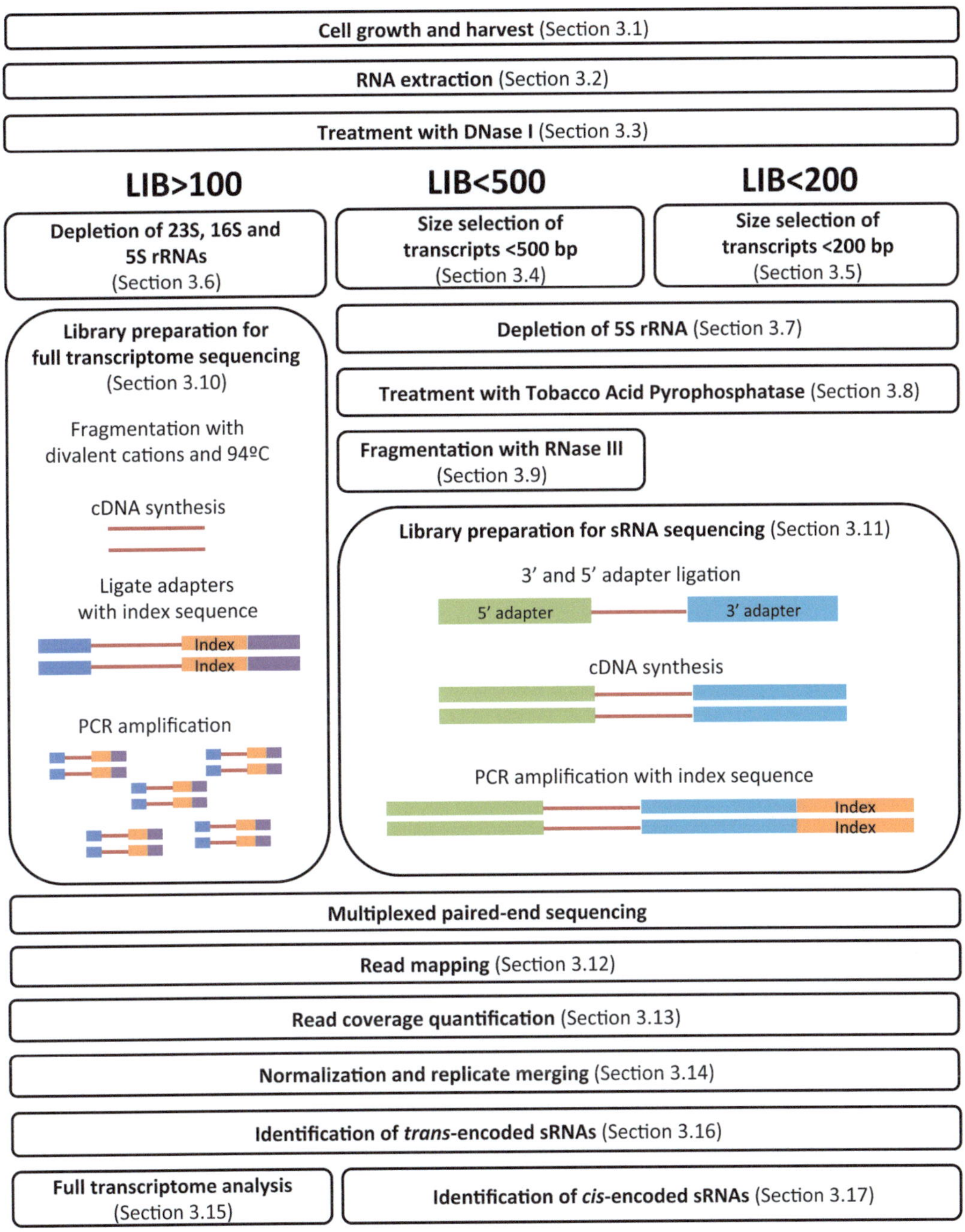

Fig. 1 Library preparation scheme summarizing the steps involved in the preparation of the different sequencing libraries (LIB > 100, LIB < 500, and LIB < 200)

Table 1
Library characteristics

	Library LIB > 100	Library LIB < 500	Library LIB < 200
Length of transcripts sequenced	>100 nt	<500 nt	<200 nt
Fragmentation	Divalent cations	RNase III	No
Strand specificity	No	Yes	Yes

and LIB < 200 to indicate the transcript sizes in nucleotides that are not excluded from each, but all transcripts in these size ranges are not necessarily detected in the corresponding library.

The three libraries were prepared with different RNA populations isolated from bacterial total RNA. The LIB > 100 library contains information corresponding to all transcripts in the cell, with the exception of rRNAs and transcripts shorter than 100 nt. This library was prepared using a standard RNA-seq protocol and is suitable for analyzing full transcriptomes. Although many known sRNAs are roughly 100 nt in length, transcripts shorter than 100 nt are not represented in this library because the corresponding cDNA fragments are not retained during the protocol. Size-selected RNA shorter than 500 and 200 nt and depleted of 5S rRNA was used to prepare the LIB < 500 and LIB < 200 libraries, respectively. As these libraries were prepared from RNA that does not include longer transcripts, they are enriched for information on sRNAs and thus more sensitive for sRNA detection compared to the LIB > 100 library that contains information on the full transcriptome [6]. Fragments corresponding to transcripts shorter than 100 nt are retained in the LIB < 500 and LIB < 200 libraries because specific adapters are ligated to the 5′- and 3′-ends of RNAs prior to cDNA synthesis. In addition, they are strand-specific, which means that the information about the strand from which the transcripts are transcribed is retained. Furthermore, treatment of RNA used to prepare the LIB < 500 and LIB < 200 libraries ensures the inclusion of primary transcripts containing a 5′-triphosphate (*see* **Note 7**).

The library preparation protocols also differ in terms of whether they include fragmentation steps. In order to be compatible with the relatively short read-lengths of the Illumina sequencing technology, long RNA molecules must be cleaved into shorter fragments. Therefore, the protocols for preparation of the LIB > 100 and LIB < 500 libraries include fragmentation steps using divalent cations and RNase III, respectively. The LIB < 200 library does not have any fragmentation-associated biases because the short RNAs used to prepare the library obviate the need for a fragmentation step.

In Subheadings 3.12–3.17, we describe a high-throughput bioinformatics pipeline for the analysis of RNA-Seq. The objectives of the pipeline are to detect transcripts and to quantify their relative abundance between different samples.

Our pipeline is designed for use on organisms where an annotated reference genome exists, and as input uses the quality-filtered and de-multiplexed sequencing reads obtained as output from sequencing centers (i.e., reads that are already in or can be converted to the FASTQ-format). Specifically, the pipeline was tested on 100-bp paired-end Illumina HiSeq2000 reads obtained from cDNA-libraries prepared from cultures of *P. aeruginosa* PAO1 [6].

3.1 Cell Growth and Harvest

1. Prepare a 15-mL conical tube with 2 mL of the STOP solution (95 % ethanol, 5 % phenol). Pre-chill on ice.
2. Dilute an overnight bacterial culture in 50–100 volumes of growth medium.
3. When the desired OD_{600} is reached, transfer 10 mL of culture into the 15-mL conical tube with the STOP solution.
4. Vortex thoroughly for 15 s. Incubate at room temperature for 5 min.
5. Pellet the bacteria by centrifugation (3,500 × *g*, 10 min, 4 °C). Remove the supernatant.
6. Dissolve the pellet in 1 mL Trizol by pipetting (*see* **Notes 8** and **9**). Incubate at room temperature for 5 min.
7. Proceed to RNA extraction (Subheading 3.2) or store at −80 °C. The homogenized samples can be stored at −80 °C for at least 1 month.

3.2 RNA Extraction

1. Spin down the Phase-Lock Gel (PLG) at 12,000–16,000 × *g* for 30 s.
2. Transfer the homogenized sample to the PLG tube.
3. Add 200 μL of chloroform (work inside a fume hood). Shake vigorously for 30 s (do not vortex).
4. Incubate at room temperature for 2 min.
5. Centrifuge at 12,000–16,000 × *g* for 5 min at room temperature. The PLG will form a barrier between the aqueous and organic phases.
6. Transfer 500 μL of the aqueous phase to a clean 1.5-mL tube.
7. Add 400 μL isopropanol, mix, and incubate for 10 min at room temperature.
8. Pellet the RNA by centrifugation (14,500 × *g*, 10 min, 4 °C). Remove the supernatant.
9. Wash the RNA pellet by adding 500 μL ice-cold 75 % ethanol and vortex briefly.

10. Pellet the RNA by centrifugation (14,500 × *g*, 5 min, 4 °C). Remove the supernatant (*see* **Note 10**).
11. Repeat the last two steps for a total of two washes with ethanol. It is important to remove the ethanol completely at the end of the second wash.
12. Air-dry the pellet at room temperature for 5 min by leaving the tube lid open.
13. Dissolve the RNA pellet in 100 μL of nuclease-free water.
14. Incubate at 65 °C with shaking for 5 min.
15. Proceed to DNase I treatment (Subheading 3.3) immediately or store at −80 °C.

3.3 Treatment with DNase I

1. Add to RNA sample (100 μL):

   1 μL RiboLock RNase inhibitor (40 U/μL)

   40 μL 10× DNase I reaction buffer with $MgCl_2$

   219 μL nuclease-free water

   40 μL DNase I 1 U/μL

   400 μL total volume
2. Mix by gently pipetting the entire volume ten times.
3. Incubate at 37 °C for 30 min.
4. Spin down the Phase-Lock Gel (PLG) at 12,000–16,000 × *g* for 30 s.
5. Transfer the reaction mixture to the PLG tube.
6. Add 400 μL PCI solution.
7. Shake vigorously for 30 s (do not vortex).
8. Centrifuge at 12,000–16,000 × *g* for 5 min at room temperature.
9. Transfer the aqueous phase to a clean 1.5-mL tube (volume should be approximately 400 μL).
10. Precipitate the RNA by adding 0.02 volume glycogen (5 mg/mL), 0.1 volume 3 M NaOAc pH 5.5, and 2.5 volumes of 100 % ethanol (ice-cold). Precipitate at −20 °C for at least 1 h. At this point, the RNA can be stored at −20 or −80 °C (*see* **Note 10**). Pellet the RNA by centrifugation (14,500 × *g*, 30 min, 4 °C) and remove the supernatant. Recover the RNA as described in **steps 9–12** of Subheading 3.2.
11. Resuspend in 50 μL TE buffer.
12. Check the concentration and purity of the RNA sample using a low volume spectrophotometer (*see* **Note 11**). Check the quality of the RNA with a RNA 6000 Nano chip on the Bioanalyzer (*see* **Note 12**). The RNA profile should appear similar to that in Fig. 2a. Proceed to the next step immediately (*see* Fig. 1) or store at −80 °C.

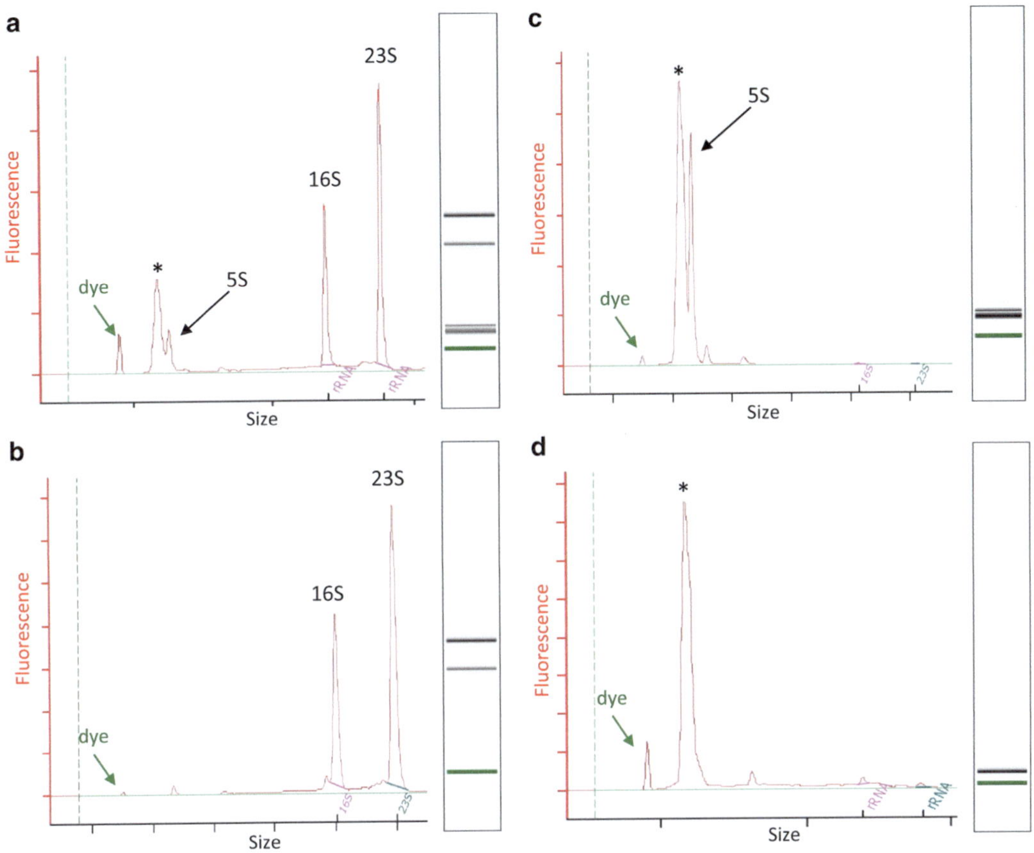

Fig. 2 Typical size profiles of RNA samples. All profiles correspond to RNA extracted from *P. aeruginosa* PAO1 cells analyzed with the RNA 6000 Nano chip on the Bioanalyzer (Agilent). (**a**) Total RNA extracted with Trizol and treated with DNase I. (**b**) Total RNA extracted with the NucleoSpin RNA II kit (Macherey Nagel) and treated with DNase I. (**c**) Size-selected RNA (<500 nt) purified on a denaturing polyacrylamide–urea gel. (**d**) An rRNA-depleted sample. The 23S (2,891 nt), 16S (1,536 nt), and 5S (120 nt) rRNAs were removed from the total RNA by subtractive hybridization as described in Subheadings 3.6 and 3.7. The peaks marked with asterisks correspond to tRNAs and sRNA species shorter than 120 nt

3.4 Size Selection of Transcripts Shorter than 500 nt (LIB < 500 Library)

Denaturing polyacrylamide–urea gels are used in order to size-select transcripts shorter than 500 nt (*see* **Note 2**).

1. Prior to sample loading the gels are pre-run in 1× TBE at 300 V for 30 min.
2. Combine on ice total RNA (up to 150 μg) in a maximum volume of 30 μL with one volume of Gel Loading Buffer II. Each sample is loaded onto two lanes of the gel.
3. For each lane containing the marker, combine on ice:

   6 μL of Low Range ssRNA Ladder (3 μg)

   4 μL nuclease-free water

   <u>10 μL Gel Loading Buffer II</u>

   20 μL total volume

4. Incubate the RNA and marker at 65 °C for 5 min just prior to loading them onto the gel.
5. Clean the wells of the gel with a syringe containing 1× TBE buffer.
6. For each RNA sample load equal volumes into two adjacent wells on the gel. Leave an empty lane next to the RNA ladder and also between different samples.
7. Run the gel at 200 V for 35 min.
8. Stain the gel by soaking it in 50 mL 1× TBE with 5 μL of SYBR Gold Nucleic Acid Gel Stain at room temperature for 30 min. Use a nuclease-free container that is protected from light.
9. Place the 0.5-mL Gel Breaker tube into a 2-mL microcentrifuge tube.
10. Visualize the RNA and cut out the gel containing RNA between 50 and 500 nt by comparison with the bands of the RNA ladder.
11. Place the gel pieces corresponding to the same sample into the 0.5-mL Gel Breaker tube.
12. Centrifuge the stacked tubes (20,000 × *g*, 2 min, room temperature) to move the gel through the holes and into the 2-mL tube. Ensure that all of the gel has moved into the bottom tube.
13. Add 400 μL 0.4 M NaCl solution to the gel debris.
14. Elute the RNA by rotating or shaking the tube overnight at room temperature.
15. Transfer the eluate and the gel debris to the top of a 5-μm filter tube.
16. Centrifuge the filter at 600 × *g* for 10 s. Discard the filter and retain the eluate which contains the RNA.
17. Precipitate the RNA by following **step 10** under Subheading 3.3.
18. Resuspend in 20 μL TE buffer.
19. Quantitate the yield of the rRNA-depleted RNA using the Qubit® RNA Assay Kit on the Qubit® Fluorometer (Invitrogen), or a low-volume spectrophotometer (*see* **Note 11**). Check the quality of the RNA with a RNA 6000 Nano chip on the Bioanalyzer (*see* **Note 12**). The profile of the RNA obtained should be similar to that depicted in Fig. 2c.
20. Proceed to the 5S rRNA depletion (Subheading 3.7) immediately or store at −80 °C.

3.5 Size Selection of Transcripts Shorter than 200 nt (LIB < 200 Library)

The *mir*Vana™ miRNA Isolation Kit from Ambion (Part Number 1560M Rev. C January 2011/http://tools.invitrogen.com/content/sfs/manuals/fm_1560.pdf) can be used to select for sRNAs shorter than 200 nt (*see* **Note 13**).

1. Follow the protocol described in section *IV. Additional Procedures. A. Isolation of Small RNAs from Total RNA Samples* of the kit manual to obtain a 100 μL solution containing the sRNAs ≤ 200 nt.
2. Precipitate the RNA by following **step 10** from Subheading 3.3.
3. Resuspend the RNA in 20 μL TE buffer.
4. Quantitate the yield of the rRNA-depleted RNA using the Qubit® RNA Assay Kit on the Qubit® Fluorometer (Invitrogen), or a low-volume spectrophotometer (*see* **Note 11**). Check the quality of the RNA with a RNA 6000 Nano chip on the Bioanalyzer (*see* **Note 12**). The profile of the RNA obtained should be similar to that depicted in Fig. 2c.
5. Proceed to the 5S rRNA depletion (Subheading 3.7) immediately or store at −80 °C.

3.6 Depletion of 23S, 16S, and 5S rRNAs (LIB > 100 Library)

After obtaining total RNA, the next step is to deplete abundant rRNAs from the sample. Bacterial total RNA consists of approximately 86 % rRNA [7]. Removal of rRNA significantly increases the fraction of mRNAs and sRNAs in the sample, allowing higher sequencing coverage and more sensitive detection of novel transcripts. Although the MICROB*Express*™ Bacterial mRNA Enrichment Kit from Ambion is used here (Part Number 1905M Rev. C January 2011/http://tools.invitrogen.com/content/sfs/manuals/fm_1905.pdf), there are other commercially available kits to deplete rRNA (*see* **Note 14**). The MICROB*Express*™ Kit is designed to enrich bacterial mRNA (including sRNAs) from purified total RNA by removing the 16S and 23S rRNAs via a subtractive hybridization approach.

The MICROB*Express*™ Kit is followed with the following two modifications: an additional oligo is employed to remove 5S rRNA and 25 % more magnetic beads are used. The modifications needed to remove the 23S, 16S, and 5S rRNAs in a single step are described below.

1. Using the concentration obtained by the spectrophotometer, calculate the volume of sample that contains 2–10 μg of RNA (*see* **Note 15**). The recommended maximum amount of RNA per reaction is 10 μg and the recommended maximum volume is 15 μL. If the recommended maximum amount of total bacterial RNA is exceeded, then rRNA removal will be incomplete.

2. Follow the protocol detailed in *Section B Anneal RNA and Capture Oligonucleotide Mix* of the kit manual.
3. After **step B.2**, add 1 μL of the HPLC-purified 5S oligo (20 μM). Continue the protocol from **step B.3**.
4. In **step C.1**, remove 62.5 μL of Oligo MagBeads for each RNA sample, corresponding to 25 % more beads per sample. The reason for this is that adding the HPLC-purified 5S oligo results in a 25 % increase in capture oligos. We have successfully processed Oligo MagBeads for up to ten samples (625 μL) in a single 1.5-mL tube. Continue the protocol from **step C.2**.
5. In **step D.2**, add 62.5 μL of the prepared Oligo MagBeads to the RNA/Capture Oligo Mix and incubate at 37 °C for 15 min. Continue the protocol from **step D.3**.
6. In **step E.2**, resuspend the RNA pellet in 25 μL of nuclease-free water. Do not resuspend in TE buffer. Continue the protocol from **step E.3** (but do not perform **step E.4**, as this removes sRNAs).
7. Quantitate the yield of the rRNA-depleted RNA using the Qubit® RNA Assay Kit on the Qubit® Fluorometer (Invitrogen). Check the quality of the RNA with a RNA 6000 Nano chip on the Bioanalyzer. The profile of the RNA obtained should be similar to that depicted in Fig. 2d.
8. Proceed to the LIB > 100 library preparation immediately (Subheading 3.10) or store at −80 °C.

3.7 Depletion of 5S rRNA After Size Selection (LIB < 500 and LIB < 200 Libraries)

Although the 23S and 16S rRNAs are depleted from the samples after the size selection step, a significant fraction of the sample now consists of 5S rRNA. In order to deplete 5S rRNA, the MICROB*Express*™ Kit manual (Part Number 1905M Rev. C January 2011/http://tools.invitrogen.com/content/sfs/manuals/fm_1905.pdf) is followed with the modification that the HPLC-purified 5S oligo is added as the only Capture Oligo as described below.

1. Using the concentration obtained by the spectrophotometer, calculate the volume of sample that contains 2–10 μg of RNA (*see* **Note 15**). The recommended maximum amount of RNA per reaction is 10 μg and the recommended maximum volume is 15 μL. If the recommended maximum amount of total bacterial RNA is exceeded, then rRNA removal will be incomplete.
2. Follow the protocol detailed in *Section B Anneal RNA and Capture Oligonucleotide Mix* of the kit manual. In **step B.2**, add 4 μL of the HPLC-purified 5S oligo (20 μM), but do not add 4 μL of Capture Oligo Mix. Continue the protocol from **step B.3**. Do not add the Capture Oligo Mix included in the kit, as this contains oligos complementary to 23S and 16S rRNAs.

3. In **step E.2**, resuspend the RNA pellet in 25 μL of nuclease-free water. Do not resuspend in TE buffer. Continue the protocol from **step E.3** (but do not perform **step E.4**, as this removes sRNAs).
4. Quantitate the yield of the rRNA-depleted RNA using the Qubit® RNA Assay Kit on the Qubit® Fluorometer (Invitrogen). Check the quality of the RNA with a RNA 6000 Nano chip on the Bioanalyzer. The profile of the RNA obtained should be similar to that depicted in Fig. 2d.
5. Proceed to the Tobacco Acid Pyrophosphatase treatment (Subheading 3.8) immediately or store at −80 °C.

3.8 Treatment with Tobacco Acid Pyrophosphatase (LIB < 500 and LIB < 200 Libraries)

After size-selection of RNA (used to prepare LIB < 500 or LIB < 200 libraries), the RNA is treated with Tobacco Acid Pyrophosphatase (TAP). TAP is used to convert 5′-triphosphate RNA into 5′-monophosphate RNA. This step is necessary to ensure that primary transcripts (containing a 5′-triphosphate) are ligated to adapters during the library preparation.

1. Assemble the reaction in a PCR tube on ice in the order given below:

   17 μL Size-selected RNA (up to 10 μg) diluted in nuclease-free water

   2 μL 10× TAP Reaction Buffer

   1 μL TAP (10 U/μL)

   20 μL total volume
2. Pipet up and down a few times to mix. Incubate at 37 °C for 1 h.
3. Spin down the Phase-Lock Gel (PLG) at 12,000–16,000 × *g* for 30 s.
4. Transfer the sample to the PLG tube.
5. Add 130 μL nuclease-free water.
6. Add 150 μL PCI solution.
7. Shake vigorously for 30 s (do not vortex).
8. Centrifuge at 12,000–16,000 × *g* at room temperature for 5 min.
9. Transfer the aqueous phase to a clean 1.5-mL tube (the volume should be approximately 150 μL).
10. Precipitate the RNA by following **step 10** from Subheading 3.3.
11. Resuspend in 15 μL nuclease-free water.
12. Quantitate the yield of the RNA using the Qubit® RNA Assay Kit on the Qubit® Fluorometer (Invitrogen). Check the quality of the RNA with a RNA 6000 Nano chip on the Bioanalyzer.

The profile of the RNA obtained should be similar to that depicted in Fig. 2d.

13. Proceed to the next step immediately (*see* Fig. 1) or store at −80 °C.

3.9 Fragmentation with RNase III (LIB < 500 Library)

During preparation of the LIB < 500 library, the size-selected RNA needs to be fragmented after the TAP treatment. The fragmentation is performed with RNase III, which cleaves long double-stranded RNA (dsRNA) into shorter dsRNAs containing 5′-PO_4, and 3′-OH termini, and a 3′ dinucleotide overhang. The fragmentation procedure should produce a distribution of RNA fragment sizes from 35 nt to several hundred nucleotides, where the average size is 100–200 nt.

1. Assemble the reaction in a microcentrifuge tube on ice in the order given below:

   32 μL RNA (2 μg, from **step 13**, Subheading 3.8) diluted in nuclease-free water

   4 μL 10× RNase III Buffer

   4 μL RNase III (1 U/μL)

   40 μL total volume
2. Pipet up and down a few times to mix. Incubate at 37 °C for 10 min.
3. Immediately after the incubation, add 110 μL of nuclease-free water and place the fragmented RNA on ice.
4. Spin down the Phase-Lock Gel (PLG) at 12,000–16,000 × *g* for 30 s.
5. Transfer the sample to a PLG tube.
6. Add 150 μL PCI solution.
7. Shake vigorously for 30 s (do not vortex).
8. Centrifuge at 12,000–16,000 × *g* for 5 min at room temperature.
9. Transfer the aqueous phase to a clean 1.5-mL tube (the volume should be approximately 150 μL).
10. Precipitate the RNA by following **step 10** from Subheading 3.3.
11. Resuspend in 15 μL nuclease-free water.
12. Quantitate the yield of the fragmented RNA using the Qubit® RNA Assay Kit on the Qubit® Fluorometer (Invitrogen). Check the quality of the RNA with a RNA 6000 Nano chip on the Bioanalyzer. The profile of the RNA obtained should be similar to that depicted in Fig. 2d. Proceed to the LIB < 500 library preparation (Subheading 3.11) immediately, or store at −80 °C.

3.10 Library Preparation for Full Transcriptome Sequencing (LIB < 100 Library)

The LIB > 100 library has been prepared with the Illumina TruSeq™ RNA Sample Preparation Kit v2 (Part Number 15026495 Rev. D September 2012/http://support.illumina.com/documents/MyIllumina/b386d5c9-c919-48db-bdc1-8a687ba2a101/TruSeq_RNA_SamplePrep_v2_Guide_15026495_D.pdf) (*see* **Notes 3** and **4**) with some modifications that are explained below. Read the *Best Practices* section of the manual carefully. We perform the Low Sample (LS) protocol. If only two to four samples are to be pooled, read the *Pooling Guidelines* section of the manual.

1. Start the protocol with 100–400 ng of rRNA-depleted RNA, diluted in a maximum volume of 5 μL nuclease-free water (*see* **Note 15**).
2. Add 13 μL Elute, Prime, Fragment Mix to the sample.
3. Continue the protocol from step *Incubate RFP* (included in section *Purify and Fragment RNA*). Perform all the steps included in sections *Incubate RFP, Synthesize First Strand cDNA, Synthesize Second Strand cDNA, Perform End Repair, Adenylate 3′ Ends, Ligate Adapters* and *Enrich DNA Fragments* (*see* **Note 16**). At this point the amplified libraries are in a 30 μL volume.
4. Quantify the libraries using the Qubit® dsDNA HS or BR Assay Kit on the Qubit® Fluorometer (Invitrogen) (*see* **Note 17**).
5. Check the size and purity of the sample by running a DNA 1000 chip on the Bioanalyzer. The final product should be a broad band with an average size of approximately 260 bp (*see* Fig. 3a).
6. Adjust the concentration of the libraries to 10 nM using a 10 mM Tris–Cl (pH 8.5) solution containing 0.1 % Tween 20.
7. Combine 10 μL of each normalized sample library to be pooled in a 1.5-mL microcentrifuge tube (*see* **Note 3**). Each library in the pooled mixture must have a different index or barcode.
8. Send an aliquot of the pooled libraries to be sequenced (*see* **Note 6**) or store at −20 °C. It is recommended that the libraries are shipped on dry ice.

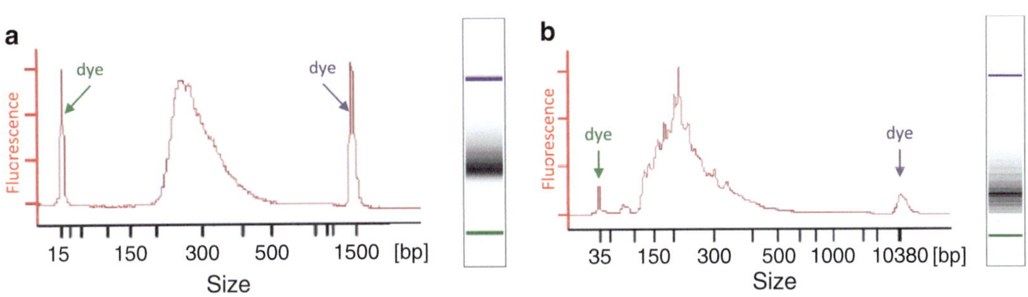

Fig. 3 Typical size profiles of amplified libraries. (**a**) The LIB > 100 library analyzed with a DNA 1000 chip on the Bioanalyzer (Agilent). (**b**) The LIB < 500 library analyzed with a High Sensitivity DNA chip on the Bioanalyzer (Agilent). Libraries prepared using the LIB < 200 protocol will have a very similar profile

3.11 Library Preparation for sRNA Sequencing (LIB < 500 and LIB < 200 Libraries)

The LIB < 500 and LIB < 200 libraries are prepared with the Illumina TruSeq™ Small RNA Sample Preparation Kit (Part Number 15004197 Rev. D May 2012/http://support.illumina.com/documents/MyIllumina/b76a55b0-4ac7-4ee7-aa23-0e993cf7f467/TruSeq_SmallRNA_SamplePrep_Guide_15004197_D.pdf) (*see* **Notes 3** and **5**) with some modifications that are explained below. Read the *Best Practices* section from the manual carefully. If more than eight samples are to be prepared, then we recommend using 250-μL PCR plates and caps for PCR strips that are compatible with the plates to perform the protocol.

1. Start the protocol at the section *Ligate 3′ and 5′ Adapters* with 100–400 ng of size-selected RNA, diluted in a maximum volume of 5 μL nuclease-free water (*see* **Note 15**).
2. Perform all steps from sections *Ligate 3′ and 5′ Adapters* and *Reverse Transcribe and Amplify*. At this point the amplified libraries are in a 50 μL volume.
3. Allow the AMPure XP Beads to come to room temperature prior to use. Vortex the beads until they are well dispersed, and then add 50 μL of the mixed AMPure XP Beads to each PCR amplified library. Gently pipette the entire volume up and down ten times to mix thoroughly.
4. Incubate the PCR tubes or plate at room temperature for 15 min.
5. Place the PCR tubes or plate on the magnetic stand at room temperature for 5 min or until the liquid appears clear.
6. Remove and discard 95 μL of the supernatant from each PCR tube or plate well.
7. With the PCR tubes or plate remaining on the magnetic stand, add 200 μL of freshly prepared 80 % EtOH to each tube or well without disturbing the beads.
8. Incubate at room temperature for 30 s, then remove and discard all of the supernatant from each tube or well.
9. Repeat the last two steps once for a total of two 80 % EtOH washes.
10. While keeping the PCR tubes or plate on the magnetic stand, let the samples air-dry at room temperature for 15 min and then remove the tubes or plate from the magnetic stand.
11. Resuspend the dried pellet with 22.5 μL TE buffer. Gently pipette the entire volume up and down ten times to mix thoroughly.
12. Incubate at room temperature for 2 min.
13. Place the PCR tubes or plate on the magnetic stand at room temperature for 5 min or until the liquid appears clear.

14. Transfer 20 μL of the clear supernatant to a clean PCR tube or plate well.
15. Quantify the libraries using the Qubit® dsDNA HS BR Assay Kit on the Qubit® Fluorometer (Invitrogen) (*see* **Note 17**).
16. Check the size and purity of the sample by running a High Sensitivity DNA chip on the Bioanalyzer. The final product should be a broad band with an average size of approximately 200 bp (*see* Fig. 3b).
17. Adjust the concentration of the libraries to 10 nM using a 10 mM Tris–Cl (pH 8.5) solution containing 0.1 % Tween 20.
18. Combine 10 μL of each normalized sample library to be pooled in a 1.5-mL microcentrifuge tube (*see* **Note 3**). Each library in the pooled mixture must have a different index or barcode.
19. Send an aliquot of the pooled libraries to be sequenced (*see* **Notes 5** and **6**) or store at −20 °C. It is recommended that the libraries are shipped on dry ice.

3.12 Read Mapping

The first step of the data analysis pipeline is to map the reads against the reference genome. Note that cDNA libraries sequenced using a paired-end protocol will yield two FASTQ files per library, one containing the forward reads, and the other the reverse reads. In the LIB < 500 and LIB < 200 libraries, information corresponding to short transcripts (<100 nucleotides) is retained in the library construction. If the native read lengths are longer than the shortest cDNA fragments, then the read sequence will run through the cDNA fragment and extend into the adapter sequence. The 3′-end adapter sequences should therefore be trimmed away before mapping. There are several tools to trim reads, including the Flexbar tool that allows for flexible barcode detection and adapter removal. We map the adapter-trimmed reads onto the reference genome using Bowtie 2, which is a fast and memory-efficient tool for aligning short sequence reads onto reference genomes. The output of Bowtie 2 is the alignment of all the reads onto the reference genome in the generic Sequence Alignment/Map (SAM) format. The SAM format is a tab-delimited text format containing generic alignment information. For faster processing the SAM format can be converted into the equivalent binary format BAM.

3.13 Read Coverage Quantification

In order to calculate the transcription level across the genome at a single base resolution, the number of aligned reads for each position in the genome is counted. Thus, the global alignment stored in the SAM format needs to be converted into a list of the genomic positions together with the number of covering bases.

We use SAMtools for manipulating alignments in the SAM format. The SAM-formatted alignments are then converted into the BAM format, which is sorted to create a pileup file. Pileup format is a text-based format for summarizing the base calls of aligned reads at each covered position in the reference sequence. In the pileup file the information about the genomic location (chromosome and position) and the count of covering reads is stored in columns one, two and eight, respectively. Note that the pileup format only contains information about genomic positions that are covered by reads. It is therefore recommended to amend the pileup file to include uncovered positions designating the number of covering reads to be 0. If there is only one chromosome, the information about coverage depth along the genome can be stored in two columns, with the genomic position specified in the first column, and the number of covering reads given in the second column. This type of file will now be referred to as the coverage file.

3.14 Normalization and Replicate Merging

After determining the read coverage across the genome, the read counts in the coverage file are then normalized according to the total number of reads in the cDNA library. This is done in order to compare the level of transcription across different libraries. After normalization the normalized coverage files from replicate samples are merged by calculating the average read count at each position.

3.15 Full Transcriptome Analysis (Library LIB > 100)

If the library LIB > 100 protocol has been followed, it is now possible to calculate the transcription levels of all already-annotated genes.

1. Using your normalized coverage files (not merged), calculate the average expression of each gene by dividing the normalized read coverage depth (the sum of coverage at each base of the gene) by the length of the transcript. It is necessary to have an annotation file containing all the annotated genes of your organism, their coordinates and length.
2. Analyze differential gene expression between the different samples, using ANOVA for example.

3.16 Identification of Trans-encoded sRNAs (All Libraries)

Most sRNAs characterized to date regulate gene expression by base pairing with mRNAs. These sRNAs fall into two broad classes: those encoded in *cis* on the DNA strand opposite the target RNA that share extended regions of perfect complementarity (*cis*-encoded antisense sRNAs), and those encoded in intergenic regions that share limited complementarity with their targets (*trans*-encoded sRNAs). In this section, we explain how to identify *trans*-encoded sRNAs. Although this type of analysis can be performed using all three types of libraries described here, a larger number of sRNAs will be detected in the LIB < 500 and LIB < 200 libraries because they are prepared from RNA that is size-selected and enriched for sRNAs [6]. For the purpose of our analysis we define *trans*-encoded sRNAs as transcripts encoded entirely in the intergenic regions. Nonetheless, some *trans*-encoded sRNAs might extend into their neighboring genes. Also, some of the identified sRNAs might represent 5′-UTRs truncated before the start codon of their respective genes. In order to detect *trans*-encoded sRNAs in a high-throughput manner, we have employed an automated algorithm that detects transcripts starting and ending in intergenic regions. The algorithm detects the sudden increases in read coverage depths in the intergenic regions by the following criteria:

(a) A transcript is detected if the coverage depth increases above a cutoff that is set to discriminate noise caused by sequenced mRNA. Our cutoff was set based on the average expression intensities of 44 previously validated sRNAs in *P. aeruginosa* [6]. (b) The average transcriptional level (calculated as the normalized read coverage depth divided by the length of the transcript) of the detected transcript must be twice the previously defined cutoff to be considered further as an sRNA candidate. The two cutoff values are necessary and used together to identify sRNAs in that the second cutoff (b) ensures detection of very abrupt increases in read coverage depths whereas the first cutoff (a) enables the length of the transcripts to be defined more accurately.

After detecting transcripts, their transcriptional levels can be compared between samples as described in the previous section. The detected *trans*-encoded sRNAs can be visualized using a global alignment viewer such as Tablet [8] that may be helpful in more accurately defining the lengths of the sRNAs. Tablet visualizes the alignment using the sorted BAM format described in Subheading 3.13. Tablet also requires an indexed BAM formatted alignment to be indexed. Indexing enables a more rapid alignment retrieval, and can be performed with SAMtools.

3.17 Identification of Cis-encoded sRNAs (LIB < 500 and LIB < 200 Libraries)

The next objective is detection of *cis*-encoded sRNAs. As these sRNAs are transcribed from the DNA strand opposite the target RNA, they can only be directly identified in strand-specific libraries like LIB < 500 or LIB < 200. The *cis*-encoded sRNAs are detected using an algorithm similar to the one used for detection of *trans*-encoded sRNAs. As the *cis*-encoded sRNAs are transcribed opposite to their respective genes, the analysis should be constrained to reads being in the opposite direction to the genes onto which they align.

Information on the direction of the aligned reads is contained in the second column of the SAM format. A number of different flags are used to describe the orientation of the single read and its corresponding mate, and flags "99," "147," "83," and "163" designate the read and its mate to be properly aligned within a distance as defined by the insert size. Read pairs with flags 99 or 147 originate from transcripts encoded on the plus strand, while read pairs with flags 163 or 83 originate from transcripts encoded on the minus strand. This allows us to only extract reads that map to a known gene, but have the opposite direction of the gene. The reads that originate from a gene encoded on the plus strand will have the flags 99 or 147. Thus, the reads that map to this gene and have the flags 163 or 83 need to be extracted. The procedure is likewise for genes encoded on the minus strand, where the reads with flags 99 or 147 need to be extracted.

The extracted reads are placed into SAM formatted files that are processed into new pileup format files which can then be normalized and merged as previously described in Subheadings 3.13 and 3.14. The algorithm described in Subheading 3.16 is used in order to detect transcripts with sudden increases in read coverage depths. The difference in this case is that only transcripts that overlap genes are evaluated. The ends of mRNA molecules that do not encode proteins are termed untranslated regions (UTRs). These regions can contain regulatory elements for controlling gene expression and extend into the neighboring genes. Be aware of the fact that some *cis*-encoded sRNAs detected in your libraries might be UTRs from neighboring genes. Calculations of transcriptional levels, comparisons between samples, and visualization of *cis*-encoded sRNAs can be performed as previously described in Subheadings 3.13–3.16.

4 Notes

1. RNA is highly susceptible to degradation by ribonucleases. These enzymes are present on skin, laboratory equipment, and in dust. They are very stable and difficult to inactivate. For these reasons, it is important to follow best laboratory practices while preparing and handling RNA samples. Wear gloves and use sterile technique at all times. Reserve a set of pipettes for RNA work. Use sterile RNase-free filter pipette tips to prevent cross-contamination. Use RNase-free disposable plasticware. All reagents and solutions should be sterile and RNase-free where possible. Use RNase decontamination solution (such as RNase-ZAP) to decontaminate work surfaces and equipment prior to starting this protocol.
2. Although we have used precast gels to size-select transcripts shorter than 500 nt, 1-mm thick, 10 % polyacrylamide TBE-Urea minigels can also be prepared. For each gel, mix 1 mL 10× TBE with 3.33 mL acrylamide: bis-acrylamide (37.5:1), 2.14 mL water, 4.2 g urea, 330 μL 1.6 % ammonium persulfate, and 5 μL tetramethylethylenediamine (TEMED). The glass plates, combs, and chambers used should be compatible with the Mini-PROTEAN or other suitable electrophoresis system.
3. It is highly recommended to work with at least two biological replicates of every sample to be sequenced. This will enable a

better comparison of the transcript expression levels between different samples. Thus, if there are 12 samples or conditions, and there are two replicates per sample, then there are 24 libraries to be sequenced. If multiplexed sequencing (sequencing different samples in the same lane) is to be performed, then each library needs to be prepared using a different index or barcode. This will allow the different libraries to be pooled together and sequenced in one lane. Therefore, the number of indexes or barcodes needed is the same as the total number of libraries to be sequenced in the same lane. We have successfully sequenced up to 24 different RNA libraries in the same lane.

4. There are two different versions of the TruSeq™ RNA Sample Prep Kit v2 (Illumina), where each contains 12 different indexes or barcodes (24 unique indexes or barcodes in total).
5. There are four different versions of the TruSeq™ Small RNA Sample Preparation Kit (Illumina), where each contains 12 different indexes or barcodes (48 unique indexes or barcodes in total). This kit is designed for sequencing of eukaryotic microRNAs (miRNAs), which are around 22 nt in length. The libraries obtained following the kit manual will contain DNA fragments of around 150 bp, including the adapters. We describe modifications that allow using the TruSeq™ Small RNA Sample Preparation Kit for preparing libraries that enable sequencing of bacterial small RNAs, which normally range between 70 and 500 nt in length. The libraries prepared following our protocol will contain fragments that range from 100 to 400 bp, including the adapters. Make sure to inform your sequencing provider about these modifications.
6. The libraries described here were sequenced using the Illumina HiSeq2000 platform with a paired-end protocol and read lengths of 100 nt. On average, 16.7 million sequence reads were generated from each library, and of these 12.3 million were of sufficient quality to be mapped onto our reference genome [6]. However, as sequencing technologies are rapidly evolving, sequencing using the latest available technology as well as the longest possible read length is recommended.
7. The RNA used to prepare the LIB < 500 and LIB < 200 libraries was treated with Tobacco Acid Pyrophosphatase (TAP) to ensure the inclusion of primary transcripts containing a 5′-triphosphate. Although primary and processed transcripts are not distinguishable in these libraries, it is possible to modify the approach with additional libraries to obtain this information. One approach is to discriminate primary from processed 5′ ends by preparing two libraries from the same RNA sample. One library is prepared with untreated RNA and the other library is enriched for primary transcripts by treatment with

Terminator 5′-Phosphate-Dependent Exonuclease (Epicentre), which degrades RNA with 5′-monophosphate but not 5′-triphosphate ends [9].

8. There are a number of protocols and kits for RNA isolation, but many of them do not quantitatively retain the smaller RNAs that are of interest here. This is the case for most of the kits that use spin columns. We use Trizol (Invitrogen) to isolate RNA because it is a quick method that allows retention of all RNA species. The protocol described here can also be used to extract RNA from mucoid cells. If another method is used to extract RNA, it is important to ensure that small RNA species are retained by checking the sample with a Bioanalyzer. The RNA profile obtained after extraction with Trizol is shown in Fig. 2a, and that obtained with a spin column kit (in this example the NucleoSpin RNA II kit from Macherey Nagel) is shown in Fig. 2b.
9. One milliliter of Trizol is sufficient to extract RNA from 10 mL of *P. aeruginosa* culture harvested at OD_{600} of 0.5–3, where the typical yields of total RNA obtained are 40–150 μg. For other growth/harvest conditions or for bacteria other than *P. aeruginosa*, optimization of the RNA extraction method may be necessary.
10. The RNA can be stored as pellets in 75 % ethanol at −20 or −80 °C for at least 1 year, or at 4 °C for at least 1 week. RNA storage in ethanol is highly recommended for long-term storage (weeks or months).
11. It is important that both the A_{260}/A_{280} and the A_{260}/A_{230} ratios of RNA solutions are close to 2.0. RNA has its absorbance maximum at 260 nm and this absorbance is not dependent on the pH of the solution. However, the absorbance of some contaminants (like proteins) is pH-dependent. This means that although the A_{260} reading of the RNA solution will remain the same at different pH values, the A_{280} reading will differ in a pH-dependent manner. Adjusting the pH of the RNA solution from approximately 5.5 to a slightly alkaline pH of 7.5–8.5 significantly increases RNA A_{260}/A_{280} ratios from approximately 1.5 to 2.0 [10]. This should be taken into account when measuring the absorbance ratio of RNA dissolved in nuclease-free water. When isolating low amounts of RNA the A_{260}/A_{230} ratio tends to be lower. This indicates contamination of the RNA solution with chaotropic salts, phenol, or protein.
12. The Bioanalyzer (Agilent) is a microfluidics-based platform for sizing, quantification and quality control of DNA, RNA, proteins and cells. Here it is used for checking the quality of RNA, and the size of DNA libraries. We recommend quantifying

RNA and DNA using other technologies such as a low-volume spectrophotometer or Qubit® Fluorometer (Invitrogen). The Bioanalyzer allows for rapid analysis of RNA and DNA samples (30–40 min) with minimal sample consumption (1 μL). The higher sensitivity afforded by the Bioanalyzer is the main advantage over gel analysis, where larger sample amounts are required.

13. Denaturing polyacrylamide–urea gels can also be used to size-select RNAs shorter than 200 nt (*see* **Note 2**). However, we find it easier and faster to use the *mir*Vana™ miRNA Isolation Kit for this purpose.

14. The MICROB*Express*™ Bacterial mRNA Enrichment Kit (Ambion) has been used because it employs subtractive hybridization to specifically deplete rRNAs from total RNA. In our libraries, the average percentage of read coverage corresponding to 23S, 16S, and 5S rRNAs was 7.1 %, 3.8 %, and 0.15 %, respectively [6]. For bacteria other than *P. aeruginosa* and *E. coli* the oligos used to deplete rRNAs may need to be redesigned. There are also other commercially available kits to deplete rRNA. One of them uses exonuclease to preferentially degrade processed RNAs with 5′ monophosphate ends (mRNA-ONLY Prokaryotic mRNA Isolation kit, from Epicentre). Exonuclease degrades all processed RNAs and, although the majority of the processed RNAs are believed to be rRNAs and tRNAs, there may be other transcripts including sRNAs that are also processed. He et al. compared the performance of the MICROB*Express*™ and mRNA-ONLY kits and found that only the former based on subtractive hybridization adequately preserved relative transcript abundance for quantitative analyses, whereas the latter based on exonuclease treatment greatly compromised mRNA abundance fidelity [11]. Another commercially available kit based on subtractive hybridization contains capture oligonucleotides to remove 23S, 16S, and 5S rRNAs (Ribo-Zero rRNA Removal Kit, Epicentre).

15. Concentration of RNA samples can be achieved via an ethanol precipitation. Follow **step 10** from Subheading 3.3 and resuspend in the appropriate volume of TE buffer or nuclease-free water.

16. Illumina recommends using Microseal "B" adhesive seals (Bio-Rad) to seal the PCR plates. However, we find that sealing the PCR plates using caps for compatible PCR strips is easier and safer.

17. In our experience the Qubit® Flurometer (Invitrogen) is a fast and easy alternative for library quantification relative to qPCR, which is recommended by Illumina.

Acknowledgments

This work was supported by the Novo Nordisk Foundation, Lundbeck Foundation, and the Danish National Research Foundation.

References

1. Waters LS, Storz G (2009) Regulatory RNAs in bacteria. Cell 136:615–628
2. Gottesman S, Storz G (2011) Bacterial small RNA regulators: versatile roles and rapidly evolving variations. Cold Spring Harb Perspect Biol 3:a00379
3. Vogel J, Sharma CM (2005) How to find small non-coding RNAs in bacteria. Biol Chem 386:1219–1238
4. Sharma C, Vogel J (2009) Experimental approaches for the discovery and characterization of regulatory small RNA. Curr Opin Microbiol 12:536–546
5. Storz G, Vogel J, Wassarman KM (2011) Regulation by small RNAs in bacteria: expanding frontiers. Mol Cell 43:880–891
6. Gómez-Lozano M, Marvig RL, Molin S, Long KS (2012) Genome-wide identification of novel small RNAs in *Pseudomonas aeruginosa*. Environ Microbiol 14:2006–2016
7. Karpinets TV, Greenwood DJ, Sams CE, Ammons JT (2006) RNA:protein ratio of the unicellular organism as a characteristic of phosphorous and nitrogen stoichiometry and of the cellular requirement of ribosomes for protein synthesis. BMC Biol 4:30
8. Milne I, Bayer M, Cardle L, Shaw P, Stephen G, Wright F, Marshall D (2010) Tablet—next generation sequence assembly visualization. Bioinformatics 26:401–402
9. Sharma CM, Hoffmann S, Darfeuille F, Reignier J, Findeiss S, Sittka A, Chabas S, Reiche K, Hackermüller J, Reinhardt R, Stadler PF, Vogel J (2010) The primary transcriptome of the major human pathogen *Helicobacter pylori*. Nature 464:250–255
10. Wilfinger W, Mackey K, Chomczynski P (1997) Effect of pH and ionic strength on the spectrophotometric assessment of nucleic acid purity. Biotechniques 22:474–481
11. He S, Wurtzel O, Singh K, Froula JL, Yilmaz S, Tringe SG, Wang Z, Chen F, Lindquist EA, Sorek R, Hugenholtz P (2010) Validation of two ribosomal RNA removal methods for microbial metatranscriptomics. Nat Methods 7:807–812

Chapter 35

Gene Amplification and qRT-PCR

Cerith Jones and Alain Filloux

Abstract

This chapter includes methods for the use of the polymerase chain reaction (PCR) with *Pseudomonas*, and several specific tips for their successful application in this organism. The first part of the chapter includes methods for purifying genomic DNA from, and amplifying genes from, *Pseudomonas*, in addition to methods which describe how to prepare a cell lysate from *Pseudomonas* species for colony PCR reactions. The chapter continues with a switch in focus from DNA to RNA, describing methods for RNA isolation from *Pseudomonas*, cDNA generation, and finally q-RT-PCR to investigate relative changes in gene expression.

Key words PCR, Cloning, Gene amplification, Colony PCR, RNA, cDNA, RT-PCR, qRT-PCR

1 Introduction

This chapter focuses on the applications of the polymerase chain reaction with *Pseudomonas* species. Methods are described for the amplification of both DNA and RNA templates to gain insights into the function and activity of *Pseudomonas* genes. The methods contained in this chapter are aimed at those familiar with these techniques, but contains specific tips for their successful application with *Pseudomonas*.

The genome sequences of many *Pseudomonas* species are freely available online [1], providing a wealth of information to those commencing work with these organisms. This is an excellent resource for primer design and for designing strategies for genomic manipulation. The methods described should allow the reader to clone genes from *Pseudomonas* and to generate genetic material for downstream applications. Additionally, colony PCR methods are described to allow for probing of *Pseudomonas* chromosomal loci to confirm genetic changes made.

Cloning may be an important step in understanding the function of a gene, but provides little information about gene expression. The polymerase chain reaction can be useful for investigating gene

Alain Filloux and Juan-Luis Ramos (eds.), *Pseudomonas Methods and Protocols*, Methods in Molecular Biology, vol. 1149, DOI 10.1007/978-1-4939-0473-0_35, © Springer Science+Business Media New York 2014

expression simply by changing the focus from genomic DNA to RNA transcripts. While single-stranded RNA itself cannot be used as a template for PCR, it can readily be converted into double-stranded cDNA. To achieve this, methods of RNA isolation and cDNA generation are included in this chapter. Having produced suitable cDNA gene expression can be investigated qualitatively or quantitatively. RT (reverse transcriptase)-PCR methods are described which will show if genes are expressed depending on the presence or absence of specific RNA transcripts. Finally, a quantitative (q)-RT-PCR method is described allowing investigation of relative changes in transcript abundance, revealing how gene expression changes under different conditions.

2 Materials

2.1 Visualization of Nucleic Acid Products

Genetic material from *Pseudomonas* spp. may be visualized on agarose gels made using TAE buffer. Standard DNA visualization protocols and agarose gel conditions should be employed.

2.2 Recommended Kits

The methods described use commercial kits which are noted in the sections below. Alternative kits can be used if manufacturer's instructions are followed, and should provide comparable results.

2.3 Thermal Cyclers

To perform most of the reactions detailed below a thermal cycler is required. For quantitative qRT-PCR experiments a cycler capable of fluorescent signal measurements will be required. The method described uses the ABI 7300 Real-Time PCR System machine for this application.

3 Methods

3.1 Purification of *P. aeruginosa* Genomic DNA

For cloning experiments, template DNA will need to be purified from the target *Pseudomonas* strain. This can be achieved by using a genomic DNA purification kit, for example, the PureLink genomic DNA kit (Invitrogen). For successful DNA purification, follow the protocol for gram-negative bacterial cell lysate, included with the kit.

1. 1 ml of an overnight culture of *Pseudomonas* is sufficient for successful purification, and elution in 100 μl of molecular biology grade water is recommended.
2. For increased yield of DNA perform multiple preparations in multiple spin columns, and reuse the eluate from the first column in subsequent columns. This will lead to an increase in DNA concentration without changing the overall volume.

3. Following purification, check for the presence of DNA by running a sample on agarose gel and measure the concentration using a NanoDrop Spectrophotometer. A genomic DNA concentration of 50–100 ng/μl will be sufficient.

3.2 Cloning Genes from *P. aeruginosa*

P. aeruginosa genes can be cloned using commercial high-fidelity DNA polymerases, following manufactures' instructions (*see* **Note 1**). Due to the high G:C content of the genome it is advisable to add betaine to a final concentration of 1 M to disrupt the formation of secondary structures [2] during the reaction. Genomic DNA purified from *Pseudomonas* strains in Subheading 3.1 should be used as a template for reactions.

1. An example reaction setup is shown in Table 1 using the Expand High Fidelity plus PCR system from Roche.
2. Example PCR reaction conditions for this enzyme are shown in Table 2. Annealing temperature and extension time may need to be optimized for each reaction.
3. To avoid accumulation of errors in PCR products the number of cycles for a cloning PCR should be kept low (e.g., no more than 20 cycles).
4. Following PCR mix 5 μl of the reaction mixture with 1 μl of DNA loading dye and visualize using an agarose gel (*see* **Note 2**). Run on a gel with a suitable molecular weight marker, check that the product is the correct molecular weight, and free from nonspecific PCR products.
5. Once the DNA product is confirmed, the remaining reaction mixture can be purified using a QIAgen DNA purification kit, and used for downstream cloning applications.

Table 1
Typical cloning PCR reaction setup

	Amount per reaction	Notes
Genomic DNA	3 μl	*See* Subheading 3.1
5′ Primer	2 μl	From 10 μM stock
3′ Primer	2 μl	From 10 μM stock
dNTPs	1 μl	From 20 mM stock
5× Buffer	10 μl	–
Betaine	10 μl	5 M stock, 1 M final concentration
HiFi PLUS polymerase	0.5 μl	Roche
Molecular Biology Water	21.5 μl	–
Total volume	50 μl	

Table 2
Typical cloning PCR reaction conditions

	Temperature	Time
Initial denaturing step	95 °C	5 min
Perform for 20 cycles:		
Denaturing	95 °C	30 s
Annealing[a]	55 °C	30 s
Extension	72 °C	1 min/kb
End cycle		
Final extension	72 °C	10 min
Hold	10 °C	∞

[a]Annealing temperature should be optimized for each primer pair

3.3 Colony PCR

Colony PCR is a useful technique for interrogation of chromosomal or plasmidic loci without having to purify genomic DNA from each clone. It is useful for screening colonies for the presence of a gene, or to screen for chromosomal changes from cloning (e.g., chromosomal insertions or gene deletions). Taq DNA polymerase can be used for colony PCR. The following protocol describes how to prepare DNA from a *Pseudomonas* colony using a freeze–thaw method (*see* **Note 3**):

1. Resuspend an isolated colony in 100 μl of molecular biology grade water.
2. Freeze the resuspended colony at −80 °C for 10 min.
3. Transfer to a heat block set at 95 °C for 10 min.
4. Return to −80 °C for 15 min, then to 95 °C for 10 min, performing two freeze–thaw cycles in total (*see* **Note 4**).
5. Spin this lysate in a benchtop centrifuge at maximum speed (~12,000 × *g*) for 1 min to pellet insoluble cell debris. The resulting supernatant will contain DNA and can be used as a template for PCR.
6. Set up a reaction as detailed in the literature provided with the TaqDNA polymerase enzyme to be used (a typical PCR setup can be seen in Table 3). To prevent the formation of secondary structures, especially between G:C rich regions of the *Pseudomonas* genome, add DMSO to the reaction mix.
7. Perform the reaction using the conditions detailed in the manufacturer's instructions. (Typical colony PCR reaction settings can be seen in Table 4). To improve detection of the product the number of PCR cycles can be increased to around

Table 3
Typical colony PCR reaction setup

	Amount per reaction	Notes
Colony DNA	5 μl	*See* Subheading 3.3
10× Taq buffer	3 μl	
5′ Primer	2 μl	From 10 μM stock
3′ Primer	2 μl	From 10 μM stock
dNTPs	2 μl	From 20 mM stock
Molecular biology water	14.3 μl	–
Taq polymerase	0.7 μl	–
DMSO	1 μl	–
Total volume	30 μl	

Table 4
Typical colony PCR reaction conditions

	Temperature	Time
Initial denaturing step	95 °C	5 min
Perform for 30 cycles:		
Denaturing	95 °C	30 s
Annealing[a]	55 °C	30 s
Extension	72 °C	1 min/kb
End cycle		
Final extension	72 °C	10 min
Hold	10 °C	∞

[a]Annealing temperature should be optimized for each primer pair

30 (fidelity of the PCR product is not important if it is not being used in downstream applications).

8. Following the reaction, add 6 μl of DNA loading dye to each 30 μl reaction and run on an agarose gel to visualize the product.
9. Clones identified as positive can then be cultured and used in subsequent experiments.

3.4 Isolation of RNA

A prerequisite to performing gene expression experiments is the production of sufficient amounts of high-quality total RNA from bacterial cells. The method below describes how to perform total RNA extraction using the QIAgen RNeasy kit. Work with RNA must be carried out under strict RNase-free conditions (*see* **Note 5**).

1. Incubate cultures under desired growth conditions.
2. Bacteria should be harvested and mixed with RNA*later* (Ambion) solution as per manufacturer's instructions. Pellet 2 ml of logarithmic phase cell culture and resuspend in 0.5–1 ml of RNA*later*, taking care to note the optical density of the culture. This preserves the RNA prior to extraction (*see* **Note 6**).
3. RNA can be extracted from *Pseudomonas* species following the "Enzymatic lysis of Bacteria" protocol in the QIAgen RNAprotect® Bacteria Reagent Handbook. For the purposes of a qPCR experiment, extraction of RNA from a volume of cells corresponding to 1 optical density unit of bacteria is sufficient.
4. Following isolation of the RNA by enzymatic lysis it should immediately be purified following the "Purification of Total RNA from Bacterial Lysate" protocol in the handbook.

3.5 DNase Treatment of RNA

Following elution of RNA the sample should be treated with DNAse to remove contaminating DNA. The following protocol describes the use of Turbo DNase (Ambion).

1. Adjust RNA sample volume to 100 μl.
2. Add 0.1 volumes of 10× Turbo DNase buffer, and 1.5 μl of Turbo DNase. Incubate at 37 °C for 30 min.
3. Add an additional 1.5 μl of turbo DNase to the sample and return to 37 °C for a further 30 min (*see* **Note 7**).
4. To stop the reaction, add 0.2 volumes of resuspended RNase inactivation agent and mix well.
5. Incubate for 5 min at room temperature, mixing occasionally by flicking the tube.
6. Centrifuge at ~6,000 × *g* for 1.5 min.
7. Transfer the RNA to a fresh tube, leaving the white pellet behind.

3.6 RNA Quality Control

Following DNase treatment the quality and quantity of the RNA should be tested to ensure it is adequate for subsequent experiments. 10 μl of each RNA sample should be removed for quality testing, and the remainder stored at −80 °C. For more detailed information on the quality of RNA required for publication, please refer to the MIQE guidelines [3].

1. The RNA concentration can be measured using a NanoDrop Spectrophotometer. A concentration of 20 ng/μl is the minimum

requirement for qPCR. The 260/280 ratio of pure RNA should be above 1.8.

2. It should be confirmed that each RNA sample is DNA free by using the RNA as a template in a PCR reaction. Primers known to amplify a region of genomic DNA should be used. A positive PCR results is indicative of DNA contamination in the RNA sample.

3.7 cDNA Generation

This step in the protocol is required to convert single-stranded RNA into first strand cDNA for use in subsequent reactions as described by Burr et al. [4]. The protocol below is for use with Invitrogen SuperScript™ II RNase H-Reverse Transcriptase.

1. For each RNA sample a reverse transcriptase reaction should be set up. Additionally, a control reaction should be performed in which reverse transcriptase is not added. Prepare two tubes per RNA sample, each with 200 ng RNA, and add RNAse-free water to a total volume of 10 μl.
2. To generate "no template" controls for use in later experiments, the reverse transcriptase reactions should also be set up with 10 μl of RNase-free water and no genetic material.
3. Heat these tubes of RNA or water to 65 °C for 5 min to disrupt secondary structures in the RNA, and then chill on ice.
4. Set up a mastermix for each reverse transcriptase reaction and controls as detailed in Table 5 (adding all components except the reverse transcriptase at this stage).
5. Add 9.0 μl of the master mix to reverse transcriptase negative samples and mix gently by pipetting (*see* **Note 8**).
6. Complete the mastermix by adding 1.0 μl of reverse transcriptase per reaction remaining.

Table 5
Reverse transcription mastermix setup

	Amount per reaction	Notes
First Strand Buffer	4.0 μl	Provided with RT kit (Invitrogen)
0.1 M DTT	1.0 μl	Provided with RT kit (Invitrogen)
dNTPs	1.0 μl	10 mM each
RNAguard	0.5 μl	From Roche
Random hexamer oligonucleotides	1.0 μl	Amersham
Nuclease-free water	1.5 μl	–
Total volume	9 μl	

Table 6
Reverse transcription reaction

Temperature	Time
25 °C	5 min
42 °C	50 min
70 °C	15 min
20 °C	∞

7. Add 10.0 μl of the complete master mix to the reverse transcriptase positive samples and water controls and mix gently by pipetting.
8. Run a thermal cycler program as detailed in Table 6.
9. Following the reaction, dilute the cDNA samples 1:5 to a total volume of 100 μl and store at −20 °C.

3.8 RT-PCR

The cDNA generated in Subheading 3.7 can be used as template for a reverse transcriptase polymerase chain reaction (RT-PCR). RT-PCR can be used as a nonquantitative approach to investigate gene expression. In order to have confidence in this reaction both reverse transcriptase plus (RT+) and reverse transcriptase negative (RT−) samples should be tested. Positive bands in RT− samples are indicative of contamination of samples by genomic DNA. Positive bands in RT+ samples (while RT− reactions are negative) indicate that the target RNA is present in the original sample, and expression of the gene is active under the conditions tested.

The method for RT-PCR is identical to that previously described for colony PCR (*see* Subheading 3.3), but DNA prepared from cell lysate is replaced with 5 μl of cDNA from RT+ and RT− reactions. All other steps should be carried out as described for colony DNA amplification.

3.9 qRT-PCR

Quantitative reverse transcriptase PCR (qRT-PCR) is a powerful technique allowing the abundance of transcripts from a particular gene to be measured. Comparison of gene expression in bacteria grown under different conditions can then be compared. qRT-PCR involves a simple PCR reaction with the addition of a fluorescent reporter system to measure the accumulation of PCR products in real time. Primers for qRT-PCR need to be designed to amplify ~100 base pair amplicons specific to the gene of interest (*see* **Note 9**). qRT-PCR reactions can be performed in an ABI 7300 Real-Time PCR System machine. The method below describes the use of the CYBR Green reporter system which binds to double-stranded DNA as it accumulates (*see* **Note 10**). The method describes relative comparison of RNA abundance using the comparative C_T method.

Table 7
q-RT-PCR reaction setup

	Amount per reaction	Notes
cDNA	5 µl	–
ABI SYBR Green PCR Master Mix	12.5 µl	Applied Biosystems
5′ Primer	1.25 µl	From 10 µM stock
3′ Primer	1.25 µl	From 10 µM stock
Nuclease-free water	5.0 µl	–
Total volume	25 µl	

1. Set up reactions in a ABI PRISM™ 96-Well Optical Reaction Plate, in technical triplicates, as described in Table 7, for each primer pair used.
2. For normalization of samples primers targeted towards a reference gene, expression of which is expected to remain constant under the growth conditions used, should be included in the reaction (*see* **Note 11**). RNA samples incubated with and without reverse transcriptase should be included for the reference control to confirm the absence of contaminating DNA in the RT– samples, as well as the no template control. RT– controls do not routinely need to be included for each of the other transcripts probed.
3. Following the reaction setup, plates should be sealed with ABI PRISM™ Optical Adhesive Covers to prevent evaporation of samples.
4. A standard curve protocol should be run on the ABI 7300 Real-Time PCR machine for the detection of CYBR green signal accumulation after each PCR cycle performed (*see* **Note 12**).
5. A dissociation curve step should be added to the protocol.
6. At the end of the reaction, the result of the dissociation curve for each primer pair should be checked. Each primer pair should show one single dissociation curve, indicating that one single PCR product is being produced.
7. The RT– reference gene control should be checked. The signal in this control should be absent, or significantly delayed in comparison to the RT+.
8. To extract data at the end of the experiment, a manual baseline should be set in a region of the curves before signal is detectible. A manual Ct (cycle time) threshold should be set to a region where all of the curves are roughly parallel in the exponential region. This data can then be analyzed, and Ct values exported to a spreadsheet.

9. The Ct times relate to the cycle number where the signal of PCR product accumulation has passed the threshold. This should be converted to the number of DNA fragments present in the reaction, achieved by calculating 2^{Ct}. This can be converted to the number of cDNA copies present at cycle number 1, by calculating the reciprocal; $1/(2^{Ct})$, representing the cDNA copies present in the original sample at the start of the reaction.
10. Normalization can be performed by dividing the values for the gene of interest by those for the reference gene under each condition. This eliminates differences due to technical variation as long as the assumption that the reference gene is constant is true.
11. Following normalisation, expression of each gene under the different conditions tested can be expressed relative to the standard conditions, set to 1.0.

4 Notes

1. For cloning experiments, where the sequence of the prepared PCR needs to be correct, a high-fidelity DNA polymerase should be used. However, for screening reactions such as colony PCR, where only the presence or absence of a band is important, a less stringent, cheaper, Taq DNA polymerase enzyme is recommended.
2. For cloning reactions the fidelity of the PCR product is very important. After removing 5 μl of the cloning PCR mixture, the remainder should be kept on ice until required. As the PCR product will be used for downstream cloning it should be carefully stored to prevent degradation of the ends of the product prior to purification. However, for colony PCR reactions the remaining PCR product can be discarded as it will not be of further use.
3. When performing colony PCR the colonies tested must be patched so that they may be re-cultured once screening is complete. This can be achieved by patching the colony onto a fresh LB plate and incubating overnight. As colony PCR is destructive to the colony, care must be taken to ensure that the colony is reproduced.
4. One freeze–thaw cycle is sufficient for *E. coli* cells, but two cycles are advisable for efficient lysis of *Pseudomonas* spp.
5. RNA extraction work should be carefully planned to ensure that processing time is kept to a minimum in order to maintain the integrity of RNA samples. Surfaces and equipment should

be kept free from RNase to eliminate the risk of sample contamination and degradation. This can be achieved by wiping with a cleaning agent such as RNase ZAP (Sigma-Aldrich).

6. RNA later solution stabilizes bacteria and RNA within them, meaning that samples can be stored for around a week at –20 °C before RNA extraction. For best results, samples should be processed as soon as possible after harvesting.
7. The 3 μl of Turbo DNase can be added all at the same time, and samples incubated for 1 h. However, as detailed in an alternative protocol for Turbo DNase, the authors find that adding 1.5 μl Turbo DNAse initially, and a further 1.5 μl after half an hour, gives the best results.
8. Reverse transcriptase negative reactions will only have a total volume of 19 μl, but these conditions are sufficient for the control reaction.
9. qRT-PCR primers must be designed to amplify approximately 100 bp amplicons specific to the gene of interest, and must meet a set of strict criteria to ensure successful experiments. Suitable primer sequences can usually be obtained using the ABI primer design program. If this is not available, primers should meet the following specifications: 30–80 % GC, 9–40 bases (ideally 20), Tm = ~61°C, with less than 2 °C difference between Tm of the two primers. Of the five bases closest to the 3′-end of the primer, a maximum two should be G or C.
10. The qRT-PCR method described here uses a Cybr Green reporter system, which binds nonspecifically to Double-Stranded DNA as it accumulates during the reaction, emitting green light as it does so. The amount of green light is thus proportional to the amount of dsDNA. Cybr Green may also bind to primer-dimers and nonspecific reaction products. Improved specificity in q-RT-PCR can be achieved using specially designed reporter probes, but these are not covered in this chapter.
11. A common reference gene used for normalization is 16S rRNA. However, this gene is very highly expressed, and due to the fast accumulation of signal, it may not be accurate. To overcome this, the authors suggest using *rpoD*. Primers for probing *rpoD* in the PAO1 strain of *P. aeruginosa*, are defined as follows: Forward: AGGCCGTGAGCAGGGATAC and Reverse: TCCCCATGTCGTTGATCATG.
12. More details on the setup of this experiment can be found in the ABI "Relative Quantitation Using Comparative CT Getting Started Guide" available from the manufacturers' Web site.

Acknowledgement

Thanks to Dr. Helga Mikkelsen for useful discussion of RNA extraction and q-RT-PCR protocols. Cerith Jones is supported by a PhD fellowship from the Biotechnology and Biological Sciences Research Council (BBSRC). Alain Filloux laboratory is supported by grants from the Wellcome Trust, BBSRC, and the Medical research Council (MRC).

References

1. Winsor GL et al (2011) *Pseudomonas* genome database: improved comparative analysis and population genomics capability for *Pseudomonas* genomes. Nucleic Acids Res 39: D596–D600
2. Henke W et al (1997) Betaine improves the PCR amplification of GC-rich DNA sequences. Nucleic Acids Res 25:3957–3958
3. Bustin SA et al (2009) The MIQE guidelines: minimum information for publication of quantitative real-time PCR experiments. Clin Chem 55:611–622
4. Burr T et al (2006) Identification of the central quorum sensing regulator of virulence in the enteric phytopathogen, *Erwinia carotovora*: the VirR repressor. Mol Microbiol 59:113–125

Chapter 36

The Standard European Vector Architecture (SEVA) Plasmid Toolkit

Gonzalo Durante-Rodríguez, Víctor de Lorenzo, and Esteban Martínez-García

Abstract

The *Standard European Vector Architecture* (SEVA) toolkit is a simple and powerful resource for constructing optimal plasmid vectors based on a backbone and three interchangeable modules flanked by uncommon restriction sites. Functional modules encode several origins of replication, diverse antibiotic selection markers, and a variety of cargoes with different applications. The backbone and DNA modules have been minimized and edited for flaws in their sequence and/or functionality. A protocol for the utilization of the SEVA platform to construct transcriptional and translational fusions between a promoter under study (the arsenic-responsive *Pars* of *Pseudomonas putida* KT2440) and the reporter *lacZ* gene is described. The resulting plasmid collection was instrumental to measure and compare the β-galactosidase activity that report gene expression (i.e., transcription and translation) in different genetic backgrounds.

Key words Genetic tools, Promoter-probe plasmids, Gene expression, β-galactosidase

1 Introduction

The Standard European Vector Architecture (SEVA) platform is a large repository of plasmids, many of them endowed with broad host range (BHR) origins of replication which cover most common genetic engineering necessities for *Pseudomonas* strains: generation of directed genomic mutations and deletions cloning and propagation of specific cistrons, parameterization of gene expression activity and regulated transcription of native or heterologous genes. The list of vectors available is organized in a database (http://seva.cnb.csic.es) [1], which compiles all relevant information for each construct and provides users with instructions for requesting specific plasmids. This collection thus broadens the landscape of molecular tools for Gram-negative bacteria and allows a degree of genetic manipulations that were hardly feasible with the non-standardized counterparts. The general architecture of plasmids that follow the SEVA format is depicted in Fig. 1.

Alain Filloux and Juan-Luis Ramos (eds.), *Pseudomonas Methods and Protocols*, Methods in Molecular Biology, vol. 1149, DOI 10.1007/978-1-4939-0473-0_36, © Springer Science+Business Media New York 2014

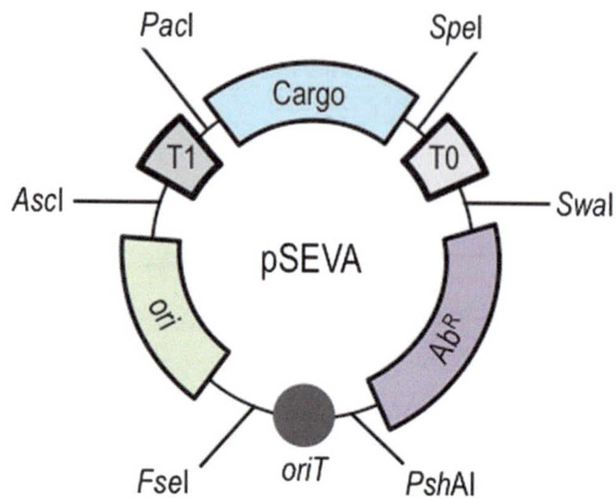

Fig. 1 pSEVA plasmid map. The figure includes the general structure and functional elements of the pSEVA plasmid architecture including the most relevant restriction sites. The cargo module flanked by *PacI* and *SpeI*; the antibiotic marker (Ab^R) punctuated with *SwaI* and *PshAI*; the origin of replication (ori) bordered with *FseI* and *AscI*; an origin of transfer region (*oriT*) and the T_0 and T_1 transcriptional terminators [1]

This resource is intended not only for description of the standard and modular structure of the plasmids that form the collection, but also to help identification of the best vectors for given purposes. The most frequent include customary cloning (for which a large number of vectors is offered in the collection) and studies on gene expression regulation. The last often relies on the construction of reporter gene fusions between given promoters and reporter genes, e.g., fusing the upstream portion of a given *cistron* to a gene encoding an easily tractable property. This sets the expression of the reporter product under the regulatory signals of the gene under examination. Reporters are structural genes that encode easily measurable and quantifiable products and must not be present in the microorganism under study. The most used ones are *lacZ*, *gfp* and *lux*. We can differentiate two types of fusions depending on the upstream DNA region linked to the reporter gene. When only the promoter region is attached to the reporter, the resulting construct is called a *transcriptional fusion*, and its output only accounts for promoter strength. However, when the promoter sequence plus the translational signals are joined in-frame to the reporter gene, the result is a translational fusion, which report both promoter strength plus any type of posttranscriptional regulation that affect eventual expression.

The modular structure and properties of the pSEVA plasmids (Fig. 1) makes them an optimal choice for these gene expression studies. The plasmid repository allows the user to choose their optimal vector depending on the antibiotic marker, origin of replication (low, medium and high copy), and reporter cassette (*lacZ*, *gfp* or *lux*). To select a pSEVA plasmid, users typically navigate through the

SEVA Web site, click the link to the plasmid list (http://seva.cnb.csic.es/SEVA/Plasmid_list.html) and scroll down until the most appropriate combination of antibiotic marker, origin of replication, and reporter cassette is found. Further links to the GenBank accession number allows downloading a file that includes the plasmid sequence with the most relevant features annotated.

Although both *gfp* and *lux*-based reporter plasmids are available in the collection, in this Chapter we describe, as a study case, a protocol to generate and analyze either transcriptional or translational fusions using *lacZ* as reporter. The *lac*Z encodes the enzyme β-galactosidase of *E. coli* that cleaves lactose into glucose and galactose, and it was the first reporter gene used. *lacZ*-based systems rely on the use of *ortho*-nitrophenyl-β-galactoside (ONPG), a colorless lactose analog, that upon hydrolysis by β-galactosidase generates the yellow compound *ortho*-nitrophenol. This color can be measured spectrophotometrically and, with this data, infer a figure of Miller units that correlate with promoter activity [2]. For a review describing the uses of *lacZ*, *see* ref. [3]. The plus and minuses of using different reporter systems is discussed in ref. [4].

Here we described a protocol to study gene regulation by creating equivalent transcriptional and translational fusions of an arsenic-responsive promoter to *lacZ*. For this purpose, we employed *Pseudomonas putida* KT2440 as the host bacterium, *Pars1* as the promoter under study, and pSEVA225 (transcriptional) and pSEVA225T (translational) plasmids as the promoter-probe vectors (Fig. 2).

The necessary steps to create a reporter fusion include [1] analysis of the DNA region at stake to identify putative promoter regions, [2] designing of primers to PCR amplify and clone the selected sequence into one of the pSEVA plasmids adopted for that purpose, [3] introduction of the constructed plasmid into the *Pseudomonas* host, and [4] quantification of the reporter gene under the selected experimental conditions. Specific details of the method are described below.

2 Materials

2.1 Strains and Vectors

See Table 1. For SEVA vectors check http://seva.cnb.csic.es.

2.2 Media and Reagents

1. Bacteria were grown in LB as rich nutrient medium. Prepare LB as follows: 10 g/l of tryptone, 5.0 g/l of yeast extract, and 5.0 g/l of NaCl, dissolve in H_2O and autoclave [2]. The LB was solidified adding 1.5 % agar (*see* **Notes 1** and **2**).
2. Kanamycin (Km): Prepare the stock at 50 mg/ml in H_2O, filter-sterilized. Store at −20 °C. Use it at a final concentration of 50 μg/ml.

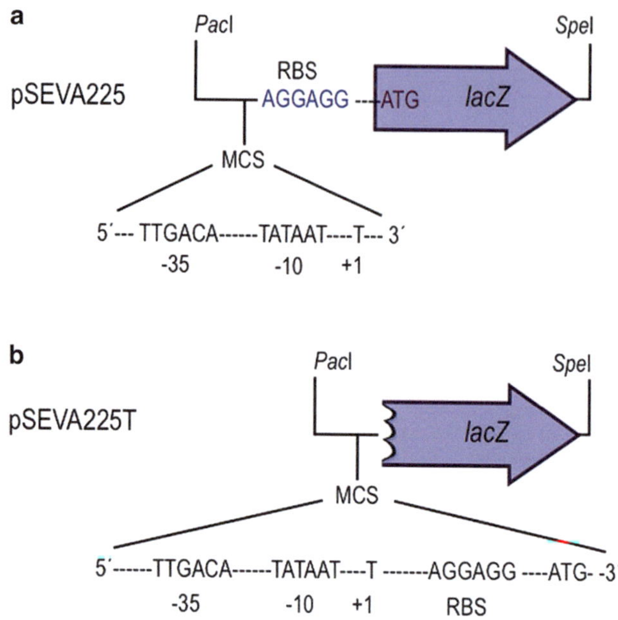

Fig. 2 Fusions to promoter-probe plasmids. (**a**) This picture shows a schematic illustration of a typical transcriptional fusion to the pSEVA225 plasmid. The general structure of this plasmid is as follows: a MCS (multiple cloning site) for cloning putative promoter sequences (represented here with −35, −10, and +1 boxes), followed by an RBS (ribosome binding site), and the *lacZ* reporter gene. (**b**) Cartoon of a translational fusion to the plasmid pSEVA225T. This plasmid is formed by a MCS attached to a *lacZ* gene lacking translational signals and the first four amino acids. Note that in both cases the whole cargo region can be excised as a *PacI* and *SpeI* fragment

Table 1
Strains

Strain	Description/relevant characteristics	Reference
Escherichia coli strains		
DH5α	*supE44*, *ΔlacU169*, (*ϕ80 lacZDM15*), *hsdR17*, (*rk-mk+*), *recA1*, *endA1*, *thi1*, *gyrA*, *relA* Recipient of SEVA plasmids	[12]
Pseudomonas putida strains		
KT2440 wt	Prototrophic, wild-type strain	[13]
KT2440 Δars1,2	KT2440, *pyrF*, *ars1* and *ars2* deletion	Lab collection
KT2440 wt-A	Prototrophic, *Pars1A::lacZ*	Lab collection
KT2440 Δars1,2-A	*pyrF*, *ars1*, and *ars2* deletion, *Pars1A::lacZ*	Lab collection
KT2440 wt-B	Prototrophic, *Pars1B::lacZ*	Lab collection
KT2440 Δars1,2-B	*pyrF*, *ars1*, and *ars2* deletion, *Pars1B::lacZ*	Lab collection

3. Sucrose solution dissolved at 300 mM in H_2O, sterilize and maintain at room temperature.
4. Prepare a stock solution of *ortho*-nitrophenyl-β-galactoside (ONPG) at 4 mg/ml, dissolved in sodium phosphate buffer 0.1 M pH 7.0, and used it at a final concentration of 500 μg/ml. Store at −20 °C.
5. The β-galactosidase assay was performed in Buffer Z [2]. Prepare as follow: 0.80 g of $Na_2HPO_4 \cdot 7H_2O$, 0.28 g of $NaH_2PO_4 \cdot 7H_2O$, 0.5 ml of KCl 1 M, 0.05 ml of $MgSO_4$ 1 M, and 0.135 ml of β-mercaptoethanol, dissolve in H_2O, adjust the pH to 7.0 and bring the volume to 50 ml. Store at 4 °C.
6. Chloroform. Add 50 μl drops in 2 ml of assay mixture (see below).
7. Sodium dodecyl sulfate (SDS). Prepare the stock at 0.1 % in H_2O. Store at room temperature. Use it at a final concentration of 0.004 %.
8. Na_2CO_3: Prepare the stock at 1 M in H_2O. Store at room temperature. Use it at a final concentration of 0.27 M.

2.3 DNA Techniques

1. For plasmid preparations use standard kits (e.g., Wizard Plus SV Minipreps kit of Promega).
2. PCR-amplified DNA and agarose DNA extractions are purified with the NucleoSpin Extract II kit (MN).
3. In DNA ligations T4 DNA ligase (NEB) is employed.
4. PCR: for general PCR reactions prepare first a PCR reaction mix in an Eppendorf tube. The values given below are calculated for a final volume of 100 μl per tube:
   - 20 μl Buffer 5×
   - 10 μl $MgCl_2$ 25 mM
   - 2 μl of upstream primer 10 μM
   - 2 μl of downstream primer 10 μM
   - 10 μl dNTPs 2.5 mM
   - 1 μl Taq Polymerase

   Then, add 55 μl of sterile H_2O into each of the PCR reaction tubes. Set up the PCR machine with the appropriate Tm (depending on primer composition) and extension time.

2.4 PCR Primers

2.4.1 Transcriptional Fusion Primers

- 5′Pars1-Eco: 5′-GCGAATTCTGATCGGTACCAAGC-3′
- 3′Pars1-Hind: 5′-GGGAAGCTTGAATGCGCGTACGGCCAATATCTG-3′

2.4.2 Translational Fusion Primers

- 5′Pars1-Eco: see above
- 3′Pars1-Bam: 5′-AGAGGATCCATCAGCAGGGTCAT-3′

3 Methods

3.1 Searching for Promoter Regions

The study of a promoter region required a previous analysis of the DNA sequence upstream the ATG codon of the gene under study (*asR1* for this particular example). This allowed us to establish the predicted common regions, like the −10 and −35 boxes with the *BPROM* program (http://xurl.es/BPROM) and palindrome regions with the program *RNAfold* (http://xurl.es/RNAfold) [5–7]. In our experimental case we predicted a consensus sequence putatively recognized by the transcriptional regulator ArsR1 (belonging to SmtB/ArsR family proteins) using the platform *EMBOSS* (http://emboss.sourceforge.net) [8]. ArsR1 is involved in the control of *ars1* operon, with the *Pars1* sequence, finding a high identity between −34 and −8 region of the promoter (Fig. 3). This information allowed us to establish the region to clone in the transcriptional and translational vectors.

3.2 Primer Design

3.2.1 Transcriptional Fusion

Select a *Pars1* region of 208 bp spanning from −182 to +26, just before the RBS sequence, and amplify by PCR using 5′Pars1-Eco and 3′Pars1-Hind primers, obtaining the fragment Pars1A, which will be cloned in the MCS, just before the RBS sequence of the *lacZ* gene of SEVA225, as described below.

3.2.2 Translational Fusion

Select a *Pars1* region of 297 bp spanning from −182 to +115 in respect to the predicted transcription initiation site including the first 26 amino acids of ArsR1. For the primer design process a key aspect to construct translational fusions is that the cloned sequence must be in frame with the *lacZ* gene of pSEVA225T. Then, amplify by PCR using 5′Pars1-Eco and 3′Pars1-Bam primers, obtaining the fragment Pars1B, which will be cloned in the MCS, entering in the same reading frame of *lacZ* gene of SEVA225T, as described below.

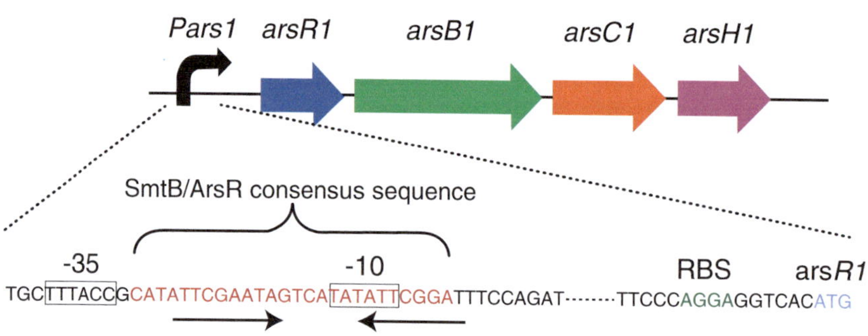

Fig. 3 The *ars1* operon and the *Pars1* promoter region. This picture shows a schematic illustration of the *ars1* operon structure in *Pseudomonas putida* KT2440, with the promoter *Pars1* followed by the transcriptional regulator *arsR1* and the metabolic genes *arsB1*, *arsC1*, and *arsH1*. Below the structure of the *Pars1* promoter is described as follows: −35 and −10 sequences in boxes, RBS (ribosome binding site) in *green letters*, first codon of *arsR1* gene (ATG) in *blue letters*, palindromic sequences *underlined* by *black arrows*, and the SmtB/ArsR consensus sequence in *red letters*

3.3 Cloning the Promoter Sequence into pSEVA225 and pSEVA225T

1. Prepare digestions of the pSEVA225 and pSEVA225T plasmids with the appropriate enzymes (*Eco*RI and *Hind*III; *Eco*RI and *Bam*HI, respectively).
2. Purify from a 0.8 % agarose gel the linear fragment of the pSEVA225 and pSEVA225T plasmid (6.9 kb, both).
3. Digest with the appropriate restriction enzymes the Pars1A, for pSEVA225, and Pars1B, for pSEVA225T, PCR fragments (*Eco*RI and *Hind*III; *Eco*RI and *Bam*HI, respectively). Example of a Pars1A digestion mixture:
   - 22 μl of Pars1A PCR fragment
   - 3 μl Buffer 10×
   - 3 μl BSA
   - 0.5 μl enzyme 1
   - 0.5 μl enzyme 2
4. Incubate for 2 h at 37 °C and inactivate restriction enzymes as required.
5. Ligate the Pars1A and Pars1B digested PCR fragments into the pSEVA225 and pSEVA225T vectors to produce plasmids 225-Pars1A and 225T-Pars1B harboring the *Pars1::lacZ* transcriptional or translational fusion, respectively. The ligation mixture comprises:
   - 11 μl of digested *Pars1* fragment
   - 6 μl of the linearized plasmid
   - 2 μl of Buffer 10×
   - 1 μl of T4 ligase
6. Incubate for 1 h at room temperature.
7. Transform DH5α competent cells [9] with all the ligation mixture (*see* **Note 3**) and plate onto LB plates supplemented with Km 50 μg/ml.
8. Select a few colonies, re-streak them, and check for the presence of the correct plasmid (225-Pars1A and 225T-Pars1B, in each case) by colony PCR using M13-reverse/3′Pars1-Hind and M13-reverse/3′Pars1-Bam primers.
9. Select a few positive clones and send them to sequence with (M13-reverse) to confirm that the target sequences did not incorporate any error due to the PCR amplification process.

3.4 Transformation of *P. putida* KT2440 Electrocompetent Cells

1. Add 200 ng of the construct plasmid (225-Pars1A or 225T-Pars1B) to a 100 μl aliquot of two different strains of *P. putida* KT2440: wt and Δars1,2 (*see* Table 1; Lab collection). Then mix gently and transfer to a 2 mm gap width electroporation cuvette (BioRad Gene Pulser® Cuvette) and proceed to electroporate [10, 11].

2. Add 800 μl of LB and incubate for 1 h at 30 °C with shaking.
3. Plate everything onto LB plates plus Km 50 μg/ml.
4. Streak a few colonies and perform a plasmid purification to check the presence of the plasmids (*see* **Note 4**), obtaining the new four strains of *P. putida* KT2440 wt-A and Δars1,2-A harboring the *Pars1A::lacZ* transcriptional fusion, and wt-B and Δars1,2-B harboring the *Pars1B'-'lacZ* translational fusion (*see* Table 1).

3.5 *Pseudomonas*-Adapted β-Galactosidase Assay

1. Grow cells overnight at 30 °C with shaking.
2. Collect the cultures and measure the OD_{600}.
3. Add 400 μl of Buffer Z (*see* **Note 5**) to 100 μl of cells (*see* **Note 6**).
4. Permeabilize cells by adding 50 μl of chloroform and 25 μl of SDS 0.1 % (*see* **Note 7**).
5. Vortex and equilibrate the tubes 5 min at room temperature.
6. Start reaction by adding 100 μl of substrate, *o*-nitrophenyl-β-D-galactoside (ONPG; 4 mg/ml). Record the time of addition precisely with a timer.
7. Incubate the cells at room temperature (*see* **Note 8**).
8. Stop the reaction after sufficient yellow color has developed by adding 250 μl of Na_2CO_3 1 M to raise the pH of the solution to 11 (this pH inactive the enzyme), and note the time of addition precisely.
9. Vortex and centrifuge for 10 min.
10. Collect 600 μl and measure the optical activity at 420 and 550 nm for each tube.
11. Calculate the Miller Units as described below.

3.6 Miller Units

The units of enzyme activity were calculated as is described in ref. 2. The absorbance at 550 nm is used to correct the light scattering produced by cell debris in the reading of *o*-nitrophenol absorbance at 420 nm, because this compound does not show absorbance at this wavelength. The light scattering at 420 nm is proportional to that at 550 nm:

$$\text{Light scattering at 420 nm} = 1.75 \times OD_{550}$$

Use the following equation to calculate Miller Units of enzyme activity:

$$\text{Miller Units} = 1{,}000 \times \left[\left(OD_{420} - 1.75 \times OD_{550}\right)\right] / \left(T \times V \times OD_{600}\right)$$

OD_{420} and OD_{550} are measured in the final reaction mixture (*see* Subheading 3.5, **step 10**).

OD_{600} shows the cell density.

T = time of the reaction measured in minutes.

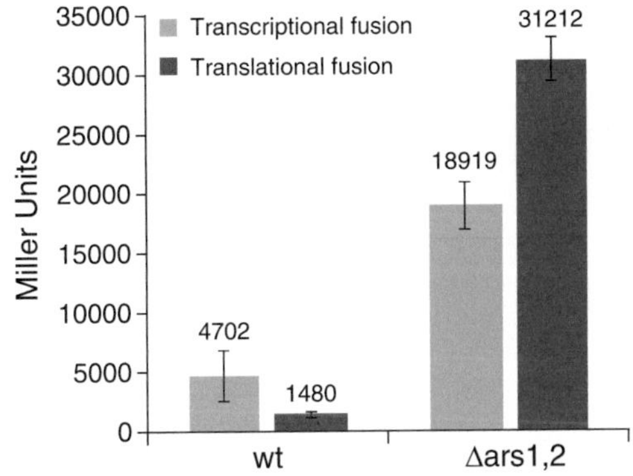

Fig. 4 β-Galactosidase activity of *P. putida* KT2440 harboring transcriptional and translational fusions. *P. putida* KT2440 wt and Δars1,2 cells harboring *Pars1A::lacZ* transcriptional fusion (*light grey*) or *Pars1B::lacZ* translational fusion (*dark grey*) were grown for 24 h in LB medium until they reached stationary phase. β-Galactosidase activity was measured as described in the text (*see* Subheading 3.5). The differences in reporter activity describe the effect of the two ArsR repressors of *P. putida* [14] in expression of the *Pars1* promoter and their effect on the product of the next ORF (*arsR*). Note the more stringent regulation of the translational fusion

V = volume of culture used in milliliters.

The Miller Units show the change in OD_{420} min^{-1} ml^{-1} OD_{600}^{-1}.

All the collected data are represented in Fig. 4.

4 Notes/Troubleshooting

1. Other experimental conditions may utilize different media instead of LB.
2. In this particular experimental set up, we used LB supplemented with uracil 0.002 % because the *P. putida* strain employed (*pyrF*, Table 1) was auxotrophic for this compound.
3. Usually, it is recommended to use the 20 μl from the ligation mixture to render a good number of colonies.
4. Broad host range plasmids can be introduced in *Pseudomonas* through either transformation or conjugative mating.
5. A general recommendation is to add fresh β-mercaptoethanol to Buffer Z each time to ensure the reaction under reductive conditions
6. Even though we recommend starting with 100 μl of cells, this total volume could vary depending on promoter strength. If the reaction mixture quickly develops to an intense yellow, scale down the method to as low as 25 μl. On the contrary, if it takes too long to render the yellow color, scale up to 250 μl.

7. The SDS 0.1 % could be added directly to the Buffer Z just prior to use
8. The original protocol suggests doing the incubation with ONPG at 28 °C.

Acknowledgments

This work was supported by the BIO and FEDER CONSOLIDER-INGENIO programs of the Spanish Ministry of Economy and Competitiveness, the MICROME, ST-FLOW and ARISYS Contracts of the EU, the ERANET-IB Program, and funding from the Autonomous Community of Madrid (PROMPT).

References

1. Silva-Rocha R, Martinez-Garcia E, Calles B, Chavarria M, Arce-Rodriguez A, de Las Heras A, Paez-Espino AD, Durante-Rodriguez G, Kim J, Nikel PI, Platero R, de Lorenzo V (2013) The Standard European Vector Architecture (SEVA): a coherent platform for the analysis and deployment of complex prokaryotic phenotypes. Nucleic Acids Res D41:D666–D675
2. Miller JH (1972) Experiments in molecular genetics. Cold Spring Harbor Laboratory, Cold Spring Harbor, NY
3. Silhavy TJ, Beckwith JR (1985) Uses of *lac* fusions for the study of biological problems. Microbiol Rev 49:398–418
4. de Las Heras A, Carreno CA, Martinez-Garcia E, de Lorenzo V (2010) Engineering input/output nodes in prokaryotic regulatory circuits. FEMS Microbiol Rev 34:842–865
5. Hofacker IL, Fontana W, Stadler PF, Bonhoeffer LS, Tacker M, Schuster P (1994) Fast folding and comparison of RNA secondary structures. Monatshefte Chem 125:167–188
6. McCaskill JS (1990) The equilibrium partition function and base pair binding probabilities for RNA secondary structure. Biopolymers 29:1105–1119
7. Zuker M, Stiegler P (1981) Optimal computer folding of large RNA sequences using thermodynamics and auxiliary information. Nucleic Acids Res 9:133–148
8. Busenlehner LS, Pennella MA, Giedroc DP (2003) The SmtB/ArsR family of metalloregulatory transcriptional repressors: structural insights into prokaryotic metal resistance. FEMS Microbiol Rev 27:131–143
9. Green MR, Sambrook J (2012) Molecular cloning: a laboratory manual. Cold Spring Harbor Laboratory Press, Cold Spring Harbor, NY
10. Choi KH, Kumar A, Schweizer HP (2006) A 10-min method for preparation of highly electrocompetent *Pseudomonas aeruginosa* cells: application for DNA fragment transfer between chromosomes and plasmid transformation. J Microb Methods 64:391–397
11. Martínez-García E, de Lorenzo V (2012) Transposon-based and plasmid-based genetic tools for editing genomes of Gram-negative bacteria. Methods Mol Biol 813:267–283
12. Hanahan D (1983) Studies on transformation of *Escherichia coli* with plasmids. J Mol Biol 166:557–580
13. Nelson KE, Weinel C, Paulsen IT, Dodson RJ, Hilbert H, Martins dos Santos VA, Fouts DE, Gill SR, Pop M, Holmes M, Brinkac L, Beanan M, DeBoy RT, Daugherty S, Kolonay J, Madupu R, Nelson W, White O, Peterson J, Khouri H, Hance I, Chris Lee P, Holtzapple E, Scanlan D, Tran K, Moazzez A, Utterback T, Rizzo M, Lee K, Kosack D, Moestl D, Wedler H, Lauber J, Stjepandic D, Hoheisel J, Straetz M, Heim S, Kiewitz C, Eisen JA, Timmis KN, Dusterhoft A, Tummler B, Fraser CM (2002) Complete genome sequence and comparative analysis of the metabolically versatile *Pseudomonas putida* KT2440. Environ Microbiol 4:799–808
14. Cánovas C, Cases I, de Lorenzo V (2003) Heavy metal tolerance and metal homeostasis in *Pseudomonas putida* as revealed by complete genome analysis. Environ Microbiol 5: 1242–1256

Chapter 37

Chromosomal Integration of Transcriptional Fusions

Rafael Silva-Rocha and Víctor de Lorenzo

Abstract

The characterization and parameterization of promoters is crucial for the study of gene regulatory networks. While a number of techniques are available for this purpose, the use of reporter fusions integrated in the chromosome of a bacterial host affords precise quantification of transcriptional responses with high reproducibility. Here, we describe the integration of green fluorescent protein (GFP) and *lacZ* reporter cassettes using either mini-Tn7-based vectors or homologous chromosomal recombination to analyze gene regulation at transcriptional and post-transcriptional levels.

Key words *Pseudomonas*, Transposon vectors, GFP, *lacZ*, Catabolic promoters

1 Introduction

Regulation of gene expression in bacteria is controlled at multiple steps and through a wide number of molecular mechanisms [1, 2]. In the last few decades, a number of reporter technologies have been developed to allow inspection of gene regulation in response to changing conditions. In general terms, these techniques are either disruptive or non-disruptive, depending on the experimental procedure used to quantify gene expression [3]. In the first case, cells are lysed to extract the reporter biomolecule to be quantified. In the second case, gene expression is analyzed through the quantification of signals from intact cells, which allows performing time course experiments in living organisms [4]. Independently of the type of technique the use of reporter genes provides a reliable way to simplify the quantification procedure. Using these systems, a target promoter along with its regulatory sequences is genetically fused to the reporter of choice and introduced into the bacterial host [5]. While a number of suitable reporter genes are currently available, two systems are of particular interest. The first and most widely used is the β-galactosidase-encoding *lacZ* gene. In this case, a modified substrate named ortho-nitrophenyl-β-D-galactopyranoside (ONPG) becomes yellow upon cleavage by the

Alain Filloux and Juan-Luis Ramos (eds.), *Pseudomonas Methods and Protocols*, Methods in Molecular Biology, vol. 1149, DOI 10.1007/978-1-4939-0473-0_37, © Springer Science+Business Media New York 2014

β-galactosidase enzyme, and thus is used to monitor the levels of this protein produced from the promoter at stake. The mainly disadvantage of this procedure is that cells must be permeabilized prior to β-galactosidase activity quantification, thus the samples cannot be used in multiple rounds of analysis. However, the enzyme is virtually the only reporter endowed with a standard unit definition (the Miller units) that allows the easy comparison of promoter strengths across different experiments in time and space [5]. Furthermore, this technique could be considered of simple implementation, as the reagents and equipment required for the assays are very easy to acquire. The other reporter of interest is the group of fluorescent proteins that allow direct monitoring of promoter activity through the quantification of a fluorescent signal [6]. While the first usable gene from this group is that of the Green Fluorescent Protein (GFP) of the jellyfish *Aquarea victoria*, several current variants with different excitation and emission spectra are available that allow the simultaneous analysis of multiple target promoters in the same cell [7]. Current optical detection devices with very high sensitivity allow the analysis of gene expression in individual cells [6] and even the activity of individual molecules [8], turning these GFP genes into a source of powerful tools for the analysis of gene expression and protein localization.

But the choice of the right reporter is not the only factor important for the reliable analysis of gene expression. Another factor to take into account regards the balance between the copy number of the fusion promoter-reporter and the regulatory elements. This is very important not only to keep the correct stoichiometry necessary for perfect regulation of the target, but also to avoid the introduction of noise in the assays [9]. This scenario is particular evident when the reporter fusion is assembled in a multi-copy plasmid while the regulatory elements are in the chromosome [10]. An alternative to avoid those interferences relies on the introduction of the reporter fusions directly into the chromosome, thus generating single-copy, stable reporter systems in the host strain [11]. Several strategies address this issue and the most popular solutions are the utilization of either mobile element such as mini-transposons [12] or through the homologous recombination (HR) of suicide vectors to the host chromosome [13]. In the case of transposons, the mini-Tn*5*, mini-Tn*10*, and mini-Tn*7* have been widely used in a number of gram-negative bacteria for decades with high efficiency [12, 14]. The two first (mini-Tn*5* and Tn*10*) are based in the random insertion of the transposon elements into the host chromosome in a process mediated by the cognate transposases. In contrast, the mini-Tn*7* based systems target the mobile element to a specific region of the chromosome, i.e., the Tn*7* attachment site (*att*Tn*7*) that is conserved in many bacterial species [14]. This last system is of particular interest as it allows the generation of isogenic reporter strains carrying different fusions [15], which enhances the accuracy of the analysis procedure. This same advantage is true for the HR-based systems [16], where the target reporter fusions are introduced into a specific position of the chromosome.

In this chapter, we describe the utilization of mini-Tn*7*-based and HR-based systems for the analysis of transcriptional fusions to the *lacZ* and GFP reporters in the gram-negative bacterium *Pseudomonas putida*. We will focus on the generation and verification of the recombinant strains, as well as in the analysis of the activity of target promoters under particular conditions in order to differentiate between transcriptional and post-transcriptional processes controlling gene expression. These protocols can be easily implemented not only for other Pseudomonads but also for a variety of gram-negative eubacteria.

2 Materials

2.1 Strains

1. *P. putida* KT2440. Laboratory strain of *P. putida* [17] used as a host for the mini-Tn*7*-based system.
2. *P. putida* MEG1. Host for the HR-based system. This strain is a mutant variant of *P. putida* KT2440 devoid of autofluorescence [16].
3. *E. coli* CC118. Laboratory strain of *E. coli* used for cloning purposes [18].
4. *E. coli* CC118*λpir*. A variant of *E. coli* CC118 with the *pir* gene integrated into the chromosome [18]. This strain is used to maintain suicide vectors based on the R6K *ori* of replication.
5. *E. coli* CC118*supF*. A variant of *E. coli* CC118 with the *supF* tRNA integrated into the chromosome [19]. This strain is used to maintain pRV1 vector variants.
6. *E. coli* HB101. Helper strain used in tripartite mating [20].

2.2 Vectors

1. pLOF-hom.fg. A mini-Tn*10*-based vector with homologue fragments for a streptomycin resistance marker and the *lacZ* gene, both in a truncated form [19].
2. pRV1. Promoter probe vector with a GFP-*lacZ* bi-cistronic reporter system [16].
3. pRV1-*Pr*. A pRV1-variant with the *Pr* promoter of the TOL system cloned as an *Eco*RI/*Bam*HI fragment [21].
4. pTn7-Z. A mini-Tn7-based promoter probe vector with a *lacZ* for translational fusions.
5. pTn7-*PrZ*. A pTn7-Z variant with a cloned fragment containing the *Pr* promoter and a 90-nt coding region of *xylR* gene [22].

6. pRK600. Helper plasmid encoding the conjugal transfer machinery used in tripartite mating [23].
7. pTnS1. Helper plasmid encoding the Tn*7* transposase required for the mobilization of the mini-Tn*7* transposon [14].

2.3 Buffers

Prepare all solutions with ultrapure water and analytical grade reagents. All reagents should be stored at room temperature (unless otherwise indicated).

1. LB media: Add 100 mL of water to a 1-L graduated flask. Weight and add to the flask 10 g bacto-tryptone, 5 g yeast extract, 10 g NaCl. Adjust the pH to 7.5 with NaOH. Make up to 1 L with water. Sterilize the solution by autoclaving.
2. PBS 10×: Add 30 mL of water to a 0.2-L graduated flask. Weight and add to the flask 8.01 g NaCl, 0.2 g KCl, 1.78 g $Na_2HPO_4{\cdot}2H_2O$, 0.27 g KH_2PO_4. Mix the solution and complete the volume to 100 mL. Sterilized the solution by autoclaving.
3. PBS 1×. Dilute 20 mL of PBS 10× solution in 90 mL water in a 0.2-L graduated flask.
4. 1 M $MgSO_4$: Dissolve 123 g $MgSO_4{\cdot}7H_2O$ in 300 mL water. Make up to 500 mL with water. Sterilize the solution by autoclaving.
5. 10 mM $MgSO_4$: Dilute 50 mL of 1 M $MgSO_4$ solution in 450 mL water in a 0.5-L graduated flask.
6. M9 media 10×: Add 100 mL of water to a 1-L graduated flask. Weight and add to the flask 128 g $Na_2HPO_4{\cdot}7H_2O$, 30 g KH_2PO_4, 5 g NaCl, 10 g NH_4Cl. Make up to 1 L. Sterilize the solution by autoclaving.
7. 20 % citrate solution: Dissolve 20 g of sodium citrate in 100 mL of water. Sterilize the solution by autoclaving.
8. 1 M succinic acid: Dissolve 11.81 g succinic acid in 100 mL water in a 0.2-L graduated flask. Sterilize the solution by autoclaving.
9. 1 M glucose: Dissolve 18.01 g glucose in 100 mL water in a 0.2-L graduated flask. Sterilize the solution by autoclaving.
10. M9 media 1×: Dilute 100 mL of M9 10× solution to a 1-L graduated flask. Add 25 mL of 1 M succinic acid solution or 25 mL of 1 M glucose solution. Add 2 mL 1 M $MgSO_4$ solution. Make up to 1 L. If required, add the proper antibiotics. Store at 4 °C.
11. M9 agar plates: Add 100 mL of M9 10× solution to a 1-L graduated flask. Add 16 g of Agar powder and mix. Add 2 mL 1 M $MgSO_4$ solution. Make up to 1 L and autoclave. After sterilization, cool down the solution to 55 °C, add 10 mL of 20 % citrate solution. If required, add the proper antibiotics.

Dispense 25 mL of the agar media into 90-mm Petri plates and wait until solidification. Store at 4 °C.

12. When required, kanamycin (Km, 50 μg/mL), chloramphenicol (Cm, 30 μg/mL), ampicillin (Ap, 150 μg/mL), gentamycin (Gm, 10 μg/mL), or streptomycin (Sm, 50 μg/mL) should be added to the media.

2.4 Primers

The primers shown below are used to check the correct insertion of the mini-Tn*7* transposon. Primers TN7-A and TN7-B are for the flaking region of the *att*Tn7 site and thus are strain specific. In case a strain other then *P. putida* KT2440 is used as host, this set of primers should be redesigned.

1. TN7-A: 5′-ATGACCAATGGCGAAGGC-3′.
2. TN7-B: 5′-GCCAGCCCATGATACTGC-3′.
3. TN7-C: 5′-ATATTTCGACCCCACGCC-3′.
4. TN7-D: 5′-CCTAGGCGGCCCCAACC-3′.

3 Methods

3.1 Construction of Host Strain for Homologue Recombination

First, it is necessary to construct a host strain bearing the homology fragment suitable to recombine with the promoter probe vector. This step is accomplished thought the insertion of a mini-Tn*10* transposon harboring this fragment, as described below Fig. 1a.

1. Inoculate single colonies of the donor strain *E. coli* CC118*λpir* (pLOF-hom.fg.), the helper strain *E. coli* HB101 (pRK600) and the receiving strain *P. putida* MEG1 in 5 mL LB + Km, 5 mL LB + Cm and 5 mL M9 media, respectively. Incubate the strains overnight with air shaking at 37 °C in the case of *E. coli* and 30 °C for *P. putida*. This step should be adapted depending on the growth requirements of the specific receiving strain.
2. After overnight growth, add 100 μL of each strain in 5 mL of 10 mM $MgSO_4$ buffer and mix vigorously.
3. Open a sterile 5 mL plastic syringe and connect to a 0.2 μm sterile filter. Load the bacterial mix in the syringe assembly and filter it. Discard the syringe and carefully remove the filter paper using sterile forceps. Place the filter paper in a LB plate supplemented with 1 mM IPTG (this compound induces the expression of the Tn*10* transposase), ensuring that the part of the filter containing the strains is face up. Incubate the plate overnight (*see* **Note 1**) at 30 °C (or the grown temperature of the receive strain, *see* **Note 2**).
4. After overnight growth, remove the filter paper form the plate and place it in 5 mL of mM $MgSO_4$ buffer, mix vigorously twice.

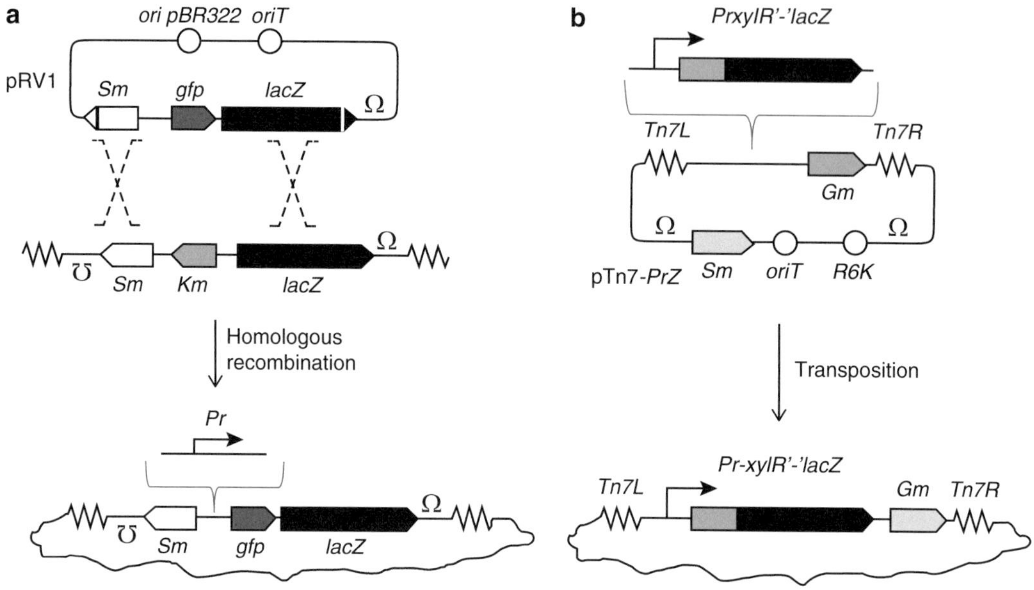

Fig. 1 Two strategies for chromosomal integration of promoter fusions. (**a**) The pRV1 vector and its main features are shown. In this system, the amber codons TGA is present in the 5′ region of the Sm maker and the *lacZ* gene (*black* and *white bars*, respectively), allowing these genes to be only functional in a strain harboring the *supF* tRNA. A GFP reporter is placed between these two genes and a target promoter can be cloned using *Eco*RI/*Bam*HI restriction sites. Additionally, the pLOF-hom.fg. mini-Tn*10* is integrated to the host chromosome and it bears 3′-truncated forms of Sm marker and *lacZ* genes, along with a Km marker. Through two steps of homologous recombination, functional genes are generated and the promoter-GFP fusion is stably integrated in the chromosome. (**b**) The mini-Tn*7* vector. In this system, a translational fusion to *lacZ* and a selection marker (Gm) are flanked by the transposition elements Tn7L and Tn7R. During transposition, these elements are mobilized to a specific region of the chromosome, the so-called *att*Tn7 site

5. Take 100 μL of the re-suspended solution and spread on a M9 agar plate supplemented with 0.2 % citrate as sole carbon source and kanamycin for selection. Incubate the plate overnight at 30 °C (*see* **Note 3**).
6. After overnight growth, pick up single colonies (at minimal eight colonies should be picked) and streak it in a fresh M9 agar plate with citrate and kanamycin to obtain isolate colonies. Incubate overnight at 30 °C.
7. After overnight growth, analyze the colonies for true transposition event. For this, replicate a colony in two new M9 agar plates, one supplemented with kanamycin and the other with ampicillin (500 μg/mL). Positive colonies should only grow on kanamycin-supplemented plates but not in those with ampicillin. After identification, store the selected strain for further analysis. In this protocol, the resulting strain is named *P. putida* MEG3 (*see* **Note 4**).

3.2 Insertion of Bi-Cistronic Reporter

This section describes the recombination between the suicide promoter probe vector and the homologue fragment placed in the chromosome of the host strain (generated beforehand in the previous section). The target promoter is previously cloned in the pRV1 vector [16], as shown in Fig. 1a.

1. Inoculate single colonies of the donor strains *E. coli* CC118*supF* (pRV1) and *E. coli* CC118*supF* (pRV1-*Pr*), the helper strain *E. coli* HB101 (pRK600) and the receiving strain *P. putida* MEG3 in 5 mL LB + Sm, 5 mL LB + Cm and 5 mL M9 media, respectively. Incubate the strains overnight with air shaking at 37 °C in the case of *E. coli* and 30 °C for *P. putida*.
2. Repeat the **steps 2–7** from Subheading 3.1 to transfer the vectors to the receive strain, with few modifications. In **step 3**, no IPTG is required. In **steps 5** and **6**, streptomycin should be used as the selection marker. The results of these steps should be single colonies grown in M9 plates.
3. Check the correct recombination of the suicide vector into the host chromosome. For this, replicate single colonies of resulting strains in one plate supplemented with streptomycin and another with kanamycin. If the recombination procedure is correct, the resulting strains should be unable to grown on kanamycin (*see* **Note 5**). Store positive strains for further analysis. In this protocol, the resulting strains are named *P. putida* MEG3-RV (resulting from the recombination of an empty pRV1 vector) and *P. putida* MEG3-*Pr* (after recombination of pRV1-*Pr* vector).

3.3 Mobilization of mini-Tn7 Transposon

This section describes the insertion of a mini-Tn7-based promoter probe system. This system uses a synthetic mini-Tn*7* vector harboring a promoter-less *lacZ* gene originated from pUJ9 vector [11]. As this particular reporter gene does not have a Shine–Dalgarno sequence, translational (or type 2) fusions can be generated and assayed. In contrast to the HR-base system, a single step is required to generate a reporter strain (Fig. 1b).

1. Inoculate single colonies of the donor strains *E. coli* CC118*λpir* (pTn7-Z) and *E. coli* CC118*λpir* (pTn7-*PrZ*), the helper strains *E. coli* HB101 (pRK600) and *E. coli* CC118*λpir* (pTnS1) and the receive strain *P. putida* KT2440 in 5 mL LB + Sm, 5 mL LB + Cm, 5 mL LB + Ap, and 5 mL M9 media, respectively. Incubate the strains overnight with air shaking at 37 °C in the case of *E. coli* and 30 °C for *P. putida*.
2. Repeat the **steps 2–7** from Subheading 3.1 to transfer the vectors to the receive strain, with few modifications. In **step 3**, no IPTG is required. In **steps 5** and **6**, gentamycin should be used as the selection marker. The results of these steps should be single colonies grown in M9 plates.

3. Check the correct insertion of the suicide vector into the host chromosome. For this, replicate single colonies of resulting strains in one plate supplemented with gentamycin and another with streptomycin. If the recombination procedure is correct, the resulting strains should be unable to grown on streptomycin. Store positive strains for further analysis. The resulting strains in this protocol are named *P. putida* KT-Z (with promoter-less *lacZ* gene) and *P. putida* KT-*PrZ* (with *Pr-xylR::lacZ* translational fusion).
4. Check the correct insertion of the mini-transposon in the *att*Tn*7* site (*see* **Note 6**). For this, perform colony PCR of few colonies using primer combinations TN7-A/TN7-B, TN7-A/TN7-C, and TN7-B/TN7-D. Positive strains should give negative amplification with the first primer combination and positive with the two last.

3.4 GFP Analysis of HR-Derivative Strains

Strains generated using the pRV1-based system harbor a transcriptional fusion to the GFP reporter stably integrated in the chromosome. These strains used to quantify the transcriptional response in a standard growth condition as shown below.

1. Inoculate single colonies of *P. putida* MEG3-RV and *P. putida* MEG3-*Pr* strains in 5 mL M9 media supplemented 25 mM of succinate. Incubate the strains with air shaking at 30 °C overnight.
2. After pre-growth, diluted the culture 1:25 in 20 mL of fresh media and incubate with air shaking at 30 °C to mid-exponential phase (about 4 h). At this point, place 500 μL with three technical replicas in 1.5 mL tubes. Pellet the cells by centrifugation at maximal speed for 5 min. Discard the supernatant and wash the pellet with 500 μL of PBS 1× solution. Repeat this wash procedure twice more. Re-suspend the cell in 500 μL of PBS 1× solution and store it in ice until analysis.
3. After samples collection, analyze the GFP distribution in the cell populations using flow cytometry and standard procedures. The results shown in Fig. 2a were obtained using a GALLIOS cytometer (Perkin Elmer).

3.5 β-Galactosidase Assay of mini-Tn7 Strains

The strains bearing the translational fusion between the target and *lacZ* genes stably integrated in the *att*Tn7 can be assayed using classical β-galactosidase assay procedures [24]. An example of experimental setup is given below.

1. Inoculate single colonies of *P. putida* KT-*PrZ* and *P. putida* KT-*PrZ* strains in 5 mL M9 media supplemented 25 mM of succinate or 25 mM of glucose. Incubate the strains with air shaking at 30 °C overnight.

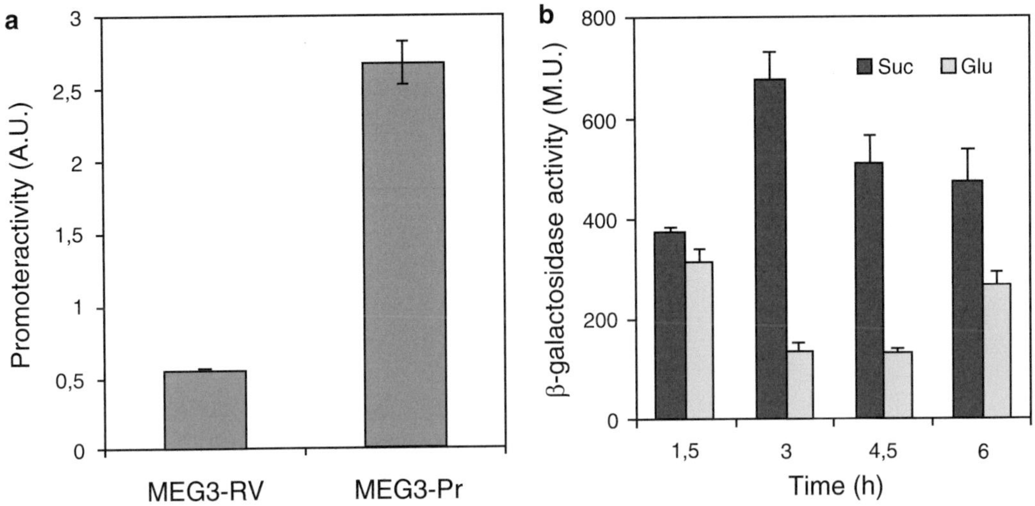

Fig. 2 Promoter activity of recombinant strains. (**a**) Assay of promoter activity for HR-generated strains using flow cytometry. Strains grown in mid-exponential phase were collected and analyzed as described [21]. Strain *P. putida* MEG-RV was generated using an empty pRV1 vector and thus only presents background GFP levels. Strain *P. putida* MEG-*Pr* was generated using pRV1-*Pr* vector with cloned *Pr* promoter cloned. Experiments were performed with three technical replicas. (**b**) β-galactosidase assays of strains with mini-Tn*10*-based systems. In this assay, *P. putida* KT-*PrZ* strain was grown in two different carbon sources (succinate and glucose). At fixed time points, samples were collected and LacZ activity measured as described previously [24]. Experiments were performed with three technical replicas

2. After pre-growth, diluted the culture 1:25 in 20 mL of fresh media and incubate with air shaking at 30 °C for 1.5 h. At this point, collect samples of 0.8 mL and measure the optical density at 600 nm. After this measurement, store the samples in ice. Repeat this process at fix time intervals (for example, every 1.5 h).
3. After sample collection, perform the β-galactosidase assay as described previously [24]. Figure 2b presents the results for the *P. putida* KT-*PrZ* strain using this specific setup. This result evidences differences in the reporter expression depending on the carbon source used, which suggest a yet unknown post-transcriptional process [22].

4 Notes

1. Transposition of mini-Tn*10* is not very efficient in Pseudomonads so the conjugation time have to be optimized. In our hands, overnight incubation for *P. putida* works best.
2. The growth condition has to be that more favorable for the receive strain of interest. All parts of the protocol involving bacterial growth have thus to fit the requirements of the particular strains.

3. If few colonies appear, simply plate a larger amount of the conjugation mix.
4. The transposition of the mini-Tn*10* element is random, so it is recommended to store at least five independent positive strains. Having many different strains can be useful to identify more suitable regions of the chromosome for promoter expression.
5. In addition to this step, correct recombination of the fragments can be assayed under a condition that allows promoter induction. In this way, positive colonies became blue in plates supplemented with 5-bromo-4-chloro-indolyl-β-D-galactopyranoside (X-gal).
6. It if not infrequent that the mini-Tn*7* element is inserted in an incorrect position in the chromosome. This is mostly true for strains with more phylogenetic distant to *E. coli* (where the system was first implemented). The introduction of the transposon in the correct site is critical for the generation of isogenic strains.

Acknowledgments

This work was supported by the BIO and FEDER CONSOLIDER-INGENIO programs of the Spanish Ministry of Economy and Competitiveness, the ARYSIS and ST-FLOW Contracts of the EU, the ERANET-IB Program and funding from the Autonomous Community of Madrid (PROMPT).

References

1. Silva-Rocha R, de Lorenzo V (2010) Noise and robustness in prokaryotic regulatory networks. Annu Rev Microbiol 64:257–275
2. Ishihama A (2010) Prokaryotic genome regulation: multifactor promoters, multitarget regulators and hierarchic networks. FEMS Microbiol Rev 34:628–645
3. Sambrook J, Fritsch EF, Maniatis T (1989) Molecular cloning: a laboratory manual. Cold Spring Harbor, New York
4. Contag CH, Bachmann MH (2002) Advances in *in vivo* bioluminescence imaging of gene expression. Annu Rev Biomed Eng 4:235–260
5. de Las Heras A, Carreno CA, Martinez-Garcia E, de Lorenzo V (2010) Engineering input/output nodes in prokaryotic regulatory circuits. FEMS Microbiol Rev 34:842–865
6. Locke JC, Elowitz MB (2009) Using movies to analyse gene circuit dynamics in single cells. Nat Rev Microbiol 7:383–392
7. Shaner NC, Steinbach PA, Tsien RY (2005) A guide to choosing fluorescent proteins. Nat Methods 2:905–909
8. Taniguchi Y, Choi PJ, Li GW, Chen H, Babu M, Hearn J, Emili A, Xie XS (2010) Quantifying *E. coli* proteome and transcriptome with single-molecule sensitivity in single cells. Science 329:533–538
9. Raj A, van Oudenaarden A (2008) Nature, nurture, or chance: stochastic gene expression and its consequences. Cell 135:216–226
10. Dublanche Y, Michalodimitrakis K, Kummerer N, Foglierini M, Serrano L (2006) Noise in transcription negative feedback loops: simulation and experimental analysis. Mol Syst Biol 2:41
11. de Lorenzo V, Herrero M, Jakubzik U, Timmis KN (1990) Mini-Tn5 transposon derivatives for insertion mutagenesis, promoter probing, and chromosomal insertion of cloned DNA in

gram-negative eubacteria. J Bacteriol 172: 6568–6572

12. de Lorenzo V, Timmis KN (1994) Analysis and construction of stable phenotypes in gram-negative bacteria with Tn*5*- and Tn*10*-derived minitransposons. Methods Enzymol 235:386–405
13. Kalogeraki VS, Winans SC (1997) Suicide plasmids containing promoterless reporter genes can simultaneously disrupt and create fusions to target genes of diverse bacteria. Gene 188:69–75
14. Choi KH, Gaynor JB, White KG, Lopez C, Bosio CM, Karkhoff-Schweizer RR, Schweizer HP (2005) A Tn*7*-based broad-range bacterial cloning and expression system. Nat Methods 2:443–448
15. de las Heras, A., Carreño, C.A. and de Lorenzo, V. (2008) Stable implantation of orthogonal sensor circuits in Gram-negative bacteria for environmental release, *Environ Microbiol* 10, 3305-3316.
16. Silva-Rocha R, de Lorenzo V (2012) A *GFP-lacZ* bicistronic reporter system for promoter analysis in environmental Gram-negative bacteria. PLoS One 7:e34675
17. Bagdasarian M, Lurz R, Ruckert B, Franklin FC, Bagdasarian MM, Frey J, Timmis KN (1981) Specific-purpose plasmid cloning vectors. II. Broad host range, high copy number, RSF1010-derived vectors, and a host-vector system for gene cloning in *Pseudomonas*. Gene 16:237–247
18. Herrero M, de Lorenzo V, Timmis KN (1990) Transposon vectors containing non-antibiotic resistance selection markers for cloning and stable chromosomal insertion of foreign genes in gram-negative bacteria. J Bacteriol 172:6557–6567
19. Kessler B, de Lorenzo V, Timmis KN (1992) A general system to integrate *lacZ* fusions into the chromosomes of gram-negative eubacteria: regulation of the *Pm* promoter of the TOL plasmid studied with all controlling elements in monocopy. Mol Gen Genet 233:293–301
20. Boyer HW, Roulland-Dussoix D (1969) A complementation analysis of the restriction and modification of DNA in *Escherichia coli*. J Mol Biol 41:459–472
21. Silva-Rocha R, de Lorenzo V (2012) Stochasticity of TOL plasmid catabolic promoters sets a bimodal expression regime in *Pseudomonas putida* mt-2 exposed to *m*-xylene. Mol Microbiol 86(1):199–211
22. Silva-Rocha R, de Lorenzo V (2011) A composite feed-forward loop I4-FFL involving IHF and Crc stabilizes expression of the XylR regulator of *Pseudomonas putida* mt-2 from growth phase perturbations. Mol Biosyst 7:2982–2990
23. Keen NT, Tamaki S, Kobayashi D, Trollinger D (1988) Improved broad-host-range plasmids for DNA cloning in Gram-negative bacteria. Gene 70:191–197
24. Miller J (ed) (1972) Experiments in molecular genetics. Cold Spring Harbor Laboratory, Cold Spring Harbor, New York

Chapter 38

A Method to Capture Large DNA Fragments from Genomic DNA

Geneviève Ball, Alain Filloux, and Romé Voulhoux

Abstract

The gene capture technique is a powerful tool that allows the cloning of large DNA regions (up to 80 kb), such as entire genomic islands, without using restriction enzymes or DNA amplification. This technique takes advantage of the high recombinant capacity of the yeast. A "capture" vector containing both ends of the target DNA region must first be constructed. The target region is then captured by co-transformation and recombination in yeast between the "capture" vector and appropriate genomic DNA. The selected recombinant plasmid can be verified by sequencing and transferred in the bacteria for multiple applications.

This chapter describes a protocol specifically adapted for *Pseudomonas aeruginosa* genomic DNA capture.

Key words Recombinational cloning, Targeting sequence, Vector capture, Gene capture, Counter-selectable marker

1 Introduction

Cloning large DNA fragment (>10 kb) is often very challenging due to the lack of suitable restriction sites and efficient DNA amplification procedure. Based on the observation that linear DNA fragment can efficiently support recombination in yeast [1], it has been demonstrated that specific DNA fragments can be directly cloned into specialized vectors by homologous recombination in yeast [2]. The fragment to be cloned must display at both ends a DNA region homologous to the target plasmid [2]. It has been shown that 40 base-pairs overlap is sufficient to ensure efficient recombination in yeast [2, 3].

The recombinational cloning in yeast was adapted to capture genomic regions in *Pseudomonas aeruginosa* and different yeast/*Escherichia coli* shuttle vectors were constructed [4–6].

Among them, Wolfgang and collaborators [6] have developed a specialized yeast/*E. coli*/*P. aeruginosa* shuttle vector, pLLX13,

Alain Filloux and Juan-Luis Ramos (eds.), *Pseudomonas Methods and Protocols*, Methods in Molecular Biology, vol. 1149, DOI 10.1007/978-1-4939-0473-0_38, © Springer Science+Business Media New York 2014

Table 1
Strains and plasmids

Strains	Genotype	Reference
Saccharomyces cerevisiae CRY1-2	Δ*ura3* Cycloheximide^R	Wolfgang et al. [6]
E. coli DH10B	*araD139* Δ*(ara,leu)7697 mcrA* Δ*(mrr-hsdRMS-mcrBC)* λ⁻	Invitrogen
Plasmids	**Features**	
pLLX8	4.9 kb plasmid used to produce by PCR the 2.95 kb *bla cyh2* amplicon. Ampicillin^R	Wolfgang et al. [6]
pLLX13	9.9 kb yeast *E. coli* shuttle vector that provides all the features for selection and replication in both yeast and bacteria. Tetracycline^R	Wolfgang et al. [6]

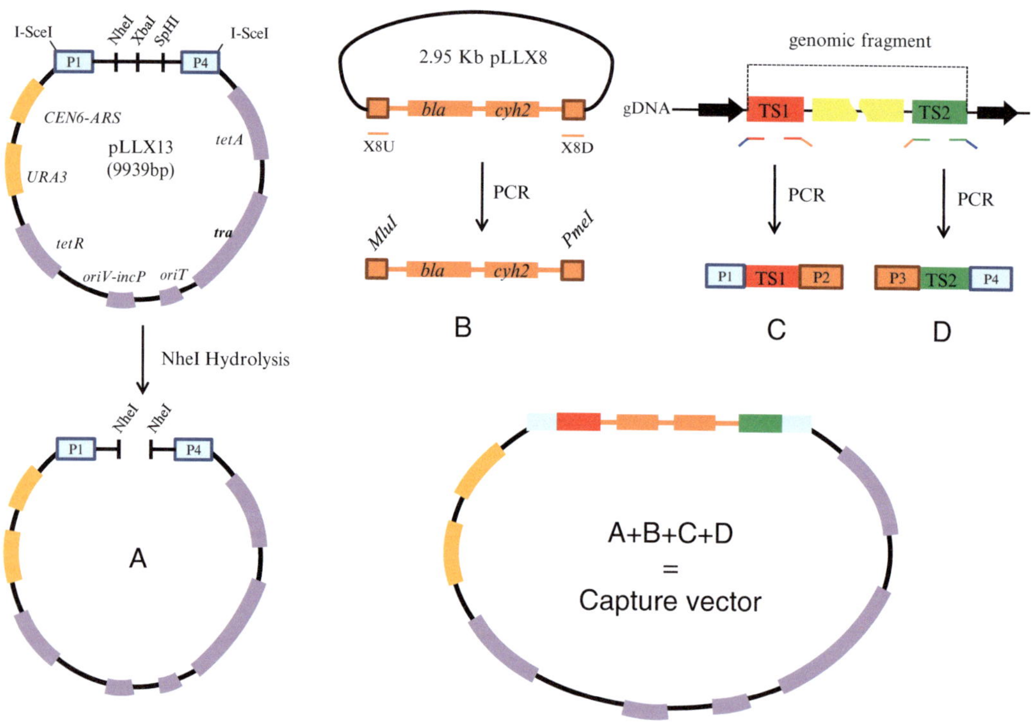

Fig. 1 Scheme depicting how the four DNA components are obtained and recombined to generate the capture vector

to target variable genomic segments in *P. aeruginosa* genomes (Table 1, Fig. 1). The backbone of pLLX13 is made with different pieces necessary for replication and genetic selection in yeast, *E. coli* and *P. aeruginosa* (Fig. 1). The *CEN6-ARSH4* genes are required for maintenance and replication of the plasmid in the yeast. The *URA3* gene is needed for plasmid selection on uracil lacking media

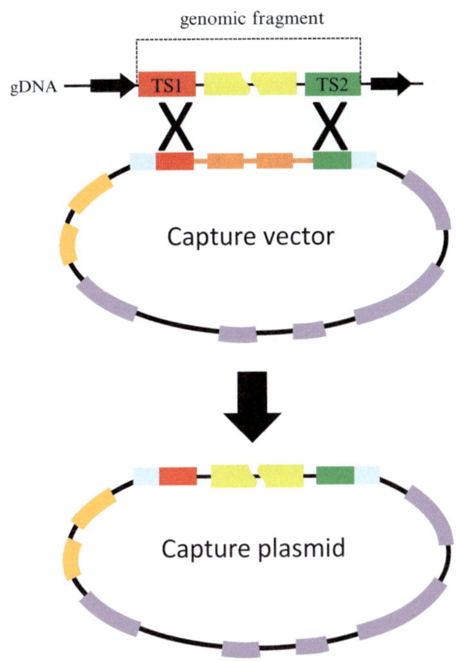

Fig. 2 Yeast recombinational cloning. Schematic representation of the recombination between the yeast capture vector and *P. aeruginosa* genomic DNA (gDNA) leading to the desired capture plasmid

(URA dropout plates). The *tetAR* gene provides tetracycline resistance for selection in *E. coli* and *P. aeruginosa*. The replication function in bacteria is provided by the origin of replication *oriV-incP*. The conjugative functions are supplied by the *tra* gene and the origin of transfer *oriT*, that allows transfer of the plasmid between *E. coli* and *P. aeruginosa* by conjugation (see corresponding chapter).

The gene capture technique follows two steps. The first one is the construction of the shuttle yeast/*E. coli* capture vector that assembles the different sources of DNA required for recombination and selection (Fig. 1). The second step is the capture of the DNA fragment by homologous recombination in yeast between the capture vector and *P. aeruginosa* genomic DNA, resulting in the desired recombinant plasmid (Fig. 2).

Four DNA sequences are assembled by recombinational cloning in *Saccharomyces cerevisiae* CRY1-2 to construct the yeast capture vector (Table 1). Two of them are universal for all capture vectors, i.e., the linearized pLLX13 vector and the pLLX8 amplicon. The later carries the *bla* gene (encoding β-lactamase) and the yeast counter-selectable marker *CYH2* gene (the introduction of the wild-type allele *CYH2* gene into the yeast cell confers sensitivity to cycloheximide). The two others components are the targeting sequences TS1 and TS2 (Fig. 1). These sequences correspond to both sides of the DNA target region, and will be the target for the recombination between the capture vector and the *P. aeruginosa* genomic DNA. Briefly, TS1 and TS2 are generated by PCR and correspond to the upstream and downstream region that flanked the DNA region to be cloned from the linearized *P. aeruginosa* genomic DNA (Fig. 2).

The gene capture technique will allow the cloning of any known large DNA fragment from characterized genomes. In addition, and combined to DNA hybridization analysis using microarrays, it can be used to identify unknown, or variable, DNA region located between two known markers and from uncharacterized clinical or environmental isolates without the need for large-scale sequencing [6, 7].

2 Materials

All solutions are prepared using ultrapure water purified with a Milli-Q system and analytical grade reagents. All reagents must be prepared, filtered and stored at room temperature (RT). Media are autoclaved 15 min at 121 °C. Antibiotics and drug solutions are filtered and stored at −20 °C. The investigator must carefully follow all waste disposal regulations.

2.1 Media

1. YPD (yeast extract–peptone–dextrose) medium (per liter): Yeast extract 1 % (w/v) (**Note 1**), Dextrose 2 %, Bacto Peptone 2 %, Adenine 0.05 %.
2. YPD agar (per liter): YPD broth plus agar 2 %.
3. Minimal SD medium (per liter): YNB (Yeast Nitrogen Base) 0.17 %, Dextrose 2 %, Ammonium sulfate 5 g. Prepare a 5× stock solution.
4. Agar solution: dissolve 20 g of agar in 800 ml water.
5. LB (Luria–Bertani) medium (per liter): Bacto Tryptone 10 g; Bacto Yeast Extract 5 g; NaCl 10 g; pH 7.4.

2.2 Solutions

1. Prepare a 1 M lithium acetate solution. Weigh 10.2 g lithium acetate in a glass beaker and add ultrapure water up to 100 ml. The solution is passed through a 0.22 μm filter.
2. 10× TE (Tris–EDTA) buffer (pH 7.5/8). Weigh 12.1 g Tris in a glass beaker and 3.7 g EDTA. Add water to a final volume of 900 ml. Mix and adjust pH with HCl. Make up to 1 L with water.
3. PEG (Polyethylene glycol) Stock solution PEG 3350 (50 %): Weigh 50 g of PEG 3350 and add water up to 100 ml. Mix and filter the solution. PEG solubilization may take about 15 min.
4. TFB (TransFormation Buffer) I and II.

TFBI (0.1 M lithium acetate in TE): Mix 1 Volume of Lithium acetate at 1 M, 1 Volume of 10× TE and 8 volumes of sterile water.

TFBII (0.1 M lithium acetate, 40 % PEG in TE): Mix 1 volume of lithium acetate at 1 M, 1 volume of 10× TE, and 8 volumes of PEG 50 %.

5. Cycloheximide 10 mg/ml: weigh 10 mg of cycloheximide and dissolve it in 1 ml of water. The stock solution can be stored several months at −20 °C.

2.3 Strains and Plasmids

For details *see* Table 1.

2.4 DNA Techniques

Genomic DNA Preparation (Pure Link Genomic DNA from Invitrogen).

DNA purification from yeast (Zymoprep from Zymo research).

DNA purification from *E. coli* (QiaSpin Miniprep from QIAGEN).

DNA restrictions are performed following provider recommendations.

PCR (polymerase chain reaction) are performed with Expand HighFidelity DNA polymerase from Roche.

3 Methods

3.1 Construction of the Capture Vector

The capture vector is assembled via yeast recombination from four different sources of DNA obtained by PCR: the vector pLLX13 (Fig. 1a), the pLLX8 fragment carrying the *bla/CYH2* cassette (Fig. 1b) and the two targeting sequences TS1 (Fig. 1c) and TS2 (Fig. 1d) that mark out the DNA region to be cloned.

1. The targeting sequences TS1 and TS2 are 1 kb long. They are generated by PCR with tailed primers (P1, P4) carrying homology to both the genomic DNA target and the vector pLLX13, and tailed primers (P2, P3) that carry homology both to TS1 and TS2 and pLLX8 (Table 2, Fig. 1c, d).
2. Amplification of the 2.95 kb *bla/CYH2* cassette is performed from pLLX8 DNA with the primer pair X8U and X8D (Table 2, Fig. 1b). During recombination in yeast, this fragment is inserted between the TS1 and TS2 sequences.
3. Linearization of the pLLX13 by cutting between P1 and P4 with *Nhe*I restriction enzyme (Fig. 1a) (*see* **Note 2**).
4. Assemble the capture vector by transforming 200 ng of each targeting sequence, 600 ng of the 2.95 kb pLLX8 fragment and 200 ng of the *Nhe*I-linearized pLLX13 vector in 200 μl of lithium acetate treated competent cells of *S. cerevisiae* CRY1-2 strain.

Table 2
Oligonucleotides sequence of the tails that are added at the 5′end of each primer

Primer names	Primers sequence 5′→3′	Description
P1	ATATTACCCTGTTATCCCTAGCGTAACTATCGATC TCGAG followed by the upstream extremity of TS1	40 bp homology to pLLX13
P2	CATATATACTTTAGATTTTAATTAAACGCGTTCTA GAAAA followed by the downstream extremity of TS1	40 bp homology to pLLX8. *Mlu*I restriction site is underlined
P3	CATTTTCACCGTTTTTTGTTTAAACGTTAACTCT AGAGGG followed by the upstream extremity of TS2	40 bp homology to pLLX8. *Pme*I restriction site is underlined
P4	TAACAGGGTAATATAGAGATCTGGTACCCTCGAG GACGTC followed by the downstream extremity of TS2	40 bp homology to pLLX13
X8U	TTTTCTAGAACGCGTTTAATTAAAATCTAAAGTATAT ATGAGTAAAC	Underlined sequence is homologous to 3′ sequence of P2
X8D	CCCTCTAGAGTTAACGTTAAACAAAAAACGGTGAA AATGGGTGATAG	Underlined sequence is homologous to the P3

The sequences of the primers P1, P2, P3, and P4 correspond to the sequence of the tail that is added at the 5′ end of each primer

3.2 Lithium Acetate Transformation

1. Streak onto an YPD plate the *S. cerevisiae* CRY1-2 strain. Grow the yeast at 30 °C until colonies could be observed.
2. Inoculate 5 ml of YPD medium with a few yeast colonies. Grow the yeast overnight at 30 °C under shaking condition (200 rpm).
3. Inoculate 50 ml of YPD medium with the overnight culture at 1:100 dilution (OD_{600} of 0.05). Let the culture grow with shaking for 5 h at 30 °C. The final OD_{600} must be around 0.4–0.5 (*see* **Note 3**).
4. Transfer the culture in a 50 ml conical tubes and centrifuge at room temperature for 5 min at 1,200 rcf (*see* **Note 4**).
5. Discard the supernatant and resuspend the pellet in 40 ml of sterile water at room temperature (RT).
6. Centrifuge for 5 min at RT at 1,200 rcf.
7. Discard the supernatant and resuspend the pellet in 250 μl of TFBI (RT).
8. Add DNA in a 15 ml falcon tube (200 ng of TS1 and TS2, 200 ng of *Nhe*I-linearized pLLX13, and 600 ng of the *bla CYH2* cassette) (*see* **Note 5**) and 200 μg of single strand herring sperm DNA denatured (*see* **Note 6**). The final volume of DNA added should not exceed 50–60 μl.

Table 3
Summary of the selection of yeast on different medias

	S. cerevisiae CRY1.2	*S. cerevisiae* CRY1.2 + pLLX13	*S. cerevisiae* CRY1.2 + Capture vector
YPD	Growth	Growth	Growth
URADO plates	No growth	Growth	Growth
URADO plates cycloheximide 2.5 μg/ml	Growth	Growth	No growth

9. Per transformation add 200 μl of the competent yeast solution in TFBI. Add 1.2 ml of TFBII and vortex the suspension until homogenization.
10. Incubate the tube at 30 °C (water bath) for 30 min then 15 min in a 42 °C (water bath) (*see* **Note 7**).
11. Pellet the cells for 2 min at 1,200 rcf and resuspend the cells in 500 μl of TE.
12. Plate 100 μl onto uracil dropout plate (*see* **Note 8**) and incubate at 30 °C. Colonies will appear after 2 or 3 days. The pLLX13 plasmid should be used as a positive control.

3.3 Selection of the Capture Vector

To obtain the desired capture vector, prepare the DNA in yeast from a pool of transformants containing closed pLLX13 plasmid and capture vector.

1. Add 2 ml of PBS in the URA DO plate and move gently the plate until you obtain the yeast colonies in suspension. Pellet 250 μl of this suspension and purify the yeast DNA with the Zymoprep yeast plasmid minikit (*see* **Note 9**).
2. Electroporate the yeast DNA in *E. coli* DH10B electroMAX competent cells (Invitrogen) (*see* **Note 10**) and select the transformants on LB agar plates supplemented with 5 μg/ml tetracycline and 50 μg/ml ampicillin. The selection on both tetracycline and ampicillin results in the specific recovery of capture vector. In general, more than 90 % of the recovered plasmids are correct.
3. When the correct capture vector is identified, perform a transformation in yeast in order to check that the cycloheximide counter-selection is efficient (Table 3). Indeed, the presence of the *CYH2* wild type gene in the capture vector confers the sensitivity to cycloheximide (*see* **Note 11**).

3.4 Capture of the DNA Target

To capture the desired *P. aeruginosa* DNA segment you have to perform recombinational cloning in yeast using the capture vector and the genomic DNA of *P. aeruginosa*.

1. Prepare genomic DNA from *P. aeruginosa* strains with the Pure Link Genomic DNA Minikit (Invitrogen). The genomic DNA is prepared from 1 ml ON culture (*see* **Note 12**).
2. Sheared the genomic DNA by passing it through a 26 1/2 gauge needle at least 40 times (*see* **Note 13**).
3. Linearized the pLLX13 by digesting it with *Pme*I or *Mlu*I restriction enzyme, the two restriction sites that have been added in the X8U and X8D primers that flanked the pLLX8 cassette (Fig. 1b).
4. Following the procedure described in Subheading 3.2, co-transform lithium acetate yeast treated cells with 500 ng of linearized capture vector and 5 μg of sheared genomic DNA. Select the transformants on URADO cycloheximide plates (*see* **Note 14**).

3.5 Screening of Yeast Colonies

1. Prepare the lysis buffer as follows: 1.2 M Sorbitol, 100 mM sodium phosphate pH 7.4, and 2.5 mg/ml of Zymolyase 20T (*see* **Note 15**). Each colony (at 1 mm diameter) is resuspended in 30 μl of lysis buffer and incubated for 1 h at 37 °C. The final mixture could be stored at −20 °C for several months and be reused. After incubation, used 1 μl of the mixture in a 25 μl of PCR reaction to select plasmids containing the desired genomic fragment.
2. To analyze the selected plasmid, grow the yeast colony in 5 ml of YPD during 12–16 h at 30 °C. Pellet the cells and prepare the plasmid DNA with the Zymoprep DNA minikit. DNA is eluted in 35 μl sterile water. Transfer the plasmid in *E. coli* DH10B by electroporation as indicated in Subheading 3.4 and select colonies on LB agar tetracycline 5 μg/ml plates (*see* **Note 16**). Prepare plasmid DNA with the QiaSpin Miniprep kit from QIAGEN and perform a restriction profile analysis in order to characterize the DNA fragment cloned (*see* **Note 17**). For a 15–20 kb insert, about 10 % of the plasmids have captured the desired genomic DNA fragment.

4 Notes

1. For every percentage (%) the ratio between the two components is weight (w) for volume (v).
2. Alternatively *Xba*I or *Sph*I restriction enzyme could also be used (Fig. 1a).
3. The quantities indicated are given for approximately 1.5 transformation.
4. 1,200 rcf correspond to 2,400 rpm (1,100 × *g*).

5. To minimize the volume of DNA to add in the 200 μl lithium acetate-treated yeast cells, perform a PEG precipitation of the PCR products as follow: prepare a solution of PEG 8000 30 %, $MgCl_2$ 30 mM. Mix 2 volumes of PEG 8000 30 %, $MgCl_2$ 30 mM in TE X1, with 1 volume PCR product, vortex, and incubate for 15 min at RT. Centrifuge 30 min at 18,625 × *g*. Discard carefully the supernatant and wash the pellet twice with 5 volumes of EtOH 70 %. Air dry the pellet 5 min at RT and resuspend the pellet in a suitable volume of ultrapure water.
6. Preparation of the denatured Single Strand herring carrier DNA (yeast maker carrier DNA BD biosciences). Before use, boil the yeast maker carrier DNA for 5 min and place on ice to ensure that you have single stranded DNA. Aliquot into appropriate volume and store at −20 °C.
7. To obtain more transformants you can increase the incubation time: up to 2 h at 30 °C, and up to 30 min at 42 °C.
8. To pour uracil deficient medium (URA DropOut plate): for 1 l add 200 ml sterile 5× Minimal SD media to 800 ml of agar and 0.77 g of uracil dropout supplement (freshly dissolved in the SD minimal medium).
9. Follow the indications of the manufacturers.
10. To perform the electroporation in *E. coli* used the ElectroMAX DH10B cells (Invitrogen) following the electroporator conditions: 2.0 kV, 200 Ω, 25 μF. You can prepare homemade electro-competent cells. Check that transformation efficiency is greater than 10^{10} transformants per μg of DNA.
11. To establish the rate of reversion for cycloheximide sensitivity, plate the transformation on URADO plates with or without cycloheximide at 2.5 μg/ml. You must not obtain more than 10 % of cycloheximide-resistant colonies.
12. Check the quality and the quantity of the genomic DNA by running a 0.4 % agarose gel. 5 μg of genomic DNA in a minimal volume of 50 μl is required to perform the yeast recombinational process.
13. Genomic DNA is well "sheared" when it runs as a smear in a 0.4 % agarose gel.
14. pLLX13 plasmid should be transformed as "positive control" and capture vector as "negative control" (Table 3). Due to the counter-selection, yeast colonies that grow on URADO cycloheximide plates should have substituted the *bla/CYH2* marker by the desired genomic DNA fragment. Re-isolate the colony on URADO cycloheximide plate to eliminate the false positives. Screen by PCR the yeast colonies to select the correct plasmids (*see* Subheading 3.5).

15. Preparation of a stock solution of zymolyase 20T at 25 mg/ml: Dissolve the lyophilized enzyme in 10 mM sodium phosphate buffer pH 7.5 and 1 M sorbitol. This enzyme transforms the intact yeast cells in spheroplasts. Aliquots of this enzyme could be stored at −80 °C for several months.
16. In the case of large DNA fragment (>30 kb), you can use a variant of strain DH10B named Genehog (Invitrogen) that allows transformation of large plasmid.
17. The entire fragment could be excised by using the I-*Sce*I restriction site present at both ends of the polylinker. The sequence of this homing endonuclease is absent in *P. aeruginosa* genome.

Acknowledgements

We gratefully thank Stephen Lory for providing the plasmids pLLX13, pLLX8 and the CRY1-2 *S. cerevisiae* strain.

The work in Romé Voulhoux's lab is supported by the "3D-Pilus" young researcher ANR grant (ANR-JC 07-183230), the "Pathomics" Era-net PATHO grant (ANR-08-PATH-004-01), and Vaincre La Mucoviscidose.

References

1. Orr-Weaver TL, Szostak JW, Rothstein RJ (1981) Yeast transformation: a model system for the study of recombination. Proc Natl Acad Sci U S A 78:6354–6358
2. Oldenburg K, Vo K, Michaelis S, Paddon C (1997) Recombination-mediated PCR-directed plasmid construction in vivo yeast. Nucleic Acids Res 25:451–452
3. Hua S, Qiu M, Chan E, Zhu L, Luo Y (1997) Minimum length of sequence homology required for *in vivo* cloning by homologous recombination in yeast. Plasmid 38:91–96
4. Raymond CK, Sims EH, Kas A, Spencer DH, Kutyavin TV, Ivey RG, Zhou Y, Kaul R, Clendenning JB, Olson MV (2002) Genetic variation of the O-antigen biosynthetic locus in *Pseudomonas aeruginosa*. J Bacteriol 184: 3614–3622
5. Smith EE, Sims EH, Spencer DH, Kaul R, Olson MV (2005) Evidence for diversifying selection at the pyoverdine locus of *Pseudomonas aeruginosa*. J Bacteriol 187:2138–2147
6. Wolfgang MC, Kulasekara BR, Liang X, Boyd D, Wu K, Yang Q, Miyada CG, Lory S (2003) Conservation of genome content and virulence determinants among clinical and environmental isolates of *Pseudomonas aeruginosa*. Proc Natl Acad Sci 100:8484–8489
7. Ball G, Viarre V, Garvis S, Voulhoux R, Filloux A (2012) Type II-dependent secretion of a *Pseudomonas aeruginosa* DING protein. Res Microbiol 163:457–469

Chapter 39

Transposon Mutagenesis

Hemantha D. Kulasekara

Abstract

Transposon-based mutagenesis of bacterial genomes is a powerful method to identify genetic elements that control specific phenotypes. The most frequently used transposon tools in *Pseudomonas aeruginosa* are based either on Himar1 mariner or Tn5 transposases, both of which have been used to generate nonredundant mutant libraries in *P. aeruginosa*. Here we present a detailed protocol for using Himar1 mariner-based transposon mutagenesis to create mutant libraries in *P. aeruginosa*.

Key words Transposon mutagenesis, *Pseudomonas aeruginosa*, Himar1 mariner, Arbitrary PCR, Mariner mutagenesis

1 Introduction

In forward genetic approaches involving bacteria, mutagenesis of chromosomal DNA is an effective experimental method to identify mechanisms responsible for a specific phenotype. An elegant technique for isolating chromosomal mutants for phenotypic analysis is chemical mutagenesis, in which DNA is briefly exposed to a mutagen such that only one locus in the chromosome is altered per cell. The resulting mutations frequently occur at individual base pair without disrupting the overall chromosomal organization. These mutant alleles may either alter gene/protein expression levels or result in production of an altered protein giving rise to such phenotypes as temperature sensitive mutants. In contrast, transposon mutagenesis employs a piece of foreign, heterologous DNA that inserts into the chromosome to disrupt genetic elements. Transposons disrupt the expression of a particular gene but rarely create proteins with altered functions. In addition to disrupting genes, transposons can alter expression of surrounding genes by positional effects or by encoding transposon-based promoters, with potentially undesirable results. Even with these shortcomings, transposon mutagenesis has become the method of choice for

Alain Filloux and Juan-Luis Ramos (eds.), *Pseudomonas Methods and Protocols*, Methods in Molecular Biology, vol. 1149, DOI 10.1007/978-1-4939-0473-0_39, © Springer Science+Business Media New York 2014

mutagenesis of bacterial genomes for the following reasons: First, transposons contain antibacterial resistance determinants that facilitate selection of mutants, a feature that is not available for chemical mutagenesis. Moreover, these selectable markers enable the subsequent transfer of mutated elements into other genetic backgrounds. Second, using semi-random arbitrary PCR sequencing methods, the site of the mutation can be easily mapped in the transposon mutant, a method not available for chemical mutagenesis in which either whole genome sequencing or more cumbersome genetic mapping would have to be carried out. Third, transposons are delivered by a suicide vector in which the transposase lies outside the transposable element, ensuring that additional transposition events will not occur after the initial event within the same chromosome. Fourth, transposons are delivered to the recipient by conjugation, transduction or transformation, techniques that are well established and relatively easy to perform. The genetics and biochemistry of bacterial transposons have been characterized in detail, and transposon tools have been used since the 1980s [1–4]. The most frequently used transposons in bacterial genetics are based on the Tn*5* and Tn*10* platforms [1–3, 5]. However, a novel transposon, the Himar1 mariner identified by Lampe et al. has recently become popular in bacteria genetics [6–8]. Transposons based on both Tn*5* and Himar1 have been used to generate nonredundant mutant libraries in *P. aeruginosa* [5, 9].

The wild-type Himar1 Mariner transposon, isolated from the horn fly *Haematobia irritans*, is a relatively simple transposon resembling an IS element and consisting of the Himar1 transposase bounded by 29 bp inverted repeats [7]. Himar1 transpose into the dinucleotide TA in a mechanism similar to that of the Tn*5* and Tn*10* transposons, in which the donor transposon DNA is cut and then pasted into the target site by the action of the transposase. A hyperactive Himar1, C9, was isolated by Lampe et al. that showed a 50-fold increase in transposition frequency in *Escherichia coli* [8]. In the C9 transposase, two mutations, Q131R and E137K, confer the hyperactivity. The transposon vectors described here, pBT20, pBTK30, and pBAM1, are based on this C9 hyperactive transposase, whereas pMAR2xT7 is based on the wild-type Himar1 mariner [10].

2 Materials

2.1 Biological Reagents Needed for Mutagenesis

2.1.1 Transposons Available for Mutagenesis

pBT20, pBT30, pMAR2XT7, or pBAM1. These should be transformed into *Escherichia coli* SM10 *lambda pir* strain (Fig. 1 and Table 1).

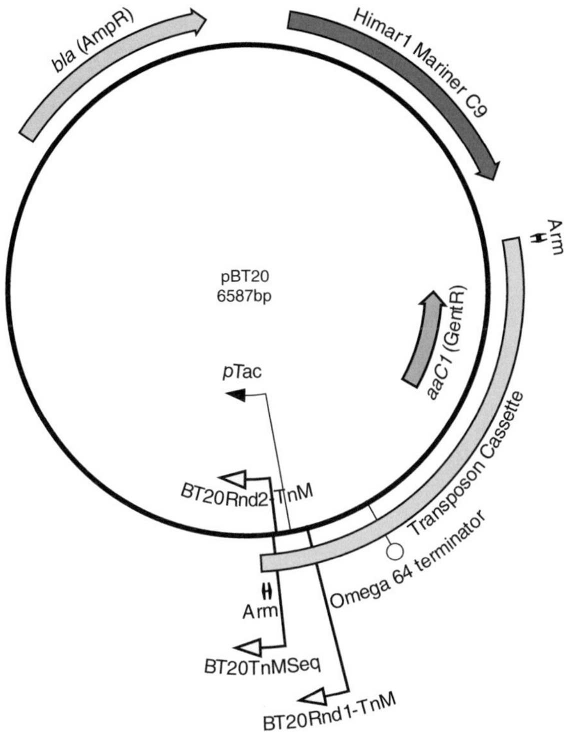

Fig. 1 Genetic organization of the transposon delivery vector pBT20

Table 1
Transposon delivery vectors

Transposase	Plasmid	Resistance marker	Antibiotic resistance	Outword facing promoter #1	Outword facing promoter #2	Reference
Himar1 C9	pBT20	*aaC1*	gentamicin	*p-tac*	*p-aaC1*	
Himar1 C9	pBT30	*aaC1*	gentamicin	None	None	
Himar1 C9	pBAM1	*tetA*	tetracycline	None	None	
Himar1 wt	pMAR2xT7	*aaC1*	gentamicin	*p-T7*	*p-T7*	

2.1.2 Recipient P. aeruginosa Isolates Available for Mutagenesis

Laboratory isolates such as PA01, PAK, or clinically derived isolates PA14 or PACS2 Early.

2.1.3 E. coli Strains Needed for Propagation or Conjugation of Transposon Vectors

For conjugation into recipient *Pseudomonas*, use the *E. coli* strain SM10 λ *pir*. Genotype*: thi thr leu tonA lacY supE recA::RP4-2-Tc::Mu Km λpir*. Because all four transposon vectors contain only the pR6K λ origin, they will only replicate in *E. coli* strains containing the *pir* gene [11] (*see* Subheading 6).

For counter selection against the donor *E. coli* strain, either irgasan (at 25 μg/ml) or chloramphenicol (10 μg/ml) can be used because wild-type *P. aeruginosa* is resistant to these antibiotics. If antibiotic-based counter selection is undesirable, minimal media including M63 salts (*see* below) can be used to counterselect for the donor SM10 *E. coli*, which is auxotrophic for leucine, thiamine (vitamin B1) and threonine. Moreover, *E. coli* K12 derivatives, including SM10, cannot transport either glutamate or citrate; therefore, VBMM media supplemented with glutamate is an alternative counterselection growth medium of choice.

2.2 Growth Media

Rich media: Any rich media such as Luria–Bertani (LB) broth (*see* **Note 1**), tryptic soy broth, brain heart infusion, Mueller Hinton broth can be used. Follow manufacturer's instruction to make the media.

Minimal media (for counter selecting donor *E. coli*): Autoclave 15 g of Bacto agar in 850 ml distilled water with a stir bar. Cool down to 60 °C and add the following: 100 ml of 10× M63 salts, 20 ml of 1 M sodium succinate, 20 ml of 1 M glutamate, 1 ml 1 M magnesium chloride, and 1 ml of 100 mM ferrous ammonium sulfate. Succinate is used here because it is the preferred carbon source for *Pseudomonas*.

10× M63 salts are made by dissolving: 70 g of K_2HPO_4, 30 g of KH_2PO_4, and 20 g of $(NH_4)_2SO_4$ in 800 ml of water, pH adjusted to 7.4 using sodium hydroxide and final volume adjusted to 1,000 ml and made sterile by autoclaving.

2.3 Antibiotics

Gentamicin stocks at 30 mg/ml or 60 mg/ml dissolved in water (*see* **Note 2**).

Irgasan (Sigma Aldrich. Cat# 72779-25G-F) stocks at 25 mg/ml dissolved in 100 ethyl alcohol.

Ampicillin stocks at 100 mg/ml in water.

Chloramphenicol stocks at 20 mg/ml dissolved in 100 % alcohol.

Tetracycline at 100 mg/ml dissolved in DMSO. Make fresh.

Ampicillin and gentamicin should be filter sterilized. Other three antimicrobials do not need to be sterilized (as the solvent used-DMSO and alcohol-are toxic to bacteria and phages).

2.4 Other Reagents and Equipment

Soda lime glass beads: VWR International, LLC (Cat#26396-508).

Petri dishes, 100 × 150 mm and 100 × 100 mm.

Cotton swabs.

Taq polymerase-based PCR kit from any vendor.

2 ml capacity 96-well deep-well plates, 96-well PCR plates, and sealing films.

PCR cleanup kit.

10 mM Tris buffer at pH 8 with 0.05 % SDS for bacterial lysis buffer.

X-Gal if the phenotype is alteration in beta-galactosidase production.

EXOSAP-IT enzyme mix (USB #78205).

2.5 Equipment

Thermocycler, incubator for growing bacteria.

3 Transposon Mutagenesis Procedure

Day 1.

3.1 Prepare Agar Plates

Autoclave four flasks of 1 L of LB agar (10 g of Tryptone, 5 g of Yeast extract and 5 g of sodium chloride and 15 g of Bacto agar per liter). Each liter should suffice for about 15 petri dishes. Pour total of at least 45 petri plates (150 mm × 100 mm diameter) of LB agar containing 25 μg/ml Irgasan and 15–60 μg/ml gentamicin (for pBT or pMAR2xT7 plasmids) or 100 μg/ml tetracycline (for pBAM1 plasmid). Dry the plates for 30–45 min in a convection incubator. Autoclave 1 L of soda lime glass beads. Pour enough glass beads (about 5 ml) into each LB agar plate onto the agar surface. Then flip the plate such that lid is on the top. This will prevent beads from making permanent indentations on the agar surface, which can be confusing when examining colonies. Store the plates on the bench top (*see* **Note 3**). Leave two LB agar plates in the incubator for drying overnight for use in conjugation.

3.2 Inoculate Bacterial Strains

The transposon vector transformed into the donor strain (in this protocol, we will be using pBT20 as the transposon vector): Using a cotton swab applicator, heavily inoculate the transposon delivery vector, pBT20 on 2 plates of ampicillin (100 μg/ml) (*see* **Note 3**; Fig. 2).

The recipient: Inoculate *P. aeruginosa* strain of choice into 2 LB agar plates and, if using PA01, leave them overnight at 42 °C. Inoculate other *P. aeruginosa* isolate similarly and incubate at 37 °C.

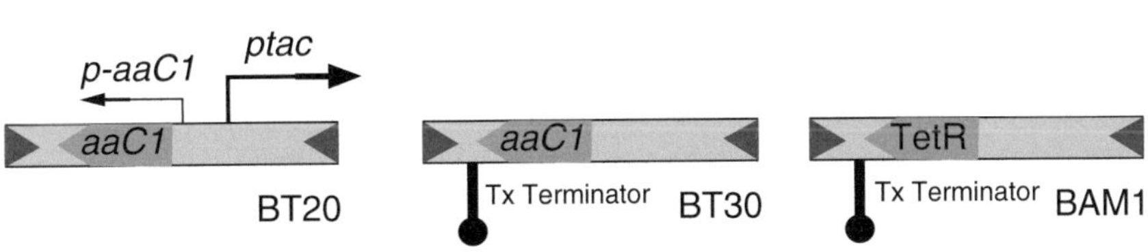

Fig. 2 Schematics of the transposon cassette of BT20 and its derivatives

Day 2.

3.3 Setting Up the Mutagenesis

If you are working with PA14 or PAK, move them into a 42 °C incubator for 1 h to further increase the recombination efficiency.

Using a cotton swab, scrape one entire plate of pBT20 colonies into a 5 ml culture tube containing 2 ml of LB broth and resuspend gently but thoroughly so that all clumps are dissolved. Make sure not to get pieces of agar during scraping. Do not vortex the donor strain but gently resuspend.

Scrape and resuspend one plate of the recipient strain in a separate 5 ml culture tube containing 2 ml of LB broth. Vortex well because plate-grown *Pseudomonas* colonies are often autoaggregate.

Add 10 μl of each suspension into 990 μl of LB broth in separate spectrophotometer cuvettes; vortex well, and measure the optical density (OD) at 600 nm. Multiply this number by 100 and this is the OD 600 nm of original suspensions.

Aliquot 500 μl of donor (*E. coli*) suspension and 500 μl recipient (*P. aeruginosa*) suspension into separate 1.5 ml microfuge tubes. And use the following equation for the OD adjustment.

$$\text{Donor}: \left(\text{OD600 of culture} \times 500 \div 40\right) - 500 = \text{“}\mathbf{p}\text{”}\left(\text{Dilution factor in } \mu\text{l}\right)$$

$$\text{Recipient}: \left(\text{OD600 of culture} \times 500 \div 20\right) - 500 = \text{“}\mathbf{q}\text{”}\left(\text{Dilution factor } \mu\text{l}\right)$$

Into the 500 μl of donor and the recipient in 1.5 microfuge tubes, add **p** microliter volume of LB broth into the donor tube and **q** microliter volume of LB broth into the recipient tube. (It is best to make your original suspensions at more than 20 and 40 OD 600 U, respectively. If they are lower, add more bacteria before performing the OD adjustment).

Gently mix and resuspend. The final OD of the recipient and the donor will be 20 and 40. Measure the optical densities of each again to confirm. If they are lower or higher, adjust them to get the desired final OD.

Combine 300 μl of the Donor (OD600 = 40) and 300 μl of the recipient (OD600 = 20) in a tube and mix thoroughly but gently. This is your mating mixture.

Pipet the following into one well-dried LB agar plate: Donor alone (control) and recipient alone (control). Into another dried LB agar plate, pipet 50 μl of the mixture containing the donor and the recipient carefully into ten different spots. In each mating spot, there are 1×10^9 cells of the donor and 2.5×10^8 cells of the recipient (*see* **Note 4**).

Incubate the matings for 2 h (*see* **Note 4**).

3.4 Plating of Mutants

Decide how many clones you need in the target petri plate. Then dilute the conjugation mixture appropriately. Dilution will depend on the amount of colonies you desire on a plate (*see* **Note 5**).

As an example, for PA14 or PAK, to obtain a 60,000 mutant library isolated as ~2,000 CFU per plate, resuspend 1 mating spot in 12 ml of 1× M63 buffer. Spread 400 μl of this culture per petridish. For PA01, resuspend all 10 spots in 12 ml of 1× M63 buffer (*see* **Note 6**).

Resuspend control spots, donor alone and recipient alone in 2 ml buffer and plate 400 μl in a single plate for each.

Technique of plating:
Stack 8 plates (containing glass beads) into one column. Turn the plates so that agar surfaces are facing up and the lid is on top, then lift each lid, starting from bottom plate, add 400 μl mating mix. Shake the column of plates using both hands until the cultures are dry. Discard the beads into a large container for future use. Incubate at the desirable temperature (example: If isolating phenotypes of cell envelop alteration or colony morphology, use room temperature or 30 °C).

Day 3. Observe each plate for expected or other desirable phenotypes, isolate these mutants and store them in a −80 °C freezer in the following manner.

3.5 Storage of Mutants

For library creation: When the mutants have grown for about 16 h at 37 °C, or when they are just visible and have a diameter about 1–2 mm, flood each plate with 2 ml rich media broth containing 20 % glycerol and scrape the bacteria using a rubber scraper or a cotton swab. Pool all the broth and mix well. Count the bacterial number by measuring the optical density at 600 nm. Make necessary adjustments to bring the concentration to 60,000 cfu/ml. Store at −80 °C. When reviving these libraries, first thaw to room temperature, then grow for 1 hour if antibiotics.

For storing mutants displaying relevant phenotypes, patch each desirable mutant into fresh phenotypic plates (*see* **Note 7**). Patching should be done in multiples of 96 in a larger plate. Mutants with dramatic phenotypes should be stored in cryovials (*see* **Note 8**).

Inoculate these patches into a deep-well 96-well plate containing 500 μl broth. Let them grow for 18 h shaking, take 50 μl of aliquot into 96-well PCR plate or PCR strip tubes. Add equal amount of 40 % glycerol + LB solution into the deep well plate, shake for 30 min, cover with adherent aluminum foil to seal, freeze at −80 °C (*see* **Note 9**).

4 Procedure for Sequencing the Transposon Insertion Site

This protocol is designed for mapping transposon insertion junctions in *P. aeruginosa* chromosome by arbitrary PCR method and subsequent sequencing of these PCR products to map the insertion junction. The "TnM" primers described here are specifically designed for mapping specific TnM-out Mariner insertions. The arbitrary primers described here can be used universally in any *P. aeruginosa* isolate. For additional information *see* Chun et al. [14]. Arbitrary PCR methods for *P. aeruginosa* are based on this publication.

4.1 Preparation of the Template Chromosomal DNA

1. Spin down the cultures that were aliquoted into 96-well PCR plates or PCR strip tubes (in Subheading 3.5). Resuspend them in 100 μl lysis buffer (10 mM Tris, pH 8 and 0.05 % SDS).
2. Run lysis program in a thermocycler (15 min at 98 °C followed by cooling to 4 °C).
3. Pellet cell debris by centrifuging at 4,000 RPM for 10 min and remove 50 μl of supernatant containing chromosomal DNA into a new plate or tubes. Seal with foil. Store at –20 °C (*see* **Note 9**).

4.2 Arbitrary PCR and Sequencing Transposon Junction

4.2.1 Round 1 PCR

Arbitrary primers (for P. aeruginosa genome): There are four arbitrary primers that have been used in this step by various research groups (refs. [4, 14, 15]) (*see* **Note 10**):

Rnd1-ARB1-Pa: 5′-GGCCACGCGTCGA CTAGTACNNNNNNNNNNAGAG-3′ (ref. [5])

Rnd1-ARB2-Pa: 5′-GGCCACGCGTCGACTAGTACNNNNNNNNNNACGCC-3′ (ref. [15])

Rnd1-ARB3-Pa: 5′-GGCCACGCGTCGACTAGTACNNNNNNNNNNGATAT-3′ (ref. [14])

Rnd1-ARB4-Pa: 5′-GGCCACGCGTCGACTAGTACSNNNNNNNSNSSSGCG-3′ (ref. [4])

Transposon specific primer for round 1 (*see refs.* [4, 10, 16]):

BT20 transposon specific primer:

Rnd1-TnM20: 5′-TATAATGTGTGGAATTGTGAGCGG-3′

BTK30 transposon specific primer:

Rnd1-TnM30: 5′-CACCGCTGCGTTCGGTCAAGGTTC-3′

MAR3XT7 transposon-specific primer:

Rnd1-TnMAR3X: 5′-TACAGTTTACGAACCGAACAGGC-3′

Compose reactions on ice:

1 μl Rnd1-TnM (20 pmol/μl) *(Rnd1-TnM20, 30, or Rnd1-TnMAR3X)*

1 μl Rnd1-ARB (50 pmol/μl each) *(Note: this could be a single primer or a mixture of Rnd1 ARB primers- see* Subheading 5.4)

1 μl bacterial lysate

2.5 μl 10× Taq PCR buffer

2.5 mM Mg^{++}

0.5 µl DNTP (10 mM stock)

1 µl DMSO

0.5 µl Taq DNA pol

De-ionized water to 25 µl

Note that a higher amount of Rnd1-ARB is used here [17]

Use the following thermocycler program

Step 1	94 °C	3 min	
Step 2	94 °C	30 s	
Step 3	49 °C	30 s	*reduce temp by 1 °C for each subsequent round
Step 4	72 °C	3 min	
Step 5	Goto Step 2: 15 rounds repeat		
Step 6	94 °C	30 s	
Step 7	60 °C	30 s	
Step 8	72 °C	3 min	
Step 9	Goto Step 6: 20 rounds repeat		
Step 10	12 °C	Hold	

Do not clean the PCR reactions from round 1 unless you have a PCR failure during the second round. In that case, cleanup round 1 PCR DNA using a commercial PCR cleanup kit.

4.2.2 Round 2 PCR

Compose reactions on ice (*see* **Notes 10–12**):

Round 2 common primer which hybridizes into the round 1 arbitrary primers:

Rnd2-ARB 5′- GGCCACGCGTCGACTAGTAC-3′ (ref. [14])

Transposon specific primers:

For pBT20 transposon:

Rnd2-TnM20 5′- ACAGGAAACAGGACTCTAGAGG-3′ (ref. [4])

For pBTK30 transposon:

Rnd2-TnM30: 5′-CGAACCGAACAGGCTTATGTCAATTC-3′ (ref. [16])

For MAR3XT7 transposon:

Rnd2-TnM-MAR3X: (5′-TGTCAACTGGGTTCGTGCCTTCAT CCG-3′) (ref. [10])

Compose reactions on ice:

1 µl Rnd2-TnM (20 pmol/µl) (choose the appropriate transposon specific primer)

1 μl Rnd2-ARB (20 pmol/μl) (the common round 2 primer)

2 μl PCR product from Round 1

10 μl 10× Taq PCR buffer

1 μl DNTP

2 μl DMSO

2 mM Mg^{++}

1 μl Taq DNA pol

Deionized water to 50 μl

Use the following thermocycler program

Step 1	94 °C	3 min
Step 2	94 °C	30 s
Step 3	60 °C	30 s
Step 4	72 °C	2 min
Step 5	Go to Step 2: 30×	
Step 6	72 °C	5 min
Step 7	12 °C	Hold

Run 5 μl of each sample on a 1.5 % agarose gel (*see* **Note 11**).

4.3 PCR Cleanup

If you have only few samples, clean positive reactions using a PCR cleanup kit. Additionally, if you have more than 12 samples, a simpler, cheaper method is given below. (*See* **Notes 12–14** if there is a PCR failure).

Compose following reaction in a PCR strip tube or PCR plate:

PCR product: 5 μl.

EXOSAP-IT enzyme mix (USB #78205): 2 μl (*see* **Note 15**).

Seal the tubes or the plate.

Use following thermocycler parameters to run the reaction:

Step 1	37°	30 min
Step 2	80°	20 min
Step 3	4°	hold

4.4 Sequencing

Sequencing primers can be added directly into the EXOSAP-IT reaction mix after the Subheading 4.3. If a commercial column-based kit was used to clean up PCR, add primer to 10 μl PCR product.

Add DMSO to 10 % for sequencing PCR DNA from *P. aeruginosa* template. This step is not necessary for bacteria with lower GC content.

Seal the tubes or the plates. Submit for sequencing.

Sequencing Primers:

For sequencing BT20 transposon junction:

BT20TnMSeq: 5′- CACCCAGCTTTCTTGTACAC

For sequencing BT30 transposon junction:

BTK30TnMSeq: 5′-TGGTGCTGACCCCGGATGAAG-3′

For sequencing MAR3XT7 transposon junction:

MAR3XT7-TnmSeq: 5′-GACCGAGATAGGGTTGAGTG-3′

(MAR3XT7-TnmSeq is originally referred to as PMFLGM. GB-4a [10])

4.5 Sequencing Result Analysis

The sequencing results should be aligned to *P. aeruginosa* genome sequence using BLAST, or using locally installed DNA analysis software. When analyzing the sequencing results, locate the transposon arm sequence 5′-AGACCGGGGACTTATCAGCCAACC TGT**TA-3**′ to map the junction. The residues TA demarcate the transposon insertion locus.

5 Relevant Information

5.1 Transposons Available for Mutagenesis

pBT20 transposon delivery vector [4]: This transposon is available as a mini-transposon cassette cloned into a vector containing the pR6K λ origin of replication [4, 6]. Therefore, propagation of the vector requires *E. coli* strains that harbor the Pir protein [20]. The backbone contains the origin of transfer from the plasmid pRP4/RK2, which enables plasmid mobilization and conjugation by RP4/RK2- based conjugation machinery [6]. The vector is delivered by conjugation via SM10 λpir *E. coli* strain into a recipient *Pseudomonas* strain [6]. pBT20 contains the C9 hyperactive derivative of Himar1 transposase conferring higher recombination frequency [8]. The transposon element contains an outward facing promoter (*ptac*) and the *aaC1* gene, which encodes gentamicin resistance. Both of these elements are flanked by mariner specific inverted repeats that are needed for recombination during transposition. The *ptac* and the *aaC1* both contribute to the utility of pBT20 for activating gene transcription. For example, *ptac* activates gene expression if the transposon inserts into the promoter region of an open reading frame. This feature is also useful if the recipient *P. aeruginosa* has a *lacZ* transcriptional fusion to a promoter of interest [4]; successful insertion in gene regulators could then increase or decrease promoter activity, which could be monitored by β-galactosidase activity using X-Gal indicator plates. The *tac* promoter, which is derived from the *trp* and *lacUV5* promoters in *E. coli*, is constitutively active in *P. aeruginosa*, which is naturally *lac* negative. The pTac can be regulated by introducing the *lacIQ* gene *in trans* via a plasmid or engineered at a heterologous site such as the *ctx* phage or the Tn7 transposon attachment sites [21, 22]. The *aaC1* gene in pBT20 faces the opposite direction to that of the *tac* promoter. The promoter of *aaC1* reads through the transposon cassette, and while it does so at lower levels than that of *pTac*, it is still strong enough to activate downstream promoters [4]. Therefore, promoter activities are present at both ends of the transposon and can activate the transcription of surrounding genes.

pBTK30 transposon delivery vector: This plasmid is derived from pBT20, but with modifications designed to not alter expression of neighboring genes. Among these alterations, the *tac* promoter is deleted in the transposon cassette [16]. Furthermore, a transcriptional terminator from bacteriophage T4, *gene 32*, has been cloned into the 3′ end of the *aaC1* gene to prevent it from extending transcription beyond the transposon and into the genomic DNA.

pBAM1 transposon delivery vector: This vector differs from pBTK30 in that *tetA* gene conferring resistance to tetracycline has been inserted into the Sph1 restriction site of *aaC1* gene [23]. This transposon is useful if the *P. aeruginosa* target isolate is already resistant to gentamicin or in other bacterial species in which gentamicin cannot be used.

pMAR2xT7 transposon delivery vector: This transposon is based on the wildtype Himar1 mariner Transposase. This transposon is unique due to the presence of T7 promoters, which can be activated if T7 polymerase is present at a heterologous site. The backbone is otherwise similar to that of pBT20 plasmid and based on the original backbone in which pBT20 was made, and contains the pR6K origin, RP4 oriT and the *aaC1* gene [10]. This transposon was used to generate a nonredundant library of *P. aeruginosa* mutants derived from the clinical strain PA14. The plasmid map and the sequence information of this transposon vector can be obtained from the PA14 Transposon Insertion Mutant Web site (http://ausubellab.mgh.harvard.edu/cgi-bin/pa14/downloads.cgi).

5.2 Recipient *P. aeruginosa* Isolates Available for Mutagenesis

Frequently used isolates of P. aeruginosa such as PA01, PA14, and PAK are all suitable for transposon mutagenesis. PA14 and PAK are excellent recipients for creating libraries because a 2-h conjugation protocol, using approximately 2.5×10^8 recipient cells and 1×10^9 donor cells, can produce a 40,000–60,000 CFU mutant library [4, 16], whereas a similar protocol using PA01 produces five- to tenfold fewer colonies (although incubating PA01 bacteria at 42 °C overnight will increase recombination frequency of foreign DNA). It is not known why PA01 is less amenable to genetic manipulation by foreign DNA compared to PA14 and PAK. Although PAK is easy to manipulate genetically, its genomic sequence is still not available, limiting its utility. For creating transposon mutant libraries in clinical isolates, good choices are PA14 isolate, obtained from a burn patient at Massachusetts General

Hospital [24], and the PACS2 Early and PACS2 Late pairs, which are two sequential, clonal isolates from the same cystic fibrosis patient. This isolate pair is maintained by the University of Washington/Seattle Children's hospital cystic fibrosis microbiology core laboratory [25]. PACS2 Early is susceptible to antibiotics used commonly in molecular biology and amenable to genetic manipulation by standard *Pseudomonas* genetic tools.

5.3 E. coli Strains Needed for Propagation or Conjugation of Transposon Vectors

For conjugation into recipient *Pseudomonas*, it is best to use the *E. coli* strain SM10 *lambda pir*. Because all four transposon vectors contain the pR6K origin, they will only replicate in *E. coli* strains containing the *pir* gene [11]. The pR6K origin is built into these plasmids so that they can be also used in enterobacteria [6]. The backbone of these vectors contains the origin of transfer from the plasmid pRP4, thus enabling plasmid mobilization and conjugation by RP4/RK2-based conjugation machinery. *E. coli* SM10 contains plasmid mobilization genes from the RP4/RK2 plasmid. Genetic manipulation of these transposon vectors requires the *E. coli* strain DH5alpha *lambda pir*. pRK2 plasmid-based helper plasmids can be also used for conjugating transposon delivery vectors into the recipient using triparental mating.

5.4 Arbitrary PCR and Sequencing Transposon Junction

Arbitrary primers (for *P. aeruginosa* genome):

Most publications describing transposon junction sequencing in *P. aeruginosa* have based their arbitrary PCR protocol using the methods published by Chun et al. [14]. Because each publication has different names for the same identical primers, we have made an effort to describe their various names in Table 2 and then assign a uniform nomenclature to them (*see* **Note 10**).

Rnd1-ARB1-Pa: 5′-GGCCACGCGTCGACTAGTACNNNNNNNNNNAGAG-3′ (ref. [5])
Rnd1-ARB2-Pa: 5′-GGCCACGCGTCGACTAGTACNNNNNNNNNNACGCC-3′ (ref. [15])
Rnd1-ARB3-Pa: 5′-GGCCACGCG TCGACTAGTACNNNNNNNNNNGATAT-3′ (ref. [14])
Rnd1-ARB4-Pa: 5′-GGCCACGCGTCGACTAGTACSNNNNNNNSNSSSGCG-3′ (ref. [4])

Table 2
Commonly used arbitrary primers used in *P. aeruginosa*

Publication	Original name	New name	Other names
Jacobs et al. [5]	CEKG2A	Rnd1-ARB1-Pa	None
O' Toole et al. [15]	ARB6	Rnd1-ARB2-Pa	CEKG2B (5), Rnd1-Pa2 (4)
Chun et al. [14]	Primer 2	Rnd1-ARB3-Pa	ARB1, CEKG2C (5), Rnd1-Pal (4)

Table 3
Frequency of the 3′ end motifs present in the PA01 genome (in both strands) for each arbitrary primer

Rnd1-ARB1-Pa: 12,447
Rnd1-ARB2-Pa: 31,119
Rnd1-ARB3-Pa: 4,212
Rnd1-ARB4-Pa: 50,000

In our hand, Rnd1-ARB1-Pa, one of the primers used to sequence PA01 mutant library, is a reliable primer for arbitrary PCR [5]. If a PCR product is not generated using ARB1-Pa, try using Rnd1-ARB2-Pa, which have more binding sites in the chromosome (*see* Table 2). On the other hand, a mixture of Rnd1-ARB1, Rnd1-ARB2 and Rnd1-ARB3 can be used during the first round [5]. Rnd1-ARB3-Pa, a popular primer that appeared in many publications [4, 5, 14–16], only binds to 4,212 sites in *P. aeruginosa* chromosome (Table 3). This is not surprising because it was first designed for amplifying yeast DNA using arbitrary PCR. In our hand, this primer is good for about 70 % of the time (figure obtained from sequencing 600 mutants). Additionally, Rnd1-ARB3-Pa has cold spots (no primer binding sites) in long stretches of DNA some reaching up to many kilobases in lengths. The 3′ end of the Rnd1-ARB3-Pa, GATAT, with 40 % GC approximately present every 600 bp of the yeast chromosome. Chun et al. designed this 3′ end sequence assuming that it will occur at least once in every 1.5 kb segment with an 90 % chance and in a 1 kb segment with an 80 % chance in the yeast genome. Therefore, there is no basis for using this primer for amplifying *P. aeruginosa* chromosome because such sequence distribution does not occur in that genome. Similar rationale should be used for designing new arbitrary primers for any genome.

The randomized portion and the 5′ end of the arbitrary primers used here were also first described by Chun et al. [15]. The basis of choosing 10 residues for randomizing the Rnd1-ARB primers is unknown but this number has been proven successful in creating large mutant banks [5, 14, 15, 17]. If the randomized portion of Rnd1-ARB1-Pa is made of all Gs and Cs, then the melting temperature of the entire primer (the 10 arbitrary residues plus the 3′ end AGAG) will be 58 °C. If the randomized portion is made of all As and Ts, then the melting temperature will be 28 °C. But if the randomized portion is about 64 % GC as found in P. aeruginosa, then the melting temperature will be 49 °C (melting temperatures were calculated using the Sci tools at the IDT Web site http://www.idtdna.com/scitools/scitools.aspx). During Round 1 PCR, annealing temperature 49 °C was chosen so that it is more biased toward high GC templates found in *P. aeruginosa*. A 15 round touch down will additionally ensure that correct annealing temperature is reached for each arbitrary primer [14, 18, 19]. The sequence GGCCAC GCGTCGACTAGTAC that is found at the 5′ end of each Rnd1-ARB was first described in Chun et al. [14]. It has a melting temperature of 60 °C with 65 % GC rich, indicating that it is suitable for PCR reactions using *P. aeruginosa* DNA, although this sequence was originally designed by Chun et al. for amplifying lower GC yeast DNA [14]. This sequence does not have secondary structures with unfavorable parameters such as stemloops or homodimers. This sequence will be the scaffold for the Rnd2-ARB primer during the second round. Therefore, when designing new arbitrary primers, it is wise to include these practically proven primer sequences.

5.5 Verification of Phenotypes Obtained from Transposon Mutants

It should be emphasized that transposons create phenotypes not only due to the disruption of genes but also due to disruption of gene expression of surrounding genes. This is especially apparent if the transposon contains outward directed promoters [4]. Therefore, complementation experiments should be carried out by cloning and expressing the wildtype gene (that was chromosomally inactivated) in the mutant background. Furthermore, complete, unmarked deletion of the genetic locus can be carried out by using the pEX-based plasmids [26]. Complementation can be carried out either using a plasmid-based system or using a chromosomal integration system. Laboratory of Dr. Herbert Schweizer has developed two systems that enable integration of DNA segments at permissive heterologous sites in the *P. aeruginosa* genome. One of the sites is located at the attachment site for the CTX phage [21]. And the other site is located at the attachment site for the Tn*7* transposon [22, 27]. If pBT20 is used for mutagenesis, whether the tac promoter is affecting a phenotype can be assessed by transforming a plasmid containing lacIQ gene and testing the phenotype by IPTG induction [4].

5.6 Primer Design for the Transposon

Primers should be nonoverlapping. Rather than using two different primers for arbitrary PCR, a single primer can be also used. The sequencing primer should be about 35–50 bp upstream of the Arm sequence (insertion site) so that the resulting sequencing read will overlaps the insertion junction.

5.7 Primer Design for Arbitrary PCR

When designing new arbitrary primers for PCR, first analyze the genomic DNA and find a repetitive 4–6 base pair sequence that appear at least once in every 1,000–1,500 base pair in each strand of the genome.

6 Notes

1. For most applications, standard LB agar is sufficient. For phenotype-based screening, it is best to use media appropriate for such phenotypes. For creating mutant libraries for cryo storage for diverse downstream applications, it is best to use rich media such as LB, tryptic soy broth, or brain heart infusion supplemented with magnesium to 5 mM, iron to 100 μM, succinate to 20 mM as a carbon source, and a 1× M63 as a buffer and a nutrient source. M63 in addition to acting as a buffer, also provides essential ions such as potassium, phosphates, sulfur and nitrogen source in the form of ammonia. This will ensure that most numbers of mutants are isolated for library creation purposes.
2. For PA01 and PAK, use gentamicin at 60 μg/ml. For PA14, 15–30 μg/ml gentamicin can be used.
3. Do not inoculate the donor strain on antibiotic plates that will select for transposon mutants in *E. coli* strain. As an example, do not use gentamicin plates to inoculate *E. coli* SM10 harboring pBT20 or tetracycline for pBAM1. Use the backbone-encoded resistance to ampicillin to select for the plasmid.
4. The pipetting should be done carefully to minimize the surface area occupied by each drop of cell suspension on the plate (to maximize cell-cell contact and mating). Filters are not needed. Let these sit for 2 h exact in the warm room. Allowing more than 2 h' incubation might result in isolation of progeny from transposon insertions producing Tn insertions in identical loci. If you are testing a new bacterium as a recipient for mutagenesis, one should first perform a growth curve for the recipient growing on the agar surface during conjugation. The recipient should divide only once during conjugation and the incubation time should be adjusted appropriately. With PAK and PA14, each mating spot should give 40,000–60,000 CFUs if 2.5×10^8 recipient cells and 1×10^9 donor cells are used. PA01 as a recipient produces five- to tenfold lower transposition frequency. Ideally, one should aim to obtain approximately 60,000 mutants during mutagenesis. Liberati et al. calculated that a gene 327 bp in length has a 95 % chance of being disrupted among a library with 60,000 random insertions [10].
5. For library creation, 5,000 colonies can be grown per plate without creating a lawn of bacteria; thus, only 12 petri plates are required to create a library of 60,000 CFUs. In contrast, for phenotype determination, each plate should contain only 1,000–2,000 CFUs per plate, which results in well-separated colonies. If you decide to obtain 2,000 CFUs per plate, 30 agar plates are needed. For certain phenotypes such as

swimming on an agar surface, which requires soft agar of 0.45 %, even less dense CFUs should be obtained per plate.

6. Test one agar plate with 400 μl LB broth to assess the dryness of the agar surface. If it is too dry, then the liquid will be absorbed too rapidly without bacteria getting spread. In that case, test again using increasing amounts of inoculation volume until agar surface is saturated. Then make appropriate adjustments to the mating culture stocks. If the agar surface is still fresh and not dry enough, 400 μl may be too much, and in that case, reduce the inoculation volume.
7. Once mutants are patched into a new plate, repeat the phenotype assays using the exact conditions that were used for transposon mutagenesis.
8. Mutants should be frozen as soon as possible after their identification. This should be done within couple of days. *P. aeruginosa* frequently undergo mutations in the *lasR* gene when growing in both rich and minimal media if grown for more than 1 week [12, 13]. These mutants can be easily identified by the presence of metallic sheen or extensive lysis on the colony surface. Do not store mutants in the cold room, again to prevent secondary mutations.
9. You can add 1 μl of the overnight culture directly into the PCR reaction for supplying template DNA. Nevertheless, making a lysate is a wise idea if there is a PCR failure and you will need to optimize PCR reactions subsequently.
10. The round 2 common primer Rnd2-ARB was first designed by Chun et al. and referred to as Primer 4 [14], and appeared as ARB2 in O'Toole et al. [15], and as Rnd2-Pa in Kulasekara et al. [4], and as CEKG4 in Jacobs et al. [5], and as ARB2A in Liberati et al. [10]. Rnd2-TnM-MAR3X is originally referred as PMFLGM.GB-2a in Liberati et al. [10].
11. Round 2 products are often less than 1,000 base pairs. You should get no product for parental (negative) control. If you run the round 1 PCR in a gel, you should see a smear or a ladder or a single band.
12. If an important mutant fails to give a PCR product, isolate genomic DNA using the Qiagen genomic DNA isolation kit and redo the random PCR steps. Use only 10 ng purified DNA for PCR reaction.
13. The starting temperature of the touch down PCR step can be also altered during round 1 if the first round does not generate a PCR product. Start with either higher temperature or lower temperature [18, 19].
14. If the above mapping procedure fails for an important mutant, digest the genomic DNA of the mutant with a frequent cutting

enzyme (that does not cut the transposon) and ligate the cut DNA into a cloning vector and select for transposon encoded antibiotic resistance. Then sequence the plasmid cloning junction.

15. EXOSAP-IT is a mixture of Exonuclease I and shrimp alkaline phosphatase. The former removes all primers and unannealed single stranded DNA. The latter removes any unincorporated dNTPs.

References

1. Ruvkun GB, Ausubel FM (1982) A general method for site-directed mutagenesis in prokaryotes. Nature 289:85–88
2. Way JC, Davis MA, Morisato D, Roberts DE, Kleckner N (1984) New Tn*10* derivatives for transposon mutagenesis and for construction of *lacZ* operon fusions by transposition. Gene 32:369–379
3. Newland JW, Green BA, Holmes RK (1984) Transposon-mediated mutagenesis and recombination in Vibrio cholerae. Infect Immun 45: 428–432
4. Kulasekara HD, Ventre I, Kulasekara BR, Lazdunski A, Filloux A, Lory S (2005) A novel two-component system controls the expression of *Pseudomonas aeruginosa* fimbrial cup genes. Mol Microbiol 55:368–380
5. Jacobs MA, Alwood A, Thaipisuttikul I, Spencer D, Haugen E, Ernst S et al (2003) Comprehensive transposon mutant library of *Pseudomonas aeruginosa*. Proc Natl Acad Sci U S A 100:14339–14344
6. Wong SM, Mekalanos JJ (2000) Genetic footprinting with mariner-based transposition in *Pseudomonas aeruginosa*. Proc Natl Acad Sci U S A 97:10191–10196
7. Rubin EJ, Akerley B, Novik VN, Lampe DJ, Husson RN, Mekalanos JJ (1999) *In vivo* transposition of mariner-based elements in enteric bacteria and mycobacteria. Proc Natl Acad Sci U S A 96:1645–1650
8. Lampe DJ, Akerley BJ, Rubin EJ, Mekalanos JJ, Robertson HM (1999) Hyperactive transposase mutants of the Himar1 mariner transposon. Proc Natl Acad Sci U S A 96:11428–11433
9. Held K, Ramage E, Jacob M, Gallaghe L, Manoil C (2012) Sequence-verified two-allele transposon mutant library for *Pseudomonas aeruginosa* PAO1. J Bacteriol 194:6387–9389
10. Liberati NT, Urbach JM, Miyata S, Lee DG, Drenkard E, Wu G et al (2006) An ordered, nonredundant library of *Pseudomonas aeruginosa* strain PA14 transposon insertion mutants. Proc Natl Acad Sci U S A 103:2833–2838
11. Kolter R, Helinski DR (1978) Construction of plasmid R6K derivatives *in vitro*: characterization of the R6K replication region. Plasmid 1:571–580
12. D'Argenio DA, Calfee MW, Rainey PB, Pesci EC (2002) Autolysis and autoaggregation in *Pseudomonas aeruginosa* colony morphology mutants. J Bacteriol 184:6481–6489
13. D'Argenio DA, Wu M, Hoffman LR, Kulasekara HD, Deziel E, Smith EE et al (2007) Growth phenotypes of *Pseudomonas aeruginosa lasR* mutants adapted to the airways of cystic fibrosis patients. Mol Microbiol 64:512–533
14. Chun K, Edenberg HJ, Kelley MR, Goebl MG (1997) Rapid amplification of uncharacterized transposon-tagged DNA sequences from genomic DNA. Yeast 13:233–240
15. O'Toole GA, Kolterm R (1998) Initiation of biofilm formation in *Pseudomonas fluorescens* WCS365 proceeds via multiple, convergent signalling pathways: a genetic analysis. Mol Microbiol 28:449–461
16. Goodman AL, Kulasekara B, Rietsch A, Boyd D, Smith RS, Lory S (2004) A signaling network reciprocally regulates genes associated with acute infection and chronic persistence in *Pseudomonas aeruginosa*. Dev Cell 7:745–754
17. Das S, Noe JC, Paik S, Kitten T (2005) An improved arbitrary primed PCR method for rapid characterization of transposon insertion sites. J Microbiol Methods 63:89–94
18. Korbie DJ, Mattick JS (2008) Touchdown PCR for increased specificity and sensitivity in PCR amplification. Nat Protoc 3:1452–1456
19. Don RH, Cox PT, Wainwright BJ, Baker K, Mattick JS (1991) "Touchdown" PCR to circumvent spurious priming during gene amplification. Nucleic Acids Res 19:4008
20. Miller VL, Mekalanos JJ (1988) A novel suicide vector and its use in construction of insertion mutations: osmoregulation of outer membrane proteins and virulence determinants in *Vibrio cholerae* requires *toxR*. J Bacteriol 170:2575–2583

21. Hoang TT, Kutchma AJ, Becher A, Schweizer HP (2000) Integration-proficient plasmids for *Pseudomonas aeruginosa*: site-specific integration and use for engineering of reporter and expression strains. Plasmid 43:59–72
22. Choi KH, Schweizer HP (2006) Mini-Tn*7* insertion in bacteria with single attTn7 sites: example *Pseudomonas aeruginosa*. Nat Protoc 1:153–161
23. Castang S, Dove SL (2012) Basis for the essentiality of H-NS family members in *Pseudomonas aeruginosa*. J Bacteriol 194:5101–5109
24. Rahme LG, Steven EJ, Wolfort SF, Shao J, Tompkins RG, Ausubel FM (1995) Common virulence factors for bacterial pathogenicity in plants and animals. Science 268: 1899–1902
25. Smith EE, Buckley DG, Wu Z, Saenphimmachak C, Hoffman R, D'Argenio DA et al (2006) Genetic adaptation by *Pseudomonas aeruginosa* to the airways of cystic fibrosis patients. Proc Natl Acad Sci U S A 103:8487–8492
26. Hoang TT, Karkhoff-Schweiz RR, Kutchma AJ, Schweizer HP (1998) A broad-host-range Flp-FRT recombination system for site-specific excision of chromosomally-located DNA sequences: application for isolation of unmarked *Pseudomonas aeruginosa* mutants. Gene 212: 77–86
27. Damron FH, McKenney ES, Schweizer HP, Goldberg JB (2013) Construction of a broad-host range Tn*7*-based vector for single copy PBAD controlled gene expression in gram-negative bacteria. Appl Environ Microbiol 79:718–721

Chapter 40

Site-Directed Mutagenesis and Gene Deletion Using Reverse Genetics

Daniela Muhl and Alain Filloux

Abstract

Understanding gene function is far easier when tools are available to engineer a bacterial strain lacking a specific gene and phenotypically compare its behavior with the corresponding parental strain. Such mutants could be selected randomly, either by natural selection under particular stress conditions or by random mutagenesis using transposon delivery as described elsewhere in this book. However, with the advent of the genomic era there are now hundreds of bacterial genomes whose sequence is available, and thus, genes can be identified, chosen, and strategies designed to specifically inactivate them. This can be done by using suicide plasmids and is most convenient when the bacterium of interest is easily amenable to genetic manipulation. The method presented here will describe the use of a suicide vector, pKNG101, which allows the selection of a double-recombination event. The first event results in the integration of the pKNG101 derivative carrying the "mutator" fragment onto the chromosome, and could be selected on plates containing appropriate antibiotics. The pKNG101 carries the *sacB* gene, which induces death when cells are grown on sucrose. Growth on sucrose plates will thus select the second homologous recombination event, which results in removing the plasmid backbone and leaving behind the mutated target gene. This method has been widely used over the last 20 years to inactivate genes in a wide range of gram-negative bacteria and in particular in *Pseudomonas aeruginosa*.

Key words Suicide vector, pKNG101, Sucrose sensitivity, Conjugation, Homologous recombination

1 Introduction

All bacterial geneticists have now become incredibly efficient at modifying genes of interest by exchanging a wild-type gene for a mutated allele directly on the bacterial chromosome. This has become easier with the ever-increasing publication of bacterial genome sequences, hundreds of which are readily available on the Web. Genetic manipulation has also become much easier with the advent of PCR, which allows to engineer and to clone any DNA fragment of interest by the appropriate design of primers.

The allelic exchange on the bacterial chromosome could be a preferred method when one wants to disrupt exclusively the gene of interest with a minimum impact on expression of other

Alain Filloux and Juan-Luis Ramos (eds.), *Pseudomonas Methods and Protocols*, Methods in Molecular Biology, vol. 1149, DOI 10.1007/978-1-4939-0473-0_40, © Springer Science+Business Media New York 2014

genes and on the genetic environment of the target gene. Random transposon insertion [2–4], or even targeted interruption of a gene by insertion of a suicide vector carrying a small region of the gene to be disrupted may have global impact on expression of downstream genes, especially when a gene is part of a cluster and included within an operon. In this case any polar effect will prevent to associate the observed phenotypic traits with the gene of interest.

In order to perform the allelic exchange one needs to direct the new allele to the target site on the genome by homologous recombination. This will be achieved by having the new allele cloned into a suicide vector, which cannot replicate in the bacterium to be genetically modified. For example the standard pUC18 or pUC19 plasmids are ColE1 replicon and do not replicate in *P. aeruginosa* except if they are specifically modified [5]. When introducing these plasmids in *P. aeruginosa*, the only way the plasmid can be maintained, together with the resistance gene they carry, is to become stably inserted into the chromosome. If the gene of interest, or part of it, is cloned into the pUC plasmids, homologous recombination will allow this integration event to occur. However, in this case there is no possibility to impose a selection pressure for the excision of the plasmid and for leaving behind the mutated allele. This step is crucial for the allelic exchange and again several plasmids have been developed that carry a counter-selectable marker. This will allow a screen for the loss of the plasmid backbone and its excision from the chromosome via a second event of homologous recombination. Such marker, which was included in the original suicide vector pKNG101 [1, 6], is the *Bacillus subtilis sacB* gene that confers sucrose sensitivity. If this plasmid is on the chromosome and the cells spread on agar plates containing 5 % sucrose, it will impose a strong selective pressure and only those bacteria, which have expelled the plasmid backbone and the *sacB* gene will survive. This occurs via a second event of recombination, which will either reconstruct the wild-type allele or will replace it with the mutated allele as precisely described in the methods presented below. Several of these counter-selectable plasmids are now available [7]. Of note another quick method to generate mutants in *P. aeruginosa* is to make use of the lambda red recombinase system [8].

The allelic exchange will allow replacing the wild-type gene with a gene, which for example contains a nucleotide(s) change resulting in the substitution of only one single amino acid in the gene product. Furthermore the method is more generally used to delete the gene of interest in a clean and precise manner. Importantly, the deletion can be obtained without leaving behind, and in the middle of the gene, an antibiotic resistance cassette. Therefore, once the deletion has been checked by PCR, additional deletions can be introduced in this new genetic background without having the problem of choosing new and different antibiotic resistance cassettes for the selection procedure. This way as many deletions as desired can be introduced in one single strain.

2 Materials

2.1 Strains and Plasmids

See Tables 1 and 2.

2.2 Amplification of the Gene of Interest and Mutator Fragment

1. Template DNA: Purified genomic *Pseudomonas aeruginosa* DNA (50–100 ng/μl).
2. Polymerase Chain Reaction (PCR) and agarose gel electrophoresis (*see* protocol in Chapter 35).
3. Purification of the PCR product: PCR purification kit.
4. NanoDrop to measure the concentration of the PCR product.
5. pCR2.1 TA cloning kit: pCR2.1 vector, 10× ligation buffer, T4 DNA ligase (4 U/μl), molecular biology water.
6. Transformation: Ice, competent Top10 *Escherichia coli* cells (50 μl competent cells/PCR reaction), 42 °C water bath, 500 μl SOC (Super Optimal broth with Catabolite repression) medium/transformation, Luria–Bertani (LB) agar plates supplemented with 50 μg/ml ampicillin (Amp50), 25 μg/ml kanamycin

Table 1
Strains used for gene deletion and site-directed mutagenesis

Strain	Genotype	Comments
Pseudomonas aeruginosa		
Escherichia coli:		
CC118 λpir	Δ(*ara-leu*) *araD* Δ*lacX74* *galE galK phoA20 thi-1 rpsE rpoB* *argE*(*Am*) *recA1 Rf*^{R} (*λpir*)	Host strain for pKNG101 replication
XL1-Blue	*recA1 endA1 gyrA96 thi-1 hsdR17* *supE44 relA1 lac* [*F′ proAB lacI*^{q}*Z*Δ*M15 Tn10* (*Tet*^{R})]	Strain used for site-directed mutagenesis
Top10	*F-mrcA* Δ(*mrr-hsdRMS-mcrBC*)Φ*80 lacZ*Δ*M15* Δ*lacX74 nupG recA1 araD139* Δ(*ara–leu*)*7697 galE15 galK16 rpsL*(*Str*^{R}) *endA1* λ-	Strain for pCR2.1 replication

Table 2
Plasmids used for gene deletion and site-directed mutagenesis

Plasmid	Resistance	Comments
pKNG101	Streptomycin (Sm)	Suicide vector in *P. aeruginosa*
pCR2.1	Kanamycin (Km), Ampicillin (Amp)	TA cloning vector for PCR products

(Kan25) and 40 μg/ml 5-bromo-4-chloro-3-indolyl-beta-D-galactopyranoside (X-gal40). Prepare a 40 mg/ml stock solution of X-gal in dimethyl sulfoxide (DMSO), *see* **Note 1**.

2.3 Site-Directed Mutagenesis

1. Template DNA: Purified methylated plasmid DNA that contains the gene/DNA fragment in which the mutation will be introduced (*see* **Notes 2** and **3**). This method uses plasmids that have a size up to 8 kb.
2. For the PCR reaction: Pfu Ultra High Fidelity^PLUS^ DNA polymerase (2.5 U/μl), 10× Pfu Ultra High Fidelity^PLUS^ DNA polymerase reaction buffer, 10 μM of primer 3′, 10 μM primer 5′, 20 mM dNTP mix, 5 M betaine solution, molecular biology water, 5–50 ng double stranded plasmid DNA (template). PCR machine with hot-top assembly.
3. 1 % agarose gel.
4. Digest of parental DNA: *Dpn*I restriction enzyme (10 U/μl), 37 °C incubator.
5. Transformation: Ice, competent XL1-Blue *E. coli* cells (80 μl competent cells/PCR reaction, *see* **Note 4**), 42 °C water bath, 500 μl SOC medium/transformation, LB agar plates supplemented with the required antibiotic.

2.4 Re-cloning into pKNG101

1. LB agar plates supplemented with 50 μg/ml streptomycin (Sm50). Make a 10 ml stock solution of 50 mg/ml Sm: Weigh 0.5 g Sm in a 15 ml falcon tube. Add water to a volume of 8 ml and vortex to resuspend. Make up to 10 ml with water. Filter-sterilize (*see* **Note 5**). For 1 plate, add 20 μl of the stock solution to 20 ml *hand warm* LB agar and pour one LB agar plate. Dry the plate next to the Bunsen burner for 30–45 min.
2. LB agar plates supplemented with 5 % sucrose. Prepare a 50 % sucrose stock solution in a total of 50 ml. Weigh 25 g of sucrose in a 50 ml falcon tube. Top up to 50 ml with water and resuspend by vortexing (*see* **Note 6**). Top up again to 50 ml with water and filter-sterilize. To make one plate mix 2 ml of the 50 % sucrose solution with 18 ml of *hand warm* LB agar. Pour the plate and leave to set. Dry next to the Bunsen burner for 30–45 min.
3. Purified sucrose-sensitive pKNG101 and pCR2.1-mutator at a concentration of 200 ng/μl at least (*see* **Notes** 7 and **8**).
4. Restriction enzymes and buffers to reclone the mutator fragment from pCR2.1 into pKNG101.
5. Purification of the digested vector and mutator: Two 1 % agarose gels, PCR and gel purification kit if required.
6. Ligation: Ligase and ligase buffer of your choice.
7. Transformation: Ice, competent CC118λ*pir* *E. coli* cells (100 μl per reaction), 42 °C water bath, 500 μl SOC medium/transformation, LB agar plates supplemented with Sm50.

2.5 Generation of the Mutant

1. LB agar plates supplemented with 2,000 μg/ml streptomycin (Sm2000). Make a stock solution of 100 mg/ml Sm.
2. LB agar plates supplemented with 5 % sucrose. Prepare fresh on the day of use.
3. Primers for colony PCR:
   (a) Primers that anneal to pKNG101: 10 μM of primer UpKn (5′—ccctggatttcactgatgag—3′), 10 μM primer RpKn (5′—catatcacaacgtgcgtgga—3′).
   (b) External primers that anneal next to the modified gene: 10 μM primer 3′, 10 μM primer 5′.
   (c) Protocol for colony PCR *see* Chapter 35.
4. PIA and LB agar plates.

3 Methods

3.1 Amplification of the Gene of Interest for Site-Directed Mutagenesis

Prior to site-directed mutagenesis, amplify a 1,000 bp DNA fragment of the gene of interest from the *Pseudomonas* genome using primers 1 and 2 (Fig. 1). The region of the desired mutation should be in the middle of the DNA fragment flanked by approximately 500 bp on each site.

1. Set up a PCR reaction following the protocols from Chapter 35.
2. Ligate the PCR product into the cloning vector pCR2.1. Mix on ice:

   2 μl of pCR2.1 vector

   1 μl of T4 DNA ligase

   1 μl of 10× ligation buffer

   1 μl of insert

   5 μl of molecular biology water to a total of 10 μl
3. Ligate overnight at 16 °C and proceed to Subheading 3.3.

3.2 Amplification of the Mutator Fragment

To engineer a clean deletion of a gene on the *Pseudomonas* chromosome, a fragment of approximately 500 bp on either side of the gene of interest (GOI) is amplified. The 3′ primer of the upstream fragment (primer 2), and the 5′ primer of the downstream fragment (primer 3) will contain a few bases of the gene, including the start and stop codons (*see* Figs. 2 and 3).

The two fragments are joined by overlapping PCR to create a mutator fragment, which is cloned into pKNG101, a *Pseudomonas* suicide vector. Subsequently, pKNG101 is conjugated into the *Pseudomonas* strain of interest. Through a series of two recombination events, the gene will be deleted.

```
TCGACGCGCAGATGCTGGCCCTGGGGCCGGGCGGACGCTCGATCGAT
ACGTCGCGTTTCGAGCGCGGCGACGTCATCGAGCCAGGCCGCTATCG      Primer 3
CCTCGACCTGCTGCTCAACAGCCGATGGCGTGGCGTCGAGGAAGTCG
AGCTGCGCCGCCAGCCGGGGCGGAAAGCGCGGTCTTCTGCTACGAC
CGGGGCCTGCTGGAGCGGGCGGGCATCGACCTGGAGAAGAGCGCGCG
TGGCCAGGACCGTTCCTCGGCTCGCGATCCTCTGCCCGAAGGTTTGC      Primer 1
ACTGCGACCCTCTCGAGCGCTATGTGCCGGGGGCCCGGGTCAAGCTC
GATATCGCCGAGCAGTCGATCTATGTCTCGGTGCCCAGCTATTACCT
GAGCCTGGATTCTTCGAAGACCTATGTCGATCCGGCGAGCTGGGACA
GCGGCATTTCCGCCGCCTTGCTCAACTACAACAGCAATCTCCACGTC
AGGGAAAACCACGGCAGGAGCGCCACCAGCGGCTATGCCGGGATGAA
CGCCGGCTTCAATTTCGGGCGGGCGCGCCTGCGCCACAACGGCACGG      500 bp
CCACCTGGTCGCGCCGCATGGGCAGCCATTACCAGCGTAGCGCAACC
TATGTGCAGACCGACCTGCCGGCCTGGCGTGCGCAGTTATTGCTGGG
AGAAAACTCCACCAGCAGCGAGTTCTTCGATGCGGTGTCCTTCCGTG
GAGTGCAGCTATCCAGCGATGACCGGATGCTGCCGGATTCGCTGCGC
TACTACGCTCCGGTGGTCCGTGGGACCGCCAGTACCAATGCGCGGGT
ATCGGTCTACCAGCGCGGCTACCTCATCTACGAAACCACGGTGGCAC      Desired SPM
CCGGGGCGTTCGCTCTCGACGAACTGCAGACCGCCAGCTATGGCGGG
GACCTGGAAGTGCGGGTGACCGAAGCCAGCGGGGAAGTCCGCAGTTT
CATCGTGCCGTTCGCCACCACCGTACAACTGCTGCGCCCCGGGACCA
CGCGCTACAGCCTGACGGCCGGGCGGCTCAACGATCCCAGCCTGGAG
CGTCGGCCGAACATGCTGCAGGGCGTCTACCAGCGCGGCCTGGGCAA      500 bp
CGACGTCACCGCATACGCGGGCGGGGCCTTCACCGGCAGCTACATGT
CCGGGTTGATGGGCGCGGCGCTGAACACGCCGGTGGGCGGATTCTCC
GGTGACGTGACGCTGGCGCGTACCGAGGTTCCCGGCGACGACCGCCT
TAGCGGCTCCAGCTACCGTCTCGCCTACAGCAAGAACCTGCCGAACA      Primer 2
CCGGCACCAACTTTTCGCTGCTCGCCTATCGTTACTCCACCGGTGGC
TATCTCGGCCTGCGCGACGCGGCCTTCATGCAGGACCGGGTAGAGCG
AGGCGAGCCGCTGGAGTCGTTCTCGCGCTTGCGCAATCGTCTCGACG
CCAACATCAGCCAGCAACTGGGCAACGGCGGCAACCTTTACCTGAAC      Primer 4
GGCTCCTCGCAGCGCTACTGGAGCGGCGGCGGGCGGGCGGTCAACTT
CTCCGTCGGCTACAGCAACCAGTGGCGCGACGTCAGTTACTCCATTT
```

Fig. 1 Primer design for the amplification of the gene of interest. Primer 1 and 2 (*red*) are used to amplify a 1,000-bp DNA fragment from the *Pseudomonas* chromosome. In the middle of this DNA fragment the point mutation (SPM) will be introduced. Primer 3 and primer 4 are external primers that bind approximately 200 bp upstream of primer 1 or downstream of primer 2, respectively. External primers will be used to check for the presence of the first and second crossover

Fig. 2 Schematic presentation of the gene of interest (GOI) and the primers used to create the mutator fragment for gene deletion. The upstream fragment is labelled up and the downstream fragment dw. ATG gene start and TAA gene stop are indicated. *Numbers in brackets* correspond to primers 1–5

1. Retrieve gene sequence of the GOI including roughly 800 bp upstream and downstream of GOI.
2. Design primers to amplify ~500 bp upstream and ~500 bp downstream of GOI (*see* Figs. 2 and 3, *see* **Note 9**). Also design

external primers that will be used to check for the deletion of the GOI on the *Pseudomonas* genome.

(a) Primer 1: Forward (Fw) primer, ca. 500 bp upstream of the ATG of the GOI.

(b) Primer 2: Reverse (Rv) primer upstream fragment; consists of 9 bases + ATG + 6 bases + 9 bases tail overlapping with Primer 3 (Fig. 4).

a

```
TGAACTGGCGCGGTAGTCCGACTCGCCGCTGGCACTGCCGTCGGCCAGGCGCCAGCGCA
GGCGCTGCGAGTCGAAGGCATGTTCGAT(5)GACCCAGCCGTTGGCGAAGTGCAGGACC   Primer 5
CGTTCGAGCCCGGCGAGATCGCCGCAGGCGGAAGGACGTGGTTGTCCGGGGCGAAGCC
GTCGGCCAGGGCGCCGACGGTGATCCATCCGGAATGCGAGT(1)GGGCGGACATAGGGT   Primer 1
TCTCCTCGTCGTGGGAAACCCATGCTGGCCTGTCTTCGAGTATCCCGATATCGACCGGA
TGGTTGATAGGTCGATATGCCACGGAAATAGCACGGAGCGGCGAAGGGAAAGTTATCGA
AGGTGAAGTCTTAGATGAGTTTTCGTTTCTTCTTTTTTTTGGTTTTTTAGTTAATGGGT   500 bp
TCCTGTTCTTCTATAAGCATTTTCTGAATTTATTCTTATGTGCTTGTTGGCTGTTGTAG
GAAATGGACTACATATATTTTCGTATGATTATTATTTTTAAATCCTGTCTTTCTTTCTA
GTATTTGGGTCTGCGTTTATCAGGGCGCAATCTTCGGCTGTCGCGTAATGGTCGGAGGC
GCTCTAGGCGGCCGTTCCGGTTGCGCGCAAAGGCTTGGCTACCTTGGCCGGTTCGTACT
GCATCGGCTCTCCAGGGAGCAGCCAGGATGTTCCACAACTCCAAAGGAAATCAGATGAA   Primer 2
CAAG(2)TTCAAACGTGCAATTCTTGCGCCCCTCGTTGCGTCGGGCACCCTTCTCGCGC
TGATGTCCCCCGTGGCCCATGCGGTCGATGGATCGTCAACTTCTCGGGCAACATCACC
GAAGTCGCGTGCGATATCAACGGCCAGGCGCCGGGTGCGGGCAACGTCACCAACGTCGA
CCTGGGTATCGTGGCGCCGACCCATTTCACCGCCGTCGGAACCAACTCCACTTTCCAGG
AGTTCGATCTGGTCCTCTCCGGTGCCGGCTGCACCAATGACAAGAAGGTTTCCATCGCC   Gene of
TTCGATCAGGTCACCAACGTCGACCGTATCAGCGGCAACCTGATGCTGATGGGGCCGTC   interest
GGAAGCCAAGGGCGTGCAGATCCAGATCTTCAACAATTCGGCGACCAACACCACGAAGA
TCCCGCTGGGCTTGGTCGAGGCCGCCCCGCAGAAGGCAACCGTCGTGAAGAACACGGCG
ACCCTGAAGTACAAGGCCTCCTATGTGTCCACCGAGAAGGCCGTGACCGCGGGTACCGG
CCTGTCCCGTATTCGCTACCTCCTGGCGTACGAGTAAGCGGCGGTC(3)GACGGACGAA   Primer 3
GAGGTGAGGCCATGGCGCCGCTAATGCATCGTTTTCATTCTTTCGTGGCGGCAAGCGCC
GTGGCCATCGCCCTGTGCGTCGGAACGGCTCACGCCGGGCTGATCGCACAGGGCACTCG
CGTCGTCTTTCCGGCCAGCGAGCGCGAGGTCACCTTGCGGGTCAGCAATACCTCCGGCA
CTCCCGTGCTGGCGCAAGCCTGGATCGACGATGGCCGGCAGGACGTTCCCCCCGAGGAA   500 bp
CTGCAGGTTCCCTTCAGTGTCACGCCGGCCGTGACGCGAGTGGAGCCCAATGGCGGCGC
TGTATTGCGCATCGCCTACCTGAAGGCCCCGTTGCCGACGGATCGCGAATCGCTGTTCT
GGCTGAATATCCTGGAAGTTCCGCCCAGGGACGAAGACGAAAACAATGCATTGCAGTTT
TCCTTTCGTTCGCGCTTCAAACTCTTCTTTCGACCCAGCCAGTTGAAGAGTGTCGATTC   Primer 4
GGCT(4)GCCGGGAAACTTCAATGGAAGTTTCTGGAGTCAGGGGGAGCAGGAAAAAAGA
CCGTAGTACAGGTAAATAATCCGACGCC(6)TTACTAC                          Primer 6
```

Fig. 3 Example strategy for amplification of the mutator fragment. *Underlined* is the gene of interest. ATG start codon and TAA stop codon are in *bold letters*. (1) Primer 1: Forward (Fw) primer ca. 500 bp upstream of the ATG of the gene to be deleted. (2) Primer 2: Reverse (Rv) primer upstream fragment; 9 bases + ATG + 6 bases + 9 bases tail overlapping with Primer 3. (3) Primer 3: Fw primer downstream fragment; 6 bases + STOP + 9 bases + 9 bases tail overlapping with primer 2. (4) Primer 4: Rv primer ca. 500 bp downstream of the TAA of the gene to be deleted. (5) Primer 5: Fw external primer, ca. 100 bp upstream of primer 1. (6) Primer 6: Rv external primer, ca. 100 bp downstream of primer 4. (**a**) Gene of interest with its upstream and downstream regions. (**b**) A PCR using primers 1 and 2 amplifies the upstream fragment, whereas the downstream fragment is amplified using primers 3 and 4 in a PCR. Both fragments are approximately 500 bp long and are used, together with primers 1 and 4, to create the mutator fragment by an overlapping PCR

b

520 bp upstream fragment:

TCCATCCGGAATGCGAGT **(1)** GGGCGGACATAGGGTTCTCCTCGTCGTGGGAAACCCATGCTGGCCT
GTCTTCGAGTATCCCGATATCGACCGGATGGTTGATAGGTCGATATGCCACGGAAATAGCACGGAGC
GGCGAAGGGAAAGTTATCGAAGGTGAAGTCTTAGATGAGTTTTCGTTTCTTCTTTTTTTTGGTTTTT
TAGTTAATGGGTTCCTGTTCTTCTATAAGCATTTTCTGAATTTATTCTTATGTGCTTGTTGGCTGTT
GTAGGAAATGGACTACATATATTTTCGTATGATTATTATTTTTAAATCCTGTCTTTCTTTCTAGTAT
TTGGGTCTGCGTTTATCAGGGCGCAATCTTCGGCTGTCGCGTAATGGTCGGAGGCGCTCTAGGCGGC
CGTTCCGGTTGCGCGCAAAGGCTTGGCTACCTTGGCCGGTTCGTACTGCATCGGCTCTCCAGGGAGC
AGCCAGGATGTTCCACAACTCCAAAGGAAATCAG**ATG**AACAAGTACGAG**TAA** **(2)**

516 bp downstream fragment:

ATGAACAAGTACGAG**TAA**GCGGCGGTC **(3)** GACGGACGAAGAGGTGAGGCCATGGCGCCGCTAATGC
ATCGTTTTCATTCTTTCGTGGCGGCAAGCGCCGTGGCCATCGCCCTGTGCGTCGGAACGGCTCACGC
CGGGCTGATCGCACAGGGCACTCGCGTCGTCTTTCCGGCCAGCGAGCGCGAGGTCACCTTGCGGGTC
AGCAATACCTCCGGCACTCCCGTGCTGGCGCAAGCCTGGATCGACGATGGCCGGCAGGACGTTCCCC
CCGAGGAACTGCAGGTTCCCTTCAGTGTCACGCCGGCCGTGACGCGAGTGGAGCCCAATGGCGGCGC
TGTATTGCGCATCGCCTACCTGAAGGCCCCGTTGCCGACGGATCGCGAATCGCTGTTCTGGCTGAAT
ATCCTGGAAGTTCCGCCCAGGGACGAAGACGAAAACAATGCATTGCAGTTTTCCTTTCGTTCGCGCT
TCAAACTCTTCTTTCGACCCAGCCAGTTGAAGAGTGTCGATTCGGCT **(4)**

1013 bp mutator fragment:

TCCATCCGGAATGCGAGT **(1)** GGGCGGACATAGGGTTCTCCTCGTCGTGGGAAACCCATGCTGGCCT
GTCTTCGAGTATCCCGATATCGACCGGATGGTTGATAGGTCGATATGCCACGGAAATAGCACGGAGC
GGCGAAGGGAAAGTTATCGAAGGTGAAGTCTTAGATGAGTTTTCGTTTCTTCTTTTTTTTGGTTTTT
TAGTTAATGGGTTCCTGTTCTTCTATAAGCATTTTCTGAATTTATTCTTATGTGCTTGTTGGCTGTT
GTAGGAAATGGACTACATATATTTTCGTATGATTATTATTTTTAAATCCTGTCTTTCTTTCTAGTAT
TTGGGTCTGCGTTTATCAGGGCGCAATCTTCGGCTGTCGCGTAATGGTCGGAGGCGCTCTAGGCGGC
CGTTCCGGTTGCGCGCAAAGGCTTGGCTACCTTGGCCGGTTCGTACTGCATCGGCTCTCCAGGGAGC
AGCCAGGATGTTCCACAACTCCAAAGGAAATCAG**ATG**AACAAGTACGAG**TAA**GCGGCGGTCGACGGA
CGAAGAGGTGAGGCCATGGCGCCGCTAATGCATCGTTTTCATTCTTTCGTGGCGGCAAGCGCCGTGG
CCATCGCCCTGTGCGTCGGAACGGCTCACGCCGGGCTGATCGCACAGGGCACTCGCGTCGTCTTTCC
GGCCAGCGAGCGCGAGGTCACCTTGCGGGTCAGCAATACCTCCGGCACTCCCGTGCTGGCGCAAGCC
TGGATCGACGATGGCCGGCAGGACGTTCCCCCCGAGGAACTGCAGGTTCCCTTCAGTGTCACGCCGG
CCGTGACGCGAGTGGAGCCCAATGGCGGCGCTGTATTGCGCATCGCCTACCTGAAGGCCCCGTTGCC
GACGGATCGCGAATCGCTGTTCTGGCTGAATATCCTGGAAGTTCCGCCCAGGGACGAAGACGAAAAC
AATGCATTGCAGTTTTCCTTTCGTTCGCGCTTCAAACTCTTCTTTCGACCCAGCCAGTTGAAGAGTG
TCGATTCGGCT **(4)**

Fig. 3 (continued)

(c) Primer 3: Fw primer downstream fragment; consists of 6 bases + STOP + 9 bases + 9 bases tail overlapping with primer 2.

(d) Primer 4: Rv primer ca. 500 bp downstream of the TAA/TGA of the GOI.

(e) Primer 5: Fw external primer, ca. 100 bp upstream of primer 1.

(f) Primer 6: Rv external primer, ca. 100 bp downstream of primer 4.

3. Check insert sequence (mutator fragment) for the presence/absence of restriction sites compatible with pKNG101.
4. Prepare on ice two separate PCR reactions to amplify the upstream (primer 1 and 2) and downstream (primer 3 and 4) DNA fragments following the protocol from Chapter 35.
5. After checking the size and purity of the PCR product by agarose gel-electrophoresis, purify the residual 45 μl PCR product with a PCR purification kit or by a method of your choice.
6. Measure DNA concentration with a NanoDrop and dilute the PCR product 1:20 in water and use 1.5 μl of the upstream and 1.5 μl of the downstream fragment for an overlapping PCR using primer 1 and 4. The overlapping PCR is designed to anneal the two fragments and create a mutator fragment for the gene knockout.

   Prepare on ice the following PCR mix and run a PCR following the conditions of Table 3:

   1.5 μl 1:20 diluted upstream fragment

   1.5 μl 1:20 diluted downstream fragment

   5 μl 10 μM primer 1

   5 μl 10 μM primer 4

   1 μl 20 mM dNTPs

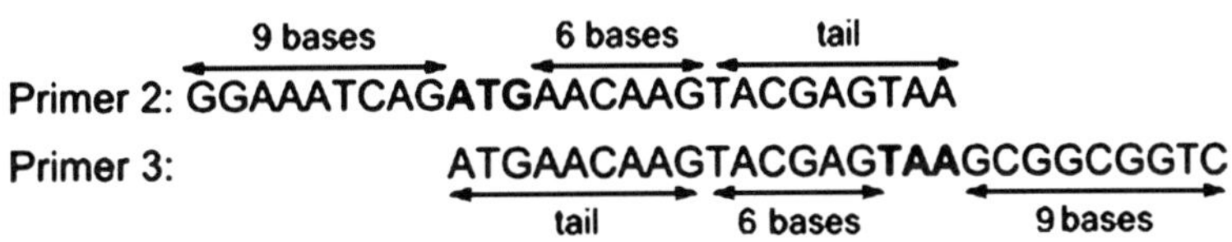

Fig. 4 Alignment of primers 2 and 3. Both primers contain 18 bases that overlap. This overlapping tail will be used in an overlapping PCR to unite both fragments and thereby forming the mutator fragment

Table 3
Reaction conditions for the overlapping PCR

	Temperature	Time
Initial denaturing step	95 °C	5 min
Start cycle: 20 cycles		
Denaturing	95 °C	30 s
Annealing	55 °C	30 s
Extension	72 °C	1 min/kb
End cycle		
Final extension	72 °C	10 min
Hold	10 °C	∞

10 μl 5× Expand High Fidelity^{PLUS} buffer

10 μl 5 M Betaine

15.5 μl molecular biology water to a final volume of 50 μl

7. After the overlapping PCR, check PCR product on an agarose gel and ligate the PCR product into pCR2.1 by following the protocol from Subheading 3.1.

3.3 Transformation into Competent Top10 E. coli Cells

Carry out all procedures on ice unless otherwise stated! Work sterile.

1. Gently thaw your competent Top10 cells on ice and transfer 50 μl into a pre-chilled 1.5 ml eppendorf tube that contains the whole ligation mixture.
2. Incubate on ice for 30 min.
3. Heat shock the transformations for 30 s at 42 °C and immediately place the reaction on ice for 1 min.
4. Add 500 μl SOC medium to each transformation and incubate the transformations at 37 °C for 1 h shaking at 200–250 rpm.
5. Spread 100 μl of the transformation on LB agar plates supplemented with Amp50, Kan25 and X-gal40, *see* **Notes 10** and **11**.
6. Spin down the residual 400 μl for 1 min in a microcentrifuge and decant the supernatant. Resuspend the pellet in the remaining 100 μl supernatant and spread on LB agar plates supplemented with Amp50, Kan25 and X-gal40.
7. Incubate the transformation plates at 37 °C for at least 16 h.
8. Positive clones that contain the insert can be identified by blue/white screening [9]. Patch positive (white) colonies on the LB Amp50, Kan25, X-gal40 agar plates and inoculate a 5 ml culture for plasmid preparation.
9. Purify plasmids and verify insert by sequencing.
10. For gene deletion, proceed directly to Subheading 3.7.

3.4 Primer Design for Site-Directed Mutagenesis

1. Both 3′ and 5′ primer have to contain the point mutation and anneal to the same sequence of DNA.
2. The optimal length of the primers is between 25 and 45 bp and the melting temperature is about ≥78 °C.
3. The point mutation should be designed in the middle of the primer, so that 10–15 bases with the correct sequence are present on either side of the primer.
4. The GC content of the primers should not be higher than 40 % and both primers should have two or three C or G bases at their ends (*see* **Note 12**).

3.5 Mutagenesis by PCR

Carry out all procedures on ice unless otherwise stated.

1. Prepare on ice the sample reaction by mixing

   5 μl of 10× Pfu Ultra High Fidelity^{PLUS} DNA polymerase buffer

   1.5 μl of plasmid DNA (*see* **Note 13**)

   1.5 μl of 3′ primer

   1.5 μl of 5′ primer

   1 μl of dNTPs

   10 μl betaine

   28.5 μl molecular biology water to a final volume of 50 μl
2. At last add 1 μl of Pfu Ultra High Fidelity^{PLUS} DNA polymerase.
3. Start the PCR reaction (*see* **Note 14**). The number of cycles depends on the type of mutation that is introduced. Point mutations require only 12 cycles, codon changes 16 cycles and multiple codons deletions or insertions 18 cycles.
4. After the PCR has finished incubate the PCR reaction on ice at least for 2 min to cool down the sample.

3.6 DpnI Digestion of the Parental DNA and Transformation into XL1-Blue Competent Cells

1. To digest the non-mutated, parental DNA add 1 μl of the restriction enzyme *Dpn*I directly into each PCR tube and mix gently by pipetting the reaction mix up and down.
2. Incubate the restriction mix for 1 h at 37 °C (*see* **Note 15**).
3. Carry out all procedures on ice! Work sterile.
4. Gently thaw your competent XL1-Blue cells on ice and aliquot 80 μl into a pre-chilled 1.5 ml eppendorf tube.
5. Add 5 μl of the *Dpn*I-digested DNA to the competent cells and gently mix by stirring with your pipette tip.
6. Incubate the reactions on ice for 30 min.
7. Heat shock the transformations for 1:15 min at 42 °C and immediately place the reaction on ice for 2 min.
8. Add 500 μl SOC medium to each transformation and incubate the transformations at 37 °C for 1 h shaking at 200–250 rpm.
9. Spread 100 μl of the transformation on LB agar plates supplemented with the required antibiotic.
10. Spin down the residual 400 μl for 1 min in a microcentrifuge and decant the supernatant. Resuspend the pellet in the remaining 100 μl supernatant and spread on a LB agar plate supplemented with the required antibiotic (*see* **Notes 10** and **11**).
11. Incubate the transformation plates at 37 °C for at least 16 h.
12. Patch the appearing colonies on the LB agar plates with antibiotic and inoculate a 5 ml culture for plasmid preparation.
13. Verify the mutation by sequencing.

3.7 Re-cloning of the Mutagenized Gene from pCR2.1 into pKNG101

1. To reclone the mutator from pCR2.1 into pKNG101, identify restriction sites which are present on the pCR2.1-mutator and pKNG101, but not in the mutator sequence. Prepare the restriction digest in 1.5 ml eppendorf tubes. If the plasmid concentration is higher than 200 ng/μl use 10 μl vector, and if the concentration is lower than 200 ng/μl use 20 μl vector.

   Restriction mix:

   10 μl/20 μl vector

   4 μl restriction enzyme buffer

   1 μl enzyme A

   1 μl enzyme B

   X μl molecular biology water to a final volume of 40 μl
2. Carry out the digest at the appropriate temperature for 2–3 h (*see* **Note 16**).
3. Take 5 μl of the restriction digest and load on a 1 % agarose gel to check for restriction. Two bands should separate for pCR2.1 and the mutator fragment and one band should be visible for pKNG101.
4. Purify the digested pKNG101 in a final volume of 30 μl water with a PCR purification kit or the method of your choice.
5. Meanwhile, load the whole volume of the pCR2.1-mutator restriction digest on a 1 % agarose gel and run the gel until vector backbone and mutator are sufficiently separated. Using a scalpel cut out the mutator band from the gel and purify with a gel purification kit or the method of your choice (*see* **Note 17**).
6. Measure the concentrations of your purified vector and mutator and ligate in a 1:3 ratio by a method of your choice.
7. Transform competent CC118*λpir E. coli* cells with the whole ligation reaction: Pre-chill the ligation reaction on ice, then add 100 μl competent cells and incubate on ice for 30 min. Heat shock the transformation reaction for 2 min at 42 °C. Place reaction on ice immediately and chill for 1 min. Add 500 μl SOC medium and incubate at 37 °C shaking at 200–250 rpm.
8. Spread 100 μl of the transformation on LB Sm50 agar plates. Spin down the residual 400 μl for 1 min in a microcentrifuge and decant the supernatant. Resuspend the pellet in the remaining 100 μl supernatant and spread on a LB Sm50 agar plate.
9. Incubate the transformation plates at 37 °C for at least 16 h.
10. Patch single colonies on LB agar plates supplemented with Sm50 and incubate at 37 °C overnight.

11. Confirm insertion of the mutator into pKNG101 by colony PCR (*see* Chapter 35) and restriction digest of the purified pKNG101.
12. Conjugate pKNG101-mutator into *Pseudomonas* by three-partner conjugation (*see* Chapter 3).

3.8 Selection of the First Recombination Event

Since pKNG101 derivatives cannot replicate in *P. aeruginosa*, a single homologous recombination event between the gene of interest on the chromosome and the mutator fragment on pKNG101 has to occur to maintain the plasmid (and corresponding antibiotic resistance), integrating the whole vector into *Pseudomonas* chromosome. The integration can take place via a crossover occurring in the pKNG101 mutator upstream or downstream of the gene of interest (for site-directed mutagenesis *see* Fig. 5A and for gene deletion *see* Fig. 6).

1. Patch appearing colonies as soon as possible on LB Sm2000 as PIA might cause mutation of *Pseudomonas*. Incubate at 37 °C overnight.
2. Confirm first event of recombination by colony PCR using external primers in combination with the plasmid primers RpKN or UpKn. Use of external primers confirms the insertion of the plasmid in the correct genomic location.
3. With a loop pick two upstream and two downstream recombinants and restreak them onto LB agar plates supplemented with Sm2000 to single colonies. Incubate at 37 °C overnight.
4. Pick a single colony of each recombinant and streak them to single colonies on LB agar plates supplemented with 5 % sucrose. Change the loop between the streaks to yield more singles colonies. Incubate the plates at *room temperature* for 48–72 h (*see* **Note 18**).

3.9 Selection of the Second Recombination Event

After 2 days on sucrose some single colonies should appear on the plates. These bacteria have ejected pKNG101 from their chromosome by a second upstream or downstream crossover event leaving behind either the mutated or the wild-type gene (*see* Fig. 5B for site-directed mutagenesis and Fig. 6 for gene deletion).

1. Patch each single colony onto an LB Sm2000 agar plate, a PIA agar plate and a LB agar plate, *in that order*, and incubate at 37 °C overnight.
2. Clones that are Sm2000 sensitive (because they have ejected pKNG101, *see* **Notes 19** and **20**), but grow on PIA and LB are checked by colony PCR using the external primers. Five µl of the PCR product are run on a 1 % agarose gel to verify purity and right size of the PCR product. For site-directed mutagenesis, residual PCR products are sequenced using primer 1 or 3 as a sequencing primer.

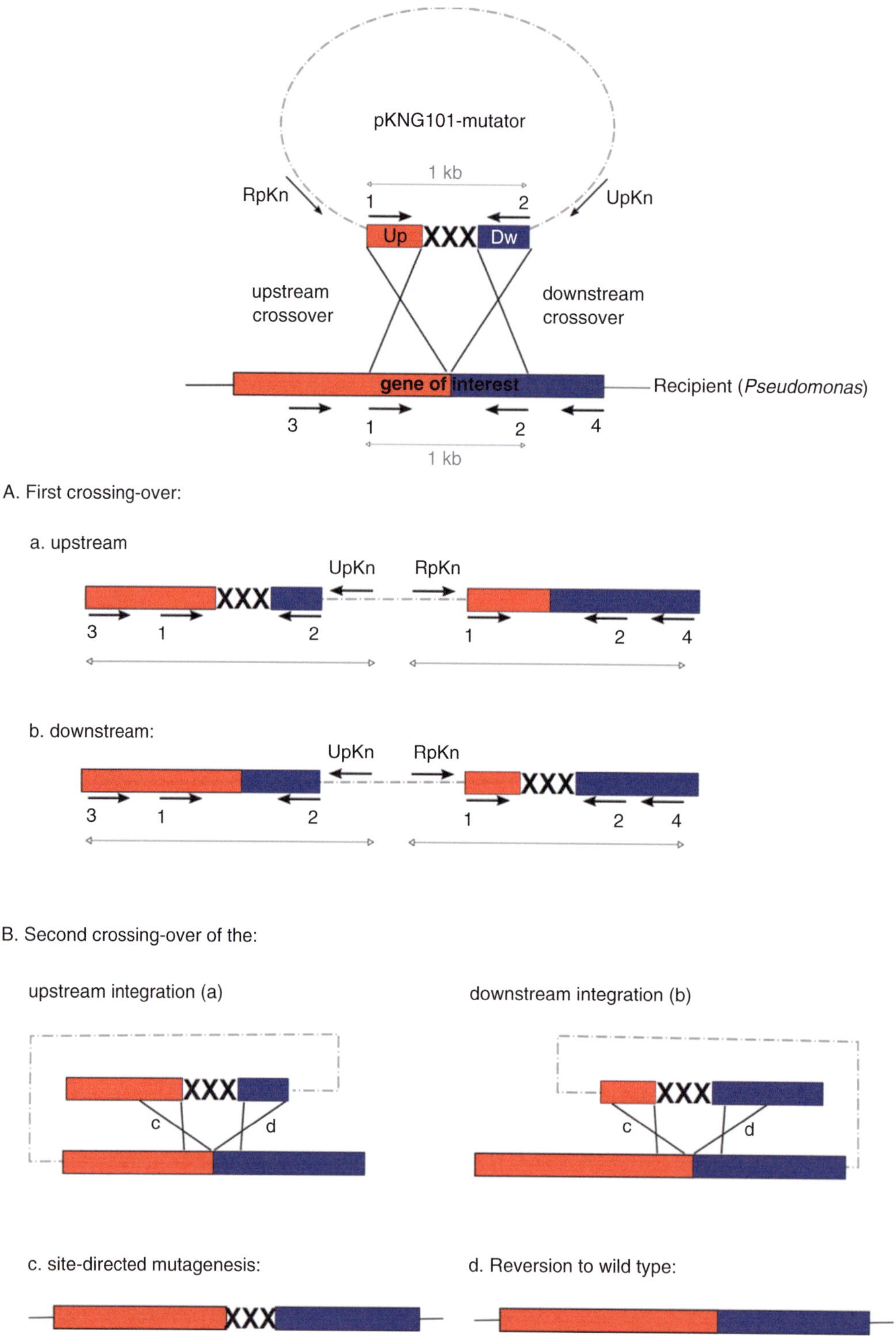

Fig. 5 Cloning strategy for site-directed mutagenesis of a gene of interest on the chromosome of *P. aeruginosa*. The pKNG101-mutator is introduced in *P. aeruginosa* by three-partner conjugation. Since pKNG101 derivatives cannot replicate in *Pseudomonas*, a single homologue recombination event between the functional GOI on the

3. Positive clones from the patch on the LB plate are streaked to single colonies on first LB Sm2000 and then on LB agar plates. Incubate at 37 °C overnight. Bacteria should not grow on Sm2000.
4. Single colonies are patched on LB agar plates, incubated at 37 °C overnight and checked again by colony PCR and sequencing. Glycerol stocks are prepared from positive clones.

4 Notes

1. The X-gal is light sensitive and should be kept in the dark. Wrap the tube in aluminum foil to protect the X-gal from light and store it in the freezer. The X-gal solution should be colorless; however, if it is yellow the X-gal has gone off.
2. Prepare plasmid template DNA from a *dam*+*E. coli* strain to ensure that the plasmid is methylated. Methylated template DNA and also hemi-methylated PCR products will be digested by the endonuclease *Dpn*I which allows selection of mutagenized plasmids only.
3. Use freshly prepared plasmid DNA as a template for the PCR, as the mutagenesis of freshly prepared DNA is more efficient than the mutagenesis of defrosted DNA.
4. XL1-Blue *E. coli* cells are resistant to tetracycline. If your mutagenized plasmid carries the tet^R resistance gene, you have to use an alternative *E. coli* strain that is sensitive to tetracycline. Alternatively, use a plasmid with a different resistance marker.
5. Sterile streptomycin stock solutions can be stored at −20 °C for several months or at 4 °C in the fridge for up to 1 week. Do not re-freeze defrosted solutions as repeated freezing and defrosting decreases the stability of the streptomycin.
6. Sucrose solutions have to be made fresh due to the low stability of sucrose in solution. When preparing a 50 % sucrose solution it will take 1–1.5 h before the sucrose will go into solution. Do *not* heat up the solution, as high temperatures decrease the

Fig. 5 (continued) chromosome and the mutator fragment on pKNG101 has to occur, integrating the whole plasmid into *Pseudomonas* chromosome. Primers are presented by *arrows labelled with a number*, e.g., Primer 1 is labelled (1). UpKn and RpKn are primers that anneal to pKNG101. (**a**) The integration can be caused by an upstream or downstream crossover. To check for pKNG101 integration primers 3 and UpKn or primers 4 and RpKn can be used. (**b**) The forced rejection of pKNG101 from the chromosome of *Pseudomonas* occurs via a second crossover, which can be again upstream or downstream. This second crossover generates either the wild-type gene or the mutated gene and is confirmed by colony PCR using the external primers 3 and 4. Both, wild-type and mutated gene, will yield the same size PCR product, and therefore, confirmation of successful site-directed mutagenesis is archived only by sequencing

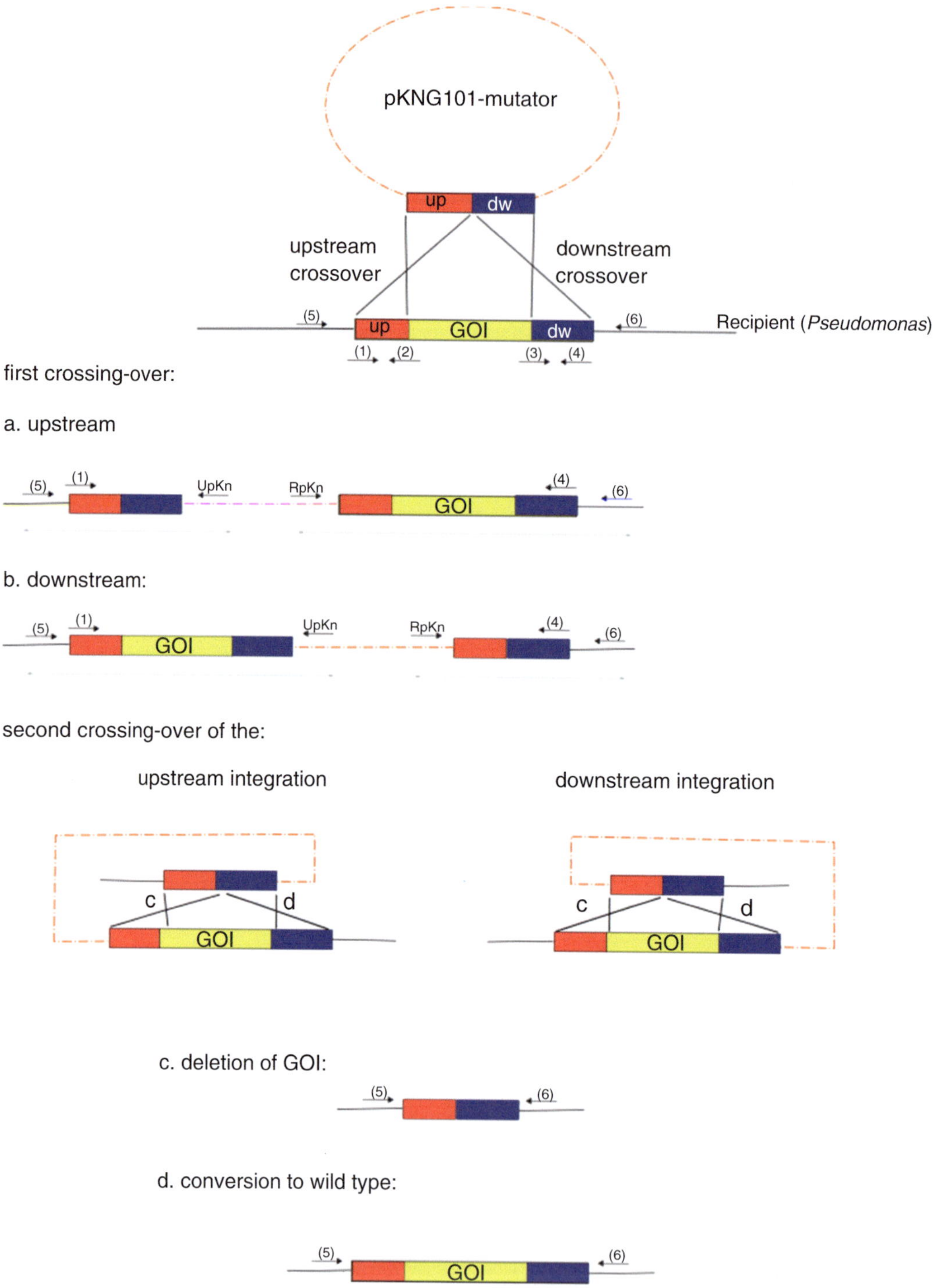

Fig. 6 Cloning strategy for gene deletion. The pKNG101-mutator is introduced in *P. aeruginosa* by three-partner conjugation and will integrate into *Pseudomonas* chromosome by homologue recombination. Primers are presented by *arrows labelled with a number*, e.g., Primer 5 is labelled (5). UpKn and RpKn are primers that anneal to pKNG101. GOI is the gene of interest. (**a**) The integration can be caused by an upstream or downstream crossover. To check for pKNG101 integration primers 5 and UpKn or primers 6 and RpKn can be used,

stability of sucrose. Also, pay attention that the LB agar has cooled down to hand temperature before adding the sucrose. You can speed down the cooling of the agar by swirling the agar bottle in a beaker with cold water.

7. Test the sucrose sensitivity of pKNG101 by streaking out a single colony of pKNG101-containing CC118*λpir E. coli* cells, first on LB agar supplemented with Sm50 and then using the same loop on LB agar supplemented with 5 % sucrose. Incubate at room temperature for 2 days. Cells should grow on Sm50 but not sucrose verifying the sucrose sensitivity of pKNG101.
8. To yield high concentrations of pKNG101, inoculate two times 50 ml LB medium with a single sucrose sensitive colony of CC118*λpir*/pKNG101, taken from a LB Sm50 agar plate, and grow overnight at 37 °C shaking. Purify pKNG101 from the total 100 ml culture by the method of your choice. The obtained concentration of purified pKNG101 in water should be approximately 800 ng/μl.
9. When you design the mutator fragment, pay attention that you (a) delete in frame, (b) do not delete the Shine–Dalgarno Sequence of downstream genes, (c) do not delete START/STOP codon of surrounding genes and (d) no not delete part of another gene when genes overlap in an operon.
10. Spread the transformation reaction only on well dried agar plates to obtain single colonies. The use of wet agar plates might result in less single colonies.
11. If your plasmid contains more than one resistance marker, use only one antibiotic for selection to facilitate the growth of bacteria.
12. To simplify the design of the primers use the free online QuikChange Primer Design program Web site from Agilent technologies (https://www.genomics.agilent.com/). You have to register with Agilent for this service, but will not be charged. Paste in the DNA sequence, choose up to 7 bases that you want to mutate and you will obtain your primer sequences along with information about their length, melting temperature and energy cost of mismatches.
13. Use a range of DNA concentrations varying from 5 to 50 ng (e.g., 5, 10, and 20 ng). Do not use more than 50 ng DNA in

Fig. 6 (continued) e.g., if an upstream crossover has occurred, primers 5 and RpKn will amplify a shorter PCR product than in the case of a downstream crossover (compare length of *grey arrows*). (**b**) The forced rejection of pKNG101 from the chromosome of *Pseudomonas* occurs via a second crossover, which can be again upstream or downstream. This second crossover generates either the wild-type gene or gene deletion and is confirmed by colony PCR using the external primers 5 and 6. The amplified PCR product for the mutant will be smaller in size than the PCR product amplified for the wild type

Table 4
Example PCR reaction for the introduction of SPM

	Temperature	Time
Initial denaturing step	95 °C	30 s
Start cycle: 12–18 cycles		
Denaturing	95 °C	30 s
Annealing	55 °C	60 s
Extension	68 °C	1 min/kb of plasmid length
End cycle		

total as high DNA concentration decrease the efficiency of the mutagenesis. As a control to check for DNA amplification, also prepare a tube that contains DNA but none Pfu Ultra High Fidelity^{PLUS} DNA polymerase. Check for amplification of the plasmid by agarose gel electrophoresis. Prepare a 1 % agarose gel and load 5 μl of the PCR reaction on the gel. Compare the strength of the DNA band of the control reaction (without polymerase) and the sample reactions (with polymerase). A much stronger band in the sample reactions indicates successful amplification of the plasmid.

14. To improve DNA amplification, increase the annealing temperature for the primers to up to 68 °C. Higher temperatures will avoid the formation of secondary structures and also increase the binding specificity of the primers (Table 4).
15. Increase the incubation time with *Dpn*I up to 1:30 h if you have a low mutagenesis efficiency to allow *Dpn*I to digest more parental DNA.
16. Do not digest for longer than 3 h as the restriction enzymes may start to cut the DNA unspecific.
17. Change the TAE buffer in the gel tank prior to running your gel to avoid any cross-contamination with DNA present in the tank. Cut out the DNA band with a clean sharp scalpel. Make sure to stay as close as possible to the DNA band.
18. The temperature should not be over 22 °C, because *Pseudomonas* would grow too quickly and eject pKNG101 too fast giving no time for the recombination event. Also high temperatures decrease the stability of the sucrose, resulting in not enough pressure on *Pseudomonas* to eject pKNG101 from the chromosome. If the temperature in the summer is too high in the lab, try to find an incubator that you can set to room temperature. In winter, at lower temperatures *Pseudomonas* might take more time to grow but will eject pKNG101 properly.
19. Some *Pseudomonas* cells might grow on the LB Sm2000 agar plates. However, the growth should not be as strong as on the

LB agar plate. If that is the case, re-streak bacteria on LB and LB Sm2000 agar plates to single colonies and incubate at 37 °C overnight. No colonies should grow on the LB Sm2000 plates. Patch colonies from the LB plate, incubate at 37 °C and carry out colony PCR to gain samples for sequencing.

20. Strong growth on the LB Sm2000 agar plate indicates that the selection did not work properly. Re-streak bacteria on LB 5 % sucrose agar plates and incubate again at room temperature for 48–72 h. Proceed as described by patching single colonies on LB Sm2000, PIA and LB agar plates.

Acknowledgements

Daniela Muhl is supported by a PhD fellowship from the Wellcome Trust N°089873/Z/08/Z. Alain Filloux laboratory is supported by grants from the Wellcome Trust, Biotechnology and Biological Sciences Research Council (BBSRC), and the Medical research Council (MRC).

References

1. Kaniga K, Delor I, Cornelis GR (1991) A wide-host-range suicide vector for improving reverse genetics in gram-negative bacteria: inactivation of the *blaA* gene of *Yersinia enterocolitica*. Gene 109:137–141
2. Kulasekara HD, Ventre I, Kulasekara BR, Lazdunski A, Filloux A, Lory S (2005) A novel two-component system controls the expression of Pseudomonas aeruginosa fimbrial cup genes. Mol Microbiol 55:368–380
3. Li R, Wang G, Shen B, Wang R, Song Y, Li S, Jiang J (2009) Random transposon vectors pUTTns for the markerless integration of exogenous genes into gram-negative eubacteria chromosomes. J Microbiol Methods 79:220–226
4. Martinez-Garcia E, de Lorenzo V (2012) Transposon-based and plasmid-based genetic tools for editing genomes of gram-negative bacteria. Methods Mol Biol 813:267–283
5. Schweizer HP (1991) Escherichia-Pseudomonas shuttle vectors derived from pUC18/19. Gene 97:109–121
6. Sarker MR, Cornelis GR (1997) An improved version of suicide vector pKNG101 for gene replacement in gram-negative bacteria. Mol Microbiol 23:410–411
7. Schweizer HD (1993) Small broad-host-range gentamycin resistance gene cassettes for site-specific insertion and deletion mutagenesis. Biotechniques 15:831–834
8. Lesic B, Rahme LG (2008) Use of the lambda Red recombinase system to rapidly generate mutants in Pseudomonas aeruginosa. BMC Mol Biol 9:20
9. Maas S (1999) Efficient and rapid procedure for blue-white screening of recombinant bacterial clones. BioTechniques 27: 1126–1128

Chapter 41

Signature-Tagged Mutagenesis

Irena Kukavica-Ibrulj and Roger C. Levesque

Abstract

Signature-tagged mutagenesis (STM) is a genome-wide functional screening assay based upon an insertional mutation technique to identify mutants having a particular phenotype from a mixed population. The selection phenotype is either negative or positive for identification of mutations in genes which will impact the host as having a diminished or enhanced adaptation to a defined environment. STM has subsequently been used in different biological systems and adapted to use different mutagens, signature tags, and detection methods. Here, STM has been modified and used for genome-wide screening of *Pseudomonas aeruginosa* to identify genes and their products essential during the infection process or in enhancing virulence in vivo.

Key words Signature-tagged mutagenesis (STM), Transposon, *Pseudomonas aeruginosa*

1 Introduction

Here, Signature-tagged mutagenesis (STM) has been modified from the original method to screen for *Salmonella* mutants and used for genome-wide screening of *Pseudomonas aeruginosa* to identify genes and their products essential during the infection process or in enhancing virulence in vivo. In STM, transposon mutants are generated and each unique bacterial clone is tagged with a specific DNA sequence (a unique bar code) part of the insertion sequence [1–7]. Each STM mutant can be rapidly identified by multiplex PCR in a pool of mutants. STM becomes an "en masse" screening technique which minimizes the number of individuals used in a specific screening assay (animal, plant, cell, or other). Previous experiments have shown that pooling of mutants can include a few STM mutants and up to 96 per assay. Technically and depending on the host used for screening, a specific STM mutant will have a higher (positive STM) or lower (negative STM as presented here) survival rate [1, 6]. The STM mutant "attenuated in virulence" or having a growth defect cannot be maintained or are significantly reduced in vivo and the attenuated mutants are selected in the host (negative selection).

Alain Filloux and Juan-Luis Ramos (eds.), *Pseudomonas Methods and Protocols*, Methods in Molecular Biology, vol. 1149, DOI 10.1007/978-1-4939-0473-0_41, © Springer Science+Business Media New York 2014

Mutants having insertional mutations can be identified by comparing the in vitro input and the in vivo output pools of mutants using multiplex PCR. Mutants identified are confirmed as definitive (enhanced or diminished in a specific phenotype) by having attenuation or enhanced virulence when compared to the wild-type strain using a competitive index (CI) analysis. For *P. aeruginosa*, the pUTminiTn5-Tc system is used and genes inactivated are identified by cloning the transposon antibiotic resistance marker (Tc). The gene inactivated is identified by Sanger DNA sequencing. An alternative method is the amplification of the DNA sequence flanking the mini-Tn5 transposon by inverse PCR which can be used not only to identify the location but also the orientation of the transposon insertion in the genome sequence [8]. The PCR-based STM that we developed and present here needs to be used in combination with bioinformatics analysis to identity DNA sequences encoding genes and their products affected by the STM insertion. Additional care must be taken for the type of operon and the gene position with an insertion and potentially causing polar mutations downstream.

To better understand the impact and the power of PCR-based STM applied to any *Pseudomonas*, we will give examples using *P. aeruginosa* strain LESB58. The "hypervirulent" LESB58 strain essentially has the core genome of *P. aeruginosa* strain PAO1 but additional DNA giving a total genome size of 6.71 Mbs. LESB58 contains six prophage gene clusters, termed here as prophages 1–6, of which four are absent from strain PAO1 and five genomic islands encoding 455 additional genes. Compared with PAO1, the highly transmissible and aggressive LESB58 displays enhanced virulence, a wider spectrum of antibiotic resistance, and presumably a better adaptation to the CF lung. The success of LES isolates in lung colonization may be due to the prior acquisition of genes or pathogenicity islands, to transcriptional variations in the level of gene expression, or to a combination of both. Such changes contribute to greater colonization and/or transmissibility of the LES strains, enhancing their ability to cause chronic infections in CF patients, and to enhanced virulence, manifesting itself in infections of non-CF parents.

The PCR-based STM method has been applied extensively to *P. aeruginosa* PAO [3, 6, 9] and LESB58 [4, 7]. The STM-based PCR STM applied to LESB58 will be used to illustrate the methods utilized for construction of the mutant libraries and the identification of mutants by multiplex PCR.

The preparation of agar beads for in vivo screening in a rat model of chronic lung, in the mouse model of acute infection and the selection of mutants attenuated for in vivo maintenance and their analysis using a competitive index are described in this book. Additional screening can be done using alternative hosts such as the worm *Caenorhabditis elegans*, the amoeba *Dictyostelium discoideum*, the wax moth *Galleria mellonella*, the model plant organism *Arabidopsis thaliana*, embryos from the zebra fish, and in the fly *Drosophila melanogaster*, in a series of host cells, and in a bacterial killing assay.

2 Materials

1. Plasmids: pUTmini-Tn5 Tc [10], pUTmini-Tn5 TcGFP [11], pTZ18R (GE Healthcare, Baie d'Urfé, Québec, Canada), pPS856, pDONR221, pEX18ApGw [12], pUCP19 [13].
2. Oligonucleotides for tag construction and universal primers for multiplex PCR are listed in Table 1.
3. Oligonucleotide buffer: 10× Medium salt buffer (10 mM Tris–HCl pH 7.5, 10 mM $MgCl_2$, 50 mM NaCl, 1 mM DTT).
4. dNTPs (dATP, dGTP, dCTP, dTTP).
5. HotStartTaq DNA polymerase (Qiagen, Mississauga, ON, Canada) and HiFi Platinum Taq (Invitrogen, Burlington, ON, Canada).
6. Restriction enzymes, restriction enzymes buffers (10× NEB 1, 2, 3, 10× BSA (1 mg/ml), T4 DNA polymerase, and T4 DNA ligase (New England Biolabs).
7. T4 DNA ligase 10× buffer (NEB, New England Biolabs, Mississauga, ON, Canada).
8. Micropure-EZ pure, microcon 30, microcon PCR (Millipore, Nepean, ON, Canada).
9. *Pseudomonas aeruginosa* strain LESB58 [7], *Escherichia coli* strains, S17-1 λ *pir*, DH5α, ElectroMax DH10B (Invitrogen), One Shot MAX Efficiency DH5α-T1^r (Invitrogen).
10. 1 mm and 2 mm electroporation gap cuvettes.
11. Bacterial growth media: Tryptic soy broth (TSB), Brain Heart Infusion (BHI), Tryptic soy agar (TSA), Mueller Hinton agar (MHA), Pseudomonas isolation agar (PIA), BHI agar.
12. Antibiotics: ampicillin (Ap), tetracycline (Tc), gentamicin (Gm), carbenicillin (Cb), chloramphenicol (Cm).
13. TE PCR buffer (10 mM Tris–HCl pH 7.4; EDTA 0.1 mM).
14. 10× HotStartTaq DNA polymerase reaction buffer with Tris–HCl, KCl, $(NH_4)_2SO_4$, 15 mM $MgCl_2$, pH 8.7 (Qiagen).
15. 10 pmoles oligonucleotide tags, universal primers listed in Table 1.
16. Mineral oil.
17. Agarose LM Nusieve GTG (FMC, Rockland, Maine).
18. Standard gel electrophoresis grade agarose, 1× Tris-borate EDTA buffer, and 0.5 μg/ml ethidium bromide solution.

Table 1
Nucleotide sequences of the oligonucleotides used for construction of signature tags (1–24) and sequences of the two universal primers for multiplex PCR-based STM

Primer	Nucleotide sequence
1	GTACCGCGCTTAA*ACGTTCA*G
2	GTACCGCGCTTAA*ATAGCCT*G
3	GTACCGCGCTTAA*AAGTCTC*G
4	GTACCGCGCTTAA*TAACGTG*G
5	GTACCGCGCTTAA*ACTGGTA*G
6	GTACCGCGCTTAA*GCATGTT*G
7	GTACCGCGCTTAA*TGTAACC*G
8	GTACCGCGCTTAA*AATCTCG*G
9	GTACCGCGCTTAA*TAGGCAA*G
10	GTACCGCGCTTAA*CAATCGT*G
11	GTACCGCGCTTAA*TCAAGAC*G
12	GTACCGCGCTTAA*CTAGTAG*G
13	CTTGCGGCGTATT*ACGTTCA*G
14	CTTGCGGCGTATT*ATAGCCT*G
15	CTTGCGGCGTATT*AAGTCTC*G
16	CTTGCGGCGTATT*TAACGTG*G
17	CTTGCGGCGTATT*ACTGGTA*G
18	CTTGCGGCGTATT*GCATGTT*G
19	CTTGCGGCGTATT*TGTAACC*G
20	CTTGCGGCGTATT*AATCTCG*G
21	CTTGCGGCGTATT*TAGGCAA*G
22	CTTGCGGCGTATT*CAATCGT*G
23	CTTGCGGCGTATT*TCAAGAC*G
24	CTTGCGGCGTATT*CTAGTAG*G
pUTgfpR2	ATCCATGCCATGTGTAATCCC
tetR1	CCATACCCACGCCGAAACAAG

Each 21-mers has a T_m of 64 °C and permits PCR amplification in one step when the 3 primer combinations are used for multiplex screening. Two sets of consensus 5′-ends comprising the first 13 nucleotides have higher ΔGs for optimizing PCR. Twelve variable 3'-ends in capital italic letters define tag specificity and allow amplification of specific DNA fragments. The set of 24 21-mers representing the complementary DNA strand in each tag is not represented and can be deduced from the sequences present. Single colonies are selected, purified, and screened by colony PCR using 10 pmol of pUTgfpR2 and tetR1 as the 3′ primers designed in the transposon resistance gene for multiplex PCR

19. MF-Millipore membrane filters 0.025 μm, 25 mm (Millipore).
20. Sterile 1× phosphate-buffered saline (PBS) (137 mM NaCl, 3 mM KCl, 10 mM Na_2HPO_4, 1.3 mM KH_2PO_4 pH 7.4).
21. 2 ml 96-well plates (QIAGEN).
22. Sprague Dawley rats 450–500 g, male.
23. Polytron homogenizer (Kinematica AG, Luzern, Switzerland).
24. QIAGEN Dneasy Tissue kit (Qiagen).
25. QIAfilter plasmid midi kit (Qiagen).
26. QIAquick gel extraction kit (Qiagen).
27. Quant-iT Picogreen ds DNA reagent and kit (Invitrogen).
28. Gateway BP Clonase II Enzyme Mix (Invitrogen).
29. Gateway LR Clonase II Enzyme Mix (Invitrogen).
30. DNA sequencing service and bioinformatics software.

3 Methods

STM is divided into two major steps: (1) the *construction of a library* of tagged mutants by transposon mutagenesis which implicates the synthesis and ligation of DNA tags into a specific site, transfer of the transposon into the recipient host, selection of transconjugants, and arraying of the mutants, and (2) a step for *screening* and comparative PCR analysis of mutants either absent or enhanced in a model used for screening (negative or positive-based STM assay). When applying STM, one must take into consideration that insertion into an essential gene gives a lethal phenotype. These genes cannot be identified by STM. All mutants selected require several rounds of screening in the system used, testing for auxotrophy and the estimation changes in the level of gene expression (diminished or enhanced and confirmed by qPCR) for a particular mutant when compared to the wild type.

The methods outlined below are as follows: (Subheading 3.1) the construction of tagged plasmids including tag annealing, plasmid preparation, plasmid and tag ligation, and electroporation; (Subheading 3.2) construction of libraries of tagged mutants by conjugation including transposon mutagenesis; (Subheading 3.3) cloning, sequencing, and analysis of disrupted genes; and (Subheading 3.4) construction of gene knockouts for selected STM mutants. Screening of STM mutants in models of chronic and acute lung infections and a CI analysis of selected mutants to estimate the level of attenuation or to enhance virulence are also provided.

3.1 Construction of Tagged Plasmids

Bar-coding for PCR-based STM involves the design of 24 primer pairs of 21-mers (Table 1) synthesized as complementary DNA strands for cloning into the mini-Tn5 plasmid vectors in Fig. 1. The two sets of plasmids used here contained 24 tags and are

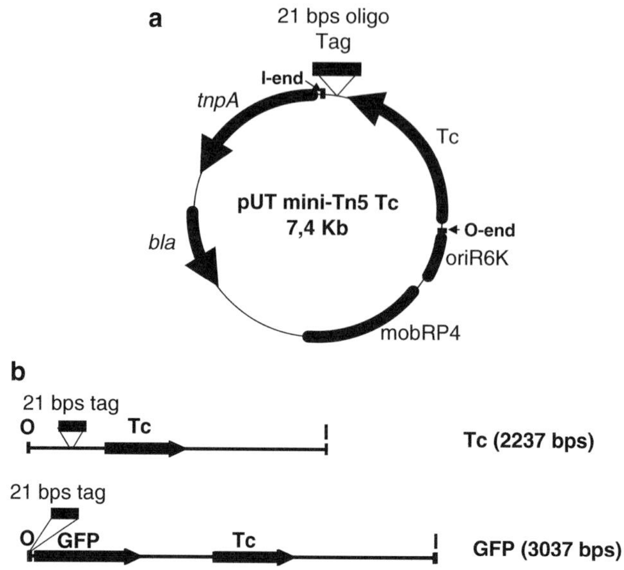

Fig. 1 (**a**) Physical and genetic maps of the pUT plasmid and the miniTn*5*Tc and miniTn*5*GFP transposons used. The transposons are located on a R6K-based suicide delivery plasmid pUT where the Pi protein is furnished by the donor cell (*E. coli* S17-1 λ pir); the pUT plasmid provides the IS50R transposase in *cis* but the *tnp* gene is external to the mobile element and whose conjugal transfer to recipients is mediated by RP4 mobilization functions in the donor [17]. (**b**) The elements are represented by *thick black lines*, inverted repeats are indicated as *vertical boxes* and genes are indicated by *arrows*. This collection of Tn5-derived minitransposons has been constructed that simplifies substantially the generation of insertion mutants, in vivo fusions with reporter genes, and the introduction of foreign DNA fragments into the chromosome of a variety of Gram negative bacteria. The miniTn*5* consists of genes specifying resistance to Tc and GFP with unique cloning sites for tag insertion flanked by 19-base-pair terminal repeats, the I and the O ends. Abbreviations: *I* and *O* inverted repeat ends; *Tc* tetracycline, *GFP* green fluorescent protein

repeatedly used to construct two sets of 24 libraries as shown in Fig. 2a giving a total of 96 STM mutants per master plate. DNA amplification using a specific tag as a PCR primer coupled to primers specific to the Tc and Tc GFP genes gives amplified DNA fragments of 980 bps and 820 bps by multiplex PCR as depicted in Fig. 2b. Multiplex PCR products obtained from arrayed bacterial clones in vitro can be compared to the amplified DNA products obtained after in vivo passage. These PCR products are visible in agarose gels as 980, 820 bps amplified products absent in defective STM mutants (Fig. 2b). The strategies used for further studies of STM mutants having no particular change in phenotype or having a specific change in phenotype is summarized in Fig. 3.

The oligonucleotides were designed as tags following three basic rules: (1) similar T_m of 64 °C to simplify tag comparisons by using one step of PCR reactions; (2) invariable 5′-ends with higher

ΔG than at the 3′-end to optimize PCR amplification reactions; (3) a variable 3′-end for an optimized yield of specific amplification product from each tag [14, 15]. The 21-mers are annealed as double stranded DNA, and are cloned into mini-Tn5 DNA. This collection of transposon can be used with any bacterial system which can conjugate with *E. coli* as a donor and is available upon request.

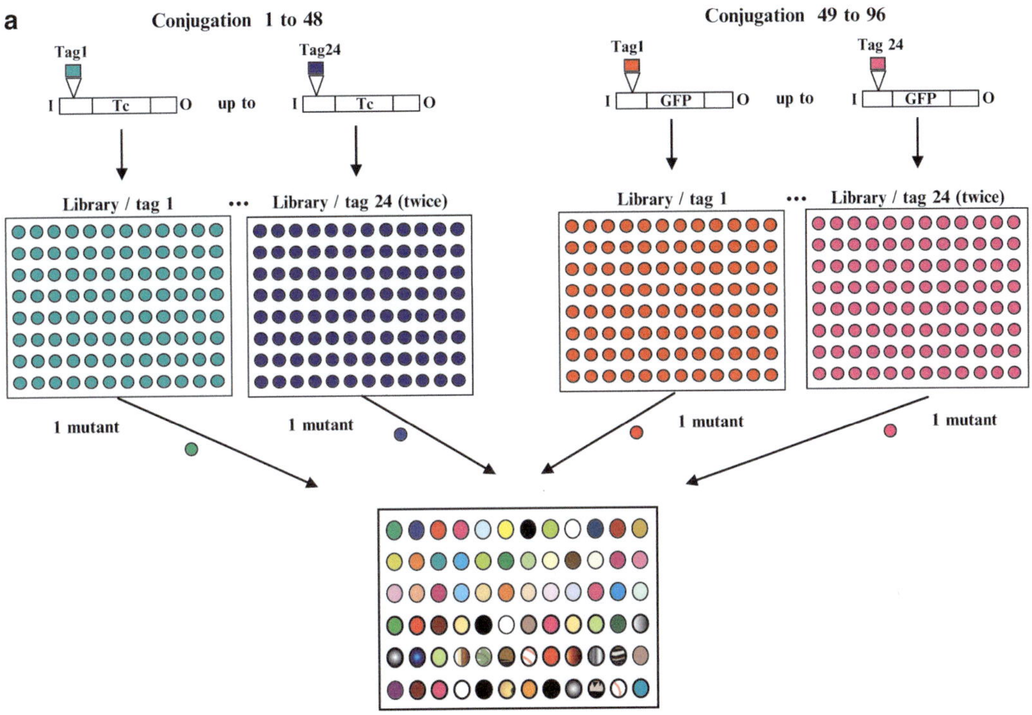

Fig. 2 (**a**) Construction of master plates of P. aeruginosa STM LESB58 mutants for in vitro and in vivo screening by PCR-based signature tagged-mutagenesis. Each master plate contains a collection of 96 mutants having unique chromosomal transposon insertions and selected from arrayed mutants obtained by conjugation. As depicted above, each conjugation set for a transposon is done using a specific marker (Tetracycline, Tc and Tc-GFP, green fluorescent protein) containing 48 tags. Selection is based upon antibiotic resistance markers and PCR for each set of specific tag. The *colors in plates* indicate a particular tag; the *color of bacteria in the master plate* represents a unique mutant with a transposon insertion. The *open-boxed lines* represent each transposon and I and O ends inverted repeats are indicated. The pUTminiTn5Tc and GFP vectors were used. (**b**) Sequence Tagged Mutagenesis: Comparison of mutants recovered from input pool with those recovered from the rat lung model. Comparative analysis between the in vitro and in vivo STM LESB58 mutant pools using multiplex PCR. An aliquot is kept as the in vitro pool and a second aliquot from the same preparation is used for passage into the rat lung for negative selection. At determined time points of infection, bacteria are recovered from the lung, and they constitute the in vivo pool. The in vitro and in vivo pools are used to prepare DNA in 48 PCR multiplex reactions using the 48 specific 21-mers tags and the Tc and Gfp-specific primers. Comparisons between in vitro and in vivo multiplex PCR products is done by agarose gel electrophoresis for identification of mutants absent in vivo (indicated by the *red halos* in lanes 5, 7, 15, and 24). The PCR products of 980 and 820 bps correspond to Tc and GFP and tag-specific PCR products, respectively. Each mutant is confirmed by a specific PCR, resistance makers are cloned and flanking regions sequenced to identify the inactivated gene

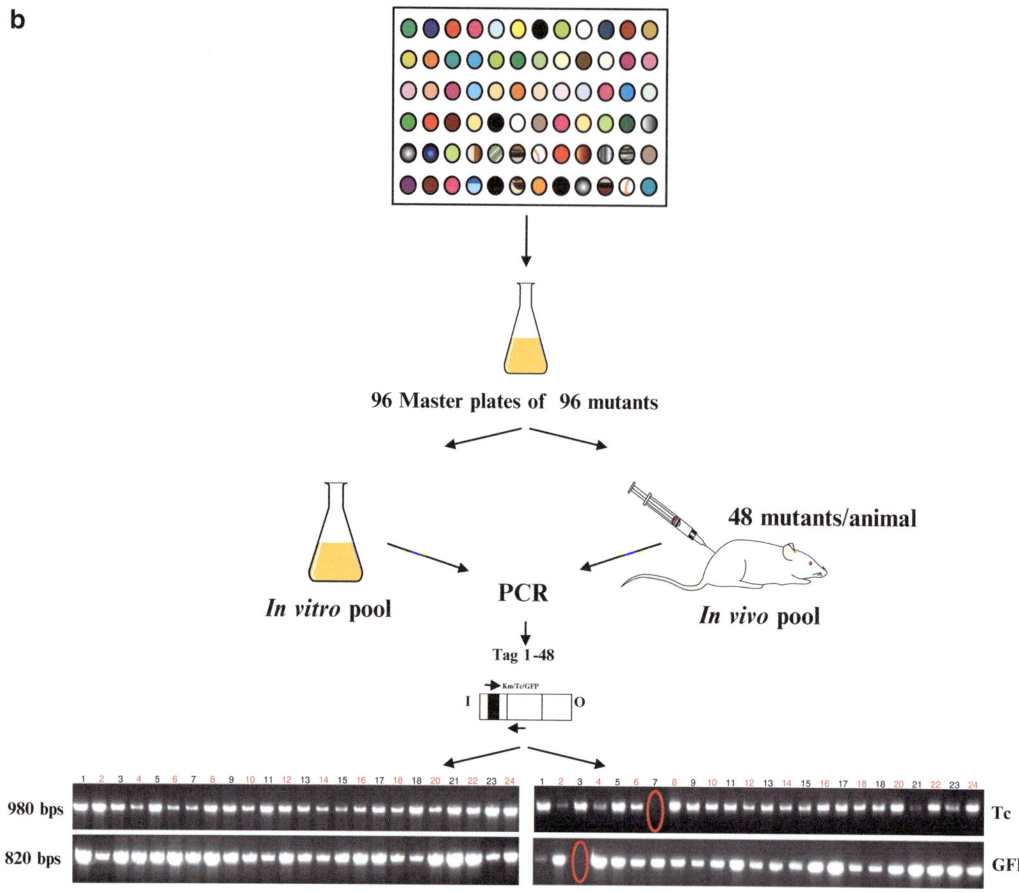

Fig. 2 (continued)

3.1.1 Tag Annealing

A collection of 24 defined 21-mers oligonucleotides should be synthesized along with their complementary DNA strands using the templates listed in Table 1. Annealing reactions contained 50 pmoles of both complementary oligonucleotides in 100 μl of 1× medium salt buffer. This oligonucleotides mixture is heated 5 min. at 95 °C, left to cool slowly at room temperature in a block heater, and kept on ice.

3.1.2 Plasmid Purification and Preparation for Tag Ligation

On a routine basis, we use the Qiagen system for plasmid preparation. DNA manipulations were performed by standard recombinant DNA procedures [16].

1. 20 μg of each pUTmini-Tn*5* plasmid DNAs is digested with 20 units of *Kpn*I in 40 μl of 1× NEB 1 buffer containing 1× BSA, incubated for 2 h at 37 °C and the enzyme is inactivated for 20 min at 65 °C.

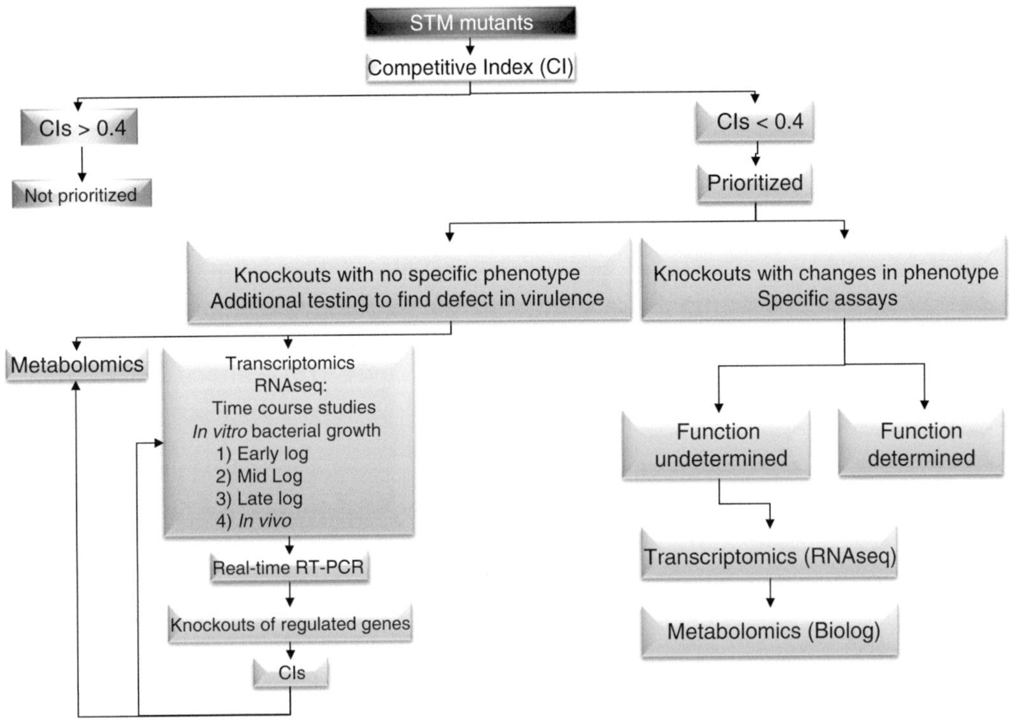

Fig. 3 General strategy for prioritization of STM mutants by construction of knockouts and in vivo screening with the competitive index (CI) analysis. Knockouts that have a CI value of <0.4 are known to be highly attenuated in vivo and will be analyzed further. Knockouts will then be prioritized for analysis of gene function using a combination of specific assays. Additional analysis will include transcriptomics profiling and other biological assays including metabolomics studies

2. Extremities are blunted with T4 DNA polymerase by adding 4 nmoles of each dNTPs and 5 units of T4 DNA polymerase.
3. Purify each blunted plasmid DNA to eliminate endonuclease and T4 DNA polymerase reactions with micropure-EZ and microcon 30 systems in a single step as described by the manufacturer's protocol.

3.1.3 Plasmid and Tag Ligation and Electroporation

1. 0.04 pmoles of each plasmid are ligated to 1 pmole of double stranded DNA tags in a final volume of 10 μl of T4 DNA ligase 1× buffer containing 400 units of T4 DNA ligase. Note that 24 ligation reactions are done for each plasmid which implies 72 single reactions, 72 electroporations, and 72 PCR analyses.
2. Ligated products are purified using microcon PCR (Millipore) as described by the manufacturer's instructions and resuspended in 5 μl of H_2O.
3. The 5 μl solution containing ligated products are transformed into *E. coli* S17-1 λ pir by electroporation using a Bio-Rad

apparatus (2.5 kV, 200 Ω, 25 μF) in a 2 mm electroporation gap cuvette. After electroporation, 0.8 ml of SOC is added to the bacterial preparation and the solution is transferred in culture tubes for incubation for 1 h at 37 °C.

4. Transformed bacteria containing tagged plasmids are selected on TSB supplemented with 50 μg/ml of Ap and 20 μg/ml of Tc by plating 100 μl of electroporated cells.
5. Single colonies are selected, purified, and screened by colony PCR in 50 μl reaction volumes containing: 10 μl of boiled bacterial colonies in 100 μl of TE PCR (10 mM Tris–HCl pH 7.4; EDTA 0.1 mM); 5 μl of 10× HotStartTaq polymerase reaction buffer; 1,5 mM $MgCl_2$; 200 μM of each dNTPs; 10 pmoles of one of the oligonucleotides used for tags as a specific 5′ primer and 10 pmoles of the pUTgfpR2 and the tetR1 (Table 1) as the universal 3′ primer; 2.5 units HotStartTaq polymerase (Invitrogen). Thermal cycling conditions are for touchdown PCR including a hot start for 15 min at 95 °C, 22 cycles at 95 °C for 1 min, decrease the temperature 70 to 60 °C after cycle 2 by 1 °C every 2 cycles, for 1 min, and at 72 °C for 1 min, followed by the 7 min final elongation step at 72 °C. Amplified products using 10 μl aliquots were analyzed by electrophoresis in a 1 % agarose gel, 1× Tris–borate EDTA buffer and stained for 10 min in 0.5 μg/ml ethidium bromide solution [16].

3.2 Construction of Libraries of Tagged Mutants

A series of suicide pUT plasmids carrying mini-Tn*5*Tc and mini-Tn*5*Tc-GFP each with a specific tag were transferred by conjugation [17] into the targeted bacteria *P. aeruginosa* giving libraries of mutants arrayed into 96-well master plates (Fig. 2a, b). Mutants from the same pool are grown separately overnight at 37 °C. Aliquots of these cultures are pooled and a sample is kept for PCR analysis (the in vitro pool). A second sample from the same pool is used for the in vivo passage.

3.2.1 Conjugation and Transposon Mutagenesis

1. *E. coli* S17-1 λ *pir* containing the pUTmini-Tn5 tagged plasmids is used as a donor for conjugal transfer into the recipient strain. The ratio of donor to recipient bacterial cells to obtain the maximum of exconjugants should be determined in preliminary experiments. For *P. aeruginosa*, we used 1 donor to 10 recipient cells. Cells are mixed and spotted as a 50 μl drop on a membrane filter placed on a non-selective BHIA plate. Plates are incubated at 30 °C for 24–48 h.
2. Filters are washed with 10 ml of PBS saline to recover bacteria.
3. Aliquots of 100 μl of the PBS solution containing exconjugants are plated on 5 BHIA plates supplemented with the appropriate antibiotic to select for the strain. For *P. aeruginosa*

LESB58, we use Tc (45 μg/ml). Plates are incubated for 24–48 h at 37 °C.

4. Tc resistant *P. aeruginosa* exconjugants are arrayed as libraries of 96 clones in 2 ml 96-wells plate in 1.5 ml of BHI supplemented with km and appropriate antibiotic. The 2 ml 96-wells plates are incubated for 24–48 h at 37 °C.
5. One mutant from each library is picked to form 96 pools of unique tagged mutants (Fig. 2a) contained in the 2 ml 96-well plates.

3.3 In Vivo Screening

3.3.1 Preparation of Arrayed Bacteria for In Vitro PCR

Pools of 96 bacteria are grown separately in the 2 ml 96-well microplates and separate in two parts: in vitro and in vivo part (*see* Fig 2b). In vitro pool is conserved for the comparative PCR and in vivo pool used to inoculate animals in vivo. The rat model of chronic lung infection is used for negative selection to identify the mutants defective for the maintenance in vivo. Using the pools of 96 bacteria per animal allows reducing considerably the number of animals for in vivo screening. *See* Chapter 58 for details for in vivo screening of tagged mutants (*see* Subheadings 3.1, 3.2, 3.4, or 3.5 of Chapter 58).

Bacteria are recovered from the homogenized lungs and used as a template in 24 distinct multiplex PCR reactions. The in vitro pool is used as control. PCR products are separated by gel electrophoresis where the presence or absence of DNA fragments and their size are compared between the in vitro and in vivo pools (*see* Fig 2b). Missing mutants after in vivo passage are incapable to be maintained and are considered as attenuated in vivo.

1. The 96 mutants from the same pool are grown separately 24–48 h at 37 °C in 200 μl of TSB containing Tc in 96-well microtiter plates.
2. Aliquots of these cultures are pooled.
3. A first sample is diluted from 10^{-1} to 10^{-4}, and plated on BHIA supplemented with Tc for transposon marker.
4. After overnight incubation at 37 °C, 10^4 colonies are recovered in 5 ml of PBS and a sample of 1 ml is removed for PCR and called the in vitro pool.
5. The 1 ml in vitro pool sample is spun down and the cell pellet is resuspended in 1 ml of TE PCR buffer.
6. The in vitro pool is boiled for 10 min, spun down, and 2 μl of supernatant is used in PCR analysis as described above (See Section 3.1.3, **step 5**).
7. A second sample from the pooled cultures is used to inoculate the in vivo system selected for screening. *See* Chapter 58 for details for in vivo screening of tagged mutants in the rat model of chronic lung infection.

3.4 Cloning and Analysis of Disrupted Genes from Attenuated Mutants

Chromosomal DNA from attenuated mutants is prepared using the QIAGEN genomic DNA extraction kit as described in the manufacturer's protocol.

1. Chromosomal DNA (1–5 μg) is digested with endonuclease (in our case *Pst*I) giving a large range of DNA fragment sizes.
2. Digested chromosomal DNA is cloned into pTZ18R (or any vector of choice) predigested with the corresponding endonuclease and ligation reactions are done as follows:
3. 1 μg of digested chromosomal DNA is mixed with 50 ng of digested pTZ18R in 20 μl of 1× T_4 DNA ligase buffer with 40 units of T_4 DNA ligase.
4. Incubate overnight at 16 °C.
5. Ligated products are purified using microcon PCR (Millipore) as described by the manufacturer's instructions and resuspended in 5 μl of H_2O.
6. The 5 μl recombinant plasmid solution is used for electroporation in *E. coli* ElectroMAX DH10B as recommended by the manufacturer.
7. All the electroporation cells are spun down and resuspended in 100 μl of BHI and plated on a selective plate. Colonies are recovered by scraping using with 5 ml of BHI.
8. Bacteria containing pTZ18R containing an insertion of genomic DNA encoding the transposon antibiotic resistance marker from miniTn5Tc and miniTn5GFP are plated on TSA with Tc (20 μg/ml), respectively.
9. Clones are kept and purified for plasmid analysis.
10. Plasmid DNA is prepared with QIAGEN midi preparation kit as described by the manufacturer.
11. These plasmids are sequenced using the complementary primer of the corresponding tagged mutant or the three conserved transposon primers encoding antibiotic resistance. Automated sequencing is done as suggested by the manufacturer.
12. DNA sequences obtained are assembled and subjected to database searches using BLAST included in the GCG Wisconsin package (version 11.0). Similarity searches with complete genomes can be performed at NCBI using the microbial genome sequences at http://www.ncbi.nlm.nih.gov or in this specific case for *P. aeruginosa* http://www.pseudomonas.com.

Since miniTn5 insertions may give polar mutations (except for insertions in genes at the end of an operon), a method is essential to construct gene knockouts in *P. aeruginosa* giving a clean genetic background. Construction of deletion alleles is most often tagged with an antibiotic resistance gene on a suicide plasmid and followed by recombination of the plasmid-borne deletions into the chromosome, usually after conjugal transfer of the suicide plasmid [18]. We use a rapid method for generation of unmarked *P. aeruginosa* deletion mutants. The method was applied to deletion of 25 *P. aeruginosa* genes encoding transcriptional regulators of the GntR family [12]. The method that we now use can be summarized as follows: Three partially overlapping DNA fragments are amplified, encoding the 5′ and 3′ regions of the target gene plus gentamicin cassette and then spliced together in vitro by overlap extension PCR. The resulting DNA fragment is cloned in vitro into the Gateway vector pDONR221 and then recombined into the Gateway-compatible gene replacement vector pEX18ApGW. The plasmid-borne deletions are next transferred to the *P. aeruginosa* chromosome by homologous recombination. Unmarked deletion mutants are finally obtained by Flp-mediated excision of the antibiotic resistance marker [12].

Acknowledgments

R. C. Levesque is a research scholar of exceptional merit from the Fond de Recherche du Québec en Santé (FRQS). His laboratory is funded by the Canadian Institute for Health Research (CIHR), a CIHR-UK team grant, the CIHR-FRQS-Québec Respiratory Health Network (RSR), the Natural Sciences and Engineering Research Council of Canada (NSERC), by Genome Québec, by the Fonds de recherche du Québec nature et technologies, and by the Alberta Innovates Bio Solutions program.

References

1. Bianconi I, Milani A, Cigana C, Paroni M, Levesque RC, Bertoni G, Bragonzi A (2012) Positive signature-tagged mutagenesis in *Pseudomonas aeruginosa*: tracking patho-adaptive mutations promoting airways chronic infection. PLoS Pathog 7:e1001270
2. Hensel M, She JE, Gleeson C, Jones MD, Dalton E, Holden DW (1995) Simultaneous identification of bacterial virulence genes by negative selection. Science 269:400–403
3. Lehoux DE, Sanschagrin F, Levesque RC (2001) Discovering essential and infection-related genes. Curr Opin Microbiol 4:515–519
4. Lemieux AK-I, Boyle B, Fothergill JL, Filion G, Dekimpe V, Laroche J, Winstanley C, Charette SJ, Déziel E, Levesque RC (2013) Genes from prophage islands of the accessory genome of the *Pseudomonas aeruginosa* LESB58 epidemic strain are essential for chronic pulmonary infection. PLoS Pathog (submitted)
5. Mazurkiewicz P, Tang CM, Boone C, Holden DW (2006) Signature-tagged mutagenesis: barcoding mutants for genome-wide screens. Nat Rev 7:929–939
6. Potvin E, Lehoux DE, Kukavica-Ibrulj I, Richard KL, Sanschagrin F, Lau GW, Levesque RC (2003) *In vivo* functional genomics of *Pseudomonas aeruginosa* for high-throughput screening of new virulence factors and antibacterial targets. Environ Microbiol 5: 1294–1308
7. Winstanley C, Langille MG, Fothergill JL, Kukavica-Ibrulj I, Paradis-Bleau C, Sanschagrin F, Thomson NR, Winsor GL, Quail MA, Lennard N, Bignell A, Clarke L, Seeger K, Saunders D, Harris D, Parkhill J, Hancock RE, Brinkman FS, Levesque RC (2009) Newly introduced genomic prophage islands are critical determinants of *in vivo* competitiveness in the Liverpool Epidemic Strain of *Pseudomonas aeruginosa*. Genome Res 19:12–23
8. Wahyudi AT (2007) Rapid and simple amplification of genomic DNA sequences flanking transposon. Microbiol Indonesia 1:4

9. Stover CK, Pham XQ, Erwin AL, Mizoguchi SD, Warrener P, Hickey MJ, Brinkman FS, Hufnagle WO, Kowalik DJ, Lagrou M, Garber RL, Goltry L, Tolentino E, Westbrock-Wadman S, Yuan Y, Brody LL, Coulter SN, Folger KR, Kas A, Larbig K, Lim R, Smith K, Spencer D, Wong GK, Wu Z, Paulsen IT, Reizer J, Saier MH, Hancock RE, Lory S, Olson MV (2000) Complete genome sequence of *Pseudomonas aeruginosa* PA01, an opportunistic pathogen. Nature 406: 959–964
10. De Lorenzo V, Herrero M, Jakubzik U, Timmis KN (1990) Mini-Tn5 transposon derivatives for insertion mutagenesis, promoter probing, and chromosomal insertion of cloned DNA in gram-negative eubacteria. J Bacteriol 172:6568–6572
11. Matthysse AG, Stretton S, Dandie C, McClure NC, Goodman AE (1996) Construction of GFP vectors for use in gram-negative bacteria other than *Escherichia coli*. FEMS Microbiol Lett 145:87–94
12. Choi KH, Schweizer HP (2005) An improved method for rapid generation of unmarked *Pseudomonas aeruginosa* deletion mutants. BMC Microbiol 5:30
13. Schweizer HP (1991) Improved broad-host-range lac-based plasmid vectors for the isolation and characterization of protein fusions in *Pseudomonas aeruginosa*. Gene 103:87–92
14. Kwok S, Kellogg DE, McKinney N, Spasic D, Goda L, Levenson C, Sninsky JJ (1990) Effects of primer-template mismatches on the polymerase chain reaction: human immunodeficiency virus type 1 model studies. Nucleic Acids Res 18:999–1005
15. Rychlik W (1993) Selection of primers for polymerase chain reaction, vol 15. Humana Press, Totowa, NJ
16. Sambrook J, Russell DW (2001) Molecular cloning: a laboratory manual, 3rd edn. Cold Spring Harbor Laboratory Press, Cold Spring Harbor, NY
17. Simon R, Priefer U, Pühler A (1983) A broad host range mobilization system for *in vivo* genetic engineering: transposon mutagenesis in gram negative bacteria. Bio/Technology 1:784–791
18. Schweizer HP (1992) Allelic exchange in *Pseudomonas aeruginosa* using novel ColE1-type vectors and a family of cassettes containing a portable *oriT* and the counter-selectable *Bacillus subtilis sacB* marker. Mol Microbiol 6:1195–1204

Chapter 42

Construction of a *Pseudomonas aeruginosa* Genomic DNA Library

Christophe Bordi

Abstract

Although the completion and annotation of the entire genomic DNA sequence of *Pseudomonas aeruginosa* PAO1 strain has been carried out, an important number of genes are still of unknown function and many genetic elements involved in various regulatory pathways like small RNA are still unrevealed. One of the strategies to identify gene function or genetic elements is the construction and utilization of DNA genomic library. Here, we describe the construction a *P. aeruginosa* DNA genomic library.

Key words DNA library, Gene function

1 Introduction

Although the genomic DNA sequence of the laboratory reference strain *Pseudomonas aeruginosa* PAO1 has been completed and annotated in 2000 [1], an important number of genes are still of unknown functions. Thus, the knowledge of DNA sequence of an open reading frame (ORF) is not enough to determine gene function and/or the biological process in which these genes are involved. PAO1 genome annotation was performed using classical bioinformatics processes based on automated identification of ORFs which encode for proteins with a minimal size of 100 amino acids (aa) followed by a function assignment based on protein sequence similarities with proteins sequences present in databases. These bioinformatic approaches are rapid and powerful for identification of genes encoding proteins but inefficient for identification of "small proteins" (size < 100 aa) and of genetic elements involved in various regulatory pathways like small RNA (sRNA). Genes encoding for sRNA, which by definition are not open reading frames, are not predictable by using classical bioinformatic approaches. Numerous biocomputational methods have been developed for predicting sRNA genes in genome; however, the

Alain Filloux and Juan-Luis Ramos (eds.), *Pseudomonas Methods and Protocols*, Methods in Molecular Biology, vol. 1149, DOI 10.1007/978-1-4939-0473-0_42, © Springer Science+Business Media New York 2014

number of sRNAs predicted by the tools is moderate and different methods that predict sRNAs often give disjointed result [2].

One of the best methods used for the identification of "small proteins" or small RNA in bacteria is thus the construction and expression of a library of randomly generate genomic DNA fragments [3–6]. A good example was provided by the identification of the small RNA RsmY in *Pseudomonas fluorescens* [3, 7]. Indeed, Rowley and collaborators reported that a DNA fragment (24.4 kb) cloned from the wild-type strain *Pseudomonas syringae pv. phaseolicola* abolishes thermoregulation of phaseolotoxin biosynthesis in the wild-type strain [3]. This DNA fragment, conserved in *P. fluorescens*, has been shown to contain a gene, unraveled during annotation genome, encoding a small RNA called RsmY, which has been described as involved in secondary metabolism control in this organism [7]. Thus, the knowledge concerning genes with unknown functions or noncoding genetic elements participating in *P. aeruginosa* physiology can be incredibly increased with genetic strategies. One of these strategies is the construction and use of a DNA genomic library for high-throughput screening. The present chapter describes the construction a DNA genomic library in *P. aeruginosa*.

2 Materials

2.1 Agarose Gel Components

1. TAE 50× Solution: dissolve 242.2 g of Tris powder in 100 mL water; add successively 57.1 mL of acetic acid and 18.5 g of EDTA (Ethylene Diamine Tetraacetic Acid). Complete to 1 L with water and adjust the pH to 8.4.
2. Loading dye 6× solution: dissolve 4 g sucrose, 25 mg bromophenol blue, and 25 mg xylene cyanol in 10 mL water. Store at 4 °C to avoid contaminants growing in the sucrose.
3. Weigh 1.2 g of agarose powder and add 15 mL of 0.5× TAE solution, and then warm the solution at 80 °C until complete dissolution of the agarose powder.
4. Pour the agarose solution in the agarose gel system and let the gel solidify for at least 30 min at room temperature.

2.2 Culture Media

1. LB (Luria-Bertani) broth: dissolve in 1 L of water 10 g of NaCl, 10 g of tryptone, 5 g of yeast extract. Adjust pH to 7 with NaOH and autoclave at 121 °C for 20 min for sterilization. LB plates are prepared by adding 1.5 % Bacto agar (Difco™ ref 214050) supplemented with ampicillin at 100 μg/mL.

2.3 Chemical Competent Bacteria

1. $CaCl_2$ 50 mM solution: dissolve 0.277 g of $CaCl_2$ powder in 100 mL water and keep the solution at 4 °C until use.
2. $CaCl_2$ 50 mM with 15 % glycerol: dissolve 0.277 g of $CaCl_2$ powder; add 18.75 mL of 80 % glycerol solution and complete

to 100 mL with water and keep the solution at 4 °C until utilization.

3. The day before, inoculate in LB the bacteria (DH5α or equivalent *E. coli* strains) and incubated under agitation at 37 °C.
4. In a 500 mL flask, place 100 mL of fresh LB medium and add 2 mL of the overnight culture. Incubate the flask at 37 °C under agitation until the OD_{600nm} reaches a value between 0.4 and 0.6.
5. Pellet the 100 mL of culture, remove the supernatant, and resuspend the pellet in 50 mL of 50 mM $CaCl_2$ solution.
6. Place the tube on ice for 20 min.
7. Pellet the bacteria, remove the supernatant, and resuspend the pellet in 5 mL of 50 mM $CaCl_2$ with 15 % glycerol solution.
8. Divide the competent bacteria into 100 μL aliquots in sterile 1.5 mL eppendorfs and store at −80 °C until use.

2.4 *E. coli* Transformation

1. Thaw the competent bacteria on ice.
2. Place 50 or 100 μL of competent bacteria in a sterile 1.5 mL eppendorf tube and add the indicated volume of ligation reaction or plasmid.
3. Incubate the eppendorf tube for 45 min on ice and then place the tube at 42 °C in a water bath for 2 min and immediately on ice for 5 min.
4. Add to the reaction 900 μL of fresh LB medium and incubate at 37 °C under agitation (around 200 rpm) during 1 h.
5. Spread the transformed bacteria on the appropriate LB plates supplemented with ampicillin.

3 Methods

3.1 Generation of *P. aeruginosa* Genome Fragment Library

The different steps of the workflow for the construction of the library are summarized in Fig. 1. Plasmid and strains used are listed in Table 1.

1. To create a representative and random DNA library, the *P. aeruginosa* chromosome has to be digested with *Sau*3AI. Add 30 μL of purified *P. aeruginosa* genomic DNA at the concentration of 150 ng/μL in a 1.5 mL eppendorf tube and add 0.3 unit of *Sau*3AI restriction enzyme with the appropriate digestion buffer in a final volume of 35 μL. Incubate the eppendorf at 37 °C for 20 min and proceed immediately to the next steps in order to rapidly stop the digestion reaction (*see* **Note 1**).
2. Pour a six-well TAE 0.65 % agarose gel, and add 6 μL of loading dye 6× in the eppendorf tube that contains the 35 μL of the digestion reaction and mix well. Load the whole digestion

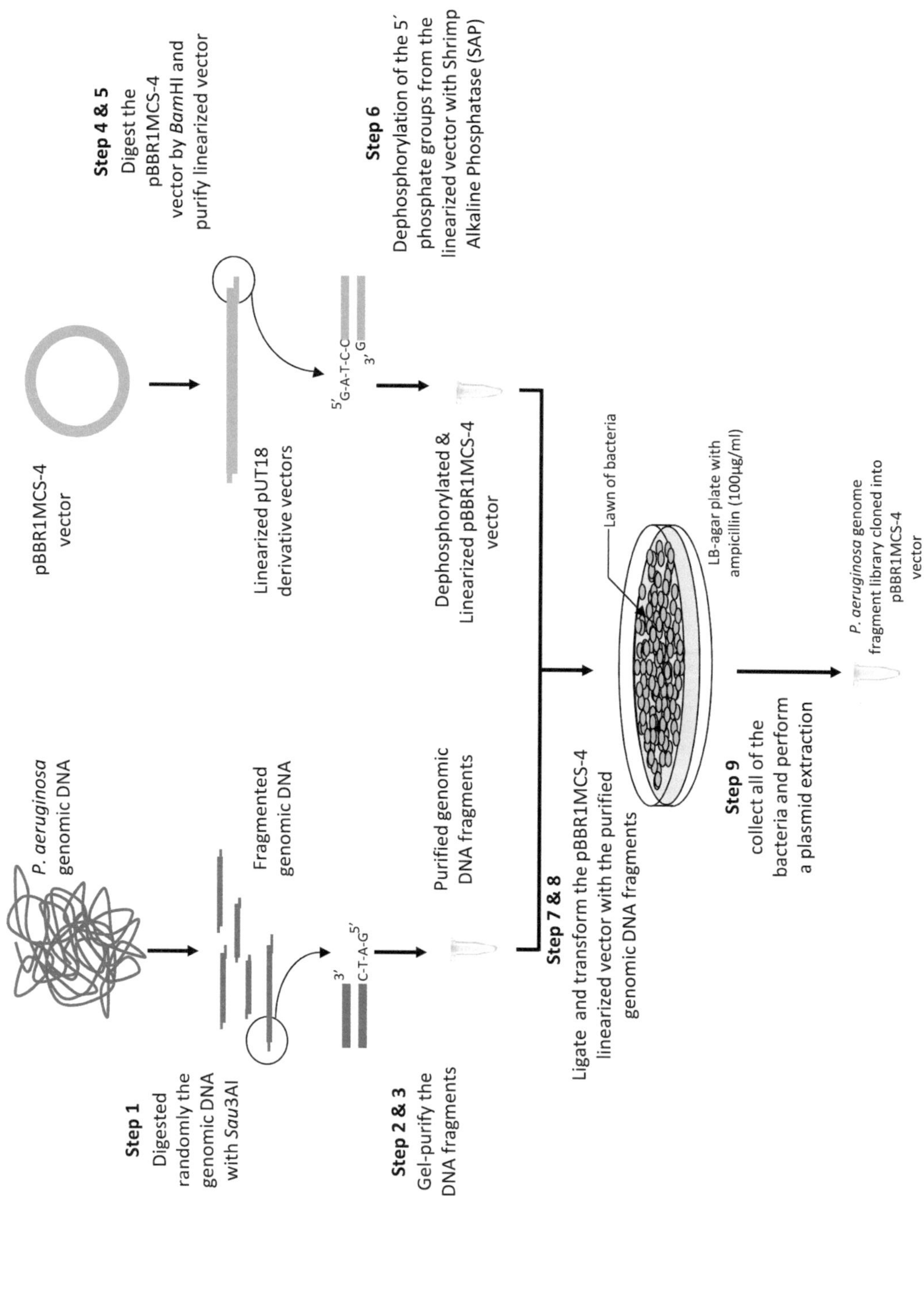

Fig. 1 Schematic representation of the different steps for construction of *P. aeruginosa* genome fragment library

Table 1
Strains and plasmids used

Strains/plasmids	Relevant characteristics	References
Strains		
DH5α	*endA1 hsdR17 supE44 thi-1 recA1 gyrA relA1 Δ(lacZYA-argF)U169 deoR (phi 80lacZ Δ M15)*	Invitrogen
Plasmids		
pBBR1MCS4	Broad-host-range plasmid, Ap^R	[9]

reaction mixed with the loading dye, and in the second well load a DNA ladder like the GeneRuler™ 1 kb DNA Ladder from Thermo Scientific-Fermentas (ref SM0313). Separate by electrophoresis in TAE 0.5× buffer at 100 V for 25 min or until the dye front has reached the bottom of the gel. Stain the gel for 10–15 min with an ethidium bromide solution at a concentration of 0.5 μg/mL and visualize DNA fragments under UV illumination. SYBR green or Gel Red dye solutions could be used as an alternative to ethidium bromide.

3. Under UV illumination, cut the loaded lane with a clean scalpel in order to collect the digested DNA fragments ranging in size from 500 to 5,000 bp (*see* **Note 2**). Gel-purify the DNA fragments with the QIAquick Gel Extraction Kit (QIAgen ref 28704) or equivalent and elute in a final volume of 40 μL water in order to concentrate the digested DNA (*see* **Note 3**). At this step, the digested genomic DNA can be stored at −20 °C for at least 6 months.
4. Add in a 1.5 mL eppendorf tube, 40 μL of the pBBR1MCS4 vector (Fig. 2) at a concentration of 200 ng/μL. The pBBR1MCS4 plasmid is a broad-host-range cloning vector replicative in *E. coli* as well as in *P. aeruginosa* strains (*see* **Note 4**). Add 1 unit of *Bam*HI restriction enzyme with the appropriate digestion buffer in a final volume of 50 μL. Incubate the reaction tube at 37 °C for 2.5 h (*see* **Note 5**).
5. Purify the digested pBBR1MCS4 vectors in order to remove the *Bam*HI enzyme and the digestion buffer by using QIAquick PCR Purification Kit (QIAgen ref 28104) or equivalent and elute in a final volume of 40 μL water in order to concentrate the linearized vector. At this step, the linearized vector can be stored at −20 °C for at least 6 months.
6. Add in a 1.5 mL eppendorf tube, 30 μL of the pBBR1MCS4 linearized vector at the concentration of 150 ng/μL and add 1 μL of Shrimp Alkaline Phosphatase (Roche ref 11 758 250 00) or equivalent with 4 μL of 10× dephosphorylation buffer in a final volume of 40 μL. Incubate the reaction tubes at 37 °C for 1 h (*see* **Note 6**).

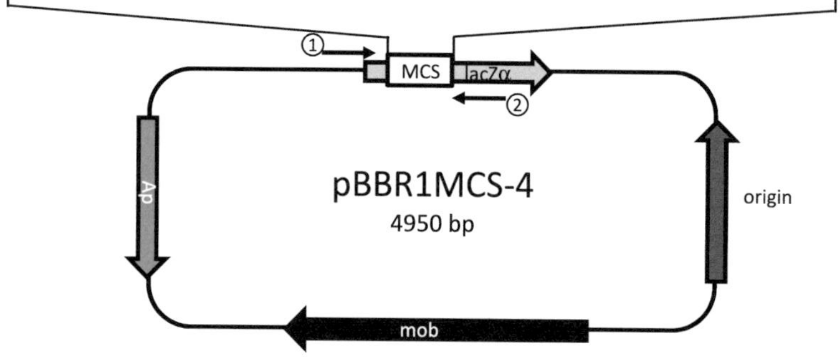

Fig. 2 Schematic representation of pBBR1MCS-4 plasmid. The *open arrow* represents the ampicillin (*Ap*[R]) resistance gene. The *grey light arrow* corresponds to the open reading frame of lacZα' sub-domain. The *open box* corresponds to the multicloning site (MCS) sequences in which only unique sites are mentioned and the origin of replication of the plasmid and *mob* genes are indicated by *grey light or black arrows. Black arrows* with numbers 1 or numbers 2 correspond respectively to the position of the Lib_pBBR1MCS4Up and Lib_ pBBR1MCS4Do primers

7. In order to remove the Shrimp Alkaline Phosphatase; purify the dephosphorylated pBBR1MCS4 vector by using the QIAquick PCR Purification Kit or equivalent.
8. Ligate the pBBR1MCS4 linearized vector with the resulting mixture of digested genomic DNA obtained at **step 3**. For this, in a 0.5 mL eppendorf tube, put 4 μL of the linearized vector at a concentration of 100 ng/μL, add 8 μL of the digested DNA at the concentration of 50 ng/μL, add 1.5 μL of T4 DNA Ligase [Roche® ref (11 635 379 001) or equivalent] with its supplied T4 DNA Ligation Buffer in a final volume of 15 μL and incubate overnight at 16 °C.
9. The whole resulting ligation mixture is transformed into *E. coli* DH5α competent cells or equivalent and spread the bacteria on two or three LB agar plates with ampicillin at 100 μg/mL. Incubate the plates overnight at 37 °C.
10. Collect all bacteria present at the surface of the different plates with a glass spreader and resuspend them in 5 mL of LB medium. Perform a plasmid extraction procedure using The Wizard® *Plus* SV Minipreps DNA Purification System (Promega ref A7100) or equivalent. At the end of this step, the plasmids collected represent the *Pseudomonas* genomic DNA library and can be stored at −20 °C (*see* **Note 7**).

3.2 Control of the Quality of the Bank and Determination of Its Size

Before using the library for any high-throughput assay, it is important to evaluate its quality. In particular, attention should be paid to the percentage of the genome covered and the percentage of the plasmids that have a fragment of *P. aeruginosa* genomic DNA inserted.

Table 2
Oligonucleotides used

Oligonucleotides	Sequence (5′ → 3′)
Library analysis	
Lib_pBBR1MCS4Up	AACAGCTATGACCATG
Lib_pBBR1MCS4Do	GTAAAACGACGGCCAGT

1. Transform 50 μL of DH5α *E. coli* chemical competent bacteria with 50 ng of plasmid library. Perform a serial dilution from 10^0 up to 10^{-5}, plate 100 μL of each dilution on LB agar plates supplemented with ampicillin (100 μg/mL) and incubate at 37 °C.
2. To determine the library size (I.E number of bacteria present in the library containing empty vector or vector with a *P. aeruginosa* DNA fragment inserted), count the number of colonies on each plate for each dilution, multiply this number by the dilution factor and calculate the mean number of colonies present per plate. Multiply this value by 10 to give the library size.
3. Determination of the percentage of plasmids that have inserted a fragment of *P. aeruginosa* genomic DNA: Randomly pick 100 clones from the dilution plates and perform a colony PCR on each clone using the oligonuleotides Lib_pBBR1MCS4Up and Lib_pBBR1MCS4Do (Table 2). This experiment will provide the percentage of empty plasmids and the mean size of the genomic fragments cloned in the different vectors.
4. The probability of having a particular fragment cloned in the library is therefore calculated using the following formula:

$$P = 1 - \left(1 - \frac{i}{G}\right)^{N}$$

where i is the mean insert size of the genomic fragment cloned, G the genome size, and N the number of clones obtained in the library [8].

4 Notes

1. Digestion of PAO1 genomic DNA is the crucial step for the construction of the library; the goal of this step is to cut the genomic DNA of *P. aeruginosa* in order to obtain a collection of DNA fragments with a range size between 500 and 5,000 kb. The quality and purity of the genomic DNA, but also the quality of the *Sau3AI* enzyme used could influence positively or negatively the efficiency of the digestion (that could lead to an insufficient or an excessive digestion of the genomic DNA).

Optimization of this step can be achieved by adding or reducing the quantity of *Sau3AI* or DNA concentration in the digestion mix or also by increasing or reducing the duration of the digestion step at 37 °C.

2. Cut the gel as fast as possible, in order to prevent DNA deterioration due to thymine dimer formation. When DNA is damaged in such a way, it cannot be replicated or transcribed and this reduces the coverage of the library.
3. At this step, monitor the DNA concentration by using a Nanodrop® system or equivalent DNA quantification method; the DNA fragment concentration after gel purification should not be below a concentration of 50 ng/μL. If the measured concentration is below this value: (1) redo the **steps 1–3** and pool with the previous fragment and/or concentrate the DNA fragment preparation, by using a vacuum concentrator like the SpeedVac® system.
4. The pBBR1MCS4 plasmid is a broad-host-range cloning vector that can be easily introduced in *P. aeruginosa* strain by mating or electroporation. Alternative vectors bearing *Bam*HI site in the multiple cloning site could be used for the construction of the library like the pBBR1MCS derivate [9], the pMMB67EH [10], or pUCP19 [11] vector.
5. Plasmid digestion by *Bam*HI will produce compatible cohesive ends (GATC) with *Sau3AI*, which will permit the correct cloning of the *P. aeruginosa* DNA fragments in the pBBR1MCS4 linearized vectors.
6. Another crucial step for the construction of the library is the dephosphorylation of the 5′ phosphate groups from the linearized vectors with SAP with the aim to prevent self relegation. This will influence the efficiency of the ligation step. A too short treatment with the SAP enzyme leads to an important percentage of plasmid self-religation. A too long treatment with the SAP enzyme heavily reduces the insertion of DNA fragments into the linearized plasmids. Optimization of this step can be achieved by increasing or reducing the quantity of SAP enzyme or DNA concentration in the digestion mix.
7. To ensure a renewal of plasmid libraries, retransform 100 μL of DH5α competent bacteria with 250 ng of plasmid library and redo the **steps** 7 and **8**.

Acknowledgements

The work of C.B. has been supported by the French Cystic Fibrosis Foundation (VLM), and CNRS institutional and ANR grant: Two-CompNet (2005–2009), as well as Europathogenomics 2005–2010 REX LSHB-CT-2005512061-EPG.

References

1. Stover CK, Pham XQ, Erwin AL et al (2000) Complete genome sequence of *Pseudomonas aeruginosa* PAO1, an opportunistic pathogen. Nature 406:959–964
2. Lu X, Goodrich-Blair H, Tjaden B (2011) Assessing computational tools for the discovery of small RNA genes in bacteria. RNA 9: 1635–1647
3. Rowley KB, Clements DE, Mandel M et al (1993) Multiple copies of a DNA sequence from *Pseudomonas syringae pathovar phaseolicola* abolish thermoregulation of phaseolotoxin production. Mol Microbiol 8:625–635
4. Kawakami K, Toma C, Honma Y (2000) Cloning, sequencing and expression of the gene encoding the extracellular metalloprotease of *Aeromonas caviae*. Microbiol Immunol 44:341–347
5. Wang Y, Ha U, Zeng L et al (2003) Regulation of membrane permeability by a two-component regulatory system in *Pseudomonas aeruginosa*. Antimicrob Agents Chemother 47:95–101
6. Ethier N, Talbot G, Sygusch J (1998) Gene cloning, DNA sequencing, and expression of thermostable beta-mannanase from Bacillus stearothermophilus. Appl Environ Microbiol 64:4428–4432
7. Valverde C, Heeb S, Keel C et al (2003) RsmY, a small regulatory RNA, is required in concert with RsmZ for GacA-dependent expression of biocontrol traits in *Pseudomonas fluorescens* CHA0. Mol Microbiol 50:1361–1379
8. Clarke L, Carbon J (1976) A colony bank containing synthetic Col El hybrid plasmids representative of the entire *E. coli* genome. Cell 9:91–99
9. Kovach ME, Elzer PH, Hill DS et al (1995) Four new derivatives of the broad-host-range cloning vector pBBR1MCS, carrying different antibiotic-resistance cassettes. Gene 166: 175–176
10. Furste JP, Pansegrau W, Frank R et al (1986) Molecular cloning of the plasmid RP4 primase region in a RT multi-host-range tacP expression vector. Gene 48:119–131
11. Schweizer HP (1991) *Escherichia-Pseudomonas* shuttle vectors derived from pUC18/19. Gene 97:109–121

Chapter 43

Strategy for Genome Sequencing Analysis and Assembly for Comparative Genomics of *Pseudomonas* Genomes

Julie Jeukens, Brian Boyle, Nicholas P. Tucker, and Roger C. Levesque

Abstract

Bacterial genome sequencing has developed rapidly in the last decade and has become a primary method for analyzing the genomic basis of differences in phenotype between strains as well as being a valuable tool for public health epidemiology. This chapter provides a comprehensive workflow for bacterial genome sequencing from experimental design to data suitable for comparative genomics analysis, while mainly focusing on the challenges associated with genome assembly. This approach was successfully applied to 19 *Pseudomonas aeruginosa* genomes from phenotypically distinct strains.

Key words Bacterial genome, Next-generation sequencing, Contig, Assembly, Finishing

1 Introduction

Much has changed in the field of bacterial genome sequencing since the first complete genome sequence was released in 1995 [1]. Almost 20 years later, owing to the advent of next-generation sequencing (NGS) technologies, time and cost for bacterial genome sequencing have decreased by multiple orders of magnitude, to the point that we might be only a few years away from the $1 bacterial genome [2] and the introduction of whole-genome sequencing into clinical microbiology [3]. This holds great promise for *Pseudomonas aeruginosa* genomics research, as the incredible versatility of this species is associated with a high level of genomic complexity and diversity [4]. Only eight strains are currently available on the Pseudomonas Genome Database [5] to represent the genomic repertoire of the *P. aeruginosa* population and little is known about the phenotypic impact of variable genomic regions already identified, let alone their significance for host response during an opportunistic infection. Thus comparative genomics analysis is a key starting point towards understanding pan-genome diversity and the genomic basis of pathogenicity in

Alain Filloux and Juan-Luis Ramos (eds.), *Pseudomonas Methods and Protocols*, Methods in Molecular Biology, vol. 1149, DOI 10.1007/978-1-4939-0473-0_43, © Springer Science+Business Media New York 2014

P. aeruginosa. In turn, it will pave the way for refined typing of clinical isolates, more personalized treatment of infection, and the identification of novel therapeutic targets.

The aim of this chapter is twofold. First, while describing detailed procedures for sequencing per se is beyond the scope of this chapter, guidelines will be provided for the reader to select the appropriate technology amongst 454 GS-FLX, Ion Torrent, and Illumina HiSeq. Information on library types and library preparation will also be provided. Second, a detailed strategy to assemble any bacterial genome into either a working draft or a finished version will be presented. Using this approach, we have successfully sequenced and assembled a total of 19 *P. aeruginosa* genomes from strains having a wide repertoire of distinct phenotypes.

2 Materials

For library preparation and sequencing, *see* **Notes 2–4. Item 2** is only relevant if a reference genome is available. **Items 4–7** are only relevant if the ultimate goal is to assemble a genome into a single finished sequence.

1. GS De Novo Assembler, version 2.6 or later (Roche) or CLC Genomics Workbench (CLC bio). Both can be run on a standard desktop or laptop. If you do not have access to these programs, a number of command line assemblers are available (e.g. Abyss [6], Mira [7]).
2. Mauve, version 2.3.1 [8].
3. Spreadsheet (e.g., Microsoft Excel and LibreOffice Calc).
4. Consed, version 20.0 or later, version 22.0 or later for Illumina mate pairs [9]. Consed can run on a standard desktop or laptop. Help from a bioinformatician might be required for software installation.
5. Thermal cycler (preferably with gradient and touchdown modes).
6. Polymerase chain reaction (PCR) kits and reagents (*see* **Note 1**).
7. GS Reference Mapper, version 2.6 or later (Roche) or CLC Genomics Workbench (CLC bio). Command line assemblers such as MAQ [10], BWA [11], and Bowtie [12] can also be used. If you are not familiar with command line bioinformatics, BWA and Bowtie can be run through Galaxy either online, in the cloud or as a local instance on your own system [13].

3 Methods

Steps 6, 7, **10**, **17**, **18**, and **20** are only relevant if a reference genome is available. **Steps 9–24** are only relevant if the ultimate goal is to assemble the genome into a single finished sequence. A schematic summary of this section is provided in Fig. 1.

1. Determine the objectives and evaluate resources (*see* **Note 2**).
2. Prepare high-quality genomic DNA (*see* **Note 3**).
3. Obtain a high-quality next-generation sequencing dataset (*see* **Note 4**).
4. Carry out quality control analysis on sequence reads in order to determine average read length, insert size (for paired-end datasets) and quality score drop off along the reads. Trim reads accordingly (*see* **Note 5**).
5. Perform de novo assembly of raw sequence data (*see* **Note 6**).
6. Align de novo contigs on a reference genome using Mauve Contig Mover [8, 14, 15] (*see* **Note** 7).

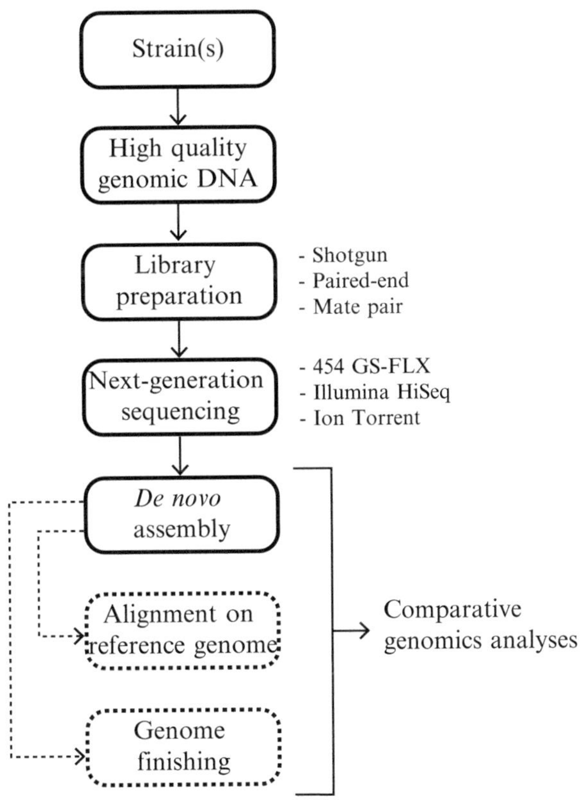

Fig. 1 Workflow for bacterial genome sequencing and assembly. *Dotted lines* represent optional steps

7. Align de novo contigs on a reference genome using BLAST (bl2seq) [16]. Export results in tabular format and filter in a spreadsheet (*see* **Note 8**).
8. In a spreadsheet, combine Mauve TAB output file and BLAST results if available, as well as the ContigGraph.txt result file from GS De Novo assembler (*see* **Note 9**). This spreadsheet will serve as a guide for **steps 10–18**.
9. Launch Consed [9] and open de novo assembly results "consed/edit_dir/*.ace.1".
10. Double-click on the appropriate contig, right-click on the consensus position that corresponds to the origin of replication and split the contig by selecting "tear contig at this consensus position".
11. Open "Assembly view". The content of this window can be modified using the "What to show" button. Namely, to view matches between contigs, use the "run crossmatch" button in the "Sequence matches" window. Sequence matches between contigs will appear in orange. Sequence matches where one of the two contig would have to be reversed and complemented will appear in black.
12. To reverse and complement contigs, go to the "Contig Arrangement/Reorient Contigs" window, select contigs, "Flip scaffold" and "Apply".
13. Begin by joining contigs that are represented only once in the genome. Select a sequence match by clicking on an orange line. A new window will appear. Select a single match and click "Show alignment" to visualize the alignment between the two contigs. To join these contigs, click "Join contigs".
14. For each contig represented more than once in the genome (according to coverage and blast results, if available), randomly parse reads into the appropriate number of new contigs (*see* **Note 10**). This can be done by selecting reads randomly across the length of the contig and selecting "Misc/Save Highlighted Read Names to File". Then, go back to the main window, open "Remove Reads," select the correct file and choose "Just Put Each Read Into Its own Contig". Finally, go back to the main window, open "Miniassembly" and reassemble the extracted reads into a new contig.
15. Join the contigs created in **step 14** with those from **step 13**.
16. Summarize contigs and remaining gaps.
17. Use the "Compare cont" button in "Aligned Reads" windows to join contigs with small overlaps that were not detected by "crossmatch" (*see* **Note 11**).
18. For remaining gaps that are not too large (expected length equivalent to 1–4 sequence reads), perform reference mapping

of the raw data in Consed. Use the "fasta2Ace.perl" command to convert your reference genome into an assembly file. Then, use the "add*Reads.perl" command to assemble all sequence reads. Go to positions where gaps in the de novo assembly are and see if sequence reads assembled onto these gaps. If so, use them to bridge gaps in the de novo assembly by finding them using the "Find reads containing" window, isolating them by selecting "Put read * into its own contig" and assembling them as in **step 9** or **step 13**. If they can't be found in the de novo assembly, they can be added with "Add new reads" in the Consed main window.

19. Export results in fasta by selecting "File/Write all contigs to fasta file".
20. Verify that there we no major errors by aligning this assembly file with the reference genome using Mauve or BLAST (*see* **Note 12**).
21. Design specific primers inside flanking contigs of any remaining gap (*see* **Note 13**).
22. Optimize amplification conditions and sequence resulting amplicons using the Sanger technology (*see* **Note 14**).
23. To add Sanger sequences in fasta format to the assembly in Consed, use the "mktrace" command and "Add new reads" in the consed main window.
24. To detect remaining minor errors, assemble raw sequence data on the newly assembled genome using GS Reference Mapper (Roche) (*see* **Note 15**).
25. Final genome version in fasta format can be annotated and used for comparative genomics analyses (*see* **Note 16**).

4 Notes

1. Multiple PCR kits will have to be tested to successfully amplify gaps in the genome sequence. We used three: HotStarTaq DNA Polymerase kit (Qiagen), Fast-Pfu DNA polymerase kit (Feldan) and Deep VentR™ DNA Polymerase kit (New England Biolabs) (*see* **Notes 9** and **10**).
2. Sequencing and assembly should only be performed with well-defined objectives. Nowadays, it can be relatively easy, fast, and inexpensive to obtain sequencing datasets. In the case of genome assemblies, obtaining a closed circular genome is by far the most difficult operation to accomplish. Bioinformatics treatment of data cannot perform miracles and the achievements of the objectives will be largely influenced by decisions taken early on in the process. A proper evaluation of existing genomic resources is also important. In the case of *P. aeruginosa*,

there are several reference genomes available, but a dataset and several reference genomes might not be sufficient to resolve all the problems encountered after the dataset is produced. The most important question that should come to mind is: "what am I looking for in this genome?"

3. High-quality DNA means free of RNA contamination with absorbance ratios 260/230 and 260/280 above 1.7. To ensure RNA-free genomic DNA, a comparison between concentrations obtained via absorbance at 260 nm using a spectrophotometer and a fluorometric method (PicoGreen or Qubit Assays from LifeTechnologies) is recommended. RNAse treatment is more efficient if it is performed early in the DNA preparation procedure, preferably at the cell lysis stage. Commercial kits like DNeasy blood & tissue kit (QIAGEN) consistently generated high-quality DNA in our hands.
4. We will try to provide tips and tricks to tailor your sequencing strategy according to objectives set in **step 1**.
   (a) Library options: Two types of libraries are commonly made for next-generation sequencing. First, the shotgun library procedure, where DNA is mechanically fragmented to size and technology specific adaptors are ligated. Second, paired-end libraries (called mate pairs in Illumina), where large DNA molecules are circularized to bring distant regions together into smaller and more manageable DNA fragments. Shotgun libraries are well suited for reference-assisted genome assembly, whereas paired-end libraries are essential for de novo genome assembly.
   (b) Sequencing options: There are numerous resources available through the net that compare/describe next-generation sequencers. These include primary research and review articles, manufacturer Web sites, service provider Web sites, forums, and much more. We will not try to describe the current state of the technologies because instruments change rapidly enough that performance reports are often outdated shortly after publication. Nonetheless, performance indicators to consider are read length, base quality, cost per base, and speed. We found that read length is an important factor to consider when the objective is to completely assemble a bacterial genome.
   (c) Coverage: Although 7–10× coverage was sufficient for high-quality assemblies using Sanger sequencing, it is recommended when using next-generation sequencers to go deeper into the sequencing and achieve at least 15–20× coverage (GS-FLX). For smaller read lengths (e.g., Illumina technology), a greater coverage, up to 70×, is recommended. However, over sequencing is not recommended

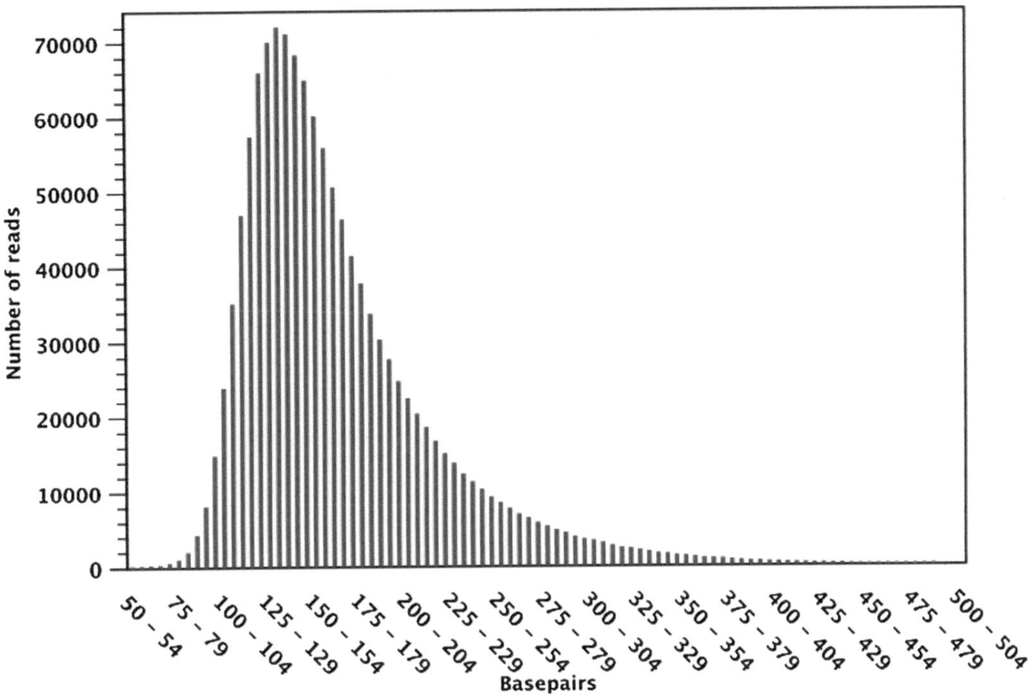

Fig. 2 Assessment of insert size determined using CLC Genomics Workbench. The data presented in this figure is one of 12 strains sequenced in a single Illumina GA2 lane with barcoded adapters

as it will increase dataset complexity, and therefore dramatically increase computation time without improving the quality of the assembly.

5. Insert size of paired-end Illumina datasets is reported in the output of a CLC Genomics Workbench reference mapping run (Fig. 2). Insert size in paired-end datasets is important in two regards. Firstly, when mapping reads to a reference genome large deletions and rearrangements can be detected because paired reads will span a region of the genome much larger than the average insert size. This can be particularly useful when studying adaptation to the CF lung. Secondly, paired reads are a powerful way of closing gaps in assemblies and extending contigs using packages such as PAGIT [17]. Quality control is heavily dependent on the sequencing platform that you are using and the aims of your experiment. In order to make an educated decision about quality control, it is important to understand the basis of sequencing quality scores, typically provided in the PHRED format. Sequencing is a degenerate process and quality score tends to decrease as reads get longer. Each base in each read is designated a score based upon the probability that there is an error at that position. In most NGS platforms, good quality scores are typically between 30 and 40,

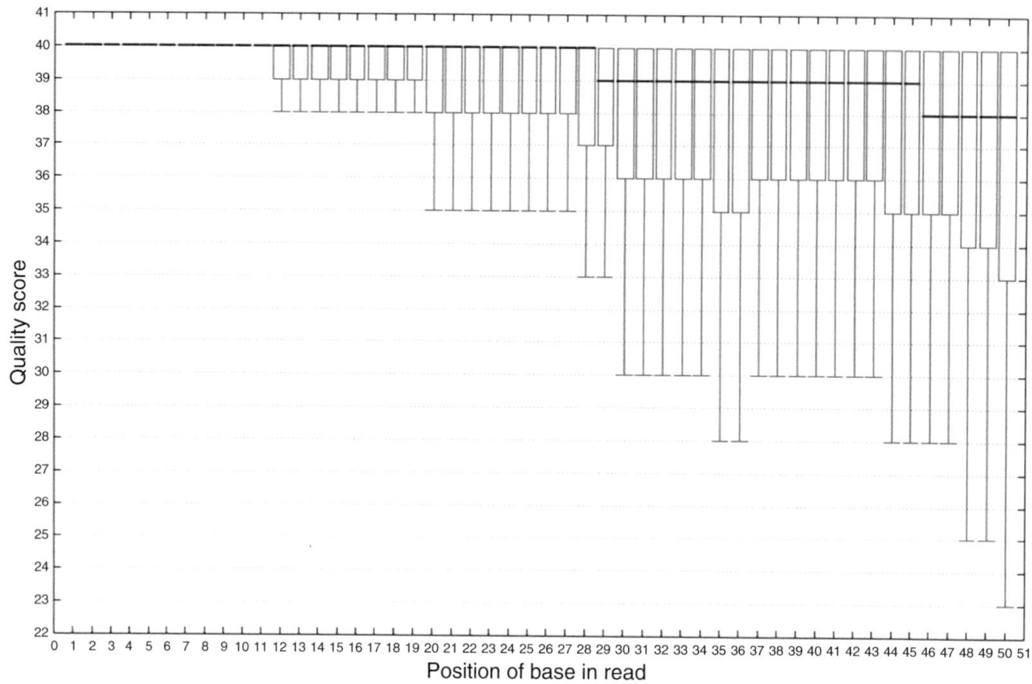

Fig. 3 Correlation between read length and quality score of a *P. aeruginosa* Illumina dataset

where the probability of an error is between 1 in 1,000 and 1 in 10,000 bases respectively. Read quality can be analyzed using the NGS QC and Manipulation feature of Galaxy by following the FastQ Manipulation tutorial [13]. As can be seen in Fig. 3, with the Illumina data, the median score tends to drop towards the end of the reads, but the range of scores rarely drops below 30. When trimming reads for de novo assembly in this example, we may simply remove the last three to five bases of every read using the trimming tools available in Galaxy or CLC Genomics Workbench.

6. Performing de novo assembly as a first step, even if a closely related reference genome is available, is a good way of letting the data "speak for themselves". For 454 data, we selected the following options in addition to default parameters: "Extend low depth overlaps," "Reads limited to one contig" and "Complete consed folder". This parameter selection is particularly relevant for genome finishing with Consed.
7. Mauve performs whole-genome alignment of two or more genome sequences that may have undergone rearrangements. The Mauve Contig Mover (MCM) tool can be used to order contigs of a draft genome. We used default parameters. *See* **Note 9** for output file use. Reordered contigs can be analyzed using Ori-Finder to check that mauve has reordered the contigs in a sensible manner (Fig. 4).

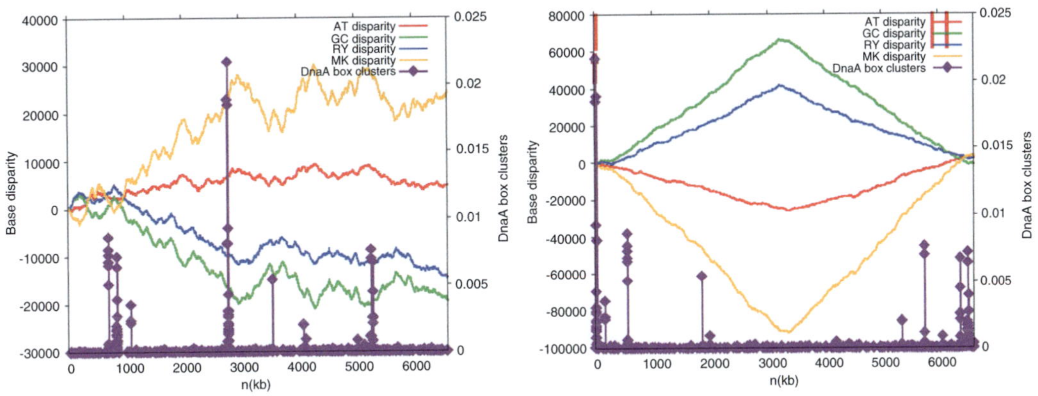

Fig. 4 Testing the contig reordering process with Ori-Finder. Ori-Finder plots the GC and AT disparity (GC skew) by base position whilst simultaneously predicting the positions of DnaA box like sequences. The *figure on the right* demonstrates effective reordering because the largest DnaA box is at the start of the sequence and the GC disparity plots match the pattern of a completed bacterial genome [29]

8. We filtered BLAST results in a spreadsheet by retaining hits with alignment length ≥ 100 bps and identity ≥ 99 %. These criteria were selected because the reference strain was very closely related to newly sequenced strains. They should be less stringent if the reference and draft genomes are more distantly related.

9. The goal of preparing this spreadsheet is to combine information that will be useful for genome finishing, namely contig order. In the Mauve TAB output file, extract ordered contigs' label and orientation. In the BLAST output, extract contig label, alignment length, query (contig) start and end, and subject (reference) start and end. Based on subject start and end (start > or < than end), create a new column with each contig's orientation. Create two other columns with 5′ and 3′ position on the reference genome sequence. Order BLAST results according to 5′ position on the reference and eliminate redundancy (e.g. very small contigs that align with regions of the reference genome that are already covered by larger contigs). Use the ContigGraph.txt result file (or any other source of per contig coverage information) to identify contigs that likely represent multiple identical regions of the genome. These contigs' coverage will be about two or more times the average. Locate and tag them in ordered BLAST results. Verify that the number of copies according to BLAST is in agreement with the coverage. Finally, combine and compare ordered BLAST results and ordered contigs from Mauve. Keep in mind that Mauve will use each contig only once. Sometimes, ContigGraph.txt indicates possible connections between pairs of contigs. This information should be validated against results from BLAST and Mauve.

10. To create multiple copies of a contig, we selected one read out of two or more across the consensus sequence. Sometimes, more than one contig will be assembled by "Miniassembly" of extracted reads. If one of these new contigs has approximately the same length as the original contig, ignore the others. Otherwise, it might be necessary to manually join two of these new contigs to mimic the original.

11. To estimate overlap length between contigs, create a column in the spreadsheet with 5′ position on the reference minus 3′ position of the previous contig. When this value is small (<100 bp), overlap may not be detected by "crossmatch."

12. What is meant here by major error is a genomic region that would be in the wrong orientation or at the wrong position. Ordering contigs with Consed can get confusing, so this type of mistake is likely to occur.

13. It is important to grasp that attempting to fill remaining gaps in a genome sequence is not straightforward and might take a significant amount of time. Recent findings clearly indicate that these regions tend to form GC-rich secondary structures that are refractory to amplification and sequencing [18]. Hence, when attempting to sequence and assemble multiple closely related strains, some problematic regions will be shared among strains. We designed primers in 2,000 bp regions on either side of gaps using Primer3 [19]. Common sets of primers were designed for gaps shared among strains. In some sequencing projects it is unreasonable to completely finish a genome. An example of this might be an Illumina resequencing project such as the one summarized in Table 1 where there are a large numbers of contigs. In these cases, de novo assembly is still very powerful but it is important to know the positions of contig boundaries in your assembled sequence. This can be achieved by adding these as features to a genbank file produced by RAST or xBASE (*see* **Note 16**) by using the "Read an entry into" feature in Artemis [20]. The entry in question is the .fna file that contains reordered contigs.

14. Hurt et al. [18] developed a strategy to circumvent problematic secondary structures using a combination of ramped extension during PCR and strand displacing DNA polymerase. We have tried to apply this strategy to the few gaps that remain in our 454 genome sequences (Table 2, last column), but were not successful. Among gaps that were resolved, some were easily amplified and sequenced on first attempt, while others, usually those shared among strains, required more effort in primer design and PCR optimization. Our largest amplicons were 3–4 kbp and could be fully sequenced with forward and reverse primers plus two internal primers.

Table 1
De novo assembly summary statistics of 12 non-scaffolded *P. aeruginosa* genomes. Strains were barcoded and sequenced in a single lane of Illumina HiSeq. De novo assembly was performed with CLC Genomics Workbench (CLCbio). Libraries were prepared for 50 bp paired-end Illumina sequencing with a target insert size of 150 bp

Strain ID	Number of reads	Total assembled bases	Average contig size	Largest contig	N50	Number of contigs
E2	4,713,048	6,657,306	38,705	335,290	105,857	172
C1426	4,608,838	6,437,521	37,211	275,912	117,451	173
C763	4,182,436	7,066,488	27,496	362,947	77,035	257
C1334	3,653,050	6,384,546	22,884	258,974	67,599	279
PA62	3,772,324	6,464,546	22,683	165,783	69,241	285
MSH10	3,469,254	6,457,887	18,611	234,172	56,273	347
PA17	3,182,194	6,356,729	13,583	207,359	39,056	468
MSH3	3,197,062	6,461,136	13,106	193,889	33,800	493
PA17SCV	3,690,844	6,363,560	9,372	94,253	23,438	679
C1433	2,724,940	6,442,798	7,753	136,157	17,044	831
J1532	2,719,066	6,546,868	6,791	305,256	14,549	964
J1385	2,734,638	6,706,156	6,646	71,358	14,646	1,009
Average	3,553,975	6,528,795	18,737	220,113	52,999	496

Table 2
De novo assembly and genome finishing summary statistics of 7 *P. aeruginosa* genomes. Shotgun libraries of the first 3 strains were sequenced individually on a quarter (LES400, LESB65) or a half (LES431) 454 plate. Shotgun libraries of the last 4 strains were barcoded and sequenced on a full 454 plate. De novo assembly was performed with GS De Novo Assembler (Roche); genome finishing was done with Consed [9]

Strain ID	Sequencing technology	Number of reads	Mean coverage in de novo assembly	Contigs after de novo assembly (>100 pb)	Contigs after finishing with Consed	Contigs after gap resolution by PCR
LES400	454 GS-FLX Titanium	316,245	19	96	9	3
LES431	454 GS-FLX Titanium	859,117	50	74	2	1
LESB65	454 GS-FLX Titanium	447,065	30	77	3	1
01-022-1	454 GS-FLX Plus	228,869	18	66	5	3
03-019-10	454 GS-FLX Plus	372,870	30	54	4	2
03-054-2	454 GS-FLX Plus	330,965	26	54	5	3
05-009-2	454 GS-FLX Plus	294,866	23	44	4	2

15. What is meant here by minor error is essentially single base error. To detect minor errors, any software that can perform reference mapping and detect polymorphism between reference and sequence reads can be used.
16. A variety of automated annotation methods are available and they each have their own merits. xBase2 [21] is particularly useful for rapid annotation of a *P. aeruginosa* strain using a previously well annotated strain as a reference. xBase uses a combination of methods to generate an annotated genbank file with familiar *P. aeruginosa* gene names and labels new genes with a strain specific identifier. Alternatively, RAST [22] can also be useful, particularly where one has identified a novel genomic island because novel genes are given predicted functions based upon homology to a range of sequence databases. IslandViewer can be used as a complement to xBase and RAST to identify genomic islands [23]. As for subsequent comparative genomics analyses, here are a few examples of software we have used: Artemis Comparison Tool [24], CGView [25] and the CGView Comparison Tool [26], Panseq [27], Mega 5 [28].

Acknowledgements

Research in R.C. Levesque's laboratory is funded by the Canadian Institute for Health Research (CIHR), the CIHR-Fonds de Recherche du Québec en Santé (FRQS)-Québec Respiratory Health Network (RSR) and by the Natural Sciences and Engineering Research Council of Canada (NSERC). Research in N.P. Tucker's laboratory has been supported through grants from the Royal Society and the University of Strathclyde. N.P.T. and R.C.L. are also grateful to the Society for General Microbiology for sponsoring N.P.T. for a research visit to the corresponding author's laboratory. J. Jeukens received a CIHR-FRQS-RSR fellowship award and was part of the Québec respiratory health training program.

References

1. Fleischmann R et al (1995) Whole-genome random sequencing and assembly of *Haemophilus influenzae* Rd. Science 269:496–512
2. Loman NJ et al (2012) High-throughput bacterial genome sequencing: an embarrassment of choice, a world of opportunity. Nat Rev Microbiol 10:599–606
3. Didelot X et al (2012) Transforming clinical microbiology with bacterial genome sequencing. Nat Rev Genet 13:601–612
4. Klockgether J et al (2011) *Pseudomonas aeruginosa* genomic structure and diversity. Front Microbiol 2:150
5. Winsor GL et al (2011) *Pseudomonas* genome database: improved comparative analysis and population genomics capability for *Pseudomonas* genomes. Nucleic Acids Res 39:D596–D600
6. Simpson JT et al (2009) ABySS: a parallel assembler for short read sequence data. Genome Res 19:1117–1123

7. Chevreux B, Wetter T, Suhai S (1999) Genome sequence assembly using trace signals and additional sequence information. Comput Sci Biol 99:45–56
8. Darling ACE et al (2004) Mauve: multiple alignment of conserved genomic sequence with rearrangements. Genome Res 14:1394–1403
9. Gordon D, Abajian C, Green P (1998) Consed: a graphical tool for sequence finishing. Genome Res 8:195–202
10. Venkata V, Prasad S, Loshma G (2011) HPC-MAQ: a parallel short-read reference assembler. Comput Sci Inform Tech 2:75–84
11. Li H, Durbin R (2009) Fast and accurate short read alignment with Burrows–Wheeler transform. Bioinformatics 25:1754–1760
12. Langmead B et al (2009) Ultrafast and memory-efficient alignment of short DNA sequences to the human genome. Genome Biol 10:R25
13. Goecks J et al (2010) Galaxy: a comprehensive approach for supporting accessible, reproducible, and transparent computational research in the life sciences. Genome Biol 11:R86
14. Darling AE, Mau B, Perna NT (2010) Progressive Mauve: multiple genome alignment with gene gain, loss and rearrangement. PLoS One 5:e11147
15. Rissman AI et al (2009) Reordering contigs of draft genomes using the Mauve aligner. Bioinformatics 25:2071–2073
16. Altschul SF et al (1990) Basic local alignment search tool. J Mol Biol 215:403–410
17. Swain M et al (2012) A post-assembly genome-improvement toolkit (PAGIT) to obtain annotated genomes from contigs. Nat Protoc 7: 1260–1284
18. Hurt RA Jr et al (2012) Sequencing intractable DNA to close microbial genomes. PLoS One 7:e41295
19. Rozen S, Skaletsky H (2000) Primer3 on the WWW for general users and for biologist programmers. Methods Mol Biol 132:365–386
20. Rutherford K et al (2000) Artemis: sequence visualization and annotation. Bioinformatics 16:944–945
21. Chaudhuri RR et al (2008) xBASE2: a comprehensive resource for comparative bacterial genomics. Nucleic Acids Res 36: D543–D546
22. Aziz R et al (2008) The RAST server: rapid annotations using subsystems technology. BMC Genomics 9:75
23. Langille MGI, Brinkman FSL (2009) IslandViewer: an integrated interface for computational identification and visualization of genomic islands. Bioinformatics 25:664–665
24. Carver TJ et al (2005) ACT: the Artemis comparison tool. Bioinformatics 21:3422–3423
25. Stothard P, Wishart DS (2005) Circular genome visualization and exploration using CGView. Bioinformatics 21:537–539
26. Grant J, Arantes A, Stothard P (2012) Comparing thousands of circular genomes using the CGView comparison tool. BMC Genomics 13:202
27. Laing C et al (2010) Pan-genome sequence analysis using Panseq: an online tool for the rapid analysis of core and accessory genomic regions. BMC Bioinformatics 11:461
28. Tamura K et al (2011) MEGA5: molecular evolutionary genetics analysis using maximum likelihood, evolutionary distance, and maximum parsimony methods. Mol Biol Evol 28:2731–2739
29. Gao F, Zhang CT (2008) Ori-Finder: a web-based system for finding oriCs in unannotated bacterial genomes. BMC Bioinformatics 9:79

Chapter 44

Promoter Fusions with Optical Outputs in Individual Cells and in Populations

Ilaria Benedetti and Victor de Lorenzo

Abstract

Reporter genes are widely used to quantify promoter activity, which controls production of mRNA through the interplay with RNA polymerases and transcription factors. Some of such reporters have either diffuse (*lux*) or focused (GFP) optical outputs that allow description of transcriptional activity in populations and in single cells. This chapter discusses the use of a dual reporter system GFP-*luxCDABE* that is placed in broad-host-range plasmids having origins of replication from *RK2* and *pBBR1*. The value of this system is shown in *Pseudomonas putida* by characterizing the activity of the *Pb* promoter, which drives an operon for benzoate biodegradation in this bacterium. To this end we compare in the same cells bioluminescence as the output signal of the whole population and single cell-bound fluorescence caused by GFP expression and revealed by flow cytometry assays.

Key words GFP, *luxCDABE*, Bacterial population, Single-cell, Flow cytometry analysis, Gram-negative bacteria

1 Introduction

Control of gene expression is the result of the interplay between a limited number of components involved in regulation of the process of RNA production [1, 2]. The molecular factors at stake can be present at low concentrations in the cytoplasm, thereby submitting gene expression to stochastic variations from cell to cell [3–6]. Collision of the molecular partners involved in the process that drives the corresponding biochemical process often makes promoters prone to a high level of noise [7–9]. The analysis of cell-to-cell variability is important for understanding both the functioning of large regulatory network and the molecular details that operate in specific circuits [10, 11]. Most available methodologies for assessing the effects of stochastic processes in single cells are based on the use of fluorescent proteins (typically GFP or similar) fused to the target promoter element [12, 13]. In contrast to other approaches based on the enzymatic quantification of reporter genes (for example

Alain Filloux and Juan-Luis Ramos (eds.), *Pseudomonas Methods and Protocols*, Methods in Molecular Biology, vol. 1149, DOI 10.1007/978-1-4939-0473-0_44, © Springer Science+Business Media New York 2014

lacZ), GFP is able to fluoresce without requirement of any cofactor [14, 15]. However, when the population as a whole is the object of analysis, fluorescent proteins fail to provide enough information [15–17] and other reporters are needed. One of such non-disruptive, real-time reporters involves the *lux* genes of marine bacteria that convert reducing power into luminescence [16, 18, 19]. However, the proper activity of *lux* is limited by their inherent dependence on energy status of the cells [18]. In order to address the issue of cell-to-cell variability versus population behavior we have merged the two products (GFP and *lux*) into a unique synthetic transcriptional unit in which the two reporters have been assembled as a single promoter-less transcriptional unit [20]. In this chapter we describe the experimental details of such a system and verify the value of the dual GFP-*luxCDABE* reporter for quantifying the activity of promoters of *Pseudomonas putida*. For the user's sake, the business DNA fragment is placed as a single *BamHI/SpeI* cassette in the frame of the synthetic, broad-host-range SEVA plasmids [21] with different origins of replication and copy numbers. As shown below, the reporter system was instrumental for characterizing the *Pb* promoter of *P. putida* that controls an operon for benzoate catabolism in this bacterium [22]. To this end, we assembled a transcriptional *Pb*→ GFP-*luxCDABE* fusion, placed it in *P. putida* and examined the kinetic characteristics of the promoter in response to benzoate. As shown below, the measure of bioluminescence as output signal afforded visualization of the population-wide landscape of *Pb* promoter activity. In addition, we could monitor single-cell activity by quantifying GFP expression in individual bacteria with flow cytometry.

2 Materials

2.1 Strains and Plasmids

See Table 1. Note that the GFP-*luxCDABE* dual reporter (Fig. 1a) is framed within two pSEVA vectors [21] each of which provides a different copy number to the resulting constructs. The high-copy number vector pGLR1 (Fig. 1a) is derived from pSEVA236 [21] and has a pBBR1 origin of replication. The low-copy number counterpart (pGLR2, Fig. 1a) stems from pSEVA221 [21], and propagates in *P. putida* through a *RK2* origin. The resulting vectors have an ample variety of restriction sites upstream of the GFP-*lux* cassette (Fig. 1b) for handler's convenience. This chapter will show only the use of pGLR1, but users can also consider running the same procedures with pGLR2 if copy number and gene dosage become an issue for the specific project.

2.2 Media and Reagents

1. To routinely grow bacteria in LB, prepare it as follows: 10 g/L of tryptone, 5.0 g/L of yeast extract, and 5.0 g/L of NaCl, dissolve in H_2O and autoclave. For LB-agar plates add 1.5 % agar.

Table 1
Bacterial strains and plasmids

	Description/relevant characteristics	References
Strain		
E. coli CC118	Δ(*ara-leu*), *araD*, Δ*lac*X174, *galE*, *galK*, *phoA*, *thi1*, *rpsE*, *rpoB*, *argE* (*Am*), *recA1*. Recipient strain of all plasmids	[30]
E. coli HB101 (pRK600)	Sm^R, *hsdR⁻M⁺*, *pro*, *leu*, *thi*, *recA*. Host of pRK600, mating helper strain	[31]
P. putida KT2440	*P. putida* mt-2-derivative cured of the pWW0 plasmid	[32]
Plasmids		
pRK600	Cm^R, *oriColE1*, *mobRK2*, *traRK2*; helper for mobilization of *oriT RK2⁺*-containing plasmids	[33]
pSEVA221	Km^R, *oriRK2*, *oriT*; standard broad-host-range plasmid for gram-negative bacteria	[21]
pSEVA236	Km^R, *oripBBR1*, *oriT*; standard broad-host-range with *luxCDABE* reporter system	[21]
pGreenTIR	Ap^R, *oriColE1*; promoterless cloning vector with *gfp tir* gene	[29]
pGLR1	Km^R, *ori pBBR1*, *oriT*; pSEVA236-derivative with dual GFP-*luxCDABE* reporter system	[20]
pGLR2	Km^R, *oriRK2*, *oriT*; pSEVA221-derivative with dual GFP-*luxCDABE* reporter system	[20]
pGLR1-*Pb*	Km^R, *ori pBBR1*, *oriT*; pGLR1-derivative with *Pb* promoter cloned as a *Eco*RI/*Bam*HI fragment	[20]

2. Minimal medium (M9; *see* **Note 1**) for *P. putida*. First prepare a 10× stock solution of M9 salts with 42.5 g $Na_2HPO_4{\cdot}2H_2O$, 15 g KH_2PO_4, 2.5 g NaCl, and 5 g NH_4Cl, dissolved in 500 ml of H_2O and sterilize. Prepare stocks of 1 M $MgSO_4$ and citrate at 20 % (as selective carbon source for *Pseudomonas*); autoclave the four components separately. After sterilization, mix components to leave them at the following concentrations 1× M9 salts, 2 mM $MgSO_4$, and 0.2 % citrate. For M9 plates, add 1.6 % agar instead of water.
3. Kanamycin (Km): Prepare the stock at 50 mg/ml with H_2O, filter-sterilize. Store at −20 °C. Use it at a final concentration of 50 μg/ml.
4. Sodium benzoate stock solution was dissolved in H_2O at a concentration of 1.0 M. Use it at final concentrations of 1 mM for flow cytometry experiments and of 1 mM, 750 μM, 500 μM, 250 μM, 100 μM, 50 μM, 25 μM, 10 μM, and 5 μM for bioluminescence experiments. Store at room temperature.
5. PBS 10×: Add NaH_2PO_4, Na_2HPO_4 and NaCl to 450 ml dH_2O. Adjust pH to 7.4 by adding diluted (1 N) HCl if necessary. Bring the final volume to 500 ml with water. Dilute 1:10 for final concentration. If necessary, filter the solution using 0.2 μm 25 mm nylon syringe filter. Store the solution at room temperature.

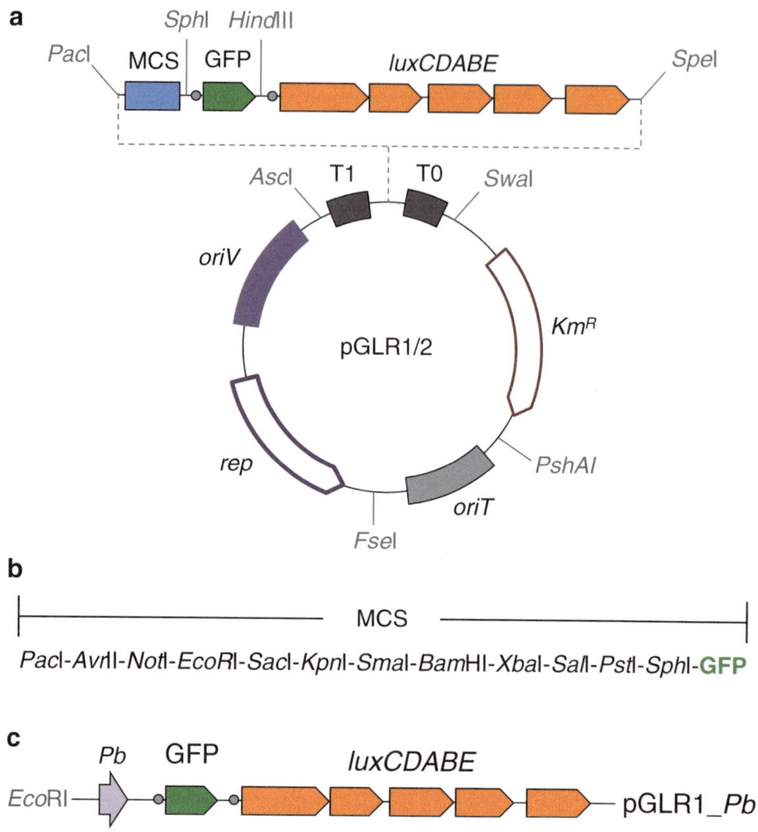

Fig. 1 Functional parts of the pGLR1/2 vectors. (**a**) The vectors each harbor a kanamycin resistance (Km^R) marker, an *oriT* for plasmid transfer through conjugation, and a broad-host-range origin of replication that consists of a vegetative origin (*oriV*) and a replication protein (*rep*). Vector pGLR1 is based on a minimal pBBR1 [34] origin, while pGLR2 is based on *ori* RK2 [35]. The GFP-*luxCDABE* reporter cassette is cloned between two strong terminators (T0 and T1) and is placed downstream of a multiple cloning site (MCS). The optimal ribosome-binding site of the TIR element [29] is represented as a *grey circle* is placed upstream of the *gfp* gene and the first gene of the promoterless *lux* operon (*luxC*). (**b**) List of enzymes included in the MCS. (**c**) Organization of the *Pb*→ GFP-*luxCDABE* fusion. The business part of the pGLR1_*Pb* plasmid assayed in the experiments with the fusion is sketched

6. Paraformaldehyde solution (*p*-formaldehyde): For 10 ml of a 4 % *p*-formaldehyde solution, weight out 0.4 g of the pure compound, transfer into 15 ml conical tube (Falcon). Add 8 ml of H_2O and 5 µl of 1 M NaOH. Heat to 70 °C and mix frequently until complete solubilization (solution should become clear). Put on ice and allow the solution to cool down to room temperature. Add 1 ml PBS 10× and mix. Filter solution using 0.2 µm 25 mm Nylon syringe filter. Store the solution at room temperature (*see* **Note 2**).

2.3 PCR Primers for Pb Promoter

5-PB *EcoRI*: 5′-TGG ATG AAT TCG ACA GTA CCC TCC-3′, which anneals with 5′ of *Pb* promoter of *P. putida* KT2440 and adds an *EcoRI* restriction site at 5′.

3-PB *BamHI*: 5′-GCG CGG ATC CGG CCA GGG TCT CCC TTG-3′, which anneals with 3′ of *Pb* promoter of *P. putida* KT2440 and adds a *BamHI* restriction site at 3′.

3 Methods

3.1 DNA Techniques

1. For plasmid preparations we routinely use the Wizard Plus SV Minipreps kit (Promega).
2. PCR-amplified DNA and agarose DNA extractions are commonly purified with the NucleoSpin Extract II kit (MN).
3. In DNA ligations the T4 ligase is employed.
4. PCR: for general PCR reactions prepare first a PCR reaction mix in an Eppendorf tube. The values given below are calculated for a final volume of 25 μl per tube:
   (a) 5 μl Buffer 5×
   (b) 1.5 μl $MgCl_2$ 25 mM
   (c) 1 μl of forward primer 5 μM
   (d) 1 μl of reverse primer 5 μM
   (e) 0.5 μl dNTPs 10 mM
   (f) 0.2 μl Taq Polymerase

Then, add 15.5 μl of sterile H_2O into each of the PCR reaction tubes. Pick fresh single colonies directly from a plate and transfer it directly to the PCR reaction tube containing the 15.5 μl of H_2O. Vortex the PCR reaction mixture and distribute 9.5 μl into PCR tubes. Set up the PCR machine with the appropriate Tm (between 55.0 and 59.0 °C, depending on the primer composition) and extension time.

3.2 Cloning and DNA Mobilization

All DNA cut-and-paste operations were made through standard protocols [23] using *E. coli* CC118 as the recipient of the resulting constructs. Broad-host-range plasmids were passed to *P. putida* KT2440 through triparental mating [24], *see* below.

3.2.1 Construction of Pb→ GFP luxCDABE Transcriptional Fusions

1. The *Pb* promoter sequence, was amplified by PCR from genomic DNA of *P. putida* using the above mentioned primers (PB-5′ *EcoRI* and PB-3′ *BamHI*), digested by *EcoRI/BamHI*, run in an agarose gel 2 %, and purified.
2. The resulting fragment of 200 bp was then cloned as *EcoRI/BamHI* fragment into pGLR1 vector, resulting in pGLR1_*Pb*, the business part of which is sketched in Fig. 1c.

3.2.2 Preparing the Mating Mixture

Plasmids pGLR1 and pGLR1_*Pb* were transferred to *P. putida* KT2440 by triparental mating:

1. Aerobically grow overnight cultures of:
   (a) Donor cells: *E. coli* CC118 (pGLR1 and pGLR1_*Pb* plasmids) in LB with Km 150 μg/ml at 37 °C.
   (b) Helper cells: *E. coli* HB101 (pRK600 plasmid) in LB with chloramphenicol (Cm) 30 μg/ml at 37 °C.
   (c) Recipient cells: *P. putida* cells in LB at 30 °C.
2. Mix the three strains in a 1:1:1 ratio (e.g., 100 μl of each strain) in an empty test tube and complete with 5 ml of 10 mM $MgSO_4$.
3. Vortex the mixture and pass through a Millipore filter disk (0.45-μm pore-size, 13-mm diameter) using a 10 ml sterile syringe.
4. Place the filter onto an LB agar plate and incubate for a maximum of 5 h at 30 °C (In this case, the optimum temperature for recipient cells).
5. With the help of flamed sterile forceps take the filter from the LB plate and introduce it into a test tube with 5.0 ml of 10 mM $MgSO_4$ and vortex to resuspend the cells.
6. Plate appropriate dilutions (*see* **Note 3**) onto M9 citrate with Km 50 μg/ml (selective media for *P. putida* cells) onto 140 mm diameter petri dishes.

3.3 Bioluminescence Assays

The methodology to assay *Pb* activity in the whole population of cultured *P. putida* was carried out by quantification of bioluminescence produced by *luxCDABE* activity. This complete operon does not need of any aldehyde addition as substrate for the activity of the luciferase enzyme encoded by *luxAB* genes [25]. As mentioned above, this benefit is apparent in experiments for determining non-disruptively the induction kinetics of the promoter under study [26, 27]. Emission of light was measured along exponential phase (early 4–5 h).

3.3.1 Expression Landscape of the Pb Promoter

The bioluminescence output signal originated by *Pb* in the pGLR1_*Pb* plasmid placed in *P. putida* was employed to generate an expression landscape of the promoter in response to various concentrations of benzoate. To this end:

1. Aerobically grow overnight cultures of *P. putida* KT2440 leading pGLR1_*Pb*. Grow in M9 citrate 0.2 % with Km 50 mg/ml.
2. Prepare separately a tube (Falcon) with 20 ml of M9 citrate 0.2 % and Km 50 mg/ml.
3. Prepare ten vials each with 2 ml of medium above mentioned plus benzoate at the following concentrations: 1 mM, 750 μM,

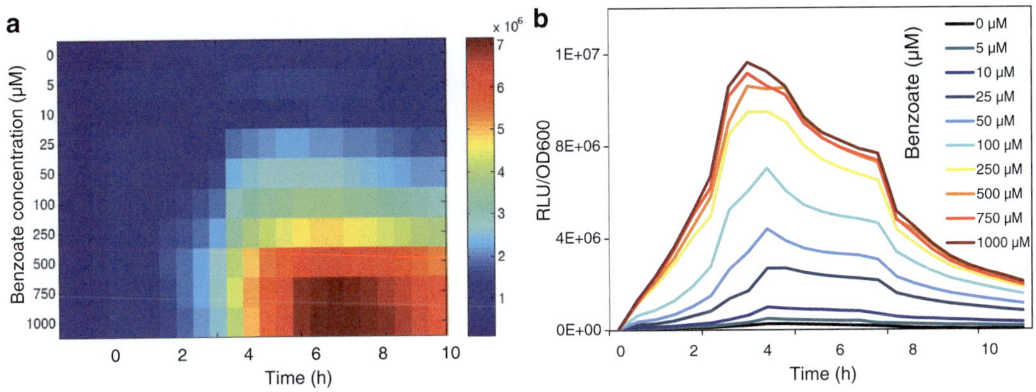

Fig. 2 Lux-based expression landscape of the *Pb* promoter in *P. putida host*. Overnight-grown *P. putida* KT2440 (pGLR1-*Pb*) was diluted 1:20 in fresh medium containing different concentrations of benzoate, as indicated. At 30 min intervals, bioluminescence and OD_{600} signals were recorded. (**a**) Level of promoter activity relative to time, where *color bars at left* indicate the scale for promoter activity. (**b**) *Pb* kinetics represented with Excel 2008 tools. In this case promoter activity is calculated by normalizing reporter signal RLU to the OD_{600} (RLU/OD_{600})

500 μM, 250 μM, 100 μM, 50 μM, 25μM, 10 μM, and 5 μM and 0 μM.

4. After pre-growth, the cells are washed twice with 10 mM $MgSO_4$ buffer and diluted 1:20 (v/v) into fresh medium with different concentrations of benzoate.
5. Take a MicrotestTM 96-well assay black-plates (Cultex), fill with 200 μl per well of diluted culture and place eight replications into a Wallac Victor 2 Microplate Reader (Perkin Elmer).
6. Plates are then incubated at 30 °C with shaking for 5–6 h (*see* **Note 4**).
7. At time intervals of 30 min, the optical density at 600 nm (OD_{600}) and the luminescence of each of the cultures are measured. Non-inoculated M9 medium is used as a blank for adjusting the baseline for measurements. Promoter activities are calculated by normalizing the reporter signals (luminescence) to the OD_{600} readings (*see* **Note 5**).
8. The resulting data was processed using MATLAB software (MathWorks) or software Excel 2008. Figure 2a (obtained by MATLAB software), represents an example of expression landscape after the induction of *Pb*, while Fig. 2b is a graphic representation of *Pb* kinetics in response to several concentrations of benzoate, obtained by Excel software.

3.4 Fluorescence Assay for Single Cell Analysis

Flow cytometry is a high resolution and quantitative methodology that offer the possibility to describe this cell–cell variability [12, 28]. Typically, the method allows detection of changes at population scale that reflect variations in the relative proportion of cells which are ON versus OFF. As indicated below, the activity of *Pb* promoter was measured in individual *P. putida* cells by monitoring the expression of the stable variant of GFP present in the dual reporter employed in this chapter [29]. This allowed us to analyze the performance of single *P. putida* KT2440 cells carrying pGLR1_*Pb* and exposed to 1 mM of benzoate.

3.4.1 Preparing Samples

1. Aerobically grow overnight cultures of *P. putida* KT2440 carrying pGLR1_Pb. Grow in M9 succinate 0.2 % with Km 50 mg/ml.
2. Overnight-grown cells are diluted 1/20 in fresh M9 media containing 0.2 % succinate and incubated for 4–5 h.
3. After this pre-incubation, at the mid-exponential phase (OD = 0.3–0.4), the cells are split into two samples: one is induced by 1 mM benzoate and the other is used as a non-induced control (*see* **Note 6**).
4. Cultures are then incubated with shaking in air at 30 °C, and each hour after induction, an aliquot of each sample (1 ml) is stored on ice until analysis (*see* **Note 7**).

3.4.2 Fixing Cells for Flow Cytometry Analysis

If cytometry is not run immediately, cells can be fixed with *p-formaldehyde* as follows (*see* **Note 8**):

1. Centrifuge 1 ml of cells at 8,000 × *g* in e.g. a tabletop Eppendorf centrifuge % $15C for 1 min.
2. Wash cells with 600 μl of sterilized and filtered PBS 1×.
3. Centrifuge cells at 8,000 × *g* for 1 min.
4. Add 300 μl of *p*-formaldehyde 0.4 % and incubate at room temperature for 10 min.
5. Centrifuge cells at 5,000 × *g* for 30 s.
6. Wash cells with 300 μl of sterilized and filtered PBS 1×.
7. Centrifuge cells at 5,000 × *g* for 30 s.
8. Discard the supernatant and resuspend cells with 600 μl of sterilized and filtered PBS 1×.
9. Keep cells at room temperature until analysis.

3.4.3 Quantification of Pb Activity

We created an induction profile of *Pb* promoter by overlapping histograms obtained by flow cytometry analysis (Fig. 3a) and calculating the fold-change that, in this case, indicated the magnitude of the induction (Fig. 3b). The fold-change was calculated as follows:

1. Divide values of GFP obtained during the hours of induction for values of correspondent uninduced samples.
2. For example, as shown in Fig. 3b, if we divide the value of GFP of sample induced with benzoate 1 mM for 3 h (16 RFU/OD) by the value of the same but non-induced sample (that is 4 RFU/OD), we will achieve 4 as the figure of fold-induction.

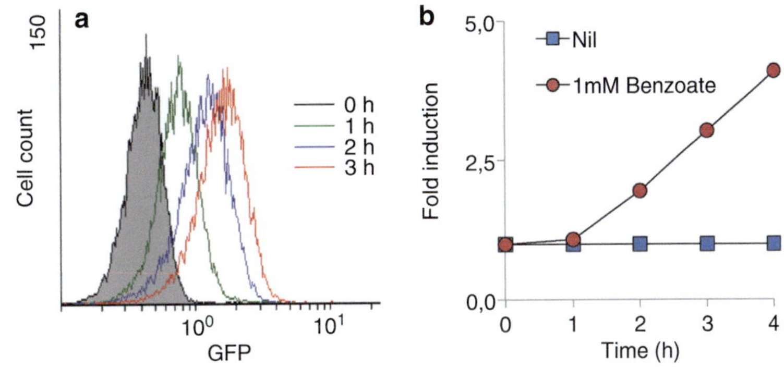

Fig. 3 Single-cell analysis of *Pb* activity. Overnight-grown *P. putida* KT2440 (pGLR1_*Pb*) was diluted 1:20 in fresh medium and allowed to grown to mid-exponential phase. At that point, 1 mM of benzoate was added to the culture. At 1 h time intervals, samples were collected and stored on ice until analysis by flow cytometry. Untreated cells were used as controls. For each assay, 15,000 cells were analyzed. (**a**) Induction profile of *P. putida* KT2440 *Pb*→ GFP-*luxCDABE* strain in response to 1 mM of benzoate. (**b**) Inducibility of the *Pb* promoter (fold-change). This parameter was calculated by normalizing the average fluorescence levels of induced populations to the corresponding signal of the control samples with no treatments

3.4.4 Flow Cytometry Data Analysis

1. Single-cell experiments were performed with a Gallios of Beckman Coulter flow cytometer, endowed with three lasers and ten fluorescence channels.
2. For the analysis of bacteria, a blue beam laser of 22 mW is used and GFP is excited at 488 nm.
3. We measured values of Forward Scatter (FSC), Side Scatter (SSC), and Fluorescence (FL1), all in a logarithmic scale.
4. Measurements of FSC were executed using the *Enhanced wide angle scatter* (w2) function.
5. Discriminator was realized on the SCS and FL1 signals and it was recovered with a filter 525(40).
6. For every aliquot of sample, we analyzed 15,000 events and data are processed by Cyflogic software (http://www.cyflogic.com/).

4 Notes

1. Instead of the M9 citrate medium, in some cases it is more convenient to use succinate as carbon source for *Pseudomonas* strains. In particular, succinate works well for flow cytometry analysis during rapid growth. We do not recommend employing LB medium because of its intrinsic auto fluorescence that can disturb GFP values in the fluorimeter (but not in flow cytometry assays).

2. Store the *p*-formaldehyde at room temperature in the dark, as it may deteriorate in the cold.
3. When all of the protocol steps are followed, we suggest to plate ≈ 100–200 μl, from the 5.0 ml mating mixture, onto 140 mm diameter petri dishes.
4. Lux activity depends on metabolic status of the cells; for this reason we suggest to consider informative the activity of *lux* as a transcriptional reporter only when the cells enter exponential phase (OD_{600} = 0.1–0.2) until early stationary phase (4–5 h after the addition of inducer).
5. It is also possible create an expression landscape using the merged GFP fluorescence of the culture (data not shown): the experimental conditions are the same of *lux*.
6. For induction assays and flow cytometry analysis, it is important to ensure that cells are in mid-exponential phase (OD_{600} = 0.3–0.4). When the culture is in a too early exponential phase, only few bacteria respond to inducer, resulting in an apparent noisy situation. For this reason, we often prefer succinate to citrate (*see* above) because of a faster growth of the cells. Additionally, in a flow cytometry experiment, it is mandatory to have a negative control (for example, we use a *P. putida* KT2440 strain containing a plasmid without GFP) together with uninduced and induced strains to ensure that the GFP signal is *bona fide*.
7. The time of induction depends on the biological system that we want to analyze. In this case, 4–5 (maximum) hours are sufficient.
8. This procedure of fixation is useful when the experiment takes a long time long (e.g., due to slow growth of the bacteria in minimal medium). The best, however, is to run the flow cytometry analysis the same day of the experiment. Never use fixed samples older than 2–3 days.

Acknowledgments

Authors are indebted to María Carmen Ortiz Navarro for help in flow cytometry assays. This work was supported by the BIO and FEDER CONSOLIDER-INGENIO programs of the Spanish Ministry of Economy and Competitiveness, the ARISYS and ST-FLOW Contracts of the EU, the ERANET-IB Program and funding from the Autonomous Community of Madrid (PROMPT). IB is the holder of an International Ph.D. Fellowship of La Caixa. All plasmids and strains described here are available upon request.

References

1. Browning DF, Busby SJ (2004) The regulation of bacterial transcription initiation. Nat Rev Microbiol 2:57–65
2. Hermsen R, Tans S, ten Wolde PR (2006) Transcriptional regulation by competing transcription factor modules. PLoS Comput Biol 2:e164
3. McAdams HH, Arkin A (1997) Stochastic mechanisms in gene expression. Proc Natl Acad Sci U S A 94:814–819
4. Kaern M, Elston TC, Blake WJ, Collins JJ (2005) Stochasticity in gene expression: from theories to phenotypes. Nat Rev Genet 6:451–464
5. Elowitz MB, Levine AJ, Siggia ED, Swain PS (2002) Stochastic gene expression in a single cell. Science 297:1183–1186
6. Elowitz MB, Leibler S (2000) A synthetic oscillatory network of transcriptional regulators. Nature 403:335–338
7. Silva-Rocha R, de Lorenzo V (2010) Noise and robustness in prokaryotic regulatory networks. Ann Rev Microbiol 64:257–275
8. Hooshangi S, Thiberge S, Weiss R (2005) Ultrasensitivity and noise propagation in a synthetic transcriptional cascade. Proc Natl Acad Sci U S A 102:3581–3586
9. Hasty J, Collins JJ (2002) Translating the noise. Nat Genet 31:13–14
10. Raj A, van Oudenaarden A (2008) Nature, nurture, or chance: stochastic gene expression and its consequences. Cell 135:216–226
11. Zaslaver A, Bren A, Ronen M, Itzkovitz S, Kikoin I, Shavit S, Liebermeister W, Surette MG, Alon U (2006) A comprehensive library of fluorescent transcriptional reporters for *Escherichia coli*. Nat Methods 3:623–628
12. Brehm-Stecher BF, Johnson EA (2004) Single-cell microbiology: tools, technologies, and applications. Microbiol Mol Biol Rev 68:538–559
13. Kelly JR, Rubin AJ, Davis JH, Ajo-Franklin CM, Cumbers J, Czar MJ, de Mora K, Glieberman AL, Monie DD, Endy D (2009) Measuring the activity of BioBrick promoters using an in vivo reference standard. J Biol Eng 3:4
14. Heim R, Cubitt AB, Tsien RY (1995) Improved green fluorescence. Nature 373:663–664
15. Kohlmeier S, Mancuso M, Tecon R, Harms H, van der Meer JR, Wells M (2007) Bioreporters: *gfp* versus *lux* revisited and single-cell response. Biosens Bioelectron 22:1578–1585
16. de Las Heras A, Carreno CA, Martinez-Garcia E, de Lorenzo V (2010) Engineering input/output nodes in prokaryotic regulatory circuits. FEMS Microbiol Rev 34:842–865
17. Keasling JD (1999) Gene-expression tools for the metabolic engineering of bacteria. Trends Biotechnol 17:452–460
18. Meighen EA (1991) Molecular biology of bacterial bioluminescence. Microbiol Rev 55:123–142
19. Meighen EA (1993) Bacterial bioluminescence: organization, regulation, and application of the *lux* genes. FASEB J 7:1016–1022
20. Benedetti IM et al (2012) Quantitative, non disruptive monitoring of transcription in single cells with a broad-host range GFP-*luxCDABE* dual reporter system. PLoS One 7:e52000
21. Silva-Rocha R, Martinez-Garcia E, Calles B, Chavarria M, Arce-Rodriguez A, de Las Heras A, Paez-Espino AD, Durante-Rodriguez G, Kim J, Nikel PI, Platero R, de Lorenzo V (2012) The Standard European Vector Architecture (SEVA): a coherent platform for the analysis and deployment of complex prokaryotic phenotypes. Nucleic Acids Res 41:D666–D675
22. Cowles CE, Nichols NN, Harwood CS (2000) BenR, a XylS homologue, regulates three different pathways of aromatic acid degradation in *Pseudomonas putida*. J Bacteriol 182:6339–6346
23. Sambrook J, Fritsch EF, Maniatis T (1989) Molecular cloning: a laboratory manual. Cold Spring Harbor Laboratory Press, Cold Spring Harbor, New York
24. de Lorenzo V, Timmis KN (1994) Analysis and construction of stable phenotypes in gram-negative bacteria with Tn*5*- and Tn*10*-derived minitransposons. Methods Enzymol 235:386–405
25. Ferri SR, Soly RR, Szittner RB, Meighen EA (1991) Structure and properties of luciferase from Photobacterium phosphoreum. Biochem Biophys Res Commun 176:541–548
26. Canton B, Labno A, Endy D (2008) Refinement and standardization of synthetic biological parts and devices. Nat Biotechnol 26:787–793
27. Young JW, Locke JC, Altinok A, Rosenfeld N, Bacarian T, Swain PS, Mjolsness E, Elowitz MB (2012) Measuring single-cell gene expression dynamics in bacteria using fluorescence time-lapse microscopy. Nat Protoc 7:80–88
28. Czechowska K, Johnson DR, van der Meer JR (2008) Use of flow cytometric methods for single-cell analysis in environmental microbiology. Curr Opin Microbiol 11:205–212
29. Miller WG, Lindow SE (1997) An improved GFP cloning cassette designed for prokaryotic transcriptional fusions. Gene 191:149–153

30. Herrero M, de Lorenzo V, Timmis KN (1990) Transposon vectors containing non-antibiotic resistance selection markers for cloning and stable chromosomal insertion of foreign genes in gram-negative bacteria. J Bacteriol 172:6557–6567
31. Boyer HW, Roulland-Dussoix D (1969) A complementation analysis of the restriction and modification of DNA in *Escherichia coli*. J Mol Biol 41:459–472
32. Bagdasarian M, Lurz R, Ruckert B, Franklin FC, Bagdasarian MM, Frey J, Timmis KN (1981) Specific-purpose plasmid cloning vectors. II. Broad host range, high copy number, RSF1010-derived vectors, and a host-vector system for gene cloning in *Pseudomonas*. Gene 16:237–247
33. Keen NT, Tamaki S, Kobayashi D, Trollinger D (1988) Improved broad-host-range plasmids for DNA cloning in gram-negative bacteria. Gene 70:191–197
34. Antoine R, Locht C (1992) Isolation and molecular characterization of a novel broad-host-range plasmid from *Bordetella bronchiseptica* with sequence similarities to plasmids from gram-positive organisms. Mol Microbiol 6: 1785–1799
35. Thomas CM, Cross MA, Hussain AA, Smith CA (1984) Analysis of copy number control elements in the region of the vegetative replication origin of the broad host range plasmid RK2. EMBO J 3:57–63

Chapter 45

Chromatin Immunoprecipitation for ChIP-chip and ChIP-seq

Sebastian Schulz and Susanne Häussler

Abstract

Bacterial adaptation to given environmental conditions is largely achieved by complex gene regulatory processes. To address the question how and to what extend single transcriptional regulators modulate gene expression, chromatin immunoprecipitation (ChIP) coupled to DNA microarrays (ChIP-chip) or to next-generation sequencing (ChIP-seq) is one of the preferred methods. Both ChIP-chip and ChIP-seq can generate genome-wide maps of protein–DNA interactions and thus identify primary regulons of transcription factors. In combination with transcriptome analyses, the obtained data can be used to compile complex regulatory networks which in terms will advance our understanding of bacterial adaptation processes to specific environmental conditions.

Key words Chromatin immunoprecipitation, ChIP-chip, ChIP-seq, Transcription factors, Protein–DNA interactions, Primary regulon, Regulatory networks

1 Introduction

Global gene expression is tightly regulated in all living cells and essential for successful adaptation to changing environmental conditions. Within the complex and dynamic interplay of numerous factors involved in gene regulatory processes, protein–DNA interactions are of particular importance. They are directly involved in the regulation of gene transcription, DNA replication, recombination, repair and structural genome arrangements [1]. Specific DNA protein complexes can be enriched by the use of primary antibodies which allow for chromatin immunoprecipitation (ChIP). This is a powerful method to selectively enrich DNA sequences to which a protein of interest has bound and if applied in combination with microarrays (ChIP-chip) or next-generation sequencing (ChIP-seq) can provide genome wide maps of protein–DNA interactions in a living cell under a given environmental condition. By the integration of ChIP-chip/seq data, global gene regulatory networks

Alain Filloux and Juan-Luis Ramos (eds.), *Pseudomonas Methods and Protocols*, Methods in Molecular Biology, vol. 1149, DOI 10.1007/978-1-4939-0473-0_45, © Springer Science+Business Media New York 2014

can be reconstructed which provide deep insight into complex adaptation processes.

In brief, the standard ChIP protocol includes the following steps: (1) cell cultivation and cross-linking of protein–DNA complexes, (2) cell lysis and DNA fragmentation, (3) protein–DNA complex enrichment by immunoprecipitation, (4) de-cross-linking and DNA purification, (5) amplification of immunoprecipitated DNA. In ChIP-chip experiments, DNA is subsequently labeled and hybridized to a DNA microarray, while in ChIP-seq approaches DNA libraries are established and subjected to high-throughput sequencing (*see* Fig. 1).

For decades, experimental approaches to characterize protein–DNA interactions suffered from constraints of traditional biochemical and genetic approaches. Unlike in ChIP, biochemical approaches, such as gel shift assays, were performed under non-physiological

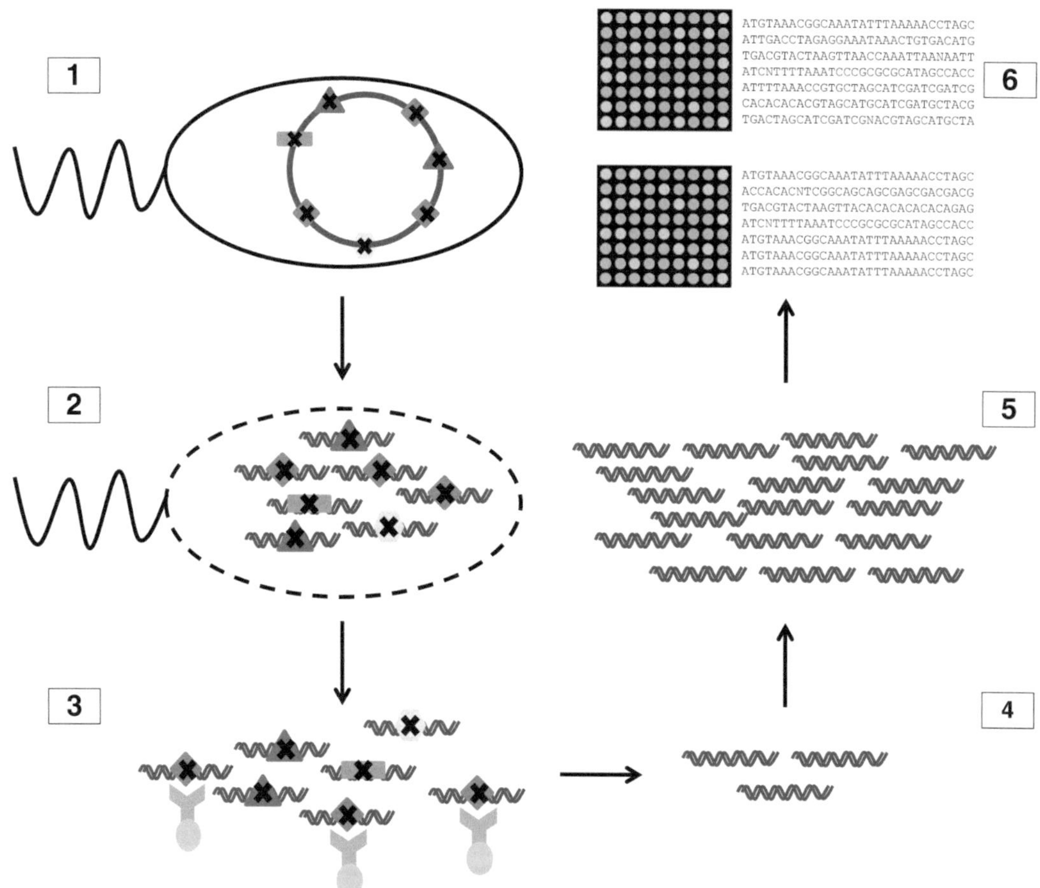

Fig. 1 Workflow of ChIP-chip and ChIP-seq. Cells are cultivated under selected condition (1) and formaldehyde is added to conserve protein–DNA interactions (2). Next, protein–DNA complexes of interest are specifically enriched by ChIP (3). After de-cross-linking protein–DNA and DNA purification (4), the immunoprecipitated DNA is amplified (5) and finally subjected either to DNA microarrays or next-generation sequencing (6)

conditions, while the genetic approach, e.g., transcriptome profiling of mutant strains vs. wild type strain, rather reflects the consequence of the binding event than the binding event itself [1]. David Gilmour and John Lis introduced ChIP in 1984, when they investigated the distribution and density of RNA-polymerase on specific bacterial genes [2]. One year later, they applied ChIP to compare interactions of RNA-polymerase II with genes of *Drosophila melanogaster* under normal and heat-shock conditions [3]. In 1988, Mark Solomon and colleagues improved the ChIP protocol by using formaldehyde as a cross-linking agent to conserve protein–DNA interactions which is commonly termed as X-CHIP protocol [4]. Since that time, formaldehyde is the most frequently used cross-linking agent as its cross-links are very tight (2 Å) and can be easily reversed with heat [5]. However, there are evidences that very fast transient protein–DNA interactions may escape the cross-linking process [6] and also the combination of formaldehyde with an additional cross-linking agent has been reported to extend the power of ChIP [7]. In addition to X-ChIP protocols, Tim Hebbes and colleagues established the native ChIP protocol (N-ChIP) without cross-linker to study the link of histone modifications and transcriptionally active chromatin in chicken cells providing a valuable technique for epigenetic analyses [8].

The success of a ChIP experiment is to a large part dependent on the selection of the right antibody [9]. Antibodies which were successfully applied in previous ChIP experiments are classified as ChIP grade or ChIP validated antibodies. These antibodies can recognize the protein of interest in the DNA-bound conformation and post formaldehyde treatment which has a large impact on the protein structure [10]. If a ChIP grade antibody is not available, the use of epitope tags is an option. However, each protein–DNA interaction is unique and the quality and functionality of each single antibody must be experimentally evaluated. In general, affinity purified antibodies are recommended and polyclonal antibodies are preferred over monoclonals to minimize the risk of epitope loss [5].

The introduction of DNA microarrays and next-generation sequencing significantly advanced the field of systems biology and genome-wide transcriptional profiling and enabled profound insights into the complexity of transcriptional regulatory networks [11]. In 2000, Bing Ren and colleagues established the first ChIP-chip protocol by coupling ChIP with DNA microarrays [12] and 7 years later David Johnson and colleagues reported the first successful ChIP-seq approach [13]. Both techniques uncover the direct regulon of specific transcriptional regulators and thus allow the discrimination of direct from indirect effects of single factors on the global gene regulatory network.

The new possibility offered by ChIP-chip and ChIP-seq gave rise to various modifications to overcome limitations like input cell number and to save both time and money [14–20]. In the end, each step was subject of extensive refinement to minimize the number of false-positive as well as false-negative results [21]. ChIP-seq offers considerable advantages over ChIP-chip such as higher resolution, better coverage, larger dynamic range and lower platform noise. Furthermore, ChIP-seq requires less immunoprecipitated DNA and samples can be multiplexed to reduce costs [22]. For these reasons, ChIP-chip becomes continuously replaced by ChIP-seq.

On the following pages, we present a standard X-ChIP protocol which was applied to enrich for *Pseudomonas aeruginosa* DNA binding proteins. The immunoprecipitated DNA was purified, amplified in a linear manner similar to the protocol recently published by Shankaranarayanan and colleagues [15] and finally subjected to Illumina sequencing.

2 Materials

2.1 Cross-linking and Cell Harvesting Components

1. Medium of choice.
2. UV/VIS spectrophotometer, e.g., Ultrospec 2000 (Pharmacia Biotech, Freiburg, Germany).
3. Formaldehyde solution 37 % (Roth, Cat. No: 4979, Karlsruhe, Germany).
4. Glycine (Roth, Cat. No: 3908.3, Karlsruhe, Germany). Dissolve in distilled water to a final concentration of 1.5 M.
5. Temperature-controlled benchtop centrifuge, e.g., Megafuge 3.0R (Heraeus Instruments, Hanau, Germany).
6. Tris buffered saline (TBS; 10×): 22.42 g/L Trizma-base, 80.06 g/L NaCl adjusted to pH 7.6 and autoclaved.

2.2 Cell Lysis and DNA Fragmentation by Sonication Components

1. Lysis buffer: 10 mM Tris–HCl pH 8, 20 % (w/v) sucrose, 50 mM NaCl, 10 mM EDTA$\times$2Na$\times$2H_2O, sterile filtrated.
2. Lysozyme (Sigma-Aldrich, Cat. No: L7651, Steinheim, Germany). Dissolve in destilled water to a final concentration of 20 mg/ml.
3. IP buffer: 50 mM Tris–HCl pH 7.5, 150 mM NaCl, 1 mM EDTA$\times$2Na$\times$2H_2O, 1 % (v/v) Triton X-100, 0.1 % (w/v) sodium deoxycholate, sterile filtrated.
4. cOmplete, EDTA-free protease inhibitor cocktail tablets (Roche Cat. No: 04693124001, Mannheim, Germany). Dissolve one tablet into 10 ml IP buffer.
5. Phenylmethanesulfonylfluoride solution (Sigma Aldrich, Cat. No: 2088311, Steinheim, Germany).
6. Sodium dodecyl sulfate (Sigma Aldrich, Cat. No: 4360.2, Karlsruhe, Germany).

7. Sonicator, e.g., Branson Sonicator S250 Analogue (Branson Ultrasonics Corporation, Danbury, USA).
8. Agarose (Biozym, Cat. No: 840004, Oldendorf, Germany).
9. TAE buffer: 40 mM Tris, 20 mM acetic acid, 1 mM EDTA × 2Na × $2H_2O$.
10. Horizon58 Horizental Gel Electrophoresis Apparatus (Biometra, Göttingen, Germany).
11. Gel documentation system (Intas, Göttingen, Germany).
12. GelStar Nucleic acid gel stain (Lonza Rockland, Cat. No: 50535, Maine, USA).
13. GeneRuler 50 bp DNA Ladder, ready-to-use (Thermo Scientific, Cat. No: SM0373).

2.3 Chromatin Immunoprecipitation (ChIP) Including Chromatin Check Components

1. Dynabeads Protein A or G (Invitrogen, Cat. No: 10003D, Oslo, Norway).
2. ChIP grade antibody (customized).
3. Rotator (Kisker Biotech, Steinfurt, Germany).
4. Magnetic stand (Qiagen, Hilden, Germany).
5. IP buffer: 50 mM Tris–HCl pH 7.5, 150 mM NaCl, 1 mM EDTA × 2Na × $2H_2O$, 1 % (v/v) Triton X-100, 0.1 % (w/v) sodium deoxycholate, sterile filtrated.
6. High salt IP buffer: 50 mM Tris–HCl pH 7.5, 500 mM NaCl, 1 mM EDTA × 2Na × $2H_2O$, 1 % (v/v) Triton X-100, 0.1 % (w/v) sodium deoxycholate, sterile filtrated.
7. Washing buffer: 10 mM Tris–HCl pH 8, 250 mM LiCl, 1 mM EDTA × 2Na × $2H_2O$, 0.5 % (v/v) Nonidet-P40, 0.5 % (w/v) sodium deoxycholate, sterile filtrated.
8. TE buffer: 10 mM Trizma-base, 1 mM EDTA × 2Na × $2H_2O$, adjusted to pH 7.5, autoclaved.
9. Thermomixer 5436 (Eppendorf, Hamburg, Germany).
10. Elution buffer: 50 mM Tris–HCl pH 7.5, 10 mM EDTA × 2Na × $2H_2O$, 1 % (w/v) SDS, sterile filtrated.

2.4 De-cross-linking, DNA Purification, and ChIP Evaluation Components

1. RNaseA (Qiagen, Cat. No: 19101, Hilden, Germany).
2. Proteinase K (Qiagen, Cat. No: 19133, Hilden, Germany).
3. Sodium acetate (Roth, Cat. No: 6773.2, Karlsruhe, Germany). Prepare a 3 M stock solution and adjust to pH 5 with glacial acetic acid.
4. QIAquick PCR purification kit (Qiagen, Cat. No: 28104, Hilden, Germany).
5. Qubit fluorometer (Invitrogen, Oslo, Norway).

2.5 Linear DNA Amplification Components

1. Alkaline Phosphatase Calf Intestinal (New England Biolabs, Art.-Nr. M0290S, Frankfurt am Main, Germany).
2. Nuclease-free water (Ambion, Cat. No: AM9937, Darmstadt, Germany).
3. MinElute Reaction Cleanup Kit (Qiagen, Cat. No: 28204, Hilden, Germany).
4. ddCTP (100 mM; GE Healthcare, Cat. No: 27-2061-01).
5. dTTP (100 mM; GE Healthcare, Cat. No: 28-4065-31).
6. Prepare T-Mix by combining 92 μl dTTP (100 μM) and 8 μl ddCTP (100 μM).
7. Terminal Transferase (New England Biolabs, Cat. No: M0315S, Frankfurt am Main, Germany).
8. 10 μM T7-promoter-BpmI-$(dA)_{15}$—primer synthesized by Eurofins mwg operon (Ebersberg, Germany): 5′CGAAATT AATACGACTCACTATAGGGCTGGAGAAAAAAAAAAA AAAA-3′.
9. 10 μM T7-promoter-BpmI-$(dA)_{18}$—primer synthesized by Eurofins mwg operon (Ebersberg, Germany): 5′CGAAATT AATACGACTCACTATAGGGCTGGAGAAAAAAAAAAA AAAAAAAA-3′.
10. LinDA buffer: 20 mM Tris acetate, 10 mM magnesium acetate, 50 mM potassium acetate, adjusted to pH 7.9.
11. dNTP mix (10 mM mix of dATP, dTTP, dCTP, and dGTP; GE Healthcare, Cat. No: 28-4065-64).
12. Klenow fragment (New England Biolabs, Cat. No: M0210S, Frankfurt am Main, Germany).
13. RNAMaxx high yield kit (Agilent Technologies, 200339, Waldbronn, Germany).
14. RNeasy MinElute Cleanup Kit (Qiagen, Cat. No: 74204, Hilden, Germany).
15. Superscript III reverse transcription kit (Invitrogen, Cat. No: 18080-093, Oslo, Norway).
16. Pfu polymerase (Promega, Cat. No: M7741, Mannheim, Germany).
17. Taq polymerase (New England Biolabs, Cat. No: M0273S, Frankfurt am Main, Germany).
18. RNase H (New England Biolabs, Cat. No: M0297S, Frankfurt am Main, Germany).
19. BpmI (New England Biolabs, Cat. No: R0565S, Frankfurt am Main, Germany).

3 Methods

Carry out all procedures at room temperature unless otherwise indicated. Please follow all safety instructions and waste disposal regulations.

Before starting a ChIP experiment, it is recommended to identify the growth conditions under which the transcription factor of interest is active and binds to its specific target DNA (*see* **Note 1**).

3.1 Cross-linking and Cell Harvesting

To achieve optimal results, please check the recommendations for cross-linking (*see* **Note 2**).

1. Inoculate pre-cultures in glass tubes with 4 ml medium and cultivate them for at least 12 h at 37 °C with shaking.
2. Start main cultures in 250 ml flasks with 50 ml medium at $OD_{600nm}=0.05$. Prepare main cultures in triplicates and cultivate them under optimal condition.
3. Check OD_{600nm} and combine the two most similar cultures. For better handling, split the culture 4×20 ml into 50 ml tubes.
4. Add 0.5 ml 37 % formaldehyde to each tube and invert gently at room temperature for 10 min (*see* **Note 3**).
5. Add 2 ml glycine (1.5 M) to each tube to quench the cross-link reaction. Invert gently at room temperature for 5 min. The following steps should be carried out at 4 °C.
6. Harvest cells in a pre-cooled centrifuge at 2,500×*g* or higher and at 4 °C for 20 min.
7. Wash cell pellets 2 times with 10 ml ice-cooled 1× TBS to remove medium and cell debris.
8. Resuspend cell pellets in 1 ml ice-cooled TBS, transfer to a new 1.5 ml tube and spin-down at 6,000×*g* for 2 min.
9. Remove supernatant and freeze cell pellets at −70 °C for at least 30 min (*see* **Note 4**).

3.2 Cell Lysis and DNA Fragmentation by Sonication

Sonication efficiency depends on several parameters. Check the guideline in advance to obtain best results (*see* **Note 5**). The efficiency of DNA fragmentation should be checked on a 2 % agarose gel (*see* Fig. 2).

1. Thaw frozen cell pellets on ice.
2. Resuspend each cell pellet in 0.5 ml lysis buffer supplemented with cOmplete EDTA-free solution to inhibit proteases.
3. Add 125 μl lysozyme (20 mg/ml) and incubate the mixture at 37 °C for 30 min.

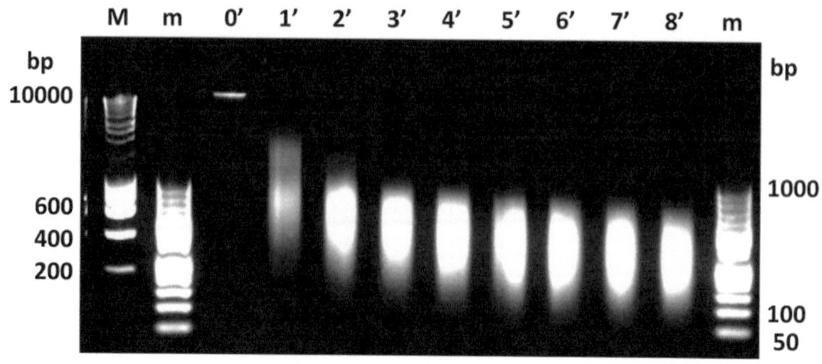

Fig. 2 DNA fragmentation pattern in correlation to the number of sonication cycles. Lane M and m are loaded with Smart Ladder and 50-bp ladder respectively. Lane 3 contains the genomic DNA of the cell lysate before sonication. Sheared DNA species upon each sonication round of 1 min are displayed in lane 4–11

4. Combine the samples in a 15 ml tube and add 1.5 ml IP-buffer supplemented with cOmplete, EDTA-free protease inhibitor cocktail, 2 mM PMSF and 0.1 % SDS.
5. Put the sample on ice and prepare an ethanol ice bath.
6. Fix the tube in the ethanol ice bath and perform sonication using a Branson Sonifier 250 at level 4, duty cycle 90 % and hold.
7. Sonicate 5 times for 45 s with a 2 min rest between the rounds of sonication (*see* **Note 6**).
8. Aliquote sample 4 × 1 ml in 1.5 ml tubes and freeze cell lysate at −70 °C.

3.3 Chromatin Immunoprecipitation (ChIP) Including Chromatin Check

Please check the recommendations and considerations for ChIP before starting your experiment (*see* **Notes 7–9**).

1. Thaw cell lysate on ice.
2. Completely resuspend magnetic beads by pipetting.
3. Transfer 25 μl magnetic beads to a 2 ml tube.
4. Separate magnetic beads from solution using a magnetic separation stand and remove solution.
5. Add the amount of cell lysate which contains 25–50 μg sheared DNA.
6. Add 5 vol or up to 1 ml IP-buffer supplemented with protease inhibitors cocktail to the cell lysate.
7. Incubate on a rotator at 4 °C and low speed for 1 h.
8. Separate supernatant from magnetic beads (*see* above) and transfer cleared supernatant to a new 2 ml tube.
9. Add 5–10 μg antibody to the supernatant and incubate on a rotator at 4 °C for at least 12 h.

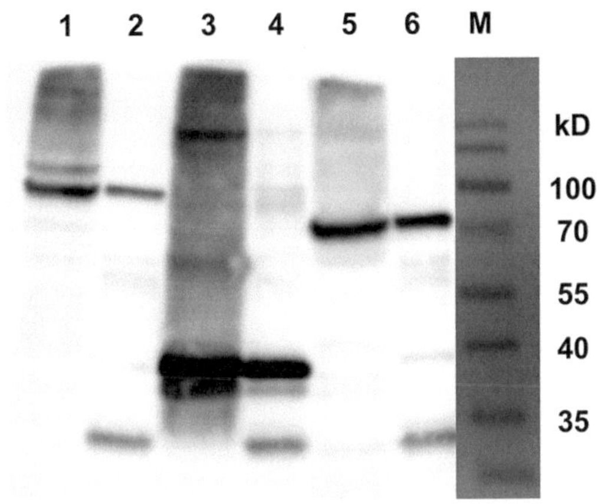

Fig. 3 Western blot analysis of octahistidine-tagged sigma factors RpoD, RpoH, and RpoN. Lane 1, 3, and 5 depict the cell lysates before ChIP. In lane 2, 4, and 6 the immunoprecipitated sigma factors are shown. Lane M contains pre-stained protein ladder

10. Add 50 μl magnetic beads and incubate on a rotator at 4 °C for 10 min up to 1 h.
11. Remove supernatant and store at −20 °C
12. Wash the magnetic bead–antibody–antigen–DNA complexes by adding 1 ml IP buffer supplemented with 1 mM PMSF and completely resuspend magnetic beads by gently inverting.
13. Separate magnetic beads from supernatant (*see* above) and store supernatant at −20 °C.
14. Repeat washing step with 1 ml IP buffer supplemented with 1 mM PMSF, 1 ml high salt IP buffer, 1 ml washing buffer and 1 ml TE-buffer (pH 7.5).
15. Resuspend magnetic bead–antibody–antigen–DNA complexes in 50 μl elution buffer and transfer mix to a new 1.5 ml tube.
16. Incubate mix at 65 °C with vigorously shaking for 15 min.
17. Separate supernatant containing antibody–antigen–DNA complexes from magnetic beads (*see* above) and transfer to new 1.5 ml tube.
18. Repeat the elution step to increase yield of immunoprecipitated protein–DNA complexes and store supernatant at −20 °C or proceed directly.

3.4 De-cross-linking, DNA Purification, and ChIP Evaluation

Before reversal of protein–DNA cross-links, it is recommended to analyze up to 30 μl eluate to check ChIP efficiency by SDS-PAGE and western blot analysis. The cell lysate can be used as a positive control (*see* Fig. 3). In the same run, ChIP supernatant and magnetic beads post elution can also be tested. *See* also recommendations for ChIP evaluation (*see* **Note 10**).

1. To de-cross-link protein–DNA complexes, transfer 70 μl eluate into a 0.2 ml PCR-tube.
2. Add 1 μl RNase A (100 mg/ml) and incubate in a PCR-cycler at 65 °C for 30 min.
3. Add 5 μl Proteinase K (20 mg/ml) and incubate at 50 °C for 1 h.
4. Shift to 65 °C for at least 6 h.
5. Add 5 μl Proteinase K (20 mg/ml) and incubate at 50 °C for 1 h.
6. Keep at 4 °C.
7. To purify DNA, adjust volume to 100 μl with nuclease-free water and add 3 μl sodium acetate (3 M, pH 5) to reduce pH value for optimal binding conditions.
8. Use the Qiagen PCR purification kit to isolate DNA according to manufacturer's instructions.
9. Elute DNA with 50 μl pre-warmed EB buffer at 55 °C to improve elution efficiency.
10. Quantify DNA concentration by the Qubit system.
11. Store DNA at −20 °C or proceed directly with Subheading 3.5.
12. To evaluate ChIP, use up to 1 ng immunoprecipitated DNA and 1 ng input DNA as template for each PCR assay.
13. Run PCR for 25–30 cycles.
14. Analyze and compare band intensity on a 2 % agarose gel to check specific DNA enrichment.

3.5 Linear DNA Amplification

Usually the amount of obtained ChIPed DNA is very low and must be amplified. Exponential amplification by PCR has been reported to be highly susceptible to bias [23]. For this reason, the linear DNA amplification (LinDA) protocol recently published by Shankaranarayanan [15] is recommended. The following procedure is a slightly modified version of this protocol.

3.5.1 DNA Dephosphorylation

1. Reduce sample volume to 26 μl in a speed vacuum machine at room temperature.
2. For DNA dephosphorylation, set up the following reaction mix:

Immunoprecipitated DNA	26 μl
NEB buffer 3 (10×)	3 μl
Shrimp alkaline phosphatase (1 U/μl)	1 μl

3. Incubate at 37 °C for 60 min.
4. Shift temperature to 70 °C for 10 min to stop reaction and cool down to 4 °C.

5. Adjust volume to 100 μl with nuclease-free water and use Qiagen MinElute reaction cleanup kit to purify DNA and elute DNA in 20 μl nuclease-free water.
6. Prepare T-Mix by mixing 92 μl dTTP (100 μM) and 8 μl ddCTP (100 μl).

3.5.2 DNA t-Tailing

1. For DNA t-tailing, set up the following reaction mix:

DNA from previous step	20 μl
Terminal transferase buffer (10×)	3 μl
2.5 mM $CoCl_2$	2 μl
T-Mix	1 μl
Terminal transferase (20 U/μl)	1 μl

2. Incubate at 37 °C for 20 min.
3. Shift temperature to 70 °C for 10 min to stop reaction and cool down to 4 °C.
4. Adjust volume to 100 μl with nuclease-free water and use Qiagen MinElute reaction cleanup kit to purify DNA and elute DNA in 20 μl nuclease-free water.

3.5.3 Primer Annealing and Extension

1. Please set up the following reaction mix:

DNA from previous step	20 μl
T7-promoter-BpmI-$(dA)_{18}$-primer (10 μM)	1 μl
10 mM dNTPs	1 μl
LinDA buffer with 1 mM DTT	4 μl

2. Incubate at 37 °C for 5 min.
3. Add 1 μl Klenow fragment and incubate at 37 °C for 55 min.
4. Shift temperature to 70 °C for 10 min to stop reaction and cool down to 4 °C.

3.5.4 Amplification via In Vitro Transcription and RNA Isolation

1. Use the RNAMaxx high yield kit and add the following components:

DNA from previous step	27 μl
RNAMaxx transcription buffer (5×)	10 μl
100 mM rUTP, rATP, rCTP, rGTP mixed	8 μl
0.75 M DTT	2 μl
yeast inorganic phosphatase (0.75 U/μl)	0.5 μl
RNase block	1 μl
T7 polymerase (200 U/μl)	1 μl

2. Incubate reaction at 37 °C for at least 16 h.
3. Adjust sample volume to 100 μl with nuclease-free water and use Qiagen RNeasy MinElute reaction cleanup kit to isolate RNA and elute RNA in 20 μl nuclease-free water.
4. Quantify RNA concentration and use maximal 500 ng RNA per reaction.

3.5.5 Reverse Transcription

1. For reverse transcription, set up the following reaction mix:

RNA	20 μl
T7-promoter-BpmI-$(dA)_{15}$-primer (10 μM)	2 μl

2. Incubate at 65 °C for 10 min to reverse secondary structure.
3. Put sample immediately on ice to maintain linear RNA.
4. Prepare master mix as follows:

1st strand buffer (5×)	8 μl
100 mM DTT	4 μl
10 mM dNTPs (each)	2 μl
RNase out	2 μl
SuperScript III (200 U/μl)	2 μl

5. Add master mix to RNA sample.
6. Incubate at 42 °C for 60 min.
7. Shift temperature to 75 °C for 15 min.
8. Cool down to 4 °C.

3.5.6 Second Strand Synthesis

1. Split cDNA sample to 2×20 μl and add the following components:

cDNA	20 μl
Pfu buffer (10×)	10 μl
BSA (100×)	1 μl
10 mM dNTPs	3 μl
RNase H (10 U/μl)	0.5 μl
Nuclease-free water	64.9 μl
Taq polymerase (5 U/μl)	0.5 μl
Pfu polymerase (2.5 U/μl)	0.1 μl

2. Incubate at 37 °C for 5 min.
3. Shift temperature to 65 °C for 1 min.

4. Shift temperature to 72 °C for 30 min.
5. Cool down to 4 °C.
6. Use for example the Qiagen PCR purification kit to isolate DNA according to manufacturer's instructions and elute in 50 μl nuclease-free water.

3.5.7 Primer Removal

1. Perform restriction assay as follows:

DNA	50 μl
NEB buffer 3 (10×)	6 μl
BpmI (2.5 U/μl)	2 μl

2. Incubate at 37 °C for 60 min.
3. Add 2 μl BpmI (2.5 U/μl).
4. Incubate at 37 °C for 30 min.
5. Use the Qiagen PCR purification kit to isolate DNA according to manufacturer's instructions and elute in 50 μl EB buffer.
6. Reduce volume in a speed vacuum machine down to 30 μl.
7. Quantify DNA concentration by the Qubit system.
8. Store DNA at −20 °C.

A minimum of 10 ng is required for ChIP-seq to proceed with the library establishment for next-generation sequencing (ChIP-seq), while ChIP-chip needs up to 5 μg immunoprecipitated DNA for labeling of the DNA for DNA microarray hybridization (ChIP-chip).

4 Notes

1. If a specific target promoter of the transcription factor of interest is known, the generation of reporter strains based on promoter-reporter gene fusions (e.g., *luxCDABE* or *lacZ*) is an option. If a promoterless reporter strain is used as a control, the reporter activity in response to selected growth conditions can be determined.
2. Formaldehyde-mediated cross-linking is a time-dependent reaction and must be optimized for each ChIP experiment. Excessive cross-linking may result in epitope masking and reduced sonication efficiency, while insufficient cross-linking decreases the yield of protein–DNA complexes. It is recommended to start with a final formaldehyde concentration of 1 % and to incubate for 10 min as a standard setting. However, a time series experiment to select the optimal conditions has to be performed by western blot analysis.

3. Formaldehyde is highly toxic and carcinogenic. Wear gloves, safety glasses and handle formaldehyde according to safety instructions. Please make sure that tubes are properly closed before inverting.
4. Cell pellets can also be processed directly. However, freezing and thawing will improve cell disruption. Be aware that depending on your sample number cell lysis and sonication may be time-consuming steps.
5. Sonication efficiency depends on several parameters such as sonication time, power output, cross-linking regime, sample volume as well as growth state of the cells and requires a time-course experiment to determine optimal conditions. For ChIP-seq the fragment size range is 100–600 bp with an optimal size of 200 bp. It is critical to avoid both heating of samples to prevent premature de-cross-linking and foam formation which reduces sonication efficiency.
6. Sonication can harm your ears. Protect your ears properly.
7. Before starting ChIP, it is recommended to check expression and presence of the transcription factor by SDS-PAGE and western blot analysis.
8. There is no consensus which kind of control sample serves best for ChIP. However, the input control is commonly used as the reference sample for subsequent enrichment calculations. The input control is a small aliquot of the cell lysate (about 10 %) that is not subjected to the antibody-mediated pull-down step. Processing the input control can simultaneously be used to check DNA fragmentation and to determine the DNA concentration to normalize DNA input for ChIP.
9. Sepharose or magnetic beads coupled to protein A or G are used to isolate antibody–antigen–DNA complexes from the cell lysate. Please check whether protein A or G shows the highest affinity to the antibody in use. Generally, magnetic beads are preferred and it is recommended to incubate the antigen–DNA complex with the antibody prior to magnetic bead addition and harvesting (indirect approach). The following protocol includes an optional step to remove unspecifically bound proteins to magnetic beads and describes the stringent washing procedure.
10. Success of ChIP should be checked by standard or quantitative PCR analysis. To this end, primer pairs are required, one pair which flanks promoter regions targeted by the transcription factor and one pair which amplifies a non-intergenic region of a housekeeping gene such as citrate synthase as control region.

Acknowledgement

This work was supported by the President's Initiative and Networking Funds of the Helmholtz Association of German Research Centers (HGF) under contract VH-GS-202.

References

1. Wade JT et al (2007) Genomic analysis of protein-DNA interactions in bacteria: insights into transcription and chromosome organization. Mol Microbiol 65(1):21–26
2. Gilmour DS, Lis JT (1984) Detecting protein-DNA interactions in vivo: distribution of RNA polymerase on specific bacterial genes. Proc Natl Acad Sci U S A 81(14):4275–4279
3. Gilmour DS, Lis JT (1985) In vivo interactions of RNA polymerase II with genes of Drosophila melanogaster. Mol Cell Biol 5(8):2009–2018
4. Solomon MJ, Larsen PL, Varshavsky A (1988) Mapping protein-DNA interactions in vivo with formaldehyde: evidence that histone H4 is retained on a highly transcribed gene. Cell 53(6):937–947
5. Orlando V (2000) Mapping chromosomal proteins in vivo by formaldehyde-crosslinked-chromatin immunoprecipitation. Trends Biochem Sci 25(3):99–104
6. Schmiedeberg L et al (2009) A temporal threshold for formaldehyde crosslinking and fixation. PLoS One 4(2):e4636
7. Zeng PY et al (2006) In vivo dual cross-linking for identification of indirect DNA-associated proteins by chromatin immunoprecipitation. Biotechniques 41(6):694, 696, 698
8. Hebbes TR, Thorne AW, Crane-Robinson C (1988) A direct link between core histone acetylation and transcriptionally active chromatin. EMBO J 7(5):1395–1402
9. Liu ET, Pott S, Huss M (2010) Q&A: ChIP-seq technologies and the study of gene regulation. BMC Biol 8:56
10. Metz B et al (2004) Identification of formaldehyde-induced modifications in proteins: reactions with model peptides. J Biol Chem 279(8):6235–6243
11. Goodman AL, Lory S (2004) Analysis of regulatory networks in Pseudomonas aeruginosa by genomewide transcriptional profiling. Curr Opin Microbiol 7(1):39–44
12. Ren B et al (2000) Genome-wide location and function of DNA binding proteins. Science 290(5500):2306–2309
13. Johnson DS et al (2007) Genome-wide mapping of in vivo protein-DNA interactions. Science 316(5830):1497–1502
14. Gilfillan GD et al (2012) Limitations and possibilities of low cell number ChIP-seq. BMC Genomics 13:645
15. Shankaranarayanan P et al (2012) Single-tube linear DNA amplification for genome-wide studies using a few thousand cells. Nat Protoc 7(2):328–338
16. O'Neill LP, VerMilyea MD, Turner BM (2006) Epigenetic characterization of the early embryo with a chromatin immunoprecipitation protocol applicable to small cell populations. Nat Genet 38(7):835–841
17. Nelson JD, Denisenko O, Bomsztyk K (2006) Protocol for the fast chromatin immunoprecipitation (ChIP) method. Nat Protoc 1(1):179–185
18. Dahl JA, Collas P (2008) A rapid micro chromatin immunoprecipitation assay (microChIP). Nat Protoc 3(6):1032–1045
19. Flanagin S et al (2008) Microplate-based chromatin immunoprecipitation method, Matrix ChIP: a platform to study signaling of complex genomic events. Nucleic Acids Res 36(3):e17
20. Lefrancois P et al (2009) Efficient yeast ChIP-Seq using multiplex short-read DNA sequencing. BMC Genomics 10:37
21. Waldminghaus T, Skarstad K (2010) ChIP on Chip: surprising results are often artifacts. BMC Genomics 11:414
22. Park PJ (2009) ChIP-seq: advantages and challenges of a maturing technology. Nat Rev Genet 10(10):669–680
23. Liu CL, Schreiber SL, Bernstein BE (2003) Development and validation of a T7 based linear amplification for genomic DNA. BMC Genomics 4(1):19

Chapter 46

Transcriptional Analysis of *Pseudomonas aeruginosa* Infected *Caenorhabditis elegans*

Ashleigh MacKenzie, Lewis Stewart, Paul A. Hoskisson, and Nicholas P. Tucker

Abstract

Here, we describe a protocol for the extraction of total RNA from *C. elegans* infected with *P. aeruginosa*. The protocol excludes *P. aeruginosa* cells that have not been ingested by the nematodes and yields total RNA that can be used for detection of transcripts from both host and pathogen.

Key words *P. aeruginosa*, RT-PCR, RNA extraction, Transcription

1 Introduction

Nonmammalian infection models for bacteria have great potential for mutant and phenotype testing and high-throughput screening of drugs and effectors. A range of invertebrate infection models for studying *Pseudomonas aeruginosa* virulence have emerged since the late 1990s, including the use of the model nematode *C. elegans* [1, 2]. The *C. elegans* infection model is particularly useful because of the organisms' small size, high reproductive rate, and ease of laboratory culture (*see* Chapter 48 in this volume). Many of the genes required for virulence in the nematode infection model are also required for virulence in mammals, suggesting that *P. aeruginosa* requires similar physiological processes in both cases. This simple assay therefore offers a simple and rapid assay for virulence gene induction to allow the investigation of regulatory mechanisms and networks. So far we have used reverse transcriptase PCR and qRT-PCR to analyze regulatory events within the nematode and we have been able to detect expression of both host and pathogen genes from a single RNA preparation. It is important to note that the nematode infection assay is carried out on a lawn of *P. aeruginosa*

Alain Filloux and Juan-Luis Ramos (eds.), *Pseudomonas Methods and Protocols*, Methods in Molecular Biology, vol. 1149, DOI 10.1007/978-1-4939-0473-0_46, © Springer Science+Business Media New York 2014

on a nematode growth medium plate. Consequently there are a significant number of *P. aeruginosa* cells on the plate that are not associated with the gut of the worms. This protocol allows for the cheap and efficient removal of *P. aeruginosa* cells that are not associated with the worms. The Tri-reagent RNA extraction method enables simultaneous lysis of both the nematodes and the *P. aeruginosa* cells so that gene expression levels can be analyzed for both organisms simultaneously. Although we have optimized this protocol for *P. aeruginosa*, the method should be applicable to a wide range of bacterial pathogens that have been studied using *C. elegans* [3, 4].

2 Materials

- Nematode Growth Medium (*see* Chapter 48).
- Sterile 1× M9 Salts (Sigma Aldrich #M6030 or homemade).
- Liquid nitrogen for freeze–thaw step.
- TriReagent (Sigma–Aldrich #93289 or equivalent).
- OneStep RT-PCR Kit (Qiagen #210212 or equivalent).
- HotStarTaq DNA Polymerase (Qiagen #203203 or equivalent).
- DNAse I (RNAse free) (NEB # M0303S or equivalent).
- RNeasy Mini Kit (Qiagen 74104).
- 25 mM EDTA, sterile.
- Ice-cold ethanol, molecular biology grade and 70 % ethanol in sterile water.
- Nanodrop or equivalent UV/Visible Spectrophotometer (Thermo).
- Qubit with RNA quantification kit and tubes (Life Technologies).
- Standard molecular biology laboratory equipment and consumables and sterile plastics including refrigerated microcentrifuge, rotating vacuum drier, thermal cycler, 1.5 ml centrifuge tubes, 0.2 ml thin-walled PCR tubes, horizontal agarose gel electrophoresis apparatus.

2.1 Standard Primers for RT-PCR

We have elected to use the following control primers in our experiments [5–7]. Recently a number of other standard genes have been evaluated for both species and these may prove useful to the reader [8, 9].

P. aeruginosa—RT-rpoD-F 5′-GATCCGGAACAGGTGGAAGAC-3′.

P. aeruginosa—RT-rpoD-R 5′-TCAGCAGTTCCACGGTACCC-3′.

C. elegans—Ama1F 5′-CAGTGGCTCATGTCGAGT-3′.

C. elegans—Ama1R 5′-CGACCTTCTTTCCATCAT-3′.

C. elegans—F55G112F 5′-TGGTTCTCCAGACGTGTTCA-3′.

C. elegans—F55G112R 5′-CAGCCTTGCCTTTACTGACA-3′.

3 Methods

3.1 Purification of RNA from Infected Worms: Tri Reagent Method

1. Infect worms as described in Chapter 48 to give ~100 worms on a 6 cm plate of NGM medium.
2. Harvest worms from plate using 1 ml 1× M9 salts per plate. Swirl gently to release worms from bacterial lawn (*see* **Note 1**).
3. Collect in a 1 ml eppendorf tube and centrifuge at low speed (400 × *g*) for 3 min (fast and long enough to pellet the worms, but too slow and short to pellet bacteria).
4. *Carefully* remove and discard supernatant and repeat wash step twice with fresh M9 salts to wash away free bacteria (*see* **Note 2**).
5. Resuspend worms in 25 μl TE and snap freeze in liquid nitrogen (wear face shield and thermal gloves). Transfer to a pre-cooled storage box in the −80 freezer. Samples can be stored until ready for RNA extraction.
6. Add 250 μl of Phenol or Tri-reagent/phenol and vortex each sample for 30 s.
7. Vortex samples intermittently at 4 °C until the worms dissolve (usually less than 20 min).
8. Centrifuge at 17,000 × *g* for 10 min at 4 °C to remove any cell debris if required.
9. Remove supernatant to a fresh tube and add 50 μl of chloroform, then vortex for 20 s.
10. Centrifuge at 14,000 rpm for 10 min at 4 °C and retain the upper aqueous phase containing the RNA.
11. Ethanol precipitate RNA with ice-cold ethanol and wash pellet with 70 % ethanol (*see* **Note 3**).
12. Dry pellet in rotating vacuum drier and resuspend in 50 μl of nuclease free water from the preferred total RNA extraction kit (e.g., RNeasy, Qiagen) (*see* **Note 4**).
13. Apply to RNA "clean-up" protocol RNeasy manual (*see* **Note 4**).
14. Quantify your RNA sample using both Qubit (quantity) and Nanodrop (purity ratio) (*see* **Note 5**).

3.2 RT-PCR: Dnase I Treatment of RNA—"In PCR" Method

Use Qiagen OneStep RT-PCR kit and Qiagen Hotstart taq (controls). The aim of this experiment is to rapidly assess the RNA prep using standard primers for both *C. elegans* and *P. aeruginosa* prior to further analysis (*see* **Note 4**). This protocol is aimed at low-throughput experiments and guarantees DNA-free template RNA. A similar adaptation can be made to quantitative RT-PCR kits from other suppliers. Remember to include HotStarTaq controls with genomic DNA and DNase I-treated RNA.

Step 1

1. To a fresh PCR tube add:
   (a) 1 μl RNA (diluted to desired [RNA]).
   (b) 0.5 μl DNase I (RNAse free).
   (c) 2 μl 10×PCR Buffer or 4 μl 5×RT buffer.
   (d) 16.5 μl RNAse free water.
2. Place in thermal cycler on manual run:
   (a) 37 °C, 30 min.
   (b) Add 1 μl 25 mM EDTA (protects RNA at 75 °C).
   (c) 75 °C, 10 min.
3. Cool to 4 °C.

While DNase I digest is being carried out, prepare master mixes for **step 2**

Step 2

1. To reactions from **step 1** add the following mixture as a master mix:
   (a) 3 μl 10×PCR buffer or 6 μl 5×RT PCR buffer.
   (b) 2 μl dNTPs (10 mM each).
   (c) 3 μl Primer 1 (10 pmol/μl).
   (d) 3 μl Primer 2 (10 pmol/μl).
   (e) 1 μl Taq or 2 μl onestep RT enzyme mix.
   (f) Water to 30 μl final volume.
2. Perform PCR reactions with the following thermal cycler conditions:

Temperature (°C)	Time (min)	Additional comments
95	15	Initial Denaturation and Hotstart Taq activation
95 55 72	1 1 1	35 Cycles
72	5	Final elongation

3. Visualize samples on a standard horizontal agarose gel (1 %).

4 Notes

1. Worms can be picked individually instead of floating as described here.

   Between 10 and 20 worms is sufficient for RT-PCR of *C. elegans* transcripts. A number of kits have become available in recent years for transcriptional analysis of very low RNA concentrations and these could easily be applied to this assay for RNA-seq experiments. We typically use ~100 worms per plate in 6 cm plates but this method can be scaled up to larger samples as required.
2. We are aware that cell strainers could be used for separating bacteria from *C. elegans* (e.g., BD Falcon #352340) although the logistics of recovering the worms from these units can be difficult. Reversible units are available, but 40 μm mesh sizes are close to the width of an L4 worm. It has also been suggested to us that a sand washing step may further reduce the number of cuticle associated bacterial cells. Although we have not tested this ourselves, readers may wish to know that this modification to our protocol could be added between **steps 3** and **4** of the RNA extraction method.
3. It is important to ethanol precipitate the sample before RNA clean up because the Tri-reagent contains guanidinium thiocyanate which has the advantage of inhibiting RNase enzymes [10], but prevents binding to commercial clean up columns.
4. We have found certain commercial "on column" DNase I digestion protocols to be unreliable (**step 14**). For this reason we have used the "In PCR" method described above for low-throughput experiments. DNase I treatment can be carried out with the entire preparation between **steps 13** and **14** and a regular control PCR reaction can be used with the standard primers described above in order to ensure that all contaminating genomic DNA has been removed.
5. This protocol should yield around 30–50 μl of high-quality RNA at a concentration of ~20 ng/μl. This yield is sufficient for most downstream applications; however, if required, this protocol can easily be scaled up to return higher yields.

Acknowledgments

NPT's lab has been funded by The Royal Society and a starter grant from the University of Strathclyde.

References

1. Tan MW, Mahajan-Miklos S, Ausubel FM (1999) Killing of *Caenorhabditis elegans* by *Pseudomonas aeruginosa* used to model mammalian bacterial pathogenesis. Proc Natl Acad Sci U S A 96:715–720
2. Mahajan-Miklos S, Tan MW, Rahme LG, Ausubel FM (1999) Molecular mechanisms of bacterial virulence elucidated using a *Pseudomonas aeruginosa–Caenorhabditis elegans* pathogenesis model. Cell 96:47–56
3. Ott L, McKenzie A, Baltazar MT, Britting S, Bischof A, Burkovski A, Hoskisson PA (2012) Evaluation of invertebrate infection models for pathogenic corynebacteria. FEMS Immunol Med Microbiol 65:413–421
4. Sifri CD, Begun J, Ausubel FM (2005) The worm has turned—microbial virulence modeled in *Caenorhabditis elegans*. Trends Microbiol 13:119–127
5. Zhang L, Mah T-F (2008) Involvement of a novel efflux system in biofilm-specific resistance to antibiotics. J Bacteriol 190: 4447–4452
6. Shapira M (2006) A conserved role for a GATA transcription factor in regulating epithelial innate immune responses. Proc Natl Acad Sci U S A 103:14086–14091
7. Addo MG, Cossard R, Pichard D, Obiri-Danso K, Rötig A, Delahodde A (2010) *Caenorhabditis elegans*, a pluricellular model organism to screen new genes involved in mitochondrial genome maintenance. Biochim Biophys Acta 1802:765–773
8. Savli H, Karadenizli A, Kolayli F, Gundes S, Ozbek U, Vahaboglu H (2003) Expression stability of six housekeeping genes: a proposal for resistance gene quantification studies of *Pseudomonas aeruginosa* by real-time quantitative RT-PCR. J Med Microbiol 52:403–408
9. Zhang Y, Chen D, Smith MA, Zhang B, Pan X (2012) Selection of reliable reference genes in *Caenorhabditis elegans* for analysis of nanotoxicity. PLoS ONE 7:e31849
10. Chomczynski P, Sacchi N (1987) Single-step method of RNA isolation by acid guanidinium thiocyanate-phenol-chloroform extraction. Anal Biochem 162:156–159

Part V

Lifestyles and Infection Models

Chapter 47

Methods for Studying Biofilm Formation: Flow Cells and Confocal Laser Scanning Microscopy

Tim Tolker-Nielsen and Claus Sternberg

Abstract

In this chapter methods for growing and analyzing biofilms under hydrodynamic conditions in flow cells are described. Use of flow cells allows for direct microscopic investigation of biofilm formation. The flow in these chambers is essentially laminar, which means that the biofilms can be grown under highly controlled conditions, and that perturbations such as addition of antibiotics or change of the growth medium can be done efficiently at a defined time point. The protocol includes construction of the flow cell and the bubble trap, assembly and sterilization of the flow cell system, inoculation of the flow cells, running of the system, confocal laser scanning microscopy and image analysis, and disassembly and cleaning of the system.

Key words Biofilm, Flow cell, Confocal laser scanning microscopy

1 Introduction

Research initiated due to the prevalent harmful fouling of industrial tubular systems encouraged the design of the first flow cells for biofilm research, e.g., the Robbins device [1]. This device has replaceable plugs with sampling surfaces that enables indirect microscopic inspection of the biofilm. The realization in the 1980s of the importance of biofilms in medical contexts, and a need for direct microscopic observation of developing biofilms, led to the development of flow cells with glass surfaces mountable on microscopes. Doug Caldwell's group at the University of Saskatchewan developed the flow cell constructed of Plexiglas and mounted with a microscope slide [2] similar to the flow cell described in this chapter. These flow cells are well suited for laminar flow conditions, whereas studies of biofilm formation under high-shear turbulent flow conditions may use other systems such as square glass tubing flow cells [3]. Biofilm growth in traditional flow cells results in relatively little biomass, but in cases where a large amount of biomass is needed, e.g., for proteome or transcriptome analysis, the biofilms can be grown in silicone tubing [4], or drip-flow reactors [5].

Alain Filloux and Juan-Luis Ramos (eds.), *Pseudomonas Methods and Protocols*, Methods in Molecular Biology, vol. 1149, DOI 10.1007/978-1-4939-0473-0_47, © Springer Science+Business Media New York 2014

Confocal laser scanning microscopy (CLSM) is particularly well suited for monitoring of the three-dimensional structure of biofilms grown in flow cells. Because CLSM is noninvasive and nondestructive it allows imaging of the structure of biofilms in their naturally hydrated form [6]. The fluorescence required for CLSM can be emitted from the bacteria if they are hybridized with fluorescently labeled nucleic acid probes, or stained with fluorescent dyes, or harbor genes encoding fluorescent proteins (e.g., GFP). Staining with fluorescently labeled lectins may visualize extracellular substances that surround and interconnect the bacteria in biofilms. A description of these staining/tagging techniques is outside the scope of this chapter, but useful information on this can be found elsewhere, e.g., in Nielsen et al. [7], Neu et al. [8], Hentzer et al. [9], Klausen et al. [10], and Pamp et al. [11].

In order to analyze structural biofilm development objectively, it is often necessary to subject the acquired micrographs to image analysis, and a number of different computer programs have been developed for this purpose. Kuehn et al. [12] developed one of the first programs for objective analysis of three-dimensional CLSM biofilm images, allowing calculation of surface coverage and biovolume. The Image Structure Analyzer (ISA3D) software package, developed at Montana State University, enables the calculation of a number of textural and areal parameters from two-dimensional biofilm images taken by light microscopy or CLSM [13, 14]. The COMSTAT program enables calculation of a variety of parameters from three-dimensional CLSM images [15, 16].

2 Materials

2.1 Components for Construction of the Flow-Cell System

1. Silicone glue (3 M Super Silicone Sealant Clear).
2. Medium appropriate for organisms and type of biofilm being grown (e.g., FAB medium [17]).
3. 70 and 96 % (v/v) ethanol.
4. 0.5 % (w/v) sodium hypochlorite.
5. H_2O, sterile.
6. 1 % hydrogen peroxide.
7. Polycarbonate sheet plastic, 6 and 35 mm thick.
8. CNC (Computer Numerical Control) tooling machine, or a drilling machine mounted on an upright stand and equipped with a milling drill-tool (3 mm).
9. Substratum: 50 × 24-mm glass coverslips or other appropriate material.
10. 2-ml syringe (Terumo).
11. Silicone tubing, 3 mm outer diameter, 1 mm inner diameter.

12. Silicone tubing, 4 mm outer diameter, 2 mm inner diameter.
13. Silicone tubing, 7 mm outer diameter, 5 mm inner diameter.
14. Marprene® tubing, 3 mm outer diameter, 1 mm inner diameter (Watson–Marlow).
15. Peristaltic Pump (Watson–Marlow, 205S).
16. Medium bottles (Schott or BIO 101).
17. Clear polypropylene plastic connectors and T-connectors (Cole Parmer, EW-06365-xx), 1/8 in. (3.175 mm) and 1/16 in. (1.588 mm). Reduction connectors 1/8–1/16 in.
18. Bubble traps (DTU Systems Biology, Technical University of Denmark, or *see* below).
19. Microscope.
20. Rolling cart for flow systems and pumps (optional).
21. Waste container.

2.2 Components for Construction of the Bubble Trap

1. 35 × 80 × 45-mm polycarbonate block.
2. CNC tooling machine.
3. 5-ml syringes with inner diameter of 12.5 mm.
4. 9 × 2-mm rubber gaskets (M-seals, 221355; http://www.mseals.dk/cms.ashx).
5. Silicone glue (3 M Super Silicone Sealant Clear).
6. Stoppers (e.g., Nordson EFD part no. 7017976; http://www.nordson.com/en-us/divisions/efd) or use the leftover needle protective cover from the inoculation needles used for inoculating the flow cells (*see* above).

Alternatively for a simple version of the bubble trap:

1. A 10 mm thick polycarbonate block, 80 × 35 mm surface area.
2. Drilling machine mounted in a vertical stand.
3. A 8 mm and a 3 mm drill suitable for drilling in plastic.
4. 2- or 5-ml syringes.
5. Silicone glue and stoppers as above.

2.3 Materials for Inoculation and Running of the Flow Cells

1. Inoculum, e.g., fresh overnight culture of the microorganisms under study.
2. 70 and 96 % (v/v) ethanol.
3. Medium (e.g., Fe-EDTA-AB (FAB) medium [17]).
4. Silicone glue (3 M Super Silicone Sealant Clear).
5. Flow-cell system (DTU Systems Biology, Technical University of Denmark, or *see* below).
6. Syringes with needles (e.g., Terumo LU-100, 27-G (0.4 × 12 mm), 0.5 ml).
7. Clamps.

2.4 Equipment for Confocal Laser Scanning Microscopy of Flow-Cell-Grown Biofilms

1. CLSM microscope (e.g., Zeiss LSM710).
2. Scalpels.
3. Computer software:

   Imaris (Bitplane; http://www.bitplane.com).

   ImageJ (http://rsb.info.nih.gov/ij).

   COMSTAT version 2 (DTU-Systems Biology, Technical University of Denmark, http://www.comstat.dk).

   Java runtime environment (needed for Comstat v. 2, http://www.java.com).

3 Methods

The flow cells are constructed from nonfluorescent material, which is tolerant to autoclaving at 115 °C. Poly carbonate (PC) is suitable for both flow cell and bubble trap construction. The components of the system are: media flask(s), a non-pulsating peristaltic pump, bubble trap(s), flow chamber(s), silicone and Marprene® tubing, waste container(s), and connectors. Flow cells are constructed from a plastic base with milled or molded channels through which the media can flow. The growth substratum is glued on top of the flow cells using silicone glue. The substratum is usually a glass coverslip, but it is possible to use other substrata, such as plastic sheets or metal slabs. The bubble trap is a simple device made from a single-use syringe mounted vertically on a plastic base. Media inlet and outlet is at the base of the syringe and any air bubbles passing the bubble trap will float to the top of the syringe and not be passed downstream of the trap. All components are assembled so media will pass from the media flask through the pump to the bubble trap, and then the flow cell. The effluent from the flow cell is diverted to the waste container.

Most components can be autoclaved after assembly, but the system should nevertheless always be sterilized using a 0.5 % sodium hypochloride solution, followed by several flushes with sterile water. It is further advisable to let the medium run in the system for several hours at the desired incubation temperature prior to inoculation.

The flow cells and bubble traps can be purchased from several sources (e.g., DTU Systems Biology, the Technical University of Denmark), or the reader can make them using the directions below. For some components advanced tooling machines are required and it is recommended to obtain assistance from trained operators.

3.1 Flow Cell Construction

Note: It is possible to construct the flow cell by molding, but it should be noted that it will require the building of a molding tool which can be costly and will require both advanced machinery for

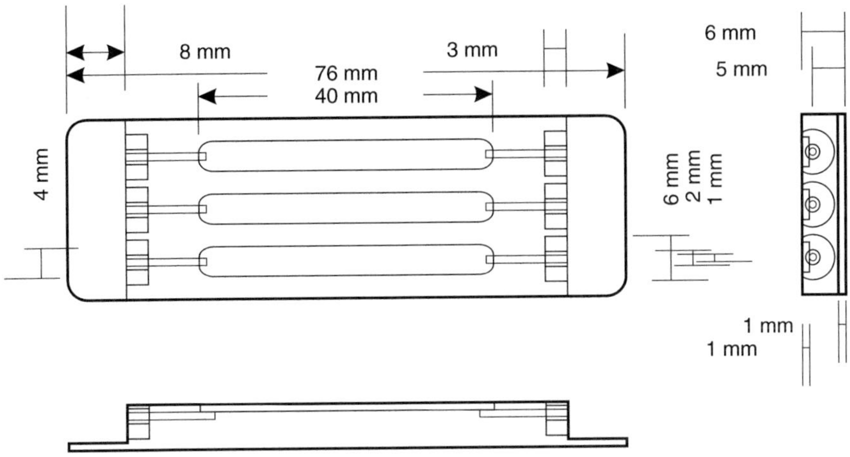

Fig. 1 Drawing of the flow cell [24]. Reproduced with permission from John Wiley and Sons

tool construction and molding. Here we will describe only the construction of flow cells using milling.

1. Use the diagram in Fig. 1 to mill the flow cell with the aid of a CNC milling machine, or an upright stand manual milling machine. The cell is made from a sheet of polycarbonate, 6×76×26 mm. This size is similar to the foot-print of a standard microscope slide which makes it easy to mount the flow cell on a microscope subsequently.
2. First remove the material in each end, leaving a 1 mm thick slab, 8 mm into each end. The remainder of the material is a central block of 60×26 mm, which will accommodate the flow channels and connectors for the silicone tubing.
3. Use a milling machine to carve three flow channels in the central block, 4 mm wide, 1 mm deep and 40 mm long. Depending on the available tools either drill 3 mm tubes in from each end to connect to the flow channels, to accommodate a silicone tube. If advanced equipment is available, use the CNC machine to carve the connector studs and drill liquid lines into the flow channels (Fig. 1).
4. Use silicone glue without added antimicrobial additives to attach a suitable substratum to the flow cell. For routine work where microscopic inspection is desired use a 50×24 mm glass coverslip as substratum. For other purposes use an appropriate substratum with the same dimensions in order to cover the area of the central block with the milled channels. Cut off the wider part of a 200 μl pipette tip and place the remaining about 1 cm tip in the out port of a 2 ml syringe where you have removed the piston. Fill about 1 ml silicone glue into the syringe from the open end and replace the piston to pump

the silicone to the tip. Use this device to make thin threads of silicone and place these between the flow channels and at each end. Make sure that all silicone threads overlap leaving no gaps to prevent subsequent leakage of medium or cross-contamination between the channels.

5. Carefully place the substratum on top of the silicone threads. Avoid using force in order not to break substratum (if fragile) but firmly press the substratum towards the flow cell. Use first a glass slide to apply even pressure to the substratum. Make sure that the glue completely seals the substratum to the base and take care to prevent gaps in the glue. Close small gaps by applying local pressure, e.g., by the handle of a syringe piston.
6. If the above procedure cannot produce a completely sealed flow cell, or if excessive silicone glue inadvertently has come into the flow channels, remove the substratum and clean the flow cell by 96 % ethanol. Then repeat the assembly **steps 4** and **5**.
7. Attach silicone tubing to the studs if present, or place cut tubing ends into the receiver holes and seal with silicone glue.

3.2 Construction of the Bubble Trap

To prevent air bubbles to disturb the biofilm growing in the flow cells it is important to remove these from the liquid lines. A bubble trap is in principle just a small vertical cylinder filled with medium where bubbles in the tubing are allowed to float atop. In practice it is constructed as a base onto which a single use syringe is placed. A removable cap allows bleeding of the bubble trap when filling or emptying the trap.

A base made with a CNC machine according to the drawing shown in Fig. 2 is easy to handle but difficult to make. A simple version where the syringes are simply glued on top of a plastic sheet with drilled holes is easy to make but can prove difficult to operate stably since the syringes easily can break off or leak.

3.2.1 Bubble Traps Made with CNC Equipment

1. Use a polycarbonate plastic block with the dimensions $35 \times 80 \times 45$-mm as base. Use the drawing in Fig. 2 to program the CNC machine for milling the base component.
2. The base has cut outs for rubber gaskets, which ensures a tight seal between the base and syringes. Place gaskets on each ring of the three columns.
3. Use 5-ml syringes (inner diameter, 12.5 mm) as liquid reservoirs. Press one new syringe without the piston onto the vertical columns. It is recommended not to fix the syringe by turning the syringe under the rails before the system is running to prevent pressure buildup in the case of a blockage downstream of the bubble trap.
4. Place a stopper on top of each syringe.

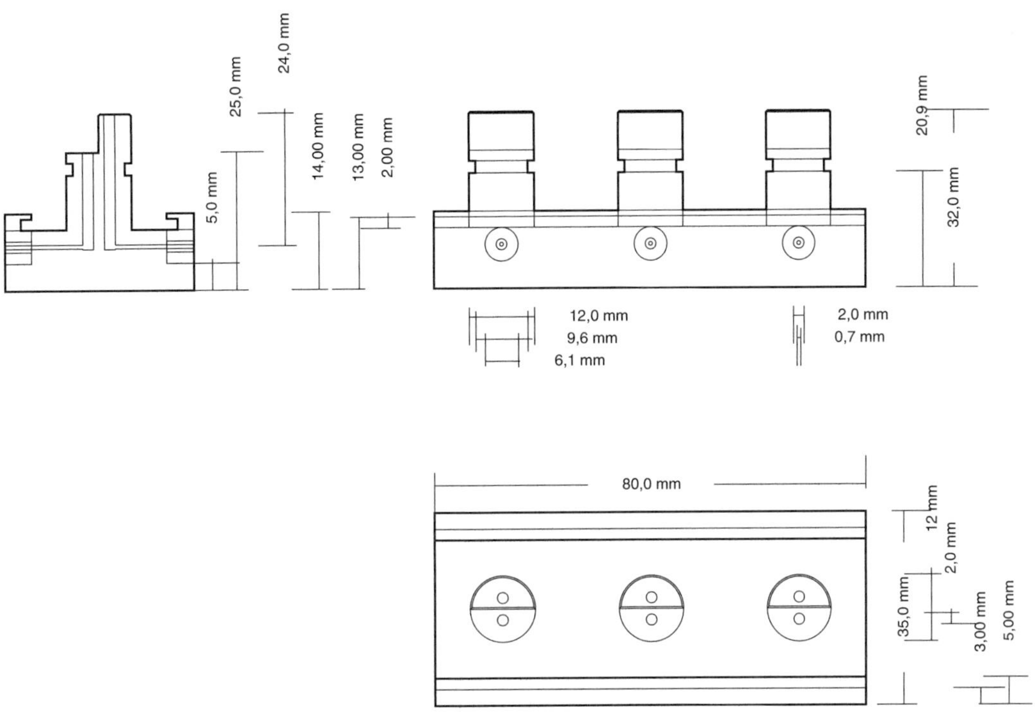

Fig. 2 Drawing of the CNC milled bubble trap [24]. Reproduced with permission from John Wiley and Sons

3.2.2 Simple Bubble Traps

This type of bubble traps can be constructed without the need for advanced equipment such as CNC machines.

1. Use a plastic (e.g., polycarbonate) slab with the dimensions 30 × 110 × 10 mm. Use a 8 mm drill to make three cavities evenly spaced approximately 8 mm into the base. Drill connecting holes from each side using a 3-mm drill. Position the drill 4 mm from the top of the base, to ensure that the channels will go into the cavities from each side.
2. Use three piece 2-ml syringes (without the piston) and glue one on top of each 8-mm hole. Use silicone tubing with 3 mm outer diameter (1 mm inner diameter) as connectors for inlet and outlet. Seal with silicone glue.
3. The assembly is quite fragile and great care must be taken to avoid breaking off the syringes when handling the bubble traps. As a preventive measure clamps can be applied to fix the syringes. Allow at least 24 h to let the glue solidify.

3.3 Assembly of the Flow-Cell System

The assembled system is shown in Fig. 3. It consists of, following the medium flow: A medium container, lead-in tubing (silicone, 1 mm inner diameter), Marprene® pump tubing, a peristaltic pump, lead-out tubing (silicone, 1 mm inner diameter), a three

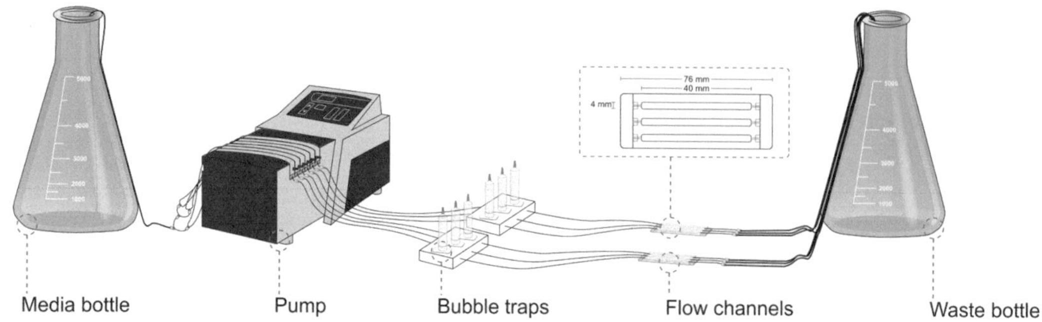

Fig. 3 Schematic of the flow-cell system [25]. Reproduced with permission from Springer Press

channel bubble trap, connecting tubing (silicone, 1 mm inner diameter), a three channel flow cell thin effluent connecting tubing (silicone 1 mm inner diameter), waste tubing (silicone, 2 mm inner diameter), and finally a waste collection container. All parts except the pump can be autoclaved and it is recommended to do so prior to final assembly.

The assembly process can be divided into a couple of subtasks: preparation of medium flask with appropriate medium, preparation of bubble traps and flow cells, and mounting of silicone tubing and connectors. The preparation of the silicone tubing can be subdivided into three tasks: Construction of a fan-out connecting part where the single medium supply line is split into the number of individual channels in the experiment. This fan-out connector will be mounted before the pump. Second task is to configure the pump Marprene® tubing (one per channel) and the final task is to cut various lengths of silicone tubing for connecting the pump tubing to bubble traps, flow cells and waste.

1. Determine how many channels you need. This will aid in calculation of the size of medium supply bottle(s). A Watson–Marlow 205S pump can accommodate up to 16 channels, which will allow for five 3-channel flow chambers.
2. Preparation of media flasks: Fit the bottle to contain the medium with a approx. 1 m long silicone tube, 2 mm inner diameter Attach a straight connector to the end of the tubing and cover it with metal foil. The connector should allow connection of the 2 mm inner diameter tubing for the medium side and 1 mm inner diameter for the pump side (i.e., a 1/8–1/16 straight reduction connector). Place the other end of the tubing inside the medium bottle and fix the tubing with autoclave tape to the neck of the flask. Important: clamp the tubing before autoclaving to prevent a siphoning effect. Fill the medium into the container, cover the opening with metal foil, and autoclave.
3. Prepare bubble traps and flow cells according to Subheadings 3.1 and 3.2.

4. (a) Prepare the tubing for the fan-out connector: Start with 50 cm silicone tubing 1 mm inner diameter and attach a 1/16 in T-connector. Cover the other end with metal foil. Then attach short pieces of silicone tubing to each free end of the T-connector and if needed attach another T-connector. Repeat the process until there are the same number of free tube ends as there are channels in the system.

   (b) Cut one piece of Marprene® tubing per channel. For the Watson–Marlow 205S pump, the length should be 25 cm. Place a 1/16 in straight connector in each end of the tubing and wrap three to four windings of autoclave tape around the tubing at one of the ends. This will prevent the pump from pulling the tubing off the connector.

   (c) For each channel make the following pieces of tubing (1 mm inner diameter):

   One piece 35 cm,

   One piece 150 cm,

   One piece 10 cm.

   One piece 2 mm inner diameter at 120 cm.

5. Assemble the tubing: Connect the ends of the fan-out to the pump tubing ends with the wound autoclave tape. Connect the other end of the pump tubing to the 35 cm pieces. Attach the bubble traps to the other end of the 35 cm pieces. *Note*: In the CNC made bubble traps the internal column of the base has two outlets. The highest should be closest to the pump. Next, attach the 150 cm pieces to the other sides of the bubble traps and connect the other ends to the flow cells. On the other side of the flow cells attach the short 1 mm inner diameter tubings and put a 1/16–1/8 reduction straight connector to the end. Finally, on the other side of the straight connector attach the 120 cm × 2 mm inner diameter waste tubing.

6. The assembled tubing with bubble traps and flow cells can optionally be autoclaved at this point.

7. Mount the flow system onto the pump: Place the system on a rolling cart or near the microscope. Make sure that you connect one line at a time in order, so it is possible to follow the line from, e.g., position 1 on the pump to the first bubble trap position 1 and on to the first channel in the first flow cell. Likewise follow the position all the way for position 2 and so on. This will make it easier to handle, e.g., leaks at a later point. Place all waste lines into an appropriate container.

3.4 Sterilization of the Flow-Cell System

1. At this point the system can be considered to be non-sterile. Despite of this it is recommended that in all steps below to use sterile distilled water.

2. Place all components at the same height level, e.g., on a table. Specifically, do not place the waste container on a higher or lower level than the flow cells. Remove the stoppers of the bubble traps and store them in a petri dish with 70 % ethanol.
3. Remove the foil from the fan-out connector and put it in a 1 l bottle of sterile water. Start the pump at maximum speed and fill the system. The bubble traps will fill slowly. When they are full, allow water to flow from the syringes for 15 s or more before replacing the stoppers. Observe that the entire system is filled with water. If any leaks are observed, these must be sealed before proceeding.
4. If any air bubbles are observed in the flow channels, remove them by tapping the flow cells on a hard surface while the pump is running.
5. Empty the system by lifting the inlet tube out of the water and allow air to be pumped into the system.
6. Remove the stoppers of the bubble traps again and fill the system as above with a solution of 0.5 % sodium hypochlorite in water. When the bubble traps are filled, replace stoppers and remove all air bubbles and continue to run disinfectant through the system for another 5–10 min. Then reduce pump speed to minimum and leave the system to sterilize for at least 2 h, but not more than 24 h.
7. From this point on the system is sterile and care should be taken when handling tubing and stoppers etc.
8. The hypochlorite is removed from the system by three consecutive fillings with air followed by complete filling with sterile water. Optionally, in the second round the sterile water can be replaced by a 1 % hydrogen peroxide solution. Conclude by passing sterile 1 l of water through the system at lower flow rate. If the experiment is not to be carried out immediately, make sure that enough water is available and leave the system running at the lowest pump speed until the experiment is ready.
9. Prior to the experiment, purge the system with air and fill it with media. When attaching the tubing observe great care not to contaminate the tube ends that are connected. If in doubt, immerse the ends briefly in 70 % ethanol before connecting.

3.5 Inoculation of the Flow Cells

1. Grow bacteria to be inoculated overnight and dilute appropriately. For *Escherichia coli* 1:100 dilution has proven suitable for most experiments, while *Pseudomonas aeruginosa* has to be diluted 1:1,000. The dilution depends on the bacterial strains used and must be determined empirically.
2. Fill a syringe with 250 μl of the diluted inoculum. Attach a thin needle to the syringe, e.g., 27-G gauge. Stop the pump and wipe inlet silicone tube with 70 % ethanol. Inject the inoculum

by perforating the tube with the needle. Be careful not to inject air bubbles in the process. Repeat the wiping with ethanol and seal the injection spot using a little silicone glue.

3. Invert the flow cells so the substratum is facing the table and leave in this position for 1 h with the flow stopped. This will allow the bacteria to make initial adhesion to the substratum.
4. Turn the flow cells over, with the substratum facing up and resume flow at a low flow rate. 1–3 ml per channel per hour has been shown to be appropriate for *Escherichia coli* and *Pseudomonas aeruginosa*, but may vary for other organisms.

3.6 Running of the Flow-Cell System

1. The system should from this point be maintenance free. It is, however, crucial to check for leaks in the initial phase of incubation, and if needed take appropriate measures to repair. Medium must be supplied at all times as an interruption will result in air entering the system with consequent destruction of the biofilm.
2. For some bacterial strains a stop of the flow will be detrimental to the biofilm integrity and should therefore be avoided. It is possible to quickly change medium bottles with the pump running, as a limited amount of introduced air will be removed in the bubble traps.
3. The system can run for several days or weeks without interruption.
4. Biological waste should be removed according to regulations, e.g., by autoclaving the waste containers before disposal.

3.7 Microscopic Inspection and Imaging of Flow-Cell-Grown Biofilms

The method of choice for microscopic inspection of flow-cell-grown biofilms is to use a confocal laser scanning microscope (CLSM). The applications possible with this equipment is beyond the scope of this paper, but in-depth information about confocal microscopy of biofilms can be found in Palmer et al. [18]. Biofilms grown in the flow cells described here can only be examined directly by microscopy if the substratum is transparent.

Here a few practical considerations will be described.

1. Place the flow cell system close to the confocal microscope.
2. Have an auxiliary waste container, such as a 0.5 l conical flask ready and place it securely on the microscope table.
3. Detach the waste line belonging to the flow cell to be examined from the waste container and place it in the auxiliary container.
4. Carefully move one flow cell to the microscope object table while observing the medium and waste lines. Firmly attach the flow cell in the specimen holder. Since the flow cell still is connected to the system it is occasionally necessary to use adhesive tape to assist mounting the sample on to the microscope.

5. Examine the sample. Be particularly careful if the microscope is adjusted to automatically set parfocality as for normal microscope slides. The flow cells are much thicker and as a consequence the microscope may position the lens wrongly and destroy the sample. It is therefore recommended to try the setup with an empty flow cell before mounting a real experiment.

3.8 Disassembly and Cleaning of the Flow-Cell System

After completion of the experiments the system must be cleaned and the components prepared for reuse.

1. Empty remaining liquid from the still assembled system by filling with air. Collect all waste and dispose of following regulations.
2. Refill once with water, using the same protocol as for preparing the system and make sure that any visible colonization is removed by massaging the tubing while the pump is still running.
3. Finalize by emptying the system one last time and detach the flow cells. All other components can be left assembled for next use.
4. The flow cells are disassembled using a scalpel to remove the substratum. If the substratum is made of fragile material such as glass, it will inevitably break in this process.
5. Remove remains of silicone glue from the flow cell using mechanical abrasion and 96 % ethanol.
6. The other components of the system can be wrapped in metal foil, autoclaved and stored for subsequent experiments.

3.9 Image Analysis

CLSM Images from biofilm experiments can be used in, in principle, two different ways: As qualitatively descriptive images or for quantitative measurements.

For qualitative analysis an appropriate software package is needed. There are several packages available. One popular commercial package is the Imaris® software suite (www.bitplane.ch) which can create three dimensional visualizations as well as animations. Simple measurements can be performed in this suite. ImageJ is a freeware alternative [19, 20] which can be supplemented by user-written plug-ins to perform several graphical visualizations of CLSM images and extensive qualitative measurements. ImageJ does require more from the user to get full benefit of its capabilities than do commercial dedicated packages.

For quantitative analysis several programs have been developed, most prominently the ISA3D [13], Comstat [15, 16] and Daime [21, 22]. All of these programs can extract basic parameters from CLSM image stacks, such as biomass, roughness, average thickness, etc. The output of quantitative analysis programs is numbers, rather than images and provides a way to directly evaluate both reproducibility of experiments and compare different biofilms which qualitatively seem similar but may have characteristic differences.

4 Notes

1. Even a small leak will render the entire flow cell useless if not sealed. It is usually very difficult to seal leaks once the system is filled with liquid, which makes it very important to be careful when assembling the system. The flow cell is vulnerable to leakage, and if the bubble traps are homemade, consisting of syringes glued to a plastic base, they can also leak. If the flow cell leaks while mounted on the microscope, it can become harmful to the equipment. Either immediately attempt to reseal the leak or remove the flow cell from the microscope. When recording time series, it is particularly important to make sure that no leaks are present. If a leak is discovered in a system under preparation, the leaking component should be emptied, dried, and the area of the leak should be covered with a layer of silicone glue and allowed to dry for at least a few hours, preferably overnight, before filling the system again. A better method is to have spare components at hand and replace the failing unit with a new one. While running an experiment, it is not possible to drain the system or even stop the flow temporarily. Clean and dry the area containing the leak, and keep a piece of paper towel at hand to repeatedly soak away leaking fluid. Then apply a large excess of silicone glue to the leak and continue to remove leaking liquid until it stops.
2. Air bubbles can destroy or alter the three-dimensional structure of biofilms, so care must be taken to prevent them from entering the flow cells. The system contains bubble traps to catch air in the medium supply. Moreover, it is recommended not to cool the medium after autoclaving, and to place it immediately at the correct temperature for the experiment. If the medium is colder than the temperature of the flow cell and tubing, air bubbles tend to emerge throughout the system. Furthermore, if running of the flow cell system at high temperatures (e.g., 37 °C) gives rise to bubble formation, a recently described [23] setup with a modified medium container may be employed.
3. Usually there will only be little fouling of the tubing upstream of the flow cells. However, because biofilm in the upstream tubing will use substrate and release waste products it may affect the biofilm in the flow cell, and therefore it should be removed. One way to reduce backgrowth is to quickly disconnect the inlet tubing from the flow cell (e.g., once a day), cut

off a few centimeters, and reconnect. It is in this case critical to clamp the tubing on the effluent side of the flow cells, so that waste medium does not flow backwards through the chamber, destroying the sample. It may also be necessary to exchange effluent tubes if they become heavily fouled.

4. Because the biofilm in the flow cell uses substrate and releases waste products, microscopic inspection and imaging of the biofilm should be done near the medium inlet to avoid uncontrolled conditions.
5. To avoid reverse siphoning from the waste container, place the end of the tubes above the surface of the waste reservoir. To avoid accidental entrance of air in case of leaks, place the effluent tubes above the level of the flow cells.
6. The biofilms formed by some microbial species will disperse in response to even short periods of flow stoppage. Hence, it may be important to ensure constant running of the peristaltic pump.
7. Microbial growth in flow cells may be influenced by selection. Biofilms are dynamic systems, and mutants may efficiently outcompete wild type organisms. Biofilm experiments should therefore always be done in replicates, and biofilm development properties should be retested to verify that the microorganisms have not undergone irreversible changes during the experiment.

Acknowledgment

Research by TT-N is supported by the Danish Research Council, the Lundbeck foundation, Novo Seeds, and Tandlægefonden. Research by CS is supported by the Danish National Advanced Technology Foundation.

References

1. McCoy WF, Bryers JD, Robbins J, Costerton JW (1981) Observations of fouling biofilm formation. Can J Microbiol 27:910–917
2. Wolfaardt GM, Lawrence JR, Robarts RD, Caldwell SJ, Caldwell DE (1994) Multicellular organization in a degradative biofilm community. Appl Environ Microbiol 60:434–446
3. Stoodley P, Lewandowski Z, Boyle JD, Lappin-Scott HM (1999) The formation of migratory ripples in a mixed species bacterial biofilm growing in turbulent flow. Environ Microbiol 1:447–455
4. Sauer K, Camper AK, Ehrlich GD, Costerton JW, Davies DG (2002) Pseudomonas aeruginosa displays multiple phenotypes during development as a biofilm. J Bacteriol 184: 1140–1154
5. Goeres DM, Hamilton MA, Beck NA, Buckingham-Meyer K, Hilyard JD, Loetterle LR, Lorenz LA, Walker DK, Stewart PS (2009) A method for growing a biofilm under low shear at the air–liquid interface using the drip flow biofilm reactor. Nat Protoc 4:783–788

6. Lawrence JR, Korber DR, Hoyle BD, Costerton JW, Caldwell DE (1991) Optical sectioning of microbial biofilms. J Bacteriol 173:6558–6567
7. Nielsen AT, Tolker-Nielsen T, Barken KB, Molin S (2000) Role of commensal relationships on the spatial structure of a surface-attached microbial consortium. Environ Microbiol 2:59–68
8. Neu T, Swerhone GD, Lawrence JR (2001) Assessment of lectin-binding analysis for in situ detection of glycoconjugates in biofilm systems. Microbiology 147:299–313
9. Hentzer M, Riedel K, Rasmussen TB, Heydorn A, Andersen JB, Parsek MR, Rice SA, Eberl L, Molin S, Hoiby N, Kjelleberg S, Givskov M (2002) Inhibition of quorum sensing in *Pseudomonas aeruginosa* biofilm bacteria by a halogenated furanone compound. Microbiology 148:87–102
10. Klausen M, Aaes-Jorgensen A, Molin S, Tolker-Nielsen T (2003) Involvement of bacterial migration in the development of complex multicellular structures in *Pseudomonas aeruginosa* biofilms. Mol Microbiol 50:61–68
11. Pamp SJ, Sternberg C, Tolker-Nielsen T (2009) Insight into the microbial multicellular lifestyle via flow-cell technology and confocal microscopy. Cytometry A 75A:90–103
12. Kuehn M, Hausner M, Bungartz HJ, Wagner M, Wilderer PA, Wuertz S (1998) Automated confocal laser scanning microscopy and semiautomated image processing for analysis of biofilms. Appl Environ Microbiol 64:4115–4127
13. Beyenal H, Donovan C, Lewandowski Z, Harkin G (2004) Three-dimensional biofilm structure quantification. J Microbiol Methods 59:395–413
14. Yang X, Beyenal H, Harkin G, Lewandowski Z (2000) Quantifying biofilm structure using image analysis. J Microbiol Methods 39: 109–119
15. Heydorn A, Nielsen AT, Hentzer M, Sternberg C, Givskov M, Ersboll BK, Molin S (2000) Quantification of biofilm structures by the novel computer program COMSTAT. Microbiology 146:2395–2407
16. http://www.comstat.dk
17. Pamp SJ, Tolker-Nielsen T (2007) Multiple roles of biosurfactants in structural biofilm development by *Pseudomonas aeruginosa*. J Bacteriol 189:2531–2539
18. Palmer RJ, Haagensen JA, Neu TR, Sternberg C (2006) Conforcal microscopy of biofilms—statiotemporal approaches. In: Pawley JB (ed) Handbook of biological confocal microscopy. Springer, New York, pp 882–900
19. Abràmoff MD, Magalhães PJ, Ram SJ (2004) Image processing with ImageJ. Biophoton Int 11:36–41
20. http://rsb.info.nih.gov/ij
21. Daims H, Lucker S, Wagner M (2006) daime, a novel image analysis program for microbial ecology and biofilm research. Environ Microbiol 8:200–213
22. http://www.microbial-ecology.net/daime
23. Crusz SA, Popat R, Rybtke MT, Camara M, Givskov M, Tolker-Nielsen T, Diggle SP, Williams P (2012) Bursting the bubble on bacterial biofilms: a flow cell methodology. Biofouling 28:835–842
24. Tolker-Nielsen T, Sternberg C (2011) Growing and analyzing biofilms in flow chambers. In: Boyle A (ed) Current protocols in microbiology. Wiley, New York, pp 1B.2.1–1B.2.17
25. Haagensen JA, Regenberg B, Sternberg C (2011) Advanced microscopy of microbial cells. Adv Biochem Eng Biotechnol 124:21–54

Chapter 48

Biofilm Formation in the 96-Well Microtiter Plate

Barbara M. Coffey and Gregory G. Anderson

Abstract

The microtiter plate (also called 96-well plate) assay for studying biofilm formation is a method which allows for the observation of bacterial adherence to an abiotic surface. In this assay, bacteria are incubated in vinyl "U"-bottom or other types of 96-well microtiter plates. Following incubation, planktonic bacteria are rinsed away, and the remaining adherent bacteria (biofilms) are stained with crystal violet dye, thus allowing visualization of the biofilm. If quantitation is desired, the stained biofilms are solubilized and transferred to a 96-well optically clear flat-bottom plate for measurement by spectrophotometry.

Key words Adhesion, Bacteria, Biofilm, Crystal violet, Microtiter, *Pseudomonas*, Surface-adhered, 96-well

1 Introduction

Biofilm formation occurs when bacteria switch from a planktonic (free-swimming) state to a surface-attached state. *Pseudomonas aeruginosa* biofilms can be found on a wide range of natural and man-made surfaces, including rocks, oil pipelines, medical catheters [1], and human tissue. To study biofilms, a number of laboratory-based biofilm formation models have been developed. The method of studying biofilm formation on plastic tissue culture plates was first described in the 1970s, and the use of 96-well tissue culture and microtiter plates was introduced in the 1980s [2, 3]. This assay tests the ability of bacteria to adhere to the plastic surface of a microtiter plate under the temperature and nutritional conditions determined by the experimenter. The method was first published in 1996 by Genevaux et al. as a way to perform rapid screens for *Escherichia coli* surface attachment [4]. Two years later, the assay was adapted for *Pseudomonas* by O'Toole et al. to support genetic studies of biofilm formation. The protocol continues to be used widely in studies of microbial biofilms [5, 6].

Today, there are a number of methods for growing and analyzing biofilms in a laboratory setting, including assays for air–liquid

Alain Filloux and Juan-Luis Ramos (eds.), *Pseudomonas Methods and Protocols*, Methods in Molecular Biology, vol. 1149, DOI 10.1007/978-1-4939-0473-0_48, © Springer Science+Business Media New York 2014

interface biofilms, colony biofilms, drip-fed biofilms, and flow-cell biofilms [7–9]. The advantages of the microtiter plate assay described here are its simplicity, the use of basic lab materials, the adaptability to small or large numbers of samples, and the variety of samples that can be tested in a single assay. This protocol has been used successfully in biology classroom settings at the high school level [10]. The microtiter plate assay is particularly useful for studying the early steps in biofilm formation, such as initial surface attachment. To examine mature biofilms, this method could be complemented with a flow system.

Various modifications have been implemented for monitoring specific activities. For instance, in a system known as the Calgary Biofilm Device, designed especially for antibiotic susceptibility testing, biofilms are grown on the surface of 96 pegs submerged in liquid medium [11]. Additionally, there are several effective stains used for biofilm evaluation, including crystal violet (described below), safranin, and Alamar blue [12]. In the last few years, safranin staining has been described in several biofilm studies of Gram-positive bacteria [13–15].

In the microtiter plate biofilm protocol, bacteria are first grown planktonically to stationary phase and then subcultured into fresh medium for the assay. 100 μL aliquots of the freshly inoculated samples are transferred to a microtiter plate and incubated. Incubation temperature is commonly 37 °C, for a length of time ranging from as little as 30 min to as much as 72 h. Assay medium, incubation temperature, and length of incubation may be varied according to the desired experimental parameters. Following incubation, the microtiter plate is rinsed in water. The water washes away non-adherent bacteria, but any biofilm that has formed will remain attached to the microtiter plate. The wells are then filled with 125 μL of 0.1 % crystal violet and allowed to sit for 10 min. The crystal violet will stain the biofilm but not the vinyl microtiter plate.

Crystal violet belongs to the family of triphenylmethane dyes, which bind to the bacterial cellular components by ionic interactions. It has been shown to bind to DNA [16], proteins [17], and apparently polysaccharides, including biofilm polysaccharides [18]. When the crystal violet is rinsed away, biofilm formation will be visible as a ring around the inside of the well (Fig. 1). If additional quantitation of the biofilm is desired, the stained biofilm should be allowed to dry for several hours. The biofilm can then be solubilized, transferred to an optically clear flat-bottom 96-well plate, and measured by spectrophotometry. This quantitation method does not necessarily correlate directly with the number of viable bacterial cells in the biofilm, since the matrix of extracellular DNA, polysaccharides, and proteins is also stained by crystal violet. Additionally, crystal violet kills bacteria, making this staining method incompatible with further analysis of viable cells.

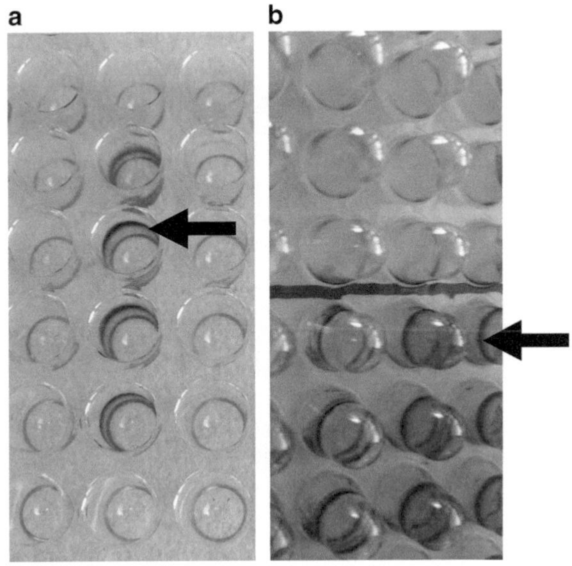

Fig. 1 Rings of crystal violet-stained biofilm are visible in the wells of a vinyl microtiter plate (indicated by *arrow*). (**a**) Looking into the wells from the *top*. (**b**) Plate is inverted. Rings of biofilm are visible in *bottom* wells, with no biofilm formation visible in *top* wells

If viable cell counts are required, a different method must be used to remove the biofilm from the surface of the microtiter plate and transfer it to another growth medium. Finally, findings discovered using microtiter plate assays should be confirmed by microscopy or another method.

2 Materials

Store all materials and solutions at room temperature.

2.1 Biofilm Assay

1. 96-well microtiter plate, not tissue culture treated (*see* **Note 1**).
2. Four flat containers of sufficient size to hold 96-well plate submerged in liquid (*see* **Note 2**).
3. 0.1 % crystal violet solution in water (*see* **Notes 3–5**).
4. Pipetter and 200 μL tips (*see* **Note 6**).
5. Bacteria of interest.
6. Media appropriate for bacterial growth and/or variations of media as desired for experimental purposes (*see* **Note 7**).
7. Lab mat or paper towels (*see* **Note 8**).
8. Lab coat and gloves.
9. Container for liquid waste (*see* **Note 9**).

10. Container for dry waste (e.g., pipette tips).
11. Funnel (*see* **Note 10**).
12. Trough or empty tip box for pipetting with multichannel pipetter.
13. 30 % glacial acetic acid solution for solubilizing biofilm (*see* **Note 11**).
14. 1.5 mL microcentrifuge tubes.
15. Water: Non-sterile tap water is fine.

2.2 Biofilm Assay Media

Some of the most commonly used media for biofilm assays are LB, Mueller–Hinton broth (MHB), M9 minimal medium, and M63 minimal medium supplemented with casamino acids and glucose [19]. LB and MHB are nutrient-rich general-purpose media for growing bacteria; they provide a nutritional environment that allows for *P. aeruginosa* biofilm growth. MHB is often used in experiments on antibiotic susceptibility testing. However, when the focus of an assay is to determine the effect of a particular nutrient on biofilm formation, it is necessary to use a chemically defined medium in which the molar concentrations of the nutrient(s) in question can be controlled. Media which can be manipulated in this manner include M9, M63 with 0.4 % arginine [20], N-minimal medium [21], and B2 [22]. A number of other media can be used, as desired.

1. M9: Per liter of water, 64 g Na_2HPO_4, 15 g KH_2PO_4, 2.5 g NaCl, 5.0 g NH_4Cl, 2 mL of 1 M $MgSO_4$, 0.1 mL of 1 M $CaCl_2$, and 20 mL of 20 % glucose solution.
2. M63: Per liter of water, 13.6 g KH_2PO_4, 2.0 g $(NH_4)_2SO_4$, 0.2 g $MgSO_4$, 0.5 mg $FeSO_4$, glucose (0.2 %), and casamino acids (0.5 %) [5]. The addition of arginine to a final concentration of 0.4 % enhances *P. aeruginosa* biofilm formation [23].
3. N-minimal: 5 mM KCl, 7.5 mM $(NH_4)_2SO_4$, 0.5 mM K_2SO_4, 1 mM KH_2PO_4, 0.1 M Tris–HCl, 0.1 % casamino acids, 38 mM glycerol, 200 μM bis-Tris [21].
4. B2: 0.1 M HEPES pH 7, 7 mM ammonium chloride, 20 mM sodium succinate pH 6.7, 2 mM magnesium sulfate, 10 μM iron sulfate, 1,600 μM phosphate buffer pH 7.2, 1.62 μM manganese sulfate, 2.45 μM calcium chloride, 13.91 μM zinc chloride, 4.69 μM boric acid, 0.67 μM cobalt chloride.

2.3 Quantitation

1. 30 % glacial acetic acid in water (*see* **Note 11**).
2. 96-well optically clear, flat-bottom plate (can be non-sterile).
3. Spectrophotometer equipped with plate reading capability.

3 Methods

3.1 Preparation of Biofilm Assay Work Area

1. *See* Figs. 2 and 3 for suggested layout of materials.
2. Bench area to be used for biofilm staining should be lined with lab mat or paper towels. A space approximately 2 ft (60 cm) wide is sufficient. It is a good idea to set up a piece of cardboard or

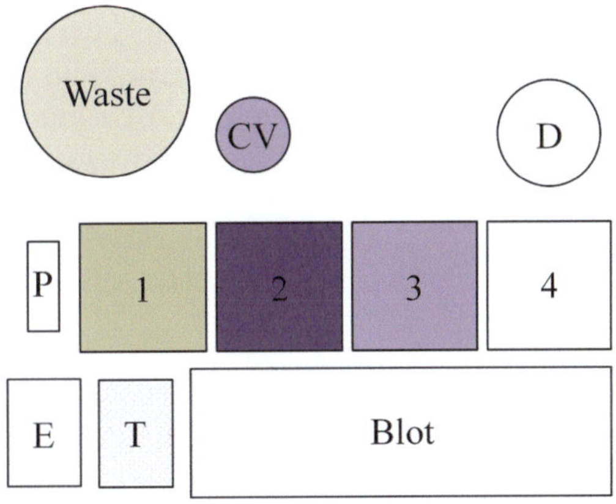

Fig. 2 Suggested benchtop layout for microtiter plate biofilm assay. *Waste* liquid waste container, *CV* bottle containing 0.1 % crystal violet, *D* disposal container for used pipette tips, *P* pipetter (single or multichannel), *1* tray 1 (waste), *2* tray 2 (crystal violet rinse 1), *3* tray 3 (crystal violet rinse 2), *4* tray 4 (rinse), *E* empty tip box or trough, *T* pipette tips, *Blot* extra layer of lab mat or paper towel

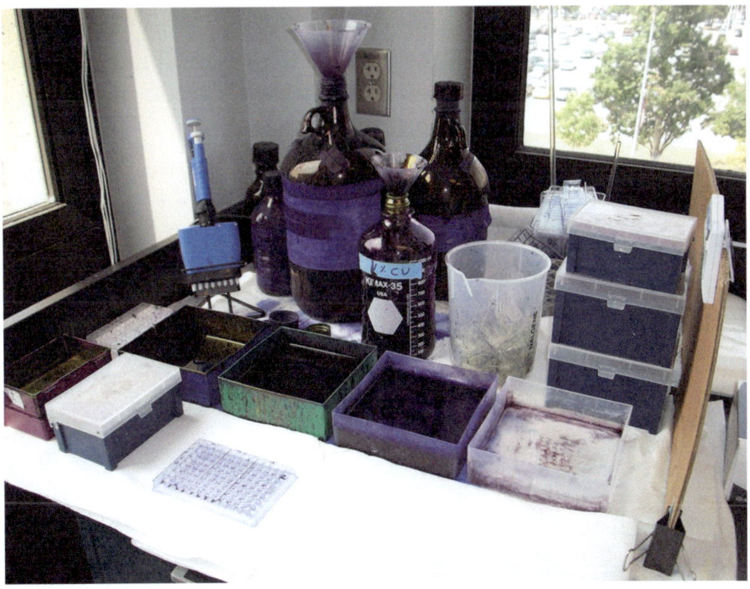

Fig. 3 Example of benchtop set up for microtiter plate biofilm assay. Note the location of materials portrayed in Fig. 2. Arrangement of materials is adaptable to fit a variety of different laboratory configurations

another material as a shield to prevent neighboring bench space from being splashed with crystal violet stain.

3. Arrange the following materials on lined bench top: Four trays, bottle of 0.1 % crystal violet solution, pipetter, pipette tips, container for liquid waste, funnel, container for used pipette tips, trough, or empty tip box (*see* Fig. 2, **Note 8**).
4. Fill trays 2, 3, and 4 with enough water to submerge microtiter plate. This water can be from the tap and need not be sterile. Water can be left in trays and reused for several assays. When the water in trays 2 and 3 becomes heavily stained with crystal violet, it should be dumped into the liquid waste containers and replaced with fresh water. The water in tray 4 should remain free of crystal violet but will become cloudy with bacteria, at which time it should be dumped into the waste container (*see* **Note 9**) and replaced with fresh water.

3.2 Preparing Bacterial Cultures

1. Inoculate bacteria into 3–5 mL appropriate growth medium in sterile culture tubes and grow to stationary phase. Typically, bacteria are grown overnight at 37 °C with shaking. If a large number of strains are to be assayed at once, such as when screening a mutant library, it may be desirable to grow cultures in 96-well plates instead of culture tubes.
2. Dilute the stationary-phase cultures 1:100 into fresh assay medium, which may vary according to the desired experimental parameters (*see* **Note 12**). It is recommended to assay all samples in triplicate (or more) so that mean values can be obtained and statistical analysis performed (*see* **Note 13**).
3. Aliquot 100 μL of diluted cultures into 96-well microtiter plate. Be sure to include a blank well containing uninoculated medium as a control.
4. Cover plate (*see* **Note 14**) and incubate at appropriate temperature for desired length of time (*see* **Notes 15** and **16**). An incubation of 8–10 h is suggested for most purposes [5].

3.3 Staining of Biofilms

1. All the remaining steps are completed at room temperature.
2. Following desired incubation time, remove planktonic cells by inverting 96-well plate over waste tray (tray 1, *see* Fig. 2) and shaking it gently.
3. Rinse the microtiter plate by submerging it in clear water in the rinsing tray (tray 4). While the plate is submerged, gently rub the surface of the plate with your gloved fingers to release bubbles and ensure that water enters all wells. After rubbing the entire surface of the plate, lift it out of the water, invert it, and shake out excess water back into the tray. Repeat this step in the same tray so that plate is rinsed two times.
4. Turn the plate face down, and pat it firmly on lab mat or paper towel to remove as much water as possible.

5. Pipette 125 μL 0.1 % crystal violet solution into wells. This volume ensures that the stain will cover the biofilm.
6. Let sit for 10 min.
7. Invert plate over waste tray (tray 1) and shake gently to remove liquid.
8. Submerge plate into tray 2, and rub the entire surface of the plate to ensure that water enters all wells. Remove plate from water, invert, and shake to remove liquid. Repeat so that plate is rinsed twice in tray 2.
9. Submerge plate into tray 3 and rinse twice as described in the previous step.
10. Invert plate over tray 3 to remove excess water.
11. With plate face down, pat firmly on lab mat or paper towel to remove as much water as possible.
12. At this point, you may be able to see purple rings where biofilms have formed at the air–liquid interface on the inner surface of the plastic wells (Fig. 1). Also, there may be biofilm on the bottom of the wells. It is important to note that crystal violet stains any substance that may be adhered to the plastic surface, including extracellular polymers or crystallized substances.
13. Allow the plate to dry for several hours, until all excess water is evaporated, prior to proceeding with quantitation.
14. If there is no visible biofilm formation in any of your samples, there are several variables you may need to adjust. Media composition, length of incubation, and incubation temperature are of primary importance. We have observed consistent *P. aeruginosa* biofilm formation (as shown in Fig. 1) in M63 medium with 0.4 % arginine, given an incubation of 24 h at 37 °C. If your assay does not produce biofilm formation under these conditions, check to make sure that the assay samples were indeed inoculated and that the assay medium did not evaporate during the incubation time.

3.4 Quantitation of Biofilms

1. Allow stained biofilm plate to air-dry for several hours or overnight.
2. Pipette 150 μL 30 % acetic acid into each well. This will solubilize the biofilm (*see* **Note 11**).
3. Allow to sit for 10 min.
4. Pipette up and down to assure that the biofilm is well solubilized, and then transfer 125 μL of each sample to a 96-well optically clear, flat-bottom plate.
5. Measure optical density of all samples in plate at 550 nm (*see* **Note 17**). Data may be graphed as shown in Fig. 4.

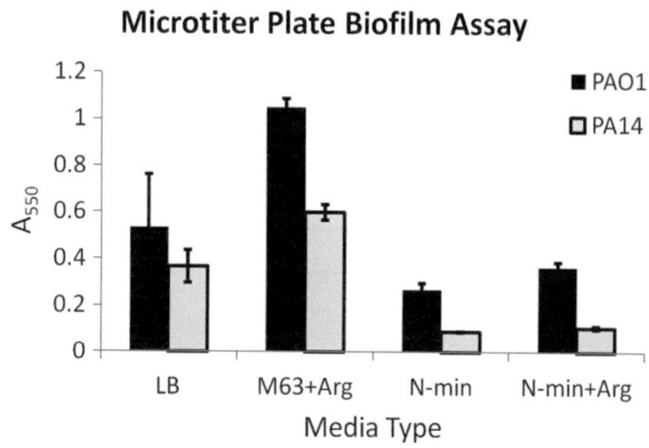

Media	PAO1	PA14
LB	0.532	0.365
M63+Arg	1.049	0.597
N-min	0.269	0.089
N-min+Arg	0.365	0.107

Fig. 4 Example of data that may be obtained by using spectrophotometry to quantitate biofilm formation. The *bar graph at left* is a graphical representation of the data shown in the table on the *right. Numbers* represent absorbance value

4 Notes

1. Several types of microtiter plates are available. We most often use vinyl U-bottom plates; however, flat- or V-bottom plates made of other materials (e.g., polystyrene or polypropylene) can also be used. Adhesion will vary depending on the properties of the surface and the *P. aeruginosa* strain (e.g., mucoid versus non-mucoid) [24].
2. Plastic food storage containers or large pipette tip boxes work well for this purpose. The container need to be just deep enough to submerge a 96-well plate. These containers will become heavily stained with crystal violet and should be used only for biofilm staining.
3. It is important to be aware of the staining properties of the crystal violet dye used in this assay. It is recommended that materials and a small portion of lab bench space be dedicated exclusively to this process.
4. When working with crystal violet, wear gloves and a lab coat, and be careful to avoid spilling the powder, which will readily stain clothes and surfaces. Stains can be removed to some degree with 70 or 95 % ethanol, but it is difficult to remove all crystal violet once it has stained a surface. It is best to dedicate a bottle and stir bar exclusively to preparation of crystal violet solution.
5. Crystal violet is classified in some countries as a hazardous substance. The use of a chemical hood while working with crystal violet is advised. Refer to your institution's chemical safety protocols for more information.
6. Multichannel pipetters greatly increase the time efficiency of these assays.

7. If the purpose of the experiment is to observe *P. aeruginosa* biofilm formation in a particular liquid substance (e.g., mouthwash, tap water, soft drinks, liquid detergent), these liquids should be used in place of the assay media described here. Thus, the stationary-phase bacteria culture would be diluted 1:100 into the substance of interest. For the first attempt at growing biofilms in liquids other than bacterial growth medium, it may be useful to include a range of dilutions, such as 1:10, 1:50, 1:100, and 1:500. This will allow a determination of the optimal dilution to use in future experiments so as to yield observable results.
8. To help prevent unwanted stain splatters in the lab, it is best to have all materials set up and ready to use before you begin biofilm staining so that you do not have to walk around the lab gathering supplies (e.g., timer, extra pipette tips, paper towels, gloves) while you are in the middle of the staining procedure.
9. Liquid waste should be placed in an autoclave-safe (e.g., glass) container. It is recommended that the liquid waste be autoclaved before disposal. It is best to autoclave when the container is no more than 2/3 full so that crystal violet waste does not spill over the top of the container due to pressure changes in the autoclave. Check with your institution's chemical safety protocol for proper disposal procedure.
10. A small funnel may be needed to transfer liquid waste from trays into waste container if the waste container is a narrow-neck bottle. This funnel will become stained with crystal violet, and so it may be desirable to dedicate a funnel specifically to biofilm assays.
11. A solution of ethanol–acetone (70:30) or 95 % ethanol may be used to solubilize stained biofilm. These solutions are particularly effective on biofilms formed by Gram-positive species [25].
12. Prior to diluting cultures, it is suggested that the density of the original cultures be adjusted to OD600 = 0.05. This will compensate for differences in cell densities of the original cultures that could potentially affect the amount of biofilm formed during the assay.
13. Before beginning to aliquot samples, particularly if working with a large number, it is a good idea to plan how the samples will be distributed in the 96-well plate and to make a written note of the layout. Permanent marker can be used to label rows and/or columns on the biofilm assay plate prior to adding samples. Labeling should not be done, however, on samples in an optically clear plate being used for spectrophotometry, as ink may interfere with readings.
14. The plate may be covered with a lid designed for 96-well plates. Placing the covered plate in a disposable plastic storage

container (such as those used for food) can reduce evaporation of media in the incubation chamber. Plate lids and storage containers may be reused by rinsing between uses with 70 % ethanol and allowing to air-dry.

15. How is incubation time determined? If the desired results of the assay are simply to determine whether or not a certain sample forms biofilm, without the need to define how long it takes or to quantitate results, then an overnight incubation is typically sufficient for *P. aeruginosa* to form biofilms. However, if the incubation is too long, the small volume of medium in the wells will become depleted of nutrients. When nutrients are unavailable, bacteria will begin to die, after which point no additional biofilm will form, and cellular debris may aggregate in the microtiter plate and produce confusing results. Studies have shown maximum biofilm formation in as little as 8–10 h [5] or as much as 72 h [23], depending on the *Pseudomonas* strain being tested, the nutrient content of the assay medium, and the type of microtiter plate. In some cases, the experimenter may wish to shorten the incubation time if, for example, the question is to see which strains or conditions produce biofilm most rapidly. In such experiments, the incubation time may be as little as 30 min, although it may take several assays and adjustments to the incubation time to optimize the results.
16. Depending on desired experimental parameters, media may or may not be changed at intervals throughout the incubation period [26]. For instance, many studies have used the microtiter plate assay to investigate the effect of antibiotics or other antimicrobial agents on *P. aeruginosa* biofilms.
17. Optical density of crystal violet-stained samples may be measured in a range of 500–600 nm [7].

Acknowledgements

This work is supported by RSFG (Indiana University Purdue University Indianapolis) and PRF (Purdue University) to Gregory G. Anderson.

References

1. Stickler DJ (2002) Susceptibility of antibiotic-resistant gram-negative bacteria to biocides: a perspective from the study of catheter biofilms. Symp Ser Soc Appl Microbiol 163S–170S
2. Christensen GD, Simpson WA, Younger JJ, Baddour LM, Barrett FF, Melton DM, Beachey EH (1985) Adherence of coagulase-negative staphylococci to plastic tissue culture plates: a quantitative model for the adherence of staphylococci to medical devices. J Clin Microbiol 22:996–1006
3. Fletcher M, Loeb GI (1979) Influence of substratum characteristics on the attachment of a marine pseudomonad to solid surfaces. Appl Environ Microbiol 37:67–72

4. Genevaux P, Muller S, Bauda P (1996) A rapid screening procedure to identify mini-Tn10 insertion mutants of Escherichia coli K-12 with altered adhesion properties. FEMS Microbiol Lett 142:27–30
5. O'Toole GA, Kolter R (1998) Initiation of biofilm formation in Pseudomonas fluorescens WCS365 proceeds via multiple, convergent signalling pathways: a genetic analysis. Mol Microbiol 28:449–461
6. O'Toole GA (2011) Microtiter dish biofilm formation assay. J Vis Exp (47):e2437. doi:10.3791/2437
7. Merritt JH, Kadouri DE, O'Toole GA (2005) Growing and analyzing static biofilms. Curr Protoc Microbiol Chapter 1, Unit 1B 1
8. Coenye T, Nelis HJ (2010) In vitro and in vivo model systems to study microbial biofilm formation. J Microbiol Methods 83:89–105
9. McBain AJ (2009) Chapter 4: in vitro biofilm models: an overview. Adv Appl Microbiol 69: 99–132
10. Redelman CV, Hawkins MA, Drumwright FR, Ransdell B, Marrs K, Anderson GG (2012) Inquiry-based examination of chemical disruption of bacterial biofilms. Biochem Mol Biol Educ 40:191–197
11. Ceri H, Olson ME, Stremick C, Read RR, Morck D, Buret A (1999) The Calgary Biofilm Device: new technology for rapid determination of antibiotic susceptibilities of bacterial biofilms. J Clin Microbiol 37:1771–1776
12. Pettit RK, Weber CA, Pettit GR (2009) Application of a high throughput Alamar blue biofilm susceptibility assay to Staphylococcus aureus biofilms. Ann Clin Microbiol Antimicrob 8:28
13. Lembke C, Podbielski A, Hidalgo-Grass C, Jonas L, Hanski E, Kreikemeyer B (2006) Characterization of biofilm formation by clinically relevant serotypes of group A streptococci. Appl Environ Microbiol 72:2864–2875
14. Lin MH, Shu JC, Huang HY, Cheng YC (2012) Involvement of iron in biofilm formation by Staphylococcus aureus. PLoS One 7:e34388
15. Standar K, Kreikemeyer B, Redanz S, Munter WL, Laue M, Podbielski A (2010) Setup of an in vitro test system for basic studies on biofilm behavior of mixed-species cultures with dental and periodontal pathogens. PLoS One 5
16. Yang Y, Jung DW, Bai DG, Yoo GS, Choi JK (2001) Counterion-dye staining method for DNA in agarose gels using crystal violet and methyl orange. Electrophoresis 22:855–859
17. Bonnekoh B, Wevers A, Jugert F, Merk H, Mahrle G (1989) Colorimetric growth assay for epidermal cell cultures by their crystal violet binding capacity. Arch Dermatol Res 281:487–490
18. Colvin KM, Gordon VD, Murakami K, Borlee BR, Wozniak DJ, Wong GC, Parsek MR (2011) The pel polysaccharide can serve a structural and protective role in the biofilm matrix of Pseudomonas aeruginosa. PLoS Pathog 7:e1001264
19. O'Toole GA, Kolter R (1998) Flagellar and twitching motility are necessary for Pseudomonas aeruginosa biofilm development. Mol Microbiol 30:295–304
20. Pardee AB, Jacob F, Monod J (1959) The genetic control and cytoplasmic expression of "inducibility" in the synthesis of B-galactosidase by E. coli. J Mol Biol 1:165–178
21. Miao EA, Freeman JA, Miller SI (2002) Transcription of the SsrAB regulon is repressed by alkaline pH and is independent of PhoPQ and magnesium concentration. J Bacteriol 184: 1493–1497
22. Mulcahy H, Charron-Mazenod L, Lewenza S (2010) Pseudomonas aeruginosa produces an extracellular deoxyribonuclease that is required for utilization of DNA as a nutrient source. Environ Microbiol 12:1621–1629
23. Musken M, Di Fiore S, Dotsch A, Fischer R, Haussler S (2010) Genetic determinants of Pseudomonas aeruginosa biofilm establishment. Microbiology 156:431–441
24. Martinez-Martinez L, Pascual A, Perea EJ (1991) Kinetics of adherence of mucoid and non-mucoid Pseudomonas aeruginosa to plastic catheters. J Med Microbiol 34:7–12
25. Fredheim, EG, Klingenberg C, Rohde H, Frankenberger S, Gaustad P, Flaegstad T, Sollid JE (2009) Biofilm formation by Staphylococcus haemolyticus. J Clin Microbiol 47:1172–1180
26. Hubner NO, Matthes R, Koban I, Randler C, Muller G, Bender C, Kindel E, Kocher T, Kramer A (2010) Efficacy of chlorhexidine, polihexanide and tissue-tolerable plasma against Pseudomonas aeruginosa biofilms grown on polystyrene and silicone materials. Skin Pharmacol Physiol 23 Suppl 28–34

Chapter 49

Methods for Studying Biofilm Dispersal in *Pseudomonas aeruginosa*

Nicolas Barraud, Joana A. Moscoso, Jean-Marc Ghigo, and Alain Filloux

Abstract

Biofilm dispersal is the last and least understood stage of the biofilm life cycle. Several recent studies have characterized dispersal events in response to various cues and signals. Here we describe a range of methods useful for the investigation of dispersal in the biofilm model organism and opportunistic pathogen *Pseudomonas aeruginosa*.

Key words Biofilm, Dispersal, Environmental cues, Nitric oxide, Batch culture, Crystal violet staining, Continuous culture, Microfermenters

1 Introduction

In most environments bacteria adopt a biphasic life cycle, with a sessile surface-attached biofilm phase and a motile free-living planktonic phase. Biofilm bacteria, compared to their planktonic counterparts, often show increased resistance to environmental stress and antimicrobials. While the mechanisms for attachment of bacteria on surfaces and the onset of biofilm formation have been well documented in many species, the reverse transition that involves the release into the bulk liquid of cells that were once strongly attached is still poorly understood. Dispersal, which is the last stage of biofilm development, appears to be a programmed, highly regulated process that can be triggered by specific cues such as nutrients, oxygen, and nitric oxide (NO) levels as well as bacterial derived signal molecules. The population of dispersal bacteria often comprises high levels of phenotypic and genetic variants, which are thought to provide an ecological advantage by facilitating colonization of new surfaces [1].

Elucidating the mechanisms of biofilm dispersal is of primary importance in order to better understand the maintenance, evolution, and dissemination of biofilm infections. It may also lead to

Alain Filloux and Juan-Luis Ramos (eds.), *Pseudomonas Methods and Protocols*, Methods in Molecular Biology, vol. 1149, DOI 10.1007/978-1-4939-0473-0_49, © Springer Science+Business Media New York 2014

the identification of potential new targets for controlling biofilms by inducing their natural detachment in a nontoxic, nonresistant manner. The successful study of this phenomenon presents several challenges: (1) one needs to grow a functional, mature biofilm, (2) then the biofilm needs to be exposed to a dispersal cue in a controlled fashion, and (3) finally the remaining biofilm needs to be analyzed for evidence of dispersal events. Since nutrients and oxygen availability can trigger dispersal, sudden changes in growth conditions during biofilm treatment or analysis may result in an uncontrolled detachment of portions of biofilm, compromising interpretation of the data. This paper describes two complementary systems that are useful to study dispersal in *Pseudomonas aeruginosa* [2]. First, a microtiter-based batch system was designed for high throughput screening of dispersal in response to various cues. Second, a microfermenter-based system allows for varying a wide range of parameters, with minimum uncontrolled disturbance to the biofilm. These procedures can be useful to screen for compounds that may induce dispersal or for mutant strains that may be defective in dispersal. The use of microfermenters, by generating abundant biomass and allowing for the extraction of nucleic acids and proteins from the attached biofilm bacteria, can also support the examination of cellular changes that regulate dispersal. Further, because the dispersal population is readily collected from the biofilm runoff effluent, it is also possible to study phenotypic and genotypic changes associated with biofilm dispersal.

2 Materials

2.1 Microtiter Plate Assay

1. Luria Bertani (LB) medium with 1 % NaCl.
2. Modified M9 minimal medium (freshly made, *see* **Note 1**): M9 1× salts consisting of 48 mM Na_2HPO_4, 22 mM KH_2PO_4, 9 mM NaCl, 19 mM NH_4Cl, pH 7.0, supplemented with the nutrients 2 mM $MgSO_4$, 100 μM $CaCl_2$, and glucose at 2 or 20 mM for continuous or batch culture experiments, respectively.
3. Tissue culture-treated 24-well plates (BD).
4. Dispersal agents. For NO donors: sodium nitroprusside (SNP, Sigma) solution in water made fresh on the day; or NONOate disodium 1-[2-(carboxylato)pyrrolidin-1-yl]diazen-1-ium-1,2-diolate (Proli, Cayman Chemicals) solution in 10 mM NaOH and stored at –20 °C for no more than 1 week.
5. Vacuum station made using a vacuum pump (Fisherbrand) connected to a side arm flask with a venting cap/stopper linked to a Pasteur pipette.
6. Repeater pipettor (Eppendorf).
7. Phosphate-buffered saline (PBS): 137 mM NaCl, 2.7 mM KCl, 10 mM Na_2HPO_4, 1.8 mM KH_2PO_4, pH 7.4.

8. Crystal violet stain: 0.2 % crystal violet prepared by diluting with PBS a 2 % crystal violet solution in 1.9 % ethanol and 0.08 % ammonium oxalate. Alternatively, a commercial stain, Gram crystal violet 0.3 % in 1:1:18 isopropanol:ethanol/methanol:water (BD), may be used.
9. 100 % ethanol.
10. Sonication bath (e.g., 150 W, Unisonics).

2.2 Microfermenter Assay

11. Microfermenters equipped with glass spatula (Institut Pasteur, Paris).
12. Silastic silicone tubing internal diameter (ID)–outside diameter (OD) 1.0–2.6 mm (Cole Parmer) for upstream sterile medium flow; Tygon tubing ID–OD 2.4–4.0 mm (Saint Gobain) for peristaltic pump; silicone microbore tubing ID–OD 1.0–2.9 mm (Cole Parmer) for downstream biofilm effluent runoff flow.
13. Polypropylene adapters and connectors (Cole Parmer): female luer × 1/8″ hose barb; male luer × 1/8″ hose barb; barbed reducer 1/8″ × 1/16″; barbed T-connector 4 mm.
14. Polypropylene 10–20 L carboys autoclavable equipped with a filling/venting cap (Cole Parmer).
15. Peristaltic pump (e.g., model 205S, Watson Marlow).
16. Water bath consisting of a 5–6 L tank and a thermostat with circulator (Cole Parmer).
17. Source of compressed air at 40 kPa.
18. Sterile filters with 0.2 μm pore size (Millipore).

3 Methods

3.1 Batch Culture Microtiter Plate Assay

In the batch assay, *P. aeruginosa* PAO1 wild type biofilms are grown in microtiter plates and dispersal is assessed as a concomitant increase in planktonic biomass and a decrease in biofilm biomass (Fig. 1).

1. Grow a *P. aeruginosa* PAO1 culture overnight in LB medium to an OD_{600} of 2.5.
2. Make a 1:200 dilution in M9 minimal medium with 20 mM glucose and aliquot 1 ml per well of a 24-well plate.
3. Put the plate on a platform shaker at 37 °C and incubate with shaking at 180 rpm (*see* **Note 2**).
4. After 6 h growth (*see* **Note 3**), induce dispersal by adding 10 μl per well of a solution at the appropriate concentration in less than 1 min per plate (*see* **Note 4**). For instance, add 10 μl of a 10 mM solution of NO donor Proli to make a final concentration of 100 μM (Fig. 1c, d). Dispersal can also be induced by reducing the shaking speed to 60 rpm (Fig. 1a, b), which is

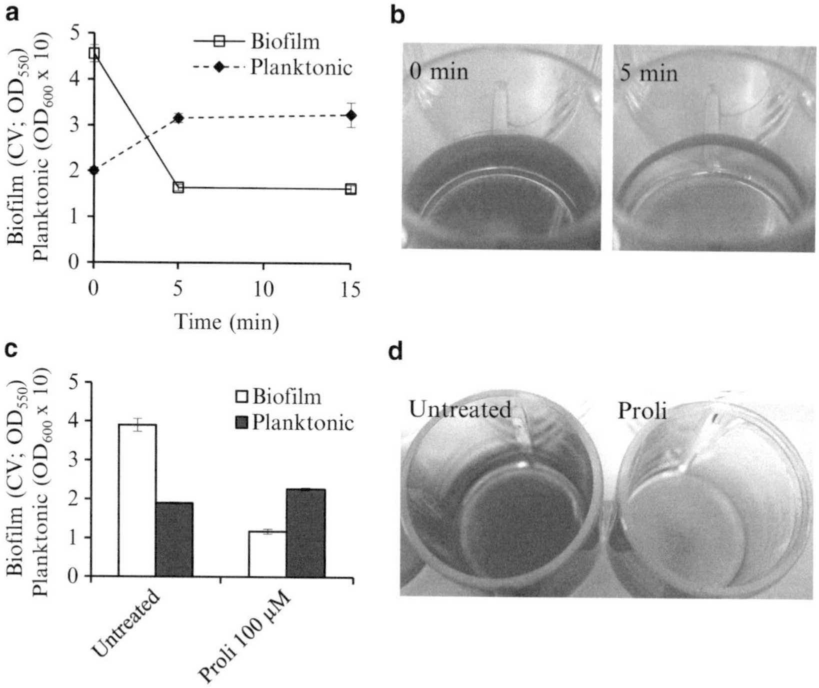

Fig. 1 Microtiter plate assay for biofilm dispersal. (**a**) Oxygen-depletion induced dispersal. *P. aeruginosa* PA01 biofilms were grown for 6 h with shaking at 200 rpm, then at $t=0$, the shaking speed was reduced to 60 rpm to reduce oxygen tension. At various time points, the planktonic biomass was assessed by measuring the OD_{600} of the supernatant and the biofilm biomass by performing crystal violet staining. (**b**) Stained biofilms before and after 5 min of incubation at a reduced shaking speed. (**c**) NO-induced dispersal. *P. aeruginosa* biofilms were grown for 6 h with shaking at 200 rpm in the absence of NO donor. Then 10 μl of a Proli solution was added to the wells to make a final concentration of 100 μM and control wells were left untreated. Plates were incubated for a further 15 min before assessing dispersal by measuring the OD_{600} of the supernatant and performing crystal violet staining of the biofilm. (**d**) Stained biofilms untreated or treated with Proli for 15 min

known to induce a decrease in oxygen tension in the culture within minutes [3].

5. Incubate for a further 15 min to 1 h.
6. The planktonic biomass is assessed by measuring the OD_{600} or colony-forming units (CFU) count of the supernatant.
7. Finally the biofilm biomass is assessed by performing either crystal violet staining [4] or CFU count of biofilm cells.

3.2 Crystal Violet Staining

Crystal violet provides a fast and highly reproducible method to quantify biofilm biomass.

1. Use the vacuum device for suction of the supernatant liquid from the wells.
2. To remove loosely attached biofilm cells, wash each well once by adding 1 ml PBS using a repeater pipettor and then removing the liquid using the vacuum device.

3. Stain with 1 ml crystal violet solution for 20 min.
4. Using the same vacuum device, remove the crystal violet dye from each well.
5. Wash twice with 1 ml PBS as in **step 2**.
6. Photograph the distribution of crystal violet dye on the walls of microtiter wells by using a digital camera (Fig. 1b, d).
7. Add 1 ml of 100 % ethanol to each well to re-elute the crystal violet dye and shake gently until all crystal violet is dissolved.
8. Measure the OD_{550}. The absorbance of an uninoculated well serves as a negative control and should be subtracted from the value of the inoculated wells.

3.3 CFU Enumeration of Biofilms

CFU count is a sensitive method that allows for detection of much lower numbers of biofilm bacteria compared to crystal violet staining. Further, CFU count is useful to assess the efficacy of biofilm control measures when dispersal treatments are combined with bactericidal agents.

1. Equip the vacuum device with a sterile Pasteur pipette and suction liquid from the cultures, briefly flaming the Pasteur pipette between each well.
2. Wash each well twice with 1 ml of sterile PBS. Drain all the wells.
3. Add 1 ml sterile PBS and resuspend biofilms on the bottom and walls of the well by swabbing with a sterile cotton bud and incubating in a sonication bath for 2 min.
4. Perform a serial dilution of resuspended biofilm cells and plate on LB agar.
5. Enumerate CFU after 24 h incubation at 37 °C. After 6 h incubation under these growth conditions, control biofilms that were not treated harbored $19.6 \pm 2.6 \times 10^6$ CFU/cm^2.

3.4 Continuous Culture Microfermenter Assay

In the continuous flow assay, *P. aeruginosa* PAO1 biofilms are grown under flow-through conditions in glass microfermenters [5] and dispersal is assessed by monitoring the biofilm effluent (Fig. 3). At the end of the assay, the remaining biofilm can be quantified or further analyzed for gene or protein expression.

1. Prepare the microfermenter system as described in Fig. 2, and sterilize all components by autoclaving. Autoclave the Tygon pump tubing separately with both extremities free of any connector and wrapped in aluminum foil.
2. Prepare a 10–20 L M9 1× salts solution and sterilize by autoclaving at 121 °C, 15 psi for 45 min instead of the usual 15 min. After cooling down, add 2 mM $MgSO_4$, 100 μM $CaCl_2$, and 2 mM glucose to the M9 1× salts solution, preferably under sterile conditions. Mix well to dissolve any calcium precipitate.

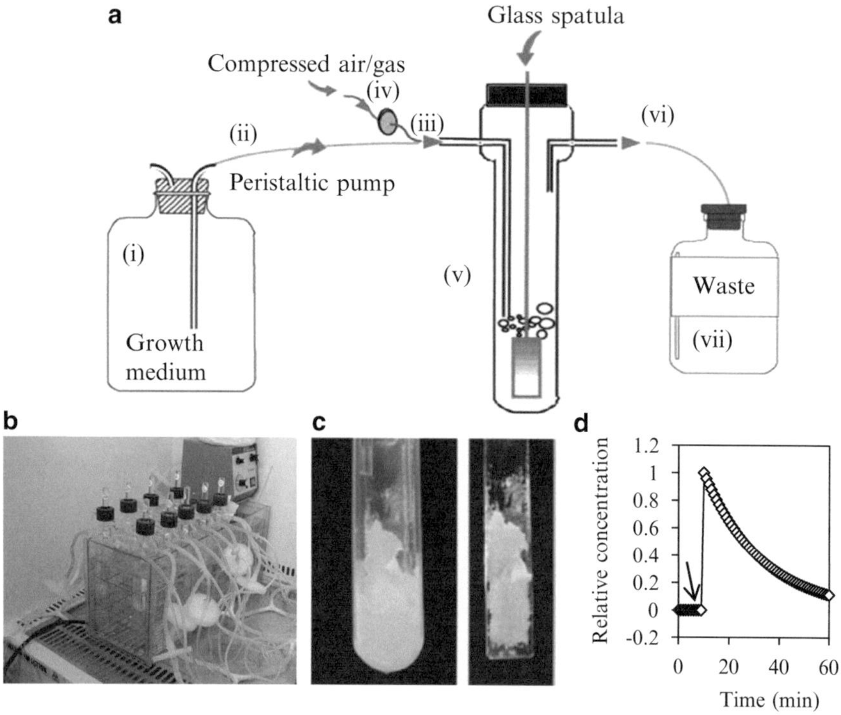

Fig. 2 (**a**) Illustration of the microfermenter flow-through culture set up: (i) 10–20 L fresh medium autoclavable carboy equipped with venting cap; (ii) 40-cm-long Silastic tubing (in a system where multiple microfermenters are used, this tubing may be diverted by using T-connectors) connected with male and female luer lock connectors to a 30-cm-long Tygon pump tubing which is pinched to the roller of a peristaltic pump and connected on the other side via a luer connection to a 20-cm-long Silastic tubing; (iii) T-connector; (iv) Silastic tubing with 0.2-μm filter connected to a source of compressed air at 40 kPa; (v) glass microfermenter with unused capillaries sealed and equipped with a removable glass spatula; (vi) 3-cm-long Silastic effluent tubing connected with reducing connector to 80-cm-long microbore effluent tubing connected with reducing connector to 3-cm-long Silastic tubing; (vii) waste collection carboy capped with aluminum foil. (**b**) Microfermenters in a water bath. (**c**) Biofilm formed on the glass surfaces of microfermenter and spatula. (**d**) Theoretical dilution curve of a solute introduced at a concentration of 1 in the microfermenter at a time indicated by the *arrow* (illustrating a sudden dispersal event) and evacuated over time with the flow of fresh medium

3. Connect the microfermenter to the medium inlet via a piece of Tygon tubing pinched to the roller of a peristaltic pump. Put the microfermenter upright in a water bath heated at 37 °C. Close the air supply line with a clamp and rinse the whole system by pumping through each microfermenter 200 ml of fresh sterile M9 medium.
4. Stop the medium flow, open the microfermenter cap under sterile conditions and in each microfermenter containing 40 ml of M9 medium inoculate 1 ml of an overnight culture of *P. aeruginosa* PAO1 grown in LB at 37 °C. Close the cap tightly.
5. Allow cells to attach with medium flow and airflow turned off for 1 h.

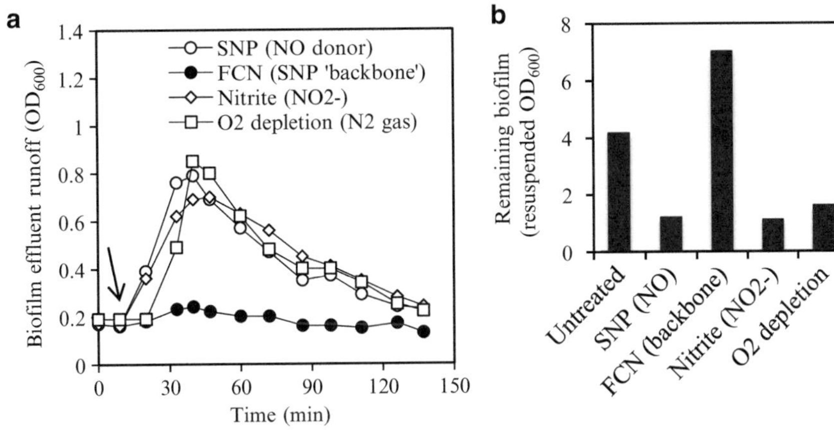

Fig. 3 Dispersal of *P. aeruginosa* PAO1 biofilms in response to NO and oxygen depletion. (**a**) One-day-old biofilms grown in microfermenters were treated (*arrow*) with the NO donor SNP at 1 mM (*white circles*), 10 mM sodium nitrite (*white diamonds*), or 1 mM potassium ferricyanide (FCN, *black circles*), the SNP "backbone" that does not release NO. To deplete the biofilm of oxygen, aeration was switched to nitrogen gas (*white squares*). To monitor dispersal events, OD_{600} of the biofilm effluent was measured. (**b**) After 2.5 h, biofilm cells remaining on the microfermenter spatula were resuspended in PBS and quantified by measuring the OD_{600}

6. Turn on liquid medium flow at a rate of 0.8 ml/min. Unclamp the air supply line and turn on the aeration at 40 kPa, 100 ml/min (*see* **Note 5**). When air and medium flows are turned on, the volume of medium inside the microfermenter is ~20 ml, thus making a residence time of 25 min.
7. Grow for 24 h without interruption or disturbance (*see* **Note 6**).
8. After 24 h, switch the medium inlet to separate containers bearing fresh media with specific treatment, e.g., NO donors SNP at 1 mM or $NaNO_2$ at 10 mM (Fig. 3), no carbon source (starvation), or 20 mM glucose (nutrient increase). For oxygen depletion-induced dispersal, switch aeration to nitrogen or argon gas (*see* **Note 7**). This procedure ensures minimal physical disruption of the biofilm growing inside the microfermenters.
9. Collect the biofilm effluent in separate 15 ml Falcon tubes (BD) every 10 min and monitor the release of dispersal cells by measuring the OD_{600} (*see* **Note 8**).
10. To assess the remaining biofilm biomass after dispersal has been induced, remove the spatula aseptically and resuspend the biofilm cells in 5 ml PBS by swabbing with a sterile cotton bud. Measure the OD_{600} of the suspension. This procedure could also be used to extract proteins or nucleic acids from biofilm cells after dispersal was induced.
11. For microscopy analysis of biofilm structures, remove the spatula and rinse once by dipping briefly in sterile PBS. Add an appropriate stain (e.g., live/dead bacterial viability kit, Invitrogen), put a cover slip on top and visualize under a microscope.

4 Notes

For the batch microtiter plate assay:

1. Complete M9 medium needs to be made fresh on the day by adding $MgSO_4$ (from 1 M stock autoclaved), glucose (from 20 % stock sterilized by passing through a 0.2 μm filter), and $CaCl_2$ (from 1 M stock autoclaved) to the 1× salts solution. Autoclave the 1× salts solution beforehand (this can be kept on the bench for long periods).
2. To minimize variations in biofilm growth due to heaters and air fans that may induce variable evaporation/aeration (edge effect), put plates on the same row/level, and preferably in the middle of the shaker. To check for possible variations, include control wells (not dispersed with any treatment) both on the edge and in the middle of the plate.
3. Incubation time and inoculum may have to be adjusted according to the strain being used. For instance, when growing *P. aeruginosa* PA14 biofilms under these conditions, a 1:1,000 initial dilution in M9 and incubation for 18 h are recommended.
4. Do not stop or open the incubator during the experiment and work quickly every time the plate is taken out of the incubator. Use only 12 wells in the two middle rows of the 24-well plate in order to keep the total number of wells to a minimum, which should allow for processing each plate quickly enough every time the plate is removed from the shaker-incubator, i.e., for addition of treatments, quantification of supernatants, washing and crystal violet staining. If assessing biofilm biomass by crystal violet analysis rather than CFU count, the system does not need to be strictly sterile at the time of adding dispersal treatments, i.e., 15 min to 1 h before the end of incubation. It is recommended to start by performing the oxygen-induced dispersal assay in order to better grasp the dynamics of dispersal that may be potentially induced when the plate is taken out of the incubator.

For the continuous flow microfermenter assay:

5. Always apply a positive pressure to medium and air supplies or the supernatant will flow back, contaminate the growth medium and block air filters. Therefore, to turn the system on, first tighten the microfermenters caps, then switch on the air pump and finally the peristaltic pump. To turn off, first switch off the peristaltic pump, then the air pump and finally loosen the caps.
6. Incubation time may have to be adjusted according to the strain being used. For instance, under these growth conditions

P. aeruginosa PA14 form visible biofilms only after 3 days. Increased volumes of media may be necessary.

7. Use three-way valves for aeration of microfermenters in order to allow for switching aeration with minimum disruption. Alternate aeration lines should be equipped with pressure gauge and flow meters as well as fine regulators to adjust the pressure quickly after switching the aeration source.
8. Using a microbore tubing (e.g., 1 mm internal diameter) for the biofilm effluent ensures minimizing biofilm growth in areas other than the microfermenter that could potentially interfere with the analysis of cells in the biofilm effluent that are expected to be exclusively released from the biofilm inside the microfermenter. A microbore tubing presents a small surface area and the flow velocity and shear stress are high.

Acknowledgement

This work was partially supported by grants from the Australian Research Council No. LX099061 and DE120101604.

References

1. McDougald D, Rice SA, Barraud N et al (2012) Should we stay or should we go: mechanisms and ecological consequences for biofilm dispersal. Nat Rev Microbiol 10:39–50
2. Barraud N, Kardak BG, Yepuri NR et al (2012) Cephalosporin-3′-diazeniumdiolates: targeted NO-donor prodrugs for dispersing bacterial biofilms. Angew Chem Int Ed Engl 51: 9057–9060
3. Härtig E, Zumft WG (1999) Kinetics of *nirS* expression (cytochrome cd_1 nitrite reductase) in *Pseudomonas stutzeri* during the transition from aerobic respiration to denitrification: evidence for a denitrification-specific nitrate- and nitrite-responsive regulatory system. J Bacteriol 181: 161–166
4. O'Toole GA, Kolter R (1998) Initiation of biofilm formation in *Pseudomonas fluorescens* WCS365 proceeds via multiple, convergent signalling pathways: a genetic analysis. Mol Microbiol 28:449–461
5. Ghigo JM (2001) Natural conjugative plasmids induce bacterial biofilm development. Nature 412:442–445

Chapter 50

Pseudomonas aeruginosa PA14 Pathogenesis in *Caenorhabditis elegans*

Natalia V. Kirienko, Brent O. Cezairliyan, Frederick M. Ausubel, and Jennifer R. Powell

Abstract

The nematode *Caenorhabditis elegans* is a simple model host for studying the interaction between bacterial pathogens such as *Pseudomonas aeruginosa* and the metazoan innate immune system. Powerful genetic and molecular tools in both *C. elegans* and *P. aeruginosa* facilitate the identification and analysis of bacterial virulence factors as well as host defense factors. Here we describe three different assays that use the *C. elegans–P. aeruginosa* strain PA14 host-pathogen system. Fast Killing is a toxin-mediated death that depends on a diffusible toxin produced by PA14 but not on live bacteria. Slow Killing is due to an active infection in which bacteria colonize the *C. elegans* intestinal lumen. Liquid Killing is designed for high-throughput screening of chemical libraries for anti-infective compounds. Each assay has unique features and, interestingly, the PA14 virulence factors involved in killing are different in each assay.

Key words *Caenorhabditis elegans*, *Pseudomonas aeruginosa*, PA14, Fast Killing, Slow Killing, Liquid Killing, Innate immunity, Pathogenesis

1 Introduction

The self-fertilizing nematode *Caenorhabditis elegans* lives in association with rotting fruit, where it feeds on microorganisms. Its small size, short generation time, powerful genetics, ease of handling, invariant developmental lineage, sequenced genome, and reverse genetic tools have formed the basis of a world-wide community of *C. elegans* researchers since its initial popularization in 1974 by Sydney Brenner as a model genetic organism [1]. It is now one of the best-understood metazoans. Only recently, however, has *C. elegans* been used as a model to study microbial pathogenesis and innate immunity.

C. elegans is typically propagated in the laboratory on agar medium spread with a lawn of *Escherichia coli* strain OP50 [1]. *C. elegans* can be infected with or intoxicated by a remarkably large number of human microbial pathogens simply by transferring the

Alain Filloux and Juan-Luis Ramos (eds.), *Pseudomonas Methods and Protocols*, Methods in Molecular Biology, vol. 1149, DOI 10.1007/978-1-4939-0473-0_50, © Springer Science+Business Media New York 2014

nematodes from a lawn of *E. coli* OP50 to a lawn of the pathogen in question. Detailed protocols for carrying out *C. elegans* killing assays with multiple bacterial and fungal pathogens have been published [2–4]. The most widely studied *C. elegans* pathogen is *Pseudomonas aeruginosa*. Different *P. aeruginosa* strains can kill *C. elegans* by a variety of mechanisms. For example, the strain PAO1 can kill via cyanide poisoning when grown on rich media or by the PQS-Fe^{3+}-mediated "red death" that is triggered by low phosphate conditions [5–7]. The strain PA14 can kill by colonizing and dividing in the intestine under low osmolarity conditions or via a low-molecular-weight diffusible toxin of the pyocyanin–phenazine class under high osmolarity conditions [8–10]. Additionally, pyoverdin, a bacterial siderophore synthesized by PA14, can disrupt host iron homeostasis, causing death in a liquid-based *C. elegans* killing assay [11].

The great potential of the *C. elegans–P. aeruginosa* model system lies not only in the multiple mechanisms of killing that can be studied but also in the powerful genetic and molecular tools available to approach this interaction from the sides of both the host and the pathogen. Indeed, several fruitful screens have found critical virulence factors in both PAO1 and PA14, and multiple genetic pathways that contribute to the *C. elegans* response have been delineated [12–18]. Adaptations that permit automated high-throughput screening in 384-well plates also permit the screening of large libraries of chemical compounds for the ability of each compound to block or cure an infection [11].

In this chapter, we will focus on three different protocols for the killing of *C. elegans* by *P. aeruginosa* strain PA14. Fast Killing is a rapid intoxication of *C. elegans* that does not depend on the accumulation of bacteria within the intestine nor even the presence of live bacteria [9, 10]. When PA14 is grown on high osmolarity-rich agar media, synchronized worms transferred to these lawns will die within 24 h. The Fast Killing assay is commonly used to compare the virulence of different PA14-derived mutants. It can also be used to compare the susceptibility of different *C. elegans* mutants to toxin-mediated death. When comparing *C. elegans* mutants, the Fast Killing assay is best suited for studying strains that are more resistant than wild type worms rather than those that are more sensitive than wild type worms, due to the extremely rapid kinetics of death.

Slow Killing is due to an active infection by live PA14 that accumulate in the lumen of the *C. elegans* intestine [8]. When synchronized worms are transferred to PA14 that has been grown on low osmolarity minimal agar media, they develop symptoms of an infection (e.g., distended intestinal lumen, reduced motility) and die over the course of several days. The Slow Killing assay has most commonly been used to compare the susceptibility of different *C. elegans* mutants to infection, but it can also be used to compare the virulence of PA14-derived mutants. Both the Fast Killing and Slow Killing assays are relatively low-throughput since worms must be transferred to the killing plates by hand, so only up to a few dozen different conditions can be realistically tested in any given assay.

Liquid Killing is a modification of the Slow Killing assay in which PA14 are grown as with Slow Killing but then the bacteria are transferred to a liquid-based medium in 384-well microtiter plates [11]. This permits automated high-throughput screening of many conditions in parallel. This assay is well-suited to screening large chemical libraries to determine the effect that different chemicals have on the infection. Different bacterial or worm strains can also be compared with this assay, but this reduces the throughput to a similar level as the Fast Killing or Slow Killing assays.

2 Materials

2.1 Equipment

1. Worm pick: Fashion a wire worm pick by inserting 2–3 cm of 32 gauge platinum wire into a holder. The holder can be a disposable glass Pasteur pipette, which is attached to one end of the wire by holding it over a flame to melt the glass around the wire. Alternatively, the wire can be screwed into the handle of a bacterial inoculating loop. If desired, flatten the end of the pick with pliers or a hammer.
2. L spreader: A glass L spreader may be made by holding a disposable glass Pasteur pipette over a flame until it bends into an L shape. Position the pipet so that the size of the L spreader is appropriate for the size of the plate you are spreading—typically about 1/2–2/3 the diameter of the plate. Seal the very end by holding the tip of the pipette just at the edge of the flame. Sterilize before each use by dipping the micro-spreader in >75 % ethanol and quickly passing it through a flame to burn off the excess ethanol. Alternatively, disposable single-use sterile plastic spreaders may be purchased.

2.2 Buffers and Media

Use ultrapure water (e.g., MilliQ ddH_2O) for the preparation of all media and buffers. Unless otherwise noted, sterilize buffers and media by autoclaving at 121 °C for 20 min for volumes less than 1 L; for larger batches, adjust autoclave time accordingly to ensure sterility. Store at room temperature unless otherwise noted.

1. M9W: Dissolve 3 g KH_2PO_4, 6 g Na_2HPO_4, and 5 g NaCl in 1 L of ddH_2O. Autoclave, cool to 60 °C, and add 1 mL of filter-sterilized 1 M $MgSO_4$.
2. S Basal: 0.1 M NaCl, 0.05 M potassium phosphate (pH 6.0). Autoclave.
3. S Basal+: Supplement 1 L of sterile, cooled S Basal with 1 mL cholesterol (5 mg/mL in ethanol) just before using.

4. Bleach Solution: Mix 20 mL NaOCl solution (available chlorine = 10–15 %), 16 mL 3 M NaOH, 64 mL H_2O. Store at 4 °C in the dark for up to 1 month.
5. 5-Fluoro-2′-deoxyuridine (FUDR): For 100×, make 10 mg/mL FUDR in ddH_2O. Filter sterilize and store at −20 °C in 1 mL aliquots.
6. Sytox Orange-S basal staining solution: Mix 4 μl Sytox Orange in DMSO with 20 mL S Basal. Make just before using.
7. LB broth: Dissolve 10 g Bacto-tryptone, 5 g Bacto-yeast extract, and 10 g NaCl in 1 L ddH_2O. Autoclave.
8. LB agar plates: Mix 10 g Bacto-tryptone, 5 g Bacto-yeast extract, 10 g NaCl, and 15 g agar in 1 L ddH_2O. Autoclave. Cool agar to 55 °C and pour into sterile Petri dishes (typically 20 mL for 10 cm plates). Dry plates at room temperature overnight, then store at 4 °C in a covered plastic box for up to 3 months.
9. Nematode Growth Medium (NGM): Mix 2.5 g Bacto-Peptone, 3 g NaCl, and 17 g Bacto-Agar in 1 L ddH_2O. Autoclave, then cool to 55 °C. Slowly add the following filter-sterilized solutions: 1 mL 1 M $MgSO_4$, 25 mL 1 M KH_2PO_4, pH 6, 1 mL 1 M $CaCl_2$, and 1 mL 5 mg/mL cholesterol in ethanol (store at 4 °C). Pour agar into sterile Petri dishes (typically 10 mL for 6 cm plates or 20 mL for 10 cm plates). Dry plates at room temperature overnight, then store at 4 °C in a covered plastic box for up to 3 months.
10. Peptone-Glucose-Sorbitol agar media (PGS): Mix 5 g Bacto proteose peptone, 5 g NaCl, 13.7 g D-sorbitol, and 8.5 g Bacto-agar in 475 mL ddH_2O. Add 25 mL of 20 % D-glucose. Autoclave, then cool to 55 °C. Pour agar into sterile Petri dishes (typically 4 mL for 3.5 cm plates). Dry plates at room temperature overnight, then store at 4 °C in a covered plastic box for up to 1 month.
11. Slow Killing (SK) agar media: Mix 3.5 g Bacto-Peptone, 3 g NaCl, and 17 g Bacto-Agar in 1 L ddH_2O. Autoclave, then cool to 55 °C. Slowly add 1 mL 1 M $MgSO_4$, 25 mL 1 M KH_2PO_4, pH 6, 1 mL 1 M $CaCl_2$, and 1 mL 5 mg/mL cholesterol in ethanol (store at 4 °C), filter sterilized. Pour agar into sterile Petri dishes (typically 4 mL for 3.5 cm plates or 20 mL for 10 cm plates). Dry plates at room temperature overnight, then store at 4 °C in a covered plastic box for up to 1 month.
12. SK broth: Prepare Slow Killing media as described above, omitting the agar. Add the $MgSO_4$, $CaCl_2$, and cholesterol just before using.
13. Liquid Kill media: Mix 1 part SK broth with 1 part S Basal+.

3 Methods

3.1 Preparation of NGM-OP50 Plates

1. *Preparation of Thin Lawn NGM-OP50 plates*: Thin lawn 6 cm NGM-OP50 plates can be prepared ahead of time and stored at 4 °C for several weeks. Inoculate a 5 mL culture of LB broth with a single colony of *E. coli* OP50 (*see* **Note 1**).
2. Incubate with agitation at 37 °C overnight.
3. Pipet about 50–200 μl of the saturated OP50 culture onto 6 cm NGM plates. If desired, spread the culture slightly by gently swirling the plates, but do not allow the bacteria to touch the sides of the plate.
4. Allow the OP50 lawn to dry and grow for 1–3 days at room temperature.
5. Store plates in a plastic box with a lid at 4 °C.
6. *Preparation of Thick Lawn NGM-OP50 plates*: Thick lawn 10 cm NGM-OP50 plates can be prepared ahead of time and stored at 4 °C for several weeks (*see* **Note 1**). Grow 1 L of *E. coli* OP50 in LB broth overnight at 37 °C with agitation to a saturated culture.
7. Spin the bacteria at 3,500 × *g* for 15 min and resuspend the pellet in 40 mL of M9W.
8. Pipet 2 mL of concentrated bacteria onto each 10 cm NGM plate.
9. Allow the OP50 lawn to dry; transfer plates to a plastic box with a lid or plastic sleeves for storage at 4 °C.
10. Pre-warm plates to room temperature before usage.

3.2 Fast Killing of C. elegans by P. aeruginosa

1. *Preparation of P. aeruginosa*: 4–5 days before the start of the assay, streak *P. aeruginosa* PA14 bacterial strains from a frozen stock on LB agar plate. Also streak *E. coli* DH5α to use as a negative control. Grow plates at 37 °C for 12–18 h and store the streaked plate at 4 °C for no longer than 2 weeks (*see* **Notes 2** and **3**).
2. Approximately 60 h prior to assay start, inoculate 3–5 mL of LB broth in a 15 mL test tube with a single colony of each bacterial strain. Grow with aeration in a tube rotator at 37 °C for 12–16 h (*see* **Note 4**).
3. Pipet 5 μl of saturated cultures onto 3.5 cm PGS agar plates. Spread the culture slightly using a sterile L spreader, leaving a border of bare agar around the edges of the plate. Assays should be run in triplicate, so prepare at least three of each type of plate (*see* **Notes 5** and **6**).
4. Allow bacteria to dry for approximately 20 min at room temperature before placing the plates upside-down in a plastic box.

5. Incubate inoculated plates at 37 °C for 24 h.
6. Transfer the plates to 23 °C and incubate for 24 h (*see* **Note 7**).
7. *Preparation of C. elegans*: 2 days before the start of the assay, pick gravid N2 (10 animals per plate) and any necessary control hermaphrodite worms to 6 cm thin lawn NGM-OP50 plates (*see* **Notes 8–10**).
8. Incubate the worms at 15 °C for 12–18 h.
9. Use a worm pick to remove all adult worms from the NGM-OP50 plates. Return the plates containing embryos to the 15 °C incubator for 8–10 h.
10. Shift the worms to a 20 °C incubator and grow 36–40 h until they are at the late L4 larval stage (*see* **Note 11**).
11. *Fast Killing assay set up*: Pick 30–40 age-matched L4 worms from the NGM-OP50 plates and transfer them to each prepared PGS agar plate (**steps 1–6**) (*see* **Notes 11** and **12**).
12. Incubate the plates at 25 °C.
13. *Assay scoring and data analysis*: The worms will initially go into shock due to the change in osmotic pressure of the media (*see* **Note 13**). They will recover in 2–3 h, so it is best to take the first time point 3–4 h after setting up the assay. Score the worms as dead or alive at various times (typically 4, 8, and 24 h), using a touch movement assay for death. Visually inspect the worm for movement. If it is not moving, then use a worm pick to touch the worm gently on the nose, stroke the worm along its side, and touch the tail. Watch for any movement, especially pumping in the pharynx, foraging behavior in the head, or curling of the tail. Worms that are not moving are scored as dead, counted, and picked off the killing plate. Worms that are still moving are scored as alive and are also counted (*see* **Note 14**).
14. Take additional time points over the next 24 h, as desired. For most assays, time points taken 4-, 8-, and 24-h after the start of the assay are sufficient to provide a nice representation of the phenotype of the worm and/or bacterial strain being tested.
15. Multiple conditions (e.g., multiple bacterial strains or multiple worm strains) can be compared at a single time point. Calculate the Percent Alive for each plate of each condition as follows:

$$\text{Percent Alive} = \frac{\text{Number alive}}{\text{Total worms}}(100)$$

16. Calculate the mean Percent Alive for all plates of each condition. Compare these means using a Student's *t*-test or ANOVA.
17. Alternatively, the best representation is to compare the entire survival curves for each condition, as described in Subheading 3.5 below.

3.3 Slow Killing of *C. elegans* by *P. aeruginosa*

1. *Preparation of P. aeruginosa*: 4–5 days before the start of the assay, streak *P. aeruginosa* PA14 and any necessary control bacteria from a frozen stock on LB agar plate. Grow at 37 °C for 12–18 h and store the streaked plate at 4 °C for no longer than a week (*see* **Notes 2** and **3**).
2. Approximately 60 h prior to infection, inoculate 3–5 mL of LB broth in a 15 mL test tube with a single colony of PA14. Grow with aeration in a tube rotator at 37 °C for 12–16 h (*see* **Note 4**).
3. Pipet 10 μL of saturated culture onto 3.5 cm SK agar plates. Spread the culture slightly using a sterile L spreader. Assays should be run in triplicate, so prepare at least three of each type of plate (*see* **Notes 5** and **6**).
4. Allow bacteria to dry for approximately 20 min at room temperature before placing them upside-down in a plastic box (*see* **Note 15**).
5. Incubate inoculated plates at 37 °C for 24 h.
6. Transfer the plates to 25 °C and incubate for 24 h (*see* **Note 16**).
7. *Preparation of C. elegans*: Three days before the start of the assay, wash gravid hermaphrodites, which contain many fertilized eggs visible in the uterus, off two to four 10 cm NGM-OP50 plates with M9W buffer into a 15 mL conical tube (*see* **Notes 8**, **10**, **17**).
8. Spin the worms in a clinical centrifuge at 1,500 × *g* for 30 s to pellet; aspirate the supernatant until approximately 1.5 mL liquid remain (Worm pellet volume ~0.1–0.5 mL).
9. Add 2.5 mL of bleach solution. Shake vigorously.
10. Monitor the lysis of the worm cuticle by laying the tube on a dissecting microscope. The worms will first break in half, releasing their eggs. Adult and larval cuticles will continue to dissolve but the unhatched progeny are partially protected from the bleach by their egg shells. Shake periodically. Within 4–5 min, most, but not all, of the adult worms should dissolve. Do not allow the reaction to proceed longer than 7 min or the eggs will begin to die.
11. Quickly add 11 mL of M9W buffer to dilute the bleach. Pellet the eggs in a clinical centrifuge at 1,500 × *g* for 30 s. Pour off or aspirate the supernatant.
12. Quickly wash the eggs three times by adding 14 mL of M9W buffer, shaking to resuspend the eggs, and spinning to pellet the eggs each time. After three washes, resuspend the eggs in 5 mL of M9W buffer.
13. Rotate 15 mL tubes on a rotisserie at room temperature for 16–24 h. The eggs will hatch and larval development will arrest in the absence of food.

14. 2 days prior to infection, pipet ~200–300 arrested L1 worms onto each 6 cm thin lawn NGM-OP50 plate. Make one plate of worms for each assay plate that will be set up (*see* **Note 18**).
15. Incubate the worms ~40–48 h at 20 °C until they reach the L4 larval stage.
16. *Slow Killing assay setup*: Pipet 40 μl of 100× 5-fluoro-2′-deoxyuridine (FUDR) onto the bare agar, distributing the volume in tiny drops as close to the edge of the plate as possible. Allow the FUDR to soak into and diffuse through the agar for about 30 min before adding worms (*see* **Note 19**).
17. Pick about 40 age-matched L4 hermaphrodite worms to the prepared Slow Killing plates. Set the worms down on the bare agar, just outside the lawn of bacteria. Try to transfer as little OP50 as possible from the maintenance plates to the killing plates (*see* **Notes 20** and **21**).
18. Incubate the killing plates at 25 °C (*see* **Note 22**).
19. *Assay scoring*: Score the worms approximately 12–24 h after setting up the assay. Score the worms as dead using a touch movement assay for death. Visually inspect the worm for movement. If it is not moving, then use a worm pick to touch the worm gently on the nose, stroke the worm along its side, and touch the tail. Watch for any movement, especially pumping in the pharynx, foraging behavior in the head, or curling of the tail. Worms that are not moving are scored as dead, counted, and picked off the killing plate. Worms that are still moving are scored as alive and are also counted (*see* **Note 14**). Return the plates to the 25 °C incubator.
20. Take additional time points, as required by the worm and/or bacterial strains being tested. The first time point should be taken when the worms have not yet begun to die (typically 12–24 h post infection). Take 2–3 time points each day for the remainder of the assay. Be sure to take at least three time points during the exponential phase of death (typically 40–72 h post infection), and at least one late time point. If possible, score the assay until all worms are dead.
21. Compare the survival curves for each condition as described in Subheading 3.5 below.

3.4 Liquid Killing of *C. elegans* by *P. aeruginosa*

1. *Preparation of P. aeruginosa*: 4–5 days before the start of the assay, streak *P. aeruginosa* PA14 bacteria from a frozen stock on LB agar plate. Grow at 37 °C for 12–18 h and store the streaked plate at 4 °C for no longer than a week (*see* **Note 2**).
2. Approximately 60 h prior to infection, inoculate 3–5 mL of LB broth with a single colony of PA14. Grow with aeration at 37 °C for 12–16 h (*see* **Note 4**).

3. Pipet 350 μL of saturated culture onto 10 cm SK plates and spread the culture using a sterile L spreader to cover the plate completely. Allow the plates to dry at room temperature for 10–15 min.
4. Incubate inoculated plates at 37 °C for 24 h.
5. Transfer the plates to 25 °C or room temperature and incubate for 24 h.
6. *Preparation of C. elegans*: Five days before the start of the assay, wash gravid *glp-4* hermaphrodites, which contain many fertilized eggs visible in the uterus, off 4–12 10 cm NGM-OP50 plates with M9W buffer into a 15 mL conical tube (*see* **Notes 8** and **23**).
7. Spin the worms in a clinical centrifuge at 1,500 × *g* for 30 s to pellet; aspirate the supernatant until approximately 1.5 mL liquid remain (Worm pellet volume ~0.5 mL).
8. Add 2.5 mL of bleach solution. Shake vigorously.
9. Monitor the lysis of the worm cuticle by laying the tube on a dissecting microscope. The worms will first break in half, releasing their eggs. Adult and larval cuticles will continue to dissolve but the unhatched progeny are partially protected from the bleach by their egg shells. Shake periodically. Within 4–5 min, most, but not all, of the adult worms should dissolve. Do not allow the reaction to proceed longer than 7 min or the eggs will begin to die.
10. Quickly add 11 mL of M9W buffer to dilute the bleach. Pellet the eggs in a clinical centrifuge at 1,500 × *g* for 30 s. Pour off or aspirate the supernatant.
11. Quickly wash the eggs three times by adding 14 mL of M9W buffer, shaking to resuspend the eggs, and spinning to pellet the eggs each time. After three washes, resuspend the eggs in 5 mL of M9W buffer.
12. Rotate 15 mL tubes on a rotisserie at 15 °C for at 40–48 h. The eggs will hatch and larval development will arrest in the absence of food.
13. 3 days (~60 h) prior to infection, spot ~5,000–6,000 worms onto each Thick lawn 10 cm NGM-OP50 plate.
14. Incubate ~16 h at 15 °C followed by ~44–48 h at 25 °C. Worms should become young adults (*see* **Note 24**).
15. *Liquid Killing assay setup*: Using a cell-scraper, scrape bacteria from one or more SK plates, and resuspend in M9. Measure OD_{600} to determine bacterial density.
16. Prepare 4× bacteria by diluting desired volume of resuspended bacteria with S Basal+ to $OD_{600} = 0.12$.

17. Combine one part diluted bacteria with three parts Liquid Kill media.
18. Use a multichannel, repeating pipette or an automated liquid-handling machine to add 50 μL/well to a cell culture-treated black wall clear bottom 384-well plate (Corning 3712).
19. Wash worms off 10–15 NGM-OP50 plates into a 50 mL conical using M9W. Allow them to gravitationally settle (*see* **Note 25**). Aspirate liquid down to ~5 mL. Wash three times with S Basal.
20. Transfer worms into sample cup of COPAS BioSort Worm Sorter. Sort 15–18 worms per well.
21. After addition of worms, seal plate with a gas-permeable membrane (BreatheEasy, Diversified Biotech) and incubate at 25 °C (without agitation), at 80–85 % humidity for 44–48 h (or a self-selected time frame).
22. To stop infection, wash plates five times using a microplate washer (e.g., BioTek ELx405). To reduce background, all but 20 μL per well can be aspirated, and the plates shaken (using MixMate plate vortexer), between the second and third washes. The worms will rapidly settle to the bottom of the wells during each wash step, and so will not be aspirated with the supernatant.
23. *Assay scoring and data analysis.* Aspirate liquid, leaving 20 μL per well.
24. Add 50 μL of Sytox Orange-S Basal solution (final dye concentration of 0.7 μM) per well. (Approximately 20 mL of S Basal is needed for each 384-well plate).
25. Stain for 16–24 h. If desired, excess background staining can be removed by washing the plates twice with S Basal.
26. Image plates in both bright field and Cy3 fluorescence channel (e.g., using a Molecular Devices IXMicro automated microscope).
27. During initial experiment design and verification, images can be scored manually by counting the number of stained worms per well. For example, if you have 5 stained worms out of 15, the fraction dead is 0.33 or 33 %.
28. The first critical step in assay validation is Z′ factor determination. For this purpose, a plate should be split evenly between negative and positive (i.e., antibiotic) controls. The Z′ factor is calculated as follows:

$$Z' \text{ factor} = 1 - \frac{|M1 - M2|}{(3)(\text{SD1}) + (3)(\text{SD2})}$$

where *M*1 is the mean of the negative control, *M*2 is the mean of the positive control. Note: you need the absolute value of the difference between the means, SD1 is the standard deviation of the negative control, SD2 is the standard deviation of the positive control.

The assay is appropriately optimized when the Z′ factor > 0.5.

29. For high-throughput experiments, unbiased, automatic-scoring procedures that can discriminate between the fluorescent and non-fluorescent worms should be developed. Both the MatLab and CellProfiler (http://www.cellprofiler.org) software packages can be used for this purpose.
30. Multiple conditions (e.g., multiple bacterial strains or multiple worm strains) can be compared at a single time point. Calculate the Percent Alive for each well of each condition as follows:

$$\text{Percent Alive} = \frac{\text{Number alive}}{\text{Total worms}}(100)$$

31. Calculate the mean Percent Alive for all wells of each condition. Compare these means using a Student's *t*-test or ANOVA.
32. Alternatively, the infection progression can be monitored along a time course. It is critical to understand that Liquid Killing is an endpoint assay because after a plate has been stained with Sytox Orange it cannot be used for subsequent time points. Therefore, each time point requires that an independent plate be set up in parallel and scored only once, at its designated time. Typically time points are taken between 24 and 72 h post infection.
33. If multiple time points are taken using parallel plates, the resulting survival curve can be analyzed as described in Subheading 3.5 below.

3.5 Statistical Analysis of Population Curves

1. Statistical analysis software such as Prism (Graphpad Software) can be used to plot the Kaplan-Meier estimate of the population survival curve. This software uses the Log-Rank method to compare survival curves. Alternately, the calculations may be performed manually with the help of a spreadsheet program such as Microsoft Excel. For each plate at each time point, calculate the percent alive:

$$\text{Total worms} = \text{Total worms dead} + \text{worms remaining alive}$$

$$\text{Number dead at time}(t) = \text{sum of all worms dead up to and including this time point}$$

$$\text{Number alive at time}(t) = \text{Total worms} - \text{Number dead at time}(t)$$

$$\text{Percent alive at time}(t) = \frac{\text{Number alive at time}(t)}{\text{Total worms}}(100)$$

2. Calculate the Mean Percent Alive for each condition at each time point. To do this, average the percent alive measurements for each of the replicate plates (*see* **Note 26**).
3. Plot the Mean Percent Alive for each condition as a function of time. This survival curve typically has a sigmoidal decay and is an

estimate of the survival function, the probability (on the *Y*-axis) of an individual worm surviving a particular length of time (along the *X*-axis). The survival curves for different conditions assayed in parallel should be plotted on the same set of axes. Importantly, however, the survival curves from assays not conducted in parallel (i.e., set up on different days) should not be compared directly on the same axes due to day-to-day variability.

4. Calculate the error for each time point. The error bars for a given time point can be presented as the standard deviation of the individual percent alive measurements across the replicates or as the standard error of the mean.
5. To compare the survival curves of two or more conditions, calculate the Mean Survival for the population of worms for each of the replicates of a given condition:

$$\text{Worm survival at time}(t) = (\text{Number dead at time}(t))(\text{time}(t))$$

$$\text{Mean Survival} = \frac{\text{Sum of Worm Survival at all time points}}{\text{Total worms}}$$

6. Calculate the overall Mean Survival for a condition by averaging the mean survival values for each of the replicates.
7. Calculate the standard deviation or standard error of the means.
8. To compare two conditions, perform a Student's *t*-test to determine whether the mean survival values for the replicates of one condition are significantly different from the mean survival values for the replicates of another condition.

4 Notes

1. Thin lawn plates are best used to grow worms that will be picked off the plates by hand, as for the Fast Killing and Slow Killing assays. Thick lawns on larger plates can support many more worms per plate than thin lawns, but the sticky OP50 bacteria is difficult to pick from. Thus, thick lawn plates are best when large numbers of worms will be washed off the plates, as for bleaching to synchronize the worms or for the Liquid Killing assay. We routinely store these plates at 4 °C for several weeks, and occasionally up to a few months, provided they are kept in a covered box so that they do not dry out. We have not tested directly whether the age of the plates affects the outcome of the killing assays, but we have not noticed a dramatic impact, if any.
2. PA14 must always be streaked from a frozen stock because the strain loses virulence with passage on plates. A streaked plate may be used for up to 1 week for Slow Killing and Liquid Killing assays or 2 weeks for Fast Killing assays.

3. When comparing different strains of bacteria in a killing assay, it is important to include wild type PA14 as a positive control and some avirulent strain as a negative control. For Fast Killing and Slow Killing assays, *E. coli* DH5α is a common choice, but other *E. coli* strains such as OP50 may be used. Alternatively, mutant PA14 strains that produce little or no phenazines such as *rhlR* or a phenazine-null mutant [19] can also be used as negative controls for Fast Killing. Attenuated mutants such as *gacA* or *lasR* can be used as negative controls for Slow Killing.
4. Under certain growth conditions, PA14 has a tendency to form phenotypic variants that are poorly characterized [20] but are less virulent in various worm killing assays. To reduce the probability of phenotypic variant emergence, grow all liquid cultures with sufficient aeration, never in static culture. It is also important not to allow the bacteria to grow in liquid culture for longer than 16 h at 37 °C, after which time the bacteria begin to lyse. If necessary, a culture may be removed from the 37 °C incubator and stored static at room temperature for a few hours before seeding the plates for a killing assay.
5. A standard lawn for Fast and Slow killing assays does not reach to the edge of the plate, but rather is surrounded by a region of bare agar. Alternatively, a big lawn can be made by entirely covering the surface of the plate with the culture. In a Slow Killing assay, comparing the phenotype of worms on standard and big lawns allows one to determine the behavioral aspects of pathogen avoidance as a component of innate immunity because worms have the ability to leave a standard lawn but not a big lawn. Standard lawns facilitate scoring of the assay because worms are less likely to crawl up the sides of the plate. Worms tend to survive slightly longer in a standard lawn Slow Killing assay than in a big lawn Slow Killing assay. The effect of big lawns on a Fast Killing assay has not been rigorously tested. In the case of the Fast Killing assay, we suspect the difference between standard and big lawns will be minimal because the mechanism of Fast Killing is via a diffusable toxin and does not require contact with live bacteria.
6. To obtain statistically significant results, assays are typically set up in triplicate. Be sure to prepare enough plates so that at least three plates can be used for each condition tested. If many conditions need to be screened, performing killing assays in triplicate can be overwhelming. In this case, assays can be carried out in duplicate and those conditions with potentially interesting results can be retested in triplicate to ascertain statistical significance.
7. The incubation temperature affects the rate of killing in the Fast Killing assay. Room temperature or up to 30 °C also provide rapid killing, but lower temperatures will result in dramatically reduced killing in this assay.

8. The simplest way to maintain worms is by chunking. Worms can survive several weeks of starvation by arresting as L1 larvae or several months of starvation by entering an alternate stress-resistant larval stage called the dauer stage. These starved plates can be wrapped in parafilm and kept until needed, although viability does diminish over time so it is best to keep freshly starved plates on hand. To resume the development of arrested starved worms and obtain gravid hermaphrodites, use a sterile spatula to cut a chunk of agar (~0.5–2 cm^2) from a starved plate and transfer it to an NGM-OP50 plate. Grow the chunked worms at 20 °C for about 48–60 h or about 72–96 h at 15 °C for temperature-sensitive strains, to obtain gravid hermaphrodites. Some mutant worm strains may require a longer growth period.
9. Killing assays are much more convenient to set up using a synchronized population of worms. The most common method of synchronization is bleaching to obtain eggs and they allowing the eggs to hatch in the absence of food so that they arrest at the L1 larval stage, as described for Slow Killing and Liquid Killing assays. However, it is critical that worms not be synchronized in this manner for a Fast Killing assay because bleaching affects the susceptibility of the worms to the toxin. This method provides reasonable synchrony without as much stress as bleaching.
10. When comparing different strains of *C. elegans* in a killing assay, it is important to include N2 wild type worms as a control. For Fast Killing assays, it is difficult to test worm mutants that are more sensitive than wild type because the killing kinetics are so rapid, but if necessary, a *pgp-1; pgp-3* mutant strain could be used as a sensitive control. For Slow Killing assays, *pmk-1(km25)* mutants are a good control strain that is more sensitive than wild type to killing. For both Fast Killing and Slow Killing, *age-1* mutants can be used as a control strain that lives longer than N2.
11. It is critical that the worms be age-matched and at the late L4 stage when a Fast Killing assay is started. Earlier L4 larvae may die more slowly, and worms that have already passed the L4 → adult mold are much less susceptible to Fast Killing. The precise temperature regimen given does not need to be followed, but in our experience this is a consistent way to time the worm growth. Alternately, worms may be grown at either 20 °C or 15 °C for the entire duration of their larval development, but the total growth time must be adjusted accordingly.
12. The worms should be carefully transferred with a worm pick to the prepared PGS plates. They may be placed anywhere on the plate. Transferring OP50 from the NGM-OP50 growth plates does not affect the outcome of a Fast Killing assay.
13. Worms will initially be paralyzed when placed on the PGS plates due to osmotic shock. Over the course of 2–3 h, they will slowly recover and regain the ability to move.

14. When scoring Fast or Slow Killing assays, be careful not to miss ghosts, worm corpses that have begun to decompose. Ghosts are frequently almost completely clear, with virtually no discernable internal structure, and can be difficult to see on the lawn of PA14. Ghosts are often easier to see if the angle of the transmitted light on the microscope is adjusted.
15. Drying time may vary, depending on the plate age and storage conditions. If the plates are very fresh or are not dry within 20 min at room temperature, place them in a laminar flow hood with their lids off until the PA14 culture has soaked in completely. Too little drying facilitates the escape of the worms onto the sides of the plate, while too much drying will cause the agar to crack.
16. This incubation allows the bacteria to produce factors required for their full virulence. The incubation time can be reduced to as little as 6 h, but the worms will die at a slower rate. Increasing the incubation time from 24 to 48 h slightly increases the rate of worm death.
17. Since the Slow Killing assay is relatively low throughput, it is also fine to prepare the worms in the same manner as described for the Fast Killing assay.
18. It is sometimes useful to pipet arrested L1s onto NGM-OP50 plates more than once during the day. This staggers the population and gives you the flexibility to choose the set of plates that are at the best stage when you want to set up the Slow Killing assay. This is particularly important when comparing worms of different genotypes because many mutations affect the growth rate of the worms. It is better to use more plates at this stage than try to set up an assay, only to find out that your worms grew slightly faster than you expected and are too old or your mutant strains are not all at the same age. Typically, we pipet L1s onto a set of plates in the morning, another set of plates in the early-mid afternoon, and a third set in the evening.
19. FUDR chemically sterilizes the worms, which is important for two reasons. First, worms are able to reproduce on SK-PA14 plates and the resulting progeny will grow up over the time course of the assay, making later time points very difficult to count. Second, reduced fertility increases the resistance of worms to PA14 in a Slow Killing assay. This is due at least in part to the fact that worms grown on PA14 retain some of their eggs, which hatch internally and kill the mother (Bag of worms phenotype). Worms that are completely or partially sterile die more slowly than worms that are fertile due to a reduction of the Bag of worms phenotype, regardless of whether the sterility is due to genetic background or to chemical sterilization by FUDR. Many mutant strains affect fertility, so to compare multiple strains of worms, it is best to control for fertility by sterilizing all animals used in a Slow Killing assay. This can be done with FUDR, as

described. Alternatively, a temperature-sensitive sterile strain such as *fer-15; fem-1* or *glp-4* can be used. These worms are 100 % sterile when grown at 25 °C. Temperature sensitive sterile strains should be maintained at 15 °C. Several days before the worms are needed for a killing assay, synchronize them as described above. Plate the synchronized L1 larvae on NGM plates with OP50 and grow to the L4 stage at 25 °C. Use these sterile worms to set up a killing assay.

20. L4 hermaphrodites die more slowly than young adult hermaphrodites. Males also may be used in Slow Killing assays, but they die more slowly on PA14 than hermaphrodites of the same age. Thus, it is critical that all worms in a given experiment be of the same sex and age.
21. Contamination of the assay plates with OP50 can be a problem for Slow Killing assays because it attenuates killing of the worms by PA14. When setting up Slow Killing assays, be particularly careful to transfer as little OP50 as possible. Use a worm pick to wipe or dab up any OP50 that is transferred after the worms crawl away from the spot where they were deposited on the agar. Note the location of their placement on the plate; if OP50 growth is visible during the course of the assay or if worms preferentially clump in this location, then the plate is contaminated and the data cannot be interpreted.
22. Changing the temperature of the killing assay slightly can change the rate of worm death. For example, incubating the killing plates at 23 °C instead of 25 °C can help resolve differences between populations with very similar phenotypes by extending the range of the assay.
23. Fertile worms used in a Liquid Killing assay retain their eggs, which hatch internally and cause a Bag of worms phenotype. Therefore, only sterile worms should be used for Liquid Killing assays. For this assay, we prefer using a temperature-sensitive sterile strain such as *glp-4(bn2)* or *fer-15(b26); fem-1(hc17)*. These worms are 100 % sterile when grown at 25 °C. Temperature sensitive sterile strains should be maintained at 15 °C and only shifted to 25 °C before a killing assay, as described.
24. Young adults are the preferred stage for Liquid Killing assays. L4s die more quickly than young adults. They also settle more slowly during wash steps, causing a greater number of worms to be lost.
25. For the wash steps, it is important to allow the worms to settle to the bottom of the wells rather than centrifuge the plate. Centrifugation will also pellet debris, which could then clog the Worm Sorter.
26. If there is a large difference in the number of worms on each of the replicate plates, averaging the percentages will give

greater weight to the survival of worms on plates with fewer individuals. This can be avoided by pooling the data for the replicate plates as if the worms were all on one plate, and then calculating a single percent alive for the pooled data. Based on our experience, however, we suspect that the plate effect, or the slightly different conditions experienced by worms on different plates, is greater than the potential skew resulting from weighting individual worms unequally. Thus, we favor calculating the percent alive for each replicate and then averaging those values.

References

1. Brenner S (1974) The genetics of *Caenorhabditis elegans*. Genetics 77:95–104
2. Powell JR, Ausubel FM (2008) Models of *Caenorhabditis elegans* infection by bacterial and fungal pathogens. Methods Mol Biol 415:403–427
3. Muhammed M, Coleman JJ, Mylonakis E (2012) *Caenorhabditis elegans*: a nematode infection model for pathogenic fungi. Methods Mol Biol 845:447–454
4. Brassinga AK, Sifri CD (2013) The *Caenorhabditis elegans* model of *Legionella* infection. Methods Mol Biol 954:439–461
5. Darby C et al (1999) Lethal paralysis of *Caenorhabditis elegans* by *Pseudomonas aeruginosa*. Proc Natl Acad Sci U S A 96:15202–15207
6. Gallagher LA, Manoil C (2001) *Pseudomonas aeruginosa* PAO1 kills *Caenorhabditis elegans* by cyanide poisoning. J Bacteriol 183:6207–6214
7. Zaborin A et al (2009) Red death in *Caenorhabditis elegans* caused by *Pseudomonas aeruginosa* PAO1. Proc Natl Acad Sci U S A 106:6327–6332
8. Tan MW, Mahajan-Miklos S, Ausubel FM (1999) Killing of *Caenorhabditis elegans* by *Pseudomonas aeruginosa* used to model mammalian bacterial pathogenesis. Proc Natl Acad Sci U S A 96:715–720
9. Mahajan-Miklos S et al (1999) Molecular mechanisms of bacterial virulence elucidated using a *Pseudomonas aeruginosa-Caenorhabditis elegans* pathogenesis model. Cell 96:47–56
10. Cezairliyan BO et al (2013) Identification of *Pseudomonas aeruginosa* phenazines that kill *Caenorhabditis elegans*. PLoS Pathog 9:e1003101
11. Kirienko NV et al (2013) *Pseudomonas aeruginosa* disrupts *Caenorhabditis elegans* iron homeostasis, causing a hypoxic response and death. Cell Host Microbe 13:406–416
12. Garvis S et al (2009) *Caenorhabditis elegans* semi-automated liquid screen reveals a specialized role for the chemotaxis gene cheB2 in *Pseudomonas aeruginosa* virulence. PLoS Pathog 5:e1000540
13. Tan MW et al (2009) *Pseudomonas aeruginosa* killing of *Caenorhabditis elegans* used to identify *P. aeruginosa* virulence factors. Proc Natl Acad Sci U S A 96:2408–2413
14. Feinbaum RL et al (2012) Genome-wide identification of Pseudomonas aeruginosa virulence-related genes using a Caenorhabditis elegans infection model. PLoS Pathog 8:e1002813
15. Kim DH et al (2002) A conserved p38 MAP kinase pathway in *Caenorhabditis elegans* innate immunity. Science 297:623–626
16. Garsin D et al (2003) Long-lived *C. elegans daf-2* mutants are resistant to bacterial pathogens. Science 300:1921
17. Powell JR, Kim DH, Ausubel FM (2009) The G protein coupled receptor FSHR-1 is required for the *Caenorhabditis elegans* innate immune response. Proc Natl Acad Sci U S A 106: 2782–2787
18. Estes KA et al (2010) bZIP transcription factor *zip-2* mediates an early response to *Pseudomonas aeruginosa* infection in *Caenorhabditis elegans*. Proc Natl Acad Sci U S A 107:2153–2158
19. Dietrich LE (2006) The phenazine pyocyanin is a terminal signaling factor in the quorum sensing network of *Pseudomonas aeruginosa*. Mol Microbiol 61:1308–1321
20. Drenkard E, Ausubel FM (2002) *Pseudomonas* biofilm formation and antibiotic resistance are linked to phenotypic variation. Nature 416: 740–743

Chapter 51

Assessing *Pseudomonas aeruginosa* Virulence Using a Nonmammalian Host: *Dictyostelium discoideum*

Geneviève Filion and Steve J. Charette

Abstract

Dictyostelium discoideum, a soil amoeba, can be used as an alternative host to study the virulence of various bacterial species, including *Pseudomonas aeruginosa*. A simple quantitative test based on the ability of *D. discoideum* to grow on a bacterial lawn has been developed using this amoeba to assay the virulence of *P. aeruginosa* strains. The assay needs to be customized for the strains to be tested in order to be able to discriminate between virulent and avirulent *P. aeruginosa* strains. These steps are described in this protocol.

Key words *Pseudomonas aeruginosa*, *Dictyostelium discoideum*, Virulence, Nonmammalian host, Protocol, Amoeba

1 Introduction

The soil amoeba *Dictyostelium discoideum* is a natural bacterial predator. It is a professional phagocyte that can be quickly and easily cultivated on inexpensive growth media at temperatures up to 25 °C with no need for any CO_2. For ethical, practical, and financial reasons, *D. discoideum* is a good alternative host for measuring the virulence of various bacterial genera [1], including *Aeromonas* [2–4], *Klebsiella* [5, 6], *Legionella* [7–10], *Mycobacterium* [11–15], *Burkholderia* [16–18], *Streptococcus* [19], and *Pseudomonas* [20–28].

In the protocol described here, the virulence of the bacterial strains is assessed by their ability to resist *D. discoideum* predation. When *D. discoideum* is grown on lawns of avirulent bacterial strains, it produces visible phagocytosis plaques. However, it is unable to grow on lawns of virulent strains and no phagocytosis plaques are visible (Fig. 1) [23]. This protocol makes it possible to quantitatively assess the virulence of bacterial strains based on the number of amoebae required to produce phagocytosis plaques on bacterial lawns. This virulence assay can be used to compare wild-type virulent strains with avirulent mutants and, as such,

Alain Filloux and Juan-Luis Ramos (eds.), *Pseudomonas Methods and Protocols*, Methods in Molecular Biology, vol. 1149, DOI 10.1007/978-1-4939-0473-0_51, © Springer Science+Business Media New York 2014

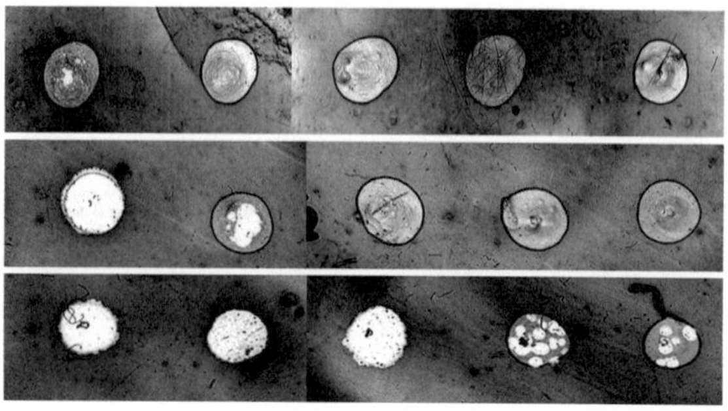

Fig. 1 Example of expected results. *D. discoideum* will not produce phagocytosis plaques (clear area in the bacterial lawn, even small) when grown on virulent strains excepted for the highest concentration after appropriate customization of the assay as it is the case here. Amoebae will, however, produce plaques at most concentrations when grown on nonpathogenic strains and avirulent mutants. In fact, for these avirulent strains and mutants, even ten *D. discoideum* cells will produce a plaque leading to multiple plaques in the footprint of the drop for the lowest amount of amoebae. Intermediate results (i.e., phagocytic plaques only visible for a fraction of amoeba drops) may also be observed. In such cases, the mutants are not considered avirulent, just less virulent than the virulent control strain

determine the contribution of mutated genes to bacterial virulence. This can be done using sets of known mutants [3, 4, 17, 19, 20] or mutants from random mutagenesis libraries to screen for new virulence genes [2, 5, 6, 18, 22]. It is also possible to compare the virulence of *Pseudomonas aeruginosa* isolates (clinical, environmental, etc.) using this assay [16, 24, 27, 28].

The virulence assay using *D. discoideum* is not only inexpensive and user-friendly, but it is also extremely versatile in that assay parameters can be easily changed to adapt it to the strains to be tested. This is very important because it is known that the virulence of *P. aeruginosa* can vary among strains from various sources [24, 28, 29] and even from laboratory to laboratory for a given strain like PAO1 [30]. The number of amoebae that is deposited on the bacteria lawn is one of the parameter that can be easily changed. In a study on *Burkholderia ambifaria* virulence, 200,000 amoeba (in 5 μL) were used as the highest concentration deposited on the bacterial lawn [16] while, in a study on *Streptococcus suis* virulence, only 3,000 amoeba (in 5 μL) were used as the highest concentration [19]. Nutrient concentrations can be also easily changed by diluting the growth medium [1, 4]. This virulence assay can be performed in Petri dishes or 24-well plates. In some cases, the type of bacterial pre-culture (liquid or agar) can influence the results (Laetitia Bonifait, Université Laval, Quebec city, QC, Canada, personal communication). Other parameters that can be easily changed include incubation time (1 day to over 7 days) and temperature (21–25 °C).

As such, when the virulence of new strains of *P. aeruginosa* is analyzed using amoebae as a host, the parameters should be varied to determine the most suitable conditions for analyzing the strains, beginning with the ideal number of amoebae and the strength of the growth medium. The ideal number of amoebae deposited on the bacterial lawn should be determined first (typically using SM agar as growth medium). SM agar is normally used to do virulence assay with *D. discoideum*. If this fails to clearly differentiate between virulent and avirulent strains, the most promising number of amoebae should then be tested on different strengths of the medium (SM, SM 1/2, SM 1/5, etc.). These two steps should give a good indication of the best conditions for analyzing the virulence of the strains. If not, other parameters such as different pre-culture media for growing the bacteria could also be tried, as could liquid and solid variants. Performing the virulence assay in 24-well plates instead of Petri dishes can be also tried (this may influence thickness of the bacterial lawns). Finally, the incubation temperature used for the virulence test could be varied from 20 °C to 25 °C, and the temperature of the bacterial pre-culture could be varied from 20 °C at 37 °C.

2 Materials

2.1 Reagents

1. Stock solution of 12.5 mg/mL of tetracycline: Dissolve 0.5 g of tetracycline in 40 mL of 75 % ethanol. Store at −20 °C.
2. Solution of 20 % of glucose: Dissolve 100 g of glucose in water. Adjust volume to 500 mL. Filter to sterilize. Store at room temperature.

2.2 Media

1. LB broth: Use as described by the manufacturer. Store at room temperature.
2. SM agar: Dissolve 10 g of bactopeptone, 1 g of yeast extract, 2.2 g of KH_2PO_4, 1 g of K_2HPO_4, and 1 g of $MgSO_4{\cdot}7H_2O$ in 900 mL of water. Adjust the volume to 950 mL. Add 20 g of agar. Autoclave for 20 min on liquid cycle. Cool to ~55 °C, and add 50 mL of 20 % glucose (filter-sterilized). Pour approximately 35 mL into each Petri dish (100 mm diameter). Store at 4 °C.
3. HL-5: Dissolve 14.3 g of bactopeptone, 7.15 g of yeast extract, 18 g of maltose monohydrate, 0.64 g of $Na_2HPO_4{\cdot}2H_2O$, and 0.49 g of KH_2PO_4 in 900 mL of water. Adjust the volume to 1 L. Autoclave for 20 min on liquid cycle. Store at room temperature.

4. HL-5 with 15 μg/mL of tetracycline: Proceed as described above, but cool the medium to room temperature before adding 1.2 mL of stock tetracycline solution per liter of medium. Store at room temperature.
5. Tryptic soy agar (TSA): Use as described by the manufacturer. Pour into Petri dishes (100 mm diameter). Store at 4 °C.

2.3 Microorganisms

1. *D. discoideum* (*see* **Note 1**): Grow in Petri dishes (100 mm diameter) containing 12 mL of HL-5 with tetracycline at 21 °C. HL-5 is the usual medium for the axenic growth of *D. discoideum*. Maintain the cultures by resuspending the amoebae in the medium using a pipette, then transfer 1–2 drops to a Petri dish containing 12 mL of fresh HL-5 with tetracycline. Never exceed a cell density of 10^6 amoebae/mL. If the cells become too confluent, they exhibit lower and less robust phagocytic activity. After 20 transfers, start a new culture from a frozen stock culture [31].
2. *P. aeruginosa*: Store stock cultures in a freezer (−80 °C) or in liquid nitrogen. To ensure reproducibility, always perform the virulence assay using bacteria grown in the same way. Streak frozen stock cultures directly on TSA agar plates. Incubate at 37 °C.

2.4 Other Materials

1. Hemacytometer: This device is used to count the concentration of amoeba. Count at least three 1 mm^2 squares to establish the average cell concentration. Please refer to the manufacturer for the usage of the hemacytometer (Hausser Scientific).

3 Methods

Since important virulence variability can be seen among *P. aeruginosa* strains, the protocol needs to be adapted to the strains being assessed. Before assessing the virulence of *P. aeruginosa* strains, the controls must be selected. A virulent *P. aeruginosa* strain (wild-type) and a known avirulent mutant should be used. If a known avirulent strain is not available, use a few suspected avirulent strains to validate protocol conditions. These controls should subsequently be used in every virulence assay to ensure reproducibility and validity of the results.

3.1 *P. aeruginosa* Manipulation

1. Streak the *P. aeruginosa* strains on TSA plates and incubate overnight at 37 °C. The agar must be completely covered by bacteria to provide sufficient cells for the assay (*see* **Note 2**).
2. Use a sterile swab to collect the bacteria. Resuspend the bacteria in 1 mL of LB broth. Light vortexing may be required to completely resuspend the bacteria (*see* **Note 3**).

3. Spread 300 μL of the bacterial suspension on an SM agar plate using an L-rod to get a uniform coverage of the plate without trying to dry it with the L-rod by excessive wiping. Dry the bacterial lawns by incubating the plates in a laminar flow hood. Drying time will vary depending on the laminar flow hood used. Do not over-dry to avoid splitting the agar (*see* **Note 4**).

3.2 *D. discoideum* Manipulation

1. Grow *D. discoideum* in Petri dishes (100 mm diameter) containing 12 mL of HL-5 with tetracycline. Incubate at 21 °C. The addition of tetracycline aims to prevent bacterial contamination of the cell culture.
2. Choose a dish in which approximately 70 % of the bottom surface is covered by *D. discoideum*.
3. Wash the amoebae with HL-5 and then resuspend in 5 mL of HL-5 (*see* **Note 5**).
4. Use a hemacytometer to count the *D. discoideum* in the suspension. Dilute the amoebae to the appropriate concentrations (*see* **Note 6**).
5. Add 5 μL of each concentration of *D. discoideum* on dried *P. aeruginosa* lawns. Let dry in a laminar flow hood. Over-drying will cause the agar to split and will prevent the amoebae from growing. The drying time here is shorter than for the bacterial lawn.
6. Invert the Petri dishes and incubate at 21–25 °C. Record the results after a few days (*see* **Note 7**).

Once the appropriate conditions have been determined to discriminate between the virulent and avirulent control strains, repeat the protocol using the same conditions for all the *P. aeruginosa* mutants and strains to be tested (*see* **Note 8**).

To ensure reproducibility, repeat the assay at least three times, especially if quantitative results are required.

This assay can also be performed by adding a single predetermined concentration of amoebae to obtain a growth/no growth result. This is especially useful when a large number of bacterial strains or mutants must be screened [22].

4 Notes

1. While DH1–10 is the most commonly used *D. discoideum* strain [20, 22, 24], others such as AX2, X22 [28], and AX3 [21] have been used successfully with *P. aeruginosa*.
2. For most *P. aeruginosa* strains, this step can be alternatively performed in liquid medium (TSB or LB). Add one bacterial

Table 1
Influence of medium concentration on the virulence assay

	SM	SM 1/5	SM 1/50
Virulent bacterial strain (CTRL)	–[a]	+	++
Mutant #1	–	+++++	+++++
Mutant #2	++	+++++	+++++
Mutant #3	+	+++	+++++

[a]Symbol explanation: – no phagocytosis plaques visible, + one phagocytosis plaque, ++ two phagocytosis plaques, +++ three phagocytosis plaques, ++++ four phagocytosis plaques, and +++++ five phagocytosis plaques (all amoeba concentrations)

colony to 3 mL of medium. Incubate overnight at 37 °C on a rotary shaker (200 rpm).

3. It is important to test strains with approximately the same growth rates. Mutants with a growth defect cannot be tested since *D. discoideum* can form phagocytic plaques more easily because of the lawn thickness, leading to incongruous results.

4. The concentration of the SM agar needs to be adjusted. Using a virulent *P. aeruginosa* strain and a well-known avirulent mutant as controls, try different dilutions of SM agar: undiluted, 1/2, 1/5, 1/10, 1/50, and 1/100. Then, select the dilution that provides a clearly visible bacterial lawn and a strong difference in the predation of the virulent and avirulent bacterial strains. Table 1 presents results related to the influence of medium concentration on the virulence assay. Five concentrations of amoebae were used: 1,000/5 μL, 300/5 μL, 100/5 μL, 30/5 μL, and 10/5 μL. In undiluted medium, all the bacterial strains, including the mutants, appeared to be virulent. *D. discoideum* was unable to grow on one mutant, while only the higher concentrations of amoebae grew on the other two mutants. With the most diluted medium, even the virulent bacterial strain appeared to be less virulent, allowing the amoebae to produce phagocytosis plaques at almost all the concentrations used. In the case of Table 1, SM 1/5 is the better choice.

5. *D. discoideum* cells adhere to the bottom of Petri dishes. To wash the amoebae, gently aspirate the medium using a pipette. Add 10 mL of fresh HL-5, without disturbing the amoebae. Gently hand-rotate the Petri dish to wash the cells, and then aspirate the medium. Repeat. Then, add 5 mL of fresh HL-5 and use a 10 mL pipette to detach the amoebae from the bottom of the Petri dish. Do not expose the amoebae to air for too long, otherwise they will stick more tightly to the bottom of the Petri dish and will be difficult to detach.

6. To begin, use the following amoeba concentrations: 10,000/5 μL, 1,000/5 μL, 100/5 μL, 10/5 μL, and 0/5 μL or

Table 2
Optimal concentrations of amoebae

	High concentrations[a]	Low concentrations	Optimal concentrations
Virulent bacterial strain	+++++[b]	–	+
Mutant #1	+++++	+++++	+++++
Mutant #2	+++++	+++++	+++++
Mutant #3	+++++	+++++	+++++

[a]High concentrations of amoebae: 32,000/5 μL, 16,000/5 μL, 8,000/5 μL, 4,000/5 μL, 2,000/5 μL. Low concentrations: 500/5 μL, 250/5 μL, 125/5 μL, 72/5 μL, 36/5 μL. Optimal concentrations: 1,000/5 μL, 300/5 μL, 100/5 μL, 30/5 μL, 10/5 μL

[b]Symbol explanation: – no phagocytosis plaques visible, + one phagocytosis plaque, ++ two phagocytosis plaques, +++ three phagocytosis plaques, ++++ four phagocytosis plaques, and +++++ five phagocytosis plaques (all amoeba concentrations)

3,000/5 μL, 1,000/5 μL, 300/5 μL, 100/5 μL, 30/5 μL, 10/5 μL, and 0/5 μL. *If no phagocytosis plaques form*: If the highest concentration of *D. discoideum* does not produce plaques on the lawns of the control virulent bacteria, increase the concentration of amoebae. One study used 200,000 amoebae/5 μL as the highest concentration [16]. *If too many phagocytosis plaques form*: If the virulent *P. aeruginosa* strain allows *D. discoideum* to grow at most concentrations, adjust the concentration of the amoeba and then the SM media (*see* **Note 5**). *Controls*: Controls (e.g., strains used in the protocol setup) must be used to validate the virulence assay. To choose the concentration of amoebae, only the higher concentration must be able to produce a visible phagocytosis plaque in the virulent control bacteria. The results in Table 2 exemplify the selection of the optimal concentrations of amoebae for the virulence assay. Select the concentrations that make it possible to discriminate between the virulent and the avirulent mutant strains. Place a drop of HL-5 medium (5 μL) on the bacterial lawn to compare the contour of the phagocytosis plaques to the footprint of the drop. This will facilitate the interpretation of the assay results.

7. When the assay is performed for the first time, we recommend verifying the Petri dishes every day. Select the day where the biggest difference between the virulent *P. aeruginosa* strain and the avirulent mutant can be observed. The results presented in Table 3 illustrate the influence of incubation time on the results. Following a 9-day incubation, even low concentrations of *D. discoideum* grew on lawns of virulent bacteria, which was not the case following a 4-day incubation. The Petri dishes usually need to be incubated for 3–7 days. However, depending on the bacterial strains and the number of amoebae used, a valid result may be observed after an overnight incubation [16].

Table 3
Influence of incubation time on the virulence assay

	Day 4	Day 9
Virulent bacterial strain	++[a]	++++
Mutant #1	+++++	+++++
Mutant #2	+++++	+++++
Mutant #3	+++++	+++++

[a]Symbol explanation: – no phagocytosis plaques visible, + one phagocytosis plaque, ++ two phagocytosis plaques, +++ three phagocytosis plaques, ++++ four phagocytosis plaques, and +++++ five phagocytosis plaques (all concentrations)

8. Small variations in the virulence assay may occur. However, if the variability is too pronounced, make sure that the amoebae have undergone more than 3 passages but no more than 20. If necessary, use a fresh amoeba stock culture. Medium used for the virulence assay should be prepared less than a month before their use. Check to make sure that all the steps have been performed correctly.

Acknowledgements

We thank Katherine H. Tanaka and Luc Trudel for critical reading of the manuscript. We are grateful to Sok Gheck E. Tan for her assistance with the protocol. This work was funded by a Discovery grant from the Natural Sciences and Engineering Research Council of Canada (NSERC), a Fonds de la Recherche du Québec en Santé (FRQS) grant, and a Young Researcher grant from the FRQS Réseau en santé respiratoire (RSR). S.J.C. is a research scholar of the FRQS.

References

1. Froquet R, Lelong E, Marchetti A, Cosson P (2009) *Dictyostelium discoideum*: a model host to measure bacterial virulence. Nat Protoc 4:25–30
2. Vilches S, Jimenez N, Merino S, Tomas JM (2011) The *Aeromonas* dsbA mutation decreased their virulence by triggering type III secretion system but not flagella production. Microb Pathog 52:130–139
3. Daher RK, Filion G, Tan SG, Dallaire-Dufresne S, Paquet VE, Charette SJ (2011) Alteration of virulence factors and rearrangement of pAsa5 plasmid caused by the growth of *Aeromonas salmonicida* in stressful conditions. Vet Microbiol 152:353–360
4. Froquet R, Cherix N, Burr SE, Frey J, Vilches S, Tomas JM, Cosson P (2007) Alternative host model to evaluate *Aeromonas* virulence. Appl Environ Microbiol 73:5657–5659
5. Benghezal M, Fauvarque MO, Tournebize R, Froquet R, Marchetti A, Bergeret E, Lardy B, Klein G, Sansonetti P, Charette SJ, Cosson P (2006) Specific host genes required for the killing of *Klebsiella* bacteria by phagocytes. Cell Microbiol 8:139–148

6. Pan YJ, Lin TL, Hsu CR, Wang JT (2011) Use of a *Dictyostelium* model for isolation of genetic loci associated with phagocytosis and virulence in *Klebsiella pneumoniae*. Infect Immun 79:997–1006
7. Hagele S, Kohler R, Merkert H, Schleicher M, Hacker J, Steinert M (2000) *Dictyostelium discoideum*: a new host model system for intracellular pathogens of the genus *Legionella*. Cell Microbiol 2:165–171
8. Solomon JM, Rupper A, Cardelli JA, Isberg RR (2000) Intracellular growth of *Legionella pneumophila* in *Dictyostelium discoideum*, a system for genetic analysis of host-pathogen interactions. Infect Immun 68:2939–2947
9. Lu H, Clarke M (2005) Dynamic properties of *Legionella*-containing phagosomes in *Dictyostelium* amoebae. Cell Microbiol 7: 995–1007
10. Shevchuk O, Steinert M (2009) Screening of virulence traits in *Legionella pneumophila* and analysis of the host susceptibility to infection by using the *Dictyostelium* host model system. Methods Mol Biol 470:47–56
11. Skriwan C, Fajardo M, Hagele S, Horn M, Wagner M, Michel R, Krohne G, Schleicher M, Hacker J, Steinert M (2002) Various bacterial pathogens and symbionts infect the amoeba *Dictyostelium discoideum*. Int J Med Microbiol 291:615–624
12. Peracino B, Wagner C, Balest A, Balbo A, Pergolizzi B, Noegel AA, Steinert M, Bozzaro S (2006) Function and mechanism of action of *Dictyostelium* Nramp1 (Slc11a1) in bacterial infection. Traffic 7:22–38
13. Hagedorn M, Soldati T (2007) Flotillin and RacH modulate the intracellular immunity of *Dictyostelium* to *Mycobacterium marinum* infection. Cell Microbiol 9:2716–2733
14. Solomon JM, Leung GS, Isberg RR (2003) Intracellular replication of *Mycobacterium marinum* within *Dictyostelium discoideum*: efficient replication in the absence of host coronin. Infect Immun 71:3578–3586
15. Hagedorn M, Rohde KH, Russell DG, Soldati T (2009) Infection by tubercular mycobacteria is spread by nonlytic ejection from their amoeba hosts. Science 323: 1729–1733
16. Vial L, Groleau MC, Lamarche MG, Filion G, Castonguay-Vanier J, Dekimpe V, Daigle F, Charette SJ, Deziel E (2010) Phase variation has a role in *Burkholderia ambifaria* niche adaptation. ISME J 4:49–60
17. Aubert DF, Flannagan RS, Valvano MA (2008) A novel sensor kinase-response regulator hybrid controls biofilm formation and type VI secretion system activity in *Burkholderia cenocepacia*. Infect Immun 76:1979–1991
18. Hasselbring BM, Patel MK, Schell MA (2011) *Dictyostelium discoideum* as a model system for identification of *Burkholderia pseudomallei* virulence factors. Infect Immun 79:2079–2088
19. Bonifait L, Charette SJ, Filion G, Gottschalk M, Grenier D (2011) Amoeba host model for evaluation of *Streptococcus suis* virulence. Appl Environ Microbiol 77:6271–6273
20. Cosson P, Zulianello L, Join-Lambert O, Faurisson F, Gebbie L, Benghezal M, Van Delden C, Curty LK, Kohler T (2002) *Pseudomonas aeruginosa* virulence analyzed in a *Dictyostelium discoideum* host system. J Bacteriol 184:3027–3033
21. Pukatzki S, Kessin RH, Mekalanos JJ (2002) The human pathogen *Pseudomonas aeruginosa* utilizes conserved virulence pathways to infect the social amoeba *Dictyostelium discoideum*. Proc Natl Acad Sci U S A 99:3159–3164
22. Alibaud L, Kohler T, Coudray A, Prigent-Combaret C, Bergeret E, Perrin J, Benghezal M, Reimmann C, Gauthier Y, van Delden C, Attree I, Fauvarque MO, Cosson P (2007) *Pseudomonas aeruginosa* virulence genes identified in a *Dictyostelium* host model. Cell Microbiol 10:729–740
23. Lima WC, Lelong E, Cosson P (2011) What can *Dictyostelium* bring to the study of *Pseudomonas* infections? Semin Cell Dev Biol 22:77–81
24. Lelong E, Marchetti A, Simon M, Burns JL, van Delden C, Kohler T, Cosson P (2010) Evolution of *Pseudomonas aeruginosa* virulence in infected patients revealed in a *Dictyostelium discoideum* host model. Clin Microbiol Infect 17:1415–1420
25. Sperandio D, Decoin V, Latour X, Mijouin L, Hillion M, Feuilloley MG, Orange N, Merieau A (2012) Virulence of the *Pseudomonas fluorescens* clinical strain MFN1032 towards *Dictyostelium discoideum* and macrophages in relation with type III secretion system. BMC Microbiol 12:223
26. Vallet-Gely I, Novikov A, Augusto L, Liehl P, Bolbach G, Pechy-Tarr M, Cosson P, Keel C, Caroff M, Lemaitre B (2010) Association of hemolytic activity of *Pseudomonas entomophila*, a versatile soil bacterium, with cyclic lipopeptide production. Appl Environ Microbiol 76:910–921
27. Janjua HA, Segata N, Bernabo P, Tamburini S, Ellen A, Jousson O (2012) Clinical populations of *Pseudomonas aeruginosa* isolated from acute infections show a wide virulence range partially correlated with population structure

and virulence gene expression. Microbiology 158:2089–2098

28. Bradbury RS, Reid DW, Inglis TJ, Champion AC (2011) Decreased virulence of cystic fibrosis *Pseudomonas aeruginosa* in *Dictyostelium discoideum*. Microbiol Immunol 55:224–230

29. Kukavica-Ibrulj I, Bragonzi A, Paroni M, Winstanley C, Sanschagrin F, O'Toole GA, Levesque RC (2008) In vivo growth of *Pseudomonas aeruginosa* strains PAO1 and PA14 and the hypervirulent strain LESB58 in a rat model of chronic lung infection. J Bacteriol 190:2804–2813

30. Klockgether J, Munder A, Neugebauer J, Davenport CF, Stanke F, Larbig KD, Heeb S, Schock U, Pohl TM, Wiehlmann L, Tummler B (2010) Genome diversity of *Pseudomonas aeruginosa* PAO1 laboratory strains. J Bacteriol 192:1113–1121

31. Fey P, Kowal AS, Gaudet P, Pilcher KE, Chisholm RL (2007) Protocols for growth and development of *Dictyostelium discoideum*. Nat Protoc 2:1307–1316

Chapter 52

Assessing *Pseudomonas* Virulence with Nonmammalian Host: *Galleria mellonella*

Gudrun Koch, Pol Nadal-Jimenez, Robbert H. Cool, and Wim J. Quax

Abstract

Pseudomonas aeruginosa is an opportunistic human pathogen responsible for severe to deadly infections in patients suffering from cystic fibrosis, AIDS, undergoing immune suppressing therapies or suffering from severe burns. In the recent years there has been an increasing interest in exploring animal infection models that, to a certain extent, could mimic human infections. Here we describe the use of the larvae of the greater wax moth *Galleria mellonella* as a non-expensive, easy-to-use, and easy-to-obtain animal model to study *P. aeruginosa* infections.

Key words *Pseudomonas aeruginosa*, *Galleria mellonella*, Virulence, Quorum sensing, Quorum quenching, Acylase

1 Introduction

Assessing the efficiency of antimicrobials is a very important issue in the battle against many pathogens. Up till now mouse models have been the primary choice of these analyses, however, these experiments are very costly and under strict severe (country-dependent) regulations. Alternative infection models are nowadays an attractive choice when assessing virulence of pathogens. The larvae of the greater wax moth *Galleria mellonella* harbor a lot of advantages when compared to other pathosystems such as *Drosophila melanogaster* or the mouse. First of all, these non-vertebrates can be purchased for a good price from animal food stores where they are regularly used to feed a large variety of reptilians, amphibians, and birds. They are excluded from ethical or legal restrictions and can be kept in a Petri dish and throughout the whole period of investigation without any feeding requirements. Secondly, the direct injection of the pathogen and/or the antimicrobial allows the administration of an accurate inoculum, thereby reducing the standard error within the experiment. Third, results can be readily obtained within 24–48 h and last but not least,

Alain Filloux and Juan-Luis Ramos (eds.), *Pseudomonas Methods and Protocols*, Methods in Molecular Biology, vol. 1149, DOI 10.1007/978-1-4939-0473-0_52, © Springer Science+Business Media New York 2014

G. mellonella infection can be followed at 37 °C, the physiologic mammalian temperature.

G. mellonella larvae as in vivo models have been investigated for a large number of pathogens, including *Pseudomonas aeruginosa* [1–5], *Burkholderia* spp. [6–9] as well as fungal pathogens such as *Aspergillus* spp. [10–12]. It has been shown that there are structural and functional similarities between the mammalian and insect immune system [13], which was further corroborated by a strong correlation between results obtained in mice and in the *G. mellonella* model, thus further promoting their use [2, 14].

Here we demonstrate that the effect of antimicrobials, such as quorum quenching enzymes can be readily assessed in a *G. mellonella* infection model. We were able to show that short incubation of the pathogen together with the antimicrobial is sufficient to monitor its effects on the survival of the wax moth.

Thus, we are able to use this pathosystem to pre-screen and also to assess the full level of efficiency of large batteries of potential therapeutics in vivo.

2 Materials

Prepare all solutions using deionized water, autoclave prior to use and store at room temperature (unless otherwise stated). *G. mellonella* larvae should be ordered shortly before the starting date of the experiment to ensure that they keep in good condition.

2.1 G. mellonella Infection Materials

1. *G. mellonella*: Order *G. mellonella* larvae at a local vendor (*see* **Note 1**). Store at room temperature but in the dark until use.
2. Injection needles: Insulin pens (*see* **Note 2**).
3. Disposable needles for the insulin pen.
4. Long tweezers (*see* **Note 3**).
5. Foam material (*see* **Note 4**).
6. Petri dishes.
7. Single-use syringes and respective needles.

2.2 Bacterial Growth Media

1. LB medium [15]: Weigh 10 g Bacto-tryptone, 5 g yeast extract, and 10 g NaCl. Add 800 ml water and adjust to pH 7.5. Add water to the final volume of 1 L. Sterilize by autoclaving.
2. Minimal media CAA medium [16]: per liter: 5 g low-iron Bacto Casamino Acids, 1.54 g $K_2HPO_4{\cdot}3H_2O$, 0.25 g $MgSO_4{\cdot}7H_2O$.
3. NaCl: 0.9 % in water.
4. $MgSO_4$: 10 mM in water.

2.3 Antimicrobials

1. Purified quorum-quenching acylase (at the concentration required for testing) solved in T50 (50 mM Tris–HCl, pH 8.8) (*see* **Note 5**).

3 Methods

All experiments are conducted at room temperature unless otherwise stated.

Here below, an experiment for three groups of larvae has been described: (1) Not treated; (2) injected with Pseudomonas, treated with the acylase PvdQ; (3) injected with Pseudomonas.

3.1 Strain Preparation

1. Grow *P. aeruginosa* overnight in LB medium at 37 °C and 250 rpm.
2. The next day, pipette under sterile conditions 10 ml of the respective growth medium (LB (rich) or CAA (minimal)) to 50 ml conical tubes.
3. Dilute the *P. aeruginosa* overnight culture 1:100 into the fresh media.
4. Incubate the strains subsequently again at 37 °C and 250 rpm until an $OD_{600} = 0.6$–0.8 is reached (*see* **Note 6**). Cool down on ice.
5. Spin down cultures ($2700 \times g$, 5 min, 4 °C) in a centrifuge and wash gently with 5 ml $MgSO_4$ (10 mM).
6. Spin down again and resuspend to a final concentration of 1×10^5 CFU/ml ($OD_{600} = 0.002$) using 10 mM $MgSO_4$ (*see* **Note 7**). Store on ice until further use.

3.2 Addition of Antimicrobials

1. Divide now your cultures into two groups, one will contain the antimicrobial (*see* **Note 8**), the other one will not be treated.
2. Add the quorum-quenching enzyme to a final concentration of 15 mg/ml to one culture tube and the same volume of buffer to the other culture tube.
3. Incubate the culture now under static conditions at 30 °C for 1 h to allow the antimicrobial to take its action. Place back on ice until further use.

3.3 Preparation of Insulin Cartridges

1. Work from now on, preferably, under a laminar flow hood to avoid potential contaminations/infections.
2. Fill up syringes with the *P. aeruginosa* samples.
3. Apply now the needle to the syringe and inject the samples into labeled insulin cartridges (*see* **Note 9**).
4. Inject a third cartridge with 10 mM $MgSO_4$ only, which will serve as an injection control.

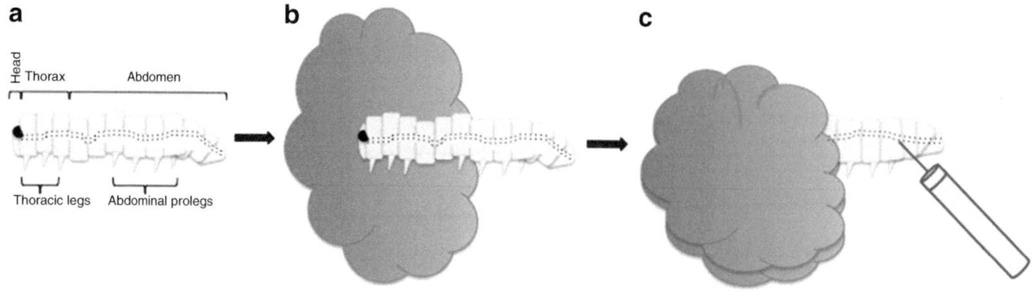

Fig. 1 Injecting *G. mellonella*. (**a**) Body parts of the larva are depicted. Pick a larva with tweezers. (**b**) Let it crawl onto the foam so that only lower abdomen is left outside. (**c**) Gently cover the larva with the foam, squeeze so that the abdomen lifts and inject the needle into the last proleg

3.4 *G. mellonella* Infection

1. Prepare 15 petri dishes and label them accordingly (5 petri dishes per group).
2. Insert now the first cartridge into the insulin pen.
3. Screw the disposable needle onto the pen and remove the safety packaging (*see* **Note 9**).
4. The first injection will serve to prime the insulin pen. Therefore, set the insulin pen to the full volume and eject this volume into the waste.
5. Set the volume to 1 which corresponds to 10 μl.
6. Grab a larva randomly from the box with the tweezers and let it crawl onto the foam material (Fig. 1a)
7. When half of its body is on the material, with only the lower part of the abdomen left outside, fold the foam around the larva and squeeze gently, so that the abdomen stands straight (Fig. 1b).
8. Now take the insulin pen and apply the needle into the last proleg of the larva (Fig. 1c).
9. Press the plunger to inject the volume (*see* **Note 10**).
10. Gently remove the pressure and subsequently the needle from the larva.
11. Place it in one of the previously prepared petri dishes.
12. Repeat this procedure now for all larvae (*see* **Note 11**).
13. After these steps we will have for each group three petri dishes with five larvae each.
14. Place the petri dishes now into a cardboard box containing air holes and incubate the larvae in the dark at 37 °C.

3.5 Scoring the Infection

1. Monitor the larvae after 24 and 48 h, respectively (*see* **Note 12**).
2. Evaluate the efficiency of the antimicrobial by scoring for life and dead larvae (*see* **Note 13**) (Fig. 2).
3. Analyze the data to produce a survival curve, which reflects the potency of your antimicrobial.

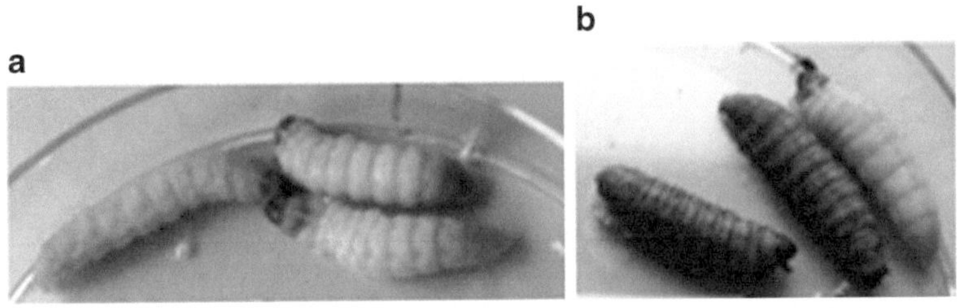

Fig. 2 Scoring life and dead larvae. (**a**) The three healthy larvae are still moving around and respond to tapping. (**b**) Two of the three larvae have already turned black because of the initiation of melanization. They are scored dead

4 Notes

1. Good experiences have been made within Europe buying from http://www.terrariumkoerier.com (to be found as "wasmotten"). 40–50 larvae are around 3 Euro. An important issue when ordering larvae from a chosen supplier is that they need to be pathogen-free upon delivery. This can easily be monitored by incubating them at 37 °C: larvae should survive up to 1 week without any problems.
2. Insulin pens are easy injection tools for accurate injections which can be bought at local pharmacies. A slight difficulty can be encountered when obtaining empty cartridges, which need to be asked for at Biotech companies, which usually distribute saline filled ones for demonstration purposes. Alternatively, Hamilton syringes can be used for injection purpose.
3. A simple trick to avoid harming the animals with the tweezers is to cut the tips of plastic Pasteur pipettes and use them to cover the endings of the tweezers.
4. Simple package fills can be used to fix the larvae in a certain position.
5. The efficiency of all kinds of antimicrobials can be assessed using this test.
6. Usually this OD is reached within 2–3 h in LB media. Growth periods may vary depending on the media used.
7. The final inoculation volume required for infection varies among *Pseudomonas* strains and has to be experimentally established prior to the target experiment.
8. Preliminary data should help to evaluate the effective range of the antimicrobial to be tested in vivo.
9. When injecting the culture into the insulin cartridges, be careful not to apply too many air bubbles, they can interfere with the injection volume. The minimal volume should be 200 μL.

10. Good precautions are strongly advised in order to avoid hurting yourselves with the syringe needle during the injection of larvae.
11. The same needle can be used for each group of larvae. Please dispose the needles afterwards in a suitable container to avoid injuries. Insulin cartridges can be reused by simply removing the rubber gum at the end by filling the cartridge with air. Subsequently, clean them with ethanol and water and allow them to dry before starting the next experiment. Experiments should be conducted independently at least three times. If more than one dead larva appears in the control group the experiment should be discarded.
12. Longer incubation times can be monitored if necessary.
13. Dead larvae do not respond anymore to tapping and will also turn black due to melanization.
14. Good experiences have been made within Europe buying from http://www.terrariumkoerier.com (to be found as "wasmotten"). 40–50 larvae are around 3 Euro. An important issue when ordering larvae from a chosen supplier is that they need to be pathogen-free upon delivery. This can easily be monitored by incubating them at 37 °C: larvae should survive up to 1 week without any problems.
15. Insulin pens are easy injection tools for accurate injections which can be bought at local pharmacies. A slight difficulty can be encountered when obtaining empty cartridges, which need to be asked for at Biotech companies, which usually distribute saline filled ones for demonstration purposes. Alternatively, Hamilton syringes can be used for injection purpose.
16. A simple trick to avoid harming the animals with the tweezers is to cut the tips of plastic Pasteur pipettes and use them to cover the endings of the tweezers.
17. Simple package fills can be used to fix the larvae in a certain position.
18. The efficiency of all kinds of antimicrobials can be assessed using this test.
19. Usually this OD is reached within 2–3 h in LB media. Growth periods may vary depending on the media used.
20. The final inoculation volume required for infection varies among *Pseudomonas* strains and has to be experimentally established prior to the target experiment.
21. Preliminary data should help to evaluate the effective range of the antimicrobial to be tested in vivo.
22. When injecting the culture into the insulin cartridges, be careful not to apply too many air bubbles, they can interfere with the injection volume. The minimal volume should be 200 μL.

23. Good precautions are strongly advised in order to avoid hurting yourselves with the syringe needle during the injection of larvae.
24. The same needle can be used for each group of larvae. Please dispose the needles afterwards in a suitable container to avoid injuries. Insulin cartridges can be reused by simply removing the rubber gum at the end by filling the cartridge with air. Subsequently, clean them with ethanol and water and allow them to dry before starting the next experiment. Experiments should be conducted independently at least three times. If more than one dead larva appears in the control group the experiment should be discarded.
25. Longer incubation times can be monitored if necessary.
26. Dead larvae do not respond anymore to tapping and will also turn black due to melanization.

Acknowledgments

We gratefully acknowledge Rien Hoge for helpful discussions regarding *Galleria* infection assays. This research was partly funded by EU grant Antibiotarget MEST-CT-2005-020278 to G.K. and P.N.J.

References

1. Hendrickson EL, Plotnikova J, Mahajan-Miklos S, Rahme LG, Ausubel FM (2001) Differential roles of the *Pseudomonas aeruginosa* PA14 rpoN gene in pathogenicity in plants, nematodes, insects, and mice. J Bacteriol 183:7126–7134
2. Jander G, Rahme LG, Ausubel FM (2000) Positive correlation between virulence of *Pseudomonas aeruginosa* mutants in mice and insects. J Bacteriol 182:3843–3845
3. Miyata S, Casey M, Frank DW, Ausubel FM, Drenkard E (2003) Use of the Galleria mellonella caterpillar as a model host to study the role of the type III secretion system in *Pseudomonas aeruginosa* pathogenesis. Infect Immun 71:2404–2413
4. Ryan RP, Lucey J, O'Donovan K, McCarthy Y, Yang L, Tolker-Nielsen T, Dow JM (2009) HD-GYP domain proteins regulate biofilm formation and virulence in *Pseudomonas aeruginosa*. Environ Microbiol 11:1126–1136
5. Inglis RF, Gardner A, Cornelis P, Buckling A (2009) Spite and virulence in the bacterium *Pseudomonas aeruginosa*. Proc Natl Acad Sci U S A 106:5703–5707
6. Mil-Homens D, Fialho AM (2012) A BCAM0223 mutant of *Burkholderia* cenocepacia is deficient in hemagglutination, serum resistance, adhesion to epithelial cells and virulence. PLoS One 7:e41747
7. Muller CM, Conejero L, Spink N, Wand ME, Bancroft GJ, Titball RW (2012) Role of RelA and SpoT in *Burkholderia pseudomallei* virulence and immunity. Infect Immun 80:3247–3255
8. Thomson EL, Dennis JJ (2012) A *Burkholderia* cepacia complex non-ribosomal peptide-synthesized toxin is hemolytic and required for full virulence. Virulence 3:286–298
9. Agnoli K, Schwager S, Uehlinger S, Vergunst A, Viteri DF, Nguyen DT, Sokol PA, Carlier A, Eberl L (2012) Exposing the third chromosome of *Burkholderia* cepacia complex strains as a virulence plasmid. Mol Microbiol 83:362–378
10. Soukup AA, Farnoodian M, Berthier E, Keller NP (2012) NosA, a transcription factor important in *Aspergillus fumigatus* stress and developmental response, rescues the germination defect of a laeA deletion. Fungal Genet Biol 49:857–865

11. Liu H, Gravelat FN, Chiang LY, Chen D, Vanier G, Ejzykowicz DE, Ibrahim AS, Nierman WC, Sheppard DC, Filler SG (2010) *Aspergillus fumigatus* AcuM regulates both iron acquisition and gluconeogenesis. Mol Microbiol 78:1038–1054
12. St Leger RJ, Screen SE, Shams-Pirzadeh B (2000) Lack of host specialization in *Aspergillus flavus*. Appl Environ Microbiol 66:320–324
13. Hoffmann JA (1995) Innate immunity of insects. Curr Opin Immunol 7:4–10
14. Brennan M, Thomas DY, Whiteway M, Kavanagh K (2002) Correlation between virulence of *Candida albicans* mutants in mice and Galleria mellonella larvae. FEMS Immunol Med Microbiol 34:153–157
15. Sambrook J, Fritsch EF, Maniatis T (2008) Molecular cloning: a laboratory manual, vol 2,1989 edn. Cold Spring Harbor Laboratory Press, Cold Spring Harbor, NY
16. Nadal Jimenez P, Koch G, Papaioannou E, Wahjudi M, Krzeslak J, Coenye T, Cool RH, Quax WJ (2010) Role of PvdQ in *Pseudomonas aeruginosa* virulence under iron-limiting conditions. Microbiology 156: 49–59

Chapter 53

Assessing *Pseudomonas* Virulence with the Nonmammalian Host Model: *Arabidopsis thaliana*

Regina L. Baldini, Melissa Starkey, and Laurence G. Rahme

Abstract

The popular plant model, *Arabidopsis thaliana*, has been used to successfully identify novel *Pseudomonas aeruginosa* genes that are involved in virulence. These genes have also been shown to be important for mammalian infection, demonstrating that this bacterium has a conserved set of virulence factors with broad range. This chapter describes using *A. thaliana* as a plant model for *P. aeruginosa* infection and describes obtaining the plants, preparing the inoculum, infecting the leaves, and collecting and interpreting the data. This protocol allows for both a qualitative assessment of symptoms and a quantitative measurement of the bacterial growth inside the leaves.

Key words *Pseudomonas aeruginosa*, *Arabidopsis thaliana*, Model host, Plant-pathogen interactions

1 Introduction

Arabidopsis thaliana has several attributes that make it an important model organism: it has a short life cycle, it does not require large spaces or expensive consumables, it is genetically tractable, and several different ecotypes are available, which allow for genetic analysis of variable phenotypic traits [1]. Since the *A. thaliana* genome was completely sequenced more than a decade ago [2], a large amount of information is available and constantly being updated in open-access databases [3]. In addition to serving as an effective model for studying plant systems, many discoveries in basic aspects of signal transduction and protein turnover and regulatory circuits, including gene silencing due to DNA methylation and small-interfering RNAs, were first made in *Arabidopsis*. These processes were later found to be conserved in humans, advancing the research in areas seemingly as distant as neurological diseases and cancer [4, 5].

Arabidopsis represents a good candidate for studying host innate immune responses to infection and virulence functions of pathogens that are related to the early or later onset of infections, as well as the underlying molecular mechanisms involved in

Alain Filloux and Juan-Luis Ramos (eds.), *Pseudomonas Methods and Protocols*, Methods in Molecular Biology, vol. 1149, DOI 10.1007/978-1-4939-0473-0_53, © Springer Science+Business Media New York 2014

host–pathogen interactions. Proteins involved in inflammatory processes in animals have homologous counterparts in plants. Examples of these conserved pathways are the NOD-like (*nucleotide oligomerization domain*) intracellular receptors in animals and the NB-LRR proteins in plants. Microbial pattern recognition receptors are also present in plants and animals [4, 6]. Moreover, bacterial pathogens, such as *Pseudomonas aeruginosa* use a subset of common virulence determinants to infect plants, invertebrates, and mammalian hosts [7–9]. Thus, both host and pathogen aspects of infection can be studied using the protocol described in this chapter. This particular protocol and model system also allows screening for disease determinants in large sets of mutants, which would not be possible to carry out in mammalian hosts due to ethical and methodological issues.

We describe all of the steps used to evaluate *P. aeruginosa* growth and ability to cause symptoms in *A. thaliana* leaves infected through the stomata, beginning with obtaining the plants through describing the analysis of data, with detailed descriptions that have not been published elsewhere [10, 11]. Despite the simplicity of the assay, some important points should be taken into account in order to obtain robust and reliable results, and these issues are presented in the corresponding notes and throughout the protocol. One important point to stress is that a positive control must always be included in each setup, to make sure the plants, the inoculum and the environmental conditions are ideal for development of the infection, since slight variations may compromise the outcome. Keeping the humidity and temperature of the incubators constant, for instance, is a key point in the success of the infection. It is also essential to run triplicates or quadruplicates of each given infection, and repeating the assays two to three times to have statistical confidence.

Specific characteristics of the plant ecotypes and the bacterial strains have also to be considered. Some *Arabidopsis* ecotypes are more susceptible than others, as noted below. Auxotrophic bacteria are not suitable for this assay, as the lower counts or less severe symptoms may be due to growth impairment *in planta* and not representative of a real attenuated in virulence phenotype. The number of cells and the stage of growth of the bacterial cultures used for infection may also greatly impact the course of the interaction.

2 Materials

2.1 Arabidopsis Growth

1. Llagostera (Ll-0) or Colombia (Col) seeds (*Arabidopsis* Biological Resource Center, http://www.biosci.ohio-state.edu/pcmb/Facilities/abrc/abrchome.htm).
2. Plastic trays with clear domed lids (e.g., T.O. Plastics, Inc., cat. no. 05023 and 05015, http://www.toplastics.com/horticulture/horticulture_products.php).

3. Plastic insert boxes for plant growth (e.g., T.O. Plastics Inc., cat. no. 01051, http://www.toplastics.com/horticulture/horticulture_products.php).
4. Potting soil, such as Metro Mix 200 or Fafard no. 2 (Griffins, cat. no. 65-3125 or 65-1523).
5. Agarose or Phytoblend agar (Caisson Labs, cat. no. PTC001) 0.1 % (w/v). Prepare the solution and autoclave to fully dissolve the agar.
6. Hoagland's nutrient solution, prepared from the stock solutions listed in Table 1 (https://amadeus.biosci.arizona.edu/~jdhall/lab_protocols.html).
7. Gnatrol WDG (Valent Biosciences, http://www.valent.com/professional/products/gnatrol/index.cfm).
8. Plant growth chamber with controlled light period and humidity (e.g., Percival Scientific model E30B).

2.2 Bacterial Culture and Infection

1. Selected strains of *Pseudomonas aeruginosa*.
2. LB medium (broth and plates, such as Fisher Scientific, cat. no. BP1427) or *Pseudomonas* Isolation Agar (Difco, cat. no. BD292710).
3. Appropriate antibiotics (e.g., rifampicin for wild-type strain PA14).
4. Sterile 10 mM $MgSO_4$ solution.

Table 1
Stock solutions and directions to prepare the Hoagland's nutrient plant solution

Component	Stock solution (g/L)	mL of stock to 1 L
Major nutrients stocks		
2 M KNO_3	202	2.5
2 M $Ca(NO_3)_2 \times 4H_2O$	472	2.5
Iron (Sprint 138 iron chelate)	15	1.5
2 M $MgSO_4 \times 7H_2O$	493	1
1 M NH_4NO_3	80	1
Minor nutrients mix		1
H_3BO_3	2.86	
$MnCl_2 \times 4H_2O$	1.81	
$ZnSO_4 \times 7H_2O$	0.22	
$Na_2MoO_4 \times 2H_2O$	0.12	
Buffer		
1 M KH_2PO_4 (pH to 6.0 with 3 M KOH)	136	0.5

Make up each major nutrient stock solution, the minor nutrients mix and the phosphate buffer and store them in separate bottles

Add the appropriate volume of each stock in 800 mL deionized water and fill to 1 L

5. 15 mL glass tubes with caps for bacterial cultures.
6. 1.5 mL centrifuge tubes.
7. 1 mL syringes without needles.
8. Paper towels.
9. Permanent markers.
10. Plastic toothpicks.
11. Safety equipment: goggles and gloves.
12. Plastic pestles for grinding and cordless grinder (Fisher Scientific, cat. no. K749521-1500 and K749521-1500).
13. No. 2 cork cutter (Fisher Scientific, cat. no. 07-865-10B).
14. Incubator at 37 °C with shaker or rotating drum for bacterial cultures.
15. Microfuge.
16. Spectrophotometer for reading absorbance at 595–600 nm.

3 Methods

3.1 *Arabidopsis* Growth

1. Soak the seeds (*see* **Note 1**) in water for 2–5 days at 4 °C to improve germination rate. Add agarose or Phytoblend agar to 0.1 % to provide the viscosity that will keep seeds in suspension.
2. Prepare the soil: fill the boxes with potting soil (such as Metromix 200), soak it in water, then Hoagland's solution. Let it drain and sprinkle with Gnatrol to avoid fungus gnat larvae (*see* **Note 2**).
3. Sow the seeds evenly on the soil using a Pasteur pipette and incubate the flats at 20 ± 2 °C, 70 % humidity and a 12 h photoperiod.
4. Cover the flats with the clear perforated domed lids.
5. Approximately 1 week later and once seeds germinate remove lids and let plants grow for another week.
6. Carefully transplant the seedlings, which will be about 5 mm tall, to new trays containing soil prepared as in **step 2** above. Each tray should contain five to six plants, 3 cm apart from each other. Incubated as in **step 3** (*see* **Note 3**).
7. Three weeks after sowing, check that leaves are green and healthy to schedule the infection. 4–6-week-old plants are used for the infection assays (*see* **Note 4**).

3.2 Preparation of the Inoculum

1. Pick a single colony from an LB-agar plate containing the appropriate antibiotics and inoculate in 3 mL of LB broth (with antibiotics if required—*see* **Note 5**). Grow overnight at 37 °C. Dilute the culture 1:100 in fresh LB and incubate with

agitation up to early stationary phase (for PA14, about 5 h at 37 °C, $OD_{600} = 3.0$) (*see* **Note 6**).

2. Pellet the cells for 5 min at 5,000 × g (*see* **Note 7**) and wash twice in 10 mM $MgSO_4$, adjusting the OD_{600} to 0.2. Dilute this suspension sequentially 1:100 and 1:1,000 in 10 mL 10 mM $MgSO_4$. The colony forming units in the 1:1,000 dilution would be around 2×10^6/mL (*see* **Notes 8** and **9**).

3.3 Infection

1. For each strain, use 12 plants (2 boxes). Make a short line at the midrib with a marker in 3–4 leaves per plant, choosing the older ones in the outside of the rosette, as they are more susceptible to infection (*see* **Note 10**).
2. Work in a well-illuminated area to keep the stomata open. A regular desk lamp may be used, with the light beam directed to the plants.
3. Fill a 1 mL syringe (without the needle) with the bacterial suspension and force it gently into the stomata in the abaxial side of a previously marked leaf, until it is completely soaked. Repeat the procedure if needed (*see* **Note 11**). Drain the excess liquid from the outside of the leaf by using filter paper, ensuring that the leaf does not drop onto the soil.
4. Support the infected leaves with plastic toothpicks to lift them from the soil (*see* **Note 12**). This is important both for keeping the leaves dry on the outside and to prevent bacterial spread in the soil and colonization of the roots.
5. Use one tray for each strain, keeping only three boxes in it to avoid cross-contamination. Put water in the bottom of the tray and cover it with a clear lid in which holes were made. This will maintain the high level of humidity needed to the symptoms to develop but also provide aeration (*see* **Note 13**).
6. Transfer the inoculated plants to a 30 °C incubator with the humidity setting as high as possible and a 12 h photoperiod.

3.4 Assessing Symptoms and Bacterial Proliferation

1. Follow the course of the infection for 5 days, assessing the symptoms daily using the parameters in Table 2 as standard (*see* **Note 14**).
2. For the bacterial count, select three to four leaves from each set of plants for each time point, starting from $t = 0$. Detach the leaves one by one and process following the steps below (*see* **Note 15**).
3. Cut two 0.28 cm^2 circles from one leaf using a #2 cork cutter, using a piece of paper towel as support to protect the finger that will be holding the leaf.
4. Grind both leaf discs in 0.3 mL 10 mM $MgSO_4$ using a plastic pestle to collect the bacteria present in the intercellular fluid.

Table 2
Symptoms observed in *Arabidopsis thaliana* Llagostera ecotype leaves infected with wild-type PA14 (severe) and isogenic mutants (moderate to none)

Symptom	Description
Severe	Water-soaked reaction zone and chlorosis around the site of inoculation 2–3 days post-inoculation. Soft rot of the entire leaf within 5 days post-inoculation
Moderate	Moderate water-soaking and chlorosis with most of the tissue softened around the inoculation site
Weak	Localized water-soaking and chlorosis of tissue around the inoculation site
None	No soft-rot symptoms, only chlorosis around the inoculation site

5. Dilute the samples serially and plate triplicates of the suspension in LB-agar. Incubate at 30 °C to count the colony forming units (*see* **Note 16**).
6. Report data as means and standard deviations of the log of cfu/cm^2. A standard t-test can be used to calculate P-values, or statistical analysis can be made using the Kaplan-Meier method.

4 Notes

1. The choice of the ecotype is important for the distinction of the symptoms or the bacterial growth *in planta*. Both Ll-O and Col have been used by us with PA14 and PA14-derived strains. Ll-O shows more accentuated symptoms with visible signs of soft rot, whereas symptoms in Col plants are less severe. Bacterial growth, however, is similar in those two ecotypes. Since the Col genome has been completely sequenced [2] and there are available tools to study its transcriptome, it may represent a better choice when one intends to study plant responses to *P. aeruginosa*.
2. The use of both Hoagland's solution and Gnatrol is optional. Hoagland's assure that the plants will grow healthy and that the results will not be affected by any nutrient limitation. Some lots of soil might be contaminated with fungus gnat eggs and the appearance of larvae will be avoided with this simple step.
3. As an alternative to transplantation, remove most seedlings and leave only the desired amount in the flat. This can save time and material in avoiding the preparation of new flats. However, transplantation is preferred because it allows for the selection and grouping of seedlings of different sizes, thus having plants ready for infection at subsequent days. This approach is particularly useful if one has many bacterial strains to test.

4. The photoperiod must be tightly controlled to avoid flowering. Plants with flower buds, even if they are still young, are more resistant to infection and should not be used.
5. An antibiotic marker is desired to avoid contamination in the plates used to count the colonies at the end of the experiment, since the plants and the environment where they are kept are not sterile. The PA14 strain widely used by many groups is originated from an isolate selected for rifampicin resistance [9]. One alternative to an antibiotic resistant strain is to use *Pseudomonas* Isolation Agar rather than LB for plating.
6. The stage of the culture is crucial for the outcome of the infection: bacterial cells have to be fully expressing their virulence factors (which occurs from late-log to early stationary phase). One way to have the culture in the right phase for starting the infection is to make serial dilutions of the inoculum in LB and incubate the whole series of tubes. Once ready to start the experiment, choose the culture with the OD_{600} closest to 3.
7. Do not over-centrifuge the cells: a loose pellet is easier to suspend with care. $8{,}000 \times g$ for 2 min can also be used, but full speed will result in a compact pellet. Do not vortex to suspend the cells, to avoid breaking appendices like flagella and thin fimbria.
8. Do not discard any viscous or insoluble material that may be present on the top of the pellet, as it may include bacterial polysaccharides or other unknown compounds important for the infection process.
9. The number of cells in the inoculum is also an important point to attend. At higher cell densities, strains that are moderately attenuated in virulence may not be distinguishable from wild-type. A lower inoculum, depending on *P. aeruginosa* strain to be used (e.g., for PA14 $1–5 \times 10^4$ cfu/mL) may be more effective in this case.
10. Using leaves in the same developmental stage helps to have consistent data. Avoid watering the plants a day or two before inoculation. This will help with the leaf inoculation. It is better to water the plants just prior to inoculation—so that the stomata are open and the bacterial inoculum goes in with less wounding to the leaves.
11. Use goggles to protect your eyes from a bacterial spray when forcing the suspension into the leaves.
12. Wooden toothpicks can absorb the moisture from the soil and transfer it to the leaves, and therefore should be avoided.
13. The right level of humidity is critical for the symptoms to develop properly. A high level is needed, but condensation cannot be excessive inside the trays. A fine layer of condensation

under the lid is what is needed. Excessive condensation leads to formation of drops inside the lid that could fall on the plants and cause rotting.

14. Note that Table 1 describes the expected symptoms in Ll-0 in ideal conditions of inoculum and incubation post-infection. Less severe symptoms are observed with Col, so these plants might show corresponding signs at later time points.
15. Before starting, have all materials ready. Cut paper towels by hand in pieces of roughly 3 × 6 cm and set marked microfuge tubes with 0.3 mL 10 mM $MgSO_4$, one for each leaf to be sampled. Do not detach all leaves at once—as they wilt, it becomes more difficult to cut them.
16. A practical way to prepare the bacteria for plating in serial dilution is to make the dilutions in 96-well plates using a multichannel pipette and plating only 10 μL of each dilution. This way, one 15 cm diameter plate can hold up to four dilutions of eight different samples. Make sure that the medium is completely dry to avoid the spreading of the drops. We use plates prepared at least 1 day in advance that were kept overnight at 37 °C. To prevent the colonies from overgrowing and coalescing, thus making them difficult to count, incubate the plates at room temperature or at a maximum of 30 °C overnight.

Acknowledgements

This work was supported by the NIH grants AI105902 and R56AI063433 to LGR. RLB is partially supported by Conselho Nacional de Desenvolvimento Científico (CNPq) and the projects in her lab are funded by São Paulo Research Foundation (FAPESP, www.fapesp.br), Brazil. We thank Jenifer Bush for her critical reading of the methods listed.

References

1. Koornneef M, Meinke D (2010) The development of Arabidopsis as a model plant. Plant J 61:909–921
2. Arabidopsis Genome Initiative (2000) Analysis of the genome sequence of the flowering plant Arabidopsis thaliana. Nature 408:796–815
3. Lamesch P, Berardini TZ, Li D, Swarbreck D, Wilks C, Sasidharan R, Muller R, Dreher K, Alexander DL, Garcia-Hernandez M, Karthikeyan AS, Lee CH, Nelson WD, Ploetz L, Singh S, Wensel A, Huala E (2012) The Arabidopsis Information Resource (TAIR): improved gene annotation and new tools. Nucleic Acids Res 40:D1202–D1210
4. Jones AM, Chory J, Dangl JL, Estelle M, Jacobsen SE, Meyerowitz EM, Nordborg M, Weigel D (2008) The impact of Arabidopsis on human health: diversifying our portfolio. Cell 133:939–943
5. Xu XM, Moller SG (2011) The value of Arabidopsis research in understanding human disease states. Curr Opin Biotechnol 22: 300–307
6. Ausubel FM (2005) Are innate immune signaling pathways in plants and animals conserved? Nat Immunol 6:973–979
7. Cao H, Baldini RL, Rahme LG (2001) Common mechanisms for pathogens of plants and animals. Annu Rev Phytopathol 39:259–284

8. Rahme LG, Ausubel FM, Cao H, Drenkard E, Goumnerov BC, Lau GW, Mahajan-Miklos S, Plotnikova J, Tan MW, Tsongalis J, Walendziewicz CL, Tompkins RG (2000) Plants and animals share functionally common bacterial virulence factors. Proc Natl Acad Sci U S A 97:8815–8821
9. Rahme LG, Stevens EJ, Wolfort SF, Shao J, Tompkins RG, Ausubel FM (1995) Common virulence factors for bacterial pathogenicity in plants and animals. Science 268: 1899–1902
10. Baldini RL, Lau GW, Rahme LG (2002) Use of plant and insect hosts to model bacterial pathogenesis. Methods Enzymol 358:3–13
11. Starkey M, Rahme LG (2009) Modeling Pseudomonas aeruginosa pathogenesis in plant hosts. Nat Protoc 4:117–124

Chapter 54

Assessing *Pseudomonas aeruginosa* Persister/Antibiotic Tolerant Cells

Ronen Hazan, Damien Maura, Yok Ai Que, and Laurence G. Rahme

Abstract

Bacterial persistence, which is observed in a broad range of microbial species, is the capacity of a bacterial cell subpopulation called "persisters" to tolerate exposure to normally lethal concentrations of bactericidal antibiotics. This ability, which is not due to antibiotic-resistant mutants, has been implicated in antibiotic treatment failures and may account for latent, chronic, and relapsing infections. Antibiotic tolerant/Persister (AT/P) cells have been notoriously difficult to study due to their low frequency and transient nature. This chapter describes the main methods used to isolate and study *Pseudomonas aeruginosa* AT/P cells and discusses new technologies that may ease research of *P. aeruginosa* persisters in the near future.

Key words Antibiotic tolerance, Bacterial persistence, Quorum sensing, Small molecules, High-throughput screen, Antibiotics, Anti-virulence, Anti-infectives, Proliferation, Virulence, Infection

1 Introduction

Bacterial persistence is the capacity of bacterial cell subpopulations to tolerate exposure to normally lethal stresses, including lethal concentrations of bactericidal antibiotics [1]. These cells represent a small subpopulation of cells that behave as dormant, latent, or viable but nonculturable (VBNC) bacterial cells and have been observed in a variety of microbial species since the 1940s [2–4]. Of particular clinical interest is multidrug tolerance, which is a phenomenon that has been implicated in treatment failures and accounts for latent, chronic, and relapsing infections that can be suppressed but not eradicated [5]. Unlike multidrug resistance, antibiotic tolerance is not due to mutations, but rather to phenotypic variations. Persisters can later resume growth when the antimicrobial agent is removed, and their progeny remains sensitive to low concentrations of antimicrobial agents [1, 6–8]. The clinical importance of antibiotic tolerance is reflected by the many cases in which antibiotics failed to clear infections despite the absence of resistant bacteria, and clinical reports suggest that the contribution

Alain Filloux and Juan-Luis Ramos (eds.), *Pseudomonas Methods and Protocols*, Methods in Molecular Biology, vol. 1149, DOI 10.1007/978-1-4939-0473-0_54, © Springer Science+Business Media New York 2014

of tolerance to treatment failure and mortality in some infections can be as significant as the contribution of antibiotic resistance.

Pseudomonas aeruginosa is one of the most well-studied bacteria, as it is a recalcitrant pathogen known to defy bactericidal eradication and exemplifies clinically problematic pathogens that cause both acute and chronic human infections [9]. This pathogen does not exhibit the canonic toxin–antitoxin systems involved in *E. coli* persisters [9, 10]. However, several other pathways, such as the metabolism of polyamines, fatty acids, phospholipids, phenylalanine, and DNA as well as global regulators, translation, quorum sensing, and antioxidant responses have been identified as having roles in the formation of *P. aeruginosa* persisters [11–23].

In this chapter, we describe the main techniques and necessary steps for isolating and quantifying *P. aeruginosa* persister cells.

2 Methods

2.1 Selection and Quantification of Antibiotic Tolerant Persister Cells

Here, we describe the protocols used to evaluate the size of the persister fraction within a bacterial culture.

1. Grow starter cells from a –80 °C stock in 5 mL LB at 37 °C overnight (*see* **Note 1**).
2. Dilute the cells 1:100 in 5 mL tubes and grow them under the desired conditions to a defined optical density.
3. Determine the concentration of viable bacteria within the culture [expressed as colony forming unit (CFU)/mL] according to Subheading 2.2 before antibiotic addition. This number will be used as a normalization reference ("normalizers") to calculate the percentage of cells that survived antibiotic treatment (surviving fraction).
4. Add antibiotic at a concentration of 10–100 times its specific minimal inhibitory concentration (MIC) as determined in Subheading 2.2. (*see* **Note 2**).
5. Incubate the cells in a shaker (200 rpm) at 37 °C for 24–48 h (*see* Fig. 1).
6. Transfer 500 μL of each sample to 1.5 μL Eppendorf tubes and centrifuge for 500 μL of each sample for 5 min at 8,000 ×*g*.
7. Remove supernatant gently without touching the pellet. Sometimes the pellet is invisible due to very low persister concentrations.
8. Wash by adding 1 mL of phosphate-buffered saline (PBS), centrifuge for 5 min at 14,000 ×*g*, and gently discard the supernatant.
9. Depending on antibiotic concentration, repeat **step 8** two to four times in order to ensure that the antibiotic has been washed out (*see* **Note 3**).

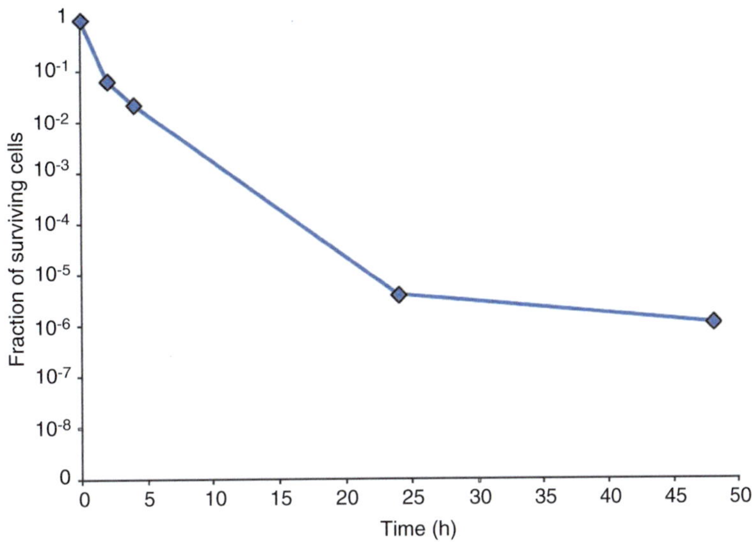

Fig. 1 Antibiotic tolerant cell assessment using the CFU counts method. The killing curve of *P. aeruginosa* strain PA14 exponential-phase cells exposed to a lethal concentration of the bactericidal antibiotic meropenem indicates that the majority of cells died quickly, showing a sharp drop-off in survival kinetics within 24 h, while a small fraction of cells (~10^{-6}) survived the treatment even after 48 h of antibiotic exposure. This surviving fraction of cells reflects the number of antibiotic tolerant cells. PA14 cells were exposed to 10 mg/L meropenem for 48 h

10. Resuspend the pellet in 500 μL of an appropriate medium.
11. Determine the surviving fraction (antibiotic tolerant persister cells) according to Subheading 2.1.1.
12. Validate that the surviving bacteria are indeed tolerant cells and not resistant mutants as described in Subheading 2.3.

2.1.1 Determination of Bacterial Number and Surviving Fractions

The gold standard method to determine bacterial concentration in a sample is the CFU plating method, which consists of plating several dilutions for each sample on LB agar plates [24]. However, this method is very time consuming and tedious in the case of high-throughput screens, for which the drop plate and *s*tart of *g*rowth *t*ime (SGT) methods are more appropriate ("Drop Plate Method" and "SGT Method") [25, 26].

CFU Plating Method (*See* **Note 4**)

1. Perform eight 1:10 serial dilutions of a bacterial sample in Eppendorf tubes. The number of dilutions can be reduced in cases where the number of bacteria can be estimated.
2. Plate triplicates of 100 μL of each dilution on agar plates containing the appropriate medium.
3. Incubate plates for 24–48 h at 37 °C.
4. Count the colonies on each plate and calculate the CFU/mL as follows: number of colonies $\times 10 \times 10^{-\text{dilution}}$.

Drop Plate Method (*See* **Note 5**)

1. Fill three columns of a 96-well plate per sample with 270 μL of LB, except the first row of wells.
2. Add 200 μL of sample in the three empty wells of the first row of wells.
3. Remove 30 μL from the wells of the first row using a multichannel pipette and place into 270 μL of LB in the wells of the second row (dilution 1:10).
4. Perform serial dilutions of the sample by repeating **step 3** down to the bottom row seven times using a multichannel pipette. This gives a range of dilutions from non-diluted (upper row) to a 10^{-7} dilution (bottom row). Use a second 96-well plate if needed to dilute samples to a factor greater than 10^{-7}.
5. Using a multichannel pipette, place drops of 10 μL on an agar plate. One plate can contain up to 6 rows of 7 dilutions. The agar plate surface should be dry enough to avoid blending of the dilution drops.
6. Let the plate dry before incubation at 37 °C for 24 h.
7. Count the colonies and calculate CFU/mL as follows: number of colonies $\times 100 \times 10^{-\text{dilution}}$.

SGT Method (*See* **Note 6**)

1. For both untreated normalizer and persister cell quantification, add 2 μL of each sample into 1 mL of LB in Eppendorf tubes (1:500 dilution).
2. Transfer 200 μL from each tube into three wells (technical triplicate) of a 96-well plate.
3. Incubate the plate in a plate reader set to measure the optical density (OD) at a 600 nm wavelength every 15 min for 24 h at 37 °C with a 10 s shaking step every 15 min.
4. Define the SGT value of each sample as the time required for the culture to reach an $OD_{600\ nm}$ threshold that is set slightly above the detectable background at the start of the logarithmic phase of growth (usually $OD_{600nm} = 0.15$ or 0.2).
5. Calculate the first ΔSGT value for each sample according to the following equation: $\Delta SGT = (SGT_{Persisters} - SGT_{Normalizers})$, where the SGT of untreated normalizer cells is subtracted from the SGT of treated persister cells.
6. Calculate the ΔΔSGT value by subtracting the ΔSGT of the reference strain/condition ("calibrator") from that of the sample as follows: $\Delta\Delta SGT = (\Delta SGT_{Sample} - \Delta SGT_{Calibrator})$.
7. Calculate the fold change between the sample and the calibrator as the following: $F = 2^{-\Delta\Delta SGT}$. Results are presented as $\log_2$ fold changes: $-\Delta\Delta SGT$ (*see* Fig. 2).

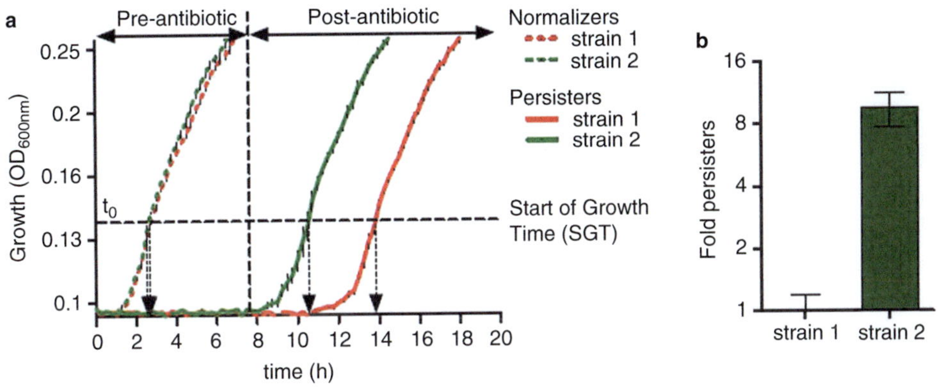

Fig. 2 Antibiotic tolerant cell assessment using the SGT method. Comparative assessment of the persister cell fraction between two strains subjected to a 24 h treatment with meropenem (10 mg/L) at 37 °C (no meropenem added to normalizers). Following a 1:500 dilution, the growth kinetics of normalizers and treated samples were recorded. (**a**) Using $OD_{600nm} = 0.15$, the ΔSGT values were calculated as the difference between treated and normalizer SGTs. The ΔΔSGT values were calculated as the difference between ΔSGTs of each strain compared to that of the calibrator. (**b**) For the SGT method, the $\log_2$-fold change was calculated as −ΔΔSGT (empty bars). For CFU counting, normalizers and treated cells were serially diluted

2.2 Determination of the Antibiotic Concentration to Be Used for Selection of Persisters

1. In order to select the persister subpopulation, the culture needs to be exposed to bactericidal levels of the antibiotic. MIC of each antibiotic could be determined according to the standard protocol [27]. However, in the case of persisters it is important to determine the killing concentration of the antibiotic in the conditions and strains of the experiment. Here we present a simple method to estimate the selective concentration of the antibiotics in specific conditions. Grow cells from a −80 °C stock in the desired conditions overnight for 18 h (usually LB at 37 °C with shaking at 200 rpm).
2. Dilute the cells 1:1,000 in the same media the next day.
3. Inoculate 100 μL of cultures in 3 rows of transparent 96-well plates that will serve as technical triplicates using a multichannel pipette (e.g., rows A–C).
4. Prepare antibiotic stock. A good starting range for the concentration of the antibiotic would be between 10 and 100 mg/mL.
5. Dilute the antibiotic stock 1:500 to 20–200 μg/mL with the media containing bacteria from **step 2**.
6. Add 100 μL of the diluted antibiotic to wells A1, B1, and C1 and mix thoroughly by pipetting up and down.
7. Replace the pipette tips and transfer 100 μL from wells A1–C1 to wells A2–C2.
8. Repeat **step 7** for all wells to A11–C11. Remove the excess 100 μL from the last row and discard.

9. Retain column 12 without antibiotic as a positive control. The plate now contains a 2,048× range of concentrations between the first well and well 11.
10. Incubate the plate for 24 h at 37 °C and read its $OD_{600\ nm}$ using a plate reader.
11. The specific MIC would be the lowest concentration of antibiotic that did not permit growth of bacteria.

2.3 Confirm that the Surviving Bacteria Are Indeed Tolerant and Not Resistant Mutants

After the first round of bacteria selection by temporary exposure to the antibiotic, the surviving colonies need to be retested to assess that they are indeed tolerant and not resistant mutants. This is done by checking the MIC and by measuring the size of the persister fraction of cultures started from surviving cells again as described in Subheadings 2.1 and 2.2.

3 Additional Methods

The major disadvantage of the culture-based persisters assessment techniques is that they indirectly determine the number of dormant cells based on the number of surviving cells by measuring the growing progeny. More direct methods to analyze dormant cells include microscopy [28], flow cytometry [21, 29], and microfluidics [19, 30] together with the use of probes that allow for the distinction between live and dead cells, even if the cells are not dividing [31]. In order to study persister "wake-up," a method using automated scans termed "ScanLag" has been described for *E. coli*, which could also be applied to *P. aeruginosa* [32].

4 Notes

1. The incubation time for the starter cultures from a −80 °C stock should always be the same duration, because bacterial cultures can accumulate persisters as incubation progresses [1]. This important precaution will help avoid having inconsistent results between experiments.
2. As a rule of thumb, the concentration of antibiotic used to select for tolerant cells is at least 10 times the MIC. It should be noted that the MIC method determines the resistance of cells to the specific antibiotic, rather than their tolerance to antibiotic-induced killing. The distinction between these two effects is whether the cells are growing in the presence of antibiotic or only surviving the antibiotic killing. Table 1 shows the MICs of commonly used antibiotics against *P. aeruginosa* with the strain PA14 grown in LB at 37 °C.

Table 1
Minimal inhibitory concentration (MIC) of commonly used antibiotics on PA14

Family	Name	MIC on PA14 (μg/mL)
Aminoglycoside	Amikacin Gentamycin	1 1.5
Beta-lactam	Imipenem Meropenem	1 0.1
Quinolone	Levofloxacin Ciprofloxacin	2 0.1

3. In order to determine the size of the antibiotic-tolerant cell fraction, the antibiotic needs to be completely removed from the culture. If the expected concentration of the fraction of bacteria surviving is greater than 1,000 cells/mL, then removal of the antibiotic used at ~100× MIC can be carried out by making ~100-fold serial dilutions. However, if the concentration of the fraction of bacteria surviving is ≤100–1,000 cells/mL, then dilutions of less than ~100-fold would result in a carryover of antibiotic, thereby restricting bacterial growth on agar plates, and >100-fold dilutions would result in bacterial concentrations that are too low for detection. Therefore, a cautious step of cell wash out needs to performed in order to remove the antibiotic without reducing the amount of detectable bacteria in samples, as described in **steps 6–10** in Subheading 2.1.
4. The advantages of this CFU plating method are that only viable bacteria are counted and the dilutions allow for any number of bacteria to be counted regardless of the starting concentration. One limitation of the CFU method is that clumps of bacterial cells can be miscounted as single colonies. In addition, CFU is usually not the method of choice for a high-throughput screen because it is a relatively time-consuming and tedious.
5. Using the drop plate method described here, the bacteria are first diluted in 96-well plates and 10 μL drops are then plated. Thus, instead of using 42 plates, an array of 6×7 drops can be easily spotted on a single standard Petri dish. One disadvantage of this method is the reduction in accuracy, but it can be overcome by plating multiple technical replicates.
6. The SGT method is a quantitative method [33]. It combines the methodology of quantitative polymerase chain reaction (qPCR) calculations [34, 35] with a previously described qualitative method of bacterial growth determination [12] to

develop an improved quantitative method [33]. The SGT method allows rapid and serial quantification of the absolute or relative number of live cells in a bacterial culture in a high-throughput manner. The SGT method is based on the regrowth time required by a growing cell culture to reach a threshold (spectrophotometrically detectable levels), and the notion that this time is proportional to the number of cells in the starting bacterial inoculum [12, 33] (*see* Fig. 2).

Acknowledgements

This work was supported by the NIH grants AI105902 and R56AI063433 to LGR.

References

1. Balaban NQ et al (2004) Bacterial persistence as a phenotypic switch. Science 305: 1622–1625
2. Lewis K (2008) Multidrug tolerance of biofilms and persister cells. Curr Top Microbiol Immunol 322:107–131
3. Oliver JD (2010) Recent findings on the viable but nonculturable state in pathogenic bacteria. FEMS Microbiol Rev 34:415–425
4. Bigger JW (1940) Treatment of *staphylococcal* infections with penicillin. Lancet 244: 497–500
5. Bizzini A et al (2010) A single mutation in enzyme I of the sugar phosphotransferase system confers penicillin tolerance to *Streptococcus gordonii*. Antimicrob Agents Chemother 54: 259–266
6. Gefen O, Balaban NQ (2009) The importance of being persistent: heterogeneity of bacterial populations under antibiotic stress. FEMS Microbiol Rev 33:704–717
7. Jayaraman R (2008) Bacterial persistence: some new insights into an old phenomenon. J Biosci 33:795–805
8. Lewis K (2007) Persister cells, dormancy and infectious disease. Nat Rev Microbiol 5:48–56
9. Fauvart M, De Groote VN, Michiels J (2011) Role of persister cells in chronic infections: clinical relevance and perspectives on anti-persister therapies. J Med Microbiol 60:699–709
10. Lewis K (2010) Persister cells. Annu Rev Microbiol 64:357–372
11. Belenky P, Collins JJ (2011) Microbiology. Antioxidant strategies to tolerate antibiotics. Science 334:915–916
12. De Groote VN et al (2009) Novel persistence genes in *Pseudomonas aeruginosa* identified by high-throughput screening. FEMS Microbiol Lett 297:73–79
13. Kayama S et al (2009) The role of *rpoS* gene and quorum-sensing system in ofloxacin tolerance in *Pseudomonas aeruginosa*. FEMS Microbiol Lett 298:184–192
14. Manuel J, Zhanel GG, de Kievit T (2010) Cadaverine suppresses persistence to carboxypenicillins in *Pseudomonas aeruginosa* PAO1. Antimicrob Agents Chemother 54: 5173–5179
15. Moker N, Dean CR, Tao J (2010) *Pseudomonas aeruginosa* increases formation of multidrug-tolerant persister cells in response to quorum-sensing signaling molecules. J Bacteriol 192: 1946–1955
16. Murakami K et al (2005) Role for *rpoS* gene of *Pseudomonas aeruginosa* in antibiotic tolerance. FEMS Microbiol Lett 242:161–167
17. Nguyen D et al (2011) Active starvation responses mediate antibiotic tolerance in biofilms and nutrient-limited bacteria. Science 334:982–986
18. Shatalin K et al (2011) H2S: a universal defense against antibiotics in bacteria. Science 334:986–990
19. Vega NM et al (2012) Signaling-mediated bacterial persister formation. Nat Chem Biol 8:431–433
20. Viducic D et al (2006) Functional analysis of *spoT*, *relA* and *dksA* genes on quinolone tolerance in *Pseudomonas aeruginosa* under non-growing condition. Microbiol Immunol 50:349–357

21. Williamson KS et al (2012) Heterogeneity in *Pseudomonas aeruginosa* biofilms includes expression of ribosome hibernation factors in the antibiotic-tolerant subpopulation and hypoxia-induced stress response in the metabolically active population. J Bacteriol 194: 2062–2073
22. Bandyopadhaya A et al (2012) The quorum sensing volatile molecule 2-amino acetophenon modulates host immune responses in a manner that promotes life with unwanted guests. PLOS Pathog 8:e1003024
23. Kesarwani M et al (2011) A quorum sensing regulated small volatile molecule reduces acute virulence and promotes chronic infection phenotypes. PLoS Pathog 7:e1002192
24. Miller JH (1972) Determination of viable cell counts: bacterial growth curves. In: Miller JH (ed) Experiments in molecular genetics. Cold Spring Harbor, New York, pp 31–36
25. Barbosa HR et al (1995) Counting of viable cluster-forming and non cluster-forming bacteria: a comparison between the drop and the spread methods. J Microbiol Methods 22:39–50
26. Hazan R et al (2012) A method for high throughput determination of viable bacteria cell counts in 96-well plates. BMC Microbiol 12:259
27. Rankin ID (2005) MIC testing. In: Coyle MB (ed) Manual of antimicrobial susceptibility testing. American Society for Microbiology, Washington, DC, pp 53–62
28. Casida LE Jr (1969) Observation of microorganisms in soil and other natural habitats. Appl Microbiol 18:1065–1071
29. Veal DA et al (2000) Fluorescence staining and flow cytometry for monitoring microbial cells. J Immunol Methods 243:191–210
30. Aldridge BB et al (2012) Asymmetry and aging of mycobacterial cells lead to variable growth and antibiotic susceptibility. Science 335: 100–104
31. Berney M et al (2007) Assessment and interpretation of bacterial viability by using the LIVE/DEAD BacLight Kit in combination with flow cytometry. Appl Environ Microbiol 73:3283–3290
32. Levin-Reisman I et al (2010) Automated imaging with ScanLag reveals previously undetectable bacterial growth phenotypes. Nat Methods 7:737–739
33. Hazan R et al (2012) A method for high throughput determination of viable bacteria cell counts in 96-well plates. BMC Microbiol 12:259
34. Heid CA et al (1996) Real time quantitative PCR. Genome Res 6:986–994
35. Nolan T, Hands RE, Bustin SA (2006) Quantification of mRNA using real-time RT-PCR. Nat Protoc 1:1559–1582

Chapter 55

Assessing *Pseudomonas* Virulence with Nonmammalian Host: Zebrafish

María A. Llamas and Astrid M. van der Sar

Abstract

In the last years, the zebrafish (*Danio rerio*) has become an important vertebrate animal model to study host–pathogen interactions, especially in its embryonic stage. The presence of a fully developed innate immune system in the first days of embryogenesis, the facility of obtaining and manipulating large numbers of embryos, the optical transparency of the embryos that allow the direct visualization of bacterial infections, a wide range of genetic tools, and extensive mutant resources and collections of transgenic reporter lines are important advantages of the zebrafish-embryo model. *Pseudomonas aeruginosa* is able to lethally infect zebrafish embryos when the amount of cells injected exceeds the phagocytic capacity of the embryo. Different studies have proved the suitability of zebrafish embryos as a model to analyze *P. aeruginosa* infection. Here we describe the detailed protocols to establish a *P. aeruginosa* infection in zebrafish embryos and to image the interaction of the bacterium with this host with fluorescent microscopy.

Key words *Danio rerio* (zebrafish), Embryo model, *Pseudomonas aeruginosa*, Bacterial infection, Fluorescence imaging, Microinjection

1 Introduction

Animal models are necessary to study microbial virulence. However, methodological, financial, and ethical issues are often a problem for those studies. These problems have been partially overcome by the use of the nonmammalian invertebrate model hosts such as the fruit fly *Drosophila melanogaster*, the nematode *Caenorhabditis elegans*, and the amoebae *Dictyostelium discoideum* [1–4]. In recent years, zebrafish (*Danio rerio*) has been increasingly used as a new model for studying host–pathogen interactions [5–8]. The zebrafish model has a number of advantages over these other models of infection. Zebrafish are vertebrates and are genetically and physically closer to humans than the invertebrate models. Zebrafish contain a fully developed immune system: the major cell types of the innate immune system, macrophages and neutrophils, already

Alain Filloux and Juan-Luis Ramos (eds.), *Pseudomonas Methods and Protocols*, Methods in Molecular Biology, vol. 1149, DOI 10.1007/978-1-4939-0473-0_55, © Springer Science+Business Media New York 2014

develop during the first days of embryogenesis [6, 9]. Furthermore, the optical transparency of the embryonic stages allows the analysis of bacterial infections in situ, in real time and at a high resolution by using fluorescent microorganisms (Fig. 1). For this reason, together with the ease of obtaining large numbers of embryos, many infectious disease studies in the zebrafish model have concentrated on the embryonic stage of development, when the advantages of the model are maximal. In addition, the availability of the zebrafish genome sequence, the efficient tools for reverse genetics and the availability of a growing library of fluorescently labelled transgenic zebrafish lines also contribute to the usefulness of the model [10, 11]. Finally, the zebrafish embryo model is highly suitable for in vivo chemical screening, with the advantage that only small amounts of compounds are needed that simply can be added to the embryo medium [12].

Infection in zebrafish embryos is most often established through microinjection of the bacteria into the embryo. The site of microinjection determines whether the infection will rapidly become systemic or will initially remain localized [6, 13, 14]. A rapid systemic infection can be established by injecting bacteria directly into the blood stream via the caudal vein near the blood island or via the Duct of Cuvier, a wide circulation channel on the yolk sac connecting the heart to the trunk vasculature. For slow-growing bacteria, like mycobacteria, systemic infections are also achieved by injection into the yolk [15]. To study directed migration of neutrophils and macrophages towards local infections, bacteria are injected into the tail muscle, otic vesicle, or hindbrain ventricle. To create an infection apparently inaccessible to phagocytes, bacteria are injected into the notochord [16].

Zebrafish embryos have been reported to be a suitable model to study *Pseudomonas aeruginosa* infections [17–19]. *P. aeruginosa* infections are normally established by injecting the bacteria into the blood stream via the caudal vein of 1 or 2 days post fertilization (dpf) old embryos (Fig. 1a). Known attenuated *P. aeruginosa* mutants, such as mutants in type III secretion or in quorum sensing, are less virulent in zebrafish embryos. Moreover, key host determinants, such as phagocytosis, play an important role in zebrafish embryos pathogenesis, and, as in humans, phagocyte depletion increases the susceptibility of the embryos to *P. aeruginosa* infection [17]. Neutrophils and macrophages rapidly phagocytose and kill *P. aeruginosa*, but if the amount of cells injected exceeds the phagocytic capacity of the embryo bacteria survive and grow causing the death of the embryo ([17, 19] and Fig. 1b). When large inocula are used, *P. aeruginosa* kills the embryos within the first 2 days post infection (dpi). Increased multiplication of the bacterium in the embryos correlates with a slower blood circulation and decreased heartbeat when compared with healthy embryos. In addition, severe damage of the tissues is observed,

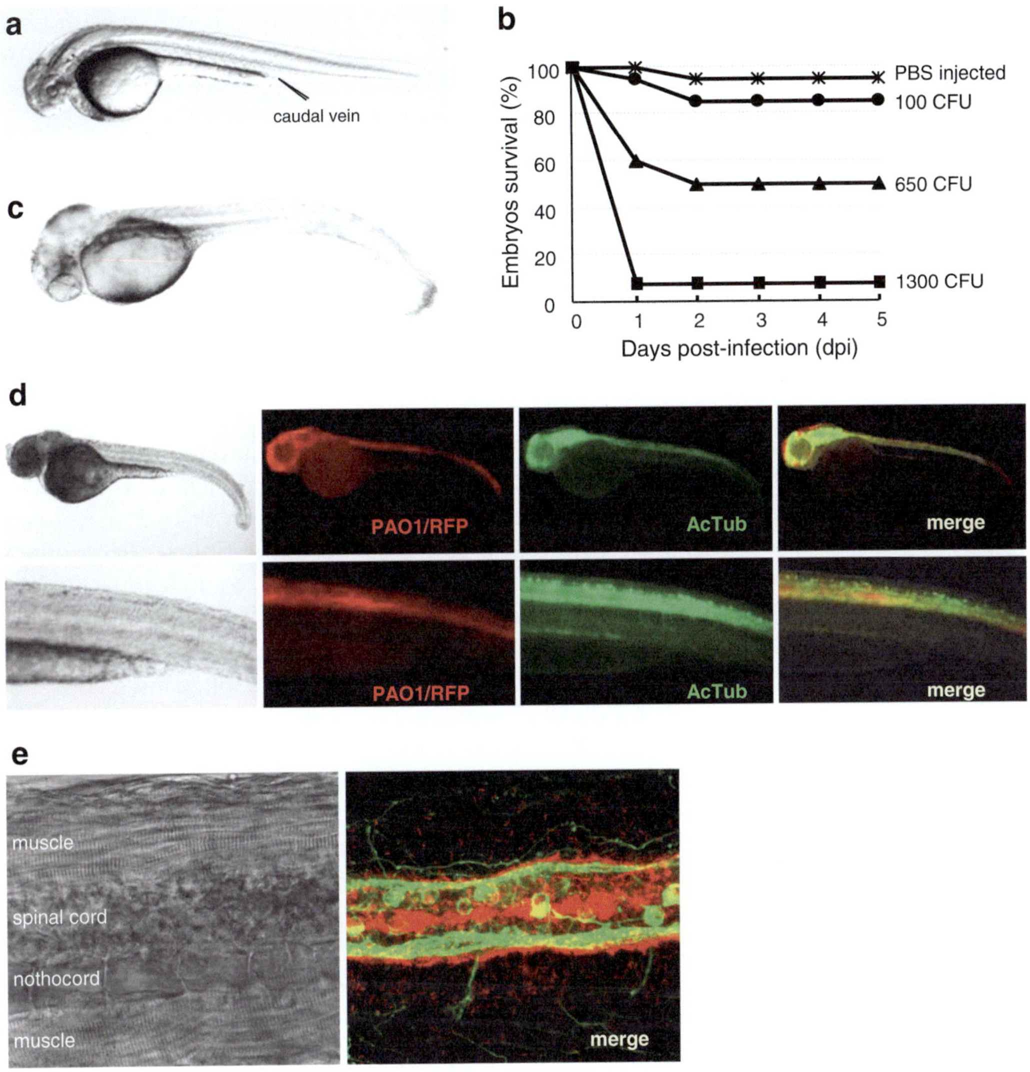

Fig. 1 Overview of the *P. aeruginosa*-zebrafish embryo model of infection. (**a**) Intravenous injections for establishing a rapid *P. aeruginosa* systemic infection is performed into the caudal vein at the posterior blood island at 1–2 dpf embryos (*arrow*). (**b**) Dose-dependent response in embryo survival to *P. aeruginosa* infection. Groups of 28–30 hpf embryos ($n = 20$ embryos/group) were inoculated with a low dose (100 CFU), an intermediate dose (650 CFU), or a high dose (1,300 CFU) of PAO1 wild-type cells. Uninfected control was injected with PBS. Embryos are resistant to low doses of bacteria, but increased mortality is observed with larger inocula. (**c**) Moribund embryo infected with a high dose of *P. aeruginosa* PAO1 cells. The bacterium produces a necrotic cell death that starts in the tail and extends to other tissues. (**d**) Fluorescent images of highly infected embryos with *P. aeruginosa* PAO1 cells labelled with an mCherry DsRed fluorescent protein variant (PAO1/RFP, *red channel*) and subjected to whole mount immunohistochemistry using an anti-acetylated tubulin (AcTub) monoclonal antibody that specifically recognizes the nerves of the embryo (*green channel*). The *last panel* shows the *red/green* overlay. (**e**) Confocal image of the embryo shown in (**d**) with the same color coding. The panel clearly shows the concentration of *P. aeruginosa* (*red channel*) in and around the spinal cord. In addition single bacteria can be seen in the muscle tissue. No co-localization of neuronal cell bodies or axons (*green channel*) with *P. aeruginosa* was seen. However, a close contact between the axon tracts in the spinal cord and *P. aeruginosa* is observed

mainly in the tail and brain from where necrotic cell death extends to other tissues (Fig. 1c). Confocal fluorescence analyses of the *P. aeruginosa* infection in zebrafish have shown that after injection in the blood stream this bacterium is able to rapidly extravasate and infect other tissues, mainly the brain and spinal cord of the embryos ([19], and Fig. 1d, e). This specific infection pattern might be the result of the fact that the amount of *P. aeruginosa* cells needed to establish an infection phenotype outnumbers the amount of phagocytic cells present in the embryo. However, this pattern seems to be specific for *P. aeruginosa* and different to the one caused by, for example, *Salmonella typhimurium*, which replicates either inside macrophages or extracellularly but always within the vascular system, or *Mycobacterium marinum*, which is located in clustered macrophages [5, 20]. Although differences in infection pattern can also be explained by the preferred niche of the bacteria, *S. typhimurium* and *M. marinum* are, in contrast to *P. aeruginosa*, intracellular pathogens. It would be interesting to determine if high doses of these intracellular pathogens would lead to a similar infection pattern as seen for *P. aeruginosa*.

In this chapter we give practical guidelines for *P. aeruginosa* infection studies in zebrafish embryos. In the following sections, we describe the detailed protocols to achieve a systemic infection of embryos with *P. aeruginosa*, and for subsequent fluorescence imaging of the interaction of the bacterium with the embryos. Additional information on the analysis of other bacterial infections in zebrafish can be found in these other methods articles [13, 14, 21].

2 Materials

2.1 Solutions/Reagents

Prepare the solutions using ultrapure water and store them at room temperature (unless indicated otherwise). Sterilize the solutions by autoclaving or by filtration through a 0.45 mm sterile filter (Millipore).

1. Phosphate-buffered saline (PBS): dissolve 5 PBS tablets (Sigma-Aldrich) in 1 L purified water. Sterilize by autoclaving.
2. Phenol red in PBS (Sigma-Aldrich).
3. 20 % (v/v) Glycerol: 20 mL glycerol (Sigma-Aldrich) in 80 mL distilled water. Sterilize by autoclaving.
4. Egg water: 60 μg/mL final concentration of commercially available Sea salts (Sigma-Aldrich) in distilled water.
5. Embryo medium stock: 5 mM NaCl, 0.17 mM KCl, 0.33 mM $CaCl_2$, and 0.33 mM $MgSO_4$ in 500 mL MilliQ water, sterile filtered.
6. Embryo medium: Add 16.67 mL of embryo medium stock and 3 mL 0.01 % (w/v) methylene-blue in MilliQ in a total volume of 1 L MilliQ water.

7. *N*-Phenylthiourea (PTU) (Sigma-Aldrich).
8. 0.02 % Buffered 3-aminobenzoic acid (Tricaine) (Sigma-Aldrich).
9. PBTx: 1 % (v/v) Triton X-100 in PBS.

2.2 Tools/Apparatus

1. Breeding tanks (Fig. 2a) (AquaSchwarz, Germany, http://www.aquaschwarz.com).
2. Injection tray mold: 5 ridges of 1 mm × 1 mm were phrased on one side of a 4 mm × 5 cm × 5 cm square Perspex plate (VU University medical center technical services) (Fig. 2e).
3. Injection tray: Prepare 1.5 % (w/v) agarose in egg water, pore the agarose in a Petri dish and place an injection tray mold with the ridges faced down on top of the agarose before it is set. The mold will form slits of 1 mm thick and 1 mm deep (Fig. 2e). The slits are just big enough for the embryos to fit in and facilitate easy injection (Fig. 2g). Store the injection trays at 4 °C.
4. Microinjector [World Precision Instruments (PV820, Pneumatic PicoPump)].
5. Micromanipulator [World Precision Instruments (M3301R, Manual)].
6. Micromanipulator stand [World Precision Instruments (M10 magnetic stand)].
7. Micropipette Holder [World Precision Instruments (5430-10, PicoNozzle Kit)].
8. Micropipette Puller [MicroData Instruments, Inc. (PMP102, Micropipette Puller)].
9. Glass Capillaries [HARVARD APPARATUS LTD. (Borosilicate glass capillaries, Harvard Part. No. 30-0038 1.0 mm O.D. × 0.78 mm I.D.)].
10. Microloader Pipettes (Eppendorf 5242956.003).
11. Microgrinder (Narishige, Inc., EG-400).
12. Fine tweezers (Fine Science Tools, Inc., Dumont #5 Forceps—InoxBiology).
13. Microscopy chambers for confocal imaging [Ibidi (1 μ-slide 8-well uncoated microscopy chamber)].
14. Stereoscopic dissecting microscope (SMZ1000, Nikon) with epi-fluorescence attachment (model P-FLA, Nikon).
15. Leica MZ16FA stereomicroscope equipped with a DFC420C digital camera and Leica microscopy suite.
16. Leica TCS SP2confocal with Leica DMIRE2 microscope and Leica confocal software suite.

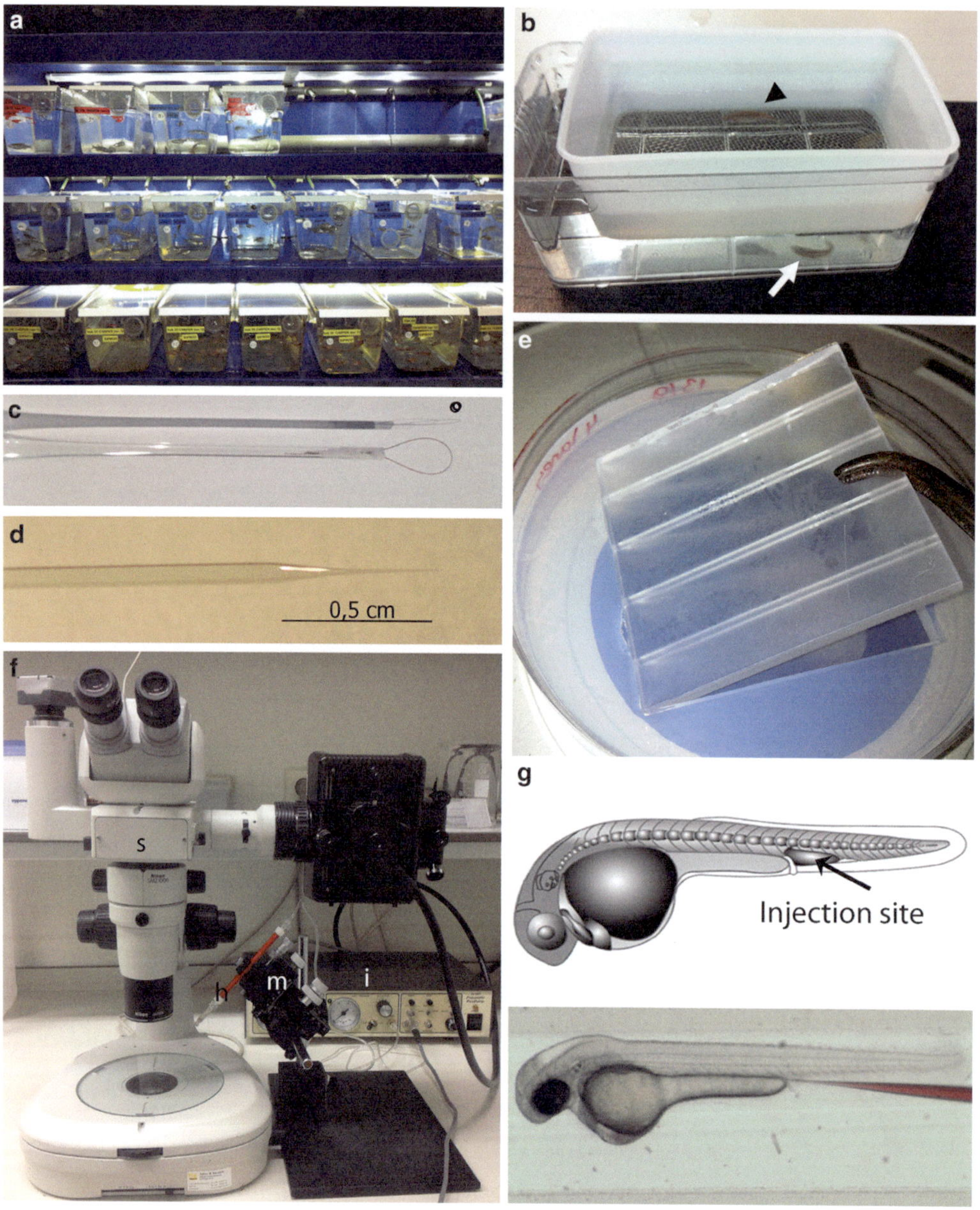

Fig. 2 Overview of tools used for infection of zebrafish embryos. (**a**) A view of zebrafish stock tanks in an Aqua Schwarz cabinet. (**b**) The breeding tank with the inset breeding cage (*in white*) in place. Overnight male zebrafish swim at the bottom part (*white arrow*) and female zebrafish swim in the inset cage (*black arrowhead*). In the morning, after the lights have switched on, the male fish are put together with the female fish and the mating can start. The eggs will fall through the mash bottom and can easily be collected. (**c**) The hairloop tool is handmade from a glass Pasteur pipette with a hair glued in the tip opening using correction fluid [Wite Out (BIC)]. The hairloop is used to move embryos without damaging them. (**d**) An example of a glass borosilicate capillary needle with a taper of 0.5 cm. (**e**) An example of an injection tray mold. The mold is removed with forceps from the 1.5 % (w/v) agarose injection tray. The mold has formed slits of 1 mm thick and 1 mm deep which are just big enough to hold embryos that are still in their chorion but also big enough to hold dechorionated embryos of 1–5 dpf. (**f**) A picture of the microinjection set up. A stereo dissection microscope (s) is shown with the micromanipulator (m) in place holding the micropipette holder (h) which is connected to the microinjector (i). A foot pedal is connected to the microinjector to regulate the injection pressure (not shown). (**g**) Shows a schematic drawing of an embryo with the injection site indicated by the *black arrow*. In addition, an embryo positioned in the injection tray with the injection needle filled with a bacterial inoculum in phenol red, ready to inject, is shown

3 Methods

3.1 Preparation of the Injection Needles

1. Prepare borosilicate glass capillary needles with a taper of approximately 0.5 cm using a micropipette pulling device (Fig. 2d).
2. Break off the needle tip under the microscope with fine forceps to create a tip opening diameter of 5–10 μm (blunt tips). In order to create sharp tips and minimize tissue damage during injection, the tip of the needles may be bevelled at an angle of 45° with a microgrinder.
3. Put a strip of double-sided foam adhesive tape (~0.5 cm wide) in a square Petri dish and stick the needles on the strip for later use. Be careful not to break the needle tips. Close the Petri dish to protect the needles from dust and breakage.

3.2 Preparation of the *P. aeruginosa* Inoculum

1. Inoculate *P. aeruginosa* cells into LB liquid medium with appropriate antibiotics to select for fluorescence expression vectors (*see* **Note 1**) and incubate overnight at 37 °C with agitation.
2. Dilute the overnight culture 1:50 in fresh LB liquid medium (with appropriate antibiotics) and incubate at 37 °C with agitation.
3. Harvest 2 mL of the bacterial culture when they are in logarithmic phase (OD_{600} around 0.6–0.8) by centrifuging and washing them two times in sterile PBS.
4. Measure the OD_{600} of the bacterial suspension in PBS and dilute the bacteria to the desired concentration (*see* protocol 3.4) in PBS optionally containing at least 0.085 % (v/v) phenol red (Sigma-Aldrich) to aid visualization during the injection process.
5. Vortex the bacterial suspension well to avoid clumping.
6. Directly use the fresh suspension for the injection or prepare glycerol stocks. To prepare glycerol stocks, spin down the freshly made injection stock with the desired concentration of bacteria and concentrate the stock by resuspending the pellet in half the starting volume in sterile 20 % (v/v) glycerol in PBS. Store the glycerol stock at −80 °C. Dilute the glycerol stock 1:1 prior to injection in sterile PBS, optionally containing 0.17 % (v/v) phenol red.
7. Load the inoculum into the microcapillary needle with a microloader tip (Eppendorf) for injection.

3.3 Zebrafish Eggs Production and Collection, and Preparation of Embryos for Injection

1. Two to three nights prior to injection, set up zebrafish breeding pairs in breeding tanks separating males and females by an inset breeding cage with a mash bottom (Fig. 2b). To increase total egg production, fish can be set up in a ratio of two females to one male if desired.

2. In the morning, after the lights turn on, put males and females together in the inset breeding cage with mash bottom and allow for approximately 20 min of mating time. Eggs will fall through the mash bottom of the inset breeding cage and can easily be collected.
3. Remove the inset breeding cage with the breeding pairs and place the zebrafish back in their tanks. Collect the eggs from the bottom of the breeding cages by pouring of the water containing the eggs through a sieve. Rinse the eggs in the sieve with egg water. Use egg water to remove the eggs from the sieve and place the eggs in a Petri dish with egg water. Remove unfertilized eggs and debris with a pipette. If required zebrafish can be regrouped in larger tanks to produce additional rounds of eggs for injection.
4. Sort embryos in a Petri dish filled with embryo medium containing approximately 60 embryos/dish and incubate at 28.5 °C.
5. If desired, add 0.003 % (w/v) PTU to the embryo medium when the embryos are about 12 h post fertilization (hpf) to prevent melanization (*see* **Note 2**).
6. Around 24 hpf, remove the embryos from their chorion with fine tweezers.
7. Keep the embryos in a Petri dish filled with embryo medium and with a layer of 1 % (w/v) agarose on the bottom to prevent embryos from sticking to the plastic surface.
8. Anesthetize the embryos with 200 μg/mL buffered Tricaine approximately 10 min prior to injections.

3.4 Injection of *P. aeruginosa* into the Caudal Vein of 1 Day Old Embryos

1. Load the microcapillary needle with 5 μl of the *P. aeruginosa* bacterium inoculum using a microloader tip.
2. Mount the loaded microcapillary needle in the micropipette holder of the PicoNozzle which fits on the micromanipulator and is connected to the microinjector. Position the needle by moving the micromanipulator stand under a stereo dissecting microscope (Nikon) (Fig. 2f). Set the injection time of the microinjector to 0.2 s and the holding pressure between 1 and 10 psi. Adjust the injection pressure between 30 and 50 psi to create a droplet with approximately the correct injection volume, this depends on the tip opening of the needle in use. Adjust the drop size further to match the desired diameter with the help of a scale bar in the ocular. Alternatively, if the ocular on the stereomicroscope does not include a scale bar, a stage micrometer slide can be used to measure the radius of the injection volume. For size determination the drop can be injected into mineral oil on a microscope slide and the diameter of the droplet can be measured. The radius of a drop of 1 nL is 0.062 mm ($V=4/3\pi r^3$).

3. Set the micromanipulator with the loaded needle into the correct position prior to injecting (i.e., approximately at a 45° angle with respect to the injection plate surface) and only use the manipulator to move the needle back and forth to inject.
4. Stage the embryos at 28 hpf. At this stage the embryos have a consistent blood circulation, beginning of pigmentation in the eye, a straight tail, and the heart is positioned just ventrally to the eye [22] (*see* **Note 3**).
5. Remove the injection tray mold from the injection tray (mold is reusable) (Fig. 2e) and add 10 mL egg water with 200 μg/mL buffered Tricaine.
6. Transfer the embryos with a pipette into the injection tray.
7. Position the embryos side by side in the slits of the injection tray using a hairloop tool (Fig. 2c) and align them with the tails oriented towards the micromanipulator and the needle tip (Fig. 2g). Move the injection tray by hand during injections to place the embryos in the preferred position for inserting the needle.
8. To start injections, place the needle tip directly above the caudal vein close to the urogenital opening (Fig. 1a) by moving the injection tray. Gently pierce the periderm with the needle tip and inject the desired dose of bacteria. The injection dose may vary depending on the desired phenotype; we use between 100 and 1,000 cfu of *P. aeruginosa* wild-type strain PAO1, containing the mCherry red fluorescent protein (RFP) expression vector pBPF-mCherry [19]. The injected bacterial suspension will follow the blood flow through the caudal vein towards the heart. Monitor if the injection was performed correctly by checking for an expanding volume of the vascular system and by the coloring of the blood flow due to injected phenol red. For dose–response experiments, two to three consecutive injections can be performed without extracting the needle. Inject the embryos in the caudal vein with ~1 nl of the bacterial suspension. Directly remove embryos that are not properly injected and discard them.
9. To ascertain the consistency of the injection volume during large experiments, frequently check and if necessary adjust the pressure to keep the injection volume similar throughout the experiment. To determine the amount of colony forming units (cfu) that are injected, inject five times a drop of bacteria directly into an eppendorf tube containing 500 μl of sterile PBS. Plate 100 μl of this "infected" PBS onto a sterile LB plate containing the appropriate antibiotics and count the cfu after incubation overnight at 37 °C.
10. Transfer the correctly infected zebrafish embryos from the injection tray into an agarose-coated Petri dish. Remove the excess amount of embryo medium with a plastic pipette and

add fresh embryo medium. Add 0.003 % (w/v) PTU to prevent melanization and improve the quality of imaging with a fluorescence stereomicroscope.

11. Incubate the infected embryos at 28.5 °C or 31 °C until further examination.

3.5 Imaging the Infection

1. Use a fluorescence stereomicroscope to observe fluorescent *P. aeruginosa* cells in the embryo and to follow the development of infection over time. Depending on the fluorescent expression levels it might be very difficult to observe individual fluorescent *P. aeruginosa* bacteria by stereo fluorescence microscopy directly after injections. However, 1 day after infection fluorescent bacteria should be visible in the zebrafish embryo.
2. Before examination, anesthetize the zebrafish embryos with 200 μg/mL buffered Tricaine.
3. Align the embryos in the desired position for imaging using the fluorescence stereo microscope and a hair loop tool. A lateral view will provide the most information about the overall infection. Therefore from 1 dpf until 5 dpf, the embryos can best be positioned on their lateral side. After 5 dpf most embryos have developed an inflated swim bladder and will start to float.
4. To prevent floating, to minimize other movements after anesthesia, or to orient embryos in a different position than the lateral view, mount the embryos in 1.5 % methyl cellulose and use the hair loop tool (Fig. 2c) for embryo manipulation.
5. After live imaging of embryos gently remove them from the Petri dishes, wash them to remove anesthetics, collect them in a Petri dish containing PTU to prevent melanization and incubate them again at 28.5 or 31 °C for further development.
6. Use a confocal microscope to analyze infections in more detail and at higher magnifications.
7. When using an inverted confocal microscope, place a drop of 1 % (w/v) Low Melting Point agarose (Boehringer Mannheim) in egg water on the glass bottom of a microscopy chambers slide (Ibidi). If an upright confocal microscope is used place a drop of Low Melting Point agarose on a single cavity depression slide (Agar Scientific, L4090).
8. Transfer the anesthetized embryos in the agarose using a minimum of embryo medium. If an inverted microscope is used for confocal imaging, mount the embryos with the region of interest flat on the bottom of the microscopy chambers, using the hair loop tool (Fig. 2c). An upright microscope can be used in combination with water immersion or long distance dry objectives

and the embryos should be positioned such that the region of interest is as close to the objective as possible.

9. Let the agarose solidify and submerge the agarose drop in egg water containing Tricaine. If an upright microscope is used, place a glass cover slip on top of the depression cavity (do not allow air bubbles to form).
10. Sequentially acquire fluorescent and transmitted light images with the Leica confocal imaging suite software.
11. After live imaging of embryos gently remove them from the agarose with fine tweezers, wash with embryo medium to remove anesthetics, and collect them in a Petri dish containing PTU to prevent melanization. Incubate the embryos again at 28.5 or 31 °C for further development and further experiments.

3.6 Whole Mount Immunostaining of Infected Embryos

1. Whole mount immune staining is used as alternative, or additive, to transgenic zebrafish lines, to visualize the interaction of bacteria with particular tissues or cells of the embryos.
2. After live analysis, fixate the embryos in 4 % (v/v) Paraformaldehyde dissolved in PBS overnight at 4 °C.
3. Rinse the embryos 5 times for 10 min with PBTx.
4. Transfer the embryos via four dehydration washing steps of 10 min from 100 % (v/v) PBTx to 75 %:25 %–50 %:50 %–25 %: 75 % (v/v) PBTx:methanol and 100 % (v/v) methanol, and store them at –20 °C.
5. Rehydrate the embryos via five times 10 min washing steps back from 100 % (v/v) methanol to 100 % (v/v) PBTx.
6. Permeate the embryos in 0.24 % (w/v) trypsin in PBS for a maximum of 5 min and wash the embryos again 5 times for 10 min with PBTx.
7. Block the embryos for 3 h in block buffer (10 % (v/v) Normal Goat Serum (NGS) in PBTx) at room temperature (RT) or overnight at 4 °C.
8. Incubate the embryos with the first antibody overnight at RT in antibody buffer (PBTx containing 1 % (v/v) NGS and 1 % (v/v) BSA). We used for example anti-Acetylated Tubulin (Sigma, 1:250 dilution) to stain cells and axons in the brain and spinal cord (Fig. 1d, e).
9. After overnight incubation wash the embryos at minimum 10 times for 10 min with PBTx.
10. Block the embryos minimal for 1 h in block buffer (10 % (v/v) NGS in PBTx).
11. Incubate the embryos with the second antibody overnight at 4 °C in antibody buffer (PBTx containing 1 % (v/v) NGS and

1 % (v/v) BSA) in the dark to prevent quenching of the conjugated fluorophore. For the acetylated tubulin staining we used Alexa 488 goat-anti-mouse as secondary antibody (Invitrogen, 1:200 solution).

12. Wash the embryos again at minimum 10 times with PBTx and store them in PBS at 4 °C, or use them directly for imaging (*see* protocol 3.5).

4 Notes

1. There are different fluorescence protein expressing vectors suitable for *P. aeruginosa* [17, 19, 23]. Given the recent progress in zebrafish genetics with regard to green (eGFP) or red (mCherry) or double fluorescent protein expressing transgenic lines, labelling the bacteria with a fluorescent protein that can be distinguished from green and red is advisable for multicolor imaging of host–pathogen interactions (Fig. 1d).
2. In most of the protocols that concern working with live zebrafish embryos, the chemicals PTU and tricaine are used. It is important to realize the chemical hazards associated with these chemicals. For up to date details on the associated chemical hazards and the proper way of waste processing of chemicals, one should always consult the manufacturer's safety sheets.
3. The time of incubation at 28 °C after fertilization gives only an approximation of the developmental age of zebrafish embryos. For this reason, staging series based on morphologic criteria can be used to precisely determine the age of the developing embryos. This is particularly important for the reproducibility of experiments performed on different days and in different laboratories. A common criterion for developmental staging is to assess the head-trunk angle of the embryo and is well explained by Kimmel et al. [22] and Westerfield [24]. At 28 hpf, the developmental stage that is used for the injections, the head-trunk angle is for example about 102–105°.

Acknowledgement

This work was supported by an EU Marie Curie CIG grant (3038130) and a Ramon&Cajal grant (RYC-2011-08874) from the Spanish Ministry of Economy. In addition support was obtained from the Smart Mix Programme of the Netherlands Ministry of Economic Affairs and the Netherlands Ministry of Education, Culture and Science.

References

1. Lemaitre B, Reichhart JM, Hoffmann JA (1997) *Drosophila* host defense: differential induction of antimicrobial peptide genes after infection by various classes of microorganisms. Proc Natl Acad Sci U S A 94:14614–14619
2. Ewbank JJ (2002) Tackling both sides of the host-pathogen equation with *Caenorhabditis elegans*. Microbes Infect 4:247–256
3. Tan MW, Mahajan-Miklos S, Ausubel FM (1999) Killing of *Caenorhabditis elegans* by *Pseudomonas aeruginosa* used to model mammalian bacterial pathogenesis. Proc Natl Acad Sci U S A 96:715–720
4. Hilbi H, Weber SS, Ragaz C, Nyfeler Y, Urwyler S (2007) Environmental predators as models for bacterial pathogenesis. Environ Microbiol 9:563–575
5. van der Sar AM, Appelmelk BJ, Vandenbroucke-Grauls CM, Bitter W (2004) A star with stripes: zebrafish as an infection model. Trends Microbiol 12:451–457
6. Meijer AH, Spaink HP (2011) Host-pathogen interactions made transparent with the zebrafish model. Curr Drug Targets 12:1000–1017
7. Phelps HA, Neely MN (2005) Evolution of the zebrafish model: from development to immunity and infectious disease. Zebrafish 2:87–103
8. Kanther M, Rawls JF (2010) Host-microbe interactions in the developing zebrafish. Curr Opin Immunol 22:10–19
9. Traver D, Herbomel P, Patton EE, Murphey RD, Yoder JA et al (2003) The zebrafish as a model organism to study development of the immune system. Adv Immunol 81:253–330
10. Moens CB, Donn TM, Wolf-Saxon ER, Ma TP (2008) Reverse genetics in zebrafish by TILLING. Brief Funct Genomic Proteomic 7:454–459
11. Wienholds E, van Eeden F, Kosters M, Mudde J, Plasterk RH et al (2003) Efficient target-selected mutagenesis in zebrafish. Genome Res 13:2700–2707
12. Bowman TV, Zon LI (2010) Swimming into the future of drug discovery: in vivo chemical screens in zebrafish. ACS Chem Biol 5:159–161
13. Cui C, Benard EL, Kanwal Z, Stockhammer OW, van der Vaart M et al (2011) Infectious disease modeling and innate immune function in zebrafish embryos. Methods Cell Biol 105:273–308
14. Benard EL, van der Sar AM, Ellett F, Lieschke GJ, Spaink HP et al (2012) Infection of zebrafish embryos with intracellular bacterial pathogens. J Vis Exp
15. Carvalho R, de Sonneville J, Stockhammer OW, Savage ND, Veneman WJ et al (2011) A high-throughput screen for tuberculosis progression. PLoS One 6:e16779
16. Alibaud L, Rombouts Y, Trivelli X, Burguiere A, Cirillo SL et al (2011) A *Mycobacterium marinum* TesA mutant defective for major cell wall-associated lipids is highly attenuated in *Dictyostelium discoideum* and zebrafish embryos. Mol Microbiol 80:919–934
17. Brannon MK, Davis JM, Mathias JR, Hall CJ, Emerson JC et al (2009) *Pseudomonas aeruginosa* Type III secretion system interacts with phagocytes to modulate systemic infection of zebrafish embryos. Cell Microbiol 11: 755–768
18. Clatworthy AE, Lee JS, Leibman M, Kostun Z, Davidson AJ et al (2009) *Pseudomonas aeruginosa* infection of zebrafish involves both host and pathogen determinants. Infect Immun 77:1293–1303
19. Llamas MA, van der Sar A, Chu BC, Sparrius M, Vogel HJ et al (2009) A novel extracytoplasmic function (ECF) sigma factor regulates virulence in *Pseudomonas aeruginosa*. PLoS Pathog 5:e1000572
20. van der Sar AM, Musters RJ, van Eeden FJ, Appelmelk BJ, Vandenbroucke-Grauls CM et al (2003) Zebrafish embryos as a model host for the real time analysis of *Salmonella typhimurium* infections. Cell Microbiol 5:601–611
21. Cosma CL, Swaim LE, Volkman H, Ramakrishnan L, Davis JM (2006) Zebrafish and frog models of *Mycobacterium marinum* infection. Curr Protoc Microbiol Chapter 10:10B.12.11–10B.12.33
22. Kimmel CB, Ballard WW, Kimmel SR, Ullmann B, Schilling TF (1995) Stages of embryonic development of the zebrafish. Dev Dyn 203:253–310
23. Singer JT, Phennicie RT, Sullivan MJ, Porter LA, Shaffer VJ et al (2010) Broad-host-range plasmids for red fluorescent protein labeling of gram-negative bacteria for use in the zebrafish model system. Appl Environ Microbiol 76:3467–3474
24. Westerfield M (2000) The zebrafish book. A guide for the laboratory use of zebrafish (*Danio rerio*), 4th edn. University of Oregon Press, Eugene

Chapter 56

Assessing *Pseudomonas* Virulence with a Nonmammalian Host: *Drosophila melanogaster*

Samantha Haller, Stefanie Limmer, and Dominique Ferrandon

Abstract

Drosophila melanogaster flies represent an interesting model to study host–pathogen interactions as: (1) they are cheap and easy to raise rapidly and do not bring up ethical issues, (2) available genetic tools are highly sophisticated, for instance allowing tissue-specific alteration of gene expression, e.g., of immune genes, (3) they have a relatively complex organization, with distinct digestive tract and body cavity in which local or systemic infections, respectively, take place, (4) a medium throughput can be achieved in genetic screens, for instance looking for *Pseudomonas aeruginosa* mutants with altered virulence. We present here the techniques used to investigate host–pathogen relationships, namely the two major models of infections as well as the relevant parameters used to monitor the infection (survival, bacterial titer, induction of host immune response).

Key words Survival assay, Bacterial titer, Host–pathogen interactions, Genetic screens for virulence factors, *Drosophila* septic injury model, *Drosophila* oral infection model

1 Introduction

A major limitation of vertebrate models is that it is difficult to screen bacterial mutant libraries for alterations in the virulence program of the pathogen. This is due to the high number of animals required to perform such screens, which are ethically questionable, costly, labor- and time-intensive. Genetic model organisms such as *Caenorhabditis elegans* and *Drosophila melanogaster* provide interesting alternatives to screen for virulence factors. Indeed, while there may be "private" virulence factors required exclusively for pathogenesis in a given model, "public" virulence factors can be identified in screens involving invertebrate or plant organisms [1–4]. To be meaningful, such screens must be performed in model systems in which the immune system is well-characterized, and *Drosophila* certainly qualifies on this account [5]. Its immune system encompasses both a humoral and a cellular response that deal with systemic infections, as well as local responses that deal with invasions

Alain Filloux and Juan-Luis Ramos (eds.), *Pseudomonas Methods and Protocols*, Methods in Molecular Biology, vol. 1149, DOI 10.1007/978-1-4939-0473-0_56, © Springer Science+Business Media New York 2014

through frontier epithelia such as that of the intestinal tract [6–8]. In addition, *Drosophila* benefits from over a century of research, which results in highly sophisticated genetic tools as well as knowledge on most major biological questions, from chromosomal structure to ecology and evolution.

Several infection models have been developed using *Pseudomonas aeruginosa* as a pathogen [9, 10]. They can be regrouped into two major categories: septic injury and oral infection models [10]. Flies succumb rapidly, within 48–72 h, to the direct introduction of *Pseudomonas* into the body cavity [11–14]. This fast infection kinetics makes it difficult to screen for virulence factors having moderate effects as the difference between survival curves between flies infected with wild-type versus mutant bacteria represents usually just a few hours. In contrast, orally infected flies succumb in 5–10 days depending on the conditions [4]. While some investigators have reported damages to the intestinal epithelium, mostly in a sensitized background (*Ras*^{V12} mutants) [15], we have failed to detect any major impact on the gut epithelium in wild-type flies [4]. Rather, we have found that some PA14 bacteria are able to escape the digestive tract and ultimately launch a systemic infection that defeats both the cellular and systemic immune responses. Of note, the use of specific conditions in which the fly immune response has been inactivated (mutants, inhibition of phagocytosis) has led to the discovery that the quorum-sensing regulator RhlR is required for PA14 to initially circumvent the cellular immune defense [4].

Thus, when performing infections, several parameters can be easily monitored, including the survival rate of flies, the bacterial titer inside the fly or tissues/compartments such as gut and hemolymph, and the activation of host antimicrobial defenses such as the induction of antimicrobial peptide genes, which is the hallmark of the systemic immune response.

Here, we present the techniques we use to probe the interactions between *P. aeruginosa* and *D. melanogaster*. These include two variations of the septic injury model (pricking and injection) and an oral infection model. Wild-type and mutant flies can be used for such assays. One way to inhibit the cellular immune response is through the saturation of the phagocytic apparatus through the injection of nondegradable latex beads. Other techniques include the determination of the bacterial titer in specific tissues or in the hemolymph. To monitor the induction of the systemic immune response, the expression levels of antimicrobial peptide genes is measured using quantitative reverse-transcription PCR, a procedure that can be performed in 96-well plates or even 384-well plates. Of note, general information on *Drosophila* can be found in several books [16, 17] as well as on the Flybase Website: http://flybase.org/.

2 Materials

2.1 Drosophila Husbandry and Culture of Bacteria

1. *Drosophila* stocks: they are kept under standardized conditions in an air-conditioned room or incubator at 25 °C and 60 % humidity. For long-term storage, stocks are kept at 18 °C. The health status of the flies has to be checked on a regular basis: tests for the presence of microsporidia, viruses, or symbionts such as *Wolbachia* are essential, since such infections can interfere with the experimental infection (*see* **Note 1**, Table 1). Common wild-type stocks are Oregon-R and Canton-S. Often, flies carrying the *white* mutation are also used as a wild-type strain. Reporter fly lines for the expression of antimicrobial peptide genes such as *Diptericin*-Tomato, *Drosomycin*-GFP, and *Diptericin*-LacZ (*see* **Note 2**) can be used to assess the spatial and temporal expression of antimicrobial peptides in the

Table 1
Primers used for the detection of microbial contamination of *Drosophila* stocks

Pathogen	Test	Primer sequences	
Wolbachia-1	Regular PCR	Fw Rv	TTGTAGCCTGCTATGGTATAACT GAATAGGTATGATTTTCATGT
Wolbachia-2	Regular PCR	Fw Rv	AAAAATTAAACGCTACTCCA TGGTCCAATAAGTGATGAAGAAAC
Nora virus	Quantitative PCR	Fw Rv	AACCTCGTAGCAATCCTCTCAAG TTCTTGTCCGGTGTATCCTGTATC
Drosophila C virus	Quantitative PCR	Fw Rv	TCATCGGTATGCACATTGCT CGCATAACCATGCTCTTCTG
Flock house virus-1	Quantitative PCR	Fw Rv	TTTAGAGCACATGCGTCCAG CGCTCACTTTCTTCGGGTTA
Flock house virus-2	Quantitative PCR	Fw Rv	CAACGTCGAACTTGATGCAG GCTTTACAGGGCATTTCCAA
Vesicular stomatitis virus	Quantitative PCR	Fw Rv	CATGATCCTGCTCTTCGTCA TGCAAGCCCGGTATCTTATC
Sindbis virus	Quantitative PCR	Fw Rv	CAAATGTGCCACAGATACCG ATACCCTGCCCTTTCAACAA
Cricket paralysis virus-1	Quantitative PCR	Fw Rv	GCTGAAACGTTCAACGCATA CCACTTGCTCCATTTGGTTT
Cricket paralysis virus-2	Quantitative PCR	Fw Rv	GGAATTTTTGGAGACGCAAA GTGAAGGGGGCAACTACAAA
Microsporidia (*Tubulinosema ratisbonensis*)	Quantitative PCR	Fw Rv	TCTCACAGTAGTGGCGAATG AACACCGTATTGGAATACAG

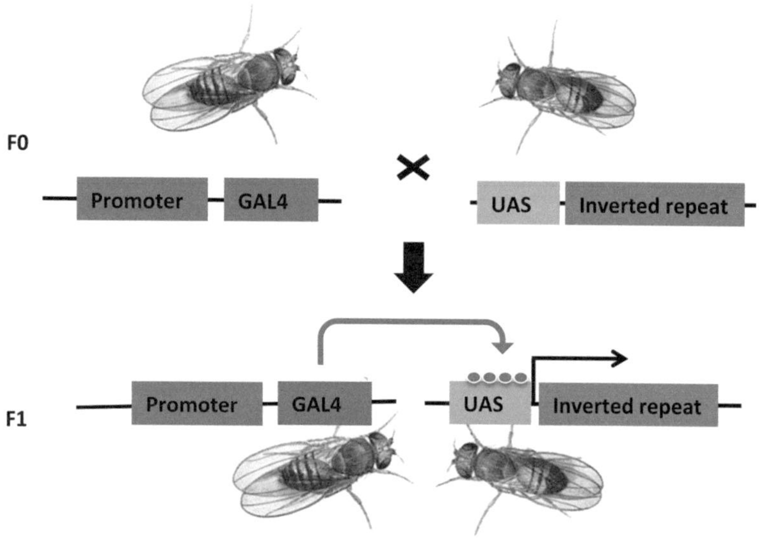

Fig. 1 The GAL4-UAS system in *Drosophila*. This system has been adapted from yeast genetics to allow the expression of any construct in a tissue-specific or cell type-specific manner. Transgenic flies containing a promoter of interest that drives the expression of the yeast transcriptional activator GAL4 are crossed to a transgenic line containing a construct that has been placed under the control of UAS enhancer sequences linked to a basal promoter. In progeny flies containing both transgenes, GAL4 will be expressed in the required tissue or cell-type and drive the expression of the construct

fly (systemic expression vs. local expression). When a gene needs to be expressed or inactivated in a given tissue, the UAS-GAL4 system is used (Fig. 1) [18]. *Drosophila* lines carrying the tissue-specific promoter of interest fused to the coding sequence of the yeast transcription activator GAL4 can either be constructed or obtained from other *Drosophila* investigators or stock centers (Vienna *Drosophila* RNAi Center: *stockcenter.vdrc.at*, Bloomington *Drosophila* Stock Center at Indiana University: *flystocks.bio.indiana.edu*, *Drosophila* Genetic Resource Center at Kyoto: *kyotofly.kit.jp*). *Drosophila* lines carrying a specific DNA sequence fused to an Upstream Activation Sequence (UAS, target sequence of GAL4) for over expression of a gene, a marker, or a hairpin construct for RNA interference mediated gene knock-down can similarly be obtained. The timing of transgene expression can be controlled by using a transgene encoding a temperature sensitive repressor of GAL4:GAL80^{ts} (*see* **Note 3**) [19].

2. *Drosophila* feed: flies are fed on a standard semi-solid cornmeal medium composed of: 6.4 % (w/v) cornmeal (Moulin des moines, France), 4.8 % (w/v) granulated sugar (Erstein, France), 1.2 % (w/v) yeast brewer's dry powder (VWR, Belgium)

(*see* **Note 4**), 0.5 % (w/v) agar (Sobigel, France), 0.004 % (w/v) 4-hydroxybenzoate sodium salt (Merck, Germany) (*see* **Note 5**). The medium is poured in plastic vials after cooking: small (Ø 25 mm), medium (Ø 32 mm), and large (Ø 50 mm). Stoppers of the adequate size are placed once the medium has cooled down. Store at 4 °C.

3. *Preparation of latex bead suspension for saturating the phagocytic apparatus*: wash the solution of beads (Invitrogen, Surfactant free red CML latex, Ø 0.30 μm, C29145) with sterile PBS 1×: centrifuge beads at 20,000×*g* for 10 min at 4 °C. Discard the supernatant and add fresh sterile PBS 1×. Repeat wash three times. After the last centrifugation, concentrate the latex bead solution fourfold by adding only a quarter of the initial volume of sterile PBS 1×.
4. *Injections (latex beads suspension)*: Use a Nanoinjector (Nanoject II ™, Drummond Scientific Company) and appropriate capillaries (3.5″ capillary, 3-000-203-G/X, Drummond Scientific Company). Injection capillaries are prepared using a needle puller that heats and subsequently pulls the capillary until it breaks in the middle (e.g., Model P-97, Flaming/Brown Micropipette puller, Sutter Instrument Co.).
5. *Bacterial culture medium (liquid)*: Brain Heart Broth (BHB; Sigma-Aldrich): Beef heart, 5 g/L, calf brains 12.5 g/L, disodium hydrogen phosphate 2.5 g/L, d(+)-glucose 2 g/L, peptone 10 g/L, sodium chloride 5 g/L. Weigh 37 g of BHB powder and dissolve in 1 L distilled water. Adjust the final pH to 7.4 ± 0.2. Sterilize the solution by autoclaving at 121 °C for 15 min. Store the prepared medium below 8 °C.
6. *Bacterial plates*: Luria Bertani broth (LB, Sigma-Aldrich) with 15 % agar in Petri dish (*see* **Note 6**) with appropriate antibiotics (*see* **Note 7**). Weigh 10 g of LB powder, 7.5 g of agar powder and add 500 mL distilled water qsp. Sterilize the solution by autoclaving at 121 °C for 15 min. Once the temperature of the medium is below 60 °C, add the appropriate antibiotics and pour the medium in Petri dishes under a laminar flow cabinet. When the medium is solid, store plates below 8 °C.
7. *Dissecting microscope*: able to enlarge until 35-fold and cold light source.
8. *Material to anesthetize the flies*: Inject+Matic Sleeper (CO_2 blow gun and porous pad) linked to a CO_2 bottle or similar (*see* http://flystocks.bio.indiana.edu/Fly_Work/supplies.htm for a list of suppliers of *Drosophila* research material) (*see* **Note 8**).
9. *Tabletop centrifuge*.
10. *Pure distilled sterile water*: for instance, Advantage A10 and RiOs™ systems, Millipore, or equivalent.

2.2 Systemic Infection

1. *Fly vials* with normal fly food and appropriate stoppers.
2. *1 M Sodium hydroxide solution* (NaOH 1 M). Weigh 4 g NaOH and add pure distilled sterile water qsp 100 mL.
3. *Tungsten needle* (Ø 0.25 mm, Sigma-Aldrich) and the material to sharpen the needle (NaOH 1 M and a low voltage generator) for an infection by pricking the flies (*see* **Note 9**, Fig. 2a). For injections of bacteria, a Nanoinjector and appropriate capillaries (*see* **item 4** of Subheading 2.1) are needed.
4. *Phosphate Buffer Saline* (*PBS 1×*): 1.06 mM potassium phosphate monobasic (KH_2PO_4), 155.17 mM sodium chloride (NaCl), 2.97 mM sodium phosphate dibasic (Na_2HPO_4–$7H_2O$). Weigh 144 mg of KH_2PO_4, 9 g of NaCl, and 795 mg of Na_2HPO_4–$7H_2O$. Add pure distilled sterile water qsp 1 L.
5. *Ethanol 70%*: Measure 700 mL pure ethanol and add pure distilled sterile water qsp 1 L.

2.3 Oral Infection

1. Medium plastic vials (Ø 32 mm, Greiner Bio-One) and appropriate stoppers.
2. *Absorbent filter*: of a size appropriate to cover the bottom of the medium plastic vial (e.g., AP1003700, Millipore).
3. *50 mM sucrose solution*: Weigh 3.42 g sucrose and add distilled water 200 mL qsp. Sterilize the solution by sterile filtration using a Ø 0.22 μm pore filter. Store at 4 °C.

2.4 Hemolymph Collection

1. *Microtubes* to collect the hemolymph: 1.5 or 900 μL volume Eppendorf or equivalent.
2. *Phosphate Buffer Saline*: *See* **item 4** of Subheading 2.2.
3. *LB agar Petri dish*: With appropriate antibiotics (*see* **Note 6**).
4. *Nanoinjector and appropriate capillaries*: *See* **item 4** of Subheading 2.1.

2.5 Analysis of Antimicrobial Peptide Gene Expression in Multiple-Well Plates

1. *Tubes to crush the flies*: specific tubes (rack with microtube strips and cap strips, Macherey-Nagel) each one containing a tungsten bead (Tungsten Carbide Beads, 3 mm diameter, Qiagen) are needed (*see* **Note 10**).
2. *Crusher* (Mixer Mill 300 MM, Retsch).
3. *Reverse transcription enzyme and reagents*: iScript™ cDNA Synthesis Kit, Bio-Rad, or equivalent.
4. *PCR tubes.*
5. *Thermocycler*: C1000™ Thermal Cycler, Bio-Rad or equivalent (*see* **Note 11**).
6. *Forward and reverse primers*: designed to detect the antimicrobial peptide mRNA sequence (*see* Table 2) at concentration of 10 μM.

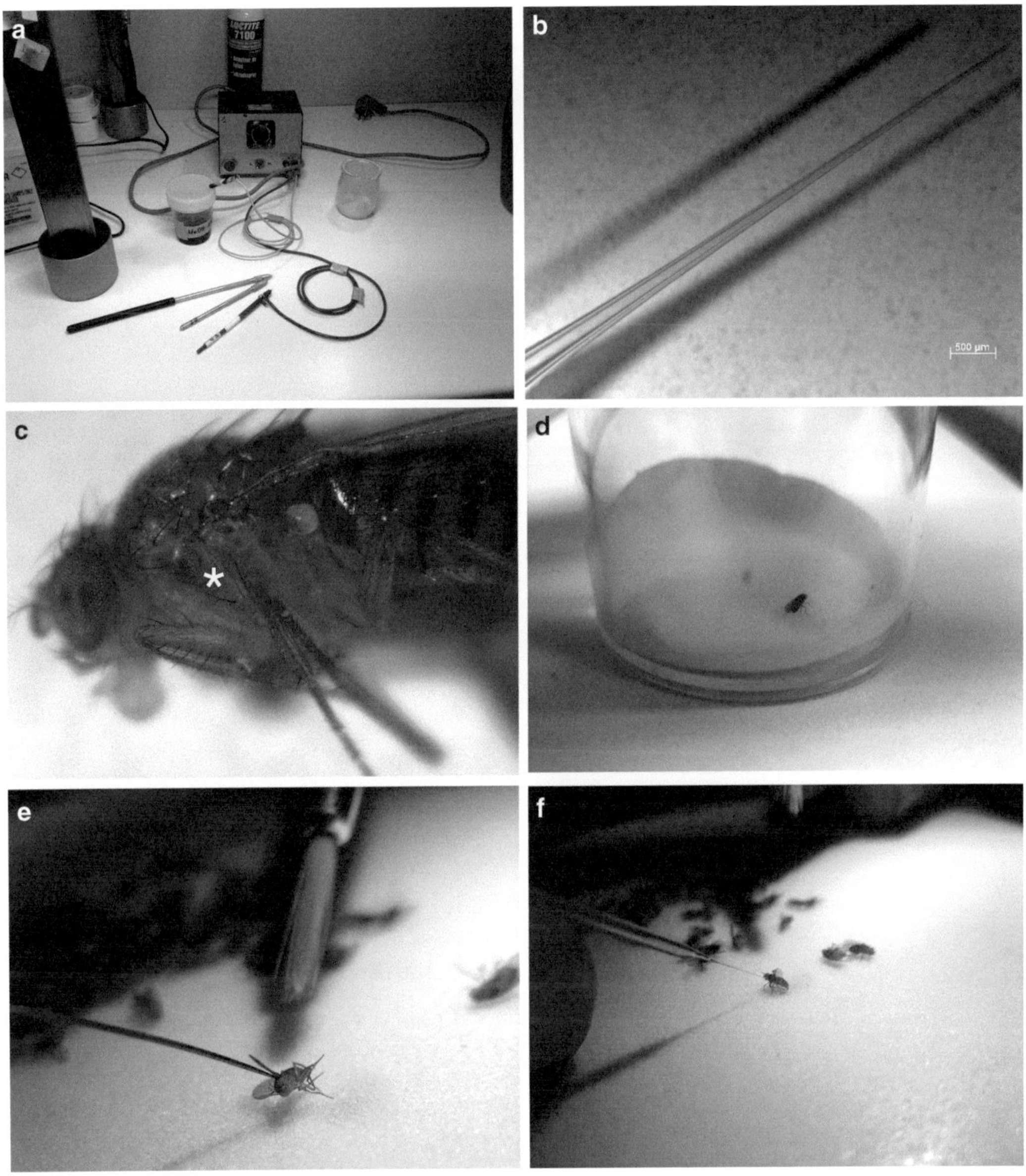

Fig. 2 Septic injury of *Drosophila*. (**a**): Setup to sharpen tungsten needles. The tungsten wire is inserted in a needle-holder. Three types of needle-holders are presented on the *left foreground* of the picture. The *middle needle-holder* has been obtained by sawing off the black plastic handle of the *left-most needle-holder*, as it is too long to be convenient. The power supply is in the background. A wire from the positive pole is connected to the *right-most needle-holder*: the wire is no longer protected by its cover and is in direct contact with the metal of the holder. The graphite electrode connected to the negative pole of the generator has been placed on *top* of the NaOH solution vessel. (**b**): Elongated capillary for injecting bacterial suspensions. The tip has been broken off with sharp tweezers. (**c**) The position where the fly should be injected is shown with an *asterisk*. Inject just above the *asterisk*, at a location where the cuticle is somewhat weaker. While this is not of utmost importance for pricking or injection, it makes the collection of hemolymph easier. (**d**) Oral infection setup. It is strongly advised to use 20 flies so that conditions remain the same when comparing different bacterial strains. (**e**) Pricked fly. When correctly pricked, the fly is held by the needle and then dropped off either on one side or directly in a vial containing food using either a brush or tweezers (need to be disinfected when changing bacterial strains). (**f**) Injected fly. Same remark as for pricked fly

Table 2
Primers used for the quantification of *Drosophila* antimicrobial peptide gene expressions

Antimicrobial peptide	Primer	
Ribosomal protein L32 (RP49)	Fw Rv	GACGCTTCAAGGGACAGTATCTG AAACGCGGTTCTGCATGAG
Drosomycin (Drs)	Fw Rv	CGTGAGAACCTTTTCCAATATGATG TCCCAGGACCACCAGCAT
Diptericin (Dpt)	Fw Rv	GCTGCGCAATCGCTTCTACT TGGTGGAGTGGGCTTCATG
Attacin-A (Att-A)	Fw Rv	GGCCCATGCCAATTTATTCA AGCAAAGACCTTGGCATCCA
Defensin (Def)	Fw Rv	GCTCAGCCAGTTTCCGATGT TCCTGGTGGGCATCCTCAT
Cecropin (Cec)	Fw Rv	ACGCGTTGGTCAGCACACT ACATTGGCGGCTTGTTGAG
Metchnikowin (Mtk)	Fw Rv	CGTCACCAGGGACCCATTT CCGGTCTTGGTTGGTTAGGA
Drosocin (Dro)	Fw Rv	CACCCATGGCAAAAACGC TGAAGTTCACCATCGTTTTCCTG

7. *Quantitative PCR reagent and enzyme*: iQ™ Sybr® Green, Bio-Rad or equivalent.
8. *384-Well plates and cover*: specific seals to close the plate that allow fluorescence detection (*see* **Note 12**).
9. *Thermocycler with a fluorescence detector*: C1000™ Thermal Cycler with CFX384 Real-Time System, Bio-Rad and the analyzer program (CFX Manager Software, Bio-Rad) or equivalent. 96-Well plates can also be used.

3 Methods

Carry out all procedures in a P2 level laboratory (except the extraction of nucleic acids from frozen flies). As *P. aeruginosa* is an opportunistic human pathogen, the experimenter must have a functional immune system, which provides already a significant barrier against infection. Always wear a laboratory coat and gloves, and work under a microbial hood when handling the bacteria, except when injecting or retrieving hemolymph from flies. If needed, a better protection against accidentally pricking one's hand with a contaminated needle can be provided by wearing a special glove (which contains quaternary amines) on the noninjecting hand (Gant BioPro, MAP203-7, MAPA professional).

3.1 Preparation of the Flies Prior to the Infection

1. If the virulence of the bacteria needs to be tested in flies in which a gene is ectopically expressed or inactivated in a given tissue, a cross needs to be performed between two specific *Drosophila* lines (Fig. 1). Collect virgin female flies from the line carrying the regulatory sequence fused to the GAL4 sequence and male flies from the line carrying the desired coding sequence under the control of the UAS sequence (or the reverse). Deposit the males and females of the appropriate genotypes in a tube containing normal fly food, which can be supplemented with a dash of live yeast. When the cross is performed at 25 °C, transfer the parents in a new tube every 3 days and when the cross is performed at 18 °C, transfer them every 7 days.
2. Collect the emerging progeny. Anesthetize these flies with a CO_2 blow gun and sort the flies carrying the genotype of interest on a CO_2 porous pad and with the help of a dissecting microscope.
3. For an oral infection only: prepare medium size tubes with two absorbent filters and 2 mL of 50 mM sucrose solution. Sort the flies that will be infected and keep them for 48 h in these tubes at 25 °C prior starting the experiment.
4. If the phagocytic ability of the flies needs to be blocked, inject latex beads in these flies 24 h before performing the infection experiment so as to allow wound closure and sufficient time for the flies to recover from the injection procedure [20]. To this end, prepare capillaries with the puller as described in **item 4** of Subheading 2.1. Break the capillary tip with tweezers using a dissecting microscope (Fig. 2b). Fill the capillary with mineral oil and assemble it onto the Nanoinjector. Discard half of the volume of oil and fill the capillary with latex beads solution (fourfold concentrated, *see* **item 3** of Subheading 2.1). Inject each fly with 92 nL latex beads suspension in the lateral part of the thorax, slightly below and anterior to the wing hinge (*see* **Note 13**; Fig. 2c, f).

3.2 Culture of Bacteria

1. Grow the bacteria on an LB agar Petri dish with the appropriate antibiotics at 37 °C.
2. Inoculate the BHB culture medium with a single colony (for instance, 10 mL BHB in a 100 mL Erlenmeyer flask). Grow the bacteria overnight at 37 °C and under agitation.

3.3 Systemic Infection Model

Flies can be infected either by pricking (septic injury) or injection (*see* **Note 14**).

Flies should be 3–10 days old when the infection starts.

3.3.1 Systemic Infection by Pricking the Flies

1. Measure the OD_{600} of the bacterial culture.
2. Adjust the OD_{600} to the appropriate concentration in sterile PBS 1× (*see* **Note 15**). Centrifuge the culture at 3,000 × *g*

(4,000 rpm on a A-4-62 rotor (5810R centrifuge, Eppendorf)) for 10 min at 4 °C in 50 mL Falcon tubes. Discard the supernatant and add the appropriate volume of sterile PBS 1×. Resuspend the bacteria by pipetting up and down.

3. Sharpen the tungsten needle by putting the negative graphite electrode in the 1 M NaOH solution and the positive electrode on the needle (Fig. 2a). Then plunge the needle into the 1 M NaOH solution and apply current until a sharp end is generated.
4. Dip the needle in the bacterial suspension.
5. Anesthetize the flies within the fly vial with a CO_2 blow gun and deposit them on a CO_2 porous pad.
6. Under a dissecting microscope, prick the flies one by one at the same location as for the injection (*see* **step 4** of Subheading 3.1; Fig. 2c, e) with the injection capillary. The tungsten needle has to be dipped again in the bacterial suspension after each fly has been pricked. Between each bacterial type (mutant or strain), sterilize the tungsten needle with ethanol 70 % and rinse with sterile water.
7. Transfer the infected flies in a fly vial containing normal fly food and place it at 25 °C with 60 % humidity.
8. For survival experiments, monitor the number of surviving flies every 1–2 h starting from 24 h after the infection. Do not take into account flies that have succumbed early to infection (before 18 h after infection). In addition, the bacterial titer in the hemolymph and the level of antimicrobial peptide expression can be measured (*see* below for protocols).

3.3.2 Systemic Infection by Injecting the Bacteria into the General Cavity (Hemocoel) of the Flies

1. Measure the OD_{600} of the bacterial culture.
2. Adjust the OD_{600} to the appropriate concentration in sterile PBS 1× (*see* **Note 15**). Centrifuge the culture at 3,000 × *g* for 10 min at 4 °C in 50 mL tubes (*see* section 3.3.2). Discard the supernatant and add the appropriate volume of sterile PBS 1×. Resuspend the bacteria by pipetting up and down.
3. Prepare capillaries with the puller as described in **item 4** of Subheading 2.1. Break the capillary tip with tweezers using a dissecting microscope. Fill the capillary with mineral oil and assemble it onto the Nanoinjector. Discard half of the volume of oil and fill the capillary with the bacterial suspension. Use a new capillary for each bacterial strain.
4. Anesthetize the flies within a fly vial with a CO_2 blow gun and deposit them on a CO_2 porous pad.
5. Under a dissecting microscope, inject into each fly one by one the appropriate volume of bacterial suspension in order to inject the desired number of bacteria (*see* **Note 16**).

6. Transfer the infected flies in a fly vial containing normal fly food and place it at 25 °C with 60 % humidity.
7. For survival experiments, monitor the number of surviving flies every 1–2 h starting 24 h after the injection. Do not take into account flies that have succumbed early to infection (before 18 h after infection). In addition, the bacterial titer in the hemolymph and the level of antimicrobial peptide expression can be measured (*see* below for protocols).

3.4 D. melanogaster Oral Infection

1. Measure the OD_{600} of the bacterial culture.
2. Adjust the OD_{600} to 2.5 per mL with fresh BHB: compute the appropriate volume of bacterial culture needed for the experiment (that is, the number of tubes needed to perform the experiment, 2 mL of bacterial suspension per tube). Centrifuge the culture at 3,000 × *g* for 10 min at 4 °C in 50 mL tubes. Discard the supernatant and add the appropriate volume of fresh BHB. Resuspend the bacteria by pipetting up and down.
3. Dilute the bacterial suspension 1/10 with 50 mM sucrose solution. The final OD_{600} of the infection solution is then 0.25 per mL (*see* **Note 17**).
4. Prepare the infection tubes by placing two absorbent filters at the bottom of the vials. Add 2 mL of the infection solution per tube.
5. Place 20 adult flies (3–10 days old) into each tube and put the vial at 25 °C with 60 % humidity.
6. For survival experiments, monitor the number of surviving flies at least each day. In addition, the bacterial titer in the hemolymph and the level of antimicrobial peptide expression can be measured (*see* below for protocols).

3.5 Determination of the Bacterial Titer from Collected Hemolymph

1. Prepare collecting tubes (e.g., 900 μL tubes) with 10 μL of sterile PBS 1×.
2. To manipulate the flies easily, anesthetize them with CO_2.
3. Prepare capillaries with the needle-puller as described in **item 4** of Subheading 2.1. In contrast to the injection procedure, *do not fill the capillary with oil before assembling it to the Nanoinjector.*
4. Prick the flies with the capillary into the lateral side of the thorax slightly in front and under the wing hinge, at a weak point of the cuticle (Fig. 2c). The hemolymph is drawn inside the capillary by capillarity force alone (*see* **Note 18**). Collect the hemolymph of at least ten flies, one by one (*see* **Note 19**). Use the Nanoinjector motor to force the hemolymph out of the capillary and into the collecting tube containing the sterile PBS 1× after each fly (*see* **Note 20**).

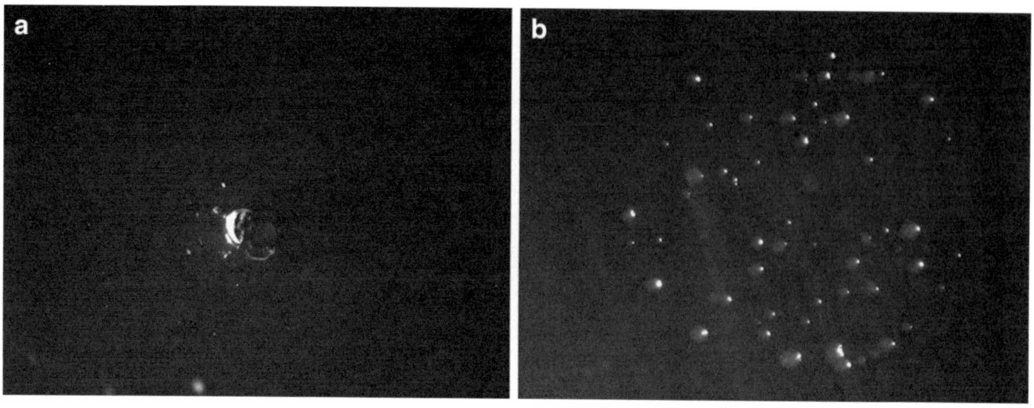

Fig. 3 Counting bacterial colonies in a drop to determine the bacterial titer. (**a**) 4 μL drop has been deposited on an agar plate containing appropriate antibiotics. Colonies are shown in (**b**), as observed under a dissecting microscope. This dilution is appropriate to count the colonies. In (**a**), no colonies have grown, and only the imprint of the tip and agar speckles can be seen. The advantage of this method is that it requires only few plates to determine the bacterial titer for many flies

5. Dilute the sample to 1/10, 1/100, 1/1,000, 1/10,000 in sterile PBS, depending for how long the flies have been fed on the bacteria (later time points tend to give higher bacterial titer).
6. Deposit two drops of 4 μL for each sample and dilutions on a single LB agar plate (*see* **Note 21**), with the appropriate antibiotics. Once the drops dried, place the plate at 37 °C for about 10 h (*see* **Note 22**) to allow bacteria to form colonies.
7. Count the number of colonies you observe for each drop using a dissecting microscope (Fig. 3). Calculate the number of colony forming units (CFU) per fly.

3.6 Analysis of Antimicrobial Peptide Gene Expression by Reverse-Transcription and Quantitative PCR (RT qPCR)

1. Freeze the flies that were used to assess the bacterial titer in the hemolymph at −80 °C (or other flies that have been infected if you are not counting CFU/fly in your experiment). Use specific tubes in which you have beforehand put a single tungsten bead in each (*see* Subheading 2).
2. Put the tubes in liquid nitrogen. Crush the flies by shaking two times for 60 s at 20 Hz using the Mixer Mill apparatus. Put the tubes in liquid nitrogen again in-between the two crushing steps. Extract total RNA with an adequate kit following the procedure recommended by the supplier (e.g., NucleoSpin 96 RNA, Macherey-Nagel). The lysis solution is added directly to the pulverized flies. Purified RNA should be kept at −80 °C.
3. Reverse transcribe the RNA into cDNA using the iScript™ cDNA Synthesis kit (BIO-RAD): 2 μL template RNA, 4 μL iScript Mix, 1 μL iScript reverse transcription enzyme, and 13 μL sterile distilled nuclease-free water (*see* **Note 23**). Use the following

program: (25 min at 65 °C, 5 min at 25 °C, 30 min at 42 °C, 5 min at 85 °C).

4. Dilute the cDNA obtained in **step 3** to 1/20 with pure distilled sterile water.
5. Prepare the RT-PCR reagent solution with 5 μL of Sybr® Green Mix, 0.3 μL of each forward and reverse primer (10 μM) (against antimicrobial peptide mRNA or other mRNA of interest: Table 2), 2 μL of cDNA (diluted to 1/20 in **step 4**). Add pure distilled sterile water to the solution to obtain a final volume of 10 μL per reaction.
6. Place the sample inside a thermocycler with the following program: 15 s at 98 °C, and then 35 times: 2 s at 95 °C and 30 s at 60 °C. Ct numbers are obtained (*see* **Note 24**). The last step is to increase the temperature by 0.5 °C each 5 s from 65 to 95 °C to check that the appropriate amplicon has been amplified (*see* **Note 25**).
7. Normalize the obtained Ct with the measured Ct on the same sample of a *Drosophila* housekeeping gene like the Ribosomal Protein L32 (RP49) (Table 2).

4 Notes

1. The physiological state and genetic background of flies is an important consideration in these assays. For instance, co-infection between *P. aeruginosa* and other pathogens or unwanted symbionts (endosymbiotic bacteria such as *Wolbachia*, viruses, or others) are best avoided, since secondary infections can influence *P. aeruginosa* infections in unpredictable manners (for instance, it has been discovered that flies harboring *Wolbachia* are protected from some viral infections). On a regular basis, flies used for experiments have to be tested for known pathogens by qPCR. Flies used for experiments should always be in the same physiological state, especially the same age. Mutant and wild-type flies should as much as possible share the same genetic background to prevent irrelevant interpretations. Ideally, the flies should have the same microbiota in their intestine, a proposition difficult to implement in practice unless axenic flies or flies mono-associated with a single commensal bacterial species are used, which has not been reported so far.
2. These reporter fly lines carry the regulatory sequences of the antimicrobial peptide gene (e.g., *Diptericin* or *Drosomycin*) fused with the coding sequences of GFP or other fluorescent proteins such as Tomato (two linked RFPs) or mCherry. β-Galactosidase was used in earlier experiments [21–23].

3. The $GAL80^{ts}$ sequence is fused to the ubiquitous promoter of *Tubulin* and will be transcribed continuously in each cell of the transgenic fly. The GAL80 protein will bind to GAL4 and prevent its binding to UAS sequences. When placed at 29 °C, the $GAL80^{ts}$ repressor is no longer functional, thus allowing GAL4-mediated gene expression. Usually, flies are left at 29 °C for a few days to allow an optimal expression of the genes under UAS control.
4. It is preferable to use dried whole yeast, as yeast extract is not nutritive enough for the flies. Check with your supplier as we have found out that suppliers sometimes change to yeast extract without even warning their customers. Of course, use pesticide-free ingredients.
5. Flies feed and reproduce on the same medium. When larvae hatch, they will work out the medium and make it semi-liquid, if enough progeny is produced (social feeding). The adults may then get stuck in the medium and will therefore be lost. Thus, one has to transfer the adults to a new tube, every 5 days for a stock kept at 25 °C.
6. Round or square Petri dishes may be used. To test large numbers of bacteria, square Petri dishes allow the use of multichannel pipettes to increase the throughput of the procedure (especially to plate hemolymph samples).
7. Depending on the strain or mutant used the resistance against antibiotics can be different. The strain UCBPP-PA14 is naturally resistant to 100 μg/mL Rifampicin.
8. If a CO_2 setup is not available, an alternative is to anesthetize flies with ether using an etherizer. This procedure requires some practice, as it is easy to overexpose the flies to ether and kill them. It is best to use a first tester batch of flies: expose them to ether for 45 s to 1 min and monitor how long they remain asleep. With the right timing of exposure, the flies remain anesthetized up to 20 min. However, too long of an exposure kills the flies. If the flies start waking up while the experiment is still underway, flies can be reexposed, very briefly, to ether. Often, a second exposure is fatal because of overexposure.
9. We sharpen the needles using a low voltage generator, a graphite electrode (taken from a 4.5 V flat battery), and a solution of 1 M NaOH. Always use gloves when manipulating NaOH. The circuit is made by plunging the graphite electrode connected to the negative pole in the NaOH solution, connecting the positive pole to a wire, which is directly attached to the metallic part of the needle holder. By plunging the tip of the needle into the NaOH solution, the circuit is closed, and one can see metal bits falling to the bottom and the production of gas bubbles. A generator for electrophoresis may also be used,

although extreme care should be taken to make sure that the voltage is really set to 6 V, otherwise there is a death hazard with DC current.

10. Individual samples can be processed in 1.5 mL collecting tubes but the quality of these tubes has to be checked otherwise there is a risk that they break during the procedure (the tungsten bead exerts a considerable stress on the frozen tubes). We use original tubes manufactured by Eppendorf.
11. Bain-Marie or hot plates at the indicated temperatures can also be used but are less convenient.
12. Other formats can be used such as 96-well plates. The use of a robot to set up the reactions in a 384-well plate is recommended.
13. We advise to check the efficiency of this treatment. One must simply inject FITC-labeled *Escherichia coli* (or any other bacterium or fungus small enough to be phagocytosed, e.g., Molecular Probes bioparticles) in latex beads-injected flies (and noninjected controls), wait 30 min, and then inject about 300 nL of Trypan blue [20], which quenches the fluorescence emitted by extracellular bacteria but not that emitted by ingested bacteria. No fluorescence should be detected under the microscope in latex beads-injected flies. An alternative is to use pH Rhodo-labeled bacteria [24], which become fluorescent when reaching the acidified phagosome (no need for Trypan blue injection).
14. The injection system compared to the pricking system allows being more accurate with respect to the number of bacteria that will effectively be introduced into the hemocoel of the flies. However, the injection system is much more time-consuming and cumbersome than the pricking system especially if high number of bacterial strains need to be tested. In addition, there is still some variability as the injection of a small volume of liquid (2.4 nL) was found to entail a 100 % variability in the quantities effectively delivered when using a radioactive solution (Marie Gottar and DF, unpublished). In practice, the bacterial dose delivered by pricking is reproducible and accurate enough for most applications.
15. The adequate concentration of bacterial suspension should be empirically determined according to the expected phenotype, that is, the time it takes for most flies to die. If the solution is too concentrated, the difference between wild-type and mutant bacterial strains may not be easily detectable.
16. Injecting a higher volume may allow to be more precise in the effective volume injected (*see* **Note 13**). Of note, the estimated volume of *Drosophila* hemolymph is 100 nL. Thus, injections

of volumes higher than 10 nL is not advised, except for specific procedures such as the injection of latex beads, in which a significant recovery period is allowed before infections.

17. Adding new fresh BHB medium is highly important for the expression of bacterial virulence, especially for the infection of wild-type flies. Without adding fresh BHB medium, wild-type PA14 are not able to kill wild-type flies but are able to kill flies with an impaired cellular immunity [4]. Using only a sucrose solution and a tenfold higher bacterial concentration as well as a 3-h starvation period of males, Mulcahy et al. have observed the *Pel*-dependent formation of biofilm in the crop, a food-storage diverticulum [25]. Of note, the *Pel* mutants we have tested in the assay we describe here did not display an altered virulence as monitored in survival assays [4]. This observation underscores the importance of experimental conditions in these assays.
18. Do NOT use the Nanoject motor to draw hemolymph. Doing so results in collecting tissue fragments and thus alters the measure as bacteria may stick to the tissues. It is not possible to crush the flies and determine the titer, as can be done in the septic injury assay, since flies contain also bacteria in their digestive tract. An alternative is to dissect the gut and the rest of the fly (and plate both separately after homogenization). In this case, the fly must first be surface-sterilized prior to dissection by dipping it for 5 s into 70 % ethanol. When using this procedure, the number of bacterial colonies retrieved is higher than the hemolymph titer by a log.
19. The higher the number of flies that are used, the closer the number of CFU obtained is representative of the number of bacteria inside the flies. Be careful not to collect tissue fragments, which would lead to altered bacterial counts. Of note, it is much easier to collect hemolymph out of females than of males, as they are bigger.
20. For beginners, it is easier to deposit the collected hemolymph into a 10 μL sterile PBS 1× drop placed on a Parafilm strip. This drop can then be transferred to a collecting tube with a micropipette. In this way, the risk of breaking the needle is alleviated.
21. Be careful to dilute the hemolymph preparation enough because the 4 μL drops will not be spread all over the plate and CFUs have to be nonconfluent to be counted.
22. 10 h incubation should be used. If the incubation time is longer, the colonies will grow too much and reach confluence. If the incubation time is shorter, the colonies will be too small to be observed.
23. The quality of the water coming from the Millipore filtration system is pure enough for that step. The water used should not contain RNases.

24. SYBR® Green is a molecule that fluoresces strongly when bound to double-stranded DNA. The Ct number represent the number of cycles needed to reach a fluorescence signal that is higher than the background. The Ct number is inversely proportional to the initial amount of mRNA copies in the sample.
25. This last step allows the measurement of the melting curve of the amplified product. The melting curve gives an indication of the purity of the amplified product.

Acknowledgements

We thank Marie-Céline Lafarge for expert technical help, Pr. Jean-Luc Imler for the sequences of the viral PCR primers and Dr. Sebastian Niehus for critical reading of the manuscript. Work in the group is funded by CNRS, Agence Nationale de la Recherche, a National Institute of Health PO1 AI085581 grant (coordinator: Pr. F. Ausubel), and Fondation pour la Recherche Médicale (Equipe FRM).

References

1. Mahajan-Miklos S, Tan MW, Rahme LG, Ausubel FM (1999) Molecular mechanisms of bacterial virulence elucidated using a Pseudomonas aeruginosa-Caenorhabditis elegans pathogenesis model. Cell 96:47–56
2. Rahme LG, Ausubel FM, Cao H, Drenkard E, Goumnerov BC, Lau GW, Mahajan-Miklos S, Plotnikova J, Tan MW, Tsongalis J, Walendziewicz CL, Tompkins RG (2000) Plants and animals share functionally common bacterial virulence factors. Proc Natl Acad Sci U S A 97:8815–8821
3. Kurz CL, Chauvet S, Andres E, Aurouze M, Vallet I, Michel GP, Uh M, Celli J, Filloux A, De Bentzmann S, Steinmetz I, Hoffmann JA, Finlay BB, Gorvel JP, Ferrandon D, Ewbank JJ (2003) Virulence factors of the human opportunistic pathogen Serratia marcescens identified by in vivo screening. EMBO J 22:1451–1460
4. Limmer S, Haller S, Drenkard E, Lee J, Yu S, Kocks C, Ausubel FM, Ferrandon D (2011) Pseudomonas aeruginosa RhlR is required to neutralize the cellular immune response in a Drosophila melanogaster oral infection model. Proc Natl Acad Sci U S A 108:17378–17383
5. Lemaitre B, Hoffmann J (2007) The Host defense of Drosophila melanogaster. Annu Rev Immunol 25:697–743
6. Ferrandon D, Imler JL, Hetru C, Hoffmann JA (2007) The Drosophila systemic immune response: sensing and signalling during bacterial and fungal infections. Nat Rev Immunol 7:862–874
7. Stuart LM, Ezekowitz RA (2008) Phagocytosis and comparative innate immunity: learning on the fly. Nat Rev Immunol 8:131–141
8. Ferrandon D (2013) The complementary facets of epithelial host defenses in the genetic model organism Drosophila melanogaster: from resistance to resilience. Curr Opin Immunol 25:59–70
9. Apidianakis Y, Rahme LG (2009) Drosophila melanogaster as a model host for studying Pseudomonas aeruginosa infection. Nat Protoc 4:1285–1294
10. Limmer S, Quintin J, Hetru C, Ferrandon D (2011) Virulence on the fly: Drosophila melanogaster as a model genetic organism to decipher host-pathogen interactions. Curr Drug Targets 12:978–999
11. Lau GW, Goumnerov BC, Walendziewicz CL, Hewitson J, Xiao W, Mahajan-Miklos S, Tompkins RG, Perkins LA, Rahme LG (2003) The Drosophila melanogaster toll pathway participates in resistance to infection by the gram-negative human pathogen Pseudomonas aeruginosa. Infect Immun 71:4059–4066
12. Apidianakis Y, Mindrinos MN, Xiao W, Lau GW, Baldini RL, Davis RW, Rahme LG (2005) Profiling early infection responses: Pseudomonas aeruginosa eludes host defenses by suppressing antimicrobial peptide gene expression. Proc Natl Acad Sci U S A 102:2573–2578

13. Fauvarque MO, Bergeret E, Chabert J, Dacheux D, Satre M, Attree I (2002) Role and activation of type III secretion system genes in Pseudomonas aeruginosa-induced Drosophila killing. Microb Pathog 32:287–295
14. Boman HG, Nilsson I, Rasmuson B (1972) Inducible antibacterial defence system in Drosophila. Nature 237:232–235
15. Apidianakis Y, Pitsouli C, Perrimon N, Rahme L (2009) Synergy between bacterial infection and genetic predisposition in intestinal dysplasia. Proc Natl Acad Sci U S A 106: 20883–20888
16. Ashburner M, Golic KG, Hawley RS (2005) Drosophila. A laboratory handbook, 2nd edn. Cold Spring Harbor Laboratory Press, Cold Spring Harbor
17. Greenspan R (2004) Fly pushing: the theory and practice of Drosophila genetics, 2nd edn. Cold Spring Harbor Laboratory Press, Cold Spring Harbor
18. Brand AH, Perrimon N (1993) Targeted gene expression as a means of altering cell fates and generating dominant phenotypes. Development 118:401–415
19. McGuire SE, Roman G, Davis RL (2004) Gene expression systems in Drosophila: a synthesis of time and space. Trends Genet 20: 384–391
20. Elrod-Erickson M, Mishra S, Schneider D (2000) Interactions between the cellular and humoral immune responses in *Drosophila*. Curr Biol 10:781–784
21. Ferrandon D, Jung AC, Criqui MC, Lemaitre B, Uttenweiler-Joseph S, Michaut L, Reichhart JM, Hoffmann JA (1998) A drosomycin-GFP reporter transgene reveals a local immune response in Drosophila that is not dependent on the Toll pathway. EMBO J 17:1217–1227
22. Tzou P, Ohresser S, Ferrandon D, Capovilla M, Reichhart JM, Lemaitre B, Hoffmann JA, Imler JL (2000) Tissue-specific inducible expression of antimicrobial peptide genes in Drosophila surface epithelia. Immunity 13: 737–748
23. Reichhart JM, Meister M, Dimarcq JL, Zachary D, Hoffmann D, Ruiz C, Richards G, Hoffmann JA (1992) Insect immunity – developmental and inducible activity of the Drosophila diptericin promoter. EMBO J 11:1469–1477
24. Cuttell L, Vaughan A, Silva E, Escaron CJ, Lavine M, Van Goethem E, Eid JP, Quirin M, Franc NC (2008) Undertaker, a Drosophila Junctophilin, links Draper-mediated phagocytosis and calcium homeostasis. Cell 135: 524–534
25. Mulcahy H, Sibley CD, Surette MG, Lewenza S (2011) Drosophila melanogaster as an animal model for the study of Pseudomonas aeruginosa biofilm infections in vivo. PLoS Pathog 7:e1002299

Chapter 57

Assessing *Pseudomonas* Virulence Using Host Cells

Iwona Bucior, Cindy Tran, and Joanne Engel

Abstract

While human or animal models are often considered the gold standard experimental system for defining virulence factors, cell culture-based infection models have proven useful for identifying important virulence factors and for examining the interactions between pathogens and the epithelial barrier. The first step in infections for most mucosal pathogens involves binding (adhesion) to the epithelial cells that line the mucosa. Successful pathogens can then penetrate the barrier by (1) inducing their uptake (i.e., "entry" or "invasion") into epithelial cells, (2) crossing the barrier by inducing epithelial cell death, and/or (3) penetrating between cells. This chapter describes growth conditions to form polarized cultures, either two-dimensional monolayers or three-dimensional cysts, of various immortalized epithelial cell lines. It describes assays to measure key early interactions between *P. aeruginosa* and host cells, including binding, invasion, and cytotoxicity. Many virulence factors defined by these criteria have been shown to be important for pathogenesis of *P. aeruginosa* infections in animals or humans. These methods are also applicable to other pathogens.

Key words Pseudomonas aeruginosa, Polarized epithelial cells, Transwells, Adhesion, Invasion, Cytotoxicity, Cysts

1 Introduction

1.1 Cell Culture Models for Polarized Epithelium

The use of polarized epithelial cell culture models to study bacterial host interactions has proven informative and often predictive of bacterial interactions with the mucosal barrier in vivo [1]. The mucosal barrier is lined by one or more layers of cells that have distinct apical and basolateral surfaces, separated by tight junctions, that form selective permeability barriers between biological compartments [5]. The apical surface of the cells faces either the outside world or the lumen of a cavity depending on the type of tissue, while the basolateral surface faces adjoining cells and the basement membrane. The apical and basolateral membrane domains are distinguished by unique assemblies of proteins, lipids, and other macromolecules and structures, such as the secreted mucous layer, and have distinct functions.

Alain Filloux and Juan-Luis Ramos (eds.), *Pseudomonas Methods and Protocols*, Methods in Molecular Biology, vol. 1149, DOI 10.1007/978-1-4939-0473-0_57, © Springer Science+Business Media New York 2014

While many investigations of host–pathogen interactions are performed on primary or immortalized tissue culture cells grown on conventional tissue culture plastic surfaces, these cells cannot completely polarize, as they are forced to mislocalize basolateral transporters to the apical surface in order to obtain nutrients. However, epithelial cells can be plated on porous filter supports, such as polycarbonate Transwell filters, so that they can be bathed in media from both the apical and basolateral sides when submerged in media in multiwall plates. Under these conditions, the cells can obtain nutrients from either the basolateral or apical surface and will form polarized monolayers when they reach confluence. Filters are available that have an average pore size of 0.4 μm (through which bacteria cannot pass) or 3.0 μm (through which bacteria can pass).

Several immortalized epithelial cell lines, including Manin-Darby Canine Kidney (MDCK, dog kidney epithelial) cells, Calu-3 cells (a human airway epithelial cell line), and 16HBE14o-cells (a human bronchial epithelial cell line), as well as primary human alveolar type II cells, can be plated at "instant" monolayer densities such that they form intact functional two-dimensional polarized monolayers within 24 h [6–9]. Importantly, these cell lines exhibit robust contact-dependent inhibition and thus do not form multilayers of cells. Another useful property is that the longer the monolayers are cultured, the more differentiated and polarized the monolayers become. We define functional polarized monolayers as cells with (1) functional tight junctions, as evidenced by lack of permeability to small molecules such as FITC-inulin, (2) an elevated transepithelial resistance, and (3) functional adherens junctions, with distinct localization of apical and basolateral markers. For MDCK cells plated at an instant monolayer density and cultured on transwells for 24 h, gp135 (podoxcalyxin) is primarily apical whereas p58 (the β-subunit of Na^+/K^+ ATPase) is primarily basolateral [10]. Basolateral localization of the epidermal growth factor receptor, however, requires longer culture times. By varying the length of time in culture, it is possible to recapitulate epithelial development and to compare "incompletely polarized" versus "fully polarized" epithelial monolayers. However, it is important to emphasize that even in "incompletely polarized" monolayers, a functional barrier is formed, with intact tight junctions and adherens junctions.

Using this system, early studies in our lab revealed a strong correlation between the virulence of various strains of *P. aeruginosa* in animal models of acute pneumonia with their ability to cause necrotic death when applied to the apical surface of Transwell-grown polarized MDCK cells, indicating that polarized cells model important aspects of human disease [7]. Studies in which a transposon mutant library of *P. aeruginosa* was screened for mutants that failed to damage polarized MDCK cells when applied apically were important in the discovery of the role of the *P. aeruginosa* type III secretion system [11], which is required for virulence in a murine pneumonia model [12–14] and whose presence correlates with outcomes in human ventilator-associated pneumonia [15, 16]. In more recent studies, this system has allowed us to compare the binding properties, signalling, and entry properties of *P. aeruginosa* when added to the apical versus basolateral surfaces. Our studies have revealed that the *N*-glycan chains at the apical surface and heparin sulfate chains at the basolateral surface are necessary and sufficient for binding, invasion, and cytotoxicity to kidney and airway cells grown at various states of polarization on Transwell filters. In incompletely polarized epithelium, heparin sulfate proteoglycan abundance is increased at the apical surface, explaining its enhanced susceptibility to *P. aeruginosa* colonization and damage [17]. In subsequent work, we found that binding to the *N*-glycans at the apical surface is primarily mediated by type IV pili, while binding to the basolaterally localized heparin sulfate proteoglycans is mediated by flagella [8].

Whereas two-dimensional cell culture systems have provided many important insights into bacterial pathogenesis, they are some limitations in the extent to which they model the complexity of an intact three-dimensional tissue. For instance, cell–cell and cell–matrix adhesion, gene expression, and orchestration of signalling pathways differ in the absence of a three-dimensional microenvironment. One practical disadvantage of using a two-dimensional MDCK monolayer is that it is difficult to infect cells from the basolateral surface, because the filter blocks access of bacteria to the cells, although this problem can be partially overcome by using filters with larger 3 μm pore size.

Recently, it has become possible to grow MDCK and some other epithelial cell lines in thick, three-dimensional gels of extracellular matrix (ECM), such as collagen I or Matrigel, a reconstituted basement membrane-like material secreted by the Engelbreth Holm Swarm tumor [18]. When MDCK cells are seeded as single cells in collagen gel, they proliferate and over 7 days form a hollow cyst lined by a monolayer of epithelial cells. The apical surfaces face inward towards a central lumen. Thus, epithelial cells can organize into a rudimentary organ-like structure. This process resembles the branching morphogenesis found during formation of many epithelial organs. These organotypic three-dimensional cultures much more closely mimic the normal environment and allow cells to respond more physiologically. Inside–out cysts, with reversed polarity, can be formed by inhibiting integrin-mediated adhesion to the ECM during cyst growth. Although much of the recent work with three-dimensional cultures has focused on cancer and related issues of differentiation [19], these three-dimensional culture systems are beginning to be applied to the study of host–pathogen interactions [17, 20]. However, due to the avid binding of *P. aeruginosa* to ECM, adhesion only be assessed by immunofluorescence microscopy [17].

1.2 Growth and Culture of *P. aeruginosa*

Most of the assays described below are carried out with commonly used well-characterized strains (PAK, PAO1, PA103) which were originally isolated from human infections but have since been passaged in the laboratory, but in theory any strain could be used. PAK expresses the major adhesins, type IV pili and flagella, and encodes and translocates into the host cell the type III secreted toxins ExoS, ExoT, and ExoY. PAO1, the first *P. aeruginosa* strain to be sequenced, also expresses type IV pili and flagella, as well as ExoS and ExoT. However, compared to PAK, lower amounts of type III secreted effectors are produced. PAO1 exhibits less cytotoxicity towards epithelial cells in culture and is less virulent in a murine model of acute pneumonia compared to PAK [7]. PA103, expresses type IV pili but not flagella, encodes and produces ExoU, a potent phospholipase, and lacks the ExoS gene [12, 21, 22]. When cocultivated with culture epithelial cells, PA103 is highly cytotoxic due to ExoU. Of the three strains, it is the most virulent in an animal pneumonia model [7].

1.3 Adhesion and Invasion Assays

The first step in establishing *P. aeruginosa* infection is receptor-mediated binding to the epithelium on the apical and/or basolateral surface, leading to bacterial internalization and/or direct host injury as well as dissemination to distant tissues and organs (reviewed in ref. [23]). *P. aeruginosa* encodes multiple adhesins that may account for its ability to bind to the multitude of host cell types that it may encounter in diverse environments, ranging from single-celled amoeba to human epithelial cells. The most important adhesins are Type IV pili, polar fimbriae that undergo extension and retraction, and flagella, which are polar organelles that also mediate swimming motility. Other minor adhesins have also been identified, including the cup fimbriae and the lectin-like proteins, LecA and LecB. A variety of host cell receptors have been identified. These include asialoGM1, mannose-containing *N*-glycoproteins, integrins, heparin sulfate proteoglycans, and the cystic fibrosis membrane receptor (reviewed in ref. [1]).

In cell culture-based models, bacterial adhesion can be monitored visually by microscopy, but quantitation is difficult and time-consuming [24]. Alternatively, adhesion can be measured by releasing surface-bound bacteria after gentle lysis of host cells under conditions that do not cause bacterial lysis (0.25 % Triton-X 100) [25]. *P. aeruginosa* binding to host cells can be measured reasonably reliably and robustly using this latter method [26]. However, given that *P. aeruginosa* also binds avidly to abiotic surfaces such as plastic, care must be taken to wash the epithelial cell monolayer thoroughly. Performing adhesion assays on confluent transwell grown cells offers several additional advantages. First, the filter can be excised and washed thoroughly, so that only binding to host cells (and not to surrounding plastic surfaces) is measured. Second, if the epithelial cells are grown as confluent monolayers on filters with 3 μm pores, bacteria can be added from either the apical or the basolateral side, allowing comparisons [8, 17, 27]. Third, since epithelial cell polarity can be controlled or perturbed when grown on porous filters, binding, invasion, or cytotoxicity as a function of cell polarity can be investigated [8, 17, 27].

In addition to assays that utilize microscopy, a significant advance in quantitating bacterial entry or invasion into cells was the development of the aminoglycoside exclusion assay [25]. This assay is similar to the adhesion assay, except that following the adhesion step (typically 1 h), an aminoglycoside antibiotic, which does not accumulate intracellularly, is added to kill extracellular bacteria for 1–2 h. Internalized bacteria are released by gentle host cell lysis. There are two important caveats to the interpretation of this assay. First, although it is possible that extracellular adherent bacteria may be protected from antibiotics, this has rarely been reported. Second, if the host cell is permeabilized during the incubation period, then aminoglycoside antibiotics may accumulate intracellularly and kill intracellular bacteria. For these reasons, it is helpful in initial experiments to perform cytotoxicity assays in parallel with the adhesion and/or invasions assays to ensure that the integrity of the host cell plasma membrane has not been affected.

Although *P. aeruginosa* is primarily considered an extracellular pathogen, many clinical and laboratory isolates demonstrate measurable internalization by various assays (reviewed in ref. [23]). Importantly, different strains and mutants vary in their internalization efficiency into cultured cells, but all strains are capable of entering into both phagocytic and non-phagocytic cells to some degree [28]. The observation that the ability to enter cells has been maintained during the evolution of *P. aeruginosa* indicates that it is likely to play a fundamental role in the pathogenesis of *P. aeruginosa* infections and/or in surviving in the environment. Both genetic and cell biologic approaches have revealed that *P. aeruginosa* internalization requires rearrangement of the actin cytoskeleton through pathways involving Abl kinase, the adaptor protein Crk, the small GTPases Rac1 and Cdc42, and p21-activated kinase [29]. The type III secreted effectors ExoS and ExoT target this pathway by interfering with GTPase function and, in the case of ExoT, by abrogating *P. aeruginosa*-induced Abl-dependent Crk phosphorylation [29]. Quantitative adhesion and invasion assays may serve as useful endpoints/readouts to assess potential inhibitors of *P. aeruginosa* infection. In addition, understanding the mechanism of bacterial entry into host cells may reveal new insights into eukaryotic cell biology.

1.4 Cytotoxicity Assays

Bacteria can induce many types of cell death in eukaryotic host cells, including necrosis, apoptosis, and pyroptosis (reviewed in ref. [30]), but most types of cell death ultimately result in loss of membrane integrity and release of cytoplasmic content (cytotoxicity).

While there are many methods for quantifying cytotoxicity, commercially available colorometric non-radioactive assays that quantify the release of the stable cytoplasmic enzyme lactate dehydrogenase (LDH) have proven very robust and sensitive. A genome-wide transposon mutant screen of PA103 in which mutants were screened for loss of cytotoxicity when applied to the apical surface of polarized MDCK cells led to the identification of type III secreted effector ExoU [26]. ExoU producing strains are highly virulent in a murine model of acute pneumonia and are associated with worse clinical outcomes in human infections [12, 15, 31, 32].

2 Materials

2.1 Cell Culture

1. MDCK clone II cells (ATCC CCL-34) (*see* **Note 1**), Calu-3 cells (ATCC HTB-55), and 16HBE14o- and CFBE14o-cells (obtained from Dr. Alan Verkman, UCSF).
2. Modified Eagle Media (MEM) with Earl's Balanced Salt Solution.
3. Trypsin.
4. Glutamine.
5. Fetal bovine serum (FBS; Thermo Scientific; product code SH30910.03).
6. 10 cm (Corning product code 430293) and 6-well plastic cell (Falcon product code 353046) culture plates.
7. 0.4 μm pore size 12 mm Transwell filters (Corning costar, product code 3101.
8. 3 μm pore size 12 mm Transwell filters (Corning Costar; product code 3402).
9. 8-well Lab-Tek Coverglass chambers (Nalge Nunc International; product code 155409).
10. 100 % Matrigel (BD Biosciences; product code 356234). Store at −20 °C.
11. Anti-β_1 integrin (AIIB2; available from Caroline Damsky through Developmental Studies Hybridoma Bank).
12. Type I collagen solution: 66 % (3 mg/ml) PureCol Vitrogen (Advanced BioMatrix; product code 5005-B), 1× MEM, 2.35 mg/ml $NaHCO_3$, and 0.025 M HEPES, pH 7.6. Store components separately at 4 °C. Prepare the solution fresh and keep on ice.

2.2 Bacterial Culture

1. 50-ml Conical tubes (Falcon; product code 352098).
2. Luria Bertani (LB) broth and plates.

2.3 Bacterial Adhesion and Invasion Assays

1. Gentamicin (Fisher Scientific; product code BP918).
2. Amikacin (Fisher Scientific; product code BP2643).

2.4 Cytotoxicity Assays

1. CytoTox 96® Non-Radioactive Cytotoxicity Assay (Promega; product code G1780).
2. Plate reader or spectrophotometer.

3 Methods

3.1 Culture of Epithelial Cells as Polarized Monolayers on Transwell Filter Supports

3.1.1 MDCK Cells

1. Grow MDCK cells in MEM supplemented with 5–10 % FBS at 37 °C with 5 % CO_2 in 10 cm plastic cell culture plates. When cells reach 100 % confluence, remove cells for splitting by adding 0.25 % Trypsin without EDTA and incubate for 30–40 min at 37 °C. After the cells come off the plastic, pellet them by centrifugation for 15 min at 220 × *g*, resuspend in MEM + FBS, and passage by diluting them 1:4–1:6. We routinely omit antibiotics for cell culture as they can inhibit the growth of *P. aeruginosa*.
2. For infection with bacteria on the apical surface, the MDCK cells should be grown on either 0.4 or 3 μm filters. For infections with bacteria on the basolateral surface, MDCK cells must be grown on 3 μm filters. If apical versus basolateral infection will be compared, grow the MDCK cells on 3 μm filters.
3. To grow incompletely polarized monolayers, add $0.5–0.7 \times 10^6$ cells (as determined by counting in a hemacytometer) in 0.5 ml MEM with 5–10 % FBS in the upper chamber of 12-mm Transwell filters, add 1.5 ml MEM with 5–10 % FBS to the lower chamber, and culture cells for 1–3 days at 37 °C with 5 % CO_2. Confluent monolayers plated on 12 mm Transwell filters contain ~2×10^6 cells.
4. To grow as polarized monolayers, add $0.4–0.6 \times 10^6$ cells in 0.5 ml MEM with 5–10 % FBS in the upper chamber of 12-mm Transwell filters placed in the well of a 6-well dish, add 1.5 ml MEM with 5–10 % FBS to the lower chamber, and culture cells for 5 days at 37 °C with 5 % CO_2. Confluent monolayers plated on 12 mm Transwell filters contain ~2×10^6 cells.

3.1.2 Calu-3 Cells: Human Airway Epithelial Cells

1. Grow Calu-3 cells in MEM supplemented with 10 % FBS and 2 mM L glutamine at 37 °C with 5 % CO_2 in plastic cell culture plates. To split cells for passaging, add 0.25 % Trypsin without EDTA for 40–60 min at 37 °C, dilute into media, passage by diluting them 1:2–1:4 (*see* **Note 2**).
2. To grow polarized monolayers, add 1×10^6 cells in 0.5 ml MEM with 10 % FBS and 2 mM L-glutamine in the upper chamber of 12-mm Transwell filters, add 1.5 ml MEM with

10 % FBS and 2 mM L-glutamine to the lower chamber, and culture cells for 3–5 days (incompletely polarized) or 9–12 days (more fully polarized) at 37 °C with 5 % CO_2. Wash once and change media every day. For the last 2 days, grow cells in an air–liquid phase by removing media from the upper chamber of the Transwell filter.

3.1.3 16hbe14o- and cfbe14o-Cells: Human Bronchial Epithelial Cells

1. Grow cells in MEM supplemented with 10 % FBS and 2 mM L-glutamine at 37 °C with 5 % CO_2 in plastic cell culture plates. To split cells for passaging, add 0.25 % Trypsin without EDTA for 40–60 min at 37 °C, centrifuge, resuspend in medium, and dilute 1:4–1:6 (*see* **Note 2**).
2. To grow polarized monolayers, seed $0.5–0.7 \times 10^6$ cells in 0.5 ml MEM with 10 % FBS and 2 mM L-glutamine in the upper chamber of 12-mm Transwell filters, add 1.5 ml MEM with 10 % FBS and 2 mM L-glutamine to the lower chamber, and culture cells for 2–3 days (incompletely polarized) or 5 days (more fully polarized) at 37 °C with 5 % CO_2. Wash once and change media every day. For the last 2 days, grow cells in an air–liquid phase by removing media from the upper chamber of the Transwell filter.

3.2 Growth of MDCK Cells as Three-Dimensional Cysts

3.2.1 Method A

1. Cover the surface of 8-well Lab-Tek Coverglass chambers with 3 μl of 100 % Matrigel. Remove a portion of 100 % Matrigel from a stock stored at −20 °C. Keep at 4 °C and store on ice during the experiment. Matrigel stored at 4 °C can be used for approximately 3 weeks.
2. Prepare a single cell suspension of $1–3 \times 10^4$ cells/ml in pre-warmed MEM supplemented with 10 % FBS and 2 % Matrigel.
3. To grow cysts with reversed polarization, add 1:100 dilution of anti-β_1 integrin (AIIB2) to a cell suspension.
4. Add 300 μl of cell suspension to each well of the coverglass chambers and grow for 2–4 days at 37 °C with 5 % CO_2.

3.2.2 Method B

1. Cover the surface of 8-well Lab-Tek Coverglass with 3 μl of 100 % Matrigel.
2. Prepare a single cell suspension of 4×10^4 cells/ml in a type I collagen solution. Keep the solution on ice before addition of cells to avoid solidification of the collagen.
3. To grow cysts with reversed polarization, add 1:100 dilution of anti-β_1 integrin (AIIB2) to a type I collagen solution.
4. Add 250 μl of a cell suspension in type I collagen to each well of coverglass chambers.
5. Allow cells to settle close to the bottom of the well by keeping chambers inside a tissue culture hood for 20 min.

6. Allow the collagen mixture to solidify fully into a gel by putting chambers inside a tissue culture incubator for 20–30 min and then add 200 μl of MEM supplemented with 10 % FBS on top.
7. Change medium every 2 days and grow cysts for 4–7 days.

3.3 Bacterial Culture

1. Inoculate one colony of bacteria from a freshly streaked Luria-Bertani (LB) plate into 5 ml LB broth in a 50-ml conical tube and grow bacteria shaking (300 RPM) overnight (~18 h) at 37 °C (*see* **Note 3**). To maintain aerobic growth conditions during overnight growth, either (1) loosen the cap and tape it to the tube or (2) remove the cap and cover the tube with parafilm.
2. If exponential phase *P. aeruginosa* are to be used, dilute overnight grown (stationary phase) bacteria 1:40 in LB, and grow in 2 ml LB shaking at 37 °C for additional 2–3 h to a final OD_{600} of 0.2–0.6 (*see* **Note 4**).

3.4 Bacterial Infections of Polarized Monolayers

3.4.1 Apical Infections of Two Dimensional Monolayers

1. Prior to addition of bacteria, gently aspirate media from the upper and lower chamber of Transwell filters using a Pasteur pipet connected to a vacuum line. Wash cells once with 0.5 ml (upper chamber) or 1.5 ml of MEM (lower chamber).
2. Dilute log phase or stationary phase bacteria in 200 μl of MEM to achieve the target multiplicity of infection (MOI) based on OD_{600} measurements (*see* **Note 5**). Typical MOIs for adhesion and invasion assays range from 5 to 30. Gently add MEM containing the bacteria inoculum to the upper chamber of the Transwell insert and, without disrupting bacterial suspension, add remaining 300 μl of MEM on top.
3. Incubate at 37 °C with 5 % CO_2 for 1 h.
4. Quantify the exact MOI by plating tenfold serial dilutions of the bacterial culture onto LB plates and incubating overnight to enumerate CFUs.

3.4.2 Basolateral Infection of Two Dimensional Monolayers

1. Prior to addition of bacteria, gently aspirate media from the upper and lower chamber of Transwell filters using a Pasteur pipet connected to a vacuum line. Wash cells once with 0.5 ml (upper chamber) or 1.5 ml of MEM (lower chamber).
2. Dilute log phase or stationary phase bacteria in 50 μl of MEM to achieve the target MOI based on OD_{600} measurements. Typical MOIs for adhesion and invasion assays range from 5 to 30.
3. Place the 50-μl drop in the middle of the well of a 6-well cell culture plate.
4. Gently place the Transwell (on which the polarized cells have been plated) on top of the drop, and cover the plate with a lid.

5. Incubate at 37 °C with 5 % CO_2 for 1 h.
6. Quantify the exact MOI by plating tenfold serial dilutions of the bacterial culture onto LB plates and incubating at 37 °C overnight to enumerate CFUs.

3.4.3 Infections of Cysts

1. For cysts grown by Method I, remove the Matrigel suspension by pipette and wash twice with 300 μl of MEM with a pipette. The majority of cysts will remain at the bottom of the well embedded in 100 % Matrigel.
2. For cysts grown by Method II, using a Pasteur pipet connected to a vacuum line, remove MEM from the top of a type I collagen solution. Add collagenase type VII at 100 U/ml in PBS to each well for 20 min at 37 °C to digest the collagen gel. Remove digested collagen with a pipette, wash twice with MEM by adding and removing 300 μl of MEM with a pipette.
3. Add 300 μl of GFP-expressing bacteria in MEM with 10 % FBS at an MOI of 100–200.
4. Incubate at 37 °C with 5 % CO_2. The length of time for the incubation will depend on whether adhesion, invasion, or cytotoxicity assays will be performed.

3.5 Apical Bacterial Adhesion and Invasion Assays of Two Dimensional Monolayers

1. Gently aspirate media from the sides of the well (to avoid disrupting the monolayers) from the upper and lower chamber using a Pasteur pipet connected to a vacuum line. Wash upper and lower chambers three times gently with 0.5 and 1.5 ml PBS (room temperature), respectively (*see* **Note 6**).
2. For adhesion assays, proceed directly to **step 4**.
3. For invasion assays, add 0.5 and 1.5 ml of MEM containing 0.4 mg/ml amikacin or 0.25 mg/ml gentamicin to the upper and lower chamber, respectively, for 2 h to kill extracellular bacteria (*see* **Note 7**). At then end of incubation with antibiotics, remove media from the upper and lower chamber using a Pasteur pipet connected to a vacuum line, and wash upper and lower chambers twice with 0.5 and 1.5 ml PBS, respectively.
4. In a sterile 50 ml conical tube, add 1 ml of 0.25 % Triton X-100 and three sterile glass beads. Using a sterile #11 scalpel and tweezers, gently excise each filter from the Transwell insert and place in the 50 ml tube. Incubate the filters in the 50 ml tubes at room temperature for 30 min. Vortex each tube every 15 min for 10–15 s to ensure thorough host cell lysis.
5. Prepare tenfold serial dilutions in 1 ml of ul LB of cell lysates. Plate 100 μl onto LB plates and incubate at 37 °C overnight to enumerate CFUs.

3.6 Basolateral Adhesion and Invasion Assays of Two Dimensional Monolayers

1. After basolateral infection of the monolayer, use sterile tweezers to transfer the Transwell filters into a 12-well plate.
2. Remove media from the upper chamber using a Pasteur pipet connected to a vacuum line, and wash upper and lower chambers three times with 0.5 and 1.5 ml PBS (room temperature), respectively.
3. For adhesion assay, proceed directly to **step 5**.
4. For invasion assays, add 0.5 and 1.5 ml of MEM containing 0.4 mg/ml amikacin or 0.25 mg/ml gentamicin to the upper and lower chamber, respectively, for 2 h to kill adherent bacteria (*see* **Note 7**). After incubation with antibiotic, remove media from the upper and lower chamber using a Pasteur pipet connected to a vacuum line, and wash upper and lower chambers twice with 0.5 and 1.5 ml PBS, respectively.
5. Remove PBS from lower chamber of Transwell. Add 500 μl of 0.25 % Triton X-100 to the upper chamber for 30 min (*see* **Notes 8** and **9**).
6. Transfer cell lysate to a sterile microcentrifuge tube. Using a sterile cell scraper, gently scrape remaining bacteria and cell debris off the upper surface of the Transwell filter. Avoid breaking the filter if possible (*see* **Note 10**). Transfer residual lysate to the sterile microcentrifuge tube.
7. Prepare tenfold serial dilutions of cell lysates 100 μl into 1 ml of LB. Plate 100 μl onto LB plates and incubate overnight to enumerate CFUs.

3.7 Bacterial Adhesion to 3D Cysts

1. After 2 h of infection, remove bacteria in MEM using a Pasteur pipet connected to a vacuum line, and wash three times with 300 μl PBS (room temperature).
2. Fix cells by adding in 300 μl of PBS containing 1 % paraformaldehyde at 37 °C for 0.5 h.
3. Wash three times with 300 μl PBS (room temperature).
4. When applicable, add primary antibodies (typically 1:300–1:600 dilution) in 300 μl PBS for 2–4 h at room temperature or overnight at 4 °C, and secondary fluorescent antibodies (typically 1:1,000 dilution) in 300 μl PBS at 4 °C overnight.
5. Wash three times with 300 μl PBS (room temperature).
6. Take images with a confocal or fluorescence microscope at a magnification of ×600 or ×1,000.
7. Perform analysis on TIFF files of images taken in the green (488 nm) channel (GFP signal) that contain only bound bacteria scored by eye with Image J: Threshold RGB images using Image > Adjust > Threshold. Score the number and size of bound bacteria (in pixels) using Analyze > Analyze Particles.

3.8 Cytotoxicity Assay

3.8.1 Measurement of Lactate Dehydrogenase (LDH) Release from the Apical Surface of 2-D Monolayers

1. Prior to bacterial infection, replace tissue culture media in the upper chamber of Transwell filter with sterile, pre-warmed MEM.
2. Infect with bacteria for desired length of time, ensuring that the total final volume of media (after bacterial inoculation) in the upper chamber of Transwell filter is 500 μl.
3. At various times after bacterial infection, remove 50 μl of media from the upper chamber of Transwell filter and transfer to a 96-well plate for the cytotoxicity assay.

3.8.2 Measurement of LDH Release from the Basolateral Surface of Cells Grown on Transwell Filters

1. If comparing directly to apical LDH release, replace tissue culture media in the lower chamber of Transwell filter with 500 μl of sterile, pre-warmed MEM. Otherwise, ensure that the lower chamber of all experimental samples contain an equivalent amount of sterile, pre-warmed MEM.
2. At various times after bacterial infection, remove 50 μl of media from the lower chamber of Transwell filter and transfer to a 96-well plate for the cytotoxicity assay.

3.8.3 Positive Control with Lysis Buffer

1. Lyse cells using either the included Lysis Buffer in the CytoTox 96 system or with 0.8 % Triton at room temperature for 45 min. An alternative method of lysis may also be used (*see* **Note 11**).
2. After lysis, remove 50 μl of media from the upper chamber of the Transwell filter and transfer to a 96-well plate for the cytotoxicity assay.

3.8.4 Cytotoxicity Assay

1. For background control, add 50 μl of MEM to three or four empty wells in the 96-well plate.
2. Add 12 ml of RT assay buffer to one vial of substrate mix (*see* **Note 12**). Add 50 μl of reconstituted substrate mix to each culture well. Incubate at room temperature for 30 min and protect from light.
3. Add 50 μl of stop solution to each culture well.
4. Record absorbance at 490 nm (*see* **Note 13**).
5. Subtract background values from the experimental sample readings and positive controls.
6. Determine percentage cell death using the formula: percentage cytotoxicity = Experimental LDH release (OD490)/Maximum LDH release (OD490).

4 Notes

1. MDCK clone I and clone II cells have somewhat distinct polarization properties.
2. Calu-3, 16HBE14o-, and CFBE14o-cells tend to clump and require repeated pipetting and vortexing before plating.

3. As there is commonly variability in these assays, it is advisable to use 3–6 replicate wells per sample and condition. In addition, it is useful to use a wild-type bacteria (such as PAK) and an isogenic mutant lacking the two major adhesins, type IV pili and flagella (PAKΔ*pilA*Δ*fliC*) as positive and negative controls. The non-adhesive mutant exhibits at least 100-fold decreased binding compared to the wild-type bacteria.
4. Either log phase or stationary phase bacteria can be used for the adhesion, invasion, or cytotoxicity assays. It is more important that the same bacterial growth conditions are used when repeating assays or when comparing apical to basolateral infections.
5. An OD_{600} of 1 is $\sim 1 \times 10^9$ CFU/ml. For infections with MOI of 20, you will need 40×10^6 bacteria (20 times 2×10^6 cells on Transwell filters). For an OD_{600} of 0.4 ($\sim 4 \times 10^8$ CFU/ml) dilute bacteria 1:10 into 200 μl of MEM.
6. To avoid disrupting the monolayer, place the Pasteur pipet at the edge of the well and slightly above the filter, and tilt the plate to aspirate the media. If any of the epithelial cells in the monolayer are injured, the bacteria will preferentially adhere to this region and the bacteria CFUs in the adhesion assays will be artificially elevated.
7. For *P. aeruginosa* strains that do not harbor antibiotic resistance, either aminoglycoside antibiotic works. If the strain is gentamicin resistant (for example, there is a transposon encoding gentamicin resistance in the genome), then use amikacin.
8. Do not use the lysis method described in Subheading 3.4 because this method would include bacteria that adhere to the bottom of the filter.
9. As the cells are lysed by 0.25 % Triton, lysate from the upper chamber may leak through the filter into the lower chamber. This lysate contains bacteria and should be collected as part of **step 6**.
10. It may be helpful to transfer the Transwell inserts to a 6-well plate for the lysis step. The larger well size allows the user to manually grasp the side of the Transwell insert while using the cell scraper. A firm grip on the Transwell insert will minimize breakage of the filter.
11. Other methods of cell lysis, such as freeze–thaw lysis or mechanical lysis by vortex, may also be used.
12. Reconstituted Substrate Mix may be stored for 6–8 weeks at –20 °C without loss of activity, per manufacturer's instructions (Promega Cytotox 96 Non-Radioactive Cytotoxicity Assay Technical Bulletin, Literature #TB163, revised 12/12).
13. If a large amount of LDH is released in your experimental conditions, positive controls and/or experimental samples may need to be serially diluted with culture medium in order to obtain measurements in the linear range of the plate reader or spectrophotometer.

References

1. Kazmierczak B, Mostov K, Engel J (2001) Interaction of bacterial pathogens with polarized epithelium. Annu Rev Microbiol 55:407–435
2. Engel J, Eran Y (2011) Subversion of mucosal barrier polarity by pseudomonas aeruginosa. Front Microbiol 2:114
3. Pizarro-Cerda J, Cossart P (2006) Bacterial adhesion and entry into host cells. Cell 124: 715–727
4. Engel J, Balachandran P (2009) Role of Pseudomonas aeruginosa type III effectors in disease. Curr Opin Microbiol 12:61–66
5. Apodaca G, Gallo LI, Bryant DM (2012) Role of membrane traffic in the generation of epithelial cell asymmetry. Nat Cell Biol 14:1235–1243
6. Breitfeld PP et al (1989) Expression and analysis of the polymeric immunoglobulin receptor in Madin-Darby canine kidney cells using retroviral vectors. Methods Cell Biol 32:329–337
7. Apodaca G et al (1995) Characterization of Pseudomonas aeruginosa-induced MDCK cell injury: glycosylation-defective host cells are resistant to bacterial killing. Infect Immun 63:1541–1551
8. Bucior I, Engel J (2012) The type IV pili and flagellar adhesins of Pseudomonas aeruginosa bind to and internalize through distinct receptors on the apical and basolateral surface of airway epithelium, PLoS Pathog (in press)
9. Fang X et al (2010) Allogeneic human mesenchymal stem cells restore epithelial protein permeability in cultured human alveolar type II cells by secretion of angiopoietin-1. J Biol Chem 285:26211–26222
10. Gassama-Diagne A et al (2006) Phosphatidylinositol-3,4,5-trisphosphate regulates the formation of the basolateral plasma membrane in epithelial cells. Nat Cell Biol 8:963
11. Kang PJ et al (1997) Identification of *Pseudomonas aeruginosa* genes required for epithelial cell injury. Mol Microbiol 24: 1249–1262
12. Hauser AR, Kang PJ, Engel JN (1998) PepA, a secreted protein of Pseudomonas aeruginosa, is necessary for cytotoxicity and virulence. Mol Microbiol 27:807–818
13. Yahr TL et al (1997) Identification of type III secreted products of the Pseudomonas aeruginosa exoenzyme S regulon. J Bacteriol 179: 7165–7168
14. Yahr TL, Goranson J, Frank DW (1996) Exoenzyme S of *Pseudomonas aeruginosa* is secreted by a type III pathway. Mol Microbiol 22:991–1003
15. Hauser AR et al (2002) Type III protein secretion is associated with poor clinical outcomes in patients with ventilator-associated pneumonia caused by *Pseudomonas aeruginosa*. Crit Care Med 30:521–528
16. Finck-Barbancon V et al (1997) ExoU expression by Pseudomonas aeruginosa correlates with acute cytotoxicity and epithelial injury. Mol Microbiol 25:547–557
17. Bucior I, Mostov K, Engel JN (2010) Pseudomonas aeruginosa-mediated damage requires distinct receptors at the apical and basolateral surfaces of the polarized epithelium. Infect Immun 78:939–953
18. Martin-Belmonte F, Mostov K (2008) Regulation of cell polarity during epithelial morphogenesis. Curr Opin Cell Biol 20: 227–234
19. Debnath J, Brugge JS (2005) Modelling glandular epithelial cancers in three-dimensional cultures. Nat Rev Cancer 5:675–688
20. Honer zu Bentrup K et al (2006) Three-dimensional organotypic models of human colonic epithelium to study the early stages of enteric salmonellosis. Microbes Infect 8:1813–1825
21. Sato H et al (2003) The mechanism of action of the *Pseudomonas aeruginosa*-encoded type III cytotoxin, ExoU. EMBO J 22:2959–2969
22. Sato H, Frank DW (2004) ExoU is a potent intracellular phospholipase. Mol Microbiol 53:1279–1290
23. Engel J (2007) Pseudomonas aeruginosa internalization by non-phagocytic cells. In: Ramos J-L, Filloux A (eds) Pseudomonas: a model system in biology. Springer, New York, pp 343–368
24. Chi E et al (1991) Interaction of *Pseudomonas aeruginosa* with A549 pneumocyte cells. Infect Immun 59:822–828
25. Isberg RR, Falkow S (1985) A single genetic locus encoded by *Yersinia pseudotuberculosis* permits invasion of cultured animal cells by *Escherichia coli* K-12. Nature 317:262–264
26. Kang PJ et al (1997) Identification of Pseudomonas aeruginosa genes required for epithelial cell injury. Mol Microbiol 24:1249–1262
27. Kazmierczak BI, Mostov K, Engel JN (2004) Epithelial cell polarity alters Rho-GTPase responses to *Pseudomonas aeruginosa*. Mol Biol Cell 15:411–419
28. Fleiszig SM et al (1997) Pseudomonas aeruginosa-mediated cytotoxicity and invasion correlate with distinct genotypes at the loci

encoding exoenzyme S. Infect Immun 65: 579–586

29. Pielage JF et al (2008) RNAi screen reveals an Abl kinase-dependent host cell pathway involved in Pseudomonas aeruginosa internalization. PLoS Pathog 4:e1000031
30. Bergsbaken T, Fink SL, Cookson BT (2009) Pyroptosis: host cell death and inflammation. Nat Rev Microbiol 7:99–109
31. Feltman H, et al. (2001) Type III secretion genotypes of clinical and environmental Pseudmonas aeruginosa isolates, ASM abstracts, D22
32. Roy-Burman A et al (2001) Type III protein secretion is associated with death in lower respiratory and systemic Pseudomonas aeruginosa infections. J Infect Dis 183: 1767–1774

Chapter 58

Assessing *Pseudomonas aeruginosa* Virulence and the Host Response Using Murine Models of Acute and Chronic Lung Infection

Irena Kukavica-Ibrulj, Marcella Facchini, Cristina Cigana, Roger C. Levesque, and Alessandra Bragonzi

Abstract

Murine models of acute and chronic lung infection have been used in studying *Pseudomonas aeruginosa* for assessing in vivo behavior and for monitoring of the host response. These models provide an important resource for studies of the initiation and maintenance of bacterial infection, identify bacterial genes essential for in vivo maintenance and for the development and testing of new therapies. The rat has been used extensively as a model of chronic lung infection, whereas the mouse has been a model of acute and chronic infection. Intratracheal administration of planktonic bacterial cells in the mouse provides a model of acute pneumonia. Bacteria enmeshed in agar beads can be used in the rat and mouse to reproduce the lung pathology of cystic fibrosis patients with advanced chronic pulmonary disease. Here, we describe the methods to assess virulence of *P. aeruginosa* using prototype and clinical strains in the Sprague-Dawley rat and the C57BL/6NCrlBR mouse by monitoring several measurable read-outs including weight loss, mortality, in vivo growth curves, the competitive index of infectivity, and the inflammatory response.

Key words Murine models, Chronic, Acute, Lung infection, Cystic fibrosis, Pneumonia, *Pseudomonas aeruginosa*

1 Introduction

Pseudomonas aeruginosa is a significant source of bacteremia in immuno-compromised patients, those in intensive care units or those connected to mechanical ventilation or other invasive devices. *P. aeruginosa* infection is also the leading cause of morbidity and mortality in individuals having cystic fibrosis (CF), whose abnormal airway epithelia allow long-term colonization of the lungs. The appropriate animal models demonstrating similar pathologies to humans are important for better understanding of the initiation and progression of the *P. aeruginosa* infection. Despite some limitations, including differences between animal and human populations, most of the models ranging from the simple models

Alain Filloux and Juan-Luis Ramos (eds.), *Pseudomonas Methods and Protocols*, Methods in Molecular Biology, vol. 1149, DOI 10.1007/978-1-4939-0473-0_58, © Springer Science+Business Media New York 2014

of septicemia to complex models of chronic pneumonia provide valuable resources to mimic the broad spectrum of infections [1].

Acute pneumonia models are useful to study non-CF-related pneumonia. This model may also be of interest for the initial phases of lung infection during *P. aeruginosa* colonization of CF airways. As infection progresses into the chronic stage in CF lungs, *P. aeruginosa* switches into a biofilm mode of growth—a population of microorganisms that aggregates on a matrix—coupled with *P. aeruginosa* phenotypic variants including mucoidy. The introduction of free-living and motile *P. aeruginosa* cells into the lung provides a model of acute lung infection; there is either rapid clearance of the organism or acute sepsis and death [2–5]. Some investigators have employed single or repeated aerosol administration of bacteria in attempts to reproduce the acquisition and initial symptoms of infection caused by *P. aeruginosa* as seen in CF patients [5, 6]. Chronic infections are obtained more readily by using *P. aeruginosa* cells enmeshed in agar, agarose or alginate beads. One of the first model of chronic bronchopulmonary infection was done in the rat and designed by Cash using *P. aeruginosa* embedded in agar beads [7]. The use of agar or agarose beads [8] is essential for the initiation and establishment of chronic lung infection. The purpose of embedding *P. aeruginosa* within beads is to retain the bacteria physically into the airways and to create an environment that mimics the bacterial biofilm and the microaerobiosis present in CF lung [9, 10]. Such retention presumably avoids physical elimination and leads to a persistent stimulation of host defenses typical of CF. Surgical implant of the beads bypasses the primary host defenses and does not take into account initial bacterial colonization. Although bacteria can migrate from the agar beads in vivo, growth is slow within the beads; this phenomenon is comparable to what has been observed in biofilm and in the planktonic state [11]. Although models using free living bacteria alone may mimic some features of CF, inoculating animals with *P. aeruginosa-laden* agar beads typically resembles the chronic lung infection of CF including histopathology, elevation in lung neutrophils and the accumulation of cytokines [12]. The "embedded bacteria model" is particularly challenging since there is extensive neutrophil influx in response to the *P. aeruginosa*-laden beads which may cause airway obstruction; effective gas exchange is not always optimal and a small percentage of animals may die. Given its simplicity and efficacy, this very first model of chronic bronchopulmonary infection with *P. aeruginosa* using agar beads is still frequently used today [13–23].

We summarize below the protocols used for both lung infection models using the rat for chronic and the mouse for chronic as well as acute infection.

2 Materials

All solutions should be prepared using ultrapure water (prepared by purifying deionized water to attain a sensitivity of 18 M Ω cm at 25 °C) and analytical grade reagents. Prepare and store all reagents at room temperature (unless indicated otherwise). Diligently follow all waste disposal regulations when disposing waste materials.

2.1 Agar-Bead Components

1. Sterile phosphate buffer saline (PBS) 1×: Mix 32 g of NaCl, 0.8 g of KCl, 5.76 g of Na_2HPO_4, 0.96 g of KH_2PO_4 in 4 l of water. Adjust the pH at 7.2 and sterilize by autoclaving for 15 min/121 °C.
2. 200 ml of sterile mineral oil with a 2 cm magnetic bar in a 250 ml Erlenmeyer (1/bead preparation). Sterilize by autoclaving 15 min/121 °C/15 lbs.
3. Sterile 1.5 % agar in Trypticase Soy Broth (TSB) for *mice* and 2 % agar in PBS 1× for *rats*. Sterilize by autoclaving 15 min/121 °C/15 lbs. Put 10 ml in 15 ml Falcon tube (1/ bead preparation) and maintain at 48 °C.
4. One 50 ml Falcon tube per bead preparation containing 9 ml of PBS and two 50 ml Falcon tubes containing sterile water and ethanol 70 % (for sterilization of the Polytron tip).
5. Prepare Petri plates with agar media containing appropriate antibiotics when requested.

2.2 Mice and Rats

Pathogen free C57BL/6NCrlBR male *mice* (22 g body weight per animal) and Sprague Dawley male *rats* (350 g body weight per animal) can be obtained from Charles River, housed in the animal care facility, allowing food and water *ad lib.* All protocols and experimentation require approval and rigorous following of the guidelines from the animal care and ethics committee.

2.3 Anesthesia and Euthanasia

For *mice* anesthesia dissolve 1 % Avertin (2,2,2-tribromethanol, 97 %) in 0.9 % NaCl and administer at a volume of 0.03 ml/g body weight by intraperitoneal injection (i.p.) using a 1 ml syringe 25 G 5/8 in. needle. CO_2 inhalation is used for euthanasia.

For *rats* combine the following for anesthesia: Ketamine (37.5 mg/ml) and Xylazine (5 mg/ml) in normal saline 0.9 % or PBS. The suggested dosage is 0.2 ml per 100 g body weight. Euthanyl (100 mg/kg) is used for euthanasia after light isoflurane anesthesia. All drugs are administrated by i.p. injection using 1 ml syringe 21–23 G 5/8 in. needle.

2.4 Intratracheal Intubation and Injection Components

The following items are necessary and should be prepared prior to the anesthesia procedure.

1. Heating blanket.
2. 70 % ethanol.

3. Cotton swabs.
4. Graefe Forceps and scissors.
5. 22 G 0.9×25 mm intravenous catheters (Becton Dickinson) for *mice* or 20 G "gavage" needle for *rats*.
6. 1 cc syringes containing agar-bead coated bacteria.

In addition only for *rat*:

7. *Rat* intubation packs (Hallowell EMC):

   Rodent tilting working stand with 40° inclination.

   Incisor loops.

   Otoscope, Welch Allyn LiIon.

   Autoclavable intubation speculum for rats.

   Rat endotracheal tube.

   Rat endotracheal tube guide wire.

   Rat lidocain applicator.

   Lidocain.

   Mirror for the verification of tracheal intubation.

2.5 Measurement of Bacterial Load in Lungs

1. Sterile PBS (for serial dilutions of bacteria).
2. 50 ml Falcon tubes containing sterile water and ethanol 70 % (for sterilization of the Polytron tip).
3. Prepare Petri plates with agar media containing appropriate antibiotics when requested.

2.6 Bronchoalveolar Lavage Fluid (BALF) Collection and Analysis

1. 70 % ethanol.
2. Cotton swabs.
3. Graefe Forceps and scissors.
4. Suture thread.
5. 22 G 0.9×25 mm intravenous catheters (Becton Dickinson) for *mice* or 20 G "gavage" needle for *rat*.
6. 1 cc syringes containing 1 ml of RPMI 1640.
7. Sterile PBS (for serial dilutions of bacteria).
8. Tuerk solution 1:2 and burker cell chamber.
9. Red Blood Cell (RBC) lysis buffer 1:10 (Biolegend).
10. Fetal Bovine Serum (Lonza).
11. Cytofunnels, superfrost ultraplus slides, and cytocentrifuge (Thermo Scientific).
12. Diff-Quik staining set (Medion Diagnostics).

2.7 Histopathology

1. 10 % Neutral buffered formalin (4 % formaldehyde) (Bio-optica).
2. Paraffin.

3. Microtome.
4. Superfrost Ultraplus Slides and cover slips (Thermo Scientific).
5. Hematoxylin and eosin solutions.
6. Brightfield microscope.

3 Methods

3.1 Preparation of Agar-Bead Embedded Bacteria for Chronic Infection of Mice and Rats

1. *Bacteria preparation.* Inoculate 5 ml of TSB, with appropriate antibiotics when requested, using one colony from a −80 °C thawed stock of a *P. aeruginosa* strain culture grown on agar plates with appropriate antibiotics when requested. Incubate overnight at 37 °C.
2. Inoculate a 250 ml Erlenmeyer containing 10 ml of TSB with 1 ml or less from an overnight culture (ON). Incubate to log phase at 37 °C/250 rpm.
3. Measure the optical density of the culture at $O.D_{600}$.
4. *For mice*: Cultures should have an O.D. between 0.8 and 1.2. Centrifuge in a centrifuge at 4,000 rpm (2,700 × *g*) for 15 min, discard the supernatant, vortexing throughly.

   For rats: Cultures should have an O.D. between 0.4 and 0.8. Wash approximately 500 μl of the log phase culture with the same volume of sterile PBS 1×. Centrifuge in a microfuge at 7,200 rpm (2,700 × *g*) for 2 min. Repeat **step 4** three times (*see* **Note 1**).
5. *Agar-bead preparation.* Prepare the setup for agar bead preparations and washing steps (*see* Fig. 1).
6. The 10 ml of sterile melted agar and the 200 ml of sterile mineral oil are pre-warmed in a water bath at 48 °C.
7. Put warm water in a glass container and place the 250 ml Erlenmeyer containing 200 ml of the mineral oil preheated at 48 °C; the agitation of the magnetic plate is adjusted at 500 rpm and must produce a visible vortex in the oil.
8. Add the individual or mixed bacterial cultures in a 15 ml Falcon tube containing 10 ml of liquid preheated (48 °C) agar 1.5–2 % in PBS or TSB. Mix quickly by vortexing.
9. Pour slowly the mixed agar-bacteria solution in the center of the Erlenmeyer containing the spinning mineral oil.
10. *For mice*: Transfer the preparation in a cold room at 4 °C to rapidly cool down and slow down the agitation at minimum speed for 20 min. Stop the agitation and maintain the preparation for 20 min in ice.

    For rats: Add rapidly ice chips to the side of the Erlenmeyer to rapidly cool down and maintain the agitation for 5 min.

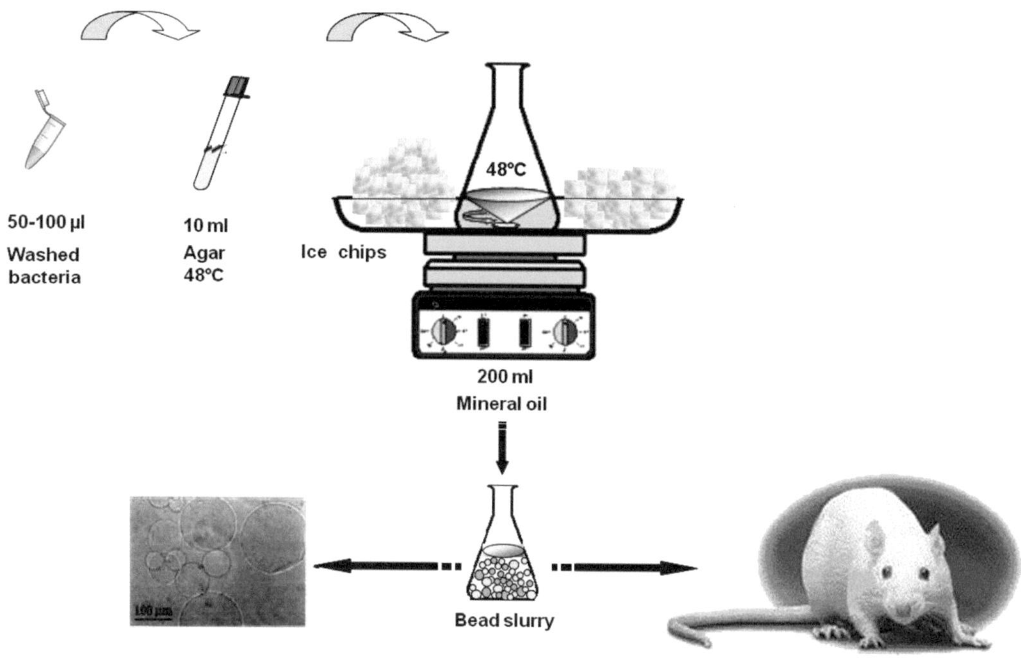

Fig. 1 Preparation of agar-bead embedded bacteria. Individual or mixed bacterial cultures are rapidly mixed with melted preheated agar and slowly poured in the center of the Erlenmeyer containing the spinning mineral oil. The preparation is rapidly cooled down and maintained in a slow stirring. Beads are then centrifuged and washed (*see* **step 5** of Subheading 3.1 for details and in accordance with using rats or mice)

Stop the agitation and maintain the preparation for 10 min at room temperature. Remove approximately a half of the mineral oil with a vacuum pump (*see* **Note 2**).

11. Pour the mixture into centrifugation tubes and complete to 200 ml with sterile PBS 1×.
12. Centrifuge at 10,000 rpm (16,274×*g*) for 20 min.
13. Aspirate the mineral oil and most of the PBS with the vacuum pump (*see* **Note 2**).
14. Complete to 200 ml with sterile PBS 1×.
15. Centrifuge at 10,000 rpm (16,274×*g*) for 20 min.
16. Repeat this wash twice (**steps 14–16**).
17. Aspirate and remove the maximum of PBS with the vacuum pump (at the end you will recover approximately 10 ml of beads in PBS) (*see* **Note 2**).
18. Recover agar beads slurry in a 50 ml sterile Falcon tube.

The agar-beads obtained can be stored for few weeks at 4 °C.

3.2 Determination of Bacterial Colony Forming Units (CFUs) in Agar Beads

1. To determine CFUs, bacteria enmeshed in beads are released using a Polytron. The Polytron probe is sterilized by immersing the probe in a 50 ml tube containing 10 ml of ethanol 70 % and rinsed in a tube containing 10 ml of sterile water.
2. Homogenize 1 ml of the bead slurry in a 50 ml Falcon tube containing 9 ml of sterile PBS with the Polytron using a burst of 30 s.
3. Prepare serial dilution (up to 10^{-6}) of the homogenized bead slurry and plate on TSA or Mueller-Hinton agar plates. If the preparation is to be used for CI, plates should contain appropriate antibiotics to select and differentiate the wild-type from the mutant strain. Both strains can be plated on Pseudomonas Isolation Agar (PIA) (*see* **Note 3**).
4. Incubate plates overnight at 37 °C.

   Continue with Subheading 3.4 for *rats*.
5. *For mice*: Determine the number of CFUs on plates and adjust the concentration of the agar beads, adding or removing PBS, to $2–4 \times 10^7$ CFUs/ml.
6. Continue with Subheading 3.5 for *mice*.

3.3 Preparation of Bacteria for Acute Infection of Mice

1. Inoculate 5 ml of TSB using one colony from a −80 °C thawed stock of a *P. aeruginosa* strain culture grown on agar plates. Incubate overnight at 37 °C.
2. Inoculate a 250 ml Erlenmeyer containing 10 ml of TSB with 1 ml or less from an overnight culture. Incubate to log phase at 37 °C/250 rpm.
3. Centrifuge at 4,000 rpm ($2{,}700 \times g$) for 15 min at 4 °C, discard the supernatant and wash with sterile PBS. Centrifuge again and resuspend the bacterial pellet in 2–3 ml of sterile PBS, vortexing throughly.
4. Measure the optical density of the culture at $O.D_{600}$. Quantify bacteria against a standard curve for the number of CFUs/OD_{600}.
5. Dilute bacteria to the desired concentration in sterile PBS.
6. Continue with Subheading 3.5.

3.4 Rat Infection via Intubation

1. Apply the ketamine/xylazine anesthesia procedure and place the rat onto the inclination table.
2. Push aside the rat's tongue.
3. Use the otoscope and visualize the trachea opening and closing.
4. Apply the lidocaine locally onto the vocal cords and wait 2–3 min.

5. Introduce the 20 G gavage needle (careful attention and practice is needed to insert into the trachea and not into the oesophage). Alternatively use an endotracheal tube and a guide wire.
6. Use a mirror to confirm that the intubation tube is well placed into the trachea.
7. Use a 1 cc syringe to inject 120 μl of beads containing between 2 and 4×10^6 CFUs/100 μl.
8. Introduce air 5–6 times with an empty syringe.
9. The animal is placed on a heating blanket until fully awake.

3.5 Mice Infection via Intratracheal Injection

1. Apply the Avertin anesthesia and place the mouse in supine position.
2. Disinfect the mouse's coat with 70 % ethanol.
3. Expose the trachea by a vertical cut of the skin and intubate the trachea with a sterile, flexible 22 G i.v. catheter.
4. Use a 1 cc syringe to inject 50 μl of beads containing between 1 and 5×10^6 CFUs of planktonic bacteria at the desired concentration, pushing gently the plunger of the syringe.
5. Close the incision by suture clips.
6. The animal is placed on a heating blanket until fully awake.

3.6 Read-Outs to Determine Virulence of P. aeruginosa During Acute and Chronic Infection

Mice can be observed daily for clinical signs including coat quality, posture, ambulation, and hydration status. Acute infection can be followed for a few days, while chronic infection is usually monitored after 7 days post infection but can be maintained up to 1 month. Animals can be monitored for several measurable readouts such as mortality, weight loss and bacterial CFUs. The host immune response is analyzed in terms of leukocyte recruitment and cytokines production from serum, BALF or directly from lung tissue. Histopathological lung studies can also be done using an additional group of animals.

3.6.1 Measurement of Bacterial Load and Cytokine Analysis in Lungs

Based on bacterial CFUs, the acute and chronic lung infection model can be used to evaluate the in vivo growth of different bacterial strains at different time points (*see* Fig. 3) or for competitive index (CI) studies between two strains (wild-type versus mutant strain into the same animal) (*see* Fig. 2b).

1. Euthanize animals by CO_2 inhalation for *mice* and Euthanyl (100 mg/kg) by i.p. injection after light isoflurane anesthesia for *rat*.
2. Place the animal in supine position. Disinfect the coat with 70 % ethanol. Expose the trachea and the thoracic cage by a vertical cut of the skin. Expose the lungs cutting the diaphragm. Excise lungs from the animal, rinse them in sterile PBS, separate lobes, put them in a round-bottom tube containing sterile PBS.

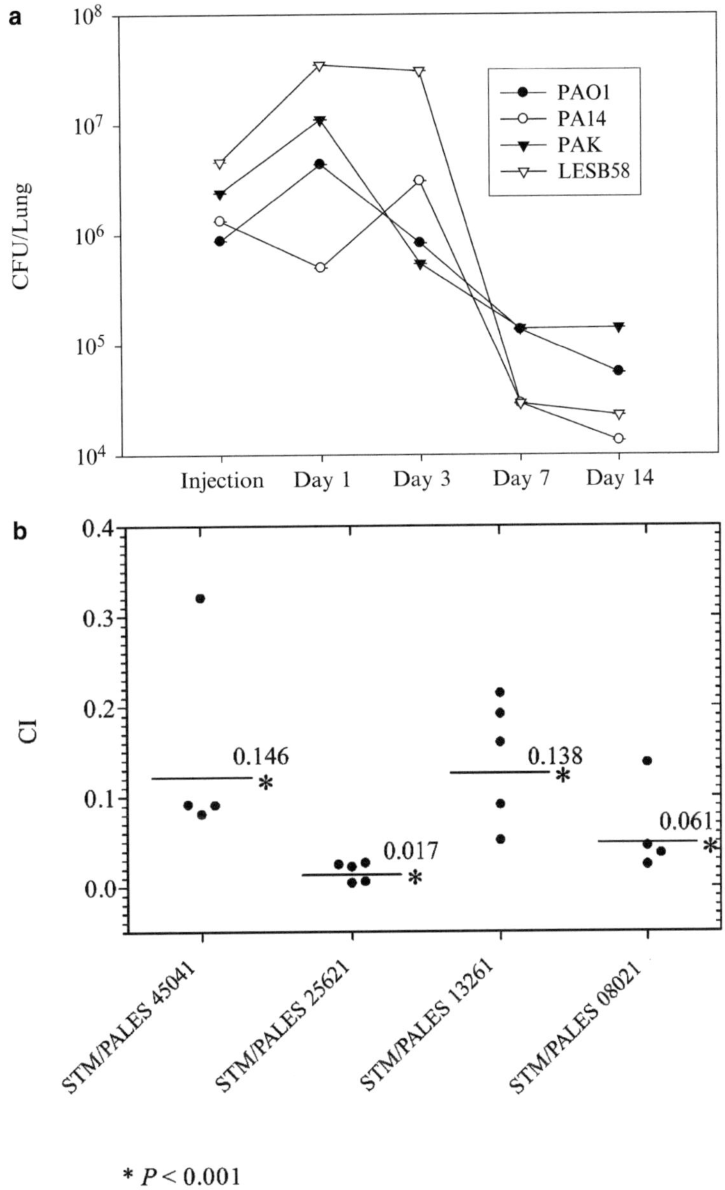

Fig. 2 (**a**) Infection kinetics of four *P. aeruginosa* strains in the rat model of chronic lung infection; (PAO1, PA14, PAK, and LESB58). Rats were infected by intubation with agar-embedded bacteria from 1×10^6–5×10^6 CFUs/lung. At different time points (1, 3, 7, and 14 days post infection), five rats were sacrificed from each group and CFUs were determined from infected lung tissues as described in Subheading 3.6.1. *Dots* represent the results of CFUs/lung for each animal, and *horizontal solid lines* represent the geometric mean value for each group [26]. (**b**) In vivo competitive index (CI) analysis between four selected *P. aeruginosa* STM mutants and the wild-type LESB58 strain in a rat model of lung infection. Equal bacterial ratios of each strain were embedded in agar beads and lungs were infected with 5×10^6 CFUs/lung by the intubation method. After 7 days post infection the lungs were recovered for CFUs determinations. Each *circle* represents the CI from a single animal in each group. A CI of less than one indicates a virulence/in vivo growth defect. The geometric mean of the CIs for all rats is shown as a *solid line*. *Asterisk* (*) indicates a *P* value: $P < 0.001$ [27]

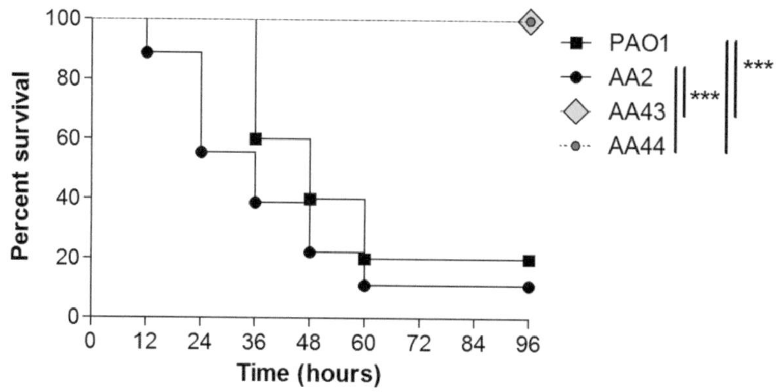

Fig. 3 Survival curve of reference strain PA01 and three clinical *P. aeruginosa* strains in the mouse model of acute infection. Mice were infected by intratracheal injection with planktonic bacterial cells (5×10^6 CFUs/lung). Survival of infected mice was followed every 12 h over a period of 4 days as described in Subheading 3.6.2. Statistical significance by Student's test is indicated: ***$P < 0.001$

If samples will be used for cytokines analysis, add protease inhibitors to PBS.

3. Homogenize the lungs under sterile conditions using a Polytron. The Polytron probe is sterilized by immersing the probe in a 50 ml tube containing 10 ml of ethanol 70 % and rinsed in a tube containing 10 ml of sterile water.
4. Plate serial dilution of the homogenized lung tissues on TSA or Mueller-Hinton agar plates. If the assay is to be used for CI, plates should contain appropriate antibiotics to select and differentiate the two strains (for example the wild-type from the mutant strain). Lungs can also be plated on selective media such as PIA.
5. Incubate plates overnight at 37 °C and count total bacterial output in the lungs.
6. *In vivo growth curves.* At different time points at least five animals are sacrificed and bacterial CFUs can be determined from infected lungs (*see* Fig. 2a and Fig. 3).
7. *CI calculations* (*see* **Note 4**) (*see* example in Fig. 2b).
8. *ELISA.* Centrifuge the remaining homogenate at 13,000 rpm ($16,000 \times g$) for 30 min at 4 °C with a microfuge. Take the supernatant for ELISA cytokine analysis and store at −80 °C.

3.6.2 In Vivo Lethality of Acute Lung Infection

The acute lung infection model can also be used to evaluate the in vivo lethality of different bacterial strains. Survival can be monitored every 12 h for up to 7–10 days. Mice that lose >25 % body weight and have evidence of severe clinical disease, such as scruffy coat, inactivity, loss of appetite, poor ambulation, dehydration or

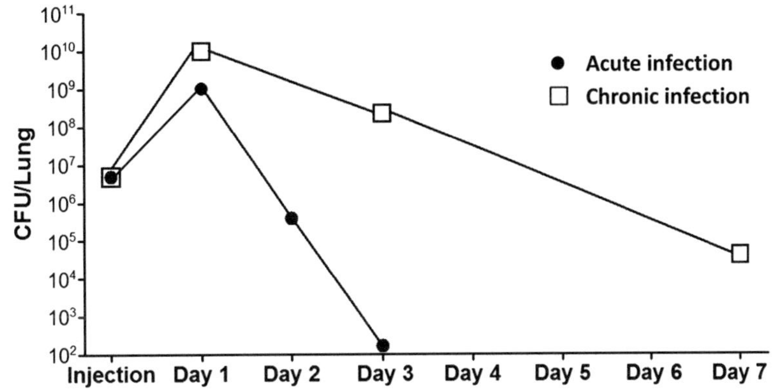

Fig. 4 Infection kinetics of *P. aeruginosa* strain PAO1 in the mouse model of acute and chronic infection. Mice were infected by intratracheal injection with 5×10^6 CFUs/lung either in planktonic suspension or embedded in agar beads. At different time points (1, 2, 3, and 7 days post infection), at least five mice were sacrificed from each group and CFUs were determined from infected lung tissues as described in Subheading 3.6.1. *Symbols* represent the geometric mean value of CFUs/lung for each group (Figure modified from Bragonzi et al. 2010 [28])

painful posture, are sacrificed before termination of the experiment as previously described. Figure 4 shows as some strains can be totally avirulent, whereas others can induce high percentage of mortality within the first 2 days from infection.

3.6.3 BALF Collection and Analysis of the Host Immune Response

Both the acute and the chronic lung infection models are powerful tools to study the host response to *P. aeruginosa* infection, by analyzing the leukocytes recruitment in the BALF and measuring cytokines/chemokines in the BALF, in the whole lung and the serum.

1. Euthanize mice as previously described, place the mouse in supine position and disinfect the coat of the mouse with 70 % ethanol. Expose the trachea and the thoracic cage by a vertical cut of the skin. Expose the lungs cutting the diaphragm. Insert a suture thread under the trachea using tweezers and intubate the trachea with a sterile, flexible 22 G 0.9×25 mm I.V. catheter. Pull up the two ends of the suture thread to bind the catheter to the trachea with a knot around the trachea.
2. Lungs are washed by pushing the plunger back and forth with 1 ml of RPMI 1640 (for cytokine analysis add protease inhibitors to RPMI medium) using a 1 ml syringe attached to a catheter. Recover immediately the liquid in a 15 ml tube. Repeat this step three times.
3. For quantification of bacteria present in the BALF, take a small aliquot, serially dilute 1:10 in PBS, plate on TSA plates, and incubate at 37 °C overnight.

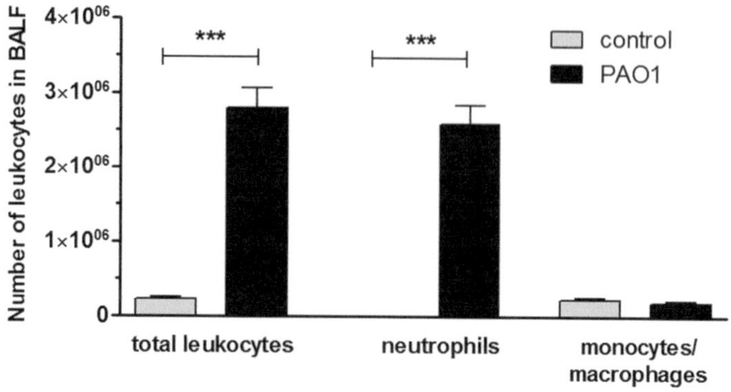

Fig. 5 Lung inflammatory response after acute *P. aeruginosa* infection in mice. Mice were infected by intratracheal injection with *P. aeruginosa* PAO1 reference strain planktonic bacterial cells (1×10^6 CFUs/lung) and euthanized after 24 h. The number of total leukocytes and in particular of neutrophils and monocytes/macrophages recruited in the airways were analyzed in BALF of infected mice in comparison with not infected mice (control), as described in Subheading 3.6.3. *Values* represent the mean ± SEM. Statistical significance by Student's test is indicated: ***$P < 0.001$

4. Count total cells using an inverted light optical microscope diluting an aliquot of the BALF 1:2 with Tuerk solution in a burker cell count chamber (see example in Fig. 4).
5. Centrifuge the remaining BALF at 1,400 rpm ($330 \times g$) for 8 min at 4 °C. Take the supernatant for ELISA cytokine analysis, storing it at −80 °C.
6. If the pellet is red, you can lyse erythrocytes resuspending pellet in 250–300 μl of RBC lysis buffer diluted 1:10 in MilliQ water for 3 min. Neutralize with 2 ml PBS and centrifuge at 1,400 rpm ($330 \times g$) for 8 min at 4 °C.
7. Discard the supernatant and resuspend pellet in RPMI 10 % fetal bovine serum (FBS). You must use a volume so as to have 1×10^6 cells/ml, basing on the total cell count.
8. Pipette 150 μl on a cytofunnel mounted with superfrost ultra plus microscope slides and centrifuge at 960 rpm for 5 min with a cytocentrifuge.
9. Stain slides by Diff-Quik staining. Differential cell count can now be performed using an inverted light optical microscope (*see* example in Fig. 5).

3.6.4 Histopathology

The host response in terms of leukocytes recruitment, tissue damage, and localization of bacteria can be also evaluated in histological samples.

1. Euthanize the mouse or the rat and recover lungs as previously described, rinse lungs in PBS, separate lobes, put them in a

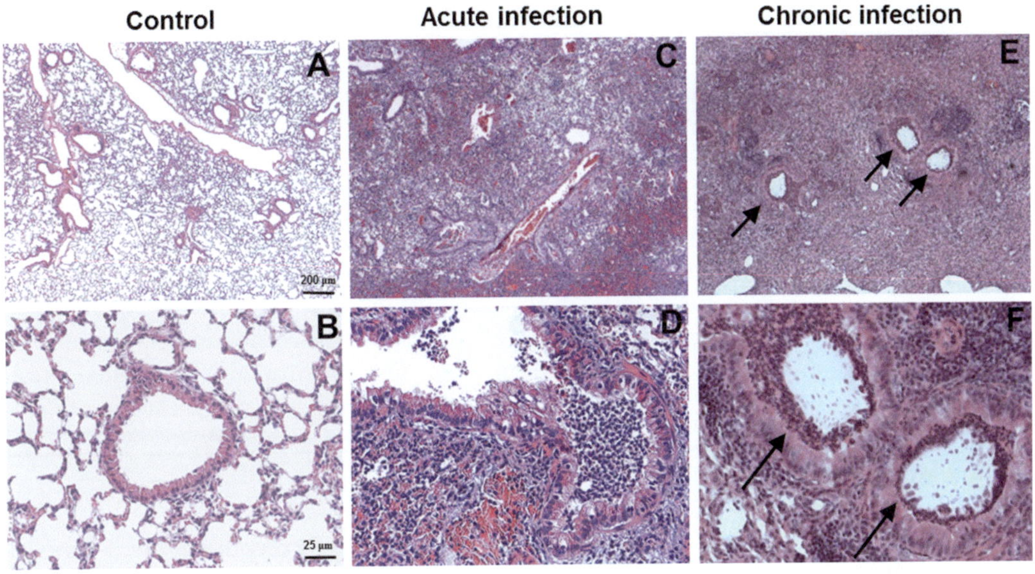

Fig. 6 Histopathology of murine lungs after acute and chronic infection with *P. aeruginosa*. Mice were infected with 1–5 × 10^6 CFUs/lung of *P. aeruginosa* planktonic bacteria for 24 h (acute infection, (**c**, **d**)) or with *P. aeruginosa* bacteria embedded in agar-beads for 14 days (chronic infection, (**e**, **f**)). Control mice were not infected (**a**, **b**). Lungs were processed and stained with H & E as described in Subheading 3.6.4. Extensive infection and inflammation are visible in bronchia and pulmonary parenchyma of infected mice. Severity of lesions and lung involvement is heterogeneous in different lobes of the same mice. In chronically infected mice, *P. aeruginosa* macrocolonies are visible in the agar-beads present in the bronchia, indicated with *arrows* in (**e**, **f**) (Figure modified from Lorè et al. 2012 [3])

tube containing 5–10 ml 10 % neutral buffered formalin (4 % formaldehyde) and store at 4 °C protected from light.

2. Embed lungs in paraffin, according to standard procedures.
3. Cut 5-µm-thick sections using a microtome.
4. Stain slides by hematoxylin and eosin (H & E) staining according to standard procedures.
5. Slides can be examined using a brightfield microscope and images can be acquired connecting it with a camera (*see* example in Fig. 6).

4 Notes

1. Please note that the concentration of bacteria to be used has to be determined by trial and error for each bacterial strain. Different *P. aeruginosa* strains may have different growth rates particularly when using clinical isolates or small colony variants. For a competitive index experiment two strains will be mixed (the wild-type and the mutant strain of interest) in the same ratio after the washing step (*see* **step 4** of Subheading 3.1).

2. This is a critical step and careful precaution should be taken not to aspirate beads on the bottom.
3. Dehydrated plates prior to plating at 37 °C for 2 h (to be dehydrated even if they are freshly prepared).
4. The competitive index (CI) is defined as the CFUs output ratio of mutant when compared to wild-type strain, divided by the CFUs input ratio (in the agar-beads) of mutant to wild-type strain [24]. The final CI was calculated as the geometric mean for animals in the same group, and experiments were done at least in triplicate [25]. Each in vivo competition was tested for statistical significance by using the Student's two-tailed *t*-test.

Acknowledgements

R. C. Levesque is a research scholar of exceptional merit from the Fond de Recherche du Québec en Santé (FRQS). His laboratory is funded by the Canadian Institute for Health Research (CIHR), a CIHR-UK team grant, the CIHR-FRQS-Québec Respiratory Health Network (RSR), the Natural Sciences and Engineering Research Council of Canada (NSERC), Genome Québec, the Fonds de recherche du Québec Nature et technologies, and the Alberta Innovates Bio Solutions program.

Research in Bragonzi's laboratory is funded by the Italian Cystic Fibrosis Foundation (CFaCore) and the European Commission (Grant NABATIVI, EU-FP7-HEALTH-2007-B, contract number 223670).

References

1. Bianconi I et al (2011) Positive signature-tagged mutagenesis in *Pseudomonas aeruginosa*: tracking patho-adaptive mutations promoting airways chronic infection. PLoS Pathog 7:e1001270
2. Dunn MM, Toews GB, Hart D, Pierce AK (1985) The effects of systemic immunization of pulmonary clearance of *Pseudomonas aeruginosa*. Am Rev Respir Dis 131:426–431
3. Lore NI et al (2012) Cystic fibrosis-niche adaptation of *Pseudomonas aeruginosa* reduces virulence in multiple infection hosts. PLoS ONE 7:e35648
4. Schook LB, Carrick L Jr, Berk RS (1977) Experimental pulmonary infection of mice by tracheal intubation of Pseudomonas aeruginosa: the use of antineoplastic agents to overcome natural resistance. Can J Microbiol 23: 823–826
5. Yu H, Hanes M, Chrisp CE, Boucher JC, Deretic V (1998) Microbial pathogenesis in cystic fibrosis: pulmonary clearance of mucoid *Pseudomonas aeruginosa* and inflammation in a mouse model of repeated respiratory challenge. Infect Immun 66:280–288
6. McCray PB Jr, Zabner J, Jia HP, Welsh MJ, Thorne PS (1999) Efficient killing of inhaled bacteria in DeltaF508 mice: role of airway surface liquid composition. Am J Physiol 277: L183–L190
7. Cash HA, Woods DE, McCullough B, Johanson WG Jr, Bass JA (1979) A rat model of chronic respiratory infection with *Pseudomonas aeruginosa*. Am Rev Respir Dis 119:453–459
8. Klinger JD, Cash HA, Wood RE, Miler JJ (1983) Protective immunization against chronic *Pseudomonas aeruginosa* pulmonary infection in rats. Infect Immun 39:1377–1384
9. Bragonzi A et al (2006) Sequence diversity of the mucABD locus in *Pseudomonas aeruginosa* isolates from patients with cystic fibrosis. Microbiology 152:3261–3269

10. Hoiby N et al (2001) *Pseudomonas aeruginosa* and the in vitro and in vivo biofilm mode of growth. Microbes Infect 3:23–35
11. Hodgson AE, Nelson SM, Brown MR, Gilbert PA (1995) Simple *in vitro* model for growth control of bacterial biofilms. J Appl Bacteriol 79:87–93
12. Rejman J et al (2010) Impact of chronic pulmonary infection with *Pseudomonas aeruginosa* on transfection mediated by viral and nonviral vectors. Hum Gene Ther 21:351–356
13. Cantin AM, Woods DE (1999) Aerosolized prolastin suppresses bacterial proliferation in a model of chronic *Pseudomonas aeruginosa* lung infection. Am J Respir Crit Care Med 160:1130–1135
14. Amano H et al (2000) Role of cytokine-induced neutrophil chemoattractant-2 (CINC-2) alpha in a rat model of chronic bronchopulmonary infections with *Pseudomonas aeruginosa*. Cytokine 12:1662–1668
15. Omri A, Suntres ZE, Shek PN (2002) Enhanced activity of liposomal polymyxin B against *Pseudomonas aeruginosa* in a rat model of lung infection. Biochem Pharmacol 64: 1407–1413
16. Potvin E et al (2003) *In vivo* functional genomics of *Pseudomonas aeruginosa* for high-throughput screening of new virulence factors and antibacterial targets. Environ Microbiol 5:1294–1308
17. Duan K, Dammel C, Stein J, Rabin H, Surette MG (2003) Modulation of *Pseudomonas aeruginosa* gene expression by host microflora through interspecies communication. Mol Microbiol 50:1477–1491
18. Lehoux DE, Sanschagrin F, Kukavica-Ibrulj I, Potvin E, Levesque RC (2004) Identification of novel pathogenicity genes by PCR signature-tagged mutagenesis and related technologies. Methods Mol Biol 266:289–304
19. Joly B et al (2005) Relative expression of *Pseudomonas aeruginosa* virulence genes analyzed by a real time RT-PCR method during lung infection in rats. FEMS Microbiol Lett 243: 271–278
20. Woods DE, Cantin A, Cooley J, Kenney DM, Remold-O'Donnell E (2005) Aerosol treatment with MNEI suppresses bacterial proliferation in a model of chronic *Pseudomonas aeruginosa* lung infection. Pediatr Pulmonol 39:141–149
21. Zhang L et al (2005) Antimicrobial peptide therapeutics for cystic fibrosis. Antimicrob Agents Chemother 49:2921–2927
22. Boyer S et al (2005) Chronic pneumonia with P*seudomonas aeruginosa* and impaired alveolar fluid clearance. Respir Res 6:17
23. Kukavica-Ibrulj I, Levesque RC (2008) Animal models of chronic lung infection with *Pseudomonas aeruginosa*: useful tools for cystic fibrosis studies. Lab Anim 42:389–412
24. Lehoux DE, Sanschagrin F, Levesque RC (2000) Genomics of the 35-kb pvd locus and analysis of novel pvdIJK genes implicated in pyoverdine biosynthesis in *Pseudomonas aeruginosa*. FEMS Microbiol Lett 190: 141–146
25. Hava DL, Camilli A (2002) Large-scale identification of serotype 4 *Streptococcus pneumoniae* virulence factors. Mol Microbiol 45: 1389–1406
26. Kukavica-Ibrulj I et al (2008) *In vivo* growth of *Pseudomonas aeruginosa* strains PAO1 and PA14 and the hypervirulent strain LESB58 in a rat model of chronic lung infection. J Bacteriol 190:2804–2813
27. Winstanley C et al (2009) Newly introduced genomic prophage islands are critical determinants of in vivo competitiveness in the Liverpool epidemic strain of *Pseudomonas aeruginosa*. Genome Res 19:12–23
28. Bragonzi A (2010) Murine models of acute and chronic lung infection with cystic fibrosis pathogens. Int J Med Microbiol 300: 584–593

Chapter 59

Assessing *Pseudomonas* Virulence Using Mammalian Models: Acute Infection Model

Antje Munder and Burkhard Tümmler

Abstract

The acute murine lung infection model monitors *Pseudomonas aeruginosa* airway infections by multiple continuous and endpoint parameters. After intratracheal or intranasal infection it characterizes the course of infection via head-out spirometry, rectal temperature, weight loss, a body condition score based on nine physiological parameters, lung bacterial numbers, organ dissemination of bacteria, and a semiquantitative assessment of lung inflammation and further analysis. The generated data allows a robust classification of virulence of mutant or wild-type *P. aeruginosa* strains and/or of the susceptibility of wild-type or engineered mouse strains to infection. If standardized, the model is applicable to the preclinical assessment of antipseudomonal prevention and intervention strategies.

Key words Mouse infection model, Noninvasive lung function, Lung inflammation, Bacterial numbers, Weight loss, Pathohistology, Body condition score

1 Introduction

The ubiquitous *Pseudomonas aeruginosa* is an opportunistic pathogen for multiple eukaryotic hosts ranging from protozoa, yeasts, and fungi to worms, insects, higher plants, and vertebrates [1]. There are anecdotic reports that among mammalian species dolphins [2] and minks [3] are particularly susceptible to infection with *P. aeruginosa*. The mouse does not exhibit such a tropism but nevertheless *Mus musculus* has become for obvious reasons the prevailing mammalian animal model for infections with *P. aeruginosa* [1]. First, the mouse is the prime model species in immunology and correspondingly an enormous body of knowledge, tools, and protocols exists to elucidate host defence mechanisms against bacterial infection. Second, hundreds of inbred mouse strains [4] with a defined genetic background are available, varying in their susceptibility towards *P. aeruginosa* and thereby suitable for all kinds of infection studies. Third, the engineering of recombinant knockout or transgenic mice has become a routine procedure so that the role

Alain Filloux and Juan-Luis Ramos (eds.), *Pseudomonas Methods and Protocols*, Methods in Molecular Biology, vol. 1149, DOI 10.1007/978-1-4939-0473-0_59, © Springer Science+Business Media New York 2014

of individual traits, epigenetic modifications, genes, and mutations can be resolved in such detail that is currently not feasible in any other mammalian species.

P. aeruginosa causes a wide range of syndromes in humans that can vary from local to systemic, subacute to chronic, and superficial and self-limiting to life-threatening. Of the severe *P. aeruginosa* infections, effective treatment measures have decreased the frequencies of eye and burn wound infections, but the incidence and prevalence of airway infections with *P. aeruginosa* have been and are still increasing. *P. aeruginosa* has become the most common Gram-negative pathogen in community-acquired pneumonia, healthcare-associated pneumonia, and ventilated-associated pneumonia in intensive care units [5, 6]. This background explains the needs and interest of health authorities, industry, and scientific community to have a robust airway infection model at hand to test novel measures of prevention and treatment.

This chapter describes an acute airway infection model in mice. *P. aeruginosa* bacteria are introduced into the murine airways by intranasal or intratracheal instillation, and the host response is then monitored by body weight, temperature, lung function, and an ethological body condition score. The longitudinal observations are supplemented by endpoint readouts selected according to the objectives of the study.

2 Materials

2.1 Media

1. Luria Bertani broth: For 1 L dissolve 15 g tryptone, 10 g NaCl, and 5 g yeast extract with stirring and heating in 950 mL deionized water. Adjust the pH of the solution to 7.0 with NaOH, and bring the volume up to 1 L. It may not be necessary to adjust pH, since pH of LB is naturally close to pH 7.0. Autoclave for 20 min at 1.21 bar; antibiotics can be added if necessary when the medium has cooled to approximately 40 °C. Medium can be stored at room temperature or at 4 °C if antibiotics are added.
2. Luria Bertani agar: Prepare LB medium as above, and add 20 g/L agar before autoclaving. After autoclaving, allow to cool to approximately 40 °C, add antibiotic if necessary, and pour onto plates. Plates can be stored for at least 1 week at +4 °C.

2.2 Mice

The model can be applied to any mouse strain; gender and age within the range of 15 to 36 g body weight (*see* **Note 1**). Susceptibility of each mouse strain has to be titrated for each *P. aeruginosa* isolate which is selected for infection. Depending on the infection dose disease manifestations can vary from almost

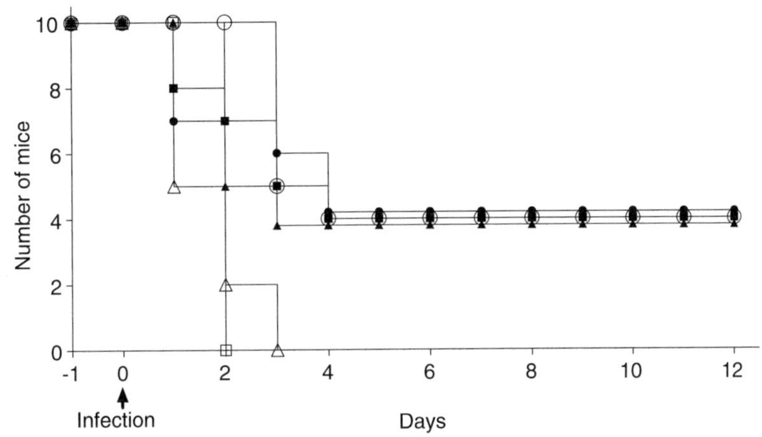

Fig. 1 Survival of C3H/HeN mice intratracheally infected with different PAO1 laboratory strains. Mice were inoculated with 10^6 or 10^7 cfu of three PAO1 laboratory strains (for strain description refer to ref. 7). Survival of mice whose airways had been inoculated with 10^6 cfu (*closed symbols*) or 10^7 cfu (*open symbols*) of PAO1-DSM (*squares*), MPAO1 (*circles*), or PAO1-UW (*triangles*). Copyright© American Society for Microbiology. J Bacteriol 192, 1113–1121, doi: 10.1128/JB.01515-09

inapparent to 100 % fatality. Infection dose therefore plays a crucial role for the target of the infection experiments (*see* Fig. 1, taken from ref. 7).

For the intratracheal (i.t.) application a body weight ≥15 g is recommended; otherwise, instillation becomes manually challenging. The device for the head-out spirometry allows lung function measurement of mice with a maximal weight of 36 g. Prior to the start of the experiments animals are acclimatized for at least 7 days (*see* **Note 2**).

2.3 Anaesthesia Components (See Table 1)

1. Systemic intraperitoneal (i.p.) anaesthesia: Mix 5 mL midazolam with 1 mL ketamine (100 mg/mL) per 10 g body weight, and fill up to 10 mL with 0.9 % (w/v) aqueous NaCl solution. Inject intraperitoneally a dose equivalent of 0.10 mL/10 g body weight.
2. Premedication: 0.5 mg atropine sulfate is filled up to 10 mL with 0.9 % (w/v) aqueous NaCl solution.
3. Heating pad to keep mice at body temperature during recovery from anaesthesia.

2.4 Materials for Intratracheal Instillation

1. Cold light lamp (e.g., KL 1500, Schott, Mainz, Germany).
2. Oblique plane (e.g., built from acrylic glass with a base of 120×100 mm; 140 mm height, 170 mm slope, and right angled; *see* Fig. 2); two bolts at a distance of approximately 40 mm on the upper site allow fixing of the rubber band.

Table 1
Compounds of anaesthesia

Systemic anaesthesia	Drug stock concentration (mg/mL)	Volume used for cocktail	Volume of cocktail administered to mouse	Dose administered to mouse
Ketamine	100 mg/mL	1.0 mL		100 mg/kg
Midazolam	1.0 mg/mL	5.0 mL		1.0 mg/kg
0.9 % (w/v) aqueous NaCl solution	Sterile, isotonic	4.0 mL		
Combination cocktail		10.0 mL total	0.10 mL/10 g	
Premedication	Drug stock concentration (mg/mL)	Volume used for cocktail	Volume of cocktail administered to mouse	Dose administered to mouse
Atropine	0.5 mg/mL	1.0 mL		0.1 mg/kg
0.9 % (w/v) aqueous NaCl solution	Sterile, isotonic	4.0 mL		
Combination cocktail		5.0 mL total	0.010 mL/10 g	

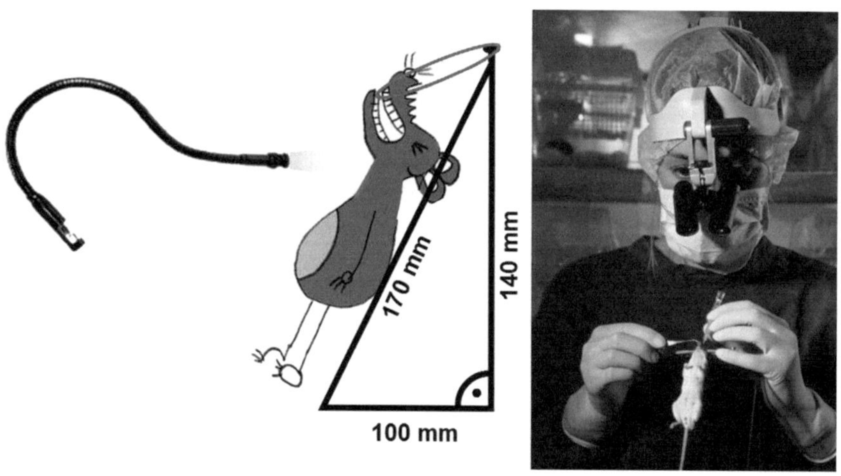

Fig. 2 *Left*: Scheme of an anaesthetized mouse positioned on the oblique plane prior to i.t. instillation, the animal is fixed via rubber band on its front teeth, its throat illuminated via a cold light lamp. *Right*: Photograph of the i.t. instillation via head-worn loupe using forceps to fix the murine tongue

3. Rubber band (like used in household application).
4. Venous catheter 24 g ¾″ (0.7 × 19 mm), with wire guide shortened (guide should be completely covered by the plastic tube).
5. Hamilton syringe with luer tip, syringe capacity 10–100 μL, needle gauge 24.

6. Spatula, forceps.
7. Head-worn loupe (Zeiss, Germany) with 8× magnification.

2.5 Material for Intranasal Application

1. Microliter pipette 10–100 μL.
2. Pipette tips, 20 μL (well suited are flexible tips which are specially designed for loading samples onto gels).

2.6 Equipment for Measuring the Rectal Temperature

1. Temperature-measuring device connected with a wired sensor.
2. Petroleum jelly to make the wired sensor more lubricated.

2.7 Equipment for Noninvasive Head-Out Spirometry

1. Sampling chambers:
   A custom-made glass corpus with four sampling chambers enables the parallel investigation of four mice (Fig. 3). The entrance end of each chamber is secured with a screwable cap with a movable glass piston. The piston enables to fix each individual mouse according to its body size. At the outlet end two membranes (see below) are fixed with rubber rings; the flexibility of the membranes allows safe positioning of the mice, and only the dense closing around the mice' necks enables measuring of the pressure variations within the chambers.

   In general, we use three different sizes of sampling chambers to hold mice of varying sizes ranging up to 36 g. The smallest inserts are 85 mm in length with an outer diameter of 30 mm and an opening at one end of 25 mm in diameter. They can accommodate young mice with a body weight up to 24 g.

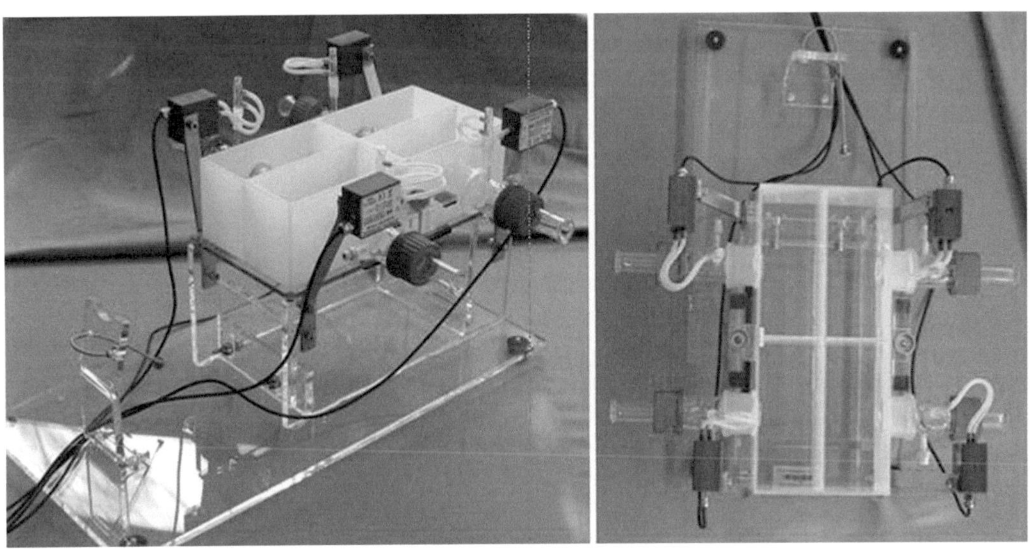

Fig. 3 The spirometry device. Photographs show the setup of the four glass sampling chambers, inserted in a rectangular box and connected via silicone tubes to the four pneumotachographs. Cables are leading from pneumotachographs to the pressure transducer (not illustrated). *Left*: side view; *right*: view from the top

For mice weighing 24–32 g inserts with a longer tube are available. They measure 110 mm in length and the same diameter and opening size as mice in this weight range increase only in length rather than diameter. For mice above 32 g inserts have a slightly larger diameter and a larger opening for easier access of the mice. These are 115 mm in length with an outer diameter of 35 mm and an opening of 35 mm. This makes the system very versatile and underlines the usability for different mouse groups/ages.

For each sampling chamber, a pneumotachograph and a pressure transducer (see below) are positioned directly above the insert. An amplifier and an analog-to-digital converter (see below) are employed for digitalization of the airflow. For data acquisition and analysis the software from Notocord Hem (see below) is used.

2. Membranes:
   Silicone and latex membranes are used to ensure a safe but flexible positioning of the mice within the sampling chambers. Membranes should be of 0.5 mm strength and have an outer circle of 5.4 cm. We use rectangular dental dam membranes which are cut with a round stencil. The inner circles are cut corresponding to the mice neck size (Table 2). Membranes are fixed with two rubber rings (outer membrane latex, inner membrane silicone on the open end of the sampling chamber).
3. Rubber rings, inner circle 21 mm, 1.5 mm size.
4. Silicone tubes (inner circle 2.0 mm, outer circle 4.0 mm) for connection of the sampling chambers to the pneumotachographs.
5. Four pneumotachographs (PTM Type 378/0.9 Hugo Sachs Elektronik—Harvard Apparatus (March-Hugstetten, Germany)).
6. Four pressure transducers (DLP10 Type 380 Hugo Sachs Elektronik—Harvard Apparatus (March-Hugstetten, Germany)).

Table 2
Diameter of membranes according to the murine body weight

Body weight (g)	Diameter in membrane (mm)
15–17	7–8
17–21	8–9
21–25	9–10
25–28	10–11
28–32	11
32+	12

7. Analog-to-digital converter (Data Translation 16 Channels, USB BNC Box, DT9800 BMC Box 16SE Hugo Sachs Elektronik—Harvard Apparatus (March-Hugstetten, Germany)).
8. Amplifier (Transducer Amplifier Module Type 705/1 Hugo Sachs Elektronik—Harvard Apparatus (March-Hugstetten, Germany)).
9. Spirometry software Notocord Hem (version 4.2.0.241, Notocord Systems, Croissy-sur-Seine, France).

2.8 Materials for Tissue Homogenization and Bacterial Counting

1. Tissue homogenizer.
2. Tubes for dilution rows.
3. Microliter pipettes, 100–1,000 μL and 10–100 μL.

3 Methods

3.1 Anaesthesia of Mice (See Table 1)

Anaesthesia for i.t. as well as for intranasal (i.n.) application is carried out via an intraperitoneal injection of 1 mg ketamine plus 10 μg midazolam per 10 g bodyweight. To reduce anaesthesia-induced salivation mice are premedicated with atropine (1 μg/animal) subcutaneously for the i.t. application. Depth of anaesthesia is tested by swallowing and foot reflex. Narcotized mice should be put on a heating pad to avoid a drop in body temperature. Artificial tears should be used to prevent eyes from drying.

3.2 Bacteria

P. aeruginosa stock cultures are maintained at –80 °C. For infection bacteria are grown in Luria broth (LB) overnight at 37 °C. The overnight culture has to be washed twice with the same volume of sterile buffer (e.g., PBS, HEPES saline) to remove cell detritus and secreted exopolysaccharides. The optical density of the bacterial suspension is determined, and the intended number of colony-forming units (cfus) is extrapolated from a standard growth curve, which should be generated for each individual *Pseudomonas* strain. Inocula with the intended infective dose are prepared by dilution with sterile buffer.

3.3 Application Techniques

The inoculation of bacteria should be performed within 30 min for the whole group of mice to exclude major changes in the viable counts of bacteria. You should be aware of the fact that the route of inoculation might have a distinct effect on the outcome of the experiment [8]. For comparison of i.t. and i.n. application *see* Fig. 4.

1. Intratracheal Instillation [9]:
   (a) Individual mice are fixed on an oblique plane, their throat illuminated with a focused cold light lamp, and the tongue of the animal pulled out of its mouth by forceps. A small spatula is used to retain the tongue.

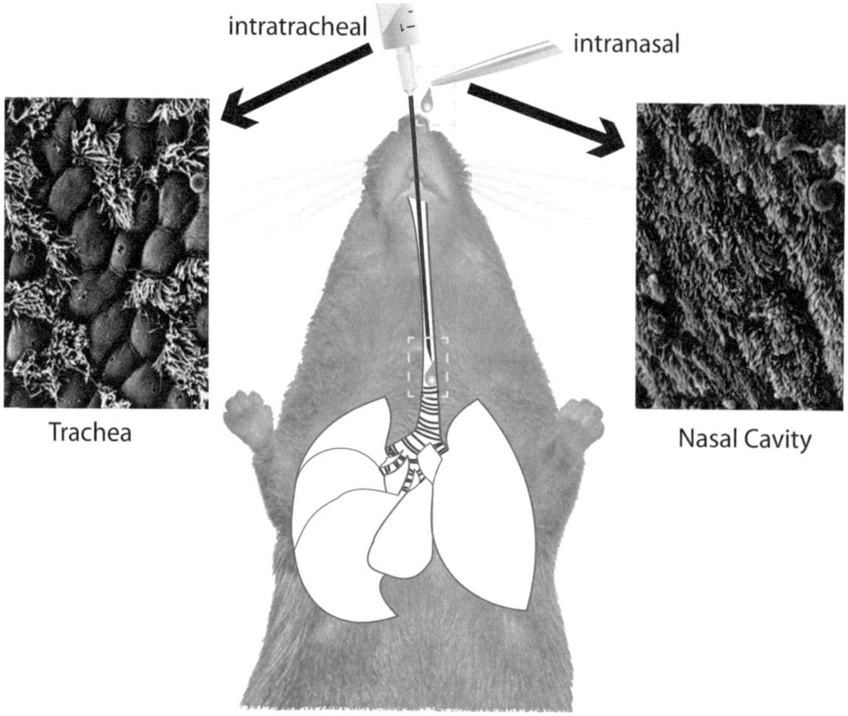

Fig. 4 Experimental lung infection—Impact of infection route. The sketch highlights the differential experimental setup for intranasal and intratracheal infection in mice. Compared to an intranasal infection route the intratracheal instillation bypasses the initial host immune response conducted by the respiratory epithelium of the upper airways and deposits more bacteria into the distal airways. Please note the differential abundance of respiratory epithelium in the upper airways (*left photograph*) and the lower conducting airways of mice (*right photograph*). Taken from ref. [8]

(b) By using an eightfold enlarging head-worn loupe instillation can be done under view, directly into the trachea (*see* **Note 3**).

(c) Insert the venous catheter into trachea, fix microliter syringe filled with bacterial suspension, and apply carefully.

(d) After instillation up to 1 mL air should be blown in the lung by the catheter to ensure that the whole volume of the bacterial suspension has reached the respiratory tract.

The maximum tolerable volume which can be applied intratracheally to the murine lung is 100 μL. Since this amount of liquid already causes dyspnea it is advisable to keep the volume below 80 μL. Instillation should be done smoothly and slowly to avoid reflux into the oropharynx.

Post-instillation mice should be kept warm and kept under surveillance till they recover from anaesthesia.

2. Intranasal Application:

(a) Mice are anaesthetized as described above (without premedication) and placed in supine position in the experimenter's hand.

(b) For injection, the tip is positioned directly on a nostril and bacteria are applied slowly and in smallest drops. Thereby the applied volume could be inhaled completely. In our model we use volumes from 20–30 μL. Up to 50 μL can be applied easily; larger amounts of liquid increase the risk of sneezing and coughing during application (*see* **Note 4**).

3.4 Assessment of Physiological Parameters

The parameters described in the following are used to characterize the course of an acute *P. aeruginosa* lung infection:

1. Body weight (Fig. 5):

   The weight of the mice is measured daily, since this parameter reflects more delayed changes. Awake mice are placed on a balance with a weighing range from 1 to 2,000 g and an accuracy of 0.1 g. It is important to use a balance with short stabilization time, since usually mice move permanently on it.

2. Rectal temperature (Fig. 6a, b):

   Temperature is assessed with an electronic temperature-measuring device connected with a wired sensor (Fig 6a). Accuracy of the measuring device should be controlled prior to the experiment with a reference thermometer. Temperature is a very sensitive and easy-to-measure parameter (*see* **Note 5**). We suggest to tape the end of the sensor at a distance of 1.5 cm and to put into petroleum jelly before insertion to avoid rectal injury.

3. Body condition score (Table 3/Fig. 7):

   The behavioral score utilizes the visual inspection of the features vocalization, piloerection, posture, movement, breathing,

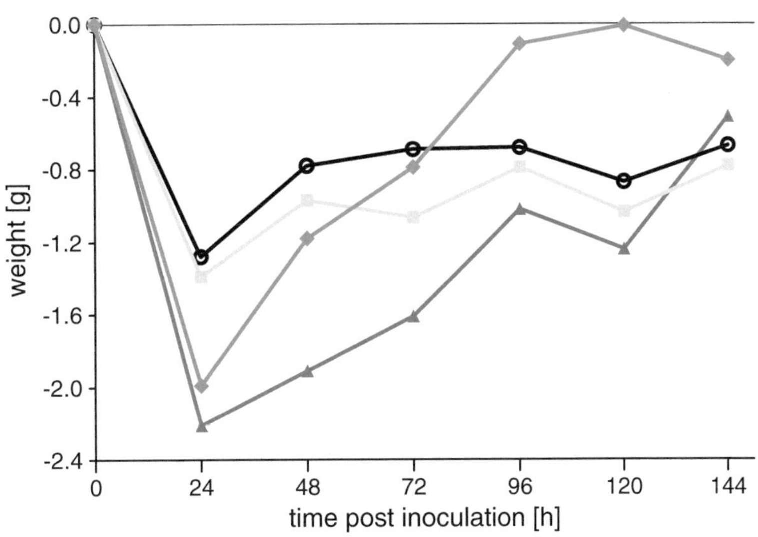

Fig. 5 Body weight. Decrease of body weight displayed in delta values in C57BL/6 mice during infection with three *P. aeruginosa* strains of different virulence. Mice were infected i.n. each with 1.5×10^6 cfu of the strains, curve with *open circles* displays the mock control (mice received PBS)

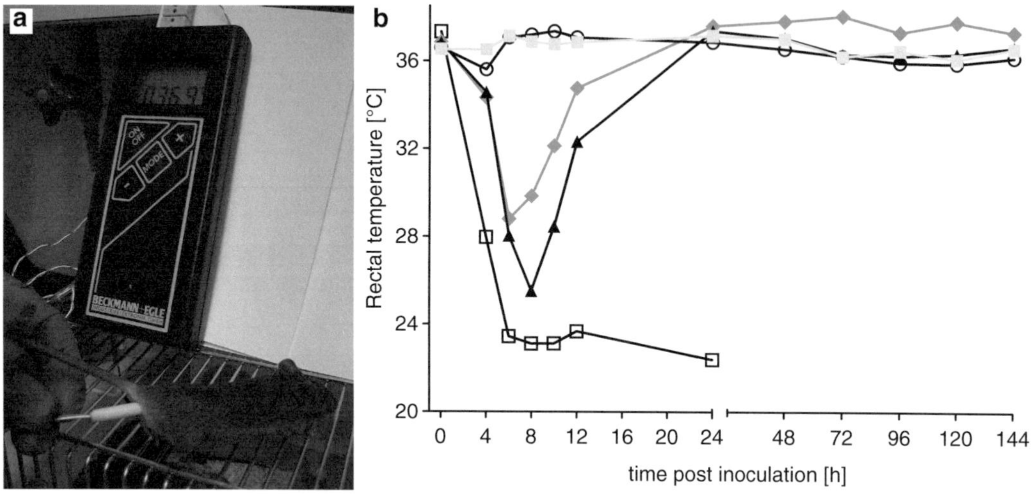

Fig. 6 Rectal temperature. (**a**) Measuring the rectal temperature in a mouse. (**b**) Body temperature shows a clear drop in mice (here C57BL/6) during infection with four *P. aeruginosa* strains of different virulence. Mice were infected i.n. each with 1.5×10^6 cfu of the strains, curve with *open circles* displays the mock control (mice received PBS). Mice infected with the most virulent isolate succumbed to death within the first 24 h p.i. (*open rectangles*)

activity, oculo-nasal secretion, grooming, and dehydration for an estimation of the health status of the mouse [10]. The sum of the nine individual scores represents the body condition score: untroubled 0–1; mildly troubled 2–4; moderately troubled 5–7; substantially troubled 8–10; moribund ≥11; and death ≥16.

The analysis scale ranges from 0 (undisturbed) to 11 (moribund). Values higher than 11 will inevitably lead to the death of the animal in due course; we therefore suggest sacrificing mice in this moribund condition for reasons of animal welfare.

4. Noninvasive head-out spirometry [11]:

   Spirometric measurements using the Notocord Hem software are performed in a custom-made head-out spirometry apparatus (*see* Fig. 3) and take about 4 min (*see* **Note 6**). Four mice can be measured in parallel. For this, mice are placed in sampling chambers with their heads sticking out through a double layer of membranes. These membranes ensure an airtight fit between the outside air and the air volume in the insert as well as restrainment of the mice (*see* Fig. 8; **Note 7**).

Prior to the start of the infection experiment mice are familiarized to the spirometric procedure for 5 days (*see* **Notes 8–10**). The first 2 days are not used for analysis due to high variability caused by movement artefacts. Of the remaining 3 days, the median is calculated for each mouse and parameter and defined as

Table 3
Parameters of the murine body condition score

Category	Quality	Score
Vocalization	No vocalization	0
	Vocalization when provoked, during handling	1
	Vocalization unprovoked, pain related	2
Piloerection	Shiny coat, close fitting	0
	Shiny coat, partial piloerection	1
	Stared, lustreless coat, marked piloerection	2
Posture	Normal	0
	Hunched intermittently	1
	Hunched permanently, partially in lateral position (mice in lateral position should be sacrificed, no recovery expectable)	2
Movement	Spontaneous locomotion without being provoked, or when sleeping, after opening of the cage and provocation	0
	Locomotion after being provoked, staggering, atactic, labored gait	1
	No locomotion	2
Respiration	Normal	0
	Tachypnea or abnormal breathing pattern	1
	Tachypnea, labored breathing, abnormal breathing pattern, occasional dropouts	2
Activity	Interaction with peers, species-typical movements and behavior, curious, responsive, normal provoked patterns of behaviour (e.g., escape reaction on approach)	0
	Interruptions in activity, subdued behavior patterns, even when provoked, reduced food and water intake	1
	Unresponsive, lethargic, no food and water intake	2
Discharges	No discharges	0
	Oculo-nasal discharge, serous secretion	1
	Oculo-nasal discharge, copious, and suppurative	2
Grooming	Normal grooming	0
	Reduced grooming, smeared anal region	1
	No grooming, smeared and incrusted anal region, facial impaction	2
Dehydration	Spontaneous urination (especially during handling), normal skin turgor	0
	No urination observable, skin turgor reduced (remaining skin fold)	1

the starting value. Post-infection the first spirometric measurements are taken after 4 h and then in the acute phase of infection at time points 6, 8, 10, 12, and 18 h post-infection. From time point 24 h post-infection until the end of a measurement series (usually we monitor infected mice for 6–8 days depending on the objective of the experiment) measurements are taken every 24 h at the same time of the day. To our experience 192 h are sufficient for a longitudinal investigation of bacterial lung infections in the acute infection model. By that time changes in the lungs are either

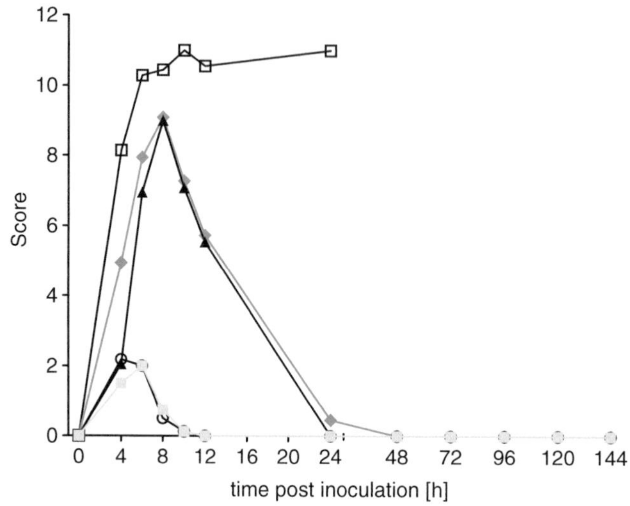

Fig. 7 Body condition score. The more the mice are disturbed in their body condition the higher are the score values (here C57BL/6). Mice were infected i.n. each with 1.5×10^6 cfu of four *P. aeruginosa* strains, the curve with *open circles* displays the mock control (mice received PBS). Mice infected with the most virulent isolate succumbed to death within the first 24 h p.i. (*open rectangles*)

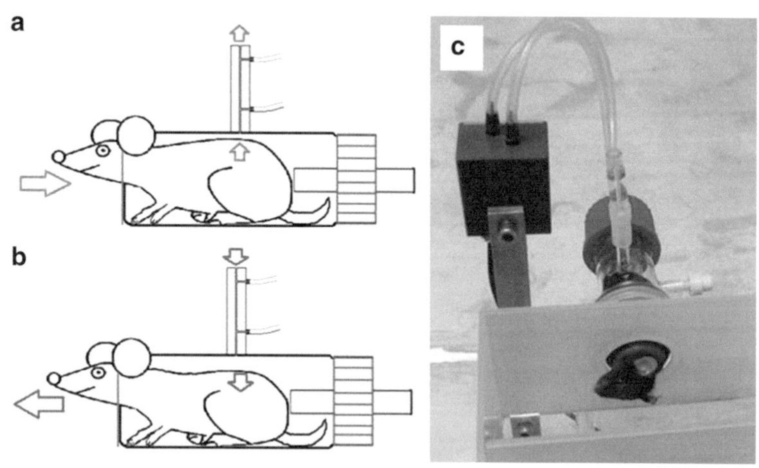

Fig. 8 Position of restrained mice in an insert of the spirometry device. Left-handed schemes of a mouse during (**a**) inspiration and (**b**) expiration, airflow shown by *blue* (**a**) and *red* (**b**) *arrows*. (**c**) Photograph of a mouse fixed during spirometric measurement, silicone tubes connecting the glass insert with a pneumotachograph (Color figure online)

irreversible, leading to the death of an animal or mice are able to eliminate *P. aeruginosa* and recover from the infection. Anyway, a further extension of the measurement period does not yield more information. A typical experimental setup is shown in Fig. 9.

The principle of the spirometry measurements is that respiration causes air to flow through a pneumotachometer positioned above

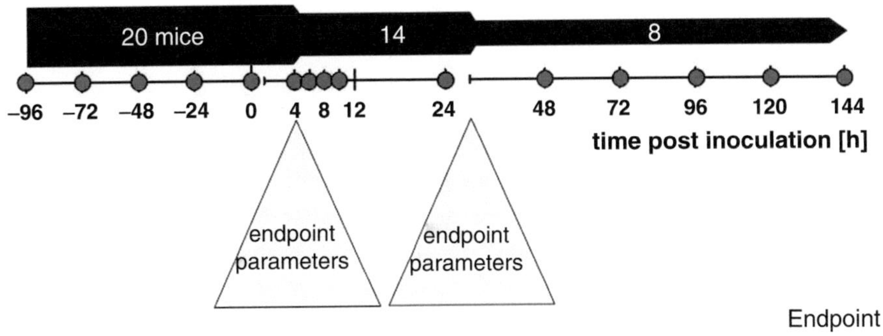

Fig. 9 Common design of an acute experimental airway infection in mice. Experiment starts with 20 mice, subgroups are used for endpoint analysis such as pathohistology or cytokine tests (time points marked by *light grey triangles*), and spirometry is symbolized by *red dots* (Color figure online)

Table 4
List of lung function parameters which are analyzed by the Notocord Hem software

Parameter	Description	Unit
Tidal volume	Total lung volume	mL
Volume	Inspiration volume	mL
Minute volume	Total volume breathed in 1 min	mL
Expiratory time	Time required for expiration	ms
Inspiratory time	Time required for inspiration	ms
Time of inspiration plus expiration	Time required for one breath	ms
IF50 (flow at 0.5 VTI)	Midtidal inspiratory flow at 50 % inspiration	mL/s
EF50 (flow at 0.5 VTE)	Midtidal expiratory flow at 50 % expiration	mL/s
PIF	Peak inspiratory flow	mL/s
PEF	Peak expiratory flow	mL/s
Respiratory rate	Breaths per minute	bpm
Relaxation time	Time required to expire (1-e-1) of tidal volume	ms
Time of pause	End expiratory pause (EEP)	ms
Time of brake	End inspiratory pause (EIP)	ms

the thorax of the mouse. The airflow is converted into an electrical signal by a pressure transducer, amplified, and digitalized before it reaches the computer for analysis. Data evaluation is done with the Notocord Hem software. For analysis, two markers are set individually for each mouse within the 4-min measurement period. In an optimum manner markers should cover a time span of approximately 1 min (*see* **Note 11**). Values of each single parameter are averaged for the marked periods, and any further analysis can be done by Excel.

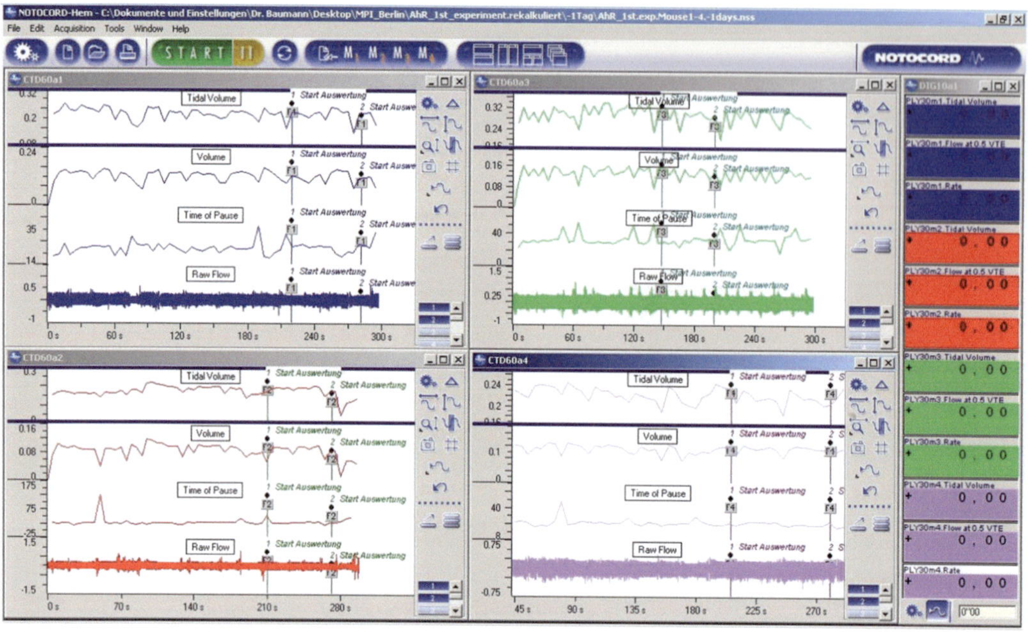

Fig. 10 Head-out body plethysmography recorded by Notocord Hem software. Four mice are measured in parallel, each animal marked by a different color. Here the parameters tidal volume, volume, time of pause, and raw flow are shown, but each of the 14 parameters can be selected to be displayed on the screen, according to the experimenter's interest. The markers are set individually for each animal and display the relevant period for analysis

The 14 parameters of lung function are shown in Table 4, and a screenshot of the software during measurement is given in Fig. 10. Exemplarily the tidal volume (total volume inspired and expired in one breath) and EF50 (expiratory flow at 50 % expiration) are shown as the most significant and robust parameters of murine lung function (Fig. 11). Please refer to the video file on springerprotocols.com for an example of how the spirometric process is performed in our laboratory (taken from ref. [11]).

3.5 Endpoint Analysis

1. Lung preparation:

   Depending on the target of the infection experiment the mouse lungs can be divided into different sections as it is shown exemplarily (Fig. 12). Lungs are removed aseptically, weighed (weight of lung in total, weight of each lobe or piece of lung tissue which is needed for a separate analysis), and divided into their different lobes.

2. Histological analysis (modified from ref. [11]):

   To investigate the signs of inflammation lungs are fixed in 4 % formalin and embedded in paraffin. The paraffin blocks are cut into 4 μm slices, stained with hematoxylin/eosin, and evaluated

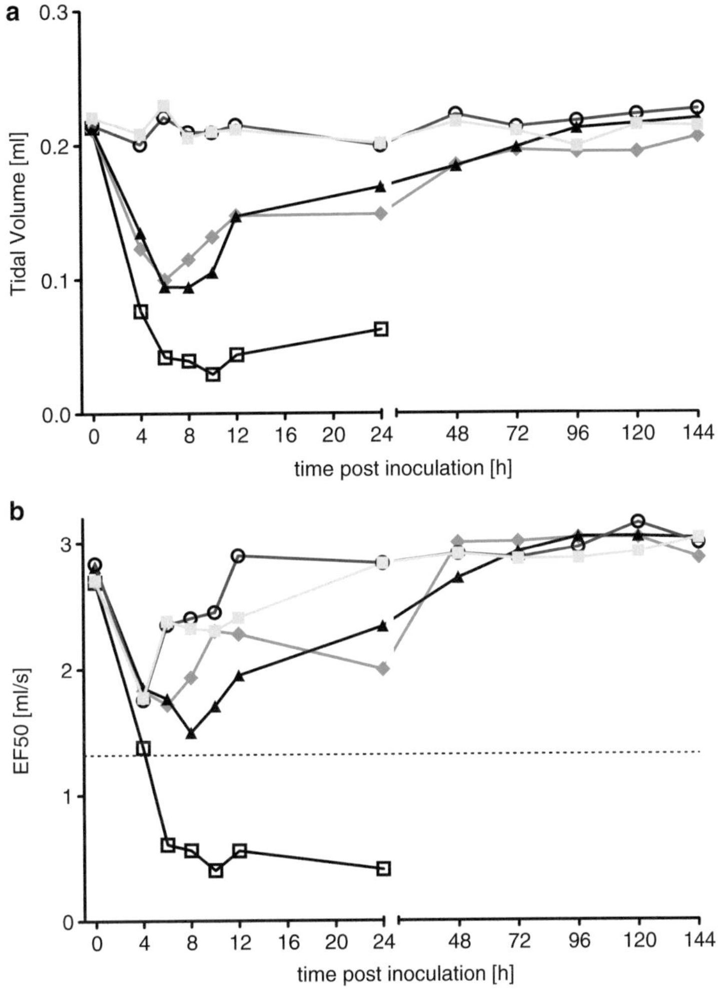

Fig. 11 Spirometry parameters. (**a**) Tidal volume, (**b**) EF50. Both parameters show a clear drop during the early phase of i.n. infection with 1.5×10^6 cfu *P. aeruginosa*. Decrease and recovery strongly depend on the virulence of the applied *P. aeruginosa* strain. Mice infected with the most virulent isolate succumbed to death within the first 24 h p.i. (*open rectangles*). Mock control: *open circles* (mice received PBS)

under a photomicroscope. Using lenses with 10- to 20-fold magnification, the inflammation of the lung parenchyma and of the bronchial areas are assessed in 20 fields of view (FOVs) per specimen slide using a three-point score. One point is given for a slight inflammation, two points for medium, and three points for a profound inflammation in each of the FOVs. A non-visible degree of airway inflammation per specimen corresponds to a total score of less than 5 points, a slight inflammation to 5–20 points, a medium inflammation to 21–40 points, and finally a profound inflammation to 41–60 points.

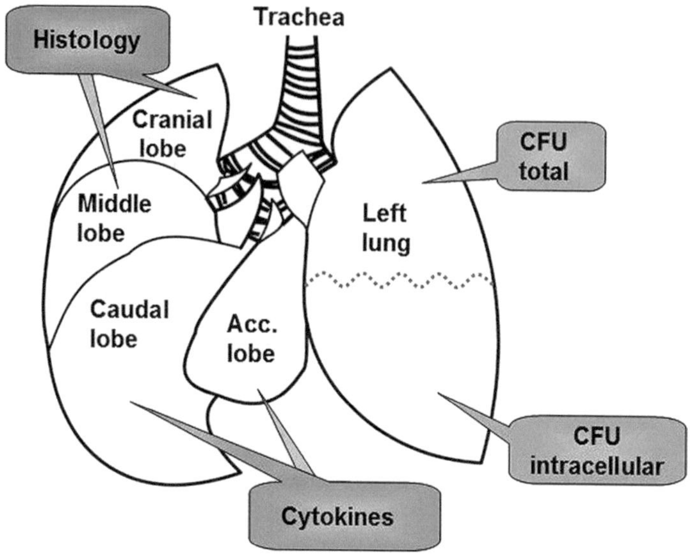

Fig. 12 Potential dissection of a murine lung. Single lobes might be differently processed depending on the aims of the experiment. The total organ and single pieces thereof have always to be weighed

3. Bacterial numbers:

   P. aeruginosa are recovered from the lungs and other organs (spleen, liver, brain, blood) by tissue homogenization and plating of aliquots.

   The cfus can be harvested from the whole lung (spleen, liver, etc.) or only parts of it. Depending on the procedure cfu values are given per organ or per gram organ. The lungs of the euthanized mice are removed aseptically and homogenized with a tissue homogenizer. The lysate is diluted sequentially in tenfold steps according to the estimates of cfu derived from infection dose, time post-inoculation, and body condition score of the individual mouse (*see* **Note 12**). Aliquots of the three last dilution steps are plated on LB agar plates. Plates have to be incubated at 37 °C for at least 12 h, and then bacterial numbers are counted.

4. Further analysis:

   The model permits the determination of numerous other endpoint parameters from tissue samples, blood, or bronchoalveolar lavage. The choice may vary from a few biomarkers of inflammation (e.g., *see* ref. 12) to a comprehensive dataset of transcriptome, proteome, and metabolome, just to quote two extreme examples. Methods are selected according to the objectives of the study.

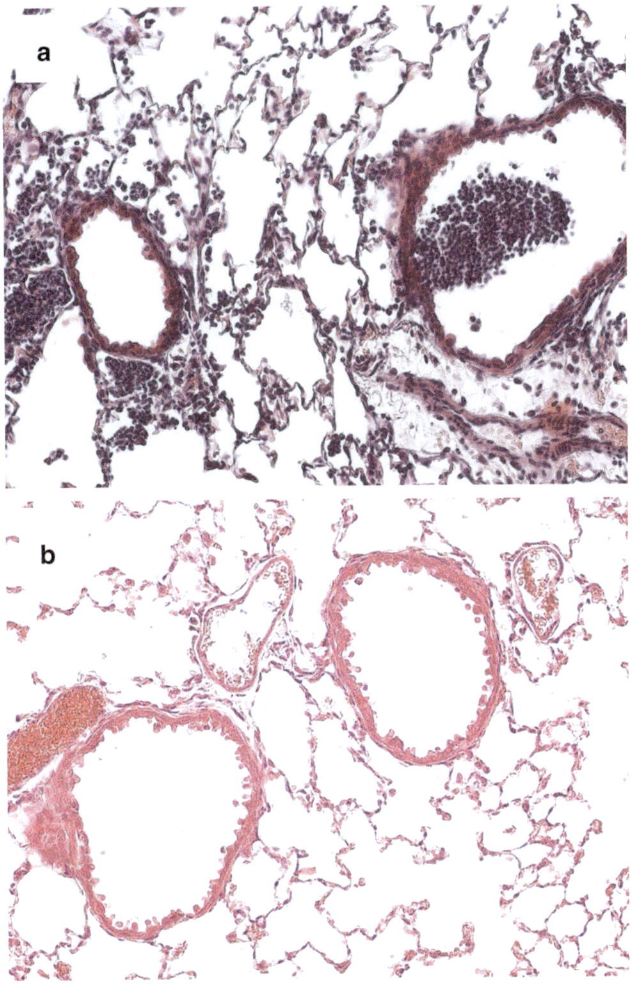

Fig. 13 Pathohistology of the murine lung on day 2 p.i. with 7.5×10^6 cfu of the virulent CF isolate *P. aeruginosa* TBCF10839 (**a**). Peri- and intrabronchiolar infiltration of neutrophils and alveolar macrophages cause numerous foci of inflammation inducing a profound purulent pneumonia. (**b**) Vehicle control instilled with 30 μL PBS. Hematoxylin–eosin staining; original magnification ×200

4 Notes

1. Groups of mice should be in the same range of age and weight. Both genders can be included into one experiment, but due to weight differences between males and females they often need to be measured with different sizes of sampling chambers.
2. Of course genders have to be separated between measurements and during the course of the experiment.

3. Sometimes prior to instillation the epiglottis has to be raised to get access to the trachea. This can easily be done with the catheter's tip.
4. The deposition of bacteria into the lungs depends on the volume of the inoculum and on the mode and/or depth of anaesthesia, respectively. According to our own experience, a sufficient depth of anaesthesia reduces defending movements as well as sneezing and coughing and therefore allows deposition of higher bacterial numbers in the lower airways. We therefore favor i.n. application on systemically anaesthetized mice (as described above) over applying bacteria to mice which are awake or only shortly put into sleep via inhaled narcotics.
5. Mice always respond to a bacterial infection with a drop of temperature; temperature measuring is therefore also suitable to quickly identify outliers, where infection does not run properly.
6. The spirometry device has to be calibrated prior to each new experimental setup. Glass inserts have to be chosen according to the murine body weight. For calibration a gauging gastight syringe with a defined volume of 200 μL (achieved via an installed Chaney adapter) is needed to control the accuracy of the measurement. Manual pumping with the syringe mimics breathing of a mouse in the chamber and must let to values of 0.2 ± 0.005 mL.
7. Density of the whole system has to be checked prior to each measurement, since even small holes in the membranes (e.g., caused by biting mice) or silicone tubes led to decrease in pressure and erroneous measurements.
8. Be cautious when handling the mice and placing them into the sampling chambers of the spirometric device. Depending on the mouse strain some animals might become excited and even aggressive, especially at the beginning of adaptation, and would try to bite. Therefore, we sometimes use a wire eye to fix the mice teeth and pull them through the membranes. A careful and safe handling protects mice from pain and injury and also prevents the experimenter from biting.
9. During closure of the sampling chamber with the screwable cap one has to take care that the mouse tail is not trapped between glass and cap.
10. It is absolutely crucial to prevent crawling of mice out of the chamber through the outlet end, since the narrowing of the membranes will make normal breathing impossible and mice will die within a very short time if not released. Instead tilting back the head into the chamber is not dangerous for a mouse, and by opening of the chamber, the animal might be repositioned even during ongoing measurement. By setting a marker

the time period of repositioning should be excluded from the analysis later on.

11. Setting of markers for subsequent analysis during measurement should reflect individual duration of acclimatization and artefacts due to movements of the mice. It is also possible to set the markers after finishing the measurement.
12. cfus are determined from plates with 20–200 colonies. Colonies should be discernable. Total bacterial numbers are then calculated according to the dilution and the applied volume of 50 μL.

Acknowledgement

This work was supported by a grant of the Deutsche Forschungsgemeinschaft to B. T. (SFB 587, project A9).

References

1. Ramos J-L (ed) (2004) Pseudomonas. Kluwer Academic/Plenum, New York
2. Diamond SS, Cadwell GA (1979) Fatal bronchopneumonia and dermatitis caused by *Pseudomonas aeruginosa* in Atlantic Bottle-Nosed Dolphin. J Am Vet Med Assoc 175: 984–987
3. Wang SZ (1987) Diagnosis and control of a disease in mink caused by *Pseudomonas aeruginosa*. Chinese J Vet Med 13:16–17
4. JAX mice and services (2014) The Jackson Laboratory, Bar Harbor, Maine. http://jaxmice.jax.org/ Accessed 03 March 2014
5. Poch DS, Ost DE (2009) What are the important risk factors for healthcare-associated pneumonia? Semin Respir Crit Care Med 30:26–35
6. American Thoracic Society; Infectious Diseases Society of America (2005) Guidelines for the management of adults with hospital-acquired, ventilator-associated, and healthcare-associated pneumonia. Am J Respir Crit Care Med 171: 388–416
7. Klockgether J et al (2010) Genome diversity of *Pseudomonas aeruginosa* PAO1 laboratory strains. J Bacteriol 192:1113–21
8. Munder A et al (2011) Acute intratracheal *Pseudomonas aeruginosa* infection in cystic fibrosis mice is age-independent. Respir Res 12:148
9. Munder A et al (2002) Pulmonary microbial infection in mice: comparison of different application methods and correlation of bacterial numbers and histopathology. Exp Toxicol Pathol 54:127–33
10. Munder A et al (2005) Murine pulmonary infection with *Listeria monocytogenes*: differential susceptibility of BALB/c, C57BL/6 and DBA/2 mice. Microbes Infect 4: 600–611
11. Wölbeling F et al (2010) Head-out spirometry accurately monitors the course of *Pseudomonas aeruginosa* lung infection in mice. Respiration 80:340–346
12. Wölbeling F et al (2011) Lung function and inflammation during murine *Pseudomonas aeruginosa* airway infection. Immunobiology 216:901–908

Chapter 60

Burn Mouse Models

Henrik Calum, Niels Høiby, and Claus Moser

Abstract

Severe thermal injury induces immunosuppression, involving all parts of the immune system, especially when large fractions of the total body surface area are affected. An animal model was established to characterize the burn-induced immunosuppression. In our novel mouse model a 6 % third-degree burn injury was induced with a hot-air blower. The third-degree burn was confirmed histologically. At 48 h, a decline in the concentration of peripheral blood leucocytes was observed in the group of mice with burn wound. The reduction was ascribed to the decline in concentration of polymorphonuclear neutrophil leucocytes and monocytes. When infecting the skin with *Pseudomonas aeruginosa*, a dissemination of bacteria was observed only in the burn wound group. Histological characterization of the skin showed an increased polymorphonuclear neutrophil granulocytes dominated inflammation in the group of mice with infected burn wound compared with the burn wound only group. The burn mouse model resembles the clinical situation and provides an opportunity to examine or develop new strategies like new antibiotics and immune therapy, in handling burn wound victims much.

Key words *Pseudomonas aeruginosa*, Burn wound infection, Animal model, Immune response, Immunosuppression

1 Introduction

In Denmark, the incidence of burn is 10,000 cases a year. Of these, almost 275 patients are admitted to the hospital and about 60 die [1]. By comparison, in the USA it is estimated by the American Burn Association that 450,000 persons experience burns requiring medical treatment each year. Furthermore, 45,000 are admitted to the hospital in the USA and 3,000–3,500 dies every year [2]. Burn injuries are also a problem in the developing countries, where 5 % of the inpatients at any given time are burn patients [3]. According to the WHO, it is estimated that 195,000 died every year of burns worldwide primarily in low- and middle-income countries [4]. The accidents are work- and home place-related. Although some burn patients die of burn shock during the first hours, later on due to pulmonary and cardiac failure. The major course of mortality is

Alain Filloux and Juan-Luis Ramos (eds.), *Pseudomonas Methods and Protocols*, Methods in Molecular Biology, vol. 1149, DOI 10.1007/978-1-4939-0473-0_60, © Springer Science+Business Media New York 2014

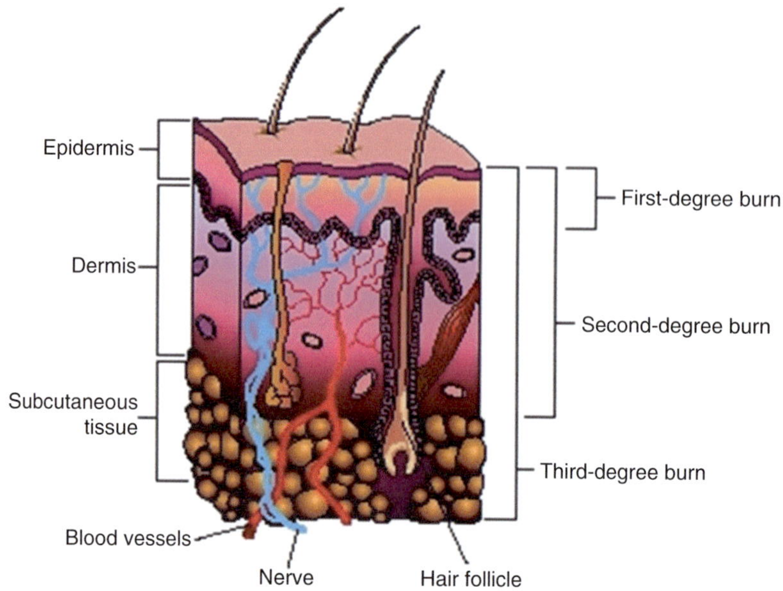

Fig. 1 From http://www.healthofchildren.com/images/gech

lung-related complications and infections. It is estimated that 75 % of all deaths following burns are related to infection [5]. Burn mouse models offer the opportunity to study different problems of the burn pathology and treatment.

1.1 Burns

Burns or ambustio can be induced by heat delivered by gaseous, liquid, or solid material. The time and temperature determine the extent of the damage. Furthermore, electricity and chemicals can also induce burns. Burn wounds of the skin can be assessed according to the classification system of first-, second-, and third-degree injury, i.e., epidermal, partial-thickness (further divided into superficial and deep), and full-thickness burn [6, 7].

First-degree burns are characterized by congestion and pain, but necrosis is not induced. The damage results in vasodilatation and increased vascular permeability. Furthermore, slight edema develops. Sunburn is an example of first-degree burns. Clinically, the burned area is red owing to the vasodilatation and painful (Fig. 1) [6–8].

Second-degree burns induce necrosis localized to the epithelium, but sparing the dermis. Clinically, these burns appear with blisters, pain, and redness. Histologically, the blister forms owing to the separation of the epidermis is separated from the dermis (Fig. 1) [6–8].

Third-degree burns damage both the epithelium and dermis so necrosis occurs. Clinically, the burn appears dry, waxy white, leathery, brown, or charred. Due to the destruction of nerves, the sensation of pain is reduced or not present. Histological, carbonization of the epidermis and dermis occurs and cells die. The potential for regeneration is limited to skin appendages, if these are spared (Fig. 1) [6–8].

1.1.1 The Microbial Flora of the Burn Wound

The microbial flora of a burn unit is dynamic over time [9]. The different microbial profiles are attributed to different policies concerning the use of antibiotics. Since colonization and infection of the burn patients often originate from the normal flora, identification of the burn wound flora is important. Furthermore, one should only extrapolate from one unit to another with precautions. For instance, Rastegar et al. reported that on admission the frequency of *Pseudomonas aeruginosa* was 50 % out of the positive isolates [10]. In comparison, Macedo reported coagulase-negative staphylococci (CoNS) (63 %) and *Staphylococcus aureus* as the most frequently recovered bacteria [11]. The change of the microbial profile over time from a Gram-positive towards a Gram-negative colonization is associated with increased morbidity and mortality. Staphylococci cause burn wound infection and failure of the skin graft but seldom lead to mortality. The increased infectivity of Gram-negative bacteria like *P. aeruginosa*, *Acinetobacter* sp., *Klebsiella* sp., and *Escherichia coli* [12] is due to the different toxins and enzymes rendering them more virulent [13]. The most frequent isolated fungus was *Candida* species, which was increased during the hospital stay and peaked within the third and fourth week [11]. The later colonizers of the burn wounds derive from the environment, the gastrointestinal tract of the patients, and the staff.

Regarding burn patients, studies have shown that the 10 % of the burn victims who become infected with *P. aeruginosa* face a mortality rate of 80 % [14]. Furthermore, *P. aeruginosa* can induce the skin graft to fail or delay the burn wound in healing. Despite improvement in the general treatment of and outcome for burn patients, those who become infected with *P. aeruginosa* have not benefited from this progress, since the mortality rate for those patients has been unchanged in the last 25 years.

Pseudomonas aeruginosa is transmitted to the burn wound by the initial hydrotherapy, which can be at the scene of the accident, where water from a river or lake can be used. Upon admission, the patients can get *P. aeruginosa*, when contaminated water is used [15]. Later on, patients can acquire bacteria from hydrotherapy, intravenous solution, aqueous solution ventilators, and nebulizers. Furthermore, some patients are harboring the bacteria rectally. However, the most important and also the most preventable transmission is the hands of healthcare workers [16].

1.1.2 Infections in the Burn Victims

Infections in burn victims can be categorized into four groups: burn wound infection, bloodstream infection, pneumonia, and urinary tract infection. Different results have been reported, in part due to the different characteristics of the burn patients.

Another explanation is the different criteria used to define the types of infections. In general, the infection rates in the burn units are comparable to other intensive care units. In some studies, pneumonia has superseded burn wound infection as a leading cause of infection-related deaths among burn victims.

The incidence of burn wound infection as a cause of sepsis-related death has been reduced due to the introduction of topical antibiotics and implementation of early excision therapy [17, 18]. From the 1950s and until 1980s, the treatment of burn patients was hydrotherapy and controlled growth of bacteria on the burn wound to allow the breakdown of the eschar, so the granulation tissue could be covered by graft [16]. Back in 1970 Janzekovic introduced early excision and grafting of burns [19]. Today most burn centers are following this strategy. The rationale for the new treatment is that early excision removes devitalized tissue and prevents colonization of the burn wound with microorganisms and furthermore removes a "site" for initiating and maintaining severe inflammatory response syndrome (SIRS). Furthermore, the excision therapy also removes the site for production of the burn toxin [20–22].

1.2 Burn Mouse Model

During time different models have been developed. One of the earliest models developed to study this condition, was performed by Mason and Walker back in 1968. They introduced the scald burn model by inflicted the thermal injury using boiling water. The anesthetized, shaved mouse was placed supine in this device and immersed in boiling water. Ten seconds of exposure caused a third-degree burn wound [23, 24].

Orenstein et al. introduced a metal plate burn model in order to examine the effect of a porphyrin mixture to eradicate *S. aureus* from burn wound. They used guinea pigs, which were anesthetized and shaved by an electrical razor. Induction of the thermal injury was conducted by placing a heated (150 °C) copper plate of 1 × 1 × 3 cm on the back for 10 s. The infection was established by the smearing of *S. aureus* suspension onto the eschar [25]. Furthermore, other animals have also been used such as swine to study treatment of infection in burn wound [26].

Inspiration to burn mouse model ad modum Rigshospitalet came from Stieritz and Holder, who back in 1975 introduced an ethanol-based burn mouse model (female CF1 mouse weighing 22–24 g) using an flame-resistant template with a window. 1 × 1.5 in. (2.52 × 3.81 cm). The burned area was estimated to be 30 %. The animals were anesthetized and prior to the burn, the hair was shaved from the back of the mice. The template is pressed against the back and 0.5 ml of ethanol is put into the open area, ignited and allowed to burn for 10 s. Immediately following the burn, the mice are given 0.5 ml saline as fluid replacement making the burn nonlethal. The infection was established by introducing bacteria (dose: 10^1–10^3 microorganisms) subcutaneous (s.c.) immediately post burn [27].

The same approach was used in our laboratory, but the use of ethanol made this procedure difficult to control, since the ethanol in our hands tended to escape the template. Due to these problems, we developed a novel model. The new burn model was developed in close collaboration with and approved by the Animal Ethics Committee of Denmark. It was crucial that the animals did not suffer unnecessarily and that the model was useful in a pathophysiological manner to help burn patients.

2 Methods

- Specified pathogen-free female 12-week-old C3H/HeN or BALB/c mice (20–25 g) are allowed to acclimatize to the animal facility for at least 1 week prior to experimentation.
- They are kept in barrier facilities with access to chow and water ad libitum.
- The animals are anesthetized s.c. with 0.25 ml 1:1 mixture of etomidate (Jansen, Birkeroed, Denmark) and midazolan (Roche, Basel, Switzerland).
- An electrical clipper is used to shave the hair on the bag of each mouse.
- The mice are placed on a sledge and covered with a fire blanket with a window (1.7×2.6 cm) corresponding to 6 % of the total body area.
- Above the fire blanket a metal plate with a window (1.7×2.6 cm) is placed.
- The sledge with the mouse, the fire blanket, and metal plate are moved into a stream of hot air with a temperature of 330 °C delivered by a hot air blower (Bosch 500-2, Germany) for 7 s (now reduced to 5 s). This procedure results in a third-degree burn confirmed by histological examination (Fig. 2).
- Immediately after the procedure, the mice are given 1.0 ml isotonic saline and 0.3 ml glucose for fluid replacement s.c. (now modified to 0.5 ml isotonic saline as in the Stierlitz and Holder).
- The mice are able to drink within 24 h after application of the thermal lesion.
- Buprenorphin (Schering-Plough, Brussels, Belgium) is given every 8 h as pain therapy during the fist 48 h after the burn procedure.
- During the first 24 h of recovery the mice are kept at electrical heating carpets to prevent cooling of the mice [28].
- The burn surface is calculated by using Meeh formula: $A = KW^{2/3}$, where A = body surface area, K = 9, and W = weight in grams [29, 30]. *See* the Fig. 3 in the appendix for induction of the thermal injury.

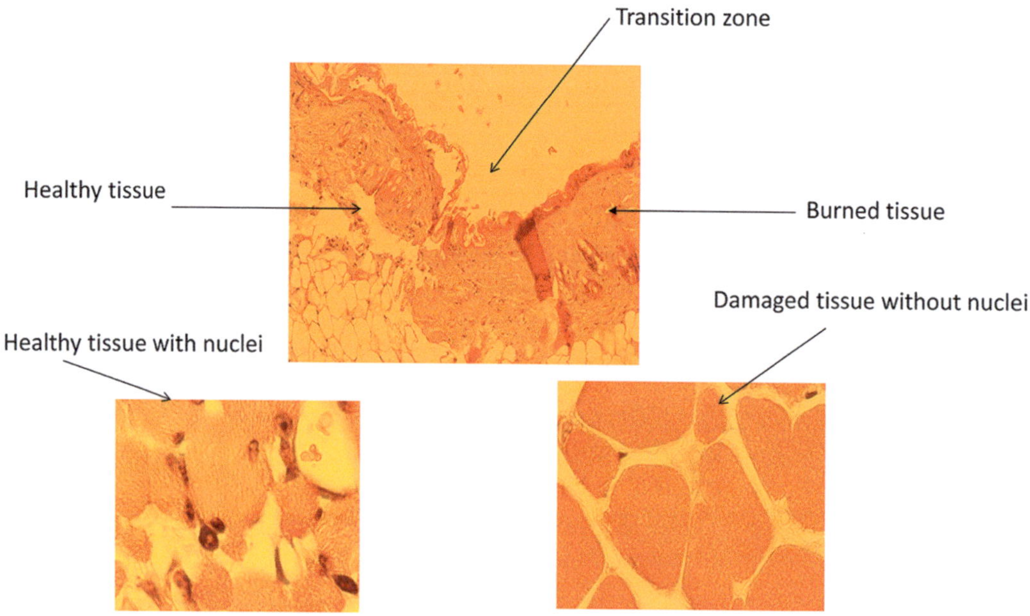

Fig. 2 Histopathology of the transition zone between burned skin and unburned tissue

- We use the wild-type *P. aeruginosa* PAO1 strain obtained from the Pseudomonas Genetic Stock Centre [31]. *P. aeruginosa* is grown in 100 ml filtered oxbroth overnight at 37 °C on a shaker.
- The mice are injected s.c. with 100 μl challenge solution (infection dose 10^7 CFU/ml) with a 27G needle (Terumo Europe NV, Leuven, Belgium) beneath the burn wound.
- Challenge dose is controlled by serial dilutions and culture overnight on selective media for Gram-negative rods (Statens Serum Institute, Copenhagen, Denmark).

The advantage of our model is the standardization and reproducibility, as well as the use of a high temperature, mimicking the temperature burn patients are exposed to. The drawback is the difficulty in comparing with other models, where the temperature often is 100 °C. The model requires two persons to perform the process.

3 Application of the Model

We used the present model to study the effect of thermal injury on the immune system.

Thermal injury was found to suppress the innate immune response. Immunosuppression was observed as a decrease in the concentration of leucocytes in peripheral blood as well as reduced function of the PMNs. By combining the burn wound with infection, the immunosuppression verified an impaired ability to control the burn wound infection. The present model, using a hot-air blower to obtain a relatively minor burn area of 6 % TBSA, was a

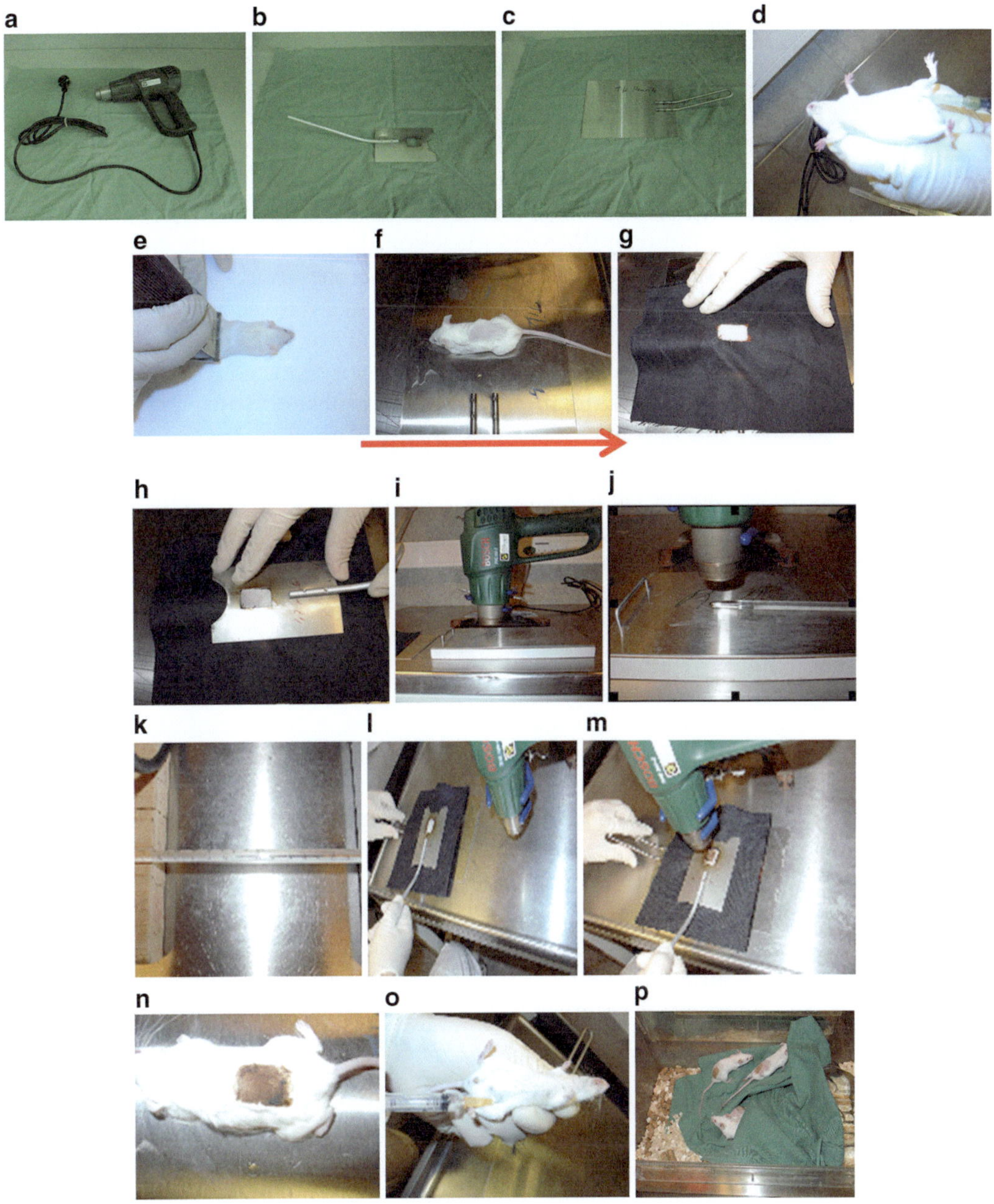

Fig. 3 The procedure. (**a**) The hot air blower, (**b**) the template with window, (**c**) the sledge, (**d**) injection of hypnorm dormicum s.c., (**e**) shaving the mouse, (**f**) placing the mouse on the sledge, (**g**) covering the mouse with fire blanket, (**h**) placing the template on the mouse, (**i**) the arrangement, (**j**) measurement of the temperature, (**k**) 330 °C, (**l**) initiating the induction of the thermal injury, (**m**) keeping the mouse for 7 s into the hot air, (**n**) clinical manifestation of the thermal injury, (**o**) injection of 1 ml of NaCl, (**p**) the mice 4–5 h after induction of the thermal injury

compromise with the Danish Animal Experimental Committee. Even with this minor thermal injury, a significant impact on the immune system was observed. Apart from temporary leucocytosis at 24 h, a reduction in peripheral blood leucocytes was observed at 48 h. The temporary peak in the concentration of leucocytes could be caused by an increased release of PMNs from the bone marrow and shifting of the marginal granulocyte pool into the circulation, as described previously [32]. The subsequent reduction in peripheral blood leucocytes has also been reported in another mouse model [33]. The reduction of PMNs could be due to diminished production the bone marrow, exhaustion of the pool of PMNs, sequestration of PMNs in the liver, lungs, spleen, or burn wound or influx of PMNs to the lesion. In our model, the influx of PMNs into the burn wound could be the cause of neutropenia. No sequestration of PMNs in the lungs was seen in this study, which may be due to the minor total body surface area burned (TBSAB). After burn injury, PMNs have been reported to display functional defects in migration, chemotaxia, phagocytosis, respiratory burst, and bacteria killing [34, 35]. Ozkan et al. have reported that an immunosuppressive glycopeptide in patients with 40 % TBSAB is capable of reducing the chemotaxia of PMNs. Furthermore, the glycopeptides was also capable of suppression of T lymphocytes [36]. It cannot be excluded that a depression of the bone marrow has occurred, as Asko-Seljavaara and coworkers investigated the bone marrow and found a depression of the DNA synthesis in the granulocytes [37]. A PMN-dominated inflammation was observed 24 h after the induction of the thermal injury. In burn patients with severe burns of 30 % or more a delayed influx of PMNs was observed [6]. Whether such a phenomenon could be observed in our model is less likely, because of the limitations of 6 % TBSA. After 48 h an influx of MNs was observed, and after 72 h the burn wound contained a mixture of PMNs and MNs. When the burn wound was superimposed with an infection the PMN-dominated response continued, probably because of the stimulus from the infection, and this response was comparable with an infection non-burn control group. Despite the first hit in terms of the thermal insult, it seems that at 96 h the mice could restore the leukocyte pool by increasing the concentration of PMNs in the blood. In contrast, it has been observed that superimposing burn wounds with *P. aeruginosa* infection leads to a reduction in the PMNs in the blood. However, these observations were obtained in mice with a much larger TBSAB (>30 %) [38].

Mice with thermal injury showed a reduced ability to mount phagocytosis and oxidative burst. Humane observations have shown that the decreased ability of PMNs to ingest and their capacity to kill bacteria occurred at day 5, independent of the burn size. When patients were discharged they had regained phagocytosis ability and killing capacity. It was also observed that, in patients with impaired killing capacity, susceptibility to infection was increased compared with patients with normal function. The explanation seemed to be burn induced reduction of myeloperoxidase, lactoferrin, and chemotrypsin-like cationic protein [39].

We examined the ability of focal infection to spread by culturing blood, liver, and spleen. We found dissemination in burn infection group but not in the infection non-burn control group. In accordance, Barnea et al. observed in their model that when using the spleen as an indicator for spread, bacterial dissemination was beginning at 6 h post infection and dissemination was seen only in the burn infection group [40].

4 Conclusion

In conclusion, the burn mouse model has been developed to study the immunological reactions in response to extreme stress such as burn injury. The burned victim is characterized by immunosuppression. The present novel burn mouse model of minor thermal lesions demonstrated a suppression of the innate immune response comparable with observations in patients suffering from burn wounds. The consequence of immunosuppression is an increased susceptibility to infections originating from colonization of the thermal injury. The burn model offers opportunity to study infections under these conditions. In our model the wound infections were examined. The present model can also be used to examine new antibiotics and immune therapy. Our animal model resembling the clinical situation is useful in developing such new treatments of burn wound vic.

References

1. Gedde H, Sparsø BH, Sheller J, Kofoed JF, Drzewiecki K, Sørensen K, Dahlstrøm K, Dahlfelt K (2006) Specialebeskrivelse I plastikkirurgi; februar:1–24
2. (2011) http://www.ameriburn.org/resources_factsheet.php
3. Askarian M, Hosseini RS, Kheirandish P, Memish ZA (2003) Incidence of urinary tract and bloodstream infections in Ghotbeddin Burn Center, Shiraz 2000–2001. Burns 29: 455–439
4. http://www.who.int/violence_injury_prevention/other_injury/burns/en/index.html
5. O'Sullivan ST, O'Connorm TPF (1997) Immunosuppression following thermal injury: the pathogenesis of immunodysfunction. Br J Plast Surg 50:615–623
6. Peng D, Huang W, Ai S, Wang S (2006) Clinical significance of leukocyte infiltrative response in deep wound of patients with major burns. Burns 3:946–50
7. Benson A (2006) ABC of wound healing. BMJ 332:649–652
8. http://www.healthofchildren.com/images/gech
9. Altoparlak U, Erol S, Akcay MN, Celebi F, Kadanali A (2004) The time-related changes of antimicrobial resistance patterns and predominant bacterial profiles of burn wounds and body flora of burned patients. Burns 30: 660–664
10. Rastegar AL, Honar HB, Alaghehbandan R (1998) *Pseudomonas* infections in Tohid Burn Centen, Iran. Burns 24:637–641
11. De Macedo JL, Santos JB (2006) Nosocomial infections in a Brazilian Burn unit. Burns 32: 477–481
12. Albrecht MA, Griffith ME, Mauray CK, Chung KK, Horvath EE, Ward JA, Hospenthal DR, Holocomb JB, Wolf SE (2006) Impact of *Acinetobacter* infection on the mortality of burn patients. J Am Coll Surg 203:546–550
13. Fogle MR, Griswold JA, Oliver JW, Hamood AN (2002) Anti-ETA IgG neutralizes the effects of *Pseudomonas aeruginosa* exotoxin A. J Surg Res 106:86–98

14. Tredget EE, Shankowsky HA, Rennie R, Burrel RE, Logsetty S (2004) *Pseudomonas aeruginosa* in the thermally injured patient. Burns 30:3–26
15. Kolmos HJ, Thuesen B, Nielsen SV, Lohmann M, Kristoffersen K, Rosdahl VT (1993) Outbreak of infection in a burns unit due to *Pseudomonas aeruginosa* originating from contaminated tubing used for irrigation of patients. J Hosp Infect 24:11–21
16. Mayhall CG (2003) The epidemiology of burn wound infections: then and now. Clin Infect Dis 37:543–550
17. Ong YS, Samuel M, Song SC (2006) Meta-analysis of early excision of burns. Burns 32: 145–150
18. Guggenheim M, Thurnheer T, Gmür R, Giovanoli P, Guggenheim B (2011) Validation of the Zürich burn-biofilm model. Burns 37:1125–1133
19. Janzekovic Z (1970) A new concept in the early excision and immediate grafting of burns. J Trauma 10:1103–1108
20. Hart DW, Wolf SE, Chinkes DL, Beaufortd RB, Mlcak RP, Heggers JP, Wolfe RR, Herndon DN (2003) Effects of early excision and aggressive enteral feeding on hypermetablism, catablism, and sepsis after severe burn. J Trauma 54:755–764
21. Chang KC, Ma H, Liao WC, Lee CK, Lin CY, Chen CC (2010) The optimal time for early burn wound excision to reduce pro-inflammatory cytokine production in a murine burn injury model. Burns 36:1059–1066
22. Alp E, Coruh A, Gunay GK, Yontar Y, Doganay M (2012) Risk factors for nosocomial infection and mortality in burn patients: 10 years of experience at a university hospital. J Burn Care Res 33:379–385
23. Walker HL, Mason AD (1968) A standard animal burn. J Trauma 8:1049–1051
24. Dai T, Kharkwal GB, Tanaka M, Huang YY, Bil de Arce VJ, Hamblin MR (2011) Animal models of external traumatic wound infections. Virulence 2:296–315
25. Orenstein A, Klein D, Kopolovic J, Winkler E, Malik Z, Keller N, Nitzan Y (1997) The use of porphyrins for eradication of *Staphylococcus aureus* in burn wound infections. FEMS Immunol Med Microbiol 19:307–314
26. Nusbaum AG, Gil J, Rippy MK, Warne B, Valdes J, Claro A, Davis SC (2012) Effective method to remove wound bacteria: comparison of various debridement modalities in an in vivo porcine model. J Surg Res 176:701–707
27. Stieritz DD, Holder LA (1975) Experimental studies of the pathogenesis of infections due to *Pseudomonas aeruginosa*: description of a burned mouse model. J Infect Dis 131: 688–691
28. Calum H, Moser C, Jensen PØ, Christophersen L, Maling DS, Van Gennip M, Bjarnholt T, Hougen HP, Givskov M, Jacobsen GK, Høiby N (2009) Thermal injury induces impaired function in polymorphonuclear neutrophil granulocytes and reduced control of burn wound infection. Clin Exp Immunol 156: 102–110
29. Gilpin DA (1996) Calculation of a new Meeh constant and experimental determination of burn size. Burns 22:607–611
30. Clancy KD, Lorenz K, Dries D, Gamelli RL, Hahn EL (2000) Chlorpromazine modulates cytokine expression in the liver and lung after burn injury and endotoxemia. J Trauma 48: 215–223
31. http://www.ecu.edu/Pseudomonas;PAO0001
32. Graddock CG (1972) Production, distribution, and fate of granulocytes. In: Williams WJ, Beutler E, Erslev AJ et al (eds) Hematology. McGraw-LliLL, New York, pp 607–618
33. Pallua N, Heimburg DV (2003) Pathogenic role of interleukin-6 in the development of sepsis. Part 1: study in a standardized contact burn murine model. Crit Care Med 31: 1490–1494
34. Babcock GF (2003) Predictive medicine: severe trauma and burns. Clin Cytom 53B: 48–53
35. Butler KL, Ambravaneswaran V, Agrawal N, Bilodeau M, Toner M, Tompkins RG, Fagan S, Irimia D (2010) Burn injury reduces neutrophil directional migration speed in microfluidic devices. PLoS One 5:e11921
36. Ozkan AX, Ninnemann JL, Sullivan J (1986) Progress in the characterization of immunosuppressive glycopeptide (SAP) from patients with major thermal injruies. J Burn Care 7: 388–397
37. Asko-Seljavaara S (1987) Granulocyte kinetics in burns. J Burn Care Rehabil 8:492–495
38. Shoup M, Weisenberger JM, Wang JL, Pyle JM, Gamelli RL, Shankar R (1998) Mechanism of neutropenia involving myeloid maturation arrest in burn sepsis. Ann Surg 228:112–122
39. Arturson G (1985) Neutrophil granulocyte functions in severely burned patients. Burns 11:309–319
40. Barnea Y, Carmeli Y, Kuzmenko B, Eyal G, Hammer-Munz O, Navon-Venezia S (2006) The establishment of a *Pseudomonas aeruginosa-infected* burn-wound sepsis model and the effect of imipenem treatment. Ann Plast Surg 56:674–679

Index

A

B

Alain Filloux and Juan-Luis Ramos (eds.), *Pseudomonas Methods and Protocols*, Methods in Molecular Biology, vol. 1149, DOI 10.1007/978-1-4939-0473-0, © Springer Science+Business Media New York 2014

C

D

E

Q

R

S

MIX
Papier aus verantwortungsvollen Quellen
Paper from responsible sources
FSC® C105338

If you have any concerns about our products,
you can contact us on
ProductSafety@springernature.com

In case Publisher is established outside the EU,
the EU authorized representative is:
Springer Nature Customer Service Center GmbH
Europaplatz 3, 69115 Heidelberg, Germany

Printed by Libri Plureos GmbH
in Hamburg, Germany